Methods in Enzymology

Volume XVIII
VITAMINS AND COENZYMES
Part B

METHODS IN ENZYMOLOGY

EDITORS-IN-CHIEF

Sidney P. Colowick   Nathan O. Kaplan

Methods in Enzymology

Volume XVIII

Vitamins and Coenzymes

Part B

EDITED BY

Donald B. McCormick and Lemuel D. Wright

GRADUATE SCHOOL OF NUTRITION AND
SECTION OF BIOCHEMISTRY AND MOLECULAR BIOLOGY
CORNELL UNIVERSITY
ITHACA, NEW YORK

1971

ACADEMIC PRESS   New York and London

ACADEMIC PRESS, INC.
111 Fifth Avenue, New York, New York 10003

United Kingdom Edition published by
ACADEMIC PRESS, INC. (LONDON) LTD.
Berkeley Square House, London W1X 6BA

LIBRARY OF CONGRESS CATALOG CARD NUMBER: 54-9110

PRINTED IN THE UNITED STATES OF AMERICA

Table of Contents

Section VII. Nicotinic Acid: Analogs and Coenzymes

Section VIII. Flavins and Derivatives

Section IX. Pteridines, Analogs, and Pterin Coenzymes

Contributors to Volume XVIII, Part B

Article numbers are shown in parentheses following the names of contributors.
Affiliations listed are current.

C. ANTHONY (201), *Department of Physiology and Biochemistry, University of Southampton, Southampton, United Kingdom*

CHARALAMPOS ARSENIS (160), *Department of Biological Chemistry, University of Illinois, Chicago, Illinois*

HERMAN BAKER (175), *Professor of Medicine and Preventive Medicine, College of Medicine and Dentistry of New Jersey, East Orange, New Jersey*

J. R. BERTINO (179), *Professor of Medicine and Pharmacology, Department of Pharmacology, Yale University School of Medicine, New Haven, Connecticut*

R. L. BLAKE (111), *The Jackson Laboratory, Bar Harbor, Maine*

J. J. BLUM (115), *Department of Physiology, Duke University, Durham, North Carolina*

GENE M. BROWN (192, 193, 194), *Department of Biology, Massachusetts Institute of Technology, Cambridge, Massachusetts*

HELEN B. BURCH (100), *Washington University School of Medicine, St. Louis, Missouri*

JAMES J. BURCHALL (196), *Burroughs Wellcome & Co. (U.S.A.) Inc., Research Triangle Park, North Carolina*

R. U. BYERRUM (110), *Michigan State University, East Lansing, Michigan*

PAOLA CAIAFA (139), *Department of Biological Chemistry, University of Rome, Rome, Italy*

D. F. CALBREATH (122), *Clinical Laboratory, Watts Hospital, Durham, North Carolina*

PAOLO CERLETTI (133, 139, 143), *Department of General Biochemistry, University of Milan, Milan, Italy*

EMMETT W. CHAPPELLE (137), *NASA, Goddard Space Flight Center, Greenbelt, Maryland*

STERLING CHAYKIN (103, 123, 128), *Department of Biochemistry and Biophysics, University of California, Davis, California*

D. R. CHRISTMAN (110), *Brookhaven National Laboratory, Upton, Long Island, New York*

ALBERT E. CHUNG (118), *Department of Biochemistry, Faculty of Arts and Sciences, University of Pittsburgh, Pittsburgh, Pennsylvania*

J. B. CLARK (102), *Department of Chemistry and Biochemistry, Medical School of St. Bartholomews Hospital, London, England*

JACK M. COOPERMAN (176), *Hematology and Nutrition Laboratory, Department of Pediatrics, New York Medical College, New York, New York*

R. F. DAWSON (110), *Rutgers University, New Brunswick, New Jersey*

A. DE KOK (156), *Department of Biochemistry Wageningen University, De Dreijen 6, Wageningen, The Netherlands*

VIRGINIA C. DEWEY (174, 190), *Biological Laboratory, Amherst College, Amherst, Massachusetts*

L. S. DIETRICH (117, 119), *Department of Biochemistry, University of Miami School of Medicine, Miami, Florida*

J. EDER (181), *Max-Planck-Institut für Biochemie, München, Germany*

A. EHRENBERG (142), *Department of Biophysics, University of Stockholm, Stockholm, Sweden*

ARPAD G. FAZEKAS (138), *Laboratorie d'Endocrinologie, Hopital Notre-Dame et Départment de Médicine, Universite de Montréal, Montreal, Canada*

RONALD L. FESLTED (128), *Biochemistry Department, University of Chicago Medical School, Chicago, Illinois*

H. S. FORREST (170), *Zoology Department, University of Texas at Austin, Austin, Texas*

WERNER FÖRY (148), *J. R. Geigy, A.G. CH-4000 Basel 21 Switzerland*

OSCAR FRANK (175), *Assistant Professor of Medicine, College of Medicine and Dentistry of New Jersey, East Orange, New Jersey*

HERBERT C. FRIEDMANN (105, 124, 125), *Department of Biochemistry, The University of Chicago, Chicago, Illinois*

SHINJI FUJIMURA (129), *Biochemistry Division, National Cancer Center Research Institute, Tsukiji, Chuo-ku, Tokyo, Japan*

JAMES A. FYFE (125), *Burroughs Wellcome Co., Research Laboratories, Research Triangle Park, North Carolina*

MARIA GRAZIA GIORDANO (133), *Department of Biological Chemistry, University of Rome, Rome, Italy*

ANN GORDON-WALKER (154), *Department of Biochemistry, Oxford University, South Parks Road, Oxford, England*

MIKI GOTO (191), *Department of Chemistry, Gakushuin University, Toshima-ku, Tokyo, Japan*

A. L. GREENBAUM (102), *Department of Biochemistry, University College, London, England*

DAVID M. GREENBERG (197, 198), *University of California Medical Center, San Francisco, California*

GORDON GUROFF (171), *Laboratory of Biomedical Sciences, National Institute of Child Health and Human Development, National Institutes of Health, Bethesda, Maryland*

W. GUTENSOHN (200), *Max-Planck-Institut für Biochemie, München, Germany*

P. HANDLER (122), *Department of Biochemistry, Duke University, Durham, North Carolina*

R. W. F. HARDY (169), *Central Research Department, Experiment Station, E. I. du Pont de Nemours and Company, Wilmington, Delaware*

R. A. HARVEY (157), *Department of Biochemistry, Rutgers Medical School, Rutgers University, New Brunswick, New Jersey*

OSAMU HAYAISHI (116, 120, 130), *Department of Medical Chemistry, Kyoto University Faculty of Medicine, Kyoto, Japan*

PETER HEMMERICH (142, 151), *Fachbereich Biologie, University of Konstanz, Konstanz, Germany*

L. M. HENDERSON (121), *Department of Biochemistry, College of Biological Sciences, University of Minnesota, St. Paul, Minnesota*

TASUKU HONJO (112, 114), *Department of Medical Chemistry, Kyoto University Faculty of Medicine, Kyoto, Japan*

S. H. HUTNER (175), *Haskins Laboratories at Pace College, New York, New York*

L. JAENICKE (172), *Institut für Biochemie, The Universität, Köln, Germany and University of Cologne, Institute of Biochemistry, Cologne, Germany*

JEAN-CLAUDE JATON (189), *Cardiac Unit, Department of Medicine, Massachusetts General Hospital, Boston, Massachusetts*

JAYANT G. JOSHI (122), *Department of Biochemistry, University of Tennessee, Knoxville, Tennessee*

VARDA KAHN (115), *Volcani Institute, Rehovoth, Israel*

ROLAND G. KALLEN (183), *Department of Biochemistry, University of Pennsylvania School of Medicine, Philadelphia, Pennsylvania*

SEYMOUR KAUFMAN (173), *Laboratory of Neurochemistry, National Institute of Mental Health, Bethesda, Maryland*

FUMIO KAWAI (158), *Faculty of Nutrition, College of Kôshien, Takarazuka, Hyôgô, Japan*

G. W. KIDDER (174, 190), *Biological Laboratory, Amherst College, Amherst, Massachusetts*

R. L. KISLIUK (180), *Department of Biochemistry, Tufts University School of Medicine, Boston, Massachusetts*

E. KNIGHT, JR. (169), *Central Research Department, Experiment Station, E. I. du Pont de Nemours and Company, Wilmington, Delaware*

EDUARD KNOBLOCH (136), *Institute for Pharmacy and Biochemistry, Prague, Czechoslovakia*

JACEK KOZIOŁ (132), *Institute of Com-Medicine, University Medical School, Szeged, Hungary*

J. F. KOSTER (155), *Department of Biochemistry Wageningen University De Dreijen 6, Wageningen, The Netherlands*

JACEK KOZIOŁ (132), *Institute of Commodity Science, Higher School of Economics, Poznań, Poland*

ERNEST KUN (111), *University of California, School of Medicine, San Francisco, California*

ERICH KUSS (104), *Laboratorium für Klinische Chemie und Biochemie, Universität Frauenklinik, München, Germany*

CARL KUTZBACH (199), *Max-Planck-Institut für Ernährungsphysiologie, Dortmund, Germany*

JOHN P. LAMBOOY (145), *Department of Biological Chemistry, School of Medicine, University of Maryland at Baltimore, Baltimore, Maryland*

DONALD B. MCCORMICK (148, 155, 159, 162, 164), *Graduate School of Nutrition and Section of Biochemistry and Molecular Biology, Cornell University, Ithaca, New York*

GIULIO MAGNI (143), *Department of Biological Chemistry, University of Camerino, Camerino, Italy*

VINCENT MASSEY (150, 151), *Department of Biological Chemistry, University of Michigan Medical School, Ann Arbor, Michigan*

F. M. MATSCHINSKY (99), *Department of Pharmacology, Washington University Medical School, St. Louis, Missouri*

KUNIO MATSUI (144), *Division of Bi-*

ology, *Research Institute for Atomic Energy, Osaka City University, Osaka, Japan*

HISATERU MITSUDA (158), *Laboratory of Nutritional Chemistry, Faculty of Agriculture, Kyoto University, Sakyo, Kyoto, Japan*

FRANZ MÜLLER (147, 150, 151), *Department of Biological Chemistry, University of Michigan Medical School, Ann Arbor, Michigan*

KEITH MURRAY (127), *Department of Biological Chemistry, M. S. Hershey Medical Center, Hershey, Pennsylvania*

G. E. NEAL (195), *Research Department, Marie Curie Memorial Foundation, Harestone Drive, Caterham Surrey, CR3 6YQ, England*

YASUTOMI NISHIZUKA (106, 113, 114, 116, 120, 126, 130), *Department of Biochemistry, Kobe University School of Medicine, Kobe, Japan*

P. F. NIXON (179), *Department of Biochemistry, Australian National University, Canberra, Australia*

EIJI OHTSU (113), *Research Division, Tenri Hospital, Mishima-cho, Nara, Japan*

HIROSHI OKAMOTO (108), *Department of Medical Chemistry, Kyoto University Faculty of Medicine, Yoshida, Sakyo-ku, Kyoto, Japan*

E. C. OWEN (135, 165, 166, 168), *Biochemistry Department, Hannah Dairy Research Institute, Ayr, Scotland*

G. R. PENZER (152), *Department of Biochemistry, Oxford University, Oxford, England*

GRACE LEE PICCIOLO (137), *NASA, Goddard Space Flight Center, Greenbelt, Maryland*

S. PINDER (102), *Research Laboratories, May and Baker Ltd., Essex, England*

G. W. E. PLAUT (157), *Department of Biochemistry, Rutgers Medical School, Rutgers University, New Brunswick, New Jersey*

A. R. PROSSER (101), *Division of Nutrition, Bureau of Foods, Pesticides and Product Safety, Food and Drug Administration, Washington, D.C.*

G. K. RADDA (152, 153, 154), *Department of Biochemistry, Oxford University, Oxford, England*

H. REMBOLD (178, 181, 200), *Max-Planck-Institut für Biochemie, München, Germany*

DAVID P. RICHEY (193), *Department of Bacteriology and Immunology, Harvard Medical School, Boston, Massachusetts*

DWIGHT R. ROBINSON (184), *Department of Medicine, Massachusetts General Hospital and, Harvard Medical School, Boston, Massachusetts*

BARBARA ROTH (196), *Burroughs Wellcome & Co. (U.S.A.) Inc., Research Triangle Park, North Carolina*

PETER B. ROWE (188), *Institute of Child Health, Royal Alexandra Hospital for Children, Comperdown, NSW, Australia*

J. SALACH (142), *Molecular Biology Division, Veterans Administration Hospital, San Francisco, California*

ANNETTE M. SANSONE (185, 186, 187), *Department of Obstetrics and Gynecology, State University of New York at Buffalo, Buffalo, New York*

THOMAS A. SCOTT (109), *Department of Biochemistry, University of Leeds, Leeds, United Kingdom*

A. J. SHEPPARD (101), *Division of Nutrition, Bureau of Foods, Pesticides and Product Safety, Food and Drug Administration, Washington, D.C.*

T. P. SINGER (142), *Molecular Biology Division, Veterans Administration Hospital, San Francisco, California and Department of Biochemistry and Biophysics, University of California, San Francisco, California*

MARGARETA SPARTHAN (103), *Department of Biochemistry and Biophysics, University of California, Davis, California*

E. R. STADTMAN (131, 163), *Laboratory of Biochemistry, National Heart and Lung Institute, National Institute of Health, Bethesda, Maryland*

B. G. STANLEY (195), *Research Department, Marie Curie Memorial Foundation, Harestone Drive, Caterham Surrey, CR3 6YQ, England*

E. L. R. STOKSTAD (199), *Department of Nutritional Sciences, University of California, Berkeley, California*

SHIRLEY SU (123), *Division of Neurological Sciences, University of British Columbia, Vancouver, British Columbia*

TAKASHI SUGIMURA (129), *Biochemistry Division, National Cancer Center Research Institute, Tsukiji, Chuo-ku, Tokyo, Japan*

KATSURA SUGIURA (191), *Department of Chemistry, Gakushuin University, Toshima-ku, Tokyo, Japan*

YUZURU SUZUKI (158), *Laboratory of Nutritional Chemistry, Faculty of Agriculture, Kyoto University, Sakyo, Kyoto, Japan*

P. B. SWAN (121), *Division of Nutrition, Institute of Agriculture, University of Minnesota, St. Paul, Minnesota*

SEI TACHIBANA (141, 161), *Department of Chemistry, Ritsumeikan University, Tojin Kitamachi, Nishijin Kyoto, Japan*

M. B. TAYLOR (153), *Department of Biochemistry, Oxford University, South Parks Road, Oxford, England*

GIULIO TESTOLIN (143), *Department of General Biochemistry, University of Milan, Milan, Italy*

L. TSAI (131, 163), *Laboratory of Biochemistry, National Heart and Lung Institute, National Institute of Health, Bethesda, Maryland*

TOHRU TSUKAHARA (177), *Laboratory of Nutrition and Microbiology, Niigata Women's College, Ebigase, Niigata, Japan*

KUNIHIRO UEDA (106, 107, 130), *Department of Medical Chemistry, Kyoto University Faculty of Medicine, Kyoto, Japan*

HANNA UNGAR-WARON (189), *Department of Chemical Immunology, The Weizmann Institute of Science, Rehovoth, Israel*

C. VEEGER (155, 156, 167), *Department*

of Biochemistry, Wageningen University, Wageningen, The Netherlands

M. VISCONTINI (182), *Professor of Organic Chemistry, University of Zürich, Zürich, Switzerland*

J. VISSER (1967), *Department of Biochemistry, Wageningen University, Wageningen, The Netherlands*

D. W. WEST (165, 166), *Biochemistry Department, Hannah Dairy Research Institute, Ayr, Scotland*

L. G. WHITBY (140), *Department of Clinical Chemistry, The Royal Infirmary, Edinburgh, Scotland*

D. C. WILLIAMS (195), *Research Department, Marie Curie Memorial Foundation, Caterham, Surrey CR 36 YQ, England*

BERNARD WITHOLT (Addendum), *Department of Chemistry, University of California, San Diego, La Jolla, California*

KUNIO YAGI (134, 146, 149), *Institute of Biochemistry, Faculty of Medicine, University of Nagoya, Nagoya, Japan*

MASAKO YAMADA (177), *Laboratory of Nutrition and Microbiology, Niigata Women's College, Ebigase, Niigata, Japan*

HIROHEI YAMAMURA (106, 107), *Department of Biochemistry, Kobe University School of Medicine, Kobe, Japan*

C. S. YANG (164), *Sterling Chemistry Laboratory, Department of Chemistry & Molecular Biophysics, Yale University, New Haven, Connecticut*

I. L. YERO (119), *University of Colorado School of Medicine, Denver, Colorado*

SIGMUND F. ZAKRZEWSKI (185, 186, 187), *Department of Experimental Therapeutics, Roswell Park Memorial Institute, Buffalo, New York*

Preface

The expansion of information with respect to vitamins and their analogs and, particularly, the advent of newer chemical and biological techniques for studying the coenzyme forms and the enzymes involved in their biosynthesis and breakdown are reflected in the scope of Volume XVIII. It soon became apparent in the preparation of this work that some subdivision of the material was essential. Consequently, Volume XVIII of *Methods in Enzymology* will appear in three parts. A somewhat arbitrary division of the subject matter had to be made, and we make no apologies for the logic or lack of logic in the way this division was made. Part A covers the vitamin and coenzyme forms of ascorbate, thiamine, lipoate, pantothenate, biotin, and pyridoxine. The present Part B covers nicotinate, flavins, and pteridines. Part C will cover the B_{12} group, ubiquinone, tocopherol, and vitamins A, K, and D.

For each vitamin–coenzyme group, detailed descriptions of current laboratory methods are given. Included are chemical and physical, enzymatic, and microbiological analyses; isolation and purification procedures for the coenzymes and derivatives and for the enzymes involved in their metabolism; chemical synthesis and reactions of natural forms, analogs, and radioactively labeled compounds; general metabolism including biosynthesis and degradation. In addition, information on properties and biochemical functions of the vitamins, coenzymes, and relevant enzymes are included.

We wish to thank the numerous contributors for their cooperation and patience. Though a few omissions of value to experimentalists in the area may have been made inadvertently, we believe that the subject has been adequately covered for the purposes intended. Even occasional overlaps, such as modifications of an assay procedure, were deliberately included to offer some flexibility in choice and to represent fairly the different researchers involved. We also wish to express our gratitude to Mrs. Patricia MacIntyre for her excellent secretarial assistance and to the numerous persons at Academic Press for their efficient and kind guidance.

DONALD B. McCORMICK
LEMUEL D. WRIGHT

July, 1970

xvii

METHODS IN ENZYMOLOGY

EDITED BY

Sidney P. Colowick and Nathan O. Kaplan

VANDERBILT UNIVERSITY
SCHOOL OF MEDICINE
NASHVILLE, TENNESSEE

DEPARTMENT OF CHEMISTRY
UNIVERSITY OF CALIFORNIA
AT SAN DIEGO
LA JOLLA, CALIFORNIA

METHODS IN ENZYMOLOGY

EDITORS-IN-CHIEF

Sidney P. Colowick   Nathan O. Kaplan

In Preparation:

VOLUME XVII. Metabolism of Amino Acids and Amines (Part B)
Edited by HERBERT TABOR AND CELIA WHITE TABOR

VOLUME XVIII. Vitamins and Coenzymes (Part C)
Edited by DONALD B. McCORMICK AND LEMUEL D. WRIGHT

VOLUME XIX. Proteolytic Enzymes
Edited by GERTRUDE E. PERLMANN AND LASZLO LORAND

VOLUME XX. Nucleic Acids and Protein Synthesis (Part C)
Edited by KIVIE MOLDAVE AND LAWRENCE GROSSMAN

VOLUME XXI. Nucleic Acids (Part D)
Edited by LAWRENCE GROSSMAN AND KIVIE MOLDAVE

VOLUME XXII. Enzyme Purification and Related Techniques
Edited by WILLIAM B. JAKOBY

VOLUME XXIII. Photosynthesis and Nitrogen Fixation (Part A)
Edited by ANTHONY SAN PIETRO

Section VII

Nicotinic Acid: Analogs and Coenzymes

[99] An Improved Catalytic Assay for DPN[1]

By F. M. MATSCHINSKY

In 1961 Lowry and co-workers introduced highly sensitive cycling assays for TPN and DPN. The system described for the analysis of DPN, though very useful, has a number of limitations. For one, the assay is linear only over a very narrow concentration range, and amplification usually cannot exceed 5,000-fold during one single cycling step. Therefore, a new catalytic assay for DPN was developed, which improves on these weak points of the older assay.

Assay Method

Principle. The pyridine nuclcotide to be measured is shared as co-enzyme by a pair of dehydrogenases. In a cyclic process the DPN alternately oxidizes the substrate of one enzyme and reduces that of the other. For each mole of DPN, more than 10,000 moles of each product can be produced in an hour. Since the nucleotide concentration is kept far below the Michaelis constants, the rate of product accumulation is proportional to pyridine nucleotide concentration. Accordingly, the amount of either product formed can be used as a measure of the pyridine nucleotide.[1a,2]

In the procedure to be described, glyceraldehyde-3-phosphate in the presence of arsenate is oxidized to 3-phosphoglycerate, and α-ketoglutarate is reduced and aminated to glutamate.

$$\text{Glyceraldehyde-3-phosphate} \rightarrow \text{3-phosphoglycerate} \tag{1}$$

$$\text{DPN}^+ \overset{\frown}{\underset{\smile}{\quad}} \text{DPNH}$$

$$\text{Glutamate} \longleftarrow \text{α-ketoglutarate} + \text{NH}_4^+ \tag{2}$$

After termination of the catalytic step, the glutamate formed is determined by using glutamic dehydrogenase in the reverse direction.

$$\text{Glutamate} + \text{DPN}^+ + \text{hydrazine} \rightarrow$$
$$\text{hydrazone of α-ketoglutarate} + \text{NH}_4^+ + \text{DPNH} \tag{3}$$

The DPNH produced in this reaction is measured by its fluorescence.

Since the catalytic step does not discriminate between DPN[+] and

[1] A preliminary report of this method has been given: F. M. Matschinsky, C. L. Rutherford, and L. Guerra, *Proc. 3rd Intern. Congr. Histochem. Cytochem. New York, 1968.*

[1a] O. H. Lowry, J. V. Passonneau, D. W. Schulz, and M. K. Rock, *J. Biol. Chem.* **236,** 2746 (1961).

[2] B. Jandorf, F. W. Klemperer, and A. B. Hastings, *J. Biol. Chem.* **138,** 34 (1941).

DPNH, the sum of the two can be determined directly. DPN$^+$ can be determined separately by first destroying DPNH with acid, whereas DPNH can be separately measured by first destroying DPN$^+$ with alkali.[3] Various procedures have been published for preparing acid or alkaline extracts of tissue suitable for measuring DPN$^+$ or DPNH, respectively.[1,4–7] The method of extraction most useful in connection with the cycling assay is that described by Burch et al.[8] For the analysis of the sum of DPN$^+$ and DPNH, a method was developed, which is based on treatment of tissue with cold, weak alkali at high tissue dilution.[8–10]

Reagents

REAGENT FOR CYCLING STEP: Phosphate buffer, 0.2 M, pH 7.1, 10 mM sodium arsenate, 10 mM EDTA, 2 mM mercaptoethanol, 0.3 mM ADP, 5 mM ammonium acetate, 0.2 mg/ml bovine serum albumin, 5 mM glyceraldehyde-3-phosphate, 5 mM α-ketoglutaric acid, 20 mM Na$_2$CO$_3$, 150 μg/ml glyceraldehyde phosphate dehydrogenase (yeast, D-glyceralde-hyde-3-phosphate: NADP oxidoreductase, EC 1.2.1.9), 300 μg/ml glutamic dehydrogenase [beef liver, L-glutamate: NAD oxidoreductase (deaminating), EC 1.4.1.2].

For convenience, a stock reagent that contains all components except the two substrates and the enzymes, is prepared and stored at $-80°$. The latter four components are added within 1 hour of use. Na$_2$CO$_3$ is added to the stock solution to prevent a drop in pH. This is necessary, since for reasons of stability, glyceraldehyde-3-phosphate is stored at pH 2–3 and α-ketoglutaric acid as free acid.

REAGENT FOR INDICATOR STEP: Hydrazine–HCl buffer, 0.2 M, pH 9.0, 0.3 mM DPN$^+$, 0.3 mM ADP, 0.2 mg/ml of bovine serum albumin, 30 μg/ml of glutamic dehydrogenase.

STANDARD SOLUTIONS: DPNH stock solutions are prepared at 5 mM concentrations in 0.08 M Na$_2$CO$_3$–0.02 M NaHCO$_3$ (pH 10.6) and heated 15 minutes at 60° to destroy any DPN$^+$ present. These are standardized by diluting 50-fold in 0.1 M phosphate, pH 7.0, containing 1 mM pyruvate. After measuring the optical density at 340 mμ, a small volume of a solution containing lactic dehydrogenase is added to give a concentration of 10 μg

[3] N. O. Kaplan, S. P. Colowick, and C. C. Barnes, J. Biol. Chem. 191, 461 (1957).

[4] O. H. Lowry, N. R. Roberts, and J. I. Kapphahn, J. Biol. Chem. 224, 1042 (1957).

[5] M. Klingenberg, Colloq. Ges. Physiol. Chem. Mosbach, Baden, 1960, p. 82. Springer-Verlag, Berlin, 1961.

[6] J. H. Bassham, L. M. Birt, R. Hems, and U. E. Loening, Biochem. J. 73, 491 (1959).

[7] J. B. Clark, A. L. Greenbaum, and P. McLean, Biochem. J. 98, 546 (1966).

[8] H. B. Burch, M. E. Bradley, and O. H. Lowry, J. Biol. Chem. 242, 4546 (1967).

[9] F. M. Matschinsky, Federation Proc. 25, 753 (1966).

[10] F. M. Matschinsky, J. Neurochem. 15, 643 (1968).

per ml. A second reading is made after the DPNH has been oxidized (5 minutes or less), and the concentration is calculated using 6.22 × 10³ cm²/ millimole as the adsorption coefficient.[11] The stock standard is stored at 4° or below −50° but not at an intermediate temperature.[12] Dilute standards are prepared daily in 0.02 N NaOH containing 0.5 mM cysteine-HCl.

DPN⁺ stock solutions are prepared in redistilled water at 5 mM or greater concentrations. For stability the pH should be 4 to 6. These solutions are standardized by diluting to about 0.1 mM concentration in imidazole-HCl buffer, pH 7.0, containing 1 mM glyceraldehyde-3-phosphate, 10 mM arsenate, 2 mM mercaptoethanol, and 1 mM EDTA. After measuring the optical density at 340 mμ, a small volume of a solution containing glyceraldehyde-3-phosphate dehydrogenase from rabbit skeletal muscle is added to give a concentration of 10 μg per ml. A second reading is made after complete reduction of the DPN⁺ and the calculation is made as above. Stock solutions of DPN⁺ are stable for months if frozen at −20° or below. Solutions more dilute than 1 mM are prepared daily.

OTHER REAGENTS: Ascorbic acid, 1.2 M, and 100 mM cysteine are prepared in deionized water and are stable for several months in the frozen state. Glutamate, which is needed to test the indicator step and to check the cycling rate, is made 5 mM in deionized water. Monosodium glutamate from Accent International, Skokie, Illinois was preferred. It is standardized spectrophotometrically at 340 mμ in 1 ml of 0.2 M hydrazine-HCl, pH 9.0, which contained 1 mM DPN⁺, 0.3 mM ADP, 0.02% bovine serum albumin, and 10 μl of 5 mM glutamate. After an initial reading, 30 μg of glutamic dehydrogenase in 50% glycerol is added. The increase of absorption due to formation of DPNH is a measure of glutamate. The assay is completed within 30 to 40 minutes.

SOURCE OF REAGENTS AND SPECIAL PREPARATIONS: All enzymes employed are commercially available from Boehringer. DPN-free crystalline glyceraldehyde-3-phosphate dehydrogenase from yeast is used rather than the muscle enzyme which cannot be crystallized after DPN⁺ has been removed. To remove ammonium sulfate, 100 μl of enzyme, containing 12 mg protein/ml is centrifuged for 15 minutes at 5000 g, the ammonium sulfate solution is pipetted off, and the precipitate dissolved in 90 μl of 0.1 M phosphate buffer (pH 7.4) containing 2 mM mercaptoethanol and 2 mM EDTA. The dissolved enzyme does not lose activity when stored for one week at 4°. Glutamic dehydrogenase from liver dissolved in glycerol is used in both the cycling and the indicator reagent. No DPN was found in various glutamic dehydrogenase preparations used. Malic dehydrogenase from skeletal mus-

---

[11] H. U. Bergmeyer, *in* "Methoden der enzymatischen Analyse" (H. U. Bergmeyer, ed.), p. 5. Verlag Chemie, Weinheim, 1962.
[12] O. H. Lowry, J. V. Passonneau, and M. K. Rock, *J. Biol. Chem.* **236,** 2756 (1961).

cle is employed. The commercial preparation is free of DPN. Glyceralde-hyde-3-phosphate is prepared as follows: 200 mg of DL-glyceraldehyde-3-phosphate (monobarium salt of the diethyl acetal, B-grade, Calbiochem) is suspended in 1 ml H_2O, then 650 μl of 1 N H_2SO_4 is added. After 90 minutes at 60°, the barium sulfate is removed by centrifugation, the pH of the supernatant solution adjusted to pH 2–3 with NaOH, and the D-glyceralde-hyde-3-phosphate that has been liberated is assayed spectrophotometrically. The assay medium for D-glyceraldehyde-3-phosphate has the following composition: 0.1 M imidazole-HCl, pH 7.3; 10 mM arsenate; 10 mM EDTA; 2 mM mercaptoethanol; 1 mM DPN$^+$; and 30–100 μM D-glycer-aldehyde-3-phosphate. After reading the absorbance at 340 mμ, 10 μg of rabbit muscle glyceraldehyde-3-phosphate dehydrogenase is added and the increase of absorption recorded. The assay is over within 20 minutes. The resulting solution contains ca. 100 mM D-glyceraldehyde-3-phosphate, which is stable for months when kept at $-20°$. α-Ketoglutarate (A-grade) is obtained from Calbiochem. Usually commercial ADP preparations are contaminated with DPN$^+$. In order to get rid of the contaminant, a stock solution of 100 mM ADP (Sigma Chemical Co.) is brought to pH 13 with 1 N NaOH. The alkaline solution is heated for 20 minutes at 60° and the pH adjusted to neutrality with 1 N HCl. DPN$^+$ is destroyed during the heat treatment with alkali.[4]

Preparation of Tissue Samples. The procedure given follows the method published by Burch *et al.*[8] A 50-mg sample of quick-frozen tissue (at $-150°$ in Freon-12, CCl_2F_2)[1] is homogenized rapidly in 5 ml of 0.04 N NaOH con-taining 0.5 mM cysteine (NaOH-cysteine) at 0°. For total DPN (DPN$^+$ and DPNH together), 200 μl are diluted at 0° with 4 ml of NaOH-cysteine. Care is taken to keep samples at 0°, and the cycling step is started within 30 min-utes. For DPNH a portion of this alkaline homogenate is heated 10 minutes at 60°. For DPN$^+$ 5 μl of 1.2 M ascorbic acid are added at 0° to 200 μl of the original homogenate which results in a concentration of 30 mM. After mixing, the sample is acidified with 2 ml of 0.02 N H_2SO_4–0.1 M Na_2SO_4 at 0° and heated 30 min at 60°. Samples are acidified within 30 minutes after homogenizing and are kept at 0° continually until heated. The addi-tion of ascorbic acid prevents DPNH oxidation to DPN$^+$ by hemoglobin and other oxidants in acid solution. The tissue must be diluted further in the homogenizing solutions before introduction into the cycling reagent. The total dilution in 50 μl of cycling reagent should be chosen in order to provide approximately 10^{-8} M DPN$^+$ or DPNH. Tissue dilutions and selection of pyridine nucleotide standards have to be adjusted according to anticipated levels of DPN$^+$ or DPNH. Regarding the analysis of tissue, the method described here has been applied to kidney only. The dilution of normal kidney at the cycling step was 50,000-fold for DPN$^+$ and 10,000-

fold for DPNH. Similarly, if the nucleotide measured is the product of a preceding reaction, appropriate dilutions have to be made.

Stability of the Sample. The frozen tissues are stable if stored at $-85°$. Once the tissues have been homogenized in the H_2SO_4–Na_2SO_4 or NaOH–cysteine it is recommended that the analysis be carried out on the same day.

Procedure

Volumes of 50 μl of the complete cycling reagent kept at 0° are pipetted into 3-ml fluorometer tubes in a rack in ice. DPN$^+$ or DPNH in a volume not larger than 2.5 μl is added to give final concentrations of the pyridine nucleotide in the range of 3×10^{-9} to 3×10^{-8} M. The rate of cycling at 0° is 10% of that at 38°. Accordingly, the time between addition of the first and the last sample to the cycling mixture should not exceed half the subsequent incubation period if the error from this source is to be kept below 5%. The rack of tubes is incubated for 1 hour in a 38° water bath, then cooled for 3 minutes in ice water. As soon as cooled, 5 μl of 3% H_2O_2 are added to each tube, and the rack of tubes is placed in a 100° water bath for 10 minutes. After the tubes have cooled to 25°, 1 ml of the reagent for the determination of glutamate is added. The tubes are incubated at 38° for 30 minutes, cooled to room temperature, and the native fluorescence of the DPNH formed is measured in the fluorometer. Farrand Models A or A-3 were used. When a tungsten-filament lamp was used, the primary filter was Corning No. 5840, and the secondary filter a combination of Corning Nos. 4303 and 3387. When a mercury arc light was employed, the primary filter was Corning No. 5860, and the secondary a combination of Corning Nos. 3387, 4308, and 5562.

Calculation. Levels of the nucleotides are calculated from enzymatically standardized DPN$^+$ or DPNH treated by precisely the same procedures as the samples. These standards of pyridine nucleotides are carried through the procedures simultaneously with the samples.

Comments on the Procedure and Useful Variations. The enzyme levels given provide a cycling rate of 6000 cycles/hour. These amounts of enzymes give activities of 300 millimoles/liter/hour, i.e., 5 I.U./ml (glyceraldehyde-3-phosphate dehydrogenase) and 150 millimoles/liter/hour, i.e., 2.5 I.U./ml (glutamic dehydrogenase). The activity of either enzyme can be assayed spectrophotometrically at 340 mμ in the complete cycling mixture, except that the second enzyme is absent and that DPN$^+$ (1 mM) or DPNH (0.1 mM), respectively, are also induced in the assay mixture at saturation levels. For this purpose the commercial enzyme preparations are diluted in 1 ml of this modified cycling mixture, to give rates of approximately 5 micromoles/liter/minute (ca. 0.1 μg/ml each of glyceraldehyde-3-phosphate dehydrogenase and glutamic dehydrogenase). The turnover rates of

DPN$^+$ and DPNH were determined directly in the fluorometer in the fashion described by Lowry et al.[1] (See below). The turnover of DPN$^+$ was 10,000 × hour^{-1}, and that of DPNH 15,700 × hour^{-1}.

It may be desirable to vary the rate of cycling. This is accomplished by increasing or decreasing both enzymes proportionally. This results in a corresponding change in rate. Thus, through the use of 10–200% of the given enzyme levels, cycling rates of 600–12,000 cycles/hour, can be achieved. In addition, it is permissible to reduce the incubation period to 30 minutes or extend it to 2 hours in order to obtain a desired amplification. Periods longer than 2 hours are not practical, since glyceraldehyde-3-phosphate is destroyed at a rate of 30% per hour at 38°. It may be necessary in order to gain in sensitivity without an excessively high blank, to perform the cycling step in a smaller volume. The use of tubes of smaller size or incubation under an oil seal in Teflon wells is then mandatory.[13,14] In such instances, the volume of H$_2$O$_2$ has to be decreased proportionately, and the heating step to terminate the cycling reaction is carried out at 75°. In such instances, the volume of the indicator step also has to be reduced but not by more than a proportionate amount (e.g., to 20 μl in case of a cycling volume of 1 μl). Of the two substrates of the cycling system, glyceraldehyde-3-phosphate may be decreased to 1 mM, provided the amount of glutamate formed is not more than 0.1 mM. The α-ketoglutarate concentration, because of the high Michaelis constant, should not be reduced. ADP is included to protect glutamic dehydrogenase. NH$_4^+$ must not exceed 5 mM, since higher levels would cause nonlinearity of the indicator assay. EDTA at the high level of 10 mM is included for chelation of any Ba^{2+}, which in variable amounts might be introduced into the cycling reagent with the glyceraldehyde-3-phosphate stock solution. Therefore, without an excess of EDTA, precipitates may form with traces of sulfate present in the glyceraldehyde phosphate dehydrogenase solution.

There may be special applications of this cycling assay for DPN, in which the use of glutamate as an indicator may lead to erroneous results. This is the case if solutions are analyzed that contain excessively high levels of glutamate. For example, brain contains ca. 10 mM glutamate. Accordingly, with this cycling system levels in the brain of 10^{-5} M of any compound would be over estimated by about 15%. The error can be minimized by increasing the cycling rate. It is also possible in such instances to replace the second reaction, Eq. (2), of the cycling assay, by another system. It was found useful to replace glutamic dehydrogenase by 150 μg/ml of crystalline malic dehydrogenase and to substitute 10 mM oxalacetate for

[13] O. H. Lowry, Harvey Lectures Ser. **58**, 1 (1963).
[14] F. M. Matschinsky, J. V. Passonneau, and O. H. Lowry, J. Histochem. Cytochem. **16**, 29 (1968).

α-ketoglutarate. As indicator of the amplification, malate is then measured similarly as described for glutamate, again exchanging glutamic dehydrogenase by 30 μg/ml of malic dehydrogenase. With the modified conditions given, the cycling rate was 4000 cycles/hour.

Sources of Blank. The overall blank, i.e., the fluorescence resulting without addition of DPN, is an indication of the useful sensitivity of the method. Under the conditions given, the overall blank need not exceed the equivalent of $5 \times 10^{-9} M$ DPN expressed as concentration in the cycling step. There are several sources of this blank. Of major importance is the fluorescence of the reagents of the indicator step and the cycling step. The use of H_2O_2, in addition to oxidizing α-ketoglutarate, prevents the formation of a fluorescent substance that otherwise develops when the cycling reagent is boiled in order to terminate the reaction. Traces of DPN ($10^{-9} M$ or less, expressed as concentration in the cycling step) may be carried into the reagent as contaminant of glyceraldehyde phosphate dehydrogenase and can contribute to the blank also. Fluorescence of blanks exceeding an equivalent of $5 \times 10^{-9} M$ DPN indicate contaminated solutions or tools. Accordingly, contamination with DPN of any analytical tools and material used for the cycling assay must be avoided. A collection of appropriate pipettes should be set aside for exclusive use in the cycling assay. These pipettes should also be cleaned specially by soaking in 0.5 N NaOH followed by rinsing with concentrated HNO_3, deionized water and acetone.[15] If blank readings are high *and* erratic, dirty tubes are usually the source of trouble. To minimize this danger, all tubes used for this method were cleaned by heating them first for 15 minutes at 100° in half concentrated HNO_3 then in distilled water, followed by thorough rinses with redistilled and deionized water.

Trouble Shooting. Difficulties may arise at various steps in the assay. The cycling reagent itself is most likely to be at fault and can be checked by means of a direct fluorometric assay.[1] At low substrate levels, the Michaelis-Menten equation reduces to a first-order equation which can be represented by $k = V_{max}/K_m$. Such an equation is applicable to DPN+ and DPNH in the present instance. At the steady state of the cycling process, the rate of reduction of DPN+ must equal the rate of oxidation of DPNH, i.e., $k_a(\text{DPN}^+) = k_b(\text{DPNH})$, where k_a and k_b represent the apparent first-order rate constants of glyceraldehyde-3-phosphate dehydrogenase and glutamic dehydrogenase, respectively. Since $k_a/k_b = \text{DPNH/DPN}^+$, the ratio of the two rate constants can be determined directly in the fluorometer by the steady-state DPNH level (Fig. 1). For this purpose, to 1 ml of the

[15] O. H. Lowry, N. R. Roberts, K. Y. Leiner, M. I. Wu, and A. L. Farr, *J. Biol. Chem.* **207**, 1 (1954).

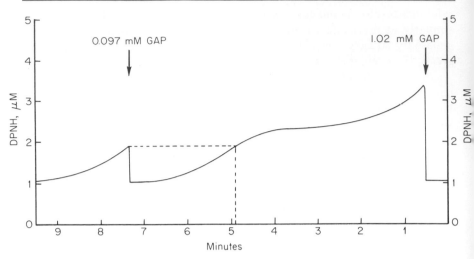

FIG. 1. Determination of cycling rate and steady state level of DPNH.[15a] Fluorescence was measured at 25° directly in the fluorometer in 1 ml of cycling reagent containing all components (see Reagents) except glyceraldehyde-3-phosphate (GAP) and in addition 3.6 μM DPN$^+$. In this example 98 micrograms of glyceraldehyde-3-phosphate dehydrogenase and 215 micrograms of glutamic dehydrogenase were added. Freshly neutralized GAP was added as indicated, and the reduction of DPN$^+$ was recorded.

cycling reagent, containing all components except glyceraldehyde-3-phosphate, is added ca. 4 μM DPN$^+$, and the blank fluorescence is recorded. After several minutes, 1 mM of freshly neutralized glyceraldehyde-3-phosphate is added, and the kinetics of the transient reduction of DPN$^+$ is observed. DPNH reaches a steady state level and then starts falling as soon as the glyceraldehyde-3-phosphate drops below the saturation concentration. When all of this substrate is used up, a second addition of a far smaller amount of glyceraldehyde-3-phosphate (0.1 mM) is made and the concentration of DPNH that is reached immediately after the addition, is observed. Accordingly, 0.9 mM (1.0–0.1 mM) glyceraldehyde-3-phosphate has been used up at one time during the first cycle, when the DPNH level equals the DPNH level obtained at the start of the second cycle. By that time, in the case described, DPN has turned over 225 times. From that, the turnover per hour can be calculated easily. In the example given, the rate found was 3450 × hour^{-1} compared to an anticipated rate of 3000 × hour^{-1}. The cycling rate obtained in this direct assay at 25° amounts to 65% of the predicted rate at 38°. If the reagent is faulty, the steady-state level of DPNH will be higher or lower than expected according to which enzyme is responsible, and the source of trouble can be pinpointed by systematic replacement of the various components. DPN contamination can be quan-

[15a] This illustrating experiment was carried out by Dr. F. Medzihradsky.

titated by comparison between blanks which have been incubated with those which have not been incubated at the cycling step.

Sensitivity and Precision of the Method

Using the conditions described here the method is capable of measuring 2×10^{-13} moles $\pm 2\%$.

The levels of pyridine nucleotides found in kidney were in terms of micromoles per kg wet weight: 725 DPN+ and 46 DPNH.[16]

The method has also been applied to the analysis of citrate in microscopic samples of exocrine and endocrine pancreas.[17] In this case the cycling step was carried out after completion of the specific enzymatic assay for citrate, which lead to the formation of stoichiometric amounts of DPN+. As little as 10^{-13} moles of citrate could be determined with a standard error of less than 5%.

The method also proved useful for the enzymatic analysis of prostaglandins. In combination with a specific DPN+ dependent prostaglandin dehydrogenase as little as 10^{-12} moles of the compound could be determined.[18]

[16] L. Guerra and F. M. Matschinsky, unpublished studies.
[17] F. M. Matschinsky, C. L. Rutherford, and J. Ellerman, *Biochem. Biophys. Res. Commun.* **33,** 855 (1968).
[18] E. Änggård, F. M. Matschinsky, and B. Samuelsson, *Science* **158,** 479 (1969).

[100] Fluorometric Measurement of Oxidized and Reduced Triphosphopyridine Nucleotides in Tissues Containing Hemoglobin[1]

By HELEN B. BURCH

Assay Method

Principle. The measurement of pyridine nucleotides in tissues has been difficult because of problems in preparing material for analysis without losses and without uncertainty as to whether oxidation or reduction may have occurred. The treatment of tissues is based on the stability characteristics of the nucleotides and means of preventing changes in redox state. NADPH is very stable in alkali, and NADP+ is stable in weak alkali for a short time at 0°.[1,2] The same alkaline homogenate can be used for measurement of NADP+ and NADPH. NADPH is measured in the alkaline

[1] H. B. Burch, M. E. Bradley, and O. H. Lowry, *J. Biol. Chem.* **242,** 4546 (1967).
[2] F. M. Matschinsky, *J. Neurochem.* **15,** 643 (1968).

homogenate after heating to destroy $NADP^+$ and interfering enzymes. $NADP^+$ is measured after destroying NADPH in acid. To prevent the oxidation of NADPH by hemoglobin during acidification, ascorbic acid is added to the homogenate beforehand.[1] Similarly, oxidation of NADPH during heating in alkali is prevented by cysteine added to the alkali used for making the homogenate.[3] Both $NADP^+$ and NADPH can be measured together in a dilute alkaline homogenate.

Measurements of the nucleotides are made by enzymatic cycling[4,5] which provides sufficient sensitivity and specificity to make deproteinization unnecessary. In the cycling process NADP is alternately oxidized and reduced in rapid succession by glutamic dehydrogenase (GDH) and glucose-6-phosphate dehydrogenase (G-6-PDH), respectively.

The system used in cycling is represented by the reactions (1) and (2).

$$NADP^+ + glucose\text{-}6\text{-}P \xrightarrow{\text{G-6-PDH}} NADPH + 6\text{-}P\text{-}gluconate + H^+ \qquad (1)$$

$$NADP^+ + glutamate \xleftarrow{\quad\text{GDH}\quad} NADPH + \alpha\text{-}ketoglutarate + NH_4^+ \qquad (2)$$

Finally 6-P-gluconate is measured by reaction (3), catalyzed by 6-phosphogluconate dehydrogenase (6-P-GDH).

$$6\text{-}P\text{-}Gluconate + NADP^+ \xrightarrow{\text{6-P-GDH}} NADPH + ribulose\text{-}5\text{-}P + CO_2 + H^+ \qquad (3)$$

Equipment for Fluorometry. The method to be described is for the Farrand fluorometer model A which is a filter instrument accommodating 1-ml samples in Pyrex test tubes (10 × 75 mm). Precautions and changes that can be made to provide increased stability and reproducibility of reading and minimal optical blanks have been described.[6] With a tungsten light source, the primary filter is Corning No. 5840; the secondary filters are Nos. 3387 and 4303 with the latter placed nearest the phototube. With a mercury arc light, the primary filter is No. 5860, and the secondary filters are Nos. 3387, 4308, and 5562.

Pyrex test tubes which have been selected for uniform size (minimal clearance in the cuvette holder) and for freedom from scratches are the cuvettes. These have the advantage of allowing the fluorescence of series of samples to be measured without intermittent cleaning of cuvettes.

Calibrated constriction pipettes are recommended for use in the cycling procedure for better precision.

[3] O. H. Lowry, J. V. Passonneau, and M. K. Rock, *J. Biol. Chem.* **236,** 2756 (1961).

[4] O. H. Lowry, J. V. Passonneau, D. W. Schulz, and M. K. Rock, *J. Biol. Chem.* **236,** 2746 (1961).

[5] O. H. Lowry and J. V. Passonneau, Vol. 6 [111].

[6] O. H. Lowry, J. V. Passonneau, F. X. Hasselberger, and D. W. Schulz, *J. Biol. Chem.* **239,** 18 (1964).

Reagents. All reagents and auxiliary enzymes must be as pure and as free from the coenzymes and from fluorescence as possible. Water is distilled and passed through a Culligan deionizer. Glassware is cleaned in nitric acid as previously described[5] and protected from contamination during storage and use.

STOCK CYCLING REAGENT

Tris-HCl buffer, 0.1 M, pH 8.0
α-Ketoglutarate, 5 mM
Glucose-6-P, 1 mM
Adenosine-5'-diphosphate, 0.1 mM
Ammonium acetate, 0.025 M
Bovine plasma albumin, 0.2 mg per ml

This reagent can be stored at $-50°$ or below. It is convenient to freeze it in aliquots large enough for one day's experiments.

Before use, glutamic dehydrogenase and glucose-6-P dehydrogenase are added at 0° to give respective concentrations of 100 and 50 μg protein per milliliter. (The latter refers to G-6-PDH preparations currently available commercially at 30% purity.) Since sulfate is inhibitory, the glucose-6-P dehydrogenase is centrifuged, as much of the supernatant $(NH_4)_2SO_4$ solution as possible is removed, and the enzyme dissolved in 2 M ammonium acetate and stored at 4°. For the same reason, it is convenient to use glutamic dehydrogenase from beef liver prepared in 50% glycerol (Boehringer & Sons, Germany).

STANDARD SOLUTIONS. NADPH stock solutions are prepared at 5 mM concentrations in 0.08 M Na_2CO_3–0.02 M $NaHCO_3$ (pH 10.6) and heated 10 minutes at 60° to destroy any $NADP^+$ present. These are standardized by diluting 50-fold in 0.1 M Tris-HCl, pH 8.0 containing 1 mM α-ketoglutarate and 0.03 mM ammonium acetate. After measuring the optical density at 340 mμ, a small volume of a solution containing glutamic dehydrogenase is added to give a concentration of 20 μg per milliliter. A second reading is made after the NADPH has been oxidized (5 minutes or less). The stock standard is stored at 4° or below $-50°$ but not at an intermediate temperature.[7] Dilute standards are prepared daily in 0.02 N NaOH containing 0.5 mM cysteine-HCl (added at 4° just before dilution).

$NADP^+$ stock solutions are prepared in redistilled water at 5 mM or greater concentrations. (For stability the pH should be 4–6.) These are standardized by diluting to about 0.1 mM concentration in Tris-HCl buffer, pH 8, containing 1 mM glucose-6-P. After measuring the optical density at 340 mμ, a small volume of a solution containing glucose-6-P

---

[7] H. W. Heldt, M. Klingenberg, and K. Pappenburg, *Biochem. Z.* **342**, 508 (1965).

dehydrogenase is added to give a concentration of 1 μg per milliliter. A second reading is made after complete reduction of the NADP$^+$ (less than 1 minute). Stock solutions of NADP$^+$ are stable indefinitely if frozen at $-20°$ or below. Solutions more dilute than 1 mM are prepared daily.

OTHER REAGENTS. Ascorbic acid, 1.2 M, and 100 mM cysteine-HCl prepared in deionized water are stable indefinitely in the frozen state.

The 6-P-gluconate dehydrogenase reagent is 0.02 M Tris-IICl buffer, pH 8, containing 0.03 M ammonium acetate, 0.1 mM EDTA, 0.03 mM NADP$^+$, 0.02% bovine plasma albumin, and 1 μg per ml of 6-P-gluconate dehydrogenase (Boehringer & Sons, Germany). This should oxidize 6-P-gluconate with a half-time of 3–6 minutes. The 6-P-gluconate is prepared as a 0.1 M solution and diluted to 1 mM. It is standardized in a spectrophotometer at 340 mμ in 500 μl of the 6-P-gluconate reagent described above with the addition of 5 μl of 100 mM NADP$^+$ and 50 μl of 1 mM 6-P-gluconate. After an initial reading, 1 μg of 6-P-gluconate dehydrogenase is added. The increase in absorption due to the formation of NADPH is a measure of 6-P-gluconate.

Procedure for Liver

Preparation of Tissue Samples. A 50-mg sample of quick-frozen liver[1] (at $-150°$ in Freon-12, CCl_2F_2) is homogenized rapidly in 5 ml of 0.04 N NaOH containing 0.5 mM cysteine (NaOH-cysteine) at $0°$.

1. For total NADP (NADP$^+$ and NADPH together), 200 μl are diluted at $0°$ with 4 ml of NaOH-cysteine. Care is taken to keep samples at $0°$ and the cycling step is started as soon as possible (within 30 minutes).

2. For NADPH, a portion of the diluted homogenate from (1) is heated 10 minutes at $60°$.

3. For NADP$^+$, 5 μl of 1.2 M ascorbic acid are added at $0°$ to 200 μl of the original homogenate to give a concentration of 30 mM. After mixing, the sample is acidified with 2 ml of 0.02 N H_2SO_4–0.1 M Na_2SO_4 (acid sulfate) at $0°$ and heated 30 minutes at $60°$. Samples are acidified within 30 minutes after homogenizing and are kept at $0°$ continually until heated. The addition of ascorbic acid prevents NADPH oxidation to NADP$^+$ by hemoglobin in acid solution.

4. Appropriate standards would usually be 20 and 40 μl of 0.1 mM NADP$^+$ or 20 and 40 μl of 1 mM NADPH added at $0°$ to 5 ml volumes of NaOH-cysteine. The additions are made concurrently with the preparation of homogenates. Thereafter aliquots of these diluted standards, together with aliquots of NaOH-cysteine solution (blanks) are carried through exactly the same diluting and heating procedures as the samples.

Cycling Procedure. A series of fluorometer tubes containing 100 μl of cycling reagent at $0°$ are placed in a rack sitting in an ice bath. For total NADP or for NADPH analyses, a 2-μl aliquot of blank (reagent identical to that containing the nucleotide), standard, or treated sample is added to a tube (duplicates are usually analyzed); for $NADP^+$ analyses, 4-μl aliquots are added. The rack of tubes is covered and incubated in a water bath at 38° for 1 hour, which permits at least 10,000-fold amplification by the cycling process. The reaction is stopped by placing the rack in a boiling water bath for 2 minutes. After cooling to room temperature, 1 ml of 6-P-gluconate dehydrogenase reagent is added and a final fluorometric reading is made as soon as the reaction is complete (20–60 minutes). Since some preparations of this dehydrogenase may contain glutamic dehydrogenase as a contaminant, a creeping blank results. If this happens, the interval between the addition of the dehydrogenase reagent and the final reading has to be kept the same for each tube. At the final step it is customary to add a 6-P-gluconate standard which has the expected concentration of the pyridine nucleotides, $2 \times 10^{-6} M$ to $5 \times 10^{-6} M$, as a check on the gluconate dehydrogenase reaction and the cycling efficiency.

The dilution of normal liver nucleotides at the cycling step is 100,000-fold for total NADP and NADPH and 25,000-fold for $NADP^+$. This results in concentrations of $5 \times 10^{-9} M$ in the former and $2 \times 10^{-9} M$ in the latter samples. After cycling 10,000 times (and after 6-P-gluconate measurement), the final concentration of NADPH in the fluorometer tube will be between 2×10^{-6} and $5 \times 10^{-6} M$. A final concentration of $10^{-5} M$ is the upper limit of this fluorometric measurement, since above this level the fluorescence of NADPH is not proportional to concentration.

Procedure for Other Tissues.[1] The only changes necessary for kidney, heart, brain, and blood are to decrease the extent of tissue dilution and to adjust the concentrations of standards according to anticipated levels of $NADP^+$ or NADPH. For kidney and heart (containing one-fourth as much total NADP as liver), the tissues are homogenized at a 1:50 dilution and after subsequent treatment as for liver, 4-μl aliquots are used for cycling. For brain (containing one-tenth as much total NADP as liver), the homogenate is a 1:20 dilution, and 4-μl aliquots are used for cycling. Blood can be diluted immediately and initially with 20 volumes of NaOH-cysteine, and the subsequent dilutions in acid–sulfate after addition of ascorbic acid, or in NaOH-cysteine can be 1:10; 4-μl aliquots are used for cycling. Thus for total NADP or NADPH the dilutions at the cycling step for kidney and heart are 25,000-fold; for brain and blood 6000-fold. The dilutions at the cycling step for $NADP^+$ are 8000-fold for kidney and heart, 2000-fold for brain, and 1000-fold for blood.

Permissible Variations. Other tissues or organisms may present different problems during treatment of the samples or they may contain substances which interfere with or enhance the cycling process. For example, in tissues high in NADPH, the chance of its oxidation to $NADP^+$ by hemoglobin during acid treatment is very great. Other tissue constituents may also oxidize the reduced coenzyme, but acidified hemoglobin appears to be the worst offender. Oxidation of NADPH by acidified hemoglobin is almost stoichiometric and can occur in both a mild acid mixture, H_2SO_4–Na_2SO_4[8] or in HCl or $HClO_4$[1] unless the oxidation is prevented by the addition of ascorbic acid. Thus any new tissue requires preliminary tests before analyses are done by this procedure.

The allowable variations in the amounts of enzymes and substrates in the cycling process and sources of error due to blanks are described by Lowry and Passonneau.[5]

In the procedure described here, the overall blank need not exceed 2×10^{-9} M NADP calculated as the concentration at cycling.

Even more sensitive procedures capable of measuring 10^{-14} moles of pyridine trinucleotides in freeze-dried sections of various tissues are described by Matschinsky[2] using similar initial treatment of samples with alkali.

Calculation. Levels of the nucleotides are calculated from enzymatically standardized NADP$^+$ or NADPH treated by precisely the same procedures as the samples and carried through the cycle simultaneously with them.

Reliability of the Method

The recovery of nucleotides added to alkaline homogenates of liver and blood averaged 96–100%. Levels in the same tissue are reproducible within ±2%. Total NADP measured directly agreed within 4% with the sum of separate analyses of liver, kidney, and heart for NADP$^+$ and NADPH, and within 7% in the case of blood. In brain, however, the value by direct analysis was 20% lower than the sum. It is likely that this results from some destruction of NADP$^+$ by "DPNase" (also active toward NADP$^+$) during the cycling step, since this enzyme is not completely destroyed in cold alkali, and since brain contains exceptionally active DPNase.

The actual levels of NADP$^+$ and NADPH found in rat liver are 63 and 486 micromoles per kilogram, respectively. The value for NADP$^+$ agrees closely with that reported by Neuhoff and Desselberger,[9] Heldt *et al.*,[7] Bücher *et al.*,[10] and Slater *et al.*[11] The NADPH level, however, is 40–100% higher.

---

[8] H. B. Burch, O. H. Lowry, and P. von Dippe, *J. Biol. Chem.* **238**, 2838 (1963).

[9] V. Neuhoff and E. Desselberger, *Arch. Exptl. Pathol. Pharmakol.* **252**, 43 (1965).

[10] T. Bücher, K. Krejei, W. Russmann, H. Schnitger, and W. Wesemann, *in* "Symposium on Rapid Mixing and Sampling Techniques in Biochemistry" (B. Chance, R. H. Eisenhardt, Q. H. Gibson, and K. K. Lonberg-Holm, eds.), p. 255. Academic Press, New York, 1964.

[11] T. F. Slater, B. Sawyer, and U. Sträuli, *Arch. Intern. Physiol. Biochim.* **72**, 427 (1964).

[101] Gas Chromatography of Niacin

By A. J. SHEPPARD and A. R. PROSSER

Assay Method

Principle. Niacin (nicotinic acid) and its analog niacinamide (nicotinic acid amide) are found in a variety of biological materials, and niacinamide is widely used in the preparation of pharmaceutical products.

Only two studies[1,2] deal with the gas chromatographic behavior of the ethyl ester derivatives. One study is limited to the ethyl ester of nicotinic acid[1] and the other[2] is concerned with the ethyl esters of niacin, niacinamide, and nonesterified niacinamide. No reports are available which indicate any application of gas chromatography (GLC) for assaying the vitamin in biological materials or pharmaceutical products. The techniques in this chapter are limited to the GLC operating parameters and the preparation of the ethyl esters of niacin and niacinamide. Application of the GLC method for assaying the vitamin has not yet been established. The techniques are provided in the hope that they may be of help to other investigators. All solvents used are ACS reagent grade unless otherwise specified.

Preparation of Packed Columns for Gas Chromatography. A Pyrex column, 183 cm × 3 mm i.d., packed with a 2.5% neopentyl glycol succinate (NPGS) and 10% SE-30 mixed immobile phase on 100/120 mesh Gas Chrom P (silanized) is the most satisfactory all-purpose column for the determination of ethyl nicotinate, niacinamide, and N-ethylniacinamide. The inert support is silanized as described in the chapter on gas chromatography of vitamin K_3. The immobile phase is coated on the silanized inert support as follows: 0.625 g of NPGS is weighed and dissolved in 50 ml of chloroform, with heat if necessary; 2.500 g of SE-30 is weighed and dissolved in 100 ml of toluene, with heat if necessary. The NPGS:chloroform and SE-30:toluene solutions are transferred to a Morton flask containing 21.875 g of the silane-treated inert support. Enough additional chloroform:toluene (1:2, v/v) is added to form a slurry. The mixture is allowed to stand for 1 hour with occasional shaking and the solvent is evaporated *in vacuo* with warming on a flash evaporator. The prepared packing is then dried at 80° for 1 hour. The finished product is stored in a dust-free container.

[1] H. Kubiezkova, V. Rezl, and N. Kucharaczyk, *Gas Chromatography Symposium, Brno, Czech., June* **1962**.

[2] A. R. Prosser and A. J. Sheppard, *J. Pharm. Sci.* **57**, 1004 (1968).

The chromatographic column is filled with the prepared packing in the manner described in the chapter on gas chromatography of vitamin K₃.

Instrumentation. Any gas-liquid chromatograph fitted with either a hydrogen flame ionization detector or a β-argon ionization detector capable of using a 183 cm glass column is satisfactory. The authors have used gas chromatographs equipped with glass injection areas and are not able to comment on the relative merits of metal injection ports compared to all-glass systems for the analysis of the ester derivatives of niacin. The electrometer outputs should not be less than 3×10^{-7} A for the β-argon detector and 1×10^{-9} A for the hydrogen flame detector as expressed in a GLC system utilizing a 5 mV recorder. The recommended chart speed is 0.33 inch per minute.

The operating parameters for ethyl nicotinate are as follows: column, 180°; detector, 230°; flash heater, 220°; carrier gas flow, 50 ml per minute. The compound appears at approximately 6.5 minutes after injection.

Another set of operating conditions is required for niacinamide and *N*-ethylnicotinamide: column, 230°; detector, 280°; flash heater, 270°; carrier gas flow, 50 ml per minute.

The column is preconditioned for 24 hours at 235° with a carrier gas flow rate of 60 ml per minute, the column outlet is connected to the detector, and the GLC operating parameters are set for the analysis desired. Niacinamide appears at approximately 7 minutes and is preceded by *N*-ethylniacinamide at about 5 minutes.

Procedure

Preparation of Ethyl Esters. A mixture of pure niacin (100 mg) or niacinamide (100.8 mg) or either compound alone, 2 ml of absolute ethanol, and 0.5 ml of concentrated H_2SO_4 is refluxed for 3 hours. The solution is then cooled and poured slowly with stirring on 20 g of crushed ice. Enough ammonium hydroxide is added to make the resulting solution strongly alkaline. Some of the ester product separates as an oil but most of it remains dissolved in the alkaline solution. The reaction solution is extracted five times with 3-ml portions of ether. The combined ether extracts are dried over anhydrous magnesium sulfate and filtered through Whatman No. 4 paper. The ether is removed from the filtrate under reduced pressure. The residue is dissolved in *n*-hexane to a final volume such that 2 μl contains 1 μg of expected final product.

Preparation of N-Ethylnicotinamide. Fifty milligrams of ethyl nicotinate is placed in a round-bottom flask with a stirring bar, 5 ml of 70% aqueous ethyl amine is added, and the mixture is stirred. The flask should be kept loosely stoppered for 3 hours, after which time the lower layer generally dissolves on shaking. The solution is saturated with the ethyl amine solution and allowed to stand for another 30 minutes; crystals of the amide begin to appear. The solution is removed under pressure and the N-ethylnicotinamide is taken up in ethanol.

Standards and Calibration. Three primary standard solutions are prepared:

1. Niacinamide: 100 mg is weighed into a 50-ml volumetric flask and diluted to the mark with n-hexane.

2. Ethyl nicotinate: 100 mg is weighed into a 50-ml volumetric flask and diluted to the mark with ethanol.

3. N-Ethylnicotinamide: 100 mg is weighed into a 50-ml volumetric flask and diluted to the mark with ethanol.

Additional working standards are prepared by diluting aliquots of each primary standard solution with the appropriate solvent to 0.5, 1.0, and 2.0 mg per milliliter.

A calibration plot covering 1 to 8 μg appears to be suitable for most gas chromatographs equipped with ionization detectors. Since instruments vary considerably in response, the calibration range must be adjusted to cover individual situations. The concentrations of the working standards should provide a minimum of three calibration points. The response for each amount of individual compounds is plotted on standard graph paper with the peak area in cm^2 as the ordinate and the amount injected as the abscissa. A straight line is fitted to the calibration data for individual compounds. It is very important that injection volumes be constant to eliminate "needle cook-out effect." A calibration plot must be determined for each compound being measured, since it has been our experience that the slopes of the calibration lines are different for each compound.

Injection Technique. This technique is described in the chapter on gas chromatography of vitamin K_3.

Application Possibilities

If one is interested only in total activity of the vitamin as niacin, the measurement of ethyl nicotinate by GLC is the most direct approach. Both niacin and niacinamide yield the same final product, i.e., ethyl nicotinate. Since the derivative yield is on an equal molar basis, any calculations are perfectly straightforward. Niacinamide is converted to niacin and then to ethyl nicotinate in the above procedure for the preparation of ethyl nicotinate.

With mixtures of niacin and niacinamide, the niacinamide can be determined directly at a GLC column temperature of 230° with the extract in ethanol. Niacin does not give a GLC peak. Then the total is esterified to form ethyl nicotinate. The GLC column temperature is lowered to 180° and the total niacin content is determined as ethyl nicotinate. The difference between the two analyses, allowing for molecular weight differences, is the nicotinic acid content.

An easier means of determining the niacin and niacinamide content, based on the experience of the authors, is to operate the GLC column temperature at 230° and determine the niacinamide content, then convert all vitamin analogs to N-ethylnicotinamide. The difference between the two measurements, allowing for molecular weight differences, gives an estimate of the nicotinic acid content. There are three advantages to this latter means of measuring niacin and/or niacinamide in a sample: (1) a constant column temperature can be used for both analogs, (2) instrument sensitivity is superior for N-ethylnicotinamide, and (3) the peaks for N-ethylnicotinamide are more symmetrical.

[102] The Assay of Intermediates and Enzymes Involved in the Synthesis of the Nicotinamide Nucleotides in Mammalian Tissues

By S. Pinder, J. B. Clark, and A. L. Greenbaum

The measurement of the activities of the enzymes involved in the biosynthesis of the nicotinamide nucleotides in animal tissues involves procedures for the preparation and estimation of each of the intermediates of the pathways that have been proposed. Methods for the preparation of several of these intermediates, labeled with ^{14}C, have been described by K. Ueda, H. Yamamura, and Y. Nishizuka, this volume [106].

The estimation of many of the intermediates involves separation and identification techniques, and these, together with methods for the deter-

[1] E. Kodicek and F. F. Reddi, *Nature* **168,** 475 (1951).

[2] N. Levitas, J. Robinson, F. Rosen, J. W. Huff, and W. A. Perlsweig, *J. Biol. Chem.* **167,** 1697 (1947).

[3] V. Pallini and C. Ricci, *Arch. Biochem. Biophys.* **112,** 282 (1965).

[4] Pabst Information Sheet OR-10.

[5] C. S. Hanes and F. A. Isherwood, *Nature* **164,** 1107 (1949).

---

[l] Place in a tank containing a freshly prepared mixture of 20 ml of 20% aqueous chloramine T suspension, 20 ml of 1 N HCl, and 10 ml of 10% KCN solution.

[m] p-Aminobenzoic acid is prepared by dissolving 2 g of p-aminobenzoic acid in 75 ml of 0.75 N HCl and making up to 100 ml with 95% ethanol.

[n] Hydrolysis: Acid—heat with 0.1 N HCl for 10 minutes at 100°. Alkaline—heat with 0.1 N NaOH for 10 minutes at 100°.

[o] R_f relative to P_i. Solvent 2 is slightly modified by changing the pH to 7.5 (see text footnote 4).

TABLE I
PROCEDURES AND DATA FOR THE SEPARATION AND IDENTIFICATION OF
INTERMEDIATES OF NAD AND NADP SYNTHESIS

| | Paper chromatography $(R_f)^a$ | | | | Thin-layer chromatography $(R_f)^b$ | | |
| | Solvent systems^c | | | | Solvent systems^c | | |
Compound	1	2	3	4	1	2	3
1. Nicotinamide^d	0.85	0.80	0.53	0.85	0.88	0.87	0.45
2. Nicotinic acid^e	0.73	0.72	0.56	0.34	0.82	0.77	0.55
3. Nicotinic acid mononucleotide^f	0.31	0.17	0.83	—	0.47	0.13	0.79
4. Deamido NAD^g	0.35	0.10	0.51	—	0.52	0.15	0.57
5. Nicotinamide mononucleotide^h	0.48	0.18	0.84	—	0.63	0.11	0.73
6. NAD^i	0.48	0.15	0.53	—	0.61	0.13	0.58
7. NADP^j	0.34	0.02	0.64	—	0.50	0.03	0.70
8. PRPP^k	—	0.15^o	—	—	—	—	—

^{a} Paper chromatography was performed using the descending technique on Whatman No. 3 MM paper for ∼16 hours.

^{b} Thin-layer chromatography was performed using ascending technique on plates coated with MN 300 G cellulose (0.25 mm thick) for 2–3 hours, greater contrast being obtained by spraying the plates with 0.2% solution of fluorescein in ethanol.

^{c} Solvent systems: (1) 66 ml isobutyric acid, 1.7 ml 0.880 NH_3, 33 ml H_2O, pH 7.4; (2) 70 ml 95% ethanol, 30 ml 1 M NH_4Ac, pH 5.0; (3) 600 g $(NH_4)_2SO_4$ dissolved in 1 liter of 0.1 M Na phosphate, pH 6.8, and 2% propanol; (4) 660 ml n-butanol, 114 ml H_2O, and 6 ml 0.880 NH_3.

^{d} Absorbs UV light; when exposed to cyanogen bromide reagent^l for 1 hour at room temperature and then sprayed with p-aminobenzoic acid,^{m} a pink coloration is observed (sensitivity 100 ng of nicotinic acid or nicotinamide; (see text footnote 1).

^{e} When subjected to treatment as above for nicotinamide, a yellow color is obtained (see text footnote 1).

^{f} When streaked with methanolic KCN, it produces a weak fluorescence (see text footnotes 2 and 3). Hydrolysis with alkali^n gives nicotinic acid (identify as above).

^{g} Reacts with methanolic KCN to give weak fluorescence; acid hydrolysis^n gives AMP, NaMN, and nicotinic acid, while alkaline hydrolysis^n gives AMP and nicotinic acid (identify as above).

^{h} Spray with ethyl methyl ketone (1:1); nicotinamide derivatives give a light blue fluorescence at 350 mμ (sensitivity 50 ng). Alkaline hydrolysis^n gives nicotinamide (identify as above).

^{i} Spray with ethyl methyl ketone (as above) to give fluorescent derivative. Acid hydrolysis^n gives NMN and nicotinamide while alkaline hydrolysis^n gives nicotinamide (identify as above).

^{j} Fluorescent compound formed when sprayed with ethyl methyl ketone (1:1) (see text footnote 1). Alkaline hydrolysis^n gives nicotinamide while acid hydrolysis^n gives NMN and nicotinamide (identify as above).

^{k} Visualized by spraying with a mixture of 60% perchloric acid, 1 M HCl, and 4% ammonium molybdate, drying, and then developing the blue spots by leaving in the presence of H_2S for a short time (for details see text footnote 5).

mination of NAD and NADP, are given in Part A below. Part B describes conditions for the assay of the enzymes involved together with some data on the intracellular location and properties of these enzymes.

PART A. IDENTIFICATION AND MEASUREMENT OF SUBSTRATES
AND INTERMEDIATES

Table I lists the procedures and data for the separation and identification of the intermediates of NAD and NADP synthesis by the use of paper or thin-layer chromatography.

These components may also be separated by ion-exchange chromatography[6]; Table II summarizes the chemical procedures for the identification of the individual compounds.

TABLE II
SPECTROSCOPIC IDENTIFICATION OF INTERMEDIATES[a]

Compound	λ_{max} in pH 7.4 buffer (mμ)	λ_{max} of secondary peak at pH 10.0 in 1 M KCN (mμ)	Results of KCN treatment
Deamido NMN	265	315	Almost complete diminution of peak at 265 mμ occurs slowly (i.e., >7 minutes), $\epsilon = 2.51 \times 10^4$ at 315 mμ (see text footnote 6)
NMN	266	325	Very quick reaction, complete disappearance of peak at 266 mμ in <20 seconds, $\epsilon = 6.2 \times 10^3$ at 325 mμ (see text footnote 7)
Deamido NAD	260	315	Slow reaction (>7 minutes), only a slight fall in peak at 260 mμ; $\epsilon = 1 \times 10^4$ at 315 mμ (see text footnote 6)
NAD	259	327	Slight decrease in peak at 259 mμ, very fast reaction (<20 seconds); $\epsilon = 5.9 \times 10^3$ at 327 mμ (see text footnote 7)
NADP	259	327	Slight decrease in peak at 259 mμ, very fast reaction (<20 seconds); $\epsilon = 6 \times 10^3$ at 327 mμ (see text footnote 7)

[a] Procedure: To 0.3–0.4 micromoles of respective intermediate add either 3 ml of 0.1 M sodium phosphate, pH 7.4, or 3 ml of 1.0 M KCN and read in a spectrophotometer.

[6] J. Preiss and P. Handler, *J. Biol. Chem.* **233**, 488 (1958).
[7] Pabst Information Sheet, OR-18 p. 16.

Nicotinamide and Nicotinic Acid

These may be measured by either chemical or microbiological methods.

Chemical Methods

The concentration of nicotinamide and nicotinic acid may be estimated by use of the König reaction[8] in which the pyridine bases are warmed with cyanogen bromide and the products so formed are coupled with an organic base (see footnote *d* of Table I for details of procedure). If sulfanilic acid is used as the organic base, then the absorption of the nicotinic acid derivative is twice that of the nicotinamide derivative when both are measured at 430 mμ.[9] It is, therefore, possible to measure the relative amounts of nicotinic acid and nicotinamide in a mixture by comparing the absorptions at 430 mμ of the mixture and of an alkaline hydrolyzate of the mixture. The hydrolysis in NaOH quantitatively hydrolyzes nicotinamide to nicotinic acid. The extinction coefficient for nicotinic acid at 430 mμ is 6.01 $\times$ 10^5.

Microbiological Method

Lactobacillus arabinosus 17-5 requires nicotinic acid as a growth factor. The rate of growth may be estimated by titrating the lactic acid formed over a specified period. For the measurement of nicotinic acid or nicotinamide, *L. arabinosus* 17-5 is grown in a medium containing all growth requirements except nicotinic acid; this medium is then supplemented with the extract containing these cofactors (A), and with an extract that has been subjected to alkaline hydrolysis (B). The effects of standard amounts of nicotinic acid are also measured. In this way, the nicotinamide content of the extract may be evaluated (B − A) and also the nicotinic acid content (A).[10,11] Modifications of this procedure are widely used in industry.[12] The basal medium for this test is as follows:[10,13]

Acid-hydrolyzed casein	0.5%
Tryptophan	0.01%
Cystine	0.01%
Glucose	1.00%
Sodium acetate	0.60%
Adenine	10 ppm
Guanine	10 ppm
Uracil	10 ppm
Thiamine	0.1 ppm
Calcium pantothenate	0.1 ppm
Vitamin B$_6$	0.1 ppm
Riboflavin	0.2 ppm
Biotin	0.4 ppb

Estimation of Nicotinic Acid Mononucleotide, Nicotinamide Mononucleotide, and Deamido-NAD Formed from Radioactive Precursors

All three intermediates may be identified and separated from each other and from other precursors and products (nicotinic acid, nicotinamide, and NAD) by paper chromatography using the solvent systems 1 and 3 described in Table I. The radioactivity associated with each intermediate may then be estimated directly either by the use of a radiochromatogram scanner or by cutting out each of the spots and counting in a liquid scintillation counter. The retardation factors (R_f) for these compounds are given in Table I.

Estimation of NAD and NADP[13a]

EXTRACTION

Although the extraction of both the oxidized and reduced forms of these nucleotides in a single neutral extraction medium has been described,[14] it is more usual to take advantage of the relative stabilities of the oxidized and reduced forms of these compounds in acid and alkali[15] so that each may be measured separately. Several extraction procedures are widely employed including procedures 1–3 described below.

Procedure 1.[16–18] For *acid extract* for NAD+ and NADP+, place 0.5–1.0 g of tissue in 10 ml of hot 0.1 N HCl. Homogenize for 30 seconds while still hot, return to a boiling water bath for 1 minute, rehomogenize for 30 seconds, reheat for 1 minute, and cool on ice.

For *alkaline extract* for NADH and NADPH, place 0.5–1.0 g of tissue in 10 ml of hot 0.1 N NaOH and then treat as for the acid extract above. To each extract add 2 ml of 0.25 M glycylglycine buffer, pH 7.6, and neu-

---

[8] W. J. König, *J. Prakt. Chem.* **69**, 105 (1904); *ibid.* **70**, 19 (1904).

[9] J. P. Sweeney and W. L. Hall, *Anal. Chem.* **23**, 983 (1957).

[10] E. E. Snell and L. D. Wright, *J. Biol. Chem.* **139**, 675 (1941).

[11] E. C. Barton-Wright, *Biochem. J.* **38**, 314 (1944).

[12] F. A. Robinson, *in* "The Vitamin Cofactors of Enzyme Systems," Macmillan (Pergamon), New York, 1966.

[13] H. McIlwain and D. A. Stanley, *J. Gen. Microbiol.* **2**, 12 (1948).

[13a] NAD and NADP refer to the nucleotide without regard to oxidoreduction state; NAD+ and NADP+ are oxidized forms; NADH and NADPH are reduced forms.

[14] M. A. Eichel and H. J. Spirtes, *Arch. Biochem. Biophys.* **53**, 309 (1954).

[15] F. Schlenk, *in* "The Enzymes" (P. D. Boyer, H. Lardy, and K. Myrbäch, eds.), Vol. 2, part 1, p. 250. Academic Press, New York, 1951.

[16] G. E. Glock and P. McLean, *Biochem. J.* **61**, 381 (1955).

[17] J. A. Bassham, L. M. Birt, R. Hems, and U. E. Loening, *Biochem. J.* **73**, 491 (1959).

[18] A. L. Greenbaum, J. B. Clark, and P. McLean, *Biochem. J.* **95**, 161 (1965).

tralize to pH 7.6. The neutralized extracts are centrifuged at 10,000 g for 10 minutes, and the supernatants are used for estimation of the nicotinamide nucleotides as described below.

When only very small amounts of tissue are available (5–10 mg wet weight), Slater, Sawyer, and Straüli[19] recommend the addition of nicotinamide (0.4 mM), EDTA (50 μM), and tryptophan (0.5 mM) as agents to prevent the destruction of the nucleotides during extraction.

Procedure 2. A modified extraction procedure has also been described by Lowry and his collaborators.[20–22] About 0.5 g of tissue is used in 10 ml of each of the extracting media.

For *acid extract*, 0.02 N H_2SO_4–0.1 M Na_2SO_4, pH 7.3, is used as extraction medium and homogenization is carried out at 0°. The extract is then heated at 60° for 45 minutes.

For *alkaline extract*, 0.02 N NaOH–0.5 mM cysteine is used as the extraction medium and homogenization is performed at 0°. The extract is then heated at 60° for 10 minutes.

The extracts are neutralized and centrifuged to remove proteins as described above.

Procedure 3.[19] For *acid extract*, 0.5–1.0 g of tissue is put in 10 ml of 1 N perchloric acid and homogenization carried out at 0°.

For *alkaline extract*, 0.5–1.0 g of tissue is put in 10 ml of 0.1 N KOH containing 0.42 mM tryptophan, and homogenization is performed at 0°. The protein should be removed by centrifugation and the extracts neutralized to pH 7.4 as before.

MEASUREMENT OF NAD AND NADP

Each of the acid extracts above contains both the NAD^+ and $NADP^+$ of the tissue; and each of the alkaline extracts, the NADH and NADPH of the tissue. The estimation of each separately is achieved by the use of specific enzymes.

Two types of assays are used: (1) those in which a stoichiometric ratio of 1:1 exists between the nucleotide being assayed and the end product being measured; (2) those in which the nucleotide is used as a rate-limiting coenzyme in a series of repetitive reactions; i.e., the nucleotide is recycled to give a yield of product which is a large integral multiple of the actual amount of nucleotide present.[23]

The type 1 assays may be used in (a) spectrophotometric or (b) fluoro-

[19] T. F. Slater, B. Sawyer, and U. Straüli, *Arch. Intern. Physiol. Biochim.* **72,** 427 (1964).

[20] H. B. Burch, O. M. Lowry, and P. Von Dippe, *J. Biol. Chem.* **238,** 2838 (1963).

[21] H. B. Burch, M. E. Bradley, and O. H. Lowry, *J. Biol. Chem.* **242,** 4546 (1967).

[22] O. H. Lowry, J. V. Passonneau, and M. K. Rock, *J. Biol. Chem.* **236,** 2756 (1961).

[23] O. H. Lowry, J. V. Passonneau, and M. K. Rock, *J. Biol. Chem.* **236,** 2746 (1961).

metric procedures. In the *spectrophotometric*[24] procedure, the levels of the nucleotides in neutralized extracts may be determined by the use of specific enzymes to distinguish between NAD (e.g., alcohol dehydrogenase, EC 1.1.1.1) and NADP (e.g., glucose-6-phosphate dehydrogenase, EC 1.1.1.49) and measuring the change in absorbance at 340 mμ, due to the production of NADH or NADPH. The sensitivity of this procedure is approximately 3 millimicromoles/ml, i.e., 3×10^{-6} M.

Determinations by *fluorometric* procedures may be made (i) by measurement of the fluorescence of addition products of the nucleotides, e.g., with methyl ethyl ketone[25]: sensitivity approximately 10^{-7} M; (ii) by measurement of the fluorescence of the reduced nicotinamide nucleotides[26]: sensitivity, approximately 10^{-7} M; or (iii) by measurement of the intense fluorescence resulting from the treatment of the nicotinamide nucleotide with alkaline solutions[27]: sensitivity approximately 10^{-7} M.

Much greater sensitivity can be achieved by using *recycling* techniques; three recycling methods are described.

Spectrophotometric Recycling Method[14, 28]

The method involves the use of NADH, generated by the oxidation of ethanol (by alcohol dehydrogenase), or of NADPH, generated by the oxidation of glucose 6-phosphate (G-6-P) (by glucose-6-phosphate dehydrogenase) for the reduction of dichlorophenolindophenol (DCPIP) to its colorless leuco base, phenazine methosulfate (PMS) being used as intermediary carrier. The rate of reduction of DCPIP, measured at 600 mμ, is proportional to the concentration of the NAD$^+$(H) or NADP$^+$(H) present when all other components of the system are in excess.

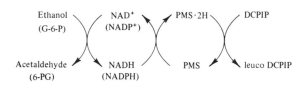

Because the overall rate of the reaction is also dependent on the turnover number of the enzyme, it is necessary to run a set of standard NAD or NADP solutions with each series of determinations. The sensitivity of the method is of the order of 10^{-7} M.

[24] E. Racker, Vol. II, p. 72.
[25] K. J. Carpenter and E. Kodicek, *Biochem. J.* **46**, 421 (1950).
[26] O. H. Lowry, N. R. Roberts, and J. I. Kapphahn, *J. Biol. Chem.* **224**, 1047 (1957).
[27] K. B. Jacobson and L. Astrachan, *Arch. Biochem. Biophys.* **71**, 69 (1957).
[28] T. F. Slater and B. Sawyer, *Nature* **193**, 454 (1962).

Reagents

NAD$^+$(H) assay:

Sodium phosphate buffer, 0.1 M, pH 7.4

Phenazine methosulfate, 1 mg/ml, freshly prepared and protected against light

Dichlorophenolindophenol, 17.5 mg/100 ml, as a stock solution. This keeps well and may be stored for weeks. It is diluted 10-fold immediately before use.

Ethanol

Alcohol dehydrogenase (EC 1.1.1.1)

NAD$^+$ solution, 1 mg/ml, in 0.02 N HCl. This stock standard solution is diluted 100-fold immediately before use. Standards should contain 1–5 μg.

NADP$^+$(H) assay:

Tris buffer, 0.1 M, pH 8.0

Glucose 6-phosphate, 0.1 M, pH 8.0

EDTA, 0.02 M, pH 8.0

Phenazine methosulfate as under NAD$^+$(H) assay

Dichlorophenolindophenol as under NAD$^+$(H) assay

Glucose-6-phosphate dehydrogenase (EC 1.1.1.49)

NADP$^+$, 1 mg/ml, in 0.02 N HCl. This stock standard solution is diluted 100-fold immediately before use. Standards should contain 1–5 μg.

Procedure. The following solutions are added to a cuvette for the NAD$^+$(H) assay: 2.5 ml of DCPIP, 0.5 ml of sodium phosphate buffer, 0.05 ml of phenazine methosulfate, 0.1 ml of ethanol, and 0.1 ml of extract or standard containing NAD$^+$(H). Some reduction of the DCPIP occurs at this stage, and the reaction is allowed to reach completion before the enzyme is added: e.g., 0.05 ml of alcohol dehydrogenase (containing 0.3 mg of protein) to the reaction cuvette and 0.05 ml of water to the blank. The rate of change of optical density at 600 mμ is obtained. This rate is then related to the rates obtained using standard amounts of NAD to obtain the concentration of NAD in the unknown.

The procedure for NADP is essentially similar except that the cuvette contains the following components for NADP$^+$(H) assay: 2.5 ml of DCPIP, 0.5 ml of Tris buffer, 0.05 ml of phenazine methosulfate, 0.1 ml of glucose 6-phosphate, 0.05 ml of EDTA, and 0.1 ml of extract or standard containing NADP$^+$(H). Again the reduction of the DCPIP is allowed to go to completion before the addition of 0.05 ml (containing 25 μg of protein) of glucose-6-phosphate dehydrogenase. As before, the rate of change in optical density at 600 mμ can be related to the rate given by standard solutions of NADP.

Fluorometric Recycling Method[23]

This procedure involves two steps. First, an amplification step by which a small amount of nucleotide is converted to a large amount of product by the use of recycling procedures; and, second, a reaction in which the product is estimated fluorometrically.

NAD Assay

The amplification phase is achieved by the following repetitive cycle of reactions which produce a large amount of pyruvate as product; the amount of pyruvate being produced is proportional to the concentration of $NAD^+(H)$ present if both enzymes and the substrates oxoglutarate and

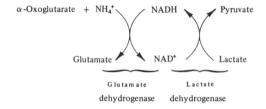

NH_4^+ are in excess. This cycle is allowed to repeat for 5000–10,000 times (30 minutes) before the reaction is stopped; the accumulated pyruvate is estimated. Standard solutions of NAD must be run simultaneously under identical conditions.

In the second phase, the pyruvate formed is estimated by measuring the increase in fluorescence at 435 mμ as a result of the reaction of pyruvate and NADH in the presence of lactate dehydrogenase. The amount of NAD present in the original extract may be calculated by comparison of the unknown extract with the standards.

Reagents. The recycling mixture contains the following components, all dissolved in 0.2 M Tris buffer, pH 8.4:

> α-Oxoglutarate, 5 mM
> Ammonium sulfate, 0.1 M
> ADP, 0.3 mM
> Sodium lactate, 100 mM
> Enzymes: 0.4 mg/ml crystalline beef liver glutamate dehydrogenase; and 0.05 mg/ml crystalline beef heart lactate dehydrogenase. Both are added immediately before use.

The pyruvate assay mixture contains:

> NaH_2PO_4, 0.65 M
> K_2HPO_4, 0.15 M

Enzyme: 1.5 μg/ml crystalline beef heart lactic dehydrogenase
NADH, 1.0 mg/ml

Other reagents required:

HCl, 5 N
NaOH, 9 N

Procedure. Between 1 and 20 μl of the extract containing NAD is added
to 100 μl of the recycling mixture and the whole is incubated at 25° for
30 minutes. The reaction is stopped by plunging the tubes into boiling
water, after which they are immediately cooled on ice.

To this mixture is now added 100 μl of the pyruvate assay mixture
plus an amount of NADH calculated as below; the mixture is incubated
at 25° for 15 minutes. The tubes are then transferred to ice, and 25 μl of
5 N HCl is added immediately. Lowry et al.[22] recommend that the amount
of NADH added at this stage be carefully regulated and should lie between
3 and 10 times the amount required by the expected pyruvate concentration. For the fluorometric measurement, 100 μl of this mixture is added to
200 μl of 9 N NaOH and heated for 10 minutes at 60°. After cooling, 1 ml
of water is added, and the fluorescence at 435 mμ is read. It is essential
that standard samples of NAD$^+$ be put through the entire procedure
simultaneously.

This is an extremely sensitive method and can measure as low as 10^{-15}
moles.

NADP Assay (see article [100])

Polarographic Recycling Method[18]

Principle. This method involves the linking of the oxidation of NAD
(by alcohol dehydrogenase) or NADP (by glucose-6-phosphate dehydrogenase) to the reduction of phenazine methosulfate (PMS). The phenazine methosulfate so reduced is reoxidized by molecular oxygen, and the
oxygen uptake is monitored on an oxygen electrode. The rate of oxygen
uptake is then proportional to the concentration of the NAD or NADP
present.

Ethanol / NAD$^+$ / PMS·2H / O$_2$
or / or
Glucose 6 phosphate / NADP$^+$

Acetaldehyde / NADH / PMS / H$_2$O$_2$
or / or
6-Phosphogluconate / NADPH

Alcohol dehydrogenase
or
Glucose-6-phosphate dehydrogenase

Reagents

NAD+(H) assay:
  Glycylglycine buffer, 0.25 M, pH 9.0
  Ethanol, absolute, RR grade
  EDTA, 0.001 M
  Phenazine methosulfate, 2 mg/ml
  Alcohol dehydrogenase
NADP+(H) assay:
  Tris buffer, 0.1 M, pH 8.0
  Glucose 6-phosphate, 0.1 M
  EDTA, 0.001 M
  Phenazine methosulfate, 2 mg/ml
  Glucose-6-phosphate dehydrogenase

Procedure. To the chamber of an oxygen electrode the following components are added for the NAD+(H) assay: 1 ml of glycylglycine buffer, 0.1 ml of ethanol, 0.1 ml of EDTA, 0.7 ml of extract and/or water, 0.1 ml of PMS.

The electrode chamber is then sealed to exclude the atmosphere and is left until a steady baseline has been achieved (1–2 minutes). Alcohol dehydrogenase solution (0.05 ml), containing 0.3 mg of enzyme protein, is then added by syringe and the oxygen uptake is recorded. The whole is then repeated with suitable standards of NAD, i.e., 1–5 μg of NAD.

For the NADP+(H) assay, the following components are added to the chamber of an oxygen electrode: 1 ml of Tris buffer, 0.1 ml of glucose 6-phosphate, 0.1 ml of EDTA, 0.7 ml of extract and/or water, 0.1 ml of PMS. The electrode is closed as for NAD and, after a steady baseline is achieved, the reaction is started by the addition of 0.05 ml of glucose-6-phosphate dehydrogenase solution, containing 5 μg of enzyme protein. The rate of oxygen uptake attained with the extracts is compared with that of suitable standards (1–5 μg) of NADP in order to estimate the amount of NADP in the extracts.

This method can measure approximately 10^{-10} mole of either NAD+(H) or NADP+(H) and has the advantage that it can be used with turbid or strongly colored suspensions and thus avoids the necessity for a high speed centrifugation.

5-Phosphoribosyl 1-Pyrophosphate (PRPP)

Two methods have been proposed for the measurement of this compound.

Method 1

The PRPP in an extract is used for the phosphoribosylation of orotic acid-^{14}C which is then decarboxylated by orotidine-5'-phosphate decarboxylase (EC 4.1.1.23) to give uridine 5'-phosphate and $^{14}CO_2$ (see footnote 29).

$$\text{Orotate-}^{14}\text{C} + \text{PRPP} \rightleftharpoons \text{orotidine 5'-phosphate} + \text{PP}_i \qquad (1)$$

$$\text{Orotidine 5'-phosphate} \rightarrow \text{uridine 5'-phosphate} + {}^{14}\text{CO}_2 \qquad (2)$$

Or, if large amounts of PRPP are present in the extract, the estimation may be done spectrophotometrically by measuring the decrease of optical density at 295 mμ as a result of the formation of the orotidine 5'-phosphate or uridine 5-phosphate (molar extinction coefficient of the latter conversion, i.e., orotate $\rightarrow$ U5P, is 3950).

Preparation of the PRPP Extract

The tissue is homogenized in 0.1 M sodium phosphate buffer, pH 7.4, to give a 1:5 tissue suspension, which is then heated at 100° for 30 seconds. The extract is cooled on ice and centrifuged to remove the protein; it may then be used directly without further treatment. Such extracts may be stored at $-20°$ for 4 weeks with no appreciable loss of PRPP content.[30]

Estimation of PRPP

Reagents

Orotate-4,7-^{14}C, 0.01 M
Tris, 1.0 M, pH 8.0
$MgCl_2$, 0.1 M
Orotidine-5'-phosphate pyrophosphorylase (EC 2.4.2.10)[29]
Orotidine-5'-phosphate decarboxylase (EC 4.1.1.23)[29]

Procedure. The following components are added to a Warburg-type vessel or a Thunberg tube with 0.5 ml of 0.2 N NaOH as absorbent: 0.02 ml of Tris buffer, 0.02 ml of orotate-^{14}C, 0.02 ml of $MgCl_2$, 0.2 ml each of orotidine-5-phosphate pyrophosphorylase[29] and orotidine-5-phosphate decarboxylase,[29] and extract containing approximately 0.03 micromole of PRPP in a final volume of 1.0 ml. The mixture is incubated at 25° for 15 minutes (with the amounts of the enzymes specified the reaction is complete in 10 minutes). The $^{14}CO_2$ yield is then counted either after precipitation as $BaCO_3$ or by scintillation counting.

Spectrophotometric Assay. A similar set-up is used except that cold orotate is used instead of the labeled compound.

[29] A. Kornberg, I. Liebermann, and E. S. Simms, *J. Biol. Chem.* **215**, 389 (1955).
[30] J. F. Henderson and M. K. Y. Khoo, *J. Biol. Chem.* **240**, 2349 (1965).

Method 2

This method uses the formation of AMP-^{14}C from adenine-^{14}C and PRPP with a measurement of the labeled AMP as the estimation of the PRPP present.[30,31]

Reagents

NaCl, 0.15 M
Tris, 0.2 M, pH 7.4
Sodium phosphate buffer, 0.1 M, pH 7.4
MgSO$_4$, 0.15 M
Adenine-^{14}C, 0.001 M
Perchloric acid, 0.2 M, containing 10% ZnSO$_4$
Adenine transphoribosylase (EC 2.4.2.7)[31a]

Procedure. The following components are added to a centrifuge tube: 0.2 ml of adenine-^{14}C, 0.1 ml of MgSO$_4$, 1.15 ml of sodium phosphate buffer, 0.25 ml of adenosine transphosphoribosylase, and PRPP extract in a final volume of 1.8 ml. The mixture is incubated at 37° for 30 minutes with shaking; the reaction is then stopped by adding 0.1 ml of the perchloric acid–ZnSO$_4$ solution and heating at 100° for 3 minutes. After centrifugation to remove the protein precipitate, 50 μl of the supernatant solution is chromatographed for 3 hours on Whatman No. 1 paper using 5% NaH$_2$PO$_4$ as a descending developing agent. The resultant spots (AMP $R_f = 0.67$) are detected under an ultraviolet lamp, cut out, and counted in a scintillation counter. Liver contains about 40 millimicromoles per gram of tissue.[32]

PART B. MEASUREMENT OF ENZYME ACTIVITIES

This section reviews the methods available for the measurement of the maximal potential activities of the enzymes involved in the biosynthesis of the nicotinamide nucleotides from nicotinamide or nicotinic acid in liver.

Conditions for the measurement of the enzymes in other tissues are

---

[31] J. D. Davidson, *J. Natl. Cancer. Inst.* **29,** 789 (1962).

[31a] Preparation of adenine transphosphoribosylase:[30] Ascites tumor cells, withdrawn from a mouse 5–6 days after inoculation, are suspended in 0.15 M NaCl and centrifuged at 0° for 3 minutes at 600 g to remove red cells. The tumor cells are then suspended in 9 volumes of Tris buffer, pH 7.4, and disrupted by sonic oscillation (2 minutes). They are then centrifuged at 2° for 20 minutes at 9000 g. The supernatant from this may be used as the enzyme source without any further preparation. The enzyme may be stored for some months at $-20°$.

[32] J. B. Clark, unpublished results (1967).

described where they are known. For other tissues it must be assumed that some modification will be needed to achieve maximal activity. In all cases the assay conditions described refer either to full tissue homogenates or to unpurified subfractions of such homogenates obtained by differential centrifugation.

Nicotinamide Deamidase

This enzyme has consistently been found to be largely associated with the microsomal fraction in rat and rabbit tissues.[33,34]

Preparation of Tissue Extract

A 1:3 tissue homogenate in 0.25 M sucrose (1 g in 3 ml) is prepared at 0–4°. To this is added Norit A charcoal (1 g/20 ml of homogenate).[35] The preparation is kept, with occasional stirring, for 30 minutes at 0°. It is then centrifuged at 8000 g for 15 minutes; the precipitate is discarded. The supernatant solution is recentrifuged at 100,000 g for 45 minutes; the microsomal pellet obtained after this treatment may be washed further with 0.25 M sucrose or resuspended in 0.25 M sucrose (at a concentration of 3 ml of sucrose per 1 g of original tissue) and used immediately. Using crude unfractionated homogenates without centrifugation gives essentially similar values.

Assay

The assay is based on the amount of nicotinic acid-^{14}C formed from nicotinamide ^{14}C under optimal conditions.[34]

Reagents

Triethanolamine, 0.25 M, pH 8.8.
Nicotinamide-^{14}C, 0.0125 M, 10 μCi/micromole
Bovine plasma albumin, crystalline
Acetic acid, 1 N, containing 12.5 mg of nicotinamide per milliliter

Procedure. The following reagents are placed in a suitable incubation tube containing 8 mg of bovine serum albumin: 0.1 ml of triethanolamine buffer, 0.2 ml of nicotinamide-^{14}C, and 0.1 ml of H_2O. The reaction is started by the addition of 0.1 ml of the enzyme preparation, and the incubation continued for 2 hours at 37° with occasional shaking. The reaction is

[33] J. Kirchner, J. G. Watson, and S. Chaykin, *J. Biol. Chem.* **241**, 953 (1966).
[34] B. Petrack, P. Greengard, A. Craston, and F. Sheppey, *J. Biol. Chem.* **240**, 1725 (1965).
[35] R. K. Gholson, I. Ueda, M. Ogasawara, and L. M. Henderson, *J. Biol. Chem.* **239**, 1208 (1964).

stopped by the addition of 0.02 ml of the acetic acid solution (the nicotinamide content of which is included as a chromatogram marker), and the tube is placed immediately in a boiling water bath for 2 minutes. After cooling on ice, the precipitate is removed by centrifugation at 4°; 0.2 ml of the supernatant solution so obtained is spotted onto Whatman 3 MM chromatographic paper, and the chromatogram is developed in solvent No. 4 (Table I).

The nicotinic acid and nicotinamide spots are then cut out of the paper and counted in a scintillation counter.

Properties and Activities of Nicotinamide Deamidase in Various Tissues

Many tissues appear to contain an endogenous inhibitor of this enzyme.[33,34] This inhibition may be removed by the action of bovine serum albumin,[33,34] although not all preparations of albumin appear to be equally effective in this property. Several albumins should be tried, and the preparation giving maximum activation should be used.

The enzyme has been partially purified by Petrack et al.[34] (38 × purification), who quote a maximal activity for the rat liver enzyme of 1 millimole per gram of purified liver enzyme per hour and a K_m for nicotinamide of 70 mM. The purified enzyme is not affected by the addition of bovine serum albumin. On the other hand, Kirchner et al.[33] reported that a purified

TABLE III

OPTIMAL ACTIVITIES OF NICOTINAMIDE DEAMIDASE FROM VARIOUS SOURCES

Source	Optical activity of enzyme (millimicromoles nicotinic acid/g wet wt/hour)		Reference[a]
	With BSA	No BSA	
Rat liver	510	80	36
Rat liver	357	67	34
Rabbit liver	155	109	33*
Rat mammary gland	<25	<25	37
Rabbit brain	7	4	33*
Rabbit kidney	14	4	33*
Rabbit lung	15	10	33*
Rabbit testis	7	2	33*
Rabbit heart	6	2	33*
Rabbit intestine	11	4	33*
Rabbit muscle	4	3	33*
Rabbit spleen	7	3	33*

[a] The numbers refer to text footnotes. Asterisk indicates that activity is expressed as millimicromoles of nicotinic acid per milligram of protein per hour.

preparation of the enzyme from rabbit liver had a maximal velocity of 0.9 millimole and 0.6 millimole per gram per hour of purified enzyme in the presence and the absence, respectively, of bovine serum albumin. The K_m for nicotinamide of this preparation was 40 mM.

Various activities for the enzyme in different tissues have been reported, all values obtained using microsomal preparations (see Table III).

Nicotinamide Mononucleotide Pyrophosphorylase (NMN-Pyrophosphate Phosphoribosyl Transferase, EC 2.4.2.12)

This enzyme is located exclusively in the cytoplasmic compartment of many tissues.[38]

Preparation of Tissue Extract

The starting material for this enzyme is the supernatant fraction prepared from a 1:10 (w/v) liver homogenate in 0.25 M sucrose by centrifugation at 105,000 g for 45 minutes. The supernatant so obtained is treated with a 1% (w/v) salmine sulfate solution, added dropwise with constant stirring, to give a final concentration of 0.7 ml of salmine sulfate per 10 ml of supernatant solution. The precipitate formed after standing at 0° for 30 minutes is removed by centrifugation at 10,000 g for 20 minutes, and the resulting supernatant solution may be used as the enzyme source. The liver enzyme is relatively stable, but extracts such as those prepared from mammary gland lose approximately 10% of their activity on storage overnight at −30° and about 30% of their activity on storage for 1 week.

Assay

The assay of NMN pyrophosphorylase is based on the amount of radioactive incorporation from ^{14}C-labeled nicotinamide into NMN.[38]

$$\text{Nicotinamide} + \text{PRPP} \xrightarrow{\text{ATP}} \text{NMN} + \text{PP}_i + \text{ADP} + \text{P}_i$$

Reagents

Tris(hydroxymethyl)aminomethane, 0.2 M, pH 7.3
MgCl$_2$, 0.1 M
5-Phosphoribosylpyrophosphate, 2.3 mM
ATP, 0.2 M, pH 7.3
Nicotinamide, 1 mM (activity approximately 10 μCi/micromole)

[36] S. Pinder, Ph.D. thesis, University of London, 1967.
[27] A. L. Greenbaum and S. Pinder, *Biochem. J.* **107**, 55 (1968).
[38] L. S. Dietrich, L. Fuller, I. L. Yero, and L. Martinez, *J. Biol. Chem.* **246**, 188 (1966).

Procedure. The following reagents are added to a suitable incubation tube: 0.25 ml of Tris buffer, 0.1 ml of $MgCl_2$, 0.1 ml of PRPP, 0.01 ml of ATP, and 0.1 ml of nicotinamide-[14]C. For tissues other than liver, the ATP and $MgCl_2$ concentrations must be varied to find an optimal concentration; for example, mammary gland requires 0.025 ml of ATP and 0.015 ml of $MgCl_2$. The reaction is started by the addition of 0.2–0.4 ml (+0.2 or 0 ml of H_2O) of the enzyme preparation and the whole is incubated for 30 minutes at 37°. The reaction is stopped by heating in a boiling water bath for 2 minutes; the tubes are cooled on ice. The precipitate is removed by centrifugation at 0–4°, and 100 μl of the resulting supernatant solution is spotted on a Whatman 3 MM chromatography paper (46 × 57 cm); the chromatogram is developed in solvent 1 (Table I) to achieve the best separation of substrate and products. The resulting spots may be identified and estimated as described above.

Properties and Activities of NMN Pyrophosphorylase in Various Tissues

A partial purification of this enzyme has been reported by Dietrich *et al.*,[38] who quote the following K_m values: ATP, $3.82 \times 10^{-4} M$; nico-

TABLE IV

OPTIMAL ACTIVITIES OF NICOTINAMIDE MONONUCLEOTIDE PYROPHOSPHORYLASE FROM VARIOUS SOURCES

Source	Optimal activity (millimicromoles/g wet wt tissue/hour)	Reference[a]
Rat liver	163	38
Rat liver	143	36
Rat liver	165	39
Rat kidney	80	38
Rat heart	60	38
Rat brain	24	38
Rat brain	16	39
Rat muscle	21	38
Rat lung	18	38
Rat mammary gland	980[b]	37
Mouse liver	146	38
Ehrlich ascites cell	90	38
Adenocarcinoma	130	38
Chick embryo liver, 12 days	126	38

[a] The numbers refer to text footnotes.

[b] Not corrected for retained milk.

[39] T. Honjo, S. Nakamura, Y. Nishizuka, and O. Hayaishi, *Biochem. Biophys. Res. Commun.* **25,** 199 (1966).

tinamide, $2.96 \times 10^{-6}\ M$; and PRPP, $3.57 \times 10^{-5}\ M$. The pH optimum is around 7.3, but the change in activity with pH shows a very broad peak. The enzyme is labile to acid, heat, and cold acetone; it is best stored at between 4° and −20°.

Values for the maximal potential activity in different tissues are given in Table IV.

Nicotinic Acid Mononucleotide Pyrophosphorylase (Nicotinate Nucleotide Pyrophosphate Phosphoribosyl Transferase, EC 2.4.2.11)

Preparation of Tissue Extract

The enzyme has been located in the cytoplasmic compartment of most tissues. A good source of the enzyme for assay is the high-speed super-natant fraction derived from a charcoal-treated homogenate as described under nicotinamide deamidase.

Assay

The assay is based on the incorporation of ^{14}C-labeled nicotinic acid into nicotinic acid mononucleotide.

Reagents

Sodium phosphate buffer, 0.17 M, pH 7.5[39a]
ATP, 0.2 M, pH 7.5[39a]
$MgCl_2$, 0.05 M
PRPP, 0.0032 M
Nicotinic acid-^{14}C, 0.001 M (10 μCi/micromole)
Acetic acid, 2 N

Procedure. The following reagents are added to an incubation tube; sodium phosphate buffer, 0.2 ml; ATP, 0.025 ml; $MgCl_2$, 0.01 ml[39b]; PRPP, 0.1 ml; nicotinic acid-^{14}C, 0.1 ml; the volume is adjusted to 1.0 ml with water. The reaction is started by the addition of the enzyme preparation (0.1–0.5 ml) and the preparation is incubated at 37° for 60 minutes. The reaction is stopped by the addition of 0.1 ml of 2 N acetic acid, and the

[39a] For mammary gland the pH should be 7.2.
[39b] For mammary gland, 0.05 ml $MgCl_2$.

tube is then heated in a boiling water bath for 2 minutes. After cooling on ice, the precipitated protein is removed by centrifugation at 4°; 100 μl of the resulting supernatant solution is spotted onto a Whatman 3 MM chromatographic paper. The chromatogram is developed in solvent system 1 (Table I) to achieve the best separation of the components. The nicotinic acid mononucleotide spot may then be cut out and counted in a scintillation counter.

Properties and Activities of Nicotinic Acid Mononucleotide Pyrophosphorylase in Various Tissues

The beef liver enzyme has been purified 500-fold by Imsande and Handler,[40] who report the following properties. The Michaelis constants for nicotinic acid and PRPP are $1 \times 10^{-6} M$ and $5 \times 10^{-5} M$, respectively. The K_i for nicotinic acid mononucleotide is $3.5 \times 10^{-5} M$. The enzyme has a pH optimum at 7.2, and its activity is stimulated by ATP, Mg^{2+}, and inorganic phosphate.

Various tissue activities, obtained with methods of estimation essentially similar to the above, have been reported (see Table V).

TABLE V

OPTIMAL ACTIVITIES OF NICOTINIC ACID MONONUCLEOTIDE PYROPHOSPHORYLASE FROM VARIOUS SOURCES

Source	Optimal activity (millimicromoles/g wet wt/hour)	Reference[a]
Rat liver	220	36
Rat liver	174	41
Rat mammary gland	40[b]	37
Rat brain	4	41

[a] The numbers refer to text footnotes.
[b] Uncorrected for retained milk.

Nicotinic Acid Mononucleotide Adenyltransferase (Nicotinate Mononucleotide-Adenylyltransferase, EC 2.7.7.18)

Preparation of Tissue Extract

This enzyme is localized exclusively in the nuclear fraction and may be the same enzyme as the nicotinamide mononucleotide adenylyltransferase. A suitable preparation for the estimation of enzyme activity is a purified nuclear fraction.

[40] J. Imsande and P. Handler, *J. Biol. Chem.* **236,** 525 (1961).
[41] J. B. Clark and S. Pinder, unpublished results (1968).

A homogenate of liver (1:10) is made in 0.25 M sucrose containing 3 mM $MgCl_2$ and 5 mM triethanolamine, pH 7.0. This homogenate should be strained through two layers of cheesecloth before being centrifuged at 900 g for 15 minutes. The sediment from the centrifugation is resuspended in a similar volume of medium and resedimented at 900 g. The sediment so obtained is then homogenized in a volume of 2.2 M sucrose equal to the original volume of the homogenate and centrifuged for 30 minutes at 50,000 g. The nuclei separate out as a clear sediment and may be freed of the dense sucrose by decantation and inversion for 30 minutes. The isolated nuclei are then suspended in the 0.25 M sucrose–$MgCl_2$–triethanolamine medium (5 ml per gram original liver) and may be used directly as the source of the enzyme.

Assay

The assay is based on the incorporation of ^{14}C-labeled nicotinic acid mononucleotide into nicotinic acid adenine dinucleotide (deamido-NAD).

Reagents

Glycylglycine, 0.25 M, pH 7.6, containing 0.075 M $MgCl_2$
ATP, 0.04 M, pH 7.6.
Nicotinic acid-^{14}C mononucleotide, 0.04 M (0.5 μCi/micromole)
Acetic acid, 2 N

Procedure. Place 0.05 ml of the glycylglycine–$MgCl_2$ mixture, 0.025 ml of ATP, 0.05 ml of nicotinic acid mononucleotide, and 0.025 ml of H_2O into a tube and add 0.1 ml of the enzyme source (either the nuclear preparation described above or a 1:10 or 1:5 homogenate of tissue in 0.25 M sucrose) to start the reaction. Incubate at 37° for 15 minutes and then stop the reaction by adding 0.1 ml of 2 N acetic acid. Place in a boiling water bath for 2 minutes. Cool. After removal of the precipitate by centrifugation, 100 μl of the supernatant solution is spotted onto a Whatman 3 MM paper and the products of the reaction are separated by development in solvent system 3 (Table I).

The deamido-NAD spot may then be identified, cut out, and counted in a scintillation counter.

Properties and Activity of Nicotinic Acid Mononucleotide Adenylyltransferase

There is substantial evidence for the identity of this enzyme with nicotinamide mononucleotide adenylyltransferase.[42,43] The properties of the nicotinic acid enzyme are therefore dealt with in the section on nicotinamide mononucleotide adenylyltransferase. It is noteworthy, however, that the enzyme is usually found to be more active when nicotinic acid mononucleotide is used as substrate. The only activity specifically quoted in the literature for nicotinic acid mononucleotide adenylyltransferase gives a value of 10.4 micromoles per gram of liver per hour.[42]

Nicotinamide Mononucleotide Adenylyltransferase (ATP-NMN Adenylyltransferase, EC 2.7.7.1)

Preparation of Tissue Extract

As this enzyme and the previous one are the same, the extracts prepared as described above will be suitable for this enzyme too. The activity of the enzyme has been found to be stable for at least 10 days if stored at 0–4° with no detectable loss of activity in the liver and mammary gland preparations.[37,46] For some tissues, e.g., liver and mammary gland, crude sucrose homogenates have the same activity as the purified nuclear fractions and may be conveniently used for the estimation of enzyme activities.[47]

Assay

This is based on the original procedure of Kornberg,[48] where the rate of formation of NAD⁺ from NMN and ATP is measured.

Reagents

Glycylglycine, 0.25 M, pH 7.6, containing 0.075 M MgCl$_2$
ATP, 0.06 M, pH 7.6
NMN, 0.04 M
Nicotinamide, 2 M
HCl, 0.15 N
Glycylglycine, 0.25 M, pH 7.6
NaOH, 1.5 N

Procedure. For the estimation of the liver enzyme the following com-

---

[42] J. Preiss and P. Handler, *J. Biol. Chem.* **233**, 493 (1958).
[43] M. R. Atkinson, J. Jackson, and R. K. Morton, *Nature* **192**, 946 (1961).
[44] W. Dahmen, B. Webb, and J. Preiss, *Arch. Biochem. Biophys.* **120**, 440 (1967).
[45] M. R. Atkinson and R. K. Morton, *Nature* **188**, 58 (1961).
[46] J. B. Clark, Ph.D. thesis, University of London, 1964.
[47] G. H. Hogeboom and W. C. Schneider, *J. Biol. Chem.* **197**, 611 (1952).
[48] A. Kornberg, *J. Biol. Chem.* **182**, 779 (1950).

ponents are placed in a tube: 0.05 ml of the glycylglycine-$MgCl_2$ mixture, 0.025 ml of ATP, 0.025 ml of nicotinamide, and 0.05 ml of NMN. (For mammary gland the appropriate quantities are 0.025 ml of the glycyl-glycine-$MgCl_2$ mixture, 0.01 ml of ATP, 0.025 ml of nicotinamide, 0.02 ml of 0.08 M NMN. For brain, the components are as for liver. For liver and mammary gland crude sucrose homogenates may be used, but for brain it is essential to use the purified nuclear preparation.) The reaction is started by the addition of 0.1 ml of the enzyme source, and the tubes are incubated at 37° for 15 minutes. The reaction is stopped by adding 0.5 ml of 0.15 N HCl and immersing the tubes in boiling water for 2 minutes, after which they are cooled on ice. Then 0.2 ml of glycylglycine buffer, pH 7.6, is added to each tube, and each is neutralized by the addition of 0.05 ml of 1.5 N NaOH. The protein precipitate is removed by centrifugation, and the NAD^+ content of the supernatant solution is estimated by any of the procedures described above.

Properties and Activities of Nicotinamide Mononucleotide Adenylyltransferase in Various Tissues

A purified semicrystalline preparation of NMN-adenylyltransferase has been obtained from pig liver nuclei.[43,45] The following constants have been reported: K_m for NMN, 0.12 mM; K_m for nicotinic acid mononu-cleotide, 0.4 mM; K_m for ATP, 0.4 mM; K_i for nicotinic acid mononu-cleotide, 0.38 mM; K_i for ITP, 1.8 mM.

Kornberg[48] has also reported the following constants for a semipurified enzyme derived from hog liver: K_m for NMN, 0.15 mM; K_m for ATP, 0.46 mM; K_i for NAD, 0.08 mM; K_i for PP_i, 0.19 mM; pH optimum, 7.6. Kornberg also reported that the equilibrium of the reaction does not favor the formation of NAD^+.

For reported activities, see Table VI.

TABLE VI

OPTIMAL ACTIVITY OF NICOTINAMIDE MONONUCLEOTIDE ADENYLYL TRANSFERASE FROM VARIOUS SOURCES

Source	Activity (millimicromoles/g wet wt/hour)	Reference[a]
Rat liver	5000	49
Rat mammary gland	1630	37
Rat brain	772	41

[a] The numbers refer to text footnotes.

[49] A. L. Greenbaum, J. B. Clark, and P. McLean, Biochem. J. 95, 167 (1965).

NAD-Synthetase [Deamido-NAD-L-Glutamine Amidoligase (AMP) EC 6.3.5.1]

Preparation of Tissue Extract

This enzyme is localized in the cytoplasm; a charcoal-treated supernatant preparation as described for nicotinamide deamidase is a convenient source, except that a 1:10 homogenate is used initially.

Assay

This is essentially according to the procedure of Preiss and Handler[42] which uses the rate of formation of NAD^+ from deamido-NAD as a measure of the enzyme activity.

$$\text{Deamido-NAD} + \text{glutamine} + \text{ATP} \rightarrow \text{NAD}^+ + \text{glutamate} + \text{AMP} + \text{PP}_i$$

Reagents

Potassium phosphate, 0.18 M, pH 7.6
Deamido-NAD, 0.005 M
L-Glutamine, 0.2 M
ATP, 0.08 M, pH 7.6
Nicotinamide, 1.64 M
$MgCl_2$, 0.05 M
HCl, 1 N
NaOH, 1 N
Glycylglycine, 0.25 M, pH 7.6

Procedure. The following reagents are placed in a centrifuge tube: 0.14 ml of phosphate buffer, 0.2 ml of deamido-NAD, 0.1 ml of L-glutamine, 0.05 ml of ATP, 0.1 ml of nicotinamide, and 0.01 ml of $MgCl_2$. The reaction is started by the addition of 0.4 ml of the enzyme source, and the mixture is incubated at 37° for 30 minutes. The reaction is stopped by the addition of 0.1 ml of 1 N HCl and then the tubes are placed in a boiling water bath for 2 minutes. After cooling, the contents of the tubes are neutralized by adding 0.1 ml of 0.25 M glycylglycine, pH 7.6, and then 0.1 ml of 1 N NaOH. The precipitated protein is removed by centrifugation, and the NAD^+ present in the supernatant is estimated by any of the procedures described above.

Properties and Activities of NAD-Synthetase from Various Tissues

The only purified preparation so far studied is one from yeast,[42] for which the following properties have been described: K_m for deamido-NAD, 1.4×10^{-4} M; K_m for ATP, 6×10^{-4} M; K_m for Mg^{2+}, 1.3×10^{-3} M; and K_m for glutamine, 3.5×10^{-3} M. Reported activities are given in Table VII.

TABLE VII
OPTIMAL ACTIVITY OF NAD SYNTHETASE FROM VARIOUS SOURCES

Source	Activity (millimicromoles/g wet wt/hour)	Reference[a]
Rat liver	2060	41
Rat liver	2180	36
Rat liver	1140	42
Rat mammary gland	<8	37
Rat brain	67	41

[a] The numbers refer to text footnotes.

NAD-Kinase (EC 2.7.1.3)

Preparation of Tissue Extract

This enzyme is located exclusively in the cytoplasm,[50] and a suitable source is a supernatant preparation from a 1:10 homogenate of tissue. The homogenate may be prepared in either 0.25 M sucrose or in 0.15 M KCl–1.6 × 10^{-4} M KHCO$_3$. For the former, centrifugation for 2 × 10^6 g-min is required; for the latter, 1.35 × 10^5 g-min suffices. Both preparations from liver give activities of similar magnitude, and both maintain their activity for several weeks when stored at −15°.[51] Other tissues give less stable extracts. Mammary gland extracts, for example, must be used within 6 hours, otherwise nonlinear kinetics become manifest.[36]

Assay

This enzyme is usually measured by the rate of formation of NADP$^+$ from NAD$^+$ but, since an unknown proportion of the NADP$^+$ formed may be reduced to NADPH by other enzymes and substrates contained in a crude extract, it is convenient to ensure that all the NADP$^+$ formed is reduced to NADPH and the activity assessed by the rate of formation of NADPH from NAD$^+$. This may be achieved by adding glucose 6-phosphate (and possibly also glucose-6-phosphate dehydrogenase where the activity of this enzyme is low in the tissue) to the assay mixture. The NADPH may be estimated by any of the procedures given above.

Reagents

Triethanolamine·HCl, 0.1 M, pH 7.6
ATP, 0.06 M, pH 7.6
Nicotinamide, 0.1 M

[50] V. Stollar and N. O. Kaplan, J. Biol. Chem. 236, 1836 (1961).
[51] T. P. Wang and N. O. Kaplan, J. Biol. Chem. 206, 311 (1954).

MgCl$_2$, 0.1 M
Glucose 6-phosphate, 0.05 M
NAD$^+$, 0.033 M
Sodium fluoride, 0.9 M, pH 7.6
NaOH, 1 N
HCl, 1 N
Glycylglycine, 0.25 M, pH 7.6

Procedure. The following reagents are added to a tube: 0.1 ml of triethanolamine buffer; 0.13 ml of ATP; 0.1 ml of nicotinamide; 0.1 ml of MgCl$_2$; 0.2 ml of NAD$^+$; 0.05 ml of glucose 6-phosphate, and water to give a final volume of 0.85 ml. To this mixture are added 0.1 ml of the enzyme preparation and, immediately afterward, 0.05 ml of sodium fluoride. The mixture is then incubated at 37° for 10 minutes; the reaction is then stopped by the addition of 0.1 ml of 1 N NaOH and heating to 100° for 2 minutes. The mixture is then cooled, 0.1 ml of the glycylglycine buffer is added, and the whole is neutralized by the addition of 0.1 ml 1 N HCl. The precipitated protein is removed, and the NADPH content of the supernatant solution is estimated as described earlier (see p. 24). For tissues other than liver, the amounts of the reagents may need to be varied. For mammary gland the components of the reaction mixture for optimal activity are 0.085 ml of ATP; 0.13 ml of MgCl$_2$; 0.14 ml of NAD$^+$, and 0.2 ml of the enzyme extract. For brain the components are the same as for liver except that brain needs the addition of 0.025 ml (containing 2.5 μg of enzyme protein) of glucose-6-phosphate dehydrogenase.

Properties and Activities of NAD$^+$-Kinase from Various Tissues

A partial purification of NAD$^+$-kinase from pigeon liver has been achieved,[52] and a complete purification and characterization of the guinea pig liver enzyme has been reported by Nemchinskaya *et al.*[52] and Apps.[53]

TABLE VIII
OPTIMAL ACTIVITIES OF NAD-KINASE FROM VARIOUS SOURCES

Source	Activity (millimicromoles/g wet wt/hour)	Reference[a]
Rat liver	3000	49
Rat mammary gland	1120	37
Rat brain	65	41

[a] The numbers refer to text footnotes.

[52] V. L. Nemchinskaya, V. P. Kushner, U. M. Bozhkov, E. I. Turcheako, and S. E. Tukachinskii, *Biochimiya* **31,** 306 (1966).
[53] D. K. Apps, *Biochem. J.* **104,** 35P (1967).

Apps quotes the following properties for his preparation: K_m for NAD^+, $2.7 \times 10^{-4}\, M$; K_m for ATP, $2.1 \times 10^{-3}\, M$. NADH inhibits competitively with respect to NAD^+ and noncompetitively with respect to ATP. The K_i for this inhibition is $9 \times 10^{-5}\, M$. ADP inhibits competitively with respect to ATP and noncompetitively with respect to NAD^+, with a K_i of $9.7 \times 10^{-3}\, M$. NADPH competes with either ATP or NAD, with a K_i of $6 \times 10^{-5}\, M$.

Reported activities are given in Table VIII.

Quinolinate Phosphoribosyl Transferase

Preparation of Tissue Extract

A convenient source of this enzyme is a charcoal-treated supernatant preparation, as described for nicotinamide deamidase.[54]

Assay

The method is based on that of Gholson et al.,[35] where the incorporation of quinolinic acid-^{14}C into nicotinic acid mononucleotide is used as an estimate of the enzyme activity.

Reagents

Potassium phosphate 0.6 M, pH 7 6
^{14}C-labeled 2,3,7,8-quinolinic acid, 0.005 M (1 μCi/micromole)
Phosphoribosyl pyrophosphate, 0.006 M
$MnCl_2$, 0.0066 M
$MgCl_2$, 0.012 M
Acetic acid, 2 N

Procedure. The following components are added to a centrifuge tube: 0.1 ml of potassium phosphate buffer, 0.1 ml of PRPP, 0.05 ml of $MnCl_2$, 0.05 ml of $MgCl_2$, and 0.2 ml of quinolinic acid-^{14}C. The reaction is started by the addition of 0.5 ml of a charcoal-treated supernatant derived from a 1:5 tissue homogenate. The reaction mixture is incubated at 37° for 1 hour; the reaction is then stopped by adding 0.1 ml of acetic acid and heating to 100° for 2 minutes. The preparation is cooled to 0°, and the precipitate is removed by centrifugation; 100 μl of the resultant supernatant are removed and chromatographed on a 3 MM Whatman filter paper using solvent system 1 (see Table I). The spots may then be identified as described above, and the radioactivity of the nicotinic acid mononucleotide spot may be estimated by cutting it out and counting in a scintillation counter.

[54] M. Ikeda, H. Tsuji, S. Nakamura, A. Ichiyama, Y. Nishizuka, and O. Hayaishi, *J. Biol. Chem.* **240**, 1395 (1965).

Properties and Activities of Quinolinate Phosphoribosyl
    Transferase from Various Tissues

Gholson et al.[35] have described a 1500-fold purification of the enzyme
from beef liver acetone powder. This preparation has a pH optimum of
6.2, a K_m for quinolinic acid of $6 \times 10^{-5}\ M$, and a K_m for PRPP of $5 \times 10^{-5}\ M$.
    Table IX lists reported activities.

TABLE IX

OPTIMAL ACTIVITIES OF QUINOLINIC ACID PHOSPHORIBOSYL TRANSFERASE
IN VARIOUS TISSUES

Tissue	Activity (millimicromoles/g wet wt/hour)	Reference[a]
Rat liver	70	36
Rat liver	83	54
Beef liver (acetone powder)	285	35
Cat liver	99	54
Rat kidney	11	54
Cat kidney	11	54

[a] The numbers refer to text footnotes.

[103] Determination of Nicotinamide N-Oxide
Using Xanthine Oxidase

By MARGARETA SPARTHAN and STERLING CHAYKIN

Assay Method

Principle. Nicotinamide N-oxide is enzymatically reduced to nicotinamide (see this volume [101]). The latter is determined by reacting it with
cyanogen bromide in the presence of ammonia to form a yellow chromophore
with a maximum absorption at 398 mμ. The method[1] is applicable over the
range 0.02–0.3 micromoles of nicotinamide N-oxide. An isotope dilution
variation on this method is used for the determination of nicotinamide
N-oxide in complex mixtures.

Reagents

Tris, 1 M, pH 9.5, containing $2.5 \times 10^{-2}\ M$ xanthine
Nicotinamide N-oxide, $10^{-2}\ M$, standard solution (see this volume
    [101])

[1] M. Sparthan and S. Chaykin, *Anal. Biochem.* **31**, 286 (1969).

Mercaptoethanol, 1.4 M

Milk xanthine oxidase[2] (xanthine:oxygen oxidoreductase, EC 1.2.3.1, 4.6 units/mg; see Vol. II [73]) 4 mg/0.2 ml of Tris buffer, 0.01 M, pH 7.0

Trichloroacetic acid (TCA), 50% w/v

NaOH, 2 N

Buffered ammonia reagent (see this volume [101])

Cyanogen bromide, 10% w/v (see this volume [101])

Procedure

The reaction mixture contains 0.1 ml of Tris-xanthine solution, 0.01 ml of mercaptoethanol, 0.2 ml of xanthine oxidase, and the solution to be assayed for nicotinamide N-oxide. The reaction mixtures are set up in Thunberg tubes with the enzyme and mercaptoethanol in the side arm. The total volume of the reaction mixture is 0.5 ml. The reaction vessels were made anaerobic by alternately evacuating and flushing with helium. Incubations are carried out at 50° for 2 hours. The reaction is initiated by tipping in the enzyme-mercaptoethanol mixture. It is terminated by the addition of 0.05 ml of 50% TCA, and precipitated protein is removed by centrifugation. A suitable portion of the deproteinized solution (representing 0.01–0.20 micromoles of nicotinamide) is neutralized with 2.0 N NaOH and analyzed for nicotinamide. The sample is diluted to 1 ml; 1.5 ml of buffered ammonia reagent and 2.5 ml of cyanogen bromide is added. Eleven minutes after the addition of the cyanogen bromide, the optical density is read at 398 mμ and the nicotinamide content is calculated from a standard curve.

Comments. The quantitative reduction of nicotinamide N-oxide, using xanthine oxidase as catalyst, can only be accomplished under the extreme reaction conditions presented above. Less than complete reduction of the N-oxide results if significantly higher or lower pH or concentrations of xanthine or mercaptoethanol are used. The enzyme is present in slight excess. Anaerobic conditions must be used because the enzyme is rapidly inactivated in the presence of oxygen. A variety of substances interfere with the attainment of the quantitative formation of nicotinamide. Since ammonium sulfate is such a compound, ammonium sulfate suspensions of the enzyme must be desalted before use.

Mercaptoethanol serves two functions in the assay procedure. During the preincubation of mercaptoethanol and enzyme at pH 7.0 in the side arm of the reaction vessel, the enzyme is activated. The tipping of mercaptoethanol into the main reaction mixture, with its high pH, causes the mer-

[2] Protein is determined by the A_{280}/A_{260} method of O. Warburg and W. Christian, Vol. III [73].

captoethanol to become very reactive with respect to oxygen and effectively completes the deoxygenation of the reaction medium. When interfering substances prevent the estimation of nicotinamide N-oxide by the foregoing direct method, the following isotope dilution method can be used.

Alternate Procedure Using Isotope Dilution. A sample of nicotinamide N-oxide-7-^{14}C of known specific activity is added to a sample containing an unknown quantity of nicotinamide N-oxide. The amount of radioactive nicotinamide N-oxide which should be added is dependent on a variety of factors including its specific activity, the amount of nicotinamide N-oxide in the unknown, the extent of reduction which can be anticipated, losses in handling, etc. Trial and error has led to the following as a suitable procedure for the determination of approximately 0.3 micromoles of nicotinamide N-oxide in 1 ml of mouse urine.

Radioactive nicotinamide N-oxide is synthesized from nicotinamide-7-^{14}C as described in article [128]. Nicotinamide N-oxide-7-^{14}C, 0.076 micromoles (specific activity 10^6 cpm/micromole) is added to 0.6 ml of mouse urine. The mixture is applied to a 7×22 inch strip of Whatman 3 MM filter paper for chromatography and developed for 18 hours in 1-butanol saturated with water. This chromatographic separation is required to remove the bulk of the materials in mouse urine which inhibit xanthine oxidase. It also accomplishes the removal of any nicotinamide which might be present and would otherwise have to be corrected for in the determination of nicotinamide derived from the N-oxide. The nicotinamide N-oxide which travels with an R_f of approximately 0.2 is located with a Geiger counter. The area of the chromatogram containing the N-oxide is cut out and the N-oxide eluted with water. The eluate is taken to dryness by lyophilization and dissolved in 0.2 ml of water. A 0.2-ml sample of this solution is treated with xanthine oxidase as described above. After the conclusion of the 2-hour incubation period, the entire reaction mixture is chromatographed under the same conditions described for the original urine sample. This time the nicotinamide-7-^{14}C containing portion of the paper is cut out and the nicotinamide eluted. Nicotinamide travels with an R_f of approximately 0.6. The amount of nicotinamide in the eluate is determined by the cyanogen bromide method used above. Radioactivity is determined by scintillation counting. These values are corrected for sampling procedures. The amount of nicotinamide N-oxide in the original sample is determined using the formula which appears below:

Micromoles of N-oxide in unknown =

$$\frac{\text{Radioactive } N\text{-oxide added (cpm)}}{\text{Specific activity of nicotinamide formed (cpm/micromole)}} -$$

Micromoles of radioactive N-oxide added

[104] Preparation and Purification of α-Amino-β-carboxy-muconic Acid-ε-semialdehyde, the Acyclic Intermediate of Nicotinic Acid Biosynthesis

By Erich Kuss

Assay Method

Principle. The intermediate 2-amino-3-carboxy-muconic acid-6-semi-aldehyde $(OCH—CH=CH—C(COOH)=C(NH_2)—COOH)$ (I) is prepared by enzymatic cleavage of 3-hydroxyanthranilic acid followed by deproteinization of the solution and extraction of the product.

Reagents

Pig liver acetone powder, prepared in the usual manner; it can be stored at $-20°$ for several months.

3-Hydroxyanthranilic acid (Fluka A.G.)

Dowex 50 X2, 200–400 mesh, H+-form

Phenol, 80% aqueous solution

n-Butanol

Ferrous sulfate

Phosphate buffer, 0.015 M, pH 7.4

Alcoholic buffer, 75 ml ethanol + 225 ml 0.015 M phosphate buffer, pH 7.4

Sodium chloride, analytical grade

Procedure

Five grams of acetone powder is extracted by stirring with 120 ml phosphate buffer for 10 minutes. After centrifuging at 10,000 g for 15 minutes the insoluble material is removed. To 100 ml of the extract, 0.5 g ferrous sulfate and finally 51 mg (0.3 millimoles) 3-hydroxyanthranilic acid are added. The solution is stirred very intensively at about 2°. After 1.5 hours the optical density of the solution at 366 mμ has reached the maximum. The reaction is then stopped by addition of an equal volume of ice-cold phenol. After shaking vigorously for 3 minutes, the layers are separated by centrifugation. The aqueous layer is washed 4 times with equal volumes of ice-cold peroxide-free ether and then stirred with 5 g of Dowex and 20 ml of butanol for 15 seconds. The resin is removed immediately by filtration. Solid sodium chloride is added to the filtrate until saturation. After withdrawing the butanol phase, the aqueous phase is extracted twice with 10 ml and once with 5 ml of butanol. The combined butanol extracts, which contain the intermediate I, are washed once with 20 ml water. If a purified aqueous solution of the intermediate is desired, 10 ml of the sodium bicarbonate solution is added to the butanol extract. After shaking, the layers are separated, and the aqueous layer is extracted twice with ether; traces of ether are removed *in vacuo*.

Determination

The determination of the intermediate I is based on the high absorbance at 360 mμ.[1,2] The molar extinction coefficient has been reported to be 4.5–4.7 × 10^4 [3,4] in water (neutral or slightly alkaline solutions). To determine the yields of the intermediate in the purification procedure, 0.05 ml of the solutions are diluted to 3 ml with buffer or alcoholic buffer solutions and extinctions are read immediately at 360 or 366 mμ against a blank. Yields in the single steps: deproteinization, at least 50%, extraction with butanol, at least 35%; extraction with bicarbonate solution, at least 70%.

Comments

To obtain nearly complete oxidation of the 3-hydroxyanthranilic acid in a concentrated solution (3 × 10^{-3} M), it is necessary to incubate in the cold; at 37° the reaction runs to completion only in dilute solutions (10^{-5} M).[5]

The intermediate is rather labile. Its stability under a number of mild conditions has been determined.[6] The extinction at 360 mμ decreases to 50% at 27° (0°, −20°) (1) in water after deproteinization within 20 minutes (6 hours, 2 days), (2) in bicarbonate solutions within 50 minutes (1 day, 2 days), and (3) in butanol solutions within 90 minutes (1 day, 14 days). The decrease of extinction in butanol solutions is partially reversible by addition of buffer.[5]

[1] A. Miyake, A. H. Bokman, and B. S. Schweigert, *J. Biol. Chem.* **211**, 391 (1954).
[2] C. L. Long, H. N. Hill, I. M. Weinstock, and L. M. Henderson, *J. Biol. Chem.* **211**, 405 (1954).
[3] O. Wiss, H. Simmer, and H. Peters, *Z. Physiol. Chem.* **304**, 221 (1956).
[4] A. Ichiyama, S. Nakamura, H. Kawai, T. Honjo, N. Nishizuka, O. Hayaishi, and S. Senoh, *J. Biol. Chem.* **240**, 740 (1965).
[5] E. Kuss, *Z. Physiol. Chem.* **348**, 1602 (1967).
[6] E. Kuss, unpublished data, 1968.

[105] Preparation of DPN+ and NMN Labeled with 14C in the Pyridine Moiety

By HERBERT C. FRIEDMANN

Principle

Nicotinic acid-7-14C is converted to 14C-labeled DPN+ by intact *Propionibacterium shermanii*. The DPN+-14C is purified from an extract of the bacteria by elution from Dowex 1 and Sephadex columns. Hydrolysis to NMN-7-14C and AMP is accomplished by incubation with snake venom phosphodiesterase. The NMN-7-14C is purified by elution from Dowex 50 and Dowex 1 columns.

Reagents

> *Propionibacterium shermanii* (ATCC 9614), grown in a medium[1] as described by Friedmann,[2] without aeration and without added cobalt, iron, or yeast. The harvested bacteria can be stored for at least 4 months at $-22°$. The glucose-fortified medium used for suspension contains 3% dextrose.[2]

Nicotinic acid-7-14C, neutralized with a molar equivalent of NaOH

Dowex 1-X2, 200–400 mesh, cycled, acetate form

Dowex 50W-X2, 200–400 mesh, cycled, H+ form

Sephadex G-10, swollen in water, deaerated

> Venom phosphodiesterase (phosphodiesterase I) from *Crotalus adamanteus*, prepared by step 1 of the method of Williams et al.,[3] bought from Worthington Biochemical Corporation. Five milligrams of lyophilized powder are dissolved in 1 ml of water and can be stored frozen for at least 3 years.

Alcohol dehydrogenase

Ethanol, absolute and 80%

Diethyl ether

Acetic acid, 0.1 N, 0.05 N, and 0.01 N

NaHCO3

MgCl2, 0.3 M

Procedure

Incubation of the Bacteria and Preparation of Crude Extract. A uniform suspension of 10 g of *P. shermanii* in 100 ml of the glucose-fortified medium

---

[1] K. Bernhauer, E. Becher, and G. Wilharm, *Arch. Biochem. Biophys.* **83**, 248 (1959).

[2] H. C. Friedmann, *J. Biol. Chem.* **243**, 2065 (1968).

[3] E. J. Williams, S. Sung, and M. Laskowski, Sr., *J. Biol. Chem.* **236**, 1130 (1961).

is prepared. Suspension is conveniently achieved[2] with a nonaerating stirrer. Then 0.1 ml of a solution containing 20 μCi of niacin-7-^{14}C is added. After an incubation for about 19 hours at 30°, the mixture is centrifuged in the cold at 17,000 g for 20 minutes. The sedimented bacteria are extracted with 60 ml of 80% ethanol at room temperature. The sediment, which centrifuges down very readily, is reextracted with 40 ml of 80% ethanol. The residue is extracted with 60 ml of water at about 95° for 6 minutes, and the mixture is centrifuged at 17,000 g for 15 minutes.

The combined centrifugates are concentrated almost to dryness *in vacuo* at about 35°. A few drops of *n*-octyl alcohol are added to prevent frothing. The material, which may be stored overnight in the refrigerator at this stage, is dissolved in about 4 ml of water. For successful fractionation on columns it is helpful to remove lipids by three extractions at room temperature with about 1.5-ml portions of diethyl ether. Phases are conveniently separated by centrifugation. The greenish-yellow aqueous phase is removed from sediment and small amounts of a white emulsion. These residues are washed twice with 1-ml portions of water. The supernatant solutions and the main extract are combined and concentrated to a small volume *in vacuo* at about 30°. The pH of the extract is about 8.

First Dowex 1 Column. The material is applied to a refrigerated Dowex 1 acetate column, 7 × 175 mm. The column is eluted with about 65 ml of water, followed by about 450 ml of 0.1 N acetic acid or preferably a smaller volume of 0.3 N acetic acid (about 280 ml) at a rate of about 0.65 ml per minute. A small amount of radioactive UV-absorbing material, eluted by the first few milliliters of water, is discarded. UV-absorbing material (continuously monitored at 254 mμ) is eluted in one or two main peaks by the remaining acetic acid, and is concentrated to near dryness *in vacuo* at about 35°. Traces of acetic acid are removed from the almost white residue by two evaporations *in vacuo* from 10-ml portions of absolute ethanol. The residue is dissolved in 9 ml of water and may be stored in the refrigerator. About 9.5% of the total radioactivity is due to nicotinic acid, the rest to DPN$^+$. The two compounds are readily distinguished by ionophoresis on Whatman No. 1 paper in 0.5 N acetic acid. While the DPN$^+$ remains near the origin half-way between the electrodes under these conditions (pH 2.8), nicotinic acid migrates toward the cathode.

Sephadex G 10 Treatment. Following the method of Olivera and Lehman,[4] the material is passed through a Sephadex G 10 column (1.5 × 25 cm), but 6% of the radioactivity in the first 90% of the void volume still consists of material with the ionophoretic behavior of nicotinic acid. The radioactivity in pooled further effluents is 70% nicotinic acid, the rest DPN$^+$.

Second Dowex 1 Column. The above 90% void volume effluent is applied

[4] B. M. Olivera and I. R. Lehman, *Proc. Natl. Acad. Sci. U.S.* **57**, 1700 (1967).

directly to a fresh refrigerated Dowex 1 acetate column, 7 × 125 mm. Elution at a rate of about 1.6 ml per minute is performed with the aid of a peristaltic pump. The first elution with about 50 ml of water removes a small amount of UV-absorbing material, which is discarded. Treatment is continued with roughly 400 ml of 0.05 N acetic acid which, following the first 200 ml or so elutes all of the nicotinic acid-7-¹⁴C without DPN⁺-¹⁴C in a small broad peak of about 140 ml volume. The DPN⁺-¹⁴C is now eluted by about 300 ml of 0.1 N acetic acid in a symmetrical peak and is concentrated *in vacuo* at about 34°. It is freed from traces of acetic acid by means of repeated evaporations from absolute ethanol, as before.

The very faintly yellow syrup is dried overnight at room temperature *in vacuo* over Linde molecular sieve type 4A.

Yield of DPN⁺-¹⁴C, based on a spectrophotometric test of aliquot with alcohol dehydrogenase (see Vol. III [128]), is 14.9 micromoles. Overall yield, based on radioactivity of administered nicotinic acid-7-¹⁴C, is 28%. Specific activity, based on counts on Whatman No. 1 paper in a thin-window 4 pi Geiger-type gas flow detector (Nuclear Chicago Actigraph III strip counter, paper stationary) is 3.48×10^5 cpm per micromole. The nicotinic acid-7-¹⁴C used (Calbiochem) had a stated specific activity of 6.5 mCi/millimole. The exact dilution with unlabeled bacterial niacin has not been determined.

Preparation of NMN-7-¹⁴C

Hydrolysis of the DPN⁺-¹⁴C to NMN-7-¹⁴C and AMP by Venom Phosphodiesterase. Dissolve the DPN⁺-¹⁴C in 1 ml of water. Add NaHCO₃ to a concentration of 1.0 M, ignoring volume change (84 mg), then 0.325 ml of 0.3 M MgCl₂, followed by 0.08 ml of the venom phosphodiesterase. Incubate 2 hours at 37°. Place in a refrigerator. The test with alcohol dehydrogenase should show that no DPN⁺ is left. This hydrolysis follows the directions of Kaplan and Stolzenbach (Vol. III [129]), who used *Crotalus adamanteus* pyrophosphatase purified essentially according to Butler (Vol. II [89]). The nucleotide pyrophosphatase from Maine potatoes prepared according to Kornberg and Pricer (see Vol. II [112]), which does not require Mg²⁺ to split DPN⁺, may also be used.

Since NMN is eluted from a Dowex 1 column by very low concentrations of acetic acid, it is not possible to obtain it free of extraneous ions by a single pass through this ion exchanger. Removal of Mg²⁺ by direct passage through Dowex 50, however, would result in high acidity due to free HCl, and in frothing due to free carbonic acid. Hence a first passage through Dowex 1 acetate is instituted to remove these anions, followed by passage through Dowex 50W to remove the Mg²⁺. A final Dowex 1 acetate treatment yields the chemically pure NMN-7-¹⁴C.

First Dowex 1 Column. The incubation mixture is applied directly to a refrigerated Dowex 1 acetate column, 7×160 mm. The fractions are collected at a rate of about 0.25 ml/min. UV-absorbing material is rapidly eluted with a water-wash, and further UV-absorbing material is eluted with a small volume of 0.01 N acetic acid. Any NMN which is eluted from Dowex 1 by water[5] is almost certainly accompanied by ions (see Vol. III [129]).

Dowex 50W Column. The eluates obtained with water and with 0.01 N acetic acid are combined, concentrated *in vacuo* at 30° to a small volume without added ethanol, and applied directly to a refrigerated Dowex 50W (H[+]) column, 7×150 mm. Elution is performed with water at the rate of about 1.25 ml/minute. UV-absorbing radioactive material, eluted between about 40 and 90 ml of water, is collected and concentrated *in vacuo* with the aid of ethanol to remove all traces of acetic acid.

Second Dowex 1 Column. The material is dissolved in 2.5 ml of water and applied to a fresh refrigerated Dowex 1 acetate column, 7×155 mm. Elution is performed at a rate of 1.25 ml per minute. In contrast to the previous Dowex 1 column, water now elutes only a slight amount of UV-absorbing material. The NMN-7-^{14}C is eluted in a sharp peak by about 30 ml of 0.01 N acetic acid after an initial volume of between 10 and 40 ml of 0.01 N acetic acid. A slight amount of yellow material is left at the top of the column.

The effluent is evaporated to dryness, freed of acetic acid as before, and dissolved in 2 ml of water. It is stored in the freezer at $-22°$ in this form.[5]

Analytical Data

Volumes of 10 μl were diluted into water and into 1.0 M KCN to final volumes of 1 ml and read against the corresponding blanks. In the absence of cyanide the wavelengths of maximum absorption (266 mμ) and of minimum absorption (249 mμ) agree exactly with standard values.[6] In the presence of cyanide this is also true for the maximum (325 mμ), but the minimum absorbance near 260 mμ was too small to be determined accurately. The inflections on either side of 266 mμ in the absence of cyanide correspond in wavelength and relative height to those recorded for the standard material. The following represent ratios of absorbance at the wavelengths indicated, compared to the corresponding values (in parentheses) given or calculated for the standard material[6]:

$A_{250}:A_{260}$	$A_{280}:A_{260}$	$A_{266}:A_{249}$	$A_{266}:A_{325(CN)}$
0.85 (0.85)	0.29 (0.27)	1.29 (1.28)	0.73 (0.74)

[5] G. W. E. Plaut and K. A. Plaut, *Arch. Biochem. Biophys.* **48**, 189 (1954).
[6] P-L Biochemicals, Inc., Circular OR-18, Milwaukee, Wisconsin, April, 1961.

Yield of NMN-7-^{14}C, based on an absorbance of 0.221 at 266 mμ of a 100-fold diluted sample, is 9.6 micromoles. Overall yield, based on recovered radioactivity, is 16.7%. Specific activity, see DPN+-^{14}C.

Other biological materials, such as *Lactobacillus arabinosus* 17-5,[7] erythrocytes,[8] *Saccharomyces cerevisiae*,[9] various bacterial, yeast, and protozoal nicotinic acid auxotrophs,[10] *Astasia longa*,[11] and doubtless others have been or may be used to prepare ^{14}C-labeled DPN+ from ^{14}C-labeled nicotinic acid. The synthesis of DPN+ from nicotinic acid by three partially purified enzymes has been described by Imsande, Preiss, and Handler (Vol. VI [44]).

In the case of *P. shermanii* the Sephadex step may be omitted. In two preparations of NMN-7-^{14}C without this step and starting with 100 μCi of niacin-7-^{14}C the yields were between 12.6 and 13.2 micromoles of NMN-7-^{14}C, and the specific activity was increased. The Sephadex treatment might possibly be of use with some other starting materials, but this has not been tested.

[7] D. E. Hughes and D. H. Williamson, *Biochem. J.* **51**, 330 (1952).
[8] J. Preiss and P. Handler, *J. Biol. Chem.* **233**, 488 (1958).
[9] T. K. Sundaram, K. V. Rajagopalan, C. V. Pichappa, and P. S. Sarma, *Biochem. J.* **77**, 145 (1960).
[10] J. Imsande, *Biochim. Biophys. Acta* **82**, 445 (1964).
[11] V. Kahn and J. J. Blum, *J. Biol. Chem.* **243**, 1441 (1968).

[106] Preparation of Labeled Pyridine Ribonucleotides and Ribonucleosides

By Kunihiro Ueda, Hirohei Yamamura, and Yasutomi Nishizuka

I. Nicotinate Ribonucleotide (NaMN) and Nicotinate Adenine Dinucleotide (Deamido-NAD)

Nicotinate ribonucleotide (NaMN) and nicotinate adenine dinucleotide (deamido-NAD) have been established to be intermediates of NAD biosynthesis both from nicotinate[1] and from tryptophan.[2] Methods for preparing the radioactive NaMN employ nicotinate phosphoribosyltransferase (EC 2.4.2.11; nicotinate nucleotide:pyrophosphate phosphoribosyltrans-

[1] J. Preiss and P. Handler, *J. Biol. Chem.* **233**, 488, 493 (1958).
[2] Y. Nishizuka and O. Hayaishi, *J. Biol. Chem.* **238**, 3369 (1963).

ferase), which catalyzes the formation of NaMN from nicotinate and 5-phosphoribosyl 1-pyrophosphate (PRPP).[3] NaMN-^{14}C may be converted further to deamido-NAD-^{14}C with deamido-NAD pyrophosphorylase.[1]

A. NaMN-NICOTINATE-^{14}C AND DEAMIDO-NAD-NICOTINATE-[14]C

Principle

Nicotinate-^{14}C ribonucleotide is prepared from commercially available nicotinate-^{14}C and PRPP with the use of nicotinate phosphoribosyltransferase. The latter enzyme is prepared from the acetone powder of human erythrocytes[1] or from bakers' yeast.[4] The crude erythrocyte extract contains also deamido-NAD pyrophosphorylase. In addition, the erythrocyte preparation contains 5-phosphoribose pyrophosphokinase. Ribose 5-phosphate and ATP may be used, therefore, to generate PRPP.

Preparation of Erythrocyte Nicotinate Phosphoribosyltransferase

The following procedures are carried out at 0–4° unless otherwise specified. Human whole blood obtained from a blood bank is centrifuged for 10 minutes at 4000 g. The precipitate is suspended in 0.85% NaCl and spun down as above. This procedure is repeated once more. The blood corpuscles thus washed are suspended and stirred briefly in about 20 volumes of cold acetone (−20°), and filtered rapidly on a Büchner funnel with Whatman No. 1 filter paper. The residue is treated twice more with cold acetone. The final residue is quickly dried at room temperature and manually reduced to powder. After being dried *in vacuo*, the powder may be stored at −15°. The nicotinate phosphoribosyltransferase activity is retained in this form for at least several months. The acetone powder is suspended in 0.05 M Tris-Cl buffer, pH 7.4 (50 ml per 10 g of powder), and stirred for 20 minutes at room temperature. The suspension is centrifuged for 20 minutes at 20,000 g. The supernatant solution (crude acetone powder extract) is employed for the preparation of NaMN and deamido-NAD without further purification.

Preparation of Yeast Nicotinate Phosphoribosyltransferase

Nicotinate phosphoribosyltransferase is purified approximately 100-fold from bakers' yeast by the method described by Honjo in this volume.[4] The preparation is free of deamido-NAD pyrophosphorylase.

[3] T. Deguchi, A. Ichiyama, Y. Nishizuka, and O. Hayaishi, *Biochim. Biophys. Acta* **158**, 382 (1968).

[4] This volume [112]. See also T. Honjo, S. Nakamura, Y. Nishizuka, and O. Hayaishi, *Biochem. Biophys. Res. Commun.* **25**, 199 (1966).

Procedure

Method A. This method produces NaMN-^{14}C as well as deamido-NAD-^{14}C in nearly the same quantities. Ten micromoles of sodium nicotinate-7-[14]C (Radiochemical Centre, 10.7 mCi/millimole) is incubated with 60 micromoles of ribose 5-phosphate, 30 micromoles of ATP, 150 micromoles of $MgCl_2$, 500 micromoles of potassium phosphate buffer, pH 7.4, and 15 ml of the acetone powder extract of human erythrocytes in a total volume of 30 ml. The incubation is carried out at 37° with shaking for 6 to 12 hours, depending upon the activity of the enzyme preparation. The reaction may be followed by measuring the formation of radioactive NaMN and deamido-NAD. An aliquot (0.05 ml) of the reaction mixture is taken at an appropriate interval and put on a small Dowex 50 II+ column (X8, 200–400 mesh, 0.5 × 2 cm). The column is washed with 2.5 ml of water. The radioactive products are not adsorbed on this column and are totally recovered in the washing, whereas the substrate, nicotinate-7-^{14}C, is completely adsorbed. The reaction is stopped by the addition of 1 ml of 60% perchloric acid. The denatured protein is removed by centrifugation for 10 minutes at 20,000 *g*. The precipitate is washed three times with 10 ml each of 2% perchloric acid. The supernatant and washings are combined and neutralized with 2 *N* KOH. Potassium perchlorate is removed by centrifugation. The clear supernatant solution is put on a Dowex 1-formate column (X2, 200–400 mesh, 0.8 × 40 cm). The column is washed with 100 ml of water, then eluted with a convex concentration gradient of formic acid; the mixing chamber initially contains 300 ml of 0.01 *N* formic acid; as the elution proceeds, 0.20 *N* formic acid (300 ml total) is introduced into the chamber. Fractions (10 ml each) are collected. Nicotinic acid which remains unaltered is eluted in this step. Then NaMN and deamido-NAD are eluted from the column by introducing 400 ml of 2.0 *N* formic acid into the chamber. The radioactive NaMN and deamido-NAD are usually recovered as sharp peaks around fractions 45 and 56, respectively. The fractions containing each nucleotide are combined and evaporated to dryness under reduced pressure at 10°–15°. The yields of NaMN and deamido-NAD are usually 40–50% and 30–40%, respectively.

Method B. This method may be employed only for the preparation of NaMN-[14]C. Ten micromoles of sodium nicotinate-7-^{14}C are incubated with 20 micromoles of PRPP, 150 micromoles of ATP, 90 micromoles of $MgCl_2$, 1.5 millimoles of potassium phosphate buffer, pH 7.8, and 0.2 unit[4] of yeast nicotinate phosphoribosyltransferase in a total volume of 15 ml. The incubation is carried out for 1 hour at 37°. The reaction is terminated by heating for a minute in a boiling water bath, and the mixture is cooled rapidly in an ice bath. After the denatured protein is removed by centrifugation, the supernatant solution is applied to a Dowex 1-formate column, and the product is eluted from the column as described in Method A. Nicotinate may be converted quantitatively to its ribonucleotide.

Properties

NaMN-^{14}C thus prepared shows no detectable radiochemical impurity upon paper chromatography and has almost the same specific activity as the nicotinate employed.

B. NaMN-RIBOSE-[14]C

Principle

NaMN labeled with ^{14}C at the ribose portion is prepared from PRPP-^{14}C and nicotinate with yeast nicotinate phosphoribosyltransferase. PRPP-^{14}C may be prepared from glucose-^{14}C by way of glucose 6-phosphate, 6-phosphogluconate, ribulose 5-phosphate, and ribose 5-phosphate.[3] This method is identical with that employed for the preparation of NAD-ribose-(NMN)-[14]C.[5] A detailed account is given in that section.

Procedure

6-Phosphogluconic dehydrogenase is purified by the method of Horecker and Smyrniotis.[6] This preparation contains hexokinase, glucose-6-phosphate dehydrogenase, and pentose phosphate isomerase in sufficient quantities. 5-Phosphoribose pyrophosphokinase and nicotinate phosphoribosyltransferase are purified according to Flaks[7] and Honjo,[4] respectively. The starting reaction mixture contains 300 micromoles of Tris-Cl buffer, pH 7.4, 50 micromoles of $MgCl_2$, 4 micromoles of NADP, 20 micromoles of ATP, 0.5 micromole of D-glucose-[14]C(U) (Radiochemical Centre, 123 mCi/millimole), and a preparation of 6-phosphogluconic dehydrogenase (about 10 units[6]) in a total volume of 5 ml. After the incubation for 60 minutes at 37°, 20 micromoles of nicotinic acid, 5-phosphoribose pyrophosphokinase (about 2 units[7]) and nicotinate phosphoribosyltransferase (about 0.1 unit[4]) are added and the incubation is continued for an additional hour at 37°. The reaction is stopped by heating the mixture in a boiling water bath for a minute. The supernatant solution is separated from denatured proteins by centrifugation in the cold. The supernatant, together with washings of insoluble material, is placed on a Dowex 1-formate column. Subsequent procedures of NaMN isolation are performed as described in the previous part. The overall conversion of glucose to NaMN is around 50%.

[5] This volume [107].
[6] See Vol. I [42].
[7] See Vol. VI [18].

Properties

The specific activity is about one-fifth that expected from glucose, suggesting that NaMN is produced from a nonradioactive precursor, the origin of which is not clear yet. A considerable amount of labeled pentose phosphate remains in the final reaction mixture. This may be employable again for the NaMN synthesis.

C. NaMN-^{32}P

Principle

NaMN-^{32}P is prepared from PRRP-^{32}P and nicotinate. PRPP-^{32}P is prepared from glucose and ATP-γ-^{32}P by the same series of enzymatic reactions as employed in the preparation of NaMN-ribose-^{14}C described above.

NaMN-^{32}P may be prepared alternatively from nicotinate ribonucleoside and ATP-γ-^{32}P by the use of nicotinate ribonucleoside kinase.[8]

Procedure

ATP-γ-^{32}P is obtained from commercial sources or prepared by photophosphorylation of ADP by the method of Jagendorf and Avron.[9] All enzymes employed are the same preparation as used for NaMN-ribose-^{14}C (see the previous part). The starting reaction mixture (5 ml) contains 300 micromoles of Tris-Cl buffer, pH 7.4, 50 micromoles of MgCl$_2$, 4 micromoles of NADP, 1 micromole each of ATP-γ-^{32}P (100 mCi/millimole) and D-glucose, and a preparation of 6-phosphogluconic dehydrogenase (about 10 units[6]). As described in the previous part, the latter preparation contains hexokinase, glucose-6-phosphate dehydrogenase, and pentose phosphate isomerase as well as 6-phosphogluconic dehydrogenase. The initial incubation is carried out for 30 minutes at 37°, the incubation period being checked beforehand to be sufficient to convert glucose quantitatively to pentose phosphate. Then, to the above mixture are added 20 micromoles of nonradioactive ATP, 20 micromoles of nicotinic acid, 5-phosphoribose pyrophosphokinase (about 2 units[7]), and nicotinate phosphoribosyltransferase (about 0.1 unit[4]); the incubation is continued for an additional hour at 37°. The reaction is terminated by heating in a boiling water bath for 1 minute. The subsequent procedure for NaMN isolation is the same as described above. The overall efficiency of a transfer of phosphate-^{32}P from ATP to NaMN is around 50%.

[8] This volume [116].
[9] A. T. Jagendorf and M. Avron, *J. Biol. Chem.* **231**, 1277 (1958).

Properties

NaMN-^{32}P thus prepared is more than 95% pure radiochemically as judged by paper chromatography. The specific activity is approximately one-third as compared with that of ATP-γ-^{32}P employed.

II. Nicotinamide Ribonucleotide (NMN)

Employing several enzymatic reactions, various labeled NMN's have been prepared.[10] NMN-^{32}P is prepared by the use of snake venom phosphodiesterase from NAD-both phosphates-^{32}P isolated from yeast cells grown in a medium containing ^{32}P.[5] NMN-nicotinamide-^{14}C is prepared from nicotinamide-^{14}C, PRPP, and ATP by nicotinamide-specific phosphoribosyltransferase from *Lactobacillus fructosus*.[11,12] NMN-ribose-^{14}C is prepared from PRPP-^{14}C, nicotinamide, and ATP by nicotinamide phosphoribosyltransferase from *L. fructosus*. PRPP-^{14}C may be synthesized by the identical method described for NaMN-ribose-^{14}C (see above).

III. Nicotinate Ribonucleoside (NaR) and Nicotinamide Ribonuceloside (NR)

Various radioactive nicotinate and nicotinamide ribonucleosides may be prepared from the respective ribonucleotides by removing the phosphate group with either snake venom 5'-nucleotidase, prostatic phosphomonoesterase or *Escherichia coli* alkaline phosphatase. The product is isolated by applying the preparation mixture on a Dowex 1-formate column (X2, 200–400 mesh) and washing the column with water. Both kinds of nucleosides are recovered in the effluents before and after washing with water, and the remaining precursors are left on the column.[13]

[10] Y. Nishizuka, K. Ueda, K. Nakazawa, and O. Hayaishi, *J. Biol. Chem.* **242**, 3164 (1967).
[11] This volume [113].
[12] NMN-nicotinamide-^{14}C may be obtained also by the phosphodiesterase digestion of NAD-nicotinamide-^{14}C prepared by an exchange reaction (see Vol. IV [34]).
[13] See Vol. III [129].

[107] Preparation of Various Labeled NAD's

By KUNIHIRO UEDA and HIROHEI YAMAMURA

NAD labeled with radioisotopes at a specified position(s) is prepared by one of the following methods: (1) the extraction of labeled NAD from microorganisms grown in a medium containing radioisotopes[1–4] or from

[1] See Vol. IV [34].

animals given an injection of a labeled compound[1]; (2) the enzymatic synthesis from ATP and NMN, either of which is labeled beforehand by some means at a specified position(s)[2,5]; (3) the enzymatic exchange of the nicotinamide moiety of NAD with externally added radioactive nicotinamide[1,6]; and (4) the enzymatic or nonenzymatic introduction of tritium into a specific position of NAD.[4,5] Method (1) can afford labeled NAD in large quantities and methods (2) or (4) can produce labeled NAD with high specific activity. Isolation of NAD from the extract or reaction mixture is performed by ion-exchange column chromatography.[7] The radiochemical purity of labeled NAD is checked by paper chromatography.[8] The concentration is determined either enzymatically or nonenzymatically.[9]

I. NAD-Both Phosphates-^{32}P

Principle

^{32}P-NAD evenly labeled at both phosphates is obtained by cultivating yeast cells in a ^{32}P-containing medium and extracting them with hot water.[2] This is a modification of the previous method,[1,3] and is capable of giving NAD with much higher specific activity. Another method employing a mouse[1] gives ^{32}P-NAD with lower specific activity.

Procedure

Saccharomyces cerevisiae (ATCC 7753) is cultured for 15 hours at 37° with constant aeration in a medium containing 75 g of glucose, 1.5 g of

---

[2] Y. Nishizuka, K. Ueda, K. Nakazawa, and O. Hayaishi, *J. Biol. Chem.* **242**, 3164 (1967).

[3] S. F. Velick, J. E. Hayes, Jr., and J. Harting, *J. Biol. Chem.* **203**, 527 (1953).

[4] S. Chaykin, *Advan. Tracer Methodology* **3**, 1 (1966).

[5] S. B. Zimmerman, J. W. Little, C. K. Oshinsky, and M. Gellert, *Proc. Natl. Acad. Sci. U.S.* **57**, 1841 (1967).

[6] L. J. Zatman, N. O. Kaplan, and S. P. Colowick, *J. Biol. Chem.* **200**, 197 (1953).

[7] The cell extract or reaction mixture, neutralized if necessary, is put on a Dowex 1-formate column (X2, 200–400 mesh; 0.8 × 40 cm). The column is eluted with a linear concentration gradient of formic acid: 200 ml of water in the mixing chamber and 200 ml of 0.15 N formic acid in the reservoir. Fractions of 8 ml each are collected at a flow rate of 20 ml per hour, and assayed for absorbancy at 260 mμ or for radioactivity. The fractions of NAD appearing around 0.1 N formic acid are concentrated to dryness either with lyophilization or flash evaporation, and the residue is dissolved in a suitable buffer solution. See also Vol. III [124].

[8] The solvent systems convenient for separation of NAD from many other nucleotides are (1) isobutyric acid–conc. NH_4OH–H_2O (66:1:33, v/v/v) containing 0.1 mM EDTA, and (2) 1 M ammonium acetate (pH 3.8)–ethanol (3:7, v/v). See also R. H. Reeder, K. Ueda, T. Honjo, Y. Nishizuka, and O. Hayaishi, *J. Biol. Chem.* **242**, 3172 (1967).

[9] See Vol. III [128].

yeast extract (Difco), 1.5 g of Bacto-Peptone (Difco) and about 10 mCi of carrier-free orthophosphate-^{32}P (Radiochemical Centre) in a total volume of 1.5 liters. Cells are collected with a Sorvall centrifuge (GSA rotor) at 5000 rpm for 10 minutes and washed twice with saline. Packed yeast cake (wet weight about 5 g) is suspended in 10 ml of hot water and stirred for 5 minutes at about 90°.[10] The hot suspension is centrifuged for 10 minutes at 5000 rpm in a Sorvall SS-34 rotor and the supernatant solution is removed. The precipitate is again suspended in 10 ml of hot water, then stirred and centrifuged as above. Both supernatant solutions are combined and subjected to the isolation procedure using a Dowex 1 column for chromatography. The final yield of NAD is usually 3–5 micromoles or more.

Properties

^{32}P-NAD prepared as above is more than 95% pure radiochemically and has a specific activity of about 5 mCi/millimole. Both phosphates in NAD are equally labeled, as judged from the fact that the NAD gives equal amounts of ^{32}P-AMP and ^{32}P-NMN upon digestion with snake venom phosphodiesterase.

Preparation of NAD-NMN-^{32}P and NAD-AMP-^{32}P^2

NAD labeled at the phosphate in the NMN portion is prepared from ^{32}P-NMN and ATP by the use of NAD pyrophosphorylase (see below). ^{32}P-NMN is prepared from the above NAD by phosphodiesterase digestion. NAD labeled at phosphate in the AMP grouping is prepared similarly from ATP-α-^{32}P and NMN with the pyrophosphorylase. ATP-α-^{32}P may be obtained commercially or prepared from ^{32}P-AMP[11] which is isolated from a hot-water extract of *S. cerevisiae*.

II. NAD-Adenine-^{14}C

Principle

NAD labeled with ^{14}C at the adenine ring is prepared from commercially available ^{14}C-ATP and NMN by the use of NAD pyrophosphorylase.[2]

Procedure

The reaction mixture contains 300 micromoles of Tris-Cl buffer (pH 7.4), 30 micromoles of MgCl$_2$, 30 micromoles of NaF, 3 micromoles of

---

[10] S. Williamson and D. E. Green, *J. Biol. Chem.* **135,** 345 (1940).
[11] The method to prepare ATP-β,γ-^{32}P from AMP is applicable to this case. See, for example, A. T. Jagendorf and M. Avron, *J. Biol. Chem.* **231,** 277 (1958).

nicotinamide, 30 micromoles of NMN, 2.4 micromoles of ATP-8-^{14}C (Radiochemical Centre, 20 mCi/millimole), and purified NAD pyrophosphorylase (about 0.5 unit) in a total volume of 3 ml. The latter enzyme is partially purified from *Lactobacillus fructosus* as described in this volume.[12] The enzyme prepared from hog liver according to Kornberg[13] can also be used. The incubation is carried out for 15 minutes at 37° and terminated by heating in a boiling water bath for a minute. The supernatant solution after a low-speed centrifugation, together with washings of the precipitate, is applied to a Dowex 1-formate column. The elution must be done with a relatively gentle gradient of formic acid (0 to 0.15 N) in order to avoid the contamination of NAD with AMP which is, more or less, produced by this enzyme preparation. The efficiency to convert ATP to NAD is about 80%.

Properties

NAD-adenine-^{14}C prepared as above is almost completely pure radiochemically. The specific activity is identical with that of the ATP employed.

Preparation of Adenosine-Labeled NAD

NAD-adenosine-^{14}C(U) and NAD-adenosine-T(G) may be prepared by the method described above from commercially available ATP-^{14}C(U) and ATP-T(G), respectively.

III. NAD-Ribose(NMN)-^{14}C

Principle

NAD carrying ribose-^{14}C in the NMN portion is prepared enzymatically from glucose-^{14}C through a number of intermediates; glucose 6-phosphate, 6-phosphogluconate, ribulose 5-phosphate, ribose 5-phosphate, 5-phosphoribosyl 1-pyrophosphate, and NMN.[2] Enzymes to catalyze each step are purified separately by the methods specified. A series of reactions involved is carried out successively in a single tube with no separation of intermediates on the way, but at a specified step the reaction mixture is heated briefly in order to inactivate the interfering enzymes which may contaminate the enzyme preparations used.

The NAD which has riboses unequally labeled and of lower specific activity is prepared more easily from ^{14}C-ribose by injection into mice as described by Colowick and Kaplan.[1]

---

[12] This volume [113].
[13] A. Kornberg, *J. Biol. Chem.* **182**, 779 (1950). See also Vol. II [116].

Procedure

6-Phosphogluconic dehydrogenase is purified from dried brewers' yeast by the method of Horecker and Smyrniotis.[14] The preparation (acid ammonium sulfate fraction) contains sufficient amounts of hexokinase, glucose-6-phosphate dehydrogenase, and pentose phosphate isomerase, and is employed as the source of these enzymes altogether. This will be referred to as yeast enzyme complex in this article. 5-Phosphoribose pyrophosphokinase is obtained from pigeon livers by the method of Flaks.[15] Nicotinamide phosphoribosyltransferase and NAD pyrophosphorylase are purified from *L. fructosus* to the stage of ammonium sulfate fractionation and protamine treatment, respectively, as described by Ohtsu and Nishizuka[12] and concentrated severalfold with a collodion bag (Sartorius Membranfilter Company, Germany). The starting mixture contains 300 micromoles of Tris-Cl buffer (pH 7.4), 80 micromoles of $MgCl_2$, 3 micromoles of NADP, 30 micromoles of ATP, 0.3 micromole of D-glucose-^{14}C(U) (Radiochemical Centre, 123 mCi/millimole), and yeast enzyme complex (about 10 units[14] of 6-phosphogluconic dehydrogenase) in a total volume of 3.7 ml. After incubation for 60 minutes at 37°, 3 micromoles of nicotinamide 5-phosphoribose pyrophosphokinase (about 2 units[15]) and nicotinamide phosphoribosyltransferase (about 0.03 unit[12]) are added. The volume increases to 5.8 ml at this stage. The incubation is continued for an additional 60 minutes at 37°. The mixture is then heated in a boiling water bath for 1 minute, cooled rapidly, and again incubated further for 60 minutes at 37° together with NAD pyrophosphorylase (about 0.3 unit[12]) and an additional 20 micromoles of ATP. The reaction is terminated by heating for 1 minute in a boiling water bath. The mixture is cooled in an ice bath, and is centrifuged to precipitate the denatured protein. The supernatant solution as well as washings of the sediment are applied to a Dowex 1 column. Since $^{14}CO_2$ is produced in this reaction, it is recommended that the entire reaction be performed in a Thunberg-type airtight tube furnished with an alkali reservoir. The overall conversion of glucose to NAD is around 50%.

Properties

NAD thus prepared has no detectable radiochemical impurity upon paper chromatography, and the label is introduced entirely into the ribose of the NMN half. The specific activity is about one-sixth that expected from glucose. A preliminary examination suggests that nonradioactive NAD is produced partly from NADP by phosphatases which contaminate the enzyme preparations employed. Other labeled products that appear

---

[14] See Vol. I [42].
[15] See Vol. VI [18].

in the final reaction mixture are mostly pentoses and pentose phosphates, as judged by Dowex 1 column and paper chromatographies. The latter compounds are again employable as the starting material for NAD synthesis. The NAD prepared from pentose phosphates once isolated has approximately twice the specific activity of that prepared directly from glucose. The yield of NAD as well as the distribution of radioactivity in the various intermediates is affected markedly by several factors, such as the enzyme activities, their quantitative combination, and the incubation period. It is, therefore, preferable to determine the reaction conditions with available enzyme species. Although hexokinase and pentose phosphate isomerase are included in the yeast enzyme complex in sufficient quantities to prepare several micromoles of NAD, the addition of these enzymes purified separately may be necessary in a large-scale preparation of the NAD.

IV. NAD-Nicotinamide-[14]C

Principle

The exchange reaction between the nicotinamide moiety of NAD and free nicotinamide catalyzed by NAD glycohydrolases from animal sources has been used for the preparation of nicotinamide-labeled NAD.[1,6] With this method, however, the yield is poor, being 20% at maximum, and the specific activity is low. An alternative method to be described consists of two enzymatic reactions, catalyzed by nicotinamide phosphoribosyltransferase and NAD pyrophosphorylase. The latter method gives a better yield of NAD with higher specific activity.

Procedure

The two enzymes are purified from L. fructosus as described by Ohtsu and Nishizuka.[12] The reaction mixture contains 500 micromoles of Tris-Cl buffer (pH 7.4), 50 micromoles of $MgCl_2$, 30 micromoles of 5-phosphoribosyl 1-pyrophosphate, 30 micromoles of ATP, 8.3 micromoles of nicotinamide-7-[14]C (Radiochemical Centre, 60 mCi/millimole), nicotinamide phosphoribosyltransferase (0.3 unit), and NAD pyrophosphorylase (0.5 unit) in a total volume of 5 ml. Incubation is carried out for 60 minutes at 37°, and is stopped by heating for 1 minute in a boiling water bath. The supernatant solution after a low-speed centrifugation and the washings of the precipitate are combined and placed on a Dowex 1 column. The yield is more than 70%.

Properties

The radiopurity of NAD thus prepared is more than 98%, and the specific activity is nearly the same as that of nicotinamide employed. The crude extract of *L. fructosus*[12] may substitute for the above two enzyme preparations when nonradioactive nicotinamide is removed by dialysis.

V. NAD-Nicotinamide-T

Principle

Tritium is introduced into the 4-position of the nicotinamide moiety of NAD from glucose-1-T through the reduction and reoxidation of pyridine nucleotides. Four consecutive enzymatic reactions are involved: Glucose-1-T is first converted to glucose-1-T 6-phosphate by hexokinase; the tritium in the latter compound is transferred by glucose-6-phosphate dehydrogenase to the 4-position of NADP on the β-side with respect to the plane of the pyridine ring; NADPT is converted to NADT by alkaline phosphatase; and finally NADT is oxidized to T-NAD by lactic dehydrogenase with the removal of hydrogen from the α-side of the 4-position, leaving the tritium on the β-side of NAD. This method is essentially identical with that of Zimmerman *et al.*[5] Nonenzymatic labeling with tritium at other position(s) in the nicotinamide portion is described by Chaykin.[4]

Procedure

Hexokinase and glucose-6-phosphate dehydrogenase are obtained from Boehringer, Mannheim, Germany. Alkaline phosphatase (type III from *E. coli*) and lactic dehydrogenase free of pyruvate kinase (type II from rabbit muscle) are the products of Sigma Chemical Company. It is convenient to carry out the whole reaction in a quartz cuvette and to follow the process of reaction by measuring the optical density at 340 mμ (OD$_{340}$). The starting reaction mixture contains 100 micromoles of Tris-Cl buffer (pH 7.6), 25 micromoles of MgCl$_2$, 2 micromoles of ATP, 0.45 millimole of NADP, 0.45 millimole of glucose-1-T (Radiochemical Centre, 250 mCi/millimole), hexokinase (30 units), and glucose-6-phosphate dehydrogenase (1.5 units) in a total volume of 3 ml. At zero time the OD$_{340}$ is 0.025. The mixture is incubated at room temperature for about 5 minutes to give an OD$_{340}$ of 0.825. Alkaline phosphatase (10 units) is then added to the mixture and the cuvette is maintained at 37° for 30 minutes. The OD$_{340}$ still increases gradually to 0.840 during this period. Finally, 2 micromoles of pyruvate and lactic dehydrogenase (40 units) are added and the mixture (3.1 ml at this stage) is incubated for 15 minutes at 37°. The OD$_{340}$ decreases to 0.060. The whole reaction mixture is then put onto a Dowex 1-formate column. The overall yield is around 50%.

Properties

T-NAD thus prepared is more than 98% pure radiochemically and has a specific activity of about 70 mCi/millimole. The specific activity remains practically constant for at least several months when stored frozen at $-10°$.

[108] Preparation and Properties of α-NAD⁺ and α-NADH

By Hiroshi Okamoto

I. Preparation of α-NAD⁺ and α-NADH

Principle[1]

α-NAD⁺ is separated from a commercial NAD⁺ preparation by employing an ion-exchange chromatographic method. α-NADH is prepared by chemical reduction of α-NAD⁺.

Procedure

All steps of the procedure are carried out at room temperature. Ten grams of NAD⁺ preparation (C. F. Boehringer & Soehne, Mannheim) are dissolved in about 60 ml of deionized H_2O. The solution is placed on a Dowex 1-formate column (200–400 mesh, 3.0 × 80 cm). The separation of adsorbed α- and β-NAD⁺ is performed by elution with 0.09 N formic acid containing 0.01 % EDTA. Fifteen-milliliter fractions are collected at the rate of one tube every 10 minutes. By measuring the optical density at 260 mμ of the fractions, two peaks of β-NAD⁺ (between about 1350 ml and 2625 ml effluent volume) and α-NAD⁺ (between about 3090 ml and 3600 ml effluent volume) are obtained. The α-NAD⁺ portion of the effluent is collected and lyophilized. The lyophilized powder is dissolved in about 20 ml of deionized H_2O and then lyophilized twice to eliminate formic acid. Thus, 10 g of the NAD⁺ yields about 160 mg of α-NAD⁺, the identity of which is established by the following criteria: (a) ultraviolet spectra in the presence and the absence of cyanide and that of the reduced form,[2,3] (b) ion-exchange chromatography with a Dowex 1-formate column,[4] (c) inactivity with alcohol dehydrogenase and muscle lactate dehydrogenase,[2,3] (d) inability to serve as substrate for a *Neurospora* NAD glycohydrolase,[2,3] and (e) optical activity.[3]

α-NADH is prepared by chemical reduction of α-NAD⁺ with hydro-

---

[1] H. Okamoto, A. Ichiyama, and O. Hayaishi, *Arch. Biochem. Biophys.* **118**, 110 (1967).

[2] See Vol. III [129].

[3] N. O. Kaplan, M. M. Ciotti, F. E. Stolzenbach, and N. R. Bachur, *J. Am. Chem. Soc.* **77**, 815 (1955).

[4] S. Suzuki, K. Suzuki, T. Imai, N. Suzuki, and S. Okuda, *J. Biol. Chem.* **240**, PC554 (1965).

sulfite and isolated as the barium salt according to the procedure of Lehninger.[5] The starting material (α-NAD$^+$) is about 90% pure, and the final product of α-NADH (Ba salt) is estimated to be approximately 60–70% pure. As described by Woenckhaus and Zumpe,[6] α-NADH, in solution, is slowly converted to the β-form; contamination by the β-isomer of the α-NADH preparation is checked by the pyruvate and lactate dehydrogenase system. The trace amounts of contaminating β-NADH in the α-NADH preparation can be eliminated by this system.

Physical Properties

The physical properties of α-NAD$^+$ and α-NADH were studied first by Kaplan *et al.*,[2,3] and later by Pfleiderer *et al.*[7] Spectrophotometric constants of α-NAD$^+$ and α-NADH are shown in Table I.

TABLE I
SPECTROPHOTOMETRIC CONSTANTS OF α-NAD$^+$ AND α-NADH[a]

	λ_{max} (mμ)	$\epsilon \times 10^{-3}$
α-NAD$^+$	260	17.9
α-NADH	259	14.9
	344	5.6

[a] From Pfleiderer *et al.*[7]

II. α-NADH Oxidizing-Enzyme System in Mammalian Cells

In contrast to the data concerning the physical properties of α-NAD$^+$, there is no available information concerning its biological significance.

There is, however, some evidence for the existence of several enzyme systems in mammals catalyzing the transfer of electrons from α-NADH to electron acceptors such as cytochrome c, 2,6-dichlorophenolindophenol (2,6-DCPIP), O$_2$,[1] etc., as described below.

Assay Method

α-NADH-Cytochrome c Reductase Activity. The assay system contains 100 mM Tris-acetate buffer, pH 7.5, 0.1 mM α-NADH,[8] 0.05 mM cytochrome c, and 1 mM NaCN, in a final volume of 3 ml. The reaction is carried out at 24° in a glass cuvette with 1-cm light path, and the reduction

[5] See Vol. III [126].

[6] C. Woenckhaus and P. Zumpe, *Biochem. Z.* **343**, 326 (1965).

[7] G. Pfleiderer, C. Woenckhaus, and M. Nelböck-Hochstetter, *Ann. Chem.* **690**, 170 (1965).

[8] Dissolved in 20 mM Tris-acetate buffer, pH 9.0, just before assay.

of cytochrome *c* is followed by measuring the increase in optical density at 550 mμ in a spectrophotometer.

α-NADH-2,6-DCPIP Reductase Activity. The assay system is the same as above except that cytochrome *c* is replaced by 0.04 mM 2,6-DCPIP. The reduction of the dyestuff is recorded at 600 mμ.

Protein Determination. Protein concentrations are estimated by the method of Lowry *et al.*,[9] using crystalline bovine plasma albumin as a standard.

Distribution of α-NADH-Cytochrome c Reductase Activity and α-NADH-2,6-DCPIP Reductase Activity in Rat Liver

As shown in Table II, about 50% of the α-NADH–cytochrome *c* reductase activity of the original homogenates is detected in the microsomal fraction, whereas the activity of the mitochondrial fraction[10] is only 12%. On the other hand, most of the α-NADH-2,6-DCPIP reductase activity is detected in the soluble fraction.

TABLE II

SUBCELLULAR DISTRIBUTION OF α-NADH-CYTOCHROME *c* REDUCTASE AND α-NADH-2,6-DCPIP REDUCTASE ACTIVITIES IN RAT LIVER

Fractions[a]	α-NADH-cytochrome *c* reductase activity (%)	α-NADH-2,6-DCPIP reductase activity (%)
Whole homogenate	100[b]	100[c]
Nuclei	20	12
Mitochondria	12	8
Microsomes	50	15
Soluble supernatant	6	85
Recovery	88	120

[a] Subcellular components were obtained by differential centrifugation by the method of Hogeboom.[11]

[b] 10.0 micromoles of cytochrome *c* are reduced per minute per gram of liver under the conditions employed.

[c] 24.4 micromoles of 2,6-DCPIP are reduced per minute per gram of liver under the conditions employed.

[9] O. H. Lowry, N. J. Rosebrough, A. L. Farr, and R. J. Randall, *J. Biol. Chem.* **193**, 265 (1951).

[10] It should be noted that α-NADH-cytochrome *c* reductase activity in mitochondria, which is devoid of microsomal contamination, is completely insensitive to amytal, rotenone, and antimycin A and is localized in the outer membrane fraction of rat liver mitochondria. Unpublished data [cf. H. Okamoto *et al.*, *Biochem. Biophys. Res. Commun.* **26**, 309 (1967); H. Okamoto and O. Hayaishi, *Arch. Biochem. Biophys.* **131**, 603 (1969)].

[11] See Vol. I [3].

Purification of "Diaphorase" (α-NADH-2,6-DCPIP Reductase) from Pig Heart

All operations of purification are carried out at 0–4°. Fresh pig heart muscle (50 g) is homogenized with 150 ml of cold 0.02 M K_2HPO_4 in a Waring blendor for 2 minutes, and the sediment is removed by centrifugation at 2000 g for 30 minutes. The supernatant fluid is brought to pH 5.4 with 1 N acetic acid and centrifuged at 2000 g for 30 minutes. The sediment containing the bulk of β-NADH-linked diaphorase activity[12] is discarded. The reddish-colored supernatant solution is brought to pH 7.5 with 1 N NaOH and passed through a Sephadex G 25 column (3 × 65 cm) which has been equilibrated with 0.005 M potassium phosphate buffer, pH 7.5. Elution is made with the same buffer. The eluate (80 ml) is passed through a DEAE-cellulose column (2.4 × 10 cm) that has previously been equilibrated with 0.005 M potassium phosphate buffer, pH 7.5, and then the column is washed with 40 ml of the same buffer. The combined washings (about 120 ml) are applied to a hydroxylapatite column (1.1 × 18.5 cm) that has been previously equilibrated with 0.005 M potassium phosphate buffer, pH 7.5. After the column is washed with 30 ml of 0.05 M potassium phosphate buffer, pH 7.5, elution is carried out with a linear concentration gradient established between 100 ml of 0.05 M potassium phosphate buffer, pH 7.5 (mixing chamber), and 100 ml of 0.20 M potassium phosphate buffer, pH 7.5 (reservoir). The active portion of the effluent, between 85 and 100 ml effluent volume, is collected. A summary of the enzyme purification is shown in Table III, in which the diaphorase activity with β-NADH is also presented. The enzyme preparation thus obtained is

TABLE III
"Diaphorase" Activities with α- and β-NADH at Each Purification Step

Purification step	ΔOD 600 mμ/min/mg protein			Yield (%)	
	α-NADH	β-NADH	$R^a = \beta/\alpha$	α-NADH	β-NADH
Homogenate	0.08	0.35	4.4	—	—
2000 g supernatant	0.28	1.11	4.0	100	100
pH 5.4 supernatant	0.45	0.83	1.8	88	40
DEAE-cellulose	6.16	9.24	1.5	60	24
Hydroxylapatite	22.2	34.3	1.5	32	12

^{a} R is the ratio of the diaphorase activity with α-NADH to that with β-NADH.

[12] See Vol. II [120].

purified about 270-fold, as judged by its diaphorase activity with α-NADH as electron donor and 2,6-DCPIP as electron acceptor. This reaction product is identified as α-NAD$^+$ by the following criteria: (a) ion-exchange chromatography on a Dowex 1 column,[4] (b) absorption spectra before and after reduction with hydrosulfite,[2,3] (c) failure of coenzyme activity of alcohol dehydrogenase as well as lactate dehydrogenase,[2,3] and (d) inability to serve as substrate for a *Neurospora* NAD glycohydrolase.[2,3] But, the purified enzyme preparation is also found to be active for β-NADH and NADPH with an oxidation rate of approximately 1.5 times that of α-NADH.

The Michaelis constants are 0.100 mM for α-NADH, 0.074 mM for β-NADH, and 0.064 mM for NADPH. At saturating concentrations, the reaction rates with α-NADH and β-NADH are not additive. Dicumarol, irrespective of the three electron donors, inhibits the enzyme 50% at a concentration of 1 mμM.

Finally, it should be mentioned that, from the data described above, it is unknown whether or not the oxidizing ability of α-NADH is truly associated with a function of some hitherto unknown enzyme(s) or is merely a manifestation of broad specificity of such an enzyme contained in pig heart.

[109] Degradation of Nicotinic Acid and Related Compounds

By THOMAS A. SCOTT

Introduction

Decarboxylation is probably the simplest degradation reaction that can be applied to nicotinic acid. Information about the distribution of radioactivity between the pyridine ring and the carboxylic acid group is sufficient for some investigations. For instance, Heidelberger *et al.*[1] showed that, in rats, the radioactivity from 3-^{14}C-labeled tryptophan is incorporated exclusively into C-7 of the urinary *N*-methyl-nicotinamide, by converting the latter into nicotinic acid, followed by decarboxylation. Similarly, decarboxylation was the only chemical process needed to demonstrate that radioactivity from 4-^{14}C-labeled aspartic acid is incorporated

[1] C. Heidelberger, M. E. Gullberg, A. F. Morgan, and S. Lepkovsky, *J. Biol. Chem.* **179**, 143 (1949).

into C-7 of nicotinic acid in *Mycobacterium tuberculosis*,[2] *Serratia marcescens*,[3] and *Clostridium butylicum*.[4]

Unfortunately, the pyridine from the decarboxylation of nicotinic acid-^{14}C has only a limited use; it is not possible to distinguish between C-2 and C-6, or between C-3 and C-5, because pyridine is symmetrical. The problem of devising a satisfactory degradation pathway lies in opening the heterocyclic ring of nicotinic acid to give a high yield of a known product. In the present scheme, this is achieved by reducing the nicotinic acid to hexahydronicotinic acid (nipecotic acid, II), followed by oxidation to 5-amino-4-carboxyvaleric acid (VII). The degradation pathway is outlined in Fig. 1. *N*-Benzoyl-5-amino-4-carboxyvaleric acid (VI) and 5-amino-4-carboxy-valeric acid (VII) are new compounds discovered in the course of this work, and the proof of their structures has been reported.[5] The remainder of the pathway employs organic reactions and compounds already known in the literature. About 100 mg of nicotinic acid-^{14}C is required for a complete analysis.

Reagents

Nicotinic acid, uniformly labeled with ^{3}H in the pyridine ring (specific activity 150–160 mCi/millimole)

Succinic acid-2,3-[3]H (specific activity 100 mCi/millimole)

Rhodium on charcoal, 5%

Hydrogen, suitable for catalytic hydrogenation

Benzoyl chloride

$KMnO_4$, solid

$KMnO_4$, 5% in N H_2SO_4

$NaNO_2$, 2% in water

Urea, solid

$Ba(OH)_2 \cdot 8 H_2O$

Sodium azide, NaN_3 solid

Chromic acid solution, $5 N$ $(= 5/3 M)$, for the Kuhn-Roth determination of methyl groups,[6] prepared by dissolving 166.77 g of chromic anhydride (reagent grade) in 1 liter of distilled water

$MgSO_4$, anhydrous solid

Sulfuric acid, 100%, prepared by mixing 1 part fuming sulfuric acid ("oleum") with 2 parts concentrated sulfuric acid

[2] E. Mothes, D. Gross, H. R. Schutte, and K. Mothes, *Naturwissenschaften* **48**, 623 (1961).

[3] T. A. Scott and H. Hussey, *Biochem. J.* **96**, 9c (1965).

[4] A. J. Isquith and A. G. Moat, *Bacteriol. Proc.* **65**, 74 (1965).

[5] T. A. Scott, *Biochem. J.* **102**, 87 (1967).

[6] N. D. Cheronis and T. S. Ma, "Organic Functional Group Analysis," p. 599. Wiley (Interscience) New York, 1964.

Hydrogen peroxide, 35%
Nitric acid, fuming (specific gravity 1.48)
Sulfuric acid, specific gravity 1.84
Bromine

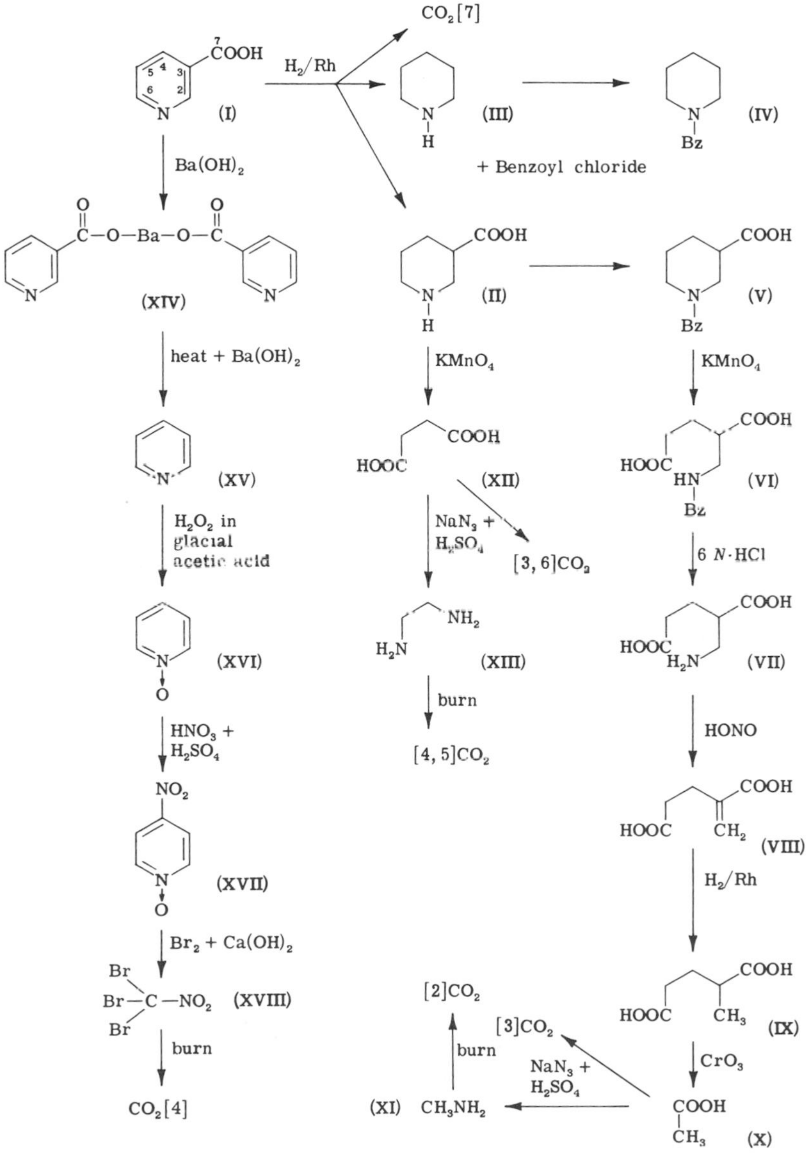

FIG. 1. Degradation scheme for ^{14}C-labeled nicotinic acid.

Calcium hydroxide

Combustion mixture of Van Slyke and Folch,[7] prepared by dissolving 25 g of chromium trioxide and 5 g of potassium iodate in 333 ml of fuming sulfuric acid (specific gravity 1.94) and adding 167 ml of syrupy phosphoric acid; the mixture is heated to 150° for a few minutes and stored in a glass-stoppered bottle, protected from dust.

HCl, concentrated

Acetic acid, glacial

$BaCl_2$, 5% in CO_2-free water

NaOH, 2 N, free from carbonate, prepared by dissolving 40 g of freshly opened "analar" grade NaOH in 100 ml of CO_2-free water to make 10 N NaOH; after standing for 24 hours, any precipitate is removed by filtration through glass wool, the solution is diluted to 2 N with CO_2-free water, and stored in a wax-lined bottle with a CO_2 trap.

Ammonia buffer (pH 8.4), made by dissolving 107 g of NH_4Cl, 87.0 g of K_2HPO_4 and 6.7 ml of ammonia solution (specific gravity 0.880) in 1 liter of distilled water.

CNBr, 10% in water, freshly prepared

Nitrogen, free from oxygen and carbon dioxide

Dowex 50 ion-exchange resin ($\times$8, 200–400 mesh)

Dowex 1 ion-exchange resin

Whatman No. 4 chromatography paper

Silicic acid for thin-layer chromatography

Scintillation mixture, consists of 5 g of 2,5-diphenyloxazole and 100 g of naphthalene dissolved in dioxane in a total volume of 1 liter

Cadmium-ninhydrin reagent; dissolve 0.05 g of cadmium acetate in 5 ml of water, followed by 1 ml of acetic acid and 50 ml of acetone. Shake until all the precipitate has dissolved; then dissolve 0.5 g of ninhydrin in the mixture.

Apparatus. A single-necked, round-bottomed flask of 25-ml capacity, with a B-14 joint is required for hydrogenation (Fig. 2) and for those stages where material is heated under reflux. For hydrogenation, the two-way head shown in Fig. 2 is required, and for heating under reflux a standard B-14 condenser, length 25 cm, is attached to the 25-ml flask.

A 50-ml pear-shaped, two necked flask with B-14 joints as shown in Figs. 3–7, is a versatile piece of apparatus, which can be used in many stages of the degradation procedure. Also required are bubblers as shown in Figs. 2 and 3. Each of the large end-bulbs of these bubblers has a capacity of 15–20 ml.

[7] D. D. Van Slyke and J. Folch, *J. Biol. Chem.* **136,** 509 (1940).

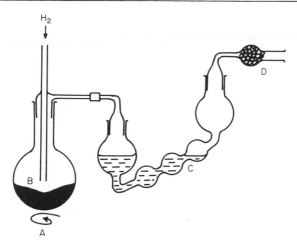

FIG. 2. Apparatus for the catalytic hydrogenation of nicotinic acid. (A) Magnetic stirrer, (B) nicotinic acid + rhodium on charcoal in 2 N H$_2$SO$_4$, (C) 2 N NaOH, (D) soda-lime.

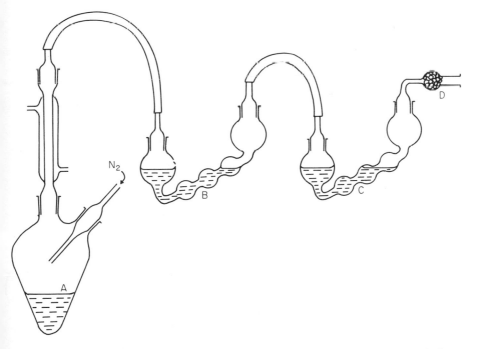

FIG. 3. Apparatus used for the total combustion of samples or for decarboxylation by the Schmidt reaction. (A) Reaction mixture, (B) 5% KMnO$_4$ in N H$_2$SO$_4$, (C) 2 N NaOH, (D) soda-lime.

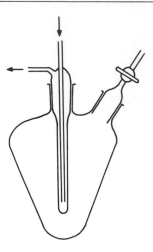

FIG. 4. Apparatus for the vacuum sublimation of nicotinic acid, or for the collection of pyridine (XV) produced by heating barium nicotinate (XIV) with barium hydroxide.

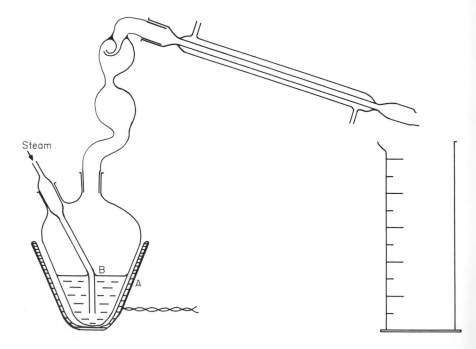

FIG. 5. Apparatus for the steam distillation of acetic acid (X) produced in the Kuhn-Roth oxidation of methylglutaric acid (IX). (A) Electric Bunsen burner, (B) chromic acid oxidation mixture with water and anhydrous MgSO₄.

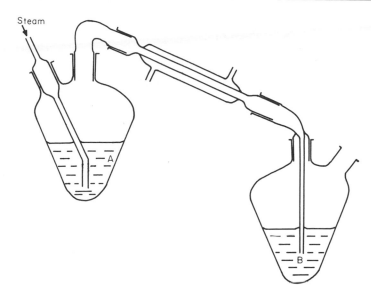

FIG. 6. Apparatus for the steam distillation of the methylamine (XI) produced in the Schmidt degradation of acetic acid (X). (A) Residue from the Schmidt decarboxylation (methylamine sulfate, Na_2SO_4, H_2SO_4) with water and excess alkali, (B) 0.1 N H_2SO_4.

The use of a water-cooled cold finger for the sublimation of nicotinic acid (1) and for the collection of pyridine (XV) from barium nicotinate (XIV) is illustrated in Fig. 4.

The apparatus for the Kuhn-Roth oxidation of 2-methylglutaric acid (IX) is minimal. There is no need for elaborate antisplash arrangements, and the single-spray trap depicted in Fig. 5 has proved adequate for the present work. By using an electric heater, which encloses most of the oxidation mixture during the steam distillation, bumping is reduced to a minimum.

Purification of Nicotinic Acid (I)

Before nicotinic acid-^{14}C is submitted to the chemical degradation, it must be radiochemically pure. The method described here has been used routinely for purifying nicotinic acid from extracts and whole cells of bacteria after incubation with radioactive substrates. It must be modified by the operator, according to the volume of the incubation mixture and the amount of nicotinic acid synthesized.

The incubation mixture is made 2 N with respect to sulfuric acid and autoclaved for 30 minutes at 15 psi and 120°. This hydrolyzes protein to

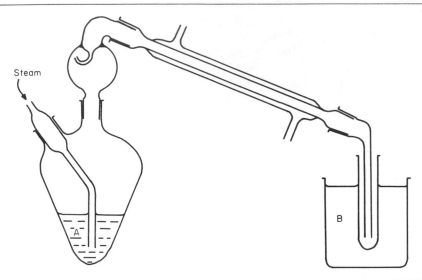

FIG. 7. Apparatus for the production and steam distillation of bromopicrin (XVIII) from 4-nitropyridine-N-oxide (XVII). (A) $Ca(OH)_2 + Br_2 + H_2O + $ 4-nitropyridine-N-oxide, (B) ice water.

amino acids, and NAD, NADP, and nicotinamide to nicotinic acid. For the eventual calculation of the specific radioactivity of the undiluted nicotinic acid, the amount of nicotinic acid in the incubation mixture must now be measured. For this purpose, the hydrolyzate is adjusted to an appropriate standard volume, and aliquots are taken for the assay of nicotinic acid.

The microbiological assay with *Lactobacillus plantarum* is well documented in the literature.[8] It has the disadvantage of being lengthy and tedious, but the great advantage of being very sensitive and requiring very little material; it is accurate in the range 0.025–0.5 μg of nicotinic acid, and for triplicate determinations at three different concentrations only 1 μg of nicotinic acid is needed.

Alternatively, the nicotinic acid can be assayed by a rapid chemical-spectrophotometric method described by Mueller and Fox,[9] and modified in the author's laboratory by Dr. E. Bellion[10]: 1 ml of ammonia buffer is pipetted into a 1-cm 3-ml cuvette, followed by 1 ml of 10% cyanogen bromide. A blank is prepared similarly. The cuvettes are placed in the carriage of an automatic recording spectrophotometer (Gilford 2000).

[8] E. C. Barton-Wright, *Biochem. J.* **38**, 314 (1944).
[9] A. Mueller and S. H. Fox, *J. Biol. Chem.* **167**, 291 (1947).
[10] E. Bellion, *Anal. Biochem.* **26**, 578 (1968).

Water, 0.5 ml, is added to the blank cuvette, and 0.5 ml of the nicotinic acid solution is added to the test cuvette; both solutions are quickly mixed. The increase in extinction at 405 mμ is measured automatically against time. The temperature of the reaction is kept at 20° by the use of a thermostatically controlled cell compartment. The maximum extinction registered on the moving chart is taken as a measure of the concentration of nicotinic acid; this value is reached in about 6 minutes, after which the color decays. With a standard solution of nicotinic acid, a linear calibration is obtained for the concentration range 0.05–40 μg of nicotinic acid per milliliter, with a standard error of less than 1%. Under the above conditions, the maximum extinction at 405 mμ represents a molar extinction coefficient of 5.904 $\times$ 10^6 cm^2/mole.

After sampling for the measurement of nicotinic acid, 100 mg of un-labeled carrier nicotinic acid are added to the remainder of the hydrolyzate. The volume is adjusted to 50 ml and the solution is again made 2 N with respect to sulfuric acid. To destroy the amino acids, 100 ml of 15% sodium nitrite are added slowly to the cooled (0–5°) solution. When the vigorous evolution of nitrogen has subsided, the solution is heated for 20 minutes at 100°, keeping the solution acid by the addition of 2 N H$_2$SO$_4$. At the end of this time, excess nitrous acid is destroyed by the gradual addition of solid urea until no more nitrogen is evolved. The solution is then evaporated to about 10 ml and placed on a column (diameter 2 cm, length 20 cm) of Dowex 50 ($\times$8) (H$^+$ form). A concave gradient of 0–4 N HCl in a total volume of 2 liters is passed through the column, and the nicotinic acid is detected in the effluent by its extinction at 260 mμ. The peak effluent volume of the nicotinic acid is about 1.5 liters when the effluent HCl is 2 N. The fractions containing nicotinic acid are evaporated to dryness.[11] The residue is redissolved in 20 ml of water and placed on a column (2 $\times$ 20 cm) of Dowex 1 (formate form). A linear gradient of 0–0.1 N formic acid in a total volume of 1.2 liters is passed through the column, and the fractions containing nicotinic acid (peak effluent volume about 1.025 liters) are evaporated to dryness as a thin layer on the inside surface of a pear-shaped sublimation flask (Fig. 4). The flask is evacuated to a pressure of 0.01 mm Hg and heated in an oil bath. Nicotinic acid sublimes onto the cold finger over the temperature range 120–180°.[12]

[11] At this stage the solution will probably contain ferric chloride, since practically all commercial HCl contains iron, and ferric ions and nicotinic acid are not separated on Dowex 50(H$^+$). If the mixture becomes too hot during evaporation, the nicotinic acid may be oxidized by the ferric chloride and thereby lost. Temperatures higher than 50° should, therefore, be avoided

[12] If traces of moisture are still present in the nicotinic acid, the sublimation will fail at first, and water droplets will appear on the cold finger. This water is removed and the

The sublimed nicotinic acid should be radiochemically pure in two-dimensional paper chromatography: first dimension, butan-1-ol–acetic acid–water (4:1:1 by volume); second dimension, ethanol–water–aqueous NH_3 (specific gravity 0.88) (18:1:1 by volume). The possibility of contamination with other radioactive materials will vary according to the organism and the substrate used in the original incubation. With any new biosynthetic system, the suitability of the above purification scheme should therefore not be taken for granted, but it should be confirmed.

Procedure

Collection of CO_2 and Conversion to $BaCO_3$. Carbon dioxide is produced by totally burning samples or by specific decarboxylation reactions. The carbon dioxide is flushed from the apparatus (Fig. 3) with CO_2-free N_2. When necessary, the effluent gases are passed through a "scrubber" of 5% $KMnO_4$ in N H_2SO_4 to remove sulfur oxides. The CO_2 is then trapped in 5 ml of carbonate-free 2 N NaOH. After completion of the combustion or decarboxylation, the apparatus is flushed for 15–20 minutes with a slow stream of CO_2-free N_2 to ensure that all the CO_2 is trapped by the NaOH. The carbonated 2 N NaOH is then transferred to a glass-stoppered 25-ml tube; the bubbler is washed with a few milliliters of CO_2-free water. $BaCO_3$ is precipitated by adding 5 ml of carbonate-free 5% $BaCl_2$ to the combined carbonated 2 N NaOH and washings.[13] The mixture is shaken thoroughly and allowed to stand in the stoppered tube for 1 hour to ensure complete precipitation of the $BaCO_3$. The $BaCO_3$ is then filtered on preweighed filter paper with a standard stainless steel or glass filter assembly and washed with water and acetone. The filter paper disk with the layer of $BaCO_3$ is dried under an infrared lamp, equilibrated for 30 minutes with the atmosphere of the balance case, then weighed to an accuracy of 0.01 mg.

Each $BaCO_3$ precipitate should be divided, and at least two weighed samples of not less than 5 mg should be prepared, i.e., the quantity of CO_2 trapped should give at least 10 mg of $BaCO_3$.

Measurement of Radioactivity. The radioactivity of the $BaCO_3$ disks can be measured by proportional counting in an end window counter or windowless gas flow counter, making the necessary correction for self absorption by the $BaCO_3$. This well-documented method[14] is, however,

---

process repeated until the preparation is dry. Alternatively, the residue of nicotinic acid can be dried *in vacuo* over P_2O_5 for 24 hours before sublimation.

[13] Large flocks of $BaCO_3$ may be precipitated almost instantaneously, or the precipitate may be fine and form slowly over a period of 30 minutes. The mode of precipitation is affected by the amount of carbonate present, temperature, and pH.

[14] S. Aronoff, "Techniques of Radiobiochemistry." The Iowa State College Press, Ames, Iowa, 1956.

best avoided if scintillation counting is available. For scintillation counting, the filter paper with the weighed sample of $BaCO_3$ is placed directly into 5–10 ml of scintillation mixture. In the author's work, the Beckman Liquid Scintillation System LS 200B was used. Under these conditions, there is a linear relationship between the weight of $BaCO_3$ and the measured counts per unit time, with a counting efficiency of 83%. Other scintillation systems will give different results, and the operator should study the efficiency of his own system for counting solid samples of $BaCO_3$.

Total Combustion of Samples. The specific radioactivity of the nicotinic acid, i.e., the average of the specific activities of C-2, 3, 4, 5, 6, and 7, is determined by totally burning a sample to CO_2 (Fig. 3). A 3–4-mg sample of nicotinic acid is totally burnt by heating with 3 ml of combustion fluid. The mixture is briefly heated with the flame of a microburner while flushing the apparatus (Fig. 3) with a slow stream of CO_2-free N_2; a rapid oxidation occurs, and in 1–2 minutes all the carbon of the nicotinic acid is converted to CO_2. For the total combustion of bromopicrin and other materials, e.g., sodium acetate and succinic acid, the procedure is exactly as for nicotinic acid.

C-7

Hydrogenation and Decarboxylation of Nicotinic Acid (I). Nicotinic acid (I), 50 mg, is dissolved in 10 ml of 2 N H_2SO_4 and 30 mg of 5% rhodium on charcoal are added to the solution. The mixture is stirred magnetically while a slow stream of CO_2-free H_2 is passed through the apparatus (Fig. 2) at room temperature and atmospheric pressure. Hydrogenation is complete after 1 hour, and it is accompanied by about 50% decarboxylation. The resulting CO_2 represents C-7 of the nicotinic acid.

C-2 and C-3

Benzoylation of Nipecotic Acid (II) and Piperidine (III).[15] The hydrogenation mixture from the determination of C-7 contains approximately equimolar proportions of nipecotic acid (II) and piperidine (III). Without removing the catalyst, the solution is made strongly alkaline by the addition of 5 N NaOH. Benzoyl chloride, 75 mg, is then added, and the mixture is stirred rapidly for 1 hour in an ice bath. During this time, the solution is kept alkaline by the addition of 5 N NaOH. The benzoylpiperidine (IV)

---

[15] Benzoylation serves two purposes: benzoyl piperidine and benzoyl nipecotic acid are easily separated by ether extraction; the oxidation of benzoyl nipecotic acid is clean, with a high yield of benzoyl-5-amino-4-carboxyvaleric acid (VI), whereas the oxidation of nipecotic acid gives many side products and a low yield of 5-amino-4-carboxyvaleric acid (VII).

is extracted from the alkaline solution with diethyl ether (3 × 20 ml) and discarded.

Oxidation of N-Benzoylnipecotic Acid (V). The remaining aqueous solution, containing N-benzoylnipecotic acid (V) is adjusted to approximately pH 7 with $2 N$ H_2SO_4. $KMnO_4$, 25 mg, is then added, and the solution is heated for 30 minutes on a boiling water bath.

After the oxidation, the resulting MnO_2 is removed by filtration and washed with hot water. The filtrate and washings are evaporated to dryness, 10 ml of 12 N HCl are added to the residue, and the solution is boiled under reflux for 12 hours. Excess HCl is removed by repeatedly evaporating the mixture to dryness under reduced pressure with the addition of extra water. The residue is dissolved in 1 ml of water and placed on a column (1 × 20 cm) of Dowex 50 (H^+ form). The column is eluted with 60 ml of 1.5 N HCl; 3-ml fractions are collected, and one spot from each fraction is analyzed by paper chromatography in the system butan-1-ol–acetic acid–water (4:1:1 by volume). The developed chromatogram is dipped in cadmium–ninhydrin reagent, dried in air, and heated at 110° for 10 minutes. 5-Amino-4-carboxyvaleric acid (VII), which has a peak effluent volume of about 39 ml, gives a red spot after ninhydrin treatment. The appropriate fractions are evaporated to dryness, the 5-amino-4-carboxyvaleric acid (VII) purified further by paper chromatography (butan-1-ol–acetic acid–water) and eluted again from Dowex 50. Later fractions from the first Dowex column contain unreacted nipecotic acid (II), which gives a characteristic deep blue color with the cadmium–ninhydrin reagent. These fractions are combined and evaporated to dryness; the nipecotic acid is oxidized to succinic acid (XII) at a later stage (see C-3,6 and C-4,5).

Oxidation of 5-Amino-4-carboxyvaleric Acid (VII) to 2-Methyleneglutaric Acid (VIII). The 5-amino-4-carboxyvaleric acid hydrochloride (VII) (about 25 mg) is dissolved in 50 ml of water, 2 ml of N H_2SO_4 and 15 ml of 2% $NaNO_2$ are added, and the mixture is heated at 100° for 15 minutes. Excess nitrous acid is destroyed by the addition of solid urea until gas evolution ceases. The reaction mixture is then extracted three times with 50 ml of diethyl ether. The pooled ether extracts are dried over anhydrous Na_2SO_4, filtered, and blown to dryness in a stream of N_2. The yield of 2-methyleneglutaric acid (VIII) is about 65%.

Reduction of 2-Methyleneglutaric Acid (VIII) to 2-Methylglutaric Acid (IX). Without further purification, the 2-methylglutaric acid from the previous stage is dissolved in 5 ml of methanol, and 5 mg of 5% rhodium on charcoal are added. The mixture is stirred magnetically, and a slow stream of H_2 is passed through the apparatus (Fig. 2) at room temperature and atmospheric pressure for 5 hours. The catalyst is removed by filtration, and the methanol is evaporated under a stream of N_2; the residue is 2-methylglutaric acid (IX) in essentially 100% yield.

Oxidation of 2-Methylglutaric Acid (IX) to Acetic Acid (X). The 2-methylglutaric acid is dissolved in approximately 1 ml of H_2SO_4 (specific gravity 1.84) with cooling. 5 N chromic acid solution, 4 ml, is added, and the mixture is boiled under reflux for 45 minutes (Fig. 5). When the mixture has cooled, the condenser is rinsed thoroughly into the flask with water and 7 g of anhydrous magnesium sulfate are added. The mixture is then steam-distilled (Fig. 5) and 50-ml fractions of the distillate are titrated to pH 9 with 0.01 N NaOH, using a pH meter.[16] Most of the acetic acid is usually collected in the first 200 ml of distillate. The sodium acetate solutions are pooled and evaporated to dryness.

Decarboxylation of Acetic Acid (X). The sodium acetate from the oxidation of 2-methylglutaric acid (IX) is dissolved in 0.1 ml of ice-cold 100% H_2SO_4, and 15 mg of NaN_3 is added. The mixture is heated at 70° for 1.5 hours, and CO_2-free N_2 is passed slowly through the apparatus (Fig. 3). The CO_2 produced represents C-3 of the nicotinic acid.

Oxidation of Methylamine (XI). The flask containing the residue from the decarboxylation of acetic acid is cooled in ice and connected via a condenser to a similar flask containing 10 ml of 0.1 N H_2SO_4 (Fig. 6). Pellets of NaOH are placed on top of the frozen residue, and about 5 ml of water is added. A vigorous reaction ensues. The alkaline mixture is then boiled down to half volume, and the distillate passes into the 0.1 N H_2SO_4. A 100-mg quantity of $KMnO_4$ is added to the resulting solution of methylamine sulfate, a trap of carbonate-free 2 N NaOH is attached to the flask, and a slow stream of CO_2-free N_2 is passed through the apparatus. The solution is boiled for 30 seconds to oxidize the methylamine to CO_2. This CO_2 represents C-2 of the nicotinic acid.

Alternative Procedure. The sodium acetate may be divided into two parts, one part decarboxylated as described above and the other totally burnt. The radioactivity of C-2 is then calculated from the values of C-3 and (C-3 + C-2)/2. This procedure for the determination of C-2 is less accurate, but it is more convenient to perform than the separate combustion of methylamine.

C-4

Decarboxylation of Nicotinic Acid (I). Nicotinic acid, 30 mg, is dissolved in 1–2 ml of water and 350 mg of $Ba(OH)_2\cdot8\ H_2O$ is added. This ensures complete conversion of the nicotinic acid to barium nicotinate (XIV) and leaves an excess of $Ba(OH)_2$. The solution of barium nicotinate (XIV) and barium hydroxide is evaporated to dryness in a 50-ml two-necked flask, fitted with a tap and water-cooled cold finger (Fig. 4). The flask is

---

[16] An indicator is not used, because its residue would give extra CO_2 in the total combustion of the sodium acetate.

heated to 140° on an oil bath to drive off the water of crystallization of the Ba(OH)$_2$. The flask is then evacuated to 0.01 mm Hg and heated above 300°, when decomposition occurs with some charring, and droplets of pyridine (XV) collect on the cold finger. Approximately 3 ml of glacial acetic acid are used to wash the pyridine from the cold finger into a 25-ml flask.

N-Oxidation and Nitration of Pyridine (XV). The solution of pyridine in glacial acetic acid (approximately 15 mg in 3 ml) is heated at 80° for 3 hours with 0.5 ml of 30% H$_2$O$_2$. A further 0.5 ml of 30% H$_2$O$_2$ is then added, and heating is continued for a further 3 hours at 80°. After concentrating the resulting mixture under reduced pressure, it is diluted with 2 ml of 2 N HCl, then evaporated to dryness. There is a 100% yield (from pyridine) of pyridine-1-oxide (XVI) as the hydrochloride.

To the pyridine-1-oxide hydrochloride are added 15 mg of fuming nitric acid (specific gravity 1.48) and 0.05 ml of H$_2$SO$_4$ (specific gravity 1.84), and the mixture is heated at 130° for 3.5 hours on an oil bath. The mixture is then diluted with sufficient (about 3 ml) ice-cold 10% NaCO$_3$ to make the solution alkaline. 4-Nitropyridine-1-oxide (XVII) is extracted from the alkaline mixture with 2 × 3 ml of cyclohexanol. The cyclohexanol solution is evaporated to dryness and the residue is dissolved in 0.2 ml of water; this is applied over a 10-cm starting line on a thin-layer plate of deactivated silicic acid (0.5 mm thickness), and the plate is developed with CHCl$_3$ for a distance of 10 cm. The band of 4-nitropyridine-1-oxide (XVII) (R_f approximately 0.3) can be detected by its UV-quenching properties.

Preparation and Combustion of Bromopicrin (XVIII). The band of 4-nitropyridine 1-oxide (XVII) is scraped from the silicic acid plate. Calcium hypobromite, freshly prepared by mixing 2.5 g of calcium hydroxide, 10 ml of water, and 1 ml of bromine, cooled to 0°, is added to the mixture of silicic acid and 4-nitropyridine 1-oxide in a 50-ml two-necked flask with a condenser and steam supply (Fig. 7). Steam is immediately passed through the flask. The first 5 ml of distillate contain oily tribromonitromethane (bromopicrin) (XVIII). The bromopicrin is washed with water by centrifugation, then burned to CO$_2$, which represents C-4 of the nicotinic acid.

C-3, 6 and C-4, 5

Oxidation of Nipecotic Acid (II). In the column purification of 5-amino-4-carboxyvaleric acid (VII) (see under C-2 and C-3), unreacted nipecotic acid is also recovered as its hydrochloride. The nipecotic acid hydrochloride is dissolved in 1 ml of 2 N H$_2$SO$_4$ and titrated with 2% KMnO$_4$ at 100°. Oxidation is complete when the red color of the permanganate persists for 1 minute. Excess permanganate is destroyed by adding a drop of dilute formic acid. The reaction mixture is placed on a column (diameter 2.5 cm, length 14 cm) of Dowex 1 (formate-form), which is then eluted with a linear gradient of 0–2 N formic acid in a total volume of 1 liter. Ten-milliliter fractions are collected. The peak effluent volume of succinic acid (XII) is about 500 ml.[17] The fractions containing succinic acid are evaporated to dryness and sublimed under reduced pressure (Fig. 4) to give a 75% yield (from nipecotic acid) of succinic anhydride. The succinic anhydride is washed from the sublimation finger with 2–3 ml of N NH$_4$OH, and the solution of ammonium succinate is evaporated to dryness.

Decarboxylation of Succinic Acid (XII). The ammonium succinate from the previous stage is dissolved in 0.1 ml of ice-cold 100% H$_2$SO$_4$. Ten milligrams of NaN$_3$ is added, and the mixture is heated to 60° for 1.5 hours, then 72° for 3 hours under a slow stream of CO$_2$-free N$_2$ (Fig. 3). The resulting CO$_2$ represents C-3, 6 of the nicotinic acid.

Oxidation of Diaminoethane (XIII). The procedure is exactly as described for the oxidation of methylamine (XI). The resulting CO$_2$ represents C-4, 5 of the nicotinic acid.

Alternative Procedure. The oxidation of diaminomethane (XIII) may be avoided and C-4,5 be calculated. The sample of ammonium succinate is divided into two: one part is decarboxylated as described above [specific ^{14}C radioactivity of the CO$_2$ is equivalent to (C-3 + C-6)/2]; the other part is totally burnt [specific ^{14}C radioactivity of the CO$_2$ is equivalent to (C-3 + C-4 + C-5 + C-6)/4].

The Use of ^{3}H:^{14}C Ratios for the Determination of ^{14}C Distribution

If nicotinic acid, uniformly labeled with tritium in the pyridine ring, is mixed with nicotinic acid-^{14}C, the ^{3}H:^{14}C ratio can be used as a basis for the determination of specific ^{14}C activity in compounds, e.g., succinic acid-^{14}C (XII) derived from the nicotinic acid-^{14}C (I). Since there is some labilization of ^{3}H during the catalytic hydrogenation of ^{3}H nicotinic acid the starting ^{3}H:^{14}C ratio is measured on the nipecotic acid (II). The specific ^{14}C activity of the starting material is measured by total combustion, and thereafter it is no longer necessary to submit known quantities of material for radioactive counting.

---

[17] If the radioactivity of the succinic acid (XII) is not high enough to permit its easy detection in the column effluent, a small quantity of high specific activity tritium-labeled succinic acid is added to the reaction mixture before chromatography. Small (about 0.1 ml) aliquots of the column fractions are then tested for succinic acid by scintillation counting. The amount of succinic acid at this stage is about 5 mg, so that up to 50 μg of additional, tritium-labeled material would introduce a maximum error of only 1% in the specific ^{14}C-activity of the sample.

Procedure. High specific activity nicotinic acid-^{3}H (150–160 mCi/milli-mole) is added to the nicotinic acid-^{14}C, so that the ^{3}H:^{14}C ratio is no lower than 20. The conversion of nicotinic acid (I) into nipecotic acid (II) and thence to succinic acid (XII) has already been described. The CO_2 (C-7) from the catalytic hydrogenation is collected, and its specific ^{14}C activity is measured as usual. For the present purposes, *N*-benzoylnipecotic acid (V) is hydrolyzed directly back to nipecotic acid (II), and the oxidation with $KMnO_4$ is omitted.

^{3}H:^{14}C ratios are measured for the nipecotic acid (II) (as the hydro-chloride) and for the succinic acid (XII) (as ammonium succinate).

Radioactive Counting. The results quoted here were obtained with the Beckman Scintillation System LS 200B, counting in two channels. The scintillation mixture described under Reagents was used; this mixture will tolerate up to 25% by volume of aqueous samples. The nipecotic acid or succinic acid is added in 0.1–0.2 ml of aqueous solution; under these conditions, quenching is practically nonexistent. Overlap from the ^{14}C channel into the ^{3}H channel is fairly high, depending on quench, while the overlap of ^{3}H into ^{14}C is so low as to be negligible. It is therefore advisable to have a very high ^{3}H:^{14}C ratio, so that interference by ^{14}C in the ^{3}H channel can be ignored. Quenching and counting efficiency are measured by the injection of an internal standard, first of ^{14}C, then of ^{3}H.

The counting data may be represented as follows:

Channel	^{14}C counts/min	^{3}H counts/min
First count	A	B
Second count, with internal ^{14}C standard	C	D
Third count, with internal ^{3}H standard	E	F

CORRECTION FOR OVERLAP OF CHANNELS

Let true ^{3}H counts/min = x
Let true ^{14}C counts/min = y
Then

$$x = B - \left[\frac{D - B}{C - A} \left(A - \frac{E - C}{F - D} x \right) \right] \qquad (1)$$

$$x - \left[\frac{D - B}{C - A} \times \frac{E - C}{F - D} \times x \right] = B - \left[\frac{D - B}{C - A} \times A \right] \qquad (2)$$

Similarly

$$y - \left[\frac{E - C}{F - D} \times \frac{D - B}{C - A} \times y \right] = A - \left[\frac{E - C}{C - A} \times B \right] \qquad (3)$$

Normally, there should be no overlap of ^3H into ^{14}C, so that $E - C/F - D = 0$. Then

$$x = B - \left[\frac{D - B}{C - A} \times A\right] \tag{4}$$

and

$$y = A \tag{5}$$

FINAL CORRECTION FOR QUENCHING AND COUNTING EFFICIENCY

The internal standard of ^{14}C contained 22,200 dpm, and the internal standard of ^3H contained 222,000 dpm.

Let true dpm of $^{14}C = Y$
Let true dpm of $^3H = X$
Then

$$Y = y \left[\frac{22,200}{C - A}\right] \tag{6}$$

and

$$X = x \left[\frac{222,000}{F - D}\right] \tag{7}$$

Calculation of Results

Let the $^3H:^{14}C$ ratio of nipecotic acid $= a$
Let the $^3H:^{14}C$ ratio of succinic acid $= b$
Then the percentage of total ^{14}C activity in C-3, 4, 5, 6 $= 50a/b\%$ and the percentage of total ^{14}C activity in C-2, 7 $= 100 - (50a/b)\%$.

The measurement of $^3H:^{14}C$ ratios thus provides a rapid method for measuring the ^{14}C activity of the C-2 of nicotinic acid. A typical application of the method is shown in Table I with the appropriate calculations.

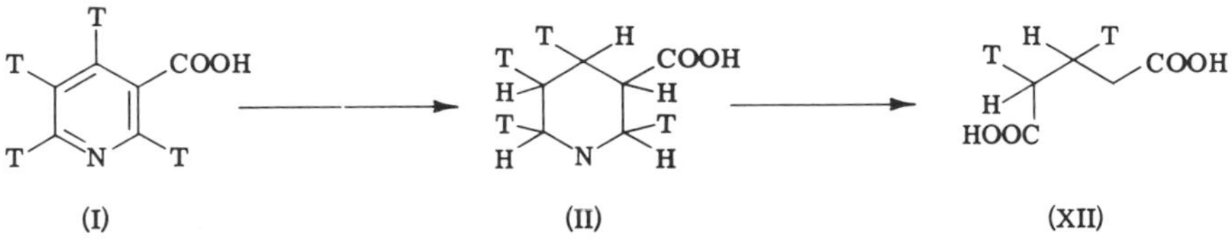

T = ^3H = Tritium

Isotopic Validity of the Degradation Pathway

Table II shows the results from the degradation of various labeled, authentic samples of nicotinic acid. For the individual carbon atoms, the

TABLE I
DEGRADATION OF NICOTINIC ACID-6-^{14}C,-2,4,5,6-^{3}H

| | Counts per min above background[a] | | | |
| | Nipecotic acid | | Succinic acid | |
Channel	^{14}C	^{3}H	^{14}C	^{3}H
First count	96.8	2158.6	51.4	572.5
Second count, with internal ^{14}C standard	14370.0	5902.0	14353.1	4335.3
Third count, with internal ^{14}C standard	14396.2	122610.2	14302.7	121478.6

[a] All activities were determined by scintillation counting, and all samples were counted to a 2 σ error of 0.3%.

^{3}H:^{14}C RATIO FOR NIPECOTIC ACID:
From Eq. (4)

$$x = 2158.6 - \left[\frac{59020 - 2158.6}{14370.7 - 96.8} \times 96.8 \right] = 2133.2$$

By inspection, there is negligible overlap of ^{3}H into the ^{14}C channel
Therefore, from Eq. (5)

$$y = 96.8$$

from Eq. (6)

$$Y = 96.8 \left[\frac{22200}{14370.7 - 96.8} \right] = 150.5$$

from Eq. (7)

$$X = 2133.2 \left[\frac{222000}{122610.2 - 5902.0} \right] = 4057.3$$

Therefore, for nipecotic acid

$$^{3}\text{H}:^{14}\text{C} = \frac{4057.3}{150.5} = 27$$

^{3}H:^{14}C RATIO FOR SUCCINIC ACID:
From Eq. (4)

$$x = 572.5 - \left[\frac{4335.3 - 572.5}{14353.1 - 51.4} \times 51.4 \right] = 559.0$$

By inspection, there is no overlap of ^{3}H into the ^{14}C channel.
Therefore, from Eq. (5)

$$y = 51.4$$

from Eq. (6)

$$Y = 51.4 \left[\frac{22200}{14353.1 - 51.4} \right] = 79.7$$

maximum error is approximately $\pm 5\%$. Comparison of the calculated and directly measured total ^{14}C activities shows an error of up to $\pm 3\%$. Another error arises from the difference calculations for C-5 and C-6; this may be as high as $\pm 3\%$ and it is mathematical in origin. Within these limits of error, the method is valid and chemical cross-contamination between carbon atoms is negligible.

TABLE II

DEGRADATION OF AUTHENTIC SAMPLES OF NICOTINIC ACID-^{14}C

	Counts per min above background per mg $BaCO_3$[a]		
Origin of CO_2	Labeling pattern of nicotinic acid (according to synthesis)		
	6-^{14}C	2,3,7-^{14}C	5-^{14}C
C-2	0.0	936.0	0.0
C-3	0.0	942.4	0.0
C-4	0.0	0.0	0.0
C-3, 6[b]	3315.9	469.4	0.0
C-4, 5[b]	0.0	0.0	571.3
C-5[c]	0.0	0.0	1138.5
C-6[c]	6653.4	−3.6	4.1
C-7	0.0	935.1	0.0
C-2, 3, 4, 5, 6, 7[b]	1098.6	471.8	186.1

[a] All activities were determined by scintillation counting, and all samples were counted to a 2 σ error of 0.3%.

[b] These represent average values, not a sum of ^{14}C activities.

[c] Calculated by difference.

Table III shows the typical results from the degradation of nicotinic acid derived metabolically from aspartate-U-^{14}C in an extract of *Clostridium butylicum*. Approximately equal ^{14}C activities were found in C-2, C-3, and C-7, and no ^{14}C was present in C-4, 5, 6. The slightly higher labeling in C-7 is the result of metabolism and does not reflect cross contamination in the degradation.

from Eq. (7)

$$X = 559.0 \left[\frac{222000}{121478.6 - 4335.3} \right] = 1062.1$$

Therefore, for succinic acid

$$^3H : ^{14}C = \frac{1062.1}{79.7} = 13.32$$

According to Eq. (9), carbon atoms 3, 4, 5, and 6 contain $(50 \times 27)/13.32 = 101.3\%$ of the total ^{14}C activity. There is therefore no activity present in C-2 (The absence of ^{14}C in C-7 is shown by decarboxylation).

TABLE III
DEGRADATION OF METABOLICALLY DERIVED NICOTINIC ACID-^{14}C[a]

Origin of CO_2	Counts/min above back-ground of $BaCO_3$[b]	Weight of $BaCO_3$ (mg)	Average counts/min per mg $BaCO_3$	Percent of total ^{14}C activity
C-2	52.7	8.65		
	42.8	7.00	6.10	31.28
C-3	48.9	7.81		
	34.2	5.41	6.29	32.25
C-4	0.0	6.42	0.00	0.0
C-3, 6[c]	19.8	6.32		
	12.8	4.10	3.12	15.10
C-4, 5[c]	0.2	5.36		
	0.0	6.02	0.01	0.05
C-5[d]	—	—	—	0.0
C-6[d]	—	—	—	0.0
C-7	28.9	3.87		
	75.6	10.20	7.43	37.95
C-2, 3, 4, 5, 6, 7[c]	31.3	9.64		
	27.0	8.32	3.25	100.00

[a] Aspartic acid-U-^{14}C was incubated with an extract of *Clostridium butylicum* in the presence of unlabeled acetate and formate under the conditions described by T. A. Scott and M. Mattey [*Biochem. J.* **107**, 606 (1968)], and the nicotinic acid was isolated and purified as described above.

[b] All activities were determined by scintillation counting, and all samples were counted to a 2 σ error of 0.3%.

[c] These represent average values, not a sum of ^{14}C activities.

[d] Calculated by difference.

Acknowledgments

The Beckman Scintillation System LS 200B used throughout this work was purchased on a grant from the British Science Research Council. The radioactive materials were purchased by Carreras Tobacco Company Ltd. The glass apparatus was made by Mr. C. Skelton. Dr. M. Mattey and Dr. E. Bellion checked the original degradation scheme in the course of their own work and made several improvements, which are included in the present paper. Mr. S. Bottomley gave valuable technical assistance in the development of the ^{3}H:^{14}C ratio method.

[110] Biosynthesis of Nicotinic Acid in Plants and Microbes

By R. F. DAWSON, D. R. CHRISTMAN, and R. U. BYERRUM

I. Introduction

Two major advances in methodology account for present knowledge of the pathways of nicotinic acid biosynthesis. The older of these, induced mutants, was employed to demonstrate the sequence tryptophan, kynurenine, 3-hydroxyanthranilic acid, quinolinic acid, nicotinic acid (I) in *Neurospora*. The other, isotopic tracers, has flourished in more recent years and has proved to be by far the more powerful of the two. By the use of isotopes it has been shown, for example, that the *Neurospora* pathway is not the sole route to nicotinic acid in nature and, indeed, that it may be replaced, or at least quantitatively overshadowed, by other pathways in certain organisms.

The present report deals with methods by which biosynthetic routes to nicotinic acid have been traced in certain higher plants. Special interest attaches to this work for the following reasons. First, the experimental plants are sufficiently complex that the array of mutants needed for detection and identification of intermediates by the older method could not have been obtained without enormous effort. Second, the nicotinic acid produced in these cases is converted to terminal metabolites, the pyridine alkaloids, which subsequently accumulate in very large quantities in the plant organs. Third, these plants do not respond to the introduction of intermediates with accelerated rates of terminal metabolite production. Fourth, classical methods of enzymology have not been applied successfully to a single step of the biosynthetic sequences. This being the case, the employment of isotopic tracer methods is both expedient and mandatory.

The methods described in this report have been applied to certain plant systems, but comparable results have been obtained with some bacteria, and these will also be discussed later when the current status of knowledge of pathways of nicotinic acid biosynthesis is described.

In the plant systems to be discussed below, it is not yet clear whether the pathways to be described exist to the exclusion of the *Neurospora* pathway and of other possible routes to nicotinic acid. The reason is that the nicotinic acid-to-alkaloid pathways have been traced by means of patterns of isotope distribution that can be observed in the terminal metabolite in question. No one has yet detected a nicotinic acid pool in these plant species of the order of magnitude of the terminal metabolite pools. Waller *et al.*[1] have shown that components of the pyridine nucleotide cycle supplied exogenously to *Ricinis communis* can contribute label to ricinine. Otherwise, little information is available on the degree of separation of the compartments within which, and the routes by means of which, the terminal metabolites, such as ricinine and nicotine on the one hand and NAD–NADP on the other, are synthesized. In consequence, although intermediates leading to the nicotinic acid precursors of, e.g., nicotine, anabasine, and ricinine, are fairly well known, it is not yet clear whether the tryptophan pathway may still be used for synthesis of NAD and NADP in the same organisms.[1]

Justification of interest in the delineation of these specialized pathways lies principally in the order of magnitude of biochemical effort that they represent. For example, nicotinic acid equivalents amounting to ca. 2% of the dry weight of green leaves may accumulate in tobacco in forms that result from replacement of the nicotinic acid carboxyl group by *N*-methyl-pyrrolidine[2,3] and other nitrogen heterocycles. Such quantities may approximate 100 times the tissue concentrations of NAD, NADP, and pool niacin combined. Analogous situations in the animal kingdom are difficult to find, but from the quantitative standpoint, at least, the excretion of dinicotinoylornithine by chicks following ingestion of excess nicotinic acid may be of some interest.

The biological system (i.e., the tobacco plant) within which nicotinic acid is converted to terminal metabolite nicotine and related alkaloids has been investigated longer and in greater detail than any other. Significant contributions to experimental methodology have been numerous, but perhaps none was more decisive than the discovery that nicotine is produced mainly in the root tips of the tobacco plant.[4,4a] With this knowledge

[1] G. R. Waller, K. S. Yang, R. K. Gholson, and L. A. Hadwiger, *J. Biol. Chem.* **241,** 4411 (1966).

[2] R. F. Dawson, D. R. Christman, A. F. D'Adamo, M. L. Solt, and A. P. Wolf, *J. Am. Chem. Soc.* **82,** 2629 (1960).

[3] S. Mizusaki, T. Kisaki, and E. Tamaki, *Plant Physiol.* **43,** 93 (1968).

[4] R. F. Dawson, *Am. J. Botany* **29,** 813 (1942).

[4a] R. F. Dawson, *in* "Advances in Enzymology" (F. F. Nord, ed.), Vol. VIII, p. 203. Wiley (Interscience), New York, 1948.

available, a number of effective means have been devised for introducing labeled organic compounds into biosynthetically active sites and for obtaining substantial radiochemical yields with respect to label in the pyridine ring. Of these methods, the least complicated biologically and the most definitive in terms of exclusion of microbial interference has been the use of the detached tobacco root in sterile culture.[5] Labeled organic compounds may be supplied to these cultures without danger of interfering microbial transformations. Furthermore, reverse transport and secondary conversions of nicotine and of its intermediates such as occur in the intact plant may be avoided altogether. However, properly handled, it is quite possible to obtain comparable results using whole plants, grown hydroponically, and much of the work with tobacco and castor plants has been done in this way (see below).

Attempts to simplify the experimental system by using cell-free preparations have produced few results. Indeed, there is evidence of the occurrence of an obligatory connection between the utilization of nicotinic acid for nicotine biosynthesis and new cell production in the apical meristems of the tobacco roots. Should nicotinic acid be converted to nicotine during some phase of the RNA or DNA cycles of the cell as is suggested by the stoichiometry of the relationship cited above, it seems unlikely that active, cell-free preparations will be found to catalyze more than a limited number of the total steps leading to alkaloid formation. As for the primary synthesis of alkaloid compartment nicotinic acid, nothing has yet been done to explore the possibilities of the enzymological approach.

Successful delineation of the pathway of nicotinic acid biosynthesis in the systems described above rested not only upon the opportunity to supply specifically labeled compounds under circumstances that favor significant precursor utilization, but also upon the availability of methods for degrading the pyridine ring, carbon atom by carbon atom, to locate the final resting place of the label. Limitations of convenience resulting from the desire to utilize small quantities of material and relatively low levels of radioactive tracers made the development of methods of stepwise degradation a slow and difficult process.

The following sections detail both biological and chemical methods insofar as they relate to the biosynthesis of the pyridine ring of nicotine or anabasine specifically and of ricinine generally. Some mention will be made later of the biosynthesis of the pyrrolidine ring of nicotine, but it is not directly concerned with the biosynthesis of the pyridine ring and, therefore, will be discussed only briefly, so as to give a more complete picture of the entire methodology involved here. Finally, biosynthetic

[5] P. R. White, *Am. J. Botany* **25,** 348 (1938).

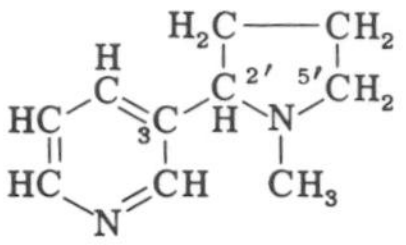

Nicotine

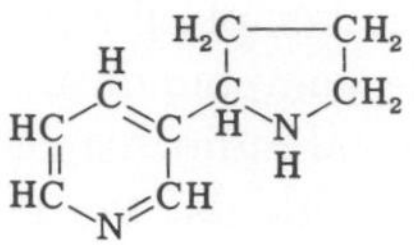

Nornicotine

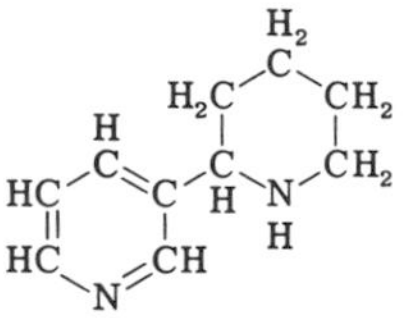

Anabasine

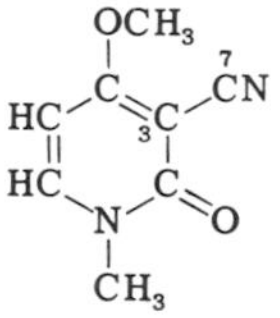

Ricinine

FIG. 1. Structures of the three major tobacco alkaloids and ricinine.

pathways leading to certain other pyridine alkaloids will be mentioned briefly, to illustrate the variety of paths that nature is apparently able to utilize to reach this rather ubiquitous ring system.

The structures of the three major tobacco alkaloids and of ricinine are shown for reference in Fig. 1.

II. Methodology

A. Tobacco Root Culture

The excised tobacco root grows at an exponential rate in sterile culture. Actually, the rate of linear extension of each root tip is arithmetic, but the rate of new tip production, as a result of branching, is exponential. Under favorable conditions, and with sufficiently frequent subdivision and transfer, it is possible to maintain these cultures in a steady state over a period of several years with respect both to growth and to nicotine yield. Due to a remarkable stoichiometric relationship between the production of new cells and nicotine, yields of nicotine also increase exponentially with elapsed culture time.

Strains of the Samsun Turkish type of *Nicotiana tabacum* L. have been most useful for biosynthetic work. It is notable that the roots of certain cigar filler-types (Pennsylvania) and Maryland-type tobaccos may be grown with equal facility in culture, and the proportionality constants between growth and nicotine yield in these cases are identical with that for the Turkish strain. This contrasts with rather divergent figures for the alkaloid content of the respective leaf crops. Strains of Virginia or Bright tobacco have never been successfully cultured in our laboratories.

Although we do not understand the cause(s), it has been our consistent experience that root cultures representing vegetatively propagated progeny of one original germinated seed (i.e., clonal material) will, after about three years of continuous subculture, suddenly and irreversibly lose growth vigor and must be discarded. We have never succeeded in restoring the vigor of a clone which has gone into growth decline. Thus, it is necessary to plan upon the introduction of a new clone approximately once every three years. It is also noteworthy that the proportionality constants relating nicotine yield to growth (dry matter production) have never differed significantly between different clones of the Turkish strain.

B. Establishing the Root Cultures

A few viable seeds of tobacco are soaked for 20 minutes in a 6% volume dilution of commercial sodium hypochlorite solution (Clorox). The seeds are transferred through several changes of sterile distilled water to remove excess alkali and are then spread over a wet filter paper contained in a sterile petri dish. After germination, ca. 1 cm of the young root of a single seedling is excised aseptically and transferred to an Erlenmeyer flask (125 ml) containing ca. 30 ml of a modified culture solution after White.[6] All subsequent cultures are made from the vegetative increase of this one rootlet.

It is to be understood throughout this work that standard methods of sterilization, aseptic transfer, and laboratory sanitation are observed. The sole innovation is the use of long-shanked surgical scissors for cutting tips from parent roots in preparation for transfer to new culture medium. These scissors are kept sterile during the cutting operations by placing them, tip down, into an Erlenmeyer flask (250-ml capacity) containing rapidly boiling distilled water while the cut tips are being transferred with a flamed Nichrome loop.

C. Maintaining the Cultures

The culture medium of White[6] has been employed routinely. This medium is too high in magnesium and in glycine for some species of *Nicotiana*, but it has supported the growth of tobacco roots well. An important criterion of a good culture medium is the absence of an appreciable initial time lag in the growth rate of a single root tip when detached from an older root and transferred to new medium.

---

[6] P. R. White, "A Handbook of Plant Tissue Culture." Cattell Press, Lancaster, Pennsylvania, 1943.

The culture medium is prepared by mixing prescribed amounts of four stock solutions (kept in the refrigerator away from light) with sucrose and yeast extract in redistilled water. The stock solution of major elements contains 7.40 g $MgSO_4 \cdot 7 H_2O$, 2.84 g $Ca(NO_3)_2 \cdot 4 H_2O$, 2.00 g Na_2SO_4, 0.80 g KNO_3, 0.65 g KCl, 0.19 g $NaH_2PO_4 \cdot H_2O$ made to 1 liter. Minor elements are supplied from a stock solution containing in 1 liter of water, 1.944 g EDTA (ethylenediaminetetracetic acid), 16 ml of N NaOH, 0.550 g of $MnSO_4 \cdot H_2O$, 0.260 g of $ZnSO_4 \cdot 7 H_2O$, 0.004 g of $CuSO_4 \cdot 5 H_2O$, 0.004 g of $(NH_4)_6Mo_7O_{24} \cdot 4 H_2O$, 0.150 g of H_3BO_3, and 0.075 g of KI, added in that order. Iron is supplied from a stock solution containing in 1 liter of water, 0.292 g of EDTA, 2.4 ml of N NaOH, and 0.250 g of $Fe_2(SO_4)_3 \cdot XH_2O$. Glycine (300 mg) and thiamine (10 mg) are dissolved in 100 ml of water. This stock solution is renewed monthly to maintain potency. One hundred milliliters of the major salts stock solution is added to ca. 750 ml of redistilled water followed by 10 ml of the minor salts stock solution, 1 ml of glycine–thiamine solution, 25 mg of Difco yeast extract, 20 g of recrystallized sucrose, and sufficient water to make to 1 liter. After thorough mixing, the resulting culture solution is dispensed in 30-ml portions into 125-ml Erlenmeyer flasks, stopped with gauze-coated cotton plugs, and autoclaved for 15 minutes at 120° and 15 psi.

On occasion, we have experienced disastrous results which have cost much lost time and effort and which resulted from the use of impure chemicals and/or distilled water, or from microbiological contamination. A precautionary note may prove valuable to biochemists who attempt to use these methods without great familiarity with microbiological culture techniques.

Distilled water may contain toxic heavy metals or pyrogens. Uncleaned still pots, resin residues, metals from laboratory piping systems, and spray carried through glass tubing of apparatus for preparing conductivity water are generally the chief sources of trouble. The widespread occurrence of organic pollutants in natural waters very readily leads to pyrogen production especially when these waters are distilled from a still pot that is lined with scale and hence tends to superheat. We have prepared water which without exception has given good results by distilling from glass through Vigreux columns (to reduce entrainment of spray). The raw water is distilled once from slightly alkaline, very dilute permanganate, once from extremely dilute sulfuric acid (a few drops in 5 liters), and once from adsorbent carbon. Heating mantles regulated by variable voltage regulators give smooth rates of distillation. Hard silicaceous boiling stones of the Hengar type prevent bumping much longer and more effectively than more porous types.

The chemicals employed to make the culture solutions are purchased in large lots and introduced one at a time into the ongoing culture program. In this manner, unsatisfactory lots are identified at once and may be discarded with minimal waste of time. This is especially important in the case of magnesium salts and of sucrose. We have found that Baker and Adamson chemicals generally offer less trouble than those of any other manufacturer. Yeast extract varies enormously in its biological potency from maker to maker. Within a very narrow range of concentrations, Difco yeast extract has proved satisfactory. Tests with specially grown (e.g., copper-low) yeasts have failed to yield advantage.

After the autoclaved medium has cooled and absorbed some air, each flask is supplied with one root tip newly excised from an older, rapidly growing culture. Best growth is secured when the inoculum represents the 2 cm-long terminus of a side branch of an older root that has many branches. Branches taken from weakly branched roots will not grow well. The tip of the principal root axis is never used, owing to its poor growth response.

To maintain inoculum for experimentation, stock cultures are kept in vigorous growth by transferring regularly every 2–3 weeks.

The cultures are incubated in the dark at ca. 29°. Tests have shown that growth rates are maximal at about this temperature. Good thermostatic control is necessary owing to the very steep slope of the thermal inactivation curve above and below ca. 30°.

Maintenance of microbial sterility is not difficult, but the presence of a fine, whitish sediment in the culture flasks, even though turbidity may be absent, is an almost positive indication of trouble. Occasional tests of spent medium for bacteria, using broth tubes, are recommended. Old cultures contaminated with sporulating fungi should never be opened, even in adjacent rooms, without first sterilizing the entire flask with contents in the autoclave. One unfortunate unstoppering of a flask containing fungus spores will contaminate a large area for months. Air conditioning introduces its own problems, often in the form of copious clouds of fungus spores brought in from outside or produced from mycelial growth on the organic binders used to coat and line air ducts. In the atmosphere of New York City, heavily contaminated with sulfur dioxide, we have never found it necessary to employ steaming or other special sanitizing devices during the transferring of root tips. On the other hand, elaborate precautions were necessary in rural Long Island (Brookhaven National Laboratory).

D. Supplying Labeled Compounds

Generally, the root cultures are grown for 3 weeks before labeled compounds are added, and harvest occurs after an additional week, owing to the fact that most of the nicotine in each culture is produced during the fourth week of growth. One culture will produce about 3 mg of product (nicotine dipicrate) in 4 weeks.

Stable intermediates are sterilized by autoclaving in 10% excess volume of redistilled water contained in cotton-stoppered flasks. Labile intermediates may be sterilized by filtration through a Pyrex UF Büchner fritted glass funnel which has been previously autoclaved (30–45 minutes) together with a small sidearm flask. The bed of the funnel is protected by inserting a large cotton plug, and the whole is wrapped in brown paper fastened with staples prior to autoclaving. The sidearm of the suction flask is fitted with two cotton air-filter plugs and a large cotton plug for the mouth. An Erlenmeyer flask with mouth opening similar to that of the suction flask is provided with a cotton plug and autoclaved simultaneously, in order to replace the filter and rubber stopper assembly after filtration has taken place. Dispensing pipettes are supplied with loose cotton plugs in the mouthpieces, wrapped in brown paper, secured with staples, and autoclaved.

Generally, it is convenient to adjust concentrations of intermediates such that ca. 0.5 ml of sterilized solution is added to each culture flask. The pipetting of the solutions should be done in a draft-free room.

Cultures are returned to the incubators and arranged in a random fashion, in the event two or more experimental variables are being tested, to avoid position effects. Air convection incubators are satisfactory provided thermostatting is both sensitive and reliable. During periods when ambient temperatures may exceed 30° for even a few hours, air conditioning may be essential to prevent serious damage to or death of the cultures.

E. Isolating Nicotine

Roots are harvested by pouring the root mass together with the spent culture fluid and flask rinsings over the plate of a large porcelain Büchner funnel resting loosely in the mouth of a sidearm flask. Grind the roots with a little sharp sand and a few crystals of trichloroacetic acid in a mortar until reduced to a cream. Add a little water, and heat gently for 10 minutes at or near boiling. Add a little diatomaceous filter aid, and filter under suction using care to avoid foaming. Reextract the pulverized cake with boiling water to which is added a few drops of N HCl for an additional 10 minutes, filter, and combine filtrates. Keep the total volume below 200 ml for 30 cultures.

Add the root extract to the accumulated spent culture fluids and flask rinsings, and acidify with a little $1\ N$ HCl to prevent loss of nicotine. Reduce to ca. 200 ml under reduced pressure in a rotating evaporator. Add 20% NaOH until the extract is strongly alkaline. Avoid overheating at this step. Add 1 drop of silicone antifoam and place in a liquid–liquid extractor. Extract with ether for at least 48 hours with a fairly rapid rate of solvent throughput.

The level of the ether in the boiling flask must always be above the tape of the heating mantle; otherwise nicotine will be lost. Glycerol is used to lubricate all ground-glass joints.

Remove the ethereal solution to a separatory funnel and shake out three times with 15, 10, and 10 ml, respectively, of 1 N HCl. Combine the extracts and warm gently on a hotplate to remove ether. Transfer to a 200-ml round-bottomed flask, and add a drop of phenolphthalein solution. Cool in an ice bath, and add 50% NaOH solution until neutrality is reached. Then add solid MgO to make the solution weakly basic, and distill the nicotine with injected steam into a saturated aqueous solution of picric acid (ca. 18 ml for 30 root cultures) contained in a 250-ml beaker. The condenser may conveniently be arranged in a vertical position and supplied with a glass tubular extension which must dip under the surface of the picric acid solution. Collect ca. 200 ml of distillate using bottom heat to reduce volume in the still pot.

Transfer the distillate quantitatively to a 1-liter round-bottomed flask and reduce to dryness in a rotating evaporator. Add just sufficient water that the solids dissolve after a few minutes of vigorous boiling in the same flask. Invert the flask into the mouth of a beaker and allow the hot liquid to drain into the beaker. Rinse the boiling flask with a very small amount of water. After standing overnight, the crystals are filtered on a sintered-glass funnel and washed with a small volume of water, then with a little methanol, and finally with ether to remove excess picric acid.

The product is usually quite pure except for a little excess picric acid. Nornicotine and anabasine do not distill appreciably under the conditions employed.

F. Whole-Plant Feeding

An alternative method of providing labeled compounds, employed by one of the authors (R.B.), is by their administration hydroponically to intact plants. When precursors of sufficiently high specific activity are available, feeding times of 2–6 hours yield nicotine containing enough ^{14}C for degradation studies. Furthermore, such feeding periods are short enough to minimize the possibility of metabolism or degradation of the precursor by the action of microorganisms.

Since nicotine is synthesized in the growing root tips, a procedure was devised to provide more actively metabolizing root tissue. *Nicotiana rustica* is grown in soil in a controlled-environment chamber for approximately 3 months to a height of about 12 cm. The plants are then removed from the soil, and their roots are removed with scissors. New roots are allowed to regenerate for about 2 weeks in 125-ml Erlenmeyer flasks containing a nutrient medium which is prepared by diluting, with 2 parts water, one part of a stock solution the composition of which is as follows: water, 1 liter; calcium nitrate, 1 g; potassium chloride, 250 mg; magnesium sulfate, 250 mg; ammonium sulfate, 250 mg; potassium dihydrogen phosphate, 250 mg; ferric chloride, 2 mg, all as the anhydrous salts. At the end of the regeneration period the nutrient medium in each flask is discarded and replaced with 1 ml of water containing 2 to 2.5×10^{-5} moles of the labeled precursors under study. After a metabolism period of 2–6 hours, plants are harvested and nicotine is isolated.[7]

This system has been used for most of the work done in this field by one of the authors (R.B.), and the results obtained thereby have been similar to those obtained via root cultures in all cases where direct comparison is possible.

G. Oxidation to Nicotinic Acid

Dissolve nicotine dipicrate in a minimum amount of 2.5 N HCl and extract with ether until the aqueous layer is colorless. Then neutralize with aqueous NaOH and add $KMnO_4$, with stirring, until the solution remains colored. Reflux for 4 hours, with additional $KMnO_4$ being added as necessary to maintain a purple color. Cool the solution, and then treat it dropwise with H_2O_2 to destroy any excess permanganate. Then add sufficient concentrated HCl to dissolve all MnO_2 present. Continuously extract the acid solution with ether for several hours; then adjust the pH to 3.0, and continuously extract for 2 days with new ether. Remove the ether and recrystallize the nicotinic acid from either ethanol or water to obtain the product in about 75% yield.

III. The Carbon-by-Carbon Degradation of Nicotinic Acid and Ricinine

A. Degradation Scheme

In this section, specific recommendations for the best methods of degrading nicotinic acid (I) carbon by carbon will be given. In addition, a brief description of the methods which have been used for the similar degradation of ricinine (V) will be given, up to the point where a common intermediate, N-methyl-2-piperidone (II), is encountered. Although ricinine has not been degraded beyond that point by the methods now recommended for nicotinic acid, the authors consider that such would now be the best way of proceeding from there in the ricinine degradation too.

[7] L. M. Henderson, J. F. Someroski, D. R. Rao, P. L. Wu, T. Griffith, and R. U. Byerrum, *J. Biol. Chem.* **234**, 93 (1959).

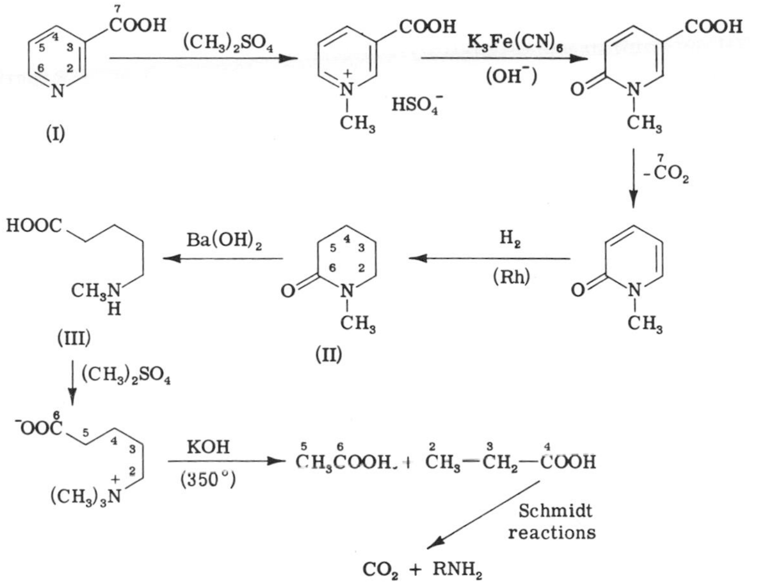

FIG. 2. Degradation scheme for nicotinic acid.

The general method is shown in Fig. 2. It is, basically, a combination of two somewhat different approaches,[8,9] combining some features of each. At several points, alternatives are possible which lead to equally satisfactory results, and there both paths are mentioned. Nicotinic acid has also been degraded in a manner whereby each carbon atom is obtained by direct attack separately from the others,[10] but the unified scheme presented here, giving each carbon in sequence through one series of reactions, would seem to be easier and more straightforward.

In general, the acid must be N-methylated, then oxidized to a pyridone which is decarboxylated, reduced, and hydrolyzed to δ-methylaminovaleric acid (III). The betaine of this acid is next formed, and heating of this betaine with solid potassium hydroxide produces acetic and propionic acids with specificity of the carbon structure being preserved.[11] The carboxyl carbon of the aminovaleric acid becomes the carboxyl carbon of acetic acid, and the 3-carbon is the carboxyl carbon of the propionic acid. These acids can be separated chromatographically[12] and then subjected to deg-

[8] D. R. Christman, R. F. Dawson, and K. I. C. Karlstrom, J. Org. Chem. 29, 2394 (1964).

[9] T. M. Jackanicz and R. U. Byerrum, J. Biol. Chem. 241, 1296 (1966).

[10] A. R. Friedman and E. Leete, J. Am. Chem. Soc. 85, 2141 (1963); 86, 1224 (1964).

[11] J. Fleeker and R. U. Byerrum, J. Biol. Chem. 240, 4099 (1965).

[12] H. E. Swim and M. F. Utter, Vol. IV, p. 585.

radation by the Schmidt reaction[13],[14] to determine the specific activity of each carbon atom in the original pyridine ring.

A method has been published[8] describing the conversion of δ-methylaminovaleric acid to valeric acid itself, via δ-bromovaleric acid, but the Schmidt reaction as applied to valeric acid gives poor yields of butyric acid for further degradation. The KOH cleavage is easier, and it gives better yields. At best, however, the overall yield of acetic and propionic acids, starting from nicotinic acid, is about 5%, so that good counting methods are required if the degradation is to be carried out from a reasonable amount of starting activity. The user of these methods must gauge the amount of starting activity needed in his specific case by the counting methods he has available, assuming the overall yield indicated above. Clearly, though, a satisfactory result can be achieved starting from 1 μCi of activity in the nicotinic acid, if modern counting methods are available.[15] If it is necessary to count samples as solid compounds, perhaps 10 times that amount of activity is desirable.

It should be pointed out that, in the case of tritium labeling on the pyridine ring, the oxidation of nicotine to nicotinic acid *must* be carried out under neutral conditions, with potassium permanganate. Use of nitric acid in this case results in considerable loss of tritium activity in the nicotinic acid by exchange with the oxidizing medium.

A detailed description of the degradation of nicotinic acid follows.

1. Trigonelline Acid Sulfate[9]

Each gram of nicotinic acid (I) is heated with 1.2 ml of dimethyl sulfate at 130° for 4 hours. After cooling, 12 ml of 0.035 N H_2SO_4 per gram of nicotinic acid is added, along with a little charcoal, then the mixture is filtered and concentrated to 1–2 ml per gram of nicotinic acid. Ethanol is added to give a concentration of 90%, after which the trigonelline acid sulfate precipitates. Filtration, reconcentration, and addition of ethanol again to 90% concentration results in additional product. A final yield of 77% is obtained. The melting point is about 200°.

2. N-Methyl-2-pyridone-5-carboxylic Acid[9]

Trigonelline acid sulfate is dissolved in 14.5 ml of 2.5 N NaOH per gram; then 11 ml of a 32% aqueous solution of $K_3Fe(CN)_6$ is added dropwise over 90 minutes. The solution is stirred for another hour, then the

[13] E. F. Phares, *Arch. Biochem. Biophys.* **33,** 173 (1951).
[14] R. C. Anderson and A. P. Wolf, Brookhaven National Laboratory Report BNL-3222, 1957.
[15] D. R. Christman, *Nucleonics* **23,** No. 12, 39 (1965).

pH is adjusted to 3.5 with 6 N H_2SO_4. The product is collected by filtration and is dissolved in a little hot water; charcoal is added, and the hot solution is filtered. The resulting product is collected by filtration and represents a yield of 64%; the melting point is 240–242°. No trace of the 2-pyridone-3-acid can be detected spectroscopically in this case.

3. N-Methyl-2-pyridone (II)[8,9]

Two methods have been used successfully for the formation of this compound. The decarboxylation is best carried out in 60% H_2SO_4 if the carbon dioxide is to be trapped and counted for study of the labeling of the pyrrolidine ring of the parent nicotine.[8] Alternatively, it can be carried out in quinoline, using copper chromite as catalyst,[9] but then the carbon dioxide is not available as such for counting. Yields are similar with these two methods, but the quinoline method is faster. Both will be described here, as each has its use.

Method a. In 15 ml of 60% H_2SO_4 is placed up to 5 millimoles of N-methyl-2-pyridone-5-carboxylic acid, in a small flask fitted with a cold finger condenser and a gas inlet tube.[8] The condenser is followed by a trap, to be cooled with liquid nitrogen. Helium or nitrogen is bubbled through the flask at a rate of 1 bubble per second, then the flask is heated to just under reflux temperature. The evolved carbon dioxide is collected in the liquid nitrogen-cooled trap and can be counted as desired. It may occasionally be necessary to add a little water, to keep the volume fairly constant. Complete decarboxylation of the pyridone takes several days under these conditions. The rate of evolution of CO_2 is about 1 mg per 4 hours. After completion of the decarboxylation, the solution is cooled, diluted to 50 ml with water, taken to pH 8 with sodium carbonate, and then continuously extracted with chloroform for 1 day. Removal of the chloroform *in vacuo* leaves the desired pyridone as an oil.

Method b. Alternatively, each gram of the above acid is dissolved in 9.5 ml of pure quinoline, along with 330 mg of freshly prepared copper chromite.[16] After refluxing for 5 hours, the quinoline is removed by steam distillation, and the aqueous portion is extracted 5 times with chloroform. Evaporation of the chloroform leaves the same oil as that described above.[9]

4. δ-Methylaminovaleric Acid (III)

The oil from the above decarboxylation is taken up in 20 ml of glacial acetic acid, then an amount of 5% rhodium on alumina (Engelhard Industries, Newark, New Jersey) about equal to the weight of the oil is added.[8]

---

[16] W. H. Lazier and H. R. Arnold, *in* "Organic Syntheses, Coll. Vol. II," p. 142. Wiley, New York, 1943.

The mixture is stirred vigorously in a hydrogen atmosphere at about 1 atmosphere, and the theoretical amount of hydrogen is taken up within 30 minutes.

The solution is filtered, the acetic acid is removed at room temperature, under reduced pressure, and then the residual oil (N-methyl-2-piperidone, II) is refluxed for 6 hours in a solution of 6 g of $Ba(OH)_2 \cdot 8\, H_2O$ in 75 ml of water.[9] The barium ions are removed by bubbling CO_2 through the solution, then filtering. Water is removed *in vacuo* and δ-methylamino-valeric acid crystallizes. The melting point is 125–126°. The overall yield from the carboxylic acid is about 30%.

5. 5-Trimethylaminovaleric Acid (IV)

This betaine has been prepared by means of methyl iodide and silver oxide,[9] and by methyl sulfate plus KOH.[8] The methyl sulfate method appears to be slightly simpler and faster and is the one described here.

In 3 ml of water is placed 3 millimoles of III plus an equivalent amount of potassium hydroxide. While the solution is stirred vigorously, 1 ml of dimethyl sulfate and a solution of 10 meq of KOH in 5 ml of water are added alternately, dropwise, over 30 minutes. The solution is refluxed for 15 minutes, cooled, taken to pH 7 with dilute H_2SO_4, then evaporated to dryness. Trituration of the solid residue with hot ethanol, evaporation of all but about 0.5 ml of the ethanol, and addition of 2–3 ml of ether gives the desired betaine (IV) in 75% yield. It crystallizes as a dihydrate, m.p. 225° (dec). If any sulfate is still present in the ethanol extract, it can be removed by adding aqueous barium hydroxide until the pH is raised to 6, then filtering and evaporating to dryness before final crystallization from ethanol and ether.

6. KOH Fusion of 5-Trimethylaminovaleric Acid[9]

Up to 10 millimoles of the betaine is mixed with crushed, solid KOH (1 g/millimole) and is heated to 350° for 10 minutes. There is rapid evolution of trimethylamine, so the reaction should be carried out under a hood. The preparation is cooled, then water is added, and alkaline and neutral contaminants are removed by steam distillation. The acetic and propionic acids are obtained by steam distillation, after acidification with 6 N H_2SO_4. The sodium salts are obtained by neutralization with NaOH and removal of the water. The acids can be separated by any desired method (e.g., vapor-phase chromatography), but are most conveniently separated by means of a Celite or silicic acid column.[12] The total yield is over 75%.

For the subsequent Schmidt reactions, the acids are best handled as their thallous salts, by neutralization of the appropriate effluent fractions with aqueous thallous hydroxide, evaporation to dryness, and recrystallization from ethanol–ether.[14] The melting point of thallous acetate is 130°, and that of thallous propionate is 188°.[9]

The carboxyl carbon of the thallous acetate is the original 6-carbon of the pyridine ring of nicotinic acid, and that of the propionate is the original 4-carbon. A total of three Schmidt reactions is required for complete degradation of the ring, the final methyl amine from each of the acids being derivatized and counted (normally as methylphenylthiourea[14]) to obtain the activity in the original 5- and 2-carbons, respectively. The activity of the 3-carbon is obtained by Schmidt reaction on thallous acetate, obtained from basic permanganate oxidation of the ethyl amine produced during the Schmidt reaction on the thallous propionate.

B. Degradation of Ricinine

Ricinine (V), existing naturally as a pyridone, is considerably easier to degrade than is nicotine itself. The methods that have been used for its carbon-by-carbon degradation have involved reactions that give specific carbon atoms one by one.[17–19] However, N-methylpiperidone (II) is reached very easily from this compound, and the best means of degrading ricinine should now proceed from that point exactly as indicated above for nicotinic acid. The requisite steps are shown in Fig. 3.

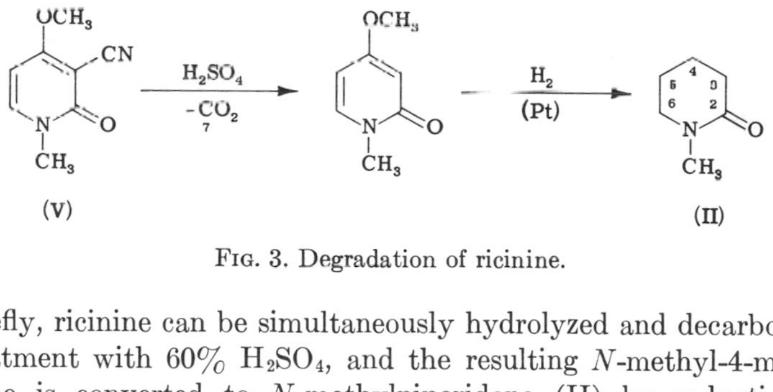

FIG. 3. Degradation of ricinine.

Briefly, ricinine can be simultaneously hydrolyzed and decarboxylated by treatment with 60% H_2SO_4, and the resulting N-methyl-4-methoxy-pyridone is converted to N-methylpiperidone (II) by reduction with hydrogen over platinum catalyst in acid medium.[18] The further degradation of (II), as given above, results in acetic acid whose carboxyl carbon is the 2-carbon of the original pyridine ring, rather than the 6-carbon as in the case of nicotinic acid itself. The carboxyl carbon of the propionic acid is still the original 4-carbon.

[17] P. F. Juby and L. Marion, Can. J. Chem. **41**, 117 (1963).
[18] J. M. Essery, P. F. Juby, L. Marion, and E. Trumbull, Can. J. Chem. **41**, 1142 (1963).
[19] K. S. Yang, R. K. Gholson, and G. R. Waller, J. Am. Chem. Soc. **87**, 4184 (1965).

IV. The Biosynthesis of Nicotinic Acid and Some Pyridine Alkaloids by Plants and Microbes

A general review of the biosyntheses of many pyridine and piperidine alkaloids was published by Leete in 1965.[20] The present section will be confined to the situation as it currently appears with respect to nicotinic acid and closely related pyridine ring compounds, plus some comments on recent results concerning the pyrrolidine ring of nicotine.

A. General Information

The actual pathway normally used by a given biological system to synthesize a given compound cannot be uniquely clarified by means of radioactive precursors, primarily for two reasons. Enzymatic alteration of the precursors may occur before their incorporation, even in a sterile medium such as the root cultures previously described (a case in point may be the results with the alanines and propionic acid mentioned below). Furthermore, the addition of any precursor perforce *may* alter the pool size (if any) of this and closely related compounds, and thus alter the normal metabolism of the system. Nevertheless, when comparable and consistent results are obtained in several different systems, and when these are based on common, ubiquitous compounds such as those which have been used in the studies described here, it is probable that the resulting picture is at least close to the way in which these systems normally *do* synthesize the compound under study. In any case, several different pathways by which plants and some lower animals *can* synthesize nicotinic acid and some other comparable pyridine ring structures have been elucidated by these methods, to a gratifyingly consistent extent.

The origins of the carboxyl group of nicotinic acid cannot be studied by using tobacco to produce nicotine or anabasine, since the carboxyl group is lost during the coupling of the two rings.[21] The origins of the 3-cyano group of ricinine must be analogous to those of the carboxyl group of nicotinic acid in tobacco, however, since the results obtained with these compounds in the case of the ring carbons are so much alike. The carboxyl group in nicotinic acid obtained by oxidation of nicotine or anabasine is actually the bridgehead carbon of the saturated ring, and this ring is biosynthesized entirely apart from the nicotinic acid.

B. Pyrrolidine Ring Biosynthesis

It was found some time ago that the pyrrolidine ring of nicotine (and the piperidine ring of anabasine) could arise from common Krebs cycle

[20] E. Leete, *Science* **147**, 1000 (1965).
[21] R. F. Dawson, D. R. Christman, and R. C. Anderson, *J. Am. Chem. Soc.* **75**, 5114 (1953).

intermediates, such as acetate. The most immediate intact precursors of the pyrrolidine ring at first appeared to be either ornithine or glutamic acid, and when either of these compounds, labeled in position 2 with ^{14}C, was administered to tobacco, nicotine was produced which was equally labeled in the 2' and the 5' positions. When ornithine was also labeled with ^{15}N, it was found that the δ-amino group was the one which was incorporated into the ring. Leete[22] and Mizusaki et al.[23] have proposed, in view of this finding, that ornithine is incorporated into the pyrrolidine ring by way of the symmetrical compound putrescine (itself a good precursor of the ring in these systems). Putrescine is believed to be converted to 4-N-methylaminobutyraldehyde, a compound which has been isolated from plants fed labeled ornithine, and which is converted to the pyrrolidine ring in high radiochemical yield when it is fed as a labeled compound. Anabasine shows a similar relationship to lysine, except that lysine-2-^{14}C appears in the anabasine with all label in the 2' position. Thus it does not pass through any symmetrical intermediate during its conversion to the piperidine ring.

Labeled carbon dioxide, when fed to photosynthesizing tobacco plants, also yields labeled nicotine. Liebman et al.,[24] using [14]CO$_2$ as a precursor, obtained labeling in the pyrrolidine ring which led them to question the concept of a symmetrical intermediate. However, more recent, similar experiments, using carbon dioxide of considerably higher specific activity, and some different degradative techniques, have shown that the pyrrolidine ring is indeed symmetrically labeled by carbon dioxide, and the labeling pattern is compatible with results from the labeled precursor studies referred to above.[25] With 6-hour exposures, in N. rustica the percentage of total nicotine activity in each of the four pyrrolidine carbons was 4.6 ± 0.2%. With N. glutinosa, it was 4.4 ± 0.3% for 6-hour exposures and 1.55 ± 0.15% with a 3-hour exposure.

The methyl group of nicotine is apparently added to the pyrrolidine ring precursors before the coupling with the pyridine ring, since nornicotine does not appear to be a direct precursor of nicotine.[26]

C. The Pyridine Ring of Nicotinic Acid

The most important results that have been obtained from the degradation of nicotine and ricinine, with regard to the pyridine ring, are sum-

[22] E. Leete, E. G. Gros, and T. J. Gilbertson, Tetrahedron Letters 11, 587 (1964).
[23] S. Mizusaki, T. Kisaki, and E. Tamaki, Plant Physiol. 43, 98 (1968).
[24] A. A. Liebman, B. P. Mundy, and H. Rapoport, J. Am. Chem. Soc. 89, 664 (1967).
[25] H. R. Zielke, R. U. Byerrum, R. M. O'Neal, L. C. Burns, and R. E. Koeppe, J. Biol. Chem., 243, 4757 (1968).
[26] S. Mizusaki, T. Kisaki, and E. Tamaki, Agr. Biol. Chem. 29, 714 (1965).

TABLE I
NICOTINE

Precursor	Percent in pyridine ring position shown[a]					Reference
	2	3	4	5	6	
Succinate-2,3-^{14}C	43	49	<4	2.7	1.9	Dawson[b]
	39	39	—	—	7	Byerrum[c]
Aspartic-3	38	57	1.1	3.4	2.4	Byerrum[d]
Malate-3	38	61	0.6	1.4	1.0	Byerrum[d]
Glycerol-1	15.7	12.1	32.3	1.8	35.1	Byerrum[e]
	17	17	36	0	31.2	Waller[f]
Glycerol-2	15	—	5	69	3.5	Dawson[b]
	2	2.4	0.1	98.5	0	Byerrum[e]
	16	16	0	75	0	Waller[f]
Acetate-2	38	34	—	—	11	Byerrum[c]
Glyceraldehyde-3	13	14	44	4.9	13	Byerrum[g]
β-Alanine-2	53	40	4.5	1.4	0.9	Dawson[b]
Alanine-3	46	47	—	—	—	Dawson[b]
Carbon dioxide-^{14}C	10	9.2	25.0	24.2	26.7	Byerrum[h]

[a] All percentages are based on total pyridine ring activity only.

[b] R. F. Dawson, D. R. Christman, and K. I. C. Karlstrom, unpublished data

[c] T. Griffith and R. U. Byerrum, *Biochem. Biophys. Res. Commun.* **10**, 293 (1963).

[d] T. M. Jackanicz and R. U. Byerrum, *J. Biol. Chem.* **241**, 1296 (1966).

[e] J. Fleeker and R. U. Byerrum, *J. Biol. Chem.* **240**, 4099 (1965).

[f] K. S. Yang, R. K. Gholson, and G. R. Waller, *J. Am. Chem. Soc.* **87**, 4184 (1965). Carbons 2 and 3 were determined together, and activity is assumed to be equally divided between the two carbons.

[g] J. Fleeker and R. U. Byerrum, *J. Biol. Chem.* **242**, 3042 (1967).

[h] H. R. Zielke, C. M. Reinke, and R. U. Byerrum, *J. Biol. Chem.*, **244**, 95 (1969).

marized in Tables I and II. None of these results has actually been obtained by the complete degradation scheme presented here, as pointed out previously. In the case of ricinine, the methyl carbons are not considered here. Where both compounds have been investigated with the same precursors, anabasine and nicotine have given essentially identical results with respect to the pyridine ring.

It seems clear that the 2- and 3-carbons of the pyridine ring can arise fairly directly from either succinic or aspartic acids. In both cases, the specific activity of these two carbons is approximately equal and accounts for about 80% of the total incorporated from these two precursors. Other compounds, such as acetic acid, which are interconvertible with succinate via the Krebs cycle, show a similar picture with respect to this part of the molecule (equal label in positions 2 and 3).

Whether or not the actual precursor is aspartic or succinic acid is open to some question. Generally, in short-term feeding to castor plants,

TABLE II
RICININE

Precursor	Percent in position shown[a]						Reference
	2	3	4	5	6	7	
Succinic-1-^{14}C	5.2	5.2	0	0	0	77.3	Waller[b]
Succinic-2,3	38.6	36.9	—	—	—	19.1	Marion[c]
	23	23	—	—	—	15.5	Waller[b]
Aspartic-3	37.2	39	0	0.8	—	18	Marion[d]
Aspartic-4	—	—	—	—	—	97	Waller[b]
	—	4	—	—	—	78	Marion[d]
Glycerol-1	16	16	13.8	0	14.9	11.0	Waller[b]
	—	—	25.8	2.2	19.7	9.4	Marion[e,f]
Glycerol-2	21	21	0	22	0	10.6	Waller[b]
	—	—	2.2	38.8	0	19.7	Marion[e,f]
Acetate-2	38.9	38.3	0	0.3	—	20.8	Marion[c]
Acetate-1	—	—	—	—	—	93	Marion[f]
β-Alanine-2	25.6	39.5	2	1	2	21	Marion[g]

[a] Percentages are based on total ricinine activity. No percentages are listed for the two methyl groups.

[b] K. S. Yang and G. R. Waller, *Phytochemistry* **4**, 881 (1965). Carbons 2 and 3 were determined together, and activity is assumed to be equally divided between the two carbons.

[c] P. F. Juby and L. Marion, *Can. J. Chem.* **41**, 117 (1963).

[d] S. R. Johns and L. Marion, *Can. J. Chem.* **44**, 23 (1966).

[e] J. M. Essery, P. F. Juby, L. Marion, and E. Trumbull, *Can. J. Chem.* **41**, 1142 (1963).

[f] P. F. Juby and L. Marion, *Biochem. Biophys. Res. Commun.* **5**, 461 (1961).

[g] P. R. Thomas, M. F. Barnes, and L. Marion, *Can. J. Chem.* **44**, 1997 (1966).

the nitrogen atom of aspartic acid may be incorporated into the ring intact.[27] When the feedings are over longer time periods, in tobacco and in the castor plant,[28,29] the nitrogen atom of aspartic acid seems to become equilibrated with the ammonia pool, and there is no excess incorporation of label into the pyridine ring when ^{15}N-labeled aspartic acid is administered under these conditions. It may be, however, that aspartic acid or a closely related unsymmetrical compound is the actual precursor. When aspartic acid-3-^{14}C or malic acid-3-^{14}C was administered to intact tobacco plants over short time periods, about 60% of the total pyridine ring activity was found in the 3-position,[9,30] and nearly all of the remainder was

---

[27] U. Schiedt and G. Boeckh-Behrens, *Z. Physiol. Chem.* **330**, 58 (1962).

[28] D. R. Christman and R. F. Dawson, *Biochemistry* **2**, 182 (1963).

[29] S. R. Johns and L. Marion, *Can. J. Chem.* **44**, 23 (1966).

[30] T. Griffith, K. P. Hellman, and R. U. Byerrum, *Biochemistry* **1**, 336 (1962).

found in the 2-position. Thus there may well be only partial equilibration with the nitrogen pool before incorporation in a short-term experiment. At least some of the material appears not to be incorporated via a symmetrical intermediate under such circumstances.

A double-label experiment with [15]N- and ^{14}C-labeled aspartic acid indicated that the compound is utilized intact in the formation of nicotinic acid by *Mycobacterium tuberculosis*,[31] with loss of label only from the carboxyl carbon in the 1-position. Johns and Marion[29] have suggested that this result may indicate a somewhat different pathway to nicotinic acid in these two systems. Another explanation could involve completely different pool sizes and kinetics in the two cases, so that nitrogen equilibration may be a much slower process in the bacterial system than with the plants.

When carboxyl labeled succinic or aspartic acid-4-^{14}C is fed, most of the activity appears in the cyano group of ricinine,[29,32] or in the carboxyl group of nicotinic acid in some bacteria.[31] No appreciable activity is found in the ring carbons after administration of these carboxyl-labeled compounds in any system studied so far.

The rest of the pyridine ring seems to be derived from glycerol or some close relative thereof. In both nicotine and ricinine, the 4- and 6-carbons are about equally labeled when glycerol-1-^{14}C is fed, while the 5-carbon is heavily labeled when glycerol-2-^{14}C is used. Of course, an appreciable part of the glycerol activity also appears in the 2- and 3-carbons, since the plants oxidize this compound to acetate and other Krebs intermediates. In ricinine,[33] furthermore, 26–35% of the activity from glycerol feedings is found in the methyl groups.

Fleeker and Byerrum[34] have presented evidence that the most immediate precursor of this portion of the pyridine ring, in tobacco plants, may be D-glyceraldehyde 3-phosphate, with the 1-carbon forming carbon 6 of the pyridine ring. The usual activity also appears in the other ring carbons, probably by oxidation to acetate and incorporation in that form. Over half of the total incorporated activity was found in the 4-carbon when D-glyceraldehyde-3-^{14}C was administered, however.

It should be noted that no other carboxyl-labeled precursors have given any appreciable incorporation into the pyridine ring of nicotinic acid in any system which has been tested. However, as would be expected [14]CO_2 is also incorporated in the pyridine ring of nicotine by several species of tobacco. Such incorporation studies provide some additional information

[31] D. Gross, H. R. Schütte, G. Hübner, and K. Mothes, *Tetrahedron Letters*, **9**, 541 (1963).

[32] K. S. Yang and G. R. Waller, *Phytochemistry* **4**, 881 (1965).

[33] P. F. Juby and L. Marion, *Biochem. Biophys. Res. Commun.* **5**, 461 (1961).

[34] J. Fleeker and R. U. Byerrum, *J. Biol. Chem.* **242**, 3042 (1967).

about the biosynthesis of the ring. In a typical experiment with *N. glutinosa* after giving $^{14}CO_2$ for 3 hours, the 2 and 3 carbons of the pyridine ring contained between 9 and 10% of the ^{14}C in the ring and the 4, 5, and 6 carbons had between 24 and 27% each[35] (Table I).

This labeling pattern was postulated to result from the condensation of two pyridine ring precursors which had different specific activities at the time of condensation. As noted above, glyceraldehyde and aspartic acid have been postulated to be these two compounds. Glyceraldehyde and aspartic acid might have different specific activities after $^{14}CO_2$ is administered because of different pool dilution rates or because of relatively slow metabolic incorporation of label into the internal carbons of aspartic acid. In any case, the labeling pattern found in this experiment is entirely consistent with the glyceraldehyde–aspartic acid hypothesis for the formation of the pyridine ring of nicotinic acid in these systems.

There are some results that cannot be fitted into this picture easily. The most apparent ones are the group consisting of alanine-3-^{14}C, β-alanine-2-^{14}C, and propionic acid-2-^{14}C. In nicotine[28] and ricinine,[36] only the one carbon indicated here as being labeled is incorporated into the pyridine ring at all, and the labeling pattern is different from that shown for any of the other compounds discussed. On the other hand, the two alanines give the highest radiochemical yield of nicotine of any precursor which has been tried with tobacco, except for nicotinic acid itself. The danger of basing conclusions strictly on relative radiochemical yields from the various compounds is obvious. Too many variables are involved from one compound to another, including differences in absorption of the precursors, different metabolism or toxicity, and different pool sizes within the plant.

D. Other Pyridine Ring Biosyntheses

Very recently, an antibiotic, pyridomycin, produced by *Streptomyces pyridomyceticus*, has been shown to involve a very similar biosynthetic pathway to the pyridine ring.[37] The compound ($C_{27}H_{32}N_4O_8$) has a complex structure containing two pyridine rings, and degradative results have indicated that these rings are formed in this case too from glycerol and aspartic acid, pyruvate being involved in the required decarboxylations and the subsequent incorporation of reduced side chains. No work was done with aspartic acid-^{15}N, so the question of direct or indirect incorporation of that compound as such still exists in this case. But there seems to be no doubt that the biosynthesis of this compound bears striking simi-

[35] H. R. Zielke, C. M. Reinke, and R. U. Byerrum, *J. Biol. Chem.*, **244**, 95(1969).

[36] P. R. Thomas, M. F. Barnes, and L. Marion, *Can. J. Chem.* **44**, 1997 (1966).

[37] H. Ogawara, K. Maeda, and H. Umezawa, *Biochemistry* **7**, 3296 (1968).

larities to that of nicotinic acid in the previously described cases. Here too, labeled precursors involved in the *Neurospora* pathway were not incorporated into the molecule at all.

In another case involving a pyridine ring (mimosine), lysine seems to be a direct precursor of the entire ring.[37a–39] This is particularly interesting because Robinson long ago suggested that lysine might be a precursor of the pyridine ring of nicotine.[40] This has proved not to be the case,[41,42] but the fact that it does appear to be the precursor of a similar ring in another alkaloid is another example of the various paths by which plants and lower animals may synthesize these basic structures. In this case, both succinic and aspartic acids also act as precursors of mimosine, but their specific incorporation is much lower than is that of lysine.[38]

Still other pyridine ring structures have been found to be basically formed from polyacetate chains. Thus fusaric acid (5-butyl-2-pyridine carboxylic acid) has been shown to be formed via a condensation of a polyacetate chain with aspartic acid, with labeled aspartic acid contributing activity only to the pyridine ring and the carboxyl group.[43] Other modes of formation of pyridine rings in various alkaloids are mentioned in Leete's review.[20]

It is quite likely that still other pathways to nicotinic acid and the pyridine ring exist in nature, and if so, degradations such as those described here will be needed for their study. For example, cinchomeronic acid has been reported to be a precursor of nicotinic acid in *Lactobacillus arabinosus*,[44] although it was almost completely inactive when applied to tobacco roots.[28]

Interesting points involved in the coupling of the pyridine and the pyrrolidine rings in nicotine are not at all understood at present. The only real clues to its nature so far are the discoveries that tritium label is lost from the 6-position of nicotinic acid during the coupling (or at least prior to it),[2] and that tritium label from the other ring positions is not only preserved, but the symmetry is also preserved during the coupling (i.e., the 2-position of the nicotinic acid is still in the 2-position in the final nicotine). The finding of retention of symmetry during the coupling has been confirmed by experiments with quinolinic acid-2,3,7,8-[14]C.[19] Leete and

[37a] J. W. Hylin, *Phytochemistry* **3**, 161 (1964).

[38] A. D. Notation and I. D. Spenser, *Can. J. Biochem.* **42**, 1803 (1964).

[39] H. P. Tiwari and I. D. Spenser, *Can. J. Biochem.* **43**, 1687 (1965).

[40] R. Robinson, *Proc. Univ. Durham Phil. Soc.* **8**, 1, 14 (1927).

[41] A. A. Bothner-by, R. F. Dawson, and D. R. Christman, *Experientia* **XII/4**, 151 (1956).

[42] E. Leete, *J. Am. Chem. Soc.* **78**, 3520 (1956)

[43] R. D. Hill, A. M. Unrau, and D. T. Canvin, *Can. J. Chem.* **44**, 2077 (1966).

[44] F. Lingens, *Angew. Chem.* **72**, 920 (1960).

others[20] have suggested possible gross mechanisms by which this coupling might occur, but the actual situation is anything but clear at present. The question still remains as to how and why the 3-carboxyl group of nicotinic acid, normally very difficult to displace chemically, becomes so labile during the biosynthesis of nicotine. These points, and others that will undoubtedly arise, should make the study of the biosynthesis of nicotinic acid and its metabolites of great interest for some time to come.

Acknowledgments

Some of the research reported here was performed under the auspices of the U.S. Atomic Energy Commission at Brookhaven National Laboratory and at Columbia University under grants G12855 and GB1129 from the National Science Foundation.

[111] Methods of Enzyme Induction by Nicotinamide

By R. L. BLAKE and ERNEST KUN[1]

Principle

The administration of nicotinamide to rats by intraperitoneal injection results in an increase in the activity of the glucocorticoid-inducible enzymes of the liver tryptophan pyrrolase (L-tryptophan:oxygen oxidoreductase, EC 1.13.1.12) and tyrosine aminotransferase (L-tyrosine:2-oxoglutarate aminotransferase, EC 2.6.1.5).[2]

Research Background

It is well known that the administration of nicotinamide to mice[3] or rats results in a rapid increase of the NAD content of the liver. During the course of investigations on the nature of tissue specific control mechanisms in the regulation of NAD biosynthesis in the rat, we observed that the activity of liver tryptophan pyrrolase, the first enzyme in the tryptophan to quinolinic acid pathway, is markedly increased after the intraperitoneal injection of nonphysiological amounts of nicotinamide. The enzyme-inducing effect of nicotinamide is not specific for tryptophan pyrrolase, however, as the activity of tyrosine aminotransferase also increases significantly. In further studies on the possible causal relationships between metabolic regulations by changes in substrate levels and direct alterations

---

[1] Recipient of a Research Career Award of the United States Public Health Service.
[2] R. L. Blake, S. L. Blake, H. H. Loh, and E. Kun, *Mol. Pharmacol.* **3**, 412 (1967).
[3] N. O. Kaplan, A. Goldin, S. R. Humphreys, M. M. Ciotti, and F. E. Stolzenbach, *J. Biol. Chem.* **219**, 287 (1956).

of rates of biosynthesis and degradation of specific enzymes, it was shown that the induction of enzyme activity by nicotinamide is inhibited by actinomycin D at doses that do not prevent the augmentation of liver NAD. It is also of significance that the injection of nicotinamide into either mice[4,5] or rats does not alter glutamate, lactate, or glucose-6-phosphate dehydrogenase content of the liver. Thus the enzyme-inducing action of nicotinamide appears to have the most prominent effect on enzymes that are rapidly induced by glucocorticoids.[6]

Methods

Reagents

Potassium chloride solution, 0.14 M KCl (adjusted to pH 8.5 with dilute KOH)
Phosphate buffer, 0.2 M NaH_2PO_4–Na_2HPO_4, pH 7.0
Trichloroacetic acid, 5%
Perchloric acid, 0.6 N $HClO_4$
Boric acid, 1.0 M H_3BO_3, pH 7.7
L-Tryptophan, 0.03 M
Hematin, 0.0004 M
Pyridoxal 5-phosphate, 0.00121 M
α-Oxoglutarate, 0.8 M (free acid, adjusted to pH 7.0)
L-Tyrosine, 0.0115 M (in borate buffer, pH 7.7)

General Procedure. Sprague-Dawley male rats weighing 200 ± 20 g are fed Wayne-Lab Blox diet ad libitum. As the extent of enzyme induction is enhanced by food deprivation, the rats are fasted for 15 hours (overnight) prior to the start of the experiment. At 0700 hours, nicotinamide is administered by intraperitoneal injection in neutralized 0.9% NaCl solution. The rats are killed by cervical dislocation at 2-hour intervals over a period of 8 hours. The livers are rapidly excised and either frozen in liquid N_2 for the NAD analyses or homogenized in a Teflon–glass homogenizer for the enzyme assays.

Methods of Enzyme Assay. The livers are rapidly excised (within 1–2 minutes after cervical dislocation), washed in an ice-cold 0.14 M KCl solution, and homogenized in a Teflon–glass homogenizer. The ratio of liver to homogenizing medium is 1:7, w/v. The homogenate is centrifuged

---

[4] E. Hirschberg, D. Snider, and M. Osmos, *in* "Advances in Enzyme Regulation" (G. Weber, ed.), Vol. II, p. 301. Macmillan, New York, 1964.
[5] The intraperitoneal injection of nonphysiological amounts of nicotinamide into mice (DBA/2J, C57BL/6J, C57BL/6J-*ob*) results in a severalfold increase in the activity of liver tyrosine aminotransferase. R. L. Blake, to be published, 1970.
[6] C. M. Berlin and R. T. Schimke, *Mol. Pharmacol.* **1**, 149 (1965).

for 20 minutes at 12,800 g in the 9RA rotor of a Servall refrigerated centrifuge at 0°. Tryptophan pyrrolase and tyrosine aminotransferase activities are determined in the centrifugal supernatant solution.

Tryptophan pyrrolase activity is determined by the spectrophotometric procedure of Knox[7] involving the measurement of kynurenine formation read at 365 mμ in a Zeiss PMQ II spectrophotometer. Each enzyme reaction mixture contains 1.0 ml of 0.2 M phosphate buffer (pH 7.0), 0.3 ml of 0.03 M L-tryptophan, 1.0 or 2.0 ml of the liver supernatant, hematin in a final concentration of 10 μM, and deionized water to make up a final volume of 4.0 ml. The blanks contain all components except L-tryptophan. Incubations are carried out in a Dubnoff shaker at 37° for 1 hour in an atmosphere of oxygen. The reaction is stopped by 2.0 ml of 5% trichloroacetic acid added to the vessels, which are chilled in ice. Aliquots of the trichloroacetate extract (after centrifugation) are assayed spectrophotometrically for kynurenine. Enzyme activity is expressed as micromoles of kynurenine accumulated per gram of liver (wet weight) per hour.

Tyrosine aminotransferase activity is determined by continuous spectrophotometric recording[8] at 310 mμ of the rate of formation of the enol–borate complex of p-hydroxyphenyl pyruvate produced at 25° in 1-cm quartz cuvettes over a period of 20 minutes. Each enzyme reaction mixture contain 0.57 M borate buffer (final pH 8.0), 12 micromoles of L-tyrosine, 30 μg pyridoxal phosphate, 0.1 ml of the liver supernatant, and 80 micromoles of α-oxoglutarate which are added last to initiate the reaction. The blanks contain all components except L-tyrosine. Enzyme activity is expressed as micromoles of p-hydroxyphenylpyruvate formed per milligram of protein per hour. Protein is determined by the modified biuret procedure of Beisenherz *et al.*[9]

Method of NAD Analysis. Tissues are rapidly removed (within 1–2 minutes after cervical dislocation), frozen in liquid N_2, weighed, and homogenized in 0.6 N $HClO_4$ at 0°. NAD is determined spectrophotometrically in neutralized extracts (after removal of ClO_4^- as K^+ salts) by the alcohol dehydrogenase[10] test using cuvettes of 5-cm light path in a Zeiss PMQ II spectrophotometer. The results are expressed as percent Δ NAD, which is the increment measured above control values (i.e., NAD content of livers of rats receiving only 0.9% NaCl) and calculated on a tissue protein basis (determined by the biuret method).

[7] W. E. Knox, Vol. II, p. 242.

[8] E. C. C. Lin and W. E. Knox. *Biochim. Biophys. Acta* **26**, 85 (1957).

[9] G. Beisenherz, H. J. Boltze, T. Bücher, R. Czok, K. H. Garbade, E. Meyer-Arendt, and G. Pfleiderer, *Z. Naturforsch.* **8b**, 555 (1953).

[10] M. Klingenberg, *in* "Methods of Enzymatic Analysis," (H. U. Bergmeyer, ed.), p. 528. Academic Press, New York, 1963.

Subcellular Distribution of NAD. The subcellular distribution of NAD in liver homogenates is determined by a simplified procedure of differential centrifugation of homogenates prepared in 0.25 M sucrose containing $5 \times 10^{-2} M$ nicotinamide (ratio of tissue to homogenizing medium is 1:4). Homogenates are separated into a pellet (containing the nuclear and mitochondrial fraction, sedimented at 12,800 g for 10 minutes at 4°) and a supernatant fraction containing soluble cell fraction and microsomes. Separation of microsomes by prolonged ultracentrifugation results in variable loss of NAD due to NAD-glycohydrolase activity of microsomes, a loss that cannot be completely prevented by nicotinamide. For this reason, the abbreviated cell fractionation was adopted. Omission of nicotinamide in the homogenizing medium results in 75% loss of NAD during centrifugation of homogenates for 10 minutes.

Properties of the Enzyme Induction System and NAD Accumulation

The results presented in Table I demonstrate an approximately 10-fold increase in tryptophan pyrrolase activity of livers of fasted rats 6 hours after the intraperitoneal injection of a dose of 5 millimoles of nicotinamide per kilogram of body weight. The induction of tryptophan pyrrolase ac-

TABLE I
TIME COURSE OF THE EFFECT OF NICOTINAMIDE ON RAT LIVER
TRYPTOPHAN PYRROLASE ACTIVITY[a]

Time (hr)	Experiments			Mean ± SE
	No. 1	No. 2	No. 3	
0	0.5	1.1	0.6	
	0.5	0.7	2.5	1.0 ± 0.3
2	1.6	2.6	2.6	
	1.9	2.3	3.5	2.4 ± 0.8
4	3.0	5.1	7.7	
	2.8	3.6	9.1	5.2 ± 1.1
6	7.8	14.0	8.9	
	8.5	18.3	9.2	11.1 ± 1.7
8	4.2	12.6	4.8	
	4.9	11.9	10.1	8.1 ± 1.6

[a] The results were determined in three separate experiments with 10 fasted rats per experiment (i.e., a total of 30 rats). In each experiment, measurements were made at 0, 2, 4, 6, and 8 hours with 2 rats at each time interval. Nicotinamide was injected intraperitoneally at a dose of 5 millimoles per kilogram body weight. Control groups (at 0 time) received only 0.9% NaCl solution (no nicotinamide). The animals were sacrificed at 2-hour intervals after the injection. Each value represents the average of duplicate determinations on individual livers. The data from the three experiments were combined as the mean ± the standard error.

TABLE II

AUGMENTATION OF TRYPTOPHAN PYRROLASE ACTIVITY BY NICOTINAMIDE
AND RELATED PYRIDINE DERIVATIVES[a]

Compound injected	Dose (millimoles per kg body wt)	Tryptophan pyrrolase activity	
		Total activity	Net increase
None	—	3.4 (4)	—
Nicotinamide	1.25	9.4 (4)	6.0
	2.50	10.2 (4)	6.8
	5.00	17.0 (4)	13.6
	10.00	11.0 (4)	7.6
Nikethamide	0.25	7.4 (4)	4.0
	0.50	9.5 (4)	6.1
	1.25	12.7 (4)	9.3
5-Fluoronicotinamide	2.50	28.5 (2)	25.1
Isonicotinic acid hydrazide	0.25	8.2 (4)	4.8
	2.50	21.3 (2)	17.9
Nicotinic acid	5.00	10.9 (2)	7.5

[a] Pyridine derivatives were administered by intraperitoneal injection to fasted rats which were sacrificed 6 hours subsequently. The livers in each experimental group, indicated by numbers in parentheses, were combined for enzyme analysis. Each value represents the average of duplicate determinations on the combined extract. Tryptophan pyrrolase activity was calculated as micromoles of kynurenine accumulated per gram of liver (wet weight) per hour. Results were also expressed as the increase above the control values (net increase).

tivity can also be demonstrated by the intraperitoneal injection of several structurally related pyridine derivatives,[11] indicating that this effect is not specific for nicotinamide (Table II). Additional characteristics of the increase of activity of rat liver adaptive enzymes by nicotinamide are presented in Table III. Both the pyrrolase and aminotransferase activities increase after injection of nicotinamide. Intraperitoneal injection of actinomycin D (2 mg/kg) 30 minutes prior to nicotinamide prevents the increase in tryptophan pyrrolase activity at 6 and 8 hours. However, if the administration of actinomycin D (2 mg/kg) is delayed 4 hours after the injection of nicotinamide, tryptophan pyrrolase activity measured at 8 hours is induced to levels comparable to those found in livers of rats treated only with nicotinamide. Liver tyrosine aminotransferase activity at 4 and 6 hours following the intraperitoneal injection of nicotinamide at a level

[11] Although optimal doses were not determined, it appears that 5-fluoronicotinamide and isonicotinic acid hydrazide are more effective than nicotinamide itself. Nikethamide was lethal in these studies at the dose of 2.5 millimoles per kilogram of body weight, and thus had to be used in smaller amounts.

TABLE III
Effect of Actinomycin D on Augmentation of Tryptophan Pyrrolase and Tryosine Transaminase by Nicotinamide[a]

| | | | Enzyme activity | | | |
| | | | TPO | | TTA | |
No.	Experimental conditions	Time of enzyme assay (hr)	No A	+A	No A	+A
1	Control (i.e., assays done prior to injections)	0 (TPO, TTA)	5.3	—	0.7	—
			4.2	—	0.4	—
			3.4	—	0.6	—
					0.4	—
2	Nicotinamide, 5 millimoles/kg, at 0 time; actinomycin at −30 min	6 (TPO) 4 (TTA)	13.8	5.1	2.7	0.9
			16.0	4.2	1.8	1.3
			27.0	3.8	2.1	0.7
3	Nicotinamide, 5 millimoles/kg, at 0 time; actinomycin at −30 min	8 (TPO) 6 (TTA)	17.4	4.9	3.8	1.1
			27.8	3.0	4.6	1.9
			22.8	—	3.6	—
			—	—	4.5	—
			—	—	2.1	—
4	Nicotinamide, 5 millimoles/kg at 0 time; actinomycin 4 hours after nicotinamide	8 (TPO)	17.4	16.8	—	—
			27.8	22.1	—	—
			22.8	—	—	—

[a] Fasted rats were injected intraperitoneally with nicotinamide at a dose of 5 millimoles per kilogram body weight. In No. 1, enzyme assays were done prior to nicotinamide injection. In No. 2 and No. 3, actinomycin (2 mg/kg) was injected 30 minutes prior to nicotinamide (−30 minutes), and in No. 4, 4 hours after nicotinamide. TPO = tryptophan pyrrolase activity (in micromoles of kynurenine per gram of liver per hour); TTA = tyrosine transaminase activity (in micromoles of p-hydroxyphenylpyruvate per milligram of protein of the centrifugal supernatant fluid per hour). A = actinomycin D. Each value is the average of duplicate assays on individual rat livers.

of 5 millimoles per kilogram is increased approximately 4- and 7-fold, respectively, above the zero time control values. Pretreatment with actinomycin D results in about 50% reduction of enzyme activity at 4 hours, and causes 60–71% diminution 6 hours after injection of nicotinamide.

Tryptophan pyrrolase activity of livers of hypophysectomized rats 6 hours following intraperitoneal injection of nicotinamide (5 millimoles/kg) or hydrocortisone (5.2×10^{-2} millimoles/kg) is shown in Table IV. Nicotinamide is not effective in hypophysectomized rats, whereas hydrocortisone causes a nearly 10-fold increase in enzyme activity during an experimental period of 6 hours.

The results presented in Tables I–IV were obtained with rats fasted overnight for approximately 15 hours prior to experiments. The effect of

TABLE IV

EFFECT OF HYPOPHYSECTOMY ON THE INCREASE OF TRYPTOPHAN PYRROLASE
BY NICOTINAMIDE AND HYDROCORTISONE[a]

Experimental conditions	Tryptophan pyrrolase activity
0 Hour, no injections	4.0
	4.3
	1.7
	2.5
	Mean = 3.1 ± 0.6
Nicotinamide, 5 millimoles/kg	2.7
	3.2
	2.7
	2.8
	Mean = 2.8 ± 0.1
Hydrocortisone, 5.2×10^{-2} millimoles/kg	21.1
	32.3
	30.1
	32.0
	Mean = 28.9 ± 2.6

[a] Hypophysectomized male rats, 1 week postoperative and fasted overnight, were injected intraperitoneally with either nicotinamide or hydrocortisone and then sacrificed 6 hours subsequently. Each value represents the average of duplicate determinations on individual rat livers. Tryptophan pyrrolase activity was calculated as micromoles of kynurenine accumulated per gram of liver (wet weight) per hour.

starvation itself on tryptophan pyrrolase activity, 6 hours following the intraperitoneal injection of either nicotinamide (5 millimoles/kg) or hydrocortisone (5.2×10^{-2} millimoles/kg) is shown in Table V. It is evident that starvation results in significant elevation of enzyme activity caused by nicotinamide injection. The absolute increase in enzyme activity caused by nicotinamide (determined by subtraction of pyrrolase activity found prior to nicotinamide injection, a value close to 4.0) was 14.3 in fasted and 7.3 in fed rats, indicating clearly that nicotinamide is far more effective in fasted rats. On the other hand, an increase in enzyme activity after injection of 5.2×10^{-2} millimoles of hydrocortisone per kilogram is uninfluenced by fasting.

A possible temporal relationship between rates of increase in pyrrolase and aminotransferase activities of rat liver and rates of NAD synthesis from pyridine-containing precursors was investigated. The time function

TABLE V

COMPARISON OF AUGMENTATION OF TRYPTOPHAN PYRROLASE ACTIVITY BY
NICOTINAMIDE AND HYDROCORTISONE IN FASTED AND FED RATS[a]

Injected substance	Tryptophan pyrrolase activity	
	Fasted	Fed
Nicotinamide, 5 millimoles/kg	22.1	9.8
	18.3	19.7
	15.8	7.0
	19.9	8.0
	17.9	11.6
	16.0	11.4
	Mean = 18.3 ± 1.0	Mean = 11.3 ± 1.8
Hydrocortisone, 5.2 × 10⁻² millimoles/kg	13.6	27.8
	23.6	20.5
	26.3	26.0
	19.9	21.3
	20.3	18.4
	17.3	12.9
	Mean = 20.2 ± 1.8	Mean = 21.2 ± 2.1

[a] For experimental details see General Procedure. Tryptophan pyrrolase activities
were determined 6 hours after injection of either nicotinamide or hydrocortisone.

of NAD formation after injection of nicotinamide, nicotinic acid, and
quinolinic acid (at a dose of 0.1 millimole per kilogram body weight) is
shown in Fig. 1. Rates of NAD augmentation as a function of nicotinamide
concentration is illustrated in Fig. 2. It is important to note that in liver
the rate of NAD augmentation after nicotinamide injection is very rapid,
even at a dose level one-tenth of that employed in experiments dealing
with the effect of nicotinamide injections on pyrrolase and aminotransferase. Maximal rates of NAD augmentation were obtained at dose levels
which cause 5- to 10-fold increase of enzymatic activities in liver tissue.
Quinolinic acid elicits an increase in NAD content of liver tissue only.
Apparent tissue-specific responses in terms of marked differences in rates
of NAD accumulation were also observed.

Since tryptophan pyrrolase and tyrosine-α-oxoglutarate aminotransferase were determined in the soluble cytoplasmic cellular fraction of liver
tissue, it was of interest to identify the subcellular localization of NAD
formed after injection of nicotinamide. As described in Methods, high
NAD glycohydrolase activity of microsomes interferes with the analysis
for NAD in subcellular fractions prepared by prolonged differential centrifugation. However, the abbreviated fractionation of liver homogenates

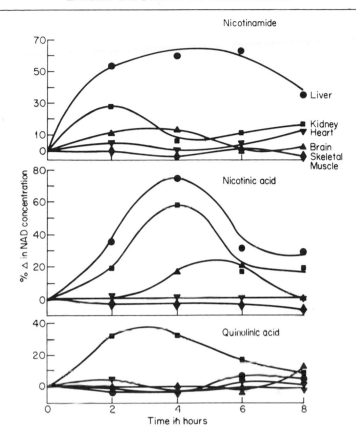

FIG. 1. The effect of nicotinamide, nicotinic acid, and quinolinic acid on NAD content of various rat tissues. The compounds were injected intraperitoneally to starved rats at a dose of 0.1 millimole/kg and analyzed at hourly intervals after injection. Each experimental point represents the NAD content of tissues pooled from 4 to 6 rats, expressed as percent change of NAD levels, compared to values obtained by analysis of pooled tissues (4 to 6 rats) of control animals, receiving only saline injections, sacrificed at time intervals indicated on abscissa.

yields useful information. Since recovery of added NAD to liver homogenates in the supernatant fraction (containing microsomes) is about 80–90%, this method gives a reasonably accurate subcellular localization of newly formed NAD, which is predominantly in the supernatant fraction (see Table VI). No definite conclusion can be drawn from these results as to the localization of NAD biosynthesis itself, since intracellular transfer processes may significantly contribute to this picture.

Actinomycin D in doses employed (2 mg/kg), injected 30 minutes prior to nicotinamide, has no influence on rates of NAD accumulation

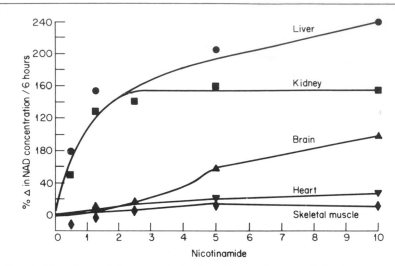

Fig. 2. The effect of the dose of nicotinamide on the rate of NAD augmentation in various rat tissues. Rates are expressed as percent change of NAD content, analyzed 6 hours after intraperitoneal injection of varying doses of nicotinamide to starved rats. Abscissa indicates the dose of nicotinamide, expressed as millimoles of nicotinamide per kilogram body weight. In obtaining experimental points, the same procedure was followed as described in the legend to Fig. 1.

caused by nicotinamide. On the other hand, fasting for 15 hours (prior to nicotinamide injection) doubles the rate of NAD accumulation which follows the injection of nicotinamide. All experiments reported here were

TABLE VI
Subcellular Distribution of NAD in Rat Liver[a]

| | NAD (micromoles/g protein) | |
Experimental conditions	Pellet	Supernatant containing microsomes
Control	1.16 ± 0.08	4.27 ± 0.2
6 Hours after injection of 5 millimoles/kg nicotinamide	2.2 ± 0.4	16.4 ± 4.0

[a] Freshly removed rat livers were homogenized in 0.25 M sucrose containing 5 × 10^{-2} M nicotinamide at 0° (ratio of tissue to homogenizing medium was 1 to 4). Homogenates were separated by centrifugation at 12,800 g for 10 minutes into a pellet (containing nuclei and mitochondria) and supernatant fluid (containing microsomes and soluble cell fraction). Omission of nicotinamide in the homogenizing medium results in 75% loss of NAD. Results shown in this table were obtained from 6 rat livers analyzed individually and expressed as the mean ± the standard error.

performed on fasted rats. It is of interest that ATP, ADP, and AMP levels of rat liver does not change during NAD accumulation following injection of nicotinamide.

[112] Nicotinate Phosphoribosyltransferase from Bakers' Yeast

By TASUKU HONJO

Nicotinate + PRPP + ATP → nicotinate ribonucleotide + PP$_i$ + ADP + P$_i$

Assay Method

Principle. The enzyme activity is assayed by measuring the formation of nicotinate ribonucleotide-^{14}C from nicotinate-^{14}C in the presence of PRPP and ATP. The ribonucleotide produced is isolated by a small Dowex 50 column, and the radioactivity is determined.[1] The alternative method based on the separation of the nucleotide by paper chromatography has been described earlier (Vol. VI [44]).

Reagents and Materials

ATP, Na salt, neutralized, 0.01 M
5-Phosphoribosyl-1-pyrophosphate (PRPP), Mg salt, 0.01 M, obtained from Sigma Chemical Company
MgCl$_2$, 0.15 M
Potassium phosphate buffer, 1.0 M, pH 7.4
Nicotinate-7-^{14}C (8700 cpm/millimicromole), 1.15 × 10^{-3} M, obtained from New England Nuclear Corporation.
Dowex 50-X8, H$^+$ form, 200–400 mesh

Procedure. The reaction mixture (0.2 ml) contains 0.02 ml of ATP, 0.01 ml each of PRPP, MgCl$_2$, and potassium phosphate buffer, 0.005 ml of nicotinate-7-^{14}C, and less than 7.5 × 10^{-5} unit of the enzyme. The mixture is incubated for 20 minutes at 37°. The reaction is terminated by heating in a boiling water bath for 1 minute. The mixture is immediately chilled, and the precipitate is removed by centrifugation. An aliquot (0.1 ml) of the supernatant is applied to a Dowex 50 column (0.5 × 2 cm). Nicotinate ribonucleotide-^{14}C produced is washed out with 5 ml of water. If necessary, nicotinate-^{14}C is subsequently eluted with 15 ml of 6 N formic

---

[1] T. Honjo, S. Nakamura, Y. Nishizuka, and O. Hayaishi, *Biochem. Biophys. Res. Commun.* **25**, 199 (1966).

acid containing 0.5 M ammonium formate. The radioactivities of the washing and eluate are determined with a Nuclear Chicago gas flow counter.

Definition of Unit and Specific Activity. One unit of enzyme activity is defined as that amount which catalyzes the formation of 1 micromole of nicotinate ribonucleotide per minute. Specific activity is expressed as units per milligram protein. Protein is determined by the method of Lowry *et al.*[2]

Purification Procedure

All procedures are operated at 0° to 4° unless otherwise specified.

Preparation of Acetone Powder. Caked bakers' yeast is obtained from Oriental Yeast Corporation, Osaka, Japan. Fresh bakers' yeast is homogenized in a Waring blendor with 5–10 volumes of cold acetone ($-10°$) and filtered on a Büchner funnel. The residue is homogenized again in cold acetone, filtered as above, and immediately dried in air. After drying completely *in vacuo*, the acetone powder may be stored for at least a year at $-20°$ without significant loss of activity.

Preparation of Crude Extract. Three hundred grams of the acetone powder is suspended in 1500 ml of 0.2 M KHCO$_3$ and stirred continuously for 30 minutes at 24°. Antifoam spray (Sigma Chemical Company) is used to prevent excess foaming. The suspension is centrifuged for 15 minutes at 12,000 g, and the slightly turbid supernatant solution is collected.

Streptomycin Treatment. Streptomycin sulfate (11.2 g) is added to the acetone powder extract (1120 ml) with continuous stirring. After 10 minutes, the precipitate is removed by centrifugation.

Ammonium Sulfate Fractionation. To the supernatant solution (1120 ml), 437 g of solid ammonium sulfate (special enzyme grade, Mann Research Laboratories, Inc.) are added to 0.60 saturation with constant stirring. The pH of the suspension is controlled above 7 by the addition of 0.5 M Tris, pH 10.7. Stirring is continued for 15 minutes. The suspension is allowed to stand for an additional 15 minutes and then centrifuged for 30 minutes at 14,000 g. The precipitate is taken up in 800 ml of 0.01 M potassium phosphate, pH 7.6. Solid ammonium sulfate (222 g, 0.45 saturation) is added to the resulting solution as described above. The suspension is centrifuged for 20 minutes at 14,000 g and the precipitate is discarded. The supernatant solution is brought to 0.60 saturation by the addition of ammonium sulfate as described above and the precipitate collected by

[2] O. H. Lowry, N. J. Rosebrough, A. L. Farr, and R. J. Randall, *J. Biol. Chem.* **193**, 265 (1951).

centrifugation is dissolved in 200 ml of 0.005 M potassium phosphate, pH 7.4. The resulting straw-colored solution is dialyzed overnight against 5 liters of 0.005 M potassium phosphate, pH 7.4. Any precipitate may be removed by centrifugation and discarded. The enzyme preparation at this stage is fairly stable when stored at $-20°$.

DEAE-Cellulose Chromatography. The dialyzed enzyme solution is applied to a DEAE-cellulose column (15 × 5 cm) which has been equilibrated with 0.005 M potassium phosphate, pH 7.4. The column is extensively washed with 500 ml of 0.005 M potassium phosphate, pH 7.4. The elution is carried out by a linear concentration gradient of potassium phosphate, pH 7.4; the mixing chamber and reservoir contain 1 liter of 0.005 M and 0.05 M potassium phosphate, pH 7.4, respectively. The buffers contain 0.002 M 2-mercaptoethanol. Fractions of 20 ml each are collected at a flow rate of 3 ml per minute. The peak of the enzyme activity is usually found at the 0.4 effluent volume fraction. To the combined eluates (420 ml), 236 g of ammonium sulfate (0.80 saturation) are added and the precipitate collected by centrifugation is dissolved in 10 ml of 0.005 M potassium phosphate, pH 7.4. The solution is dialyzed for 3 hours against 2 liters of 0.005 M potassium phosphate, pH 7.4, changing the external buffer every hour. Any insoluble material is removed by centrifugation.

Hydroxylapatite Chromatography. The enzyme solution is subjected to a hydroxylapatite column (6 × 2.6 cm) equilibrated with 0.005 M potassium phosphate, pH 7.0, containing 0.001 M 2-mercaptoethanol. The column is washed with 150 ml of 0.02 M potassium phosphate, pH 7.0, containing 0.001 M 2-mercaptoethanol. The elution is accomplished by a linear concentration gradient with 250 ml of 0.1 M potassium phosphate, pH 7.0, containing 0.001 M 2-mercaptoethanol in the reservoir flowing into the mixer (250 ml) containing 0.02 M potassium phosphate, pH 7.0, and 0.001 M 2-mercaptoethanol. Fractions of 10 ml each are collected at a flow rate of 1 ml per minute. The peak of the enzyme activity appears around the 0.5 effluent volume. Fractions 23–28 are combined and stored at $-20°$. Sometimes the enzyme solution is concentrated with a collodion bag (Sartorius Membranfilter Company, Germany). The purified preparation is unstable, especially when the protein concentration is low. A summary of the purification is given in the table.

Properties

Specificity. Nicotinate phosphoribosyltransferase is specific for nicotinate. Nicotinamide, quinolinate, purines, and pyrimidines are totally inert as substrates. The enzyme is relatively specific for ATP. The relative

PURIFICATION OF NICOTINATE PHOSPHORIBOSYLTRANSFERASE FROM BAKERS' YEAST

Fraction	Volume (ml)	Protein (mg/ml)	Activity (units/ml)	Specific activity (units/mg protein)	Yield (%)
Acetone powder extract	1120	20.9	0.0155	0.00074	100
Streptomycin treatment	1120	24.6	0.0155	0.00063	100
Ammonium sulfate (0–0.6 saturation)	800	23.0	0.0168	0.00073	77.5
Ammonium sulfate (0.45–0.6 saturation)	200	20.9	0.0408	0.00195	47.0
DEAE-cellulose	10	24.0	0.427	0.0175	24.2
Hydroxylapatite	60	0.51	0.0337	0.066	11.7

activities of the nucleotide triphosphates ($10^{-4} M$) are as follows: ATP, 100; GTP, 6.2; ITP, 15; CTP, 3.0; UTP, 3.5; dATP, 61.5; and dCTP, 0.5. Ribose 5-phosphate does not substitute for PRPP.

Stability. The enzyme preparation of high protein concentration retains at least 50% of the activity when stored at $-20°$ for a month. However, dilute preparations such as DEAE-cellulose eluate are unstable and lose 70% of the activity when dialyzed overnight at 4°. The enzyme activities remaining after heating for 5 minutes at 50° in the presence of various compounds indicated are as follows (ratio to the original activity): none, 0.28; PRPP ($2 \times 10^{-4} M$), 1.0; NAD ($2 \times 10^{-3} M$), 0.46; deamido-NAD ($1.3 \times 10^{-3} M$), 0.58; potassium phosphate, pH 7.5 ($1 \times 10^{-1} M$), 0.49; and ATP ($2 \times 10^{-3} M$), 0.60. The following compounds do not prevent the loss of activity by the heat treatment; nicotinate, nicotinamide, ADP, GTP, ribose 5-phosphate and NADP at $2 \times 10^{-3} M$.

Activators and Inhibitors. The enzyme requires magnesium absolutely with a saturating concentration of approximately $5 \times 10^{-3} M$. The enzyme is activated by orthophosphate, the maximal activity being observed at 1 to $2 \times 10^{-2} M$. In the absence of added phosphate, one-third of the maximal activity is observed. ADP acts as a competitive inhibitor against ATP with a K_i value of $1.7 \times 10^{-3} M$.

Effect of pH. The enzyme shows the maximal activity around pH 8.0.

Kinetic Properties. The Michaelis constants for nicotinate, ATP and PRPP are $1.85 \times 10^{-6} M$, $1.2 \times 10^{-4} M$, and $7.7 \times 10^{-6} M$, respectively.

Stoichiometry. When 1 mole of nicotinate ribonucleotide is synthesized, each mole of nicotinate, PRPP, and ATP disappears. The amounts of ADP and orthophosphate produced are also equivalent to that of nicotinate ribonucleotide synthesized.[1] Nicotinate ribonucleotide, however, is also produced in the absence of ATP when high concentrations of nicotinate

$(10^{-3} M)$ and of PRPP $(10^{-3} M)$ are employed. Under these conditions the activity in the absence of ATP is approximately 25% of that observed in the presence of ATP.[3]

[3] A similar observation was made with a preparation of nicotinate phosphoribosyltransferase obtained from beef liver [R. Seifert, M. Kittler, and H. Hilz, *in* "Current Aspects of Biochemical Energetics" (N. O. Kaplan and E. P. Kennedy, eds.), p. 413. Academic Press, New York, 1966; S. Nakamura, and T. Honjo, unpublished observations]. It is uncertain whether a single enzyme is responsible for both ATP-dependent and ATP-independent reactions.

[113] Nicotinamide Phosphoribosyltransferase and NAD Pyrophosphorylase[1] from *Lactobacillus fructosus*

By EIJI OHTSU and YASUTOMI NISHIZUKA

Nicotinamide + PRPP + ATP → NMN + PP + ADP + P$_i$

NMN + ATP → NAD + PP

Lactobacillus fructosus has been reported by Kodama[1a] to require nicotinamide for growth and maintenance but does not utilize nicotinate. The biosynthetic pathway of NAD in this microorganism was subsequently studied by Ohtsu, Ichiyama, Nishizuka and Hayaishi,[2] who showed that the microorganism contains a phosphoribosyltransferase and an NAD pyrophosphorylase that are specific for nicotinamide and NMN, respectively.

I. Nicotinamide Phosphoribosyltransferase

Assay Method

Principle. The activity of nicotinamide phosphoribosyltransferase is assayed by measuring the formation of NMN-^{14}C from nicotinamide-^{14}C in the presence of PRPP and ATP. The ribonucleotide produced is isolated by paper chromatography, and the radioactivity is determined.

Reagents

Tris-acetate buffer, 0.25 M, pH 7.0
5-Phosphoribosyl-1-pyrophosphate (PRPP), 4 mM, Mg salt, obtained from Sigma Chemical Company

[1] EC 2.7.7.1, ATP: NMN Adenyltransferase.
[1a] R. Kodama, *J. Agr. Chem. Soc. Japan* **30**, 219 (1956).
[2] E. Ohtsu, A. Ichiyama, Y. Nishizuka, and O. Hayaishi, *Biochem. Biophys. Res. Commun.* **29**, 635 (1967).

ATP, Na salt (neutralized), 0.02 M
$MgCl_2$, 0.25 M
Nicotinamide-7-^{14}C, 0.2 mM (9000 cpm/millimicromole)
Enzyme, diluted with 0.05 M phosphate buffer, pH 7.0, to obtain
a solution containing 0.002–0.015 unit/ml

Procedure. The assay system contains, in a total volume of 0.5 ml:
0.2 ml of Tris-acetate buffer, pH 7.0; 0.05 ml each of PRPP, ATP, $MgCl_2$,
nicotinamide-7-^{14}C, enzyme, and water. After incubation for 10 minutes
at 37°, the reaction is stopped by heating the mixture for a minute in a
boiling water bath. After centrifugation, an aliquot (0.20 ml) of the supernatant solution is chromatographed on Whatman No. 1 paper with 1 M
ammonium acetate, pH 5.0–95% ethanol (3:7) as a solvent (R_f value for
nicotinamide, 0.80; for NMN, 0.27; for NAD 0.13). The area corresponding
to authentic NMN is cut out, and the radioactivity is determined directly
with a Tri-Carb liquid scintillation spectrometer.[3]

Definition of Unit and Specific Activity. One unit of enzyme is defined
as that amount which converts 1 micromole of nicotinamide to NMN per
minute at 37°. The specific activity is defined as units per milligram of
protein. Protein is determined by the method of Lowry *et al.*[4]

Application to Crude Extracts. When crude extracts are employed, a
significant quantity of NMN produced is converted further to NAD under
the conditions. The radioactivity of NAD thus produced is also determined
and is combined with that of NMN.

Purification Procedure

Growth of Organisms. Cells of *L. fructosus* are grown in a glucose (2.0%)–
polypeptone (0.5%) medium supplemented with yeast extract (0.5%),
nicotinamide (10 mg/liter), and fructose (1.0%). The pH is adjusted to
around 7 with potassium phosphate buffer in a final concentration of
0.01 M. Growth is allowed to proceed at 26° without aeration in a 5-liter
Erlenmeyer flask containing 3 liters of the medium. After 24 hours the
cells are harvested by a Kubota continuous-flow centrifuge, and washed
twice with 0.14 M KCl. Approximately 1.5 g of the packed cells are obtained per liter of medium.

Preparation of Acetone Powder of L. fructosus. All subsequent manipulations are performed at 0–4°, and centrifugations are carried out at 12,000 g
unless otherwise stated. The washed cells are homogenized with 20 volumes
of cold acetone (−15°) for a minute in a Waring blendor. The suspension

[3] Y. Nishizuka and O. Hayaishi, *J. Biol. Chem.* **238**, 3369 (1963).
[4] O. H. Lowry, N. J. Rosebrough, A. L. Farr, and R. J. Randall, *J. Biol. Chem.* **193**,
265 (1951).

is filtered rapidly on a Büchner funnel, and the cell cake is treated again with cold acetone in the same way. The cell cake thus obtained is brought to room temperature (25°) and quickly crushed, and dried under vacuum for 2 hours. The powder may be stocked at −15° for several months under reduced pressure without significant loss of activity.

Step 1. Acetone Powder Extract. One gram of the acetone powder is suspended by stirring in 20 ml of 0.05 M potassium phosphate buffer, pH 7.0, for 15 minutes. The suspension is sonicated for 10 minutes in a Kubota 9-kc sonic oscillator followed by centrifugation for 20 minutes.

Step 2. Protamine Treatment. To the supernatant thus obtained (12 ml), an equal volume of protamine sulfate solution (0.4%, pH 7) is added. The mixture is then stirred for 30 minutes and centrifuged for 15 minutes. The supernatant solution is practically free of NAD pyrophosphorylase. The precipitate is stored to prepare the pyrophosphorylase (see below).

Step 3. Ammonium Sulfate Fractionation. To the supernatant solution (20 ml), 5.5 g of solid ammonium sulfate is slowly added, and the suspension is stirred for 30 minutes. The precipitate is discarded after centrifugation for 30 minutes. To the supernatant is added an additional 1.5 g of solid ammonium sulfate in the same way. The precipitate produced is collected by centrifugation and dissolved in 10 ml of 0.05 M potassium phosphate buffer, pH 7.0.

Step 4. Chromatography on DEAE-Cellulose. After dialysis for 12 hours against a liter of 5 mM potassium phosphate buffer, pH 7.0, the enzyme solution is applied on to a DEAE-cellulose column (1.5 × 10 cm) equilibrated with 5 mM potassium phosphate buffer, pH 7.0. The column is washed with 100 ml of 0.05 M potassium phosphate buffer, pH 7.0. Then the enzyme is eluted from the column with application of a linear concentration gradient of potassium phosphate buffer, pH 7.0 (mixing chamber, 0.05 M; reservoir, 0.25 M; 100 ml each). Fractions of 10 ml each are collected. The enzyme is eluted usually as a sharp peak around the 0.4 effluent volume. By this procedure the enzyme is purified about 25- to 30-fold with an overall recovery of 70%. Upon fractionation by ammonium sulfate, the enzyme may be recovered with more than a 100% yield. This is presumably due to the removal of some endogenous inhibitors. A summary of the purification is given in the table.

Properties

Stability. The purified enzyme in potassium phosphate buffer, pH 7.0, may be stored for at least several months at −20° without detectable loss of the activity.

pH Optimum. The pH optimum for the activity is 6 to 7.

Kinetic Constants. The K_m values for nicotinamide, PRPP, and ATP are

PURIFICATION OF NICOTINAMIDE PHOSPHORIBOSYLTRANSFERASE FROM
Lactobacillus fructosus

Step	Fraction	Volume (ml)	Total protein (mg)	Total activity (units)	Specific activity (units/mg)	Yield (%)
1.	Crude extract	12	112.7	0.0496	0.00044	100
2.	Protamine sulfate	20	31.1	0.0333	0.00107	67
3.	Ammonium sulfate	10	25.1	0.0583	0.00232	118
4.	DEAE-cellulose	30	3.0	0.0365	0.0122	73

calculated to be $2 \times 10^{-6} M$, $4 \times 10^{-5} M$, and $6.7 \times 10^{-4} M$, respectively.

Inhibitors. The enzyme activity is inhibited to 20% of the original activity by 0.1 mM of NMN, and to 60% by the same concentration of NAD. These inhibitions appear to be of competitive nature with K_i values of about $7.5 \times 10^{-4} M$.

Substrate Specificity. Nicotinamide is a specific substrate for this enzyme. Neither nicotinate phosphoribosyltransferase nor quinolinate phosphoribosyltransferase is detected in the crude extract or in the purified enzyme preparations.

Stoichiometry. The purified enzyme preparation catalyzes the conversion of nicotinamide and PRPP to stoichiometric quantities of NMN and PP. This reaction has an almost absolute specificity for ATP under the assay conditions employed. With a more purified preparation of the enzyme,[5] ATP is shown definitely to be utilized in this reaction and hydrolyzed to stoichiometric quantities of ADP and orthophosphate. However, a very small but significant quantity of NMN (approximately 5% of that produced in the presence of ATP) is produced without ATP supplemented to the reaction mixture. It is not clear at present whether this NMN formation is due to endogenous ATP which contaminates the purified enzyme preparations. ATP does not generate PRPP in this system, since the preparations are essentially free of 5-phosphoribose pyrophosphokinase. Nicotinate phosphoribosyltransferase purified from beef liver catalyzes the formation of nicotinate ribonucleotide from nicotinate and PRPP in the absence of added ATP when higher concentrations of these substrates are employed. But the latter enzyme also requires ATP in a stoichiometric quantity when lower concentrations of the substrates are used.[6] Nicotinate

---

[5] The purified preparation contains some ATPase activities. These activities may be removed by further purification of the enzyme by DEAE-Sephadex column chromatography.

[6] R. Seifert, M. Kittler, and H. Hilz, "Current Aspects of Biochemical Energetics," p. 413. Academic Press, New York, 1966. Also unpublished observations by S. Nakamura, Y. Nishizuka, and O. Hayaishi, 1965.

phosphoribosyltransferase purified from yeast shows similar properties.[7] The exact role of ATP in these reactions may be elucidated by further investigations.

II. NAD Pyrophosphorylase

Assay Method

Principle. Enzyme activity may be assayed conveniently by measuring the formation of radioactive NAD from NMN-^{14}C and ATP. The NAD-^{14}C is separated from NMN-^{14}C by paper chromatography, and the radioactivity is determined. The activity may be assayed alternatively in a larger-scale incubation with higher concentrations of nonradioactive NMN, ATP, and the enzyme preparation. The NAD produced is determined by measuring the increase in absorbance at 340 mμ upon the addition of alcohol and alcohol dehydrogenase.[8]

Reagents

Tris-acetate buffer, 0.25 M, pH 5.8
ATP, Na salt (neutralized), 0.05 M
NMN-^{14}C, 0.2 mM (9000 cpm/millimicromole)[9]
MgCl$_2$, 0.1 M
Enzyme solution (0.002–0.015 unit/ml)

Procedure. The assay system contains, in a total volume of 0.5 ml: 0.2 ml of Tris-acetate buffer, pH 5.8; 0.05 ml each of ATP, NMN-^{14}C, and MgCl$_2$; enzyme solution and water. The reaction is started by the addition of enzyme, and the incubation is carried out for 10 minutes at 37°. The reaction is stopped by heating the mixture for a minute in a boiling water bath, and the denatured protein is removed by centrifugation for 10 minutes in the cold (0–4°). An aliquot (0.02 ml) of the supernatant is chromatographed on paper, and the radioactivity of NAD-^{14}C produced is determined by direct paper strip counting as described above.

Definition of Unit and Specific Activity. One unit of enzyme is defined as that amount which converts 1 micromole of NMN to NAD per minute under the standard assay conditions. Specific activity is defined as units

---

[7] T. Honjo, S. Nakamura, Y. Nishizuka, and O. Hayaishi, *Biochem. Biophys. Res. Commun.* **25,** 199 (1966).

[8] M. M. Ciotti and N. O. Kaplan, see Vol. III [128].

[9] Either NMN-nicotinamide-7-^{14}C or NMN-ribose-^{14}C is employed as a substrate. The radioactive ribonucleotide may be prepared by the method of Ueda, Yamamura, and Nishizuka (see this volume [106]).

per milligram of protein. Protein is determined by the method of Lowry *et al.*[4]

Application to Crude Extract. The method may be employed for the crude extract, since the NAD glycohydrolase and NAD pyrophosphatase activities are negligible.

Purification Procedure

The enzyme may be partially purified as follows: The precipitate, which is obtained from the crude extract by treatment with protamine sulfate (Step 2, see above), is suspended in 20 ml of distilled water and homogenized in a glass–Teflon homogenizer. The suspension is centrifuged for 10 minutes at 12,000 *g*, and the supernatant is discarded. The precipitate is washed again in the same way. Then, the enzyme is eluted twice from the precipitate, each time with 10 ml of 0.2 *M* K$_2$HPO$_4$, using a glass–Teflon homogenizer as described above. By this procedure the enzyme may be enriched approximately 10-fold with an overall yield of 30–40%. The specific activities of the crude extract and final preparation are 0.00081 and 0.008, respectively.

Properties

The enzyme is specific for NMN and does not react with nicotinic acid ribonucleotide. The final preparation is practically free of nicotinamide phosphoribosyltransferase.

The K_m values for NMN and ATP are 6.7 × 10^{-4} *M* and 2.7 × 10^{-3} *M*, respectively. The maximum activity is observed at pH 5.8.

[114] Enzymatic Preparation of Nicotinic Acid Adenine Dinucleotide, Nicotinic Acid Ribonucleotide, and Nicotinic Acid Ribonucleoside

By Tasuku Honjo and Yasutomi Nishizuka

I. Nicotinic Acid Adenine Dinucleotide (Deamido-NAD)

Method

Principle.[1,2] The procedure is based on the observation of Ballio and Serlupi-Crescenzi[3] that beef spleen NAD(P) glycohydrolase (EC 3.2.2.6)

[1] Y. Nishizuka and O. Hayaishi, *J. Biol. Chem.* **238**, 3369 (1963).

[2] T. Honjo, M. Ikeda, A. J. Andreoli, Y. Nishizuka, and O. Hayaishi, *Biochim. Biophys. Acta* **89**, 549 (1964).

[3] A. Ballio and G. Serlupi-Crescenzi, *Nature* **180**, 1203 (1957).

catalyzes a direct exchange reaction of the nicotinamide moiety of NAD with nicotinic acid to produce deamido-NAD.

Reagents

Beef spleen NAD(P) glycohydrolase, 80 mg of protein per milliliter[4]

NAD, 1.4 millimoles (approximately 1 g) of NAD is dissolved in 20 ml of water and neutralized to about pH 7 with 4 N KOH

Nicotinate, 60 millimoles (7.38 g) of nicotinic acid is suspended in about 20 ml of water and neutralized to about pH 7 with 10 N KOH. The solution is made up to 50 ml with water.

Tris-Cl buffer, 1 M, pH 7.5

NaCN, 1.2 M

Alcohol dehydrogenase, crystalline, yeast, about 100 μg/ml[5]

NAD assay mixture, 0.1 M ethanol in 0.06 M Tris-Cl buffer, pH 8.5

Procedure. The incubation mixture contains 20 ml of NAD, 50 ml of nicotinate, 8 ml of Tris-Cl buffer, and 40 ml of beef spleen NAD(P) glycohydrolase and water in a total volume of 140 ml. The incubation is carried out in a 300-ml Erlenmeyer flask at 37° with constant shaking. To follow the progress of the reaction, 0.2-ml aliquots of the reaction mixture are taken, diluted with water to 2.0 ml, and centrifuged for 5 minutes at 24,000 g. An 0.5-ml aliquot of the supernatant solution is added to a 3-ml quartz cuvette with 1-cm light path containing 2.4 ml of the NAD assay mixture, and the amount of NAD remaining is determined by measuring the increase in absorbance at 340 mμ upon the addition of 0.1 ml of alcohol dehydrogenase. The amount of total dinucleotides, deamido-NAD plus NAD, is determined by measuring the increase in absorbance at 315 mμ upon the addition of 2.5 ml of NaCN to another 0.5-ml aliquot of the supernatant placed in a 3-ml quartz cuvette with 1-cm light path.[6] The quantity of deamido-NAD is calculated from the difference between the values mentioned above.

NAD disappears rapidly with concomitant formation of deamido-NAD. The maximum yield of deamido-NAD is obtained usually at about 90 minutes' incubation and is diminished gradually by prolonged incubation.[7] When the dehydrogenase assay reaches a minimum and the cyanide addi-

---

[4] See Vol. II [113].

[5] See Vol. I [79].

[6] This cyanide addition reaction is based on the procedure described by S. P. Colowick, N. O. Kaplan, and M. M. Ciotti [*J. Biol. Chem.* **191**, 447 (1951)]. Molar extinction coefficients at 315 mμ of deamido-NAD and NAD in 1 M NaCN are 4.6 × 10³ and 5.0 × 10³, respectively.

[7] The relative rate of hydrolysis of deamido-NAD by beef spleen NAD(P) glycohydrolase to that of NAD (100) is 40. The splitting reaction of deamido-NAD is inhibited to the extent of 20% in the presence of 0.4 M nicotinate.

tion reaction begins to decrease, the reaction is stopped by the addition of 10 ml of 20% perchloric acid. The denatured protein is removed by centrifugation, and the precipitate is washed twice with 30 ml of 2% perchloric acid. The supernatant solution and washings are combined and adjusted to pH 7 with 10 N KOH. The yield of deamido-NAD at this stage is about 45%.

The clear solution is put on a Dowex 1 X8 formate column (3 × 70 cm), and the column is washed with 0.10 N formic acid until no more UV-absorbing material is eluted from the column. Deamido-NAD is then eluted from the column by application of a formic acid concentration gradient (mixing chamber, 750 ml of 0.10 N formic acid; reservoir, 2 N formic acid). The elution of deamido-NAD is followed by measuring the absorbance at 260 mμ. Fractions (20 ml each) are collected. The dinucleotide is usually eluted in the 900–1200 ml effluent volume and the fractions are combined and evaporated to dryness under reduced pressure at 5–10°. The residue is taken up in a small volume of water and freed from as much formic acid as possible by repeated evaporation to dryness under reduced pressure. The residue is taken up in a small volume of water. The overall yield of deamido-NAD is approximately 40% based on the NAD employed as starting material. A dry, white powder may be obtained by lyophilization.

Properties of Deamido-NAD

Analysis of the product gives satisfactory results for the several criteria of deamido-NAD. Purity of the deamido-NAD is shown to be more than 90%. A single UV-absorbing spot is obtained by paper chromatography in several different solvent systems and by paper electrophoresis at pH 5.0. The product is practically free of nicotinic acid or nicotinamide as judged by bioassay with several microorganisms listed below. The absorption maximum of the ultraviolet spectrum is found at 260 mμ at pH 7.0. The cyanide addition compound of the product has an absorption maximum at 315 mμ. The aqueous cyanide addition reaction is slow in developing, and about 15 minutes is required for the maximal addition. Molar extinction coefficients of deamido-NAD at 260 mμ, pH 7.0, and its cyanide addition compound at 315 mμ (in 1 M NaCN) are 15.7 × 10³ and 4.6 × 10³, respectively.[8] When the product is heated for 10 minutes at 100° in 0.1 N NaOH, the material disappears completely, and a quantitative amount of nicotinic acid is recovered.

In microbiological growth studies, *Lactobacillus casei* (ATCC 7469)

---

[8] The cyanide addition compound of NAD has an absorption maximum at 325 mμ. The molar extinction coefficient at 325 mμ in 1 M NaCN is 6.0 × 10³. In contrast to deamido-NAD, NAD is immediately converted to the cyanide complex upon the addition of 1 M NaCN.

is found to grow on deamido-NAD to the same extent as on nicotinamide or nicotinic acid. *Hemophilus parainfluenzae* grows on the dinucleotide to an extent of 20% as compared with NAD.[9] However, *Lactobacillus arabinosus* (ATCC 8014), *Streptococcus faecalis* (ATCC 8043), *Leuconostoc mesenteroides* (ATCC 9135), *Lactobacillus fructosus* (IFO 3516), and *Pediococcus acidilactici* (ATCC 8042) are unable to utilize deamido-NAD as a growth factor.[10]

Alternative Method. Lamborg, Stolzenbach, and Kaplan[9] prepared deamido-NAD by hydrolysis of the nicotinic acid ethyl ester analog of NAD which was obtained by a pig brain NAD(P) glycohydrolase base exchange reaction. The overall yield by this procedure was about 8% based on the NAD employed as starting material, and the final product contained a small quantity of the nicotinic acid ethyl ester analog.

II. Nicotinic Acid Ribonucleotide

Method

Principle. Deamido-NAD is split by snake venom phosphodiesterase to produce 5'-AMP and nicotinate ribonucleotide. The nicotinate ribonucleotide is separated from 5'-AMP by a Dowex 1 X2 formate column.

Snake Venom Phosphodiesterase. The preparation to be employed for this preparation must be free of phosphomonoesterase; it may be prepared by the method of Butler.[11]

Reagents

Deamido-NAD, prepared as described above. Approximately 250 micromoles (about 180 mg) of deamido-NAD is dissolved in 2 ml of water, and the solution is adjusted to about pH 9 with 2 N KOH.
Tris-Cl buffer, 1 M, pH 9.2
$MgCl_2$, 0.2 M
Snake venom phosphodiesterase. The activity may be checked with NAD as substrate by measuring the decrease of NAD with the dehydrogenase assay as described above.

Procedure. The reaction mixture contains 2.0 ml of deamido-NAD, 0.2 ml of Tris-Cl buffer, 0.1 ml of $MgCl_2$, and a sufficient quantity of snake venom phosphodiesterase in a total volume of 3.0 ml. The incubation is carried out at 37° with shaking. During the course of the reaction, the

---

[9] M. Lamborg, F. E. Stolzenbach, and N. O. Kaplan, *J. Biol. Chem.* **231**, 685 (1958).
[10] The product was tested at a comparable concentration with nicotinic acid or nicotinamide. The assays were carried out by F. Tanaka, M. Nakamura, and T. Suzuki.
[11] See Vol. II [89].

pH is checked and, when necessary, KOH is added to bring the solution back to about pH 9. Since no simple and rapid assay is available to measure the decrease of deamido-NAD, it may be suggested that one carry out small-scale incubation with NAD as substrate, and to follow the cleavage of NAD using the alcohol dehydrogenase assay as described above. The addition of further enzyme may be helpful to bring the reaction to completion; the incubation period for complete cleavage of deamido-NAD may be from 10 hours to 24 hours, depending upon the quantity of snake venom phosphodiesterase employed.

Upon completion of the reaction, 0.5 ml of 20% perchloric acid is added to terminate the reaction. Any insoluble material is removed by centrifugation, and the precipitate is washed twice with 2 ml of 2% perchloric acid. The supernatant solution and washings are combined, and adjusted to pH 7 with 10 N KOH. The clear supernatant solution is applied to a Dowex 1 X2 formate (200–400 mesh) column (0.8 $\times$ 30 cm). After 5'-AMP is washed out with 0.2 N formic acid, nicotinic acid ribonucleotide is eluted by application of a formic acid concentration gradient (mixing chamber, 250 ml of 0.2 N formic acid; reservoir, 250 ml of 2.0 N formic acid). Fractions (10 ml) are collected. The ribonucleotide is detected by measuring the absorbance at 260 mμ, and is usually eluted at 100–200 ml effluent volume. The combined fractions containing nicotinic acid ribonucleotide is lyophilized and stored at $-20°$. This procedure gives more than 80% recovery.

Properties of Nicotinic Acid Ribonucleotide

Analysis of the product gives satisfactory results for the several criteria of nicotinic acid ribonucleotide. Purity of the ribonucleotide is shown to be more than 90%, and the product is practically free of nicotinic acid, 5'-AMP and deamido-NAD. The absorption spectrum at pH 7.0 shows a maximum at 275 mμ with molar extinction coefficient of 4.3 $\times$ 10^3. The cyanide addition compound of the product has an absorption maximum at 315 mμ with the molar extinction coefficient of 4.6 $\times$ 10^3 (in 1 N NaCN).[1] The aqueous cyanide addition reaction is slow in developing, and about 15 minutes is required for maximal addition. When the ribonucleotide is heated for 10 minutes at 100° in 0.1 N NaOH, a stoichiometric quantity of nicotinic acid is recovered. None of the microorganisms thus far tested is capable of utilizing the ribonucleotide as a growth factor, including *Lactobacillus arabinosus* (ATCC 8014), *Streptococcus faecalis* (ATCC 8043), *Leuconostoc mesenteroides* (ATCC 9135), *Lactobacillus casei* (ATCC 7469), *Lactobacillus fructosus* (IFO 3516), and *Pediococcus acidilactici* (ATCC 8042).[10]

III. Nicotinic Acid Ribonucleoside

Method

Principle. Nicotinic acid ribonucleotide is hydrolyzed with the prostatic phosphomonoesterase, and inorganic phosphate is adsorbed on a Dowex 1 X2 formate column. Nicotinic acid ribonucleoside is not adsorbed on the column. This procedure is essentially similar to that for the preparation of nicotinamide ribonucleoside.[12]

Reagents

Nicotinic acid ribonucleotide, prepared as described above
Sodium acetate buffer, 1 M, pH 6.0
Prostatic phosphomonoesterase[13]

Procedure. A hundred micromoles of the lyophilized nicotinic acid ribonucleotide are dissolved in 20 ml of H_2O, and the solution is neutralized to about pH 6. Sodium acetate buffer, 0.2 ml, and the prostatic phosphomonoesterase, approximately 200 Schmidt units,[13] are added, and the mixture is incubated at 37°. The amount of enzyme added should be sufficient to complete the reaction in less than 10 hours. The reaction may be followed by measuring the formation of inorganic phosphate by the method of Fiske and SubbaRow.[14] When the reaction is ended, the mixture is heated for 1 minute to inactivate the enzyme. Any insoluble material is removed by centrifugation. The supernatant solution is placed on a Dowex 1 X2 formate (200–400 mesh) column (2 X 10 cm). Effluents are collected before and after washing with water. Nicotinic acid ribonucleotide and nicotinic acid, if any, remain on this column. The ribonucleoside may be stored in the frozen state, or it can be lyophilized. This procedure gives almost a quantitative yield of the ribonucleoside.

---

[12] See Vol. III [129].
[13] See Vol. II [79].
[14] See Vol. III [115].

[115] Nicotinic Acid Mononucleotide Pyrophosphorylase[1] and Quinolinic Acid Phosphoribosyl Transferase from Astasia longa

By J. J. BLUM *and* VARDA KAHN

I. Nicotinic Acid Mononucleotide Pyrophosphorylase

Nicotinic acid + PRPP $\rightleftharpoons$ deamido-NMN + PP

Assay Method

Principle. Labeled nicotinic acid is converted to deamido-NMN in the presence of excess PRPP, and the labeled deamido-NMN is separated from nicotinic acid by paper chromatography.[1a]

Procedure

Growth of Organism. *Astasia longa* (Jahn strain), a naturally apochlorotic flagellate closely related to *Euglena gracilis*, may be obtained from the Culture Collection of Algae at the Department of Botany, Indiana University, Bloomington, Indiana. Cells grow with a doubling time of about 11 hours at 25° in a defined medium similar to that of Cramer and Myers.[2] One liter of medium contains 690 mg of sodium citrate·2 H_2O, 10 gm of $(NH_4)_2PO_4$, 10 g of KH_2PO_4, 200 mg of $CaCl_2$, 98 mg of $MgSO_4$, 1.5 mg of $FeCl_3$·6 H_2O, 1.3 mg of $CoCl_2$·6 H_2O, 1.8 mg of $MnCl_2$·4 H_2O, 0.4 mg of $ZnSO_4$·7 H_2O, 0.2 mg of Na_2MoO_4·2 H_2O, 0.02 mg of $CuSO_4$·5 H_2O, and either 3.72 g of sodium acetate·3 H_2O or 3 ml of ethanol. The pH of the medium should be about 6.8. After sterilization by autoclaving, thiamine hydrochloride (0.02 mg) and vitamin B_{12} (0.01 mg) are added aseptically. Cells need not be stirred if a sufficient surface to volume ratio is maintained (e.g., 100 ml in a 500 ml Erlenmeyer flask). A bleached strain of *Euglena gracilis* (strain SML1) can also be grown under these conditions and yields cell-free preparations with properties similar to those described below for *Astasia*.

Preparation of Sonic Lysate. Cells are harvested during exponential growth at densities of about 200,000 cells per milliliter by brief centrifugation at 1500 g in a refrigerated centrifuge and washed with 0.1 M Tris-HCl, pH 7.4, containing millimolar reduced glutathione. There is no difference in specific activity, however, if the glutathione is omitted or replaced by

[1] Nicotinate nucleotide: Pyrophosphate Phosphoribosyltransferase, EC 2.4.2.11.

[1a] J. Imsande, J. Preiss, and P. Handler, Vol. VI, p. 345.

[2] M. Cramer and J. Myers, *Arch. Mikrobiol.* **17,** 384 (1952).

millimolar mercaptoethanol. The washed cells are resuspended in this buffer, and aliquots of 5 ml (containing about 15×10^6 cells) are exposed to ultrasound from a Branson sonifier, Model LS-75, for two 30-second intervals, the temperature being maintained near 0°. The disrupted cell suspension is then centrifuged at 120 g for 10 minutes, and the supernatant solution is used as the enzyme source.

Assay Procedure. The reaction mixture contains, in a final volume of 1.0 ml: 0.2 ml of PRPP (3 micromoles), 0.1 ml of $MgCl_2$ (5.4 micromoles), 0.2 ml of Tris-HCl, pH 7.4 (12.5 micromoles), 0.2 ml of nicotinic acid 7-^{14}C (0.5 micromole, specific activity about 10^7 cpm/micromole), and 0.1–0.3 ml of the cell-free extract (corresponding to about 0.5×10^6 cells or about 150 μg of protein). The reaction is carried out at 30°. Aliquots are withdrawn at 15-minute intervals and deproteinized by heating in boiling water for 1 minute. Samples are then applied to sheets of Whatman No. 1 filter paper and subjected to ascending chromatography using molar ammonium acetate (pH 5.0)–95% ethanol (3:7, v/v) as the solvent system. The dried chromatogram is cut into serial strips of 1 cm length, and each strip is counted in a liquid scintillation spectrometer. The percent of nicotinic acid converted to deamido-NMN is computed by dividing the counts associated with this product by the total number of counts on the paper strips.

Knowledge of the specific activity of the labeled nicotinic acid used and of the percent conversion to deamido-NMN permits computation of the amount of product formed. A unit of activity is defined as the amount of enzyme required to form 1 millimicromole of deamido-NMN in 1 hour.

Properties

General. Deamido-NMN is the only product detected under the assay conditions. The rate of its formation is linear with time up to at least 1 hour and is proportional to protein concentration. The initial reaction rate is independent of PRPP concentration in the range 0.7–2.0 mM, and in this range of PRPP concentrations the nicotinic acid is quantitatively converted to deamido-NMN. No conversion occurs in the absence of added PRPP, and the requirement for PRPP cannot be circumvented by the addition of ribose 5-phosphate and ATP. ATP does not alter the rate of the reaction, but the addition of as little as 2.5×10^{-5} M ATP results in the conversion of some of the deamido-NMN into deamido-NAD by nicotinic acid adenine dinucleotide pyrophosphorylase present in the cell-free extracts.[3] The rate of formation of deamido-NMN is not altered if an ATP-trapping system is added to the reaction mixture.[4] Therefore,

[3] V. Kahn and J. J. Blum, *J. Biol. Chem.* **243**, 1448 (1968).
[4] V. Kahn and J. J. Blum, *Biochim. Biophys. Acta* **146**, 305 (1967).

contrary to the nicotinic acid mononucleotide pyrophosphorylases from several other species,[5-7] the *Astasia* enzyme is neither stimulated by nor requires ATP. The activity of this enzyme in the cell-free preparations of *Astasia* (about 300 units per milligram of protein) is much higher than the activities of crude enzyme preparations from several sources,[7] and is comparable to that of a partially purified preparation obtained from yeast.[5]

Reversibility. The reaction in the reverse direction is assayed by incubating deamido-NMN-7-^{14}C in the presence of excess pyrophosphate and measuring the rate of formation of labeled nicotinic acid.[3] Although the reactions catalyzed by the ATP-sensitive nicotinic acid mononucleotide pyrophosphorylases of yeast[5] and of bovine liver[8] are not easily reversible, the reaction catalyzed by the ATP-independent enzyme of *Astasia* has an equilibrium constant of the order of unity.[3]

Kinetics. The enzyme obeys Michaelis-Menten kinetics; the K_m for nicotinic acid is about $5 \times 10^{-5} M$ at pH 7.4. NAD and NADP at concentrations up to 5 mM do not inhibit the activity. Deamido-NMN competitively inhibits the enzyme (with respect to nicotinic acid), with a K_i value of about $2.2 \times 10^{-4} M$. Deamido-NAD is approximately as effective an inhibitor as deamido-NMN and, unlike deamido-NMN, is present *in situ* in *Astasia* at a high enough concentration to exert physiological inhibition.[9] Quinolinic acid, nicotinuric acid, nicotinamide, N-methylnicotinic acid, and picolinic acid are weak inhibitors. Deamino-NAD, 3-acetylpyridine-NAD, thionicotinamide-NAD, N-methylnicotinamide, 3-pyridine sulfonic acid, nicotinic acid methylester, and nicotinic acid-N-oxide do not alter the activity of this enzyme.

II. Quinolinic Acid Phosphoribosyl Transferase

Quinolinic acid + PRPP $\rightleftharpoons$ deamido-NMN + PP

Assay Method

Principle. Labeled quinolinic acid is converted to deamido-NMN in the presence of excess PRPP, and the ^{14}C-deamido-NMN is separated from quinolinic acid by paper chromatography.

Procedure. The conditions for growth and for preparation of the sonic lysate are exactly as described for the nicotinic acid mononucleotide pyrophosphorylase of *Astasia*. The assay conditions are also the same except

[5] T. Honjo, S. Nakamura, Y. Nishizuka, and O. Hayaishi, *Biochem. Biophys. Res. Commun.* **25**, 199 (1966).

[6] N. Ogasawara and R. K. Gholson, *Biochim. Biophys. Acta* **118**, 422 (1966).

[7] J. Imsande, *Biochim. Biophys. Acta* **82**, 445 (1964).

[8] J. Imsande and P. Handler, *J. Biol. Chem.* **236**, 525 (1961).

[9] V. Kahn and J. J. Blum, *J. Biol. Chem.* **243**, 1441 (1968).

that 0.2 ml of quinolinic acid-6-^{14}C (0.5 micromole, specific activity about 3×10^5 cpm/micromole) is used instead of labeled nicotinic acid. The paper chromatographic procedure is also as described above except that strips of 0.5 cm are cut.

Properties

Under these assay conditions the rate of conversion of quinolinic acid to deamido-NMN is linear for up to 4 hours. No product is formed if PRPP is omitted from the reaction mixture. Nicotinic acid is not an intermediate in the reaction, but millimolar nicotinic acid decreases the reaction rate by about 40%.[3] Quinolinic acid phosphoribosyl transferase activity in cell-free preparations of *Astasia* is comparable to that of nicotinic acid mononucleotide pyrophosphorylase. The former activity is much higher than that observed in a crude preparation of mycobacteria[10] but comparable to that obtained in a crude preparation of a pseudomonad.[11]

Effect of Growth Conditions

Growth of *Astasia* in the presence of 0.16 M nicotinamide for several generations considerably reduces the specific activities of nicotinic acid mononucleotide pyrophosphorylase and of quinolinic acid phosphoribosyl transferase, but growth in the presence of 1.6 mM nicotinic acid does not alter the specific activity of either of these enzymes.[3]

[10] K. Konno, K. Oizumi, and S. Oka, *Nature* **205**, 874 (1965).
[11] P. M. Packman and W. B. Jakoby, *J. Biol. Chem.* **240**, PC 4107 (1965).

[116] Mammalian Pyridine Ribonucleoside Phosphokinase

By YASUTOMI NISHIZUKA and OSAMU HAYAISHI

Nicotinate ribonucleoside + ATP → nicotinate ribonucleotide + ADP
Nicotinamide ribonucleoside + ATP → NMN + ADP

Assay Method

Principle. Enzyme activity may be assayed by measuring the formation of nicotinate ribonucleotide with nicotinate-7-^{14}C ribonucleoside and ATP as substrates. The radioactive ribonucleotide produced is isolated by a small Dowex 50 column and is determined using a Geiger-Müller gas-flow counter.

Reagents

Nicotinate-7-^{14}C ribonucleoside, 0.16 mM (2–10 μCi/millimole)[1]
ATP, 0.02 M
MgCl$_2$, 0.06 M
Tris-acetate buffer, 0.4 M pH 7.4
Enzyme solution
Dowex 50 (H$^+$ form, 200–400 mesh, X2)

Procedure. The incubation mixture contains 0.05 ml each of nicotinate-7-^{14}C ribonucleoside, ATP, MgCl$_2$, plus Tris-acetate buffer, enzyme solution, and water in a total volume of 0.40 ml. The reaction is started by the addition of enzyme. After 20 minutes at 37°, the reaction mixture is placed in a boiling water bath for 1 minute, cooled immediately in an ice bath, and centrifuged for 10 minutes at 4000 g to precipitate the denatured protein. An aliquot (0.2 ml) of the supernatant solution is put on a Dowex 50 column (0.5 × 1 cm). The column is washed with 2.5 ml of water to obtain the radioactive nicotinate ribonucleotide produced. Under these conditions nicotinate ribonucleotide passes through the column with a recovery of 98%, whereas the substrate is adsorbed. An aliquot (0.5 ml) of the washing is placed on an aluminum disk and dried; the radioactivity is determined by means of a Geiger-Müller gas-flow counter.

Definition of Unit and Specific Activity. One unit of enzyme is defined as that amount which produces 1 micromole of nicotinate ribonucleotide per minute under the assay conditions. Specific activity is expressed as units per milligram of protein. Protein is determined by the method of Lowry et al.[2]

Application of Assay Method to Crude Preparations. This method has been found to be a valid assay using crude extracts of beef liver acetone powder. Phosphate buffer should not be used, since a phosphorylase cleaves the ribosyl bond to produce nicotinic acid. If a portion of the product, nicotinate ribonucleotide, is converted further to deamido-NAD by NAD pyrophosphorylase which contaminates the preparation, the latter product is not adsorbed on the column.

Purification Procedure

The enzyme is purified from the aqueous extract of beef liver acetone powder.

Preparation of Acetone Powder. Fresh beef liver (100-g portions) is homogenized for 2 minutes in a Waring blender with 500 ml of acetone

---

[1] See this volume [106].
[2] O. H. Lowry, N. J. Rosebrough, A. L. Farr, and R. J. Randall, *J. Biol. Chem.* **193**, 265 (1951).

($-15°$), then filtered rapidly with suction on a Büchner funnel with Whatman No. 1 filter paper. The residue is resuspended in 500 ml of cold acetone and is homogenized and filtered. This procedure is repeated once more. The liver residue is quickly dried at room temperature (at $25°$). Connective tissue is removed by passing the filter pad through a wire screen. The powder is dried completely under reduced pressure. When the odor of acetone is no longer detected, the powder can be stored at $-15°$ for at least several months.

Step 1. Crude Extract. The acetone powder, 20 g, is extracted for 30 minutes at $25°$ with 200 ml of 0.01 M Tris-HCl buffer, pH 7.0, with continuous stirring. The suspension is then centrifuged for 10 minutes at 10,000 g at 0–4°, and the residue is discarded. All subsequent manipulations are carried out at 0–4°.

Step 2. Protamine Treatment. To the crude extract (170 ml), 34 ml of 1% protamine sulfate solution, pH 7.0, are added slowly with continuous stirring. The mixture is centrifuged for 10 minutes at 10,000 g, and the precipitate is discarded.

Step 3. Ammonium Sulfate Fractionation. Solid ammonium sulfate, 47.8 g, is added slowly to the protamine fraction (200 ml) with continuous stirring. After 10 minutes, the mixture is centrifuged to remove the precipitate. To the supernatant solution, 21.1 g of additional ammonium sulfate, is added slowly with stirring, and the precipitate is collected by centrifugation. The ammonium sulfate fraction, taken up in 60 ml of 0.01 M Tris-HCl buffer, pH 7.0, is dialyzed overnight against 4 liters of 0.01 M Tris-HCl buffer, pH 7.0.

Step 4. DEAE-Cellulose Column. The dialyzed enzyme solution (63 ml) is applied to a column of DEAE-cellulose (4 × 6 cm) which has been equilibrated with 0.01 M Tris-HCl, pH 7.0. The column is washed with 150 ml of the same buffer, followed by 150 ml of the buffer containing 0.1 M KCl. The enzyme is eluted from the column with 0.01 M Tris-HCl buffer, pH 7.0, containing 0.4 M KCl. Fractions from about 40–110 ml are pooled.

Step 5. Second Ammonium Sulfate Fractionation. To the enzyme solution (70 ml) 19.4 g of solid ammonium sulfate is added to give 45% saturation. After the mixture is centrifuged to remove the precipitate, the supernatant solution is brought to 63% saturation by the addition of 8.0 g of ammonium sulfate. The precipitate formed is collected by centrifugation, dissolved in 7 ml of 0.01 M Tris-HCl buffer, pH 7.0, and dialyzed overnight against a liter of the same buffer. The final preparation represents a purification of about 40-fold over the crude extract and gives an overall yield of 20%.

A typical purification is summarized in the table.

PURIFICATION OF RIBONUCLEOSIDE PHOSPHOKINASE FROM
BEEF LIVER ACETONE POWDER

Fraction	Total volume (ml)	Protein (mg/ml)	Units (ml)	Specific activity	Yield (%)
1. Crude extract	170	42.8	0.014	0.00033	100
2. Protamine sulfate	200	24.5	0.010	0.00040	83
3. First ammonium sulfate	63	19.8	0.021	0.00105	55
4. DEAE-cellulose	70	1.38	0.013	0.0094	36
5. Second ammonium sulfate	7	6.12	0.074	0.0118	20

Properties

The enzyme is stable and can be stored for at least several weeks at $-20°$ without significant loss of the activity. The enzyme has a definite requirement for magnesium ion, and $8 \times 10^{-3} M$ $MgCl_2$ saturates the reaction. The optimal pH of the reaction is found at pH 6–8. The K_m values for nicotinate ribonucleoside and for ATP are $5 \times 10^{-6} M$ and $3.3 \times 10^{-4} M$, respectively. The partially purified enzyme preparation (Step 5) also reacts with nicotinamide ribonucleoside at the rate of 45% as compared to nicotinate ribonucleoside. Although the preparation converts adenosine to 5'-AMP at a slower rate (about 4% as active as nicotinate ribonucleoside), the adenosine kinase activity can be removed completely by further purification of the enzyme upon hydroxylapatite column chromatography. Uridine and cytidine are inert as substrates.

[117] Nicotinamide Phosphoribosyltransferase[1] from Rat Liver

By L. S. DIETRICH

$$\text{Nicotinamide} + \text{PRPP} \xrightarrow[\text{Mg}^{2+}]{\text{ATP}} \text{nicotinamide mononucleotide} + \text{PP}_i$$

Assay Method

Principle. The method is based on the chromatographic separation of the reaction product, ^{14}C-labeled nicotinamide mononucleotide, from the substrate ^{14}C-labeled nicotinamide.[1a] The amount of radioactivity is then quantitated employing liquid scintillation spectrometry.

---

[1] Nicotinamid enucleotide: Pyrophosphate Phosphoribosyltransferase, EC 2.4.2.12.
[1a] L. S. Dietrich, L. Fuller, I. L. Yero, and L. Martinez, *J. Biol. Chem.* **241**, 188 (1966).

Reagents

Tris-chloride, 0.043 *M*, pH 8.0
$MgCl_2$, 0.001 *M*
ATP, 0.001 *M*
PRPP, 0.0004 *M*
Nicotinamide-^{14}C, 0.0001 *M* (10–11 μCi/micromole)

The reagents are dissolved in water. Immediately upon receipt, the ^{14}C-labeled substrate is dissolved in water and stored frozen. Employing this procedure, the nicotinamide-^{14}C can normally be used without further purification.

Procedure. The reaction mixture contains 0.4 micromole of PRPP, 1.0 micromole of ATP, 1.0 micromole of $MgCl_2$, 0.1 micromole of nicotinamide-^{14}C (10–11 μCi/micromole), 43 micromoles of Tris, pH 8.0, and enzyme in a total volume of 0.5 ml. Incubation is carried out at 37° for 30 minutes or 1 hour. The reaction is stopped by heating in a boiling water bath for 1 minute. The samples are centrifuged, and the supernatant material is chromatographed with a solvent system containing 95% ethanol (7 parts) and 1 *M* ammonium acetate (3 parts). The area corresponding to NMN is cut into small pieces and placed into counting vials; 6 ml of a toluene scintillation mixture [5 g of 2,5-diphenyloxazole and 100 mg of 1,4-bis-2′-(5′-phenyloxazolyl)benzene per liter of toluene] is then added, and the radioactivity is quantitated employing a liquid scintillation spectrometer.

The results are normally reported as a matter of convenience as counts per minute. The values can, however, be quite easily converted to millimicromoles of NMN formed, assuming that the NMN-^{14}C has the same specific activity as the nicotinamide-^{14}C which is employed as the substrate.

Purification Procedure

Holtzman rats weighing 350–450 g are killed by decapitation. The livers are immediately removed and collected on crushed ice. Pooled livers (40–80 g) are homogenized in a Teflon homogenizer in 9 volumes of 0.05 *M* Tris-chloride buffer, pH 7.3. The crude homogenate is centrifuged at 22,000 *g* for 30 minutes, and the pellet is discarded. Protamine sulfate (1%) is then added to the supernatant material (fraction 2) to a final concentration of 0.7 ml of protamine sulfate solution per 10 ml of supernatant material. After standing in the cold for 30 minutes, the material is centrifused at 22,000 *g* for 15 minutes, the pellet is discarded, and the supernatant material (fraction 3) is subjected to an ammonium sulfate fractionation. The addition of 31.77 g of solid ammonium sulfate per 100 ml of supernatant solution results in a precipitate, (collected 10 minutes after complete solution of the salt by centrifugation at 22,000 g for 15 minutes), which contains all the nicotinate mononucleotide pyrophosphorylase activity. Further addition of 7.06 g of ammonium sulfate per 100 ml results in the precipitation of a protein fraction containing all the NMN pyrophosphorylase activity (fraction 4). This material is dissolved in 0.1 M Tris-chloride, pH 7.4. The enzyme at this stage of purification contains no detectable NAD glycohydrolase, nicotinate mononucleotide pyrophosphorylase, NAD pyrophosphorylase, nicotinamide deamidase, or NAD kinase. Inorganic pyrophosphatase can be removed by heating fraction 4 at 55° for 5 minutes and then centrifuging at 21,000 g for 10 minutes. To the resulting fluid, ammonium sulfate, 40 g/100 ml, is added, and the resulting precipitate is collected 10 minutes after complete solution of the salt by centrifugation at 22,000 g for 15 minutes. The pellet is dissolved in 0.1 M Tris-chloride buffer, pH 7.3, containing 0.001 M dithioerythritol and dialyzed for 3 hours against 0.1 M Tris-chloride, pH 7.4, containing 0.001 M dithioerythritol, (fraction 5), 4.0 ml of dialyzate being used per milliliter of enzyme preparation. Fraction 5 is then passed through a Sephadex G 75 column that has been equilibrated with 0.1 M Tris-chloride containing 0.001 M dithioerythritol. The void volume is collected, and the enzymatic activity is precipitated with ammonium sulfate (40 g per 100 ml of enzyme solution). The pellet is then suspended (ca. 50 mg of protein per milliliter) in a small amount of Tris-chloride, pH 7.4 containing 0.001 M dithioerythritol. Although the increased specific activity obtained employing a Sephadex G 75 column at this stage is not great, it is included since it consistently makes the succeeding calcium phosphate step more efficient. The enzymatic material from the Sephadex G 75 column (fraction 6) is precipitated with an equal volume of calcium phosphate gel (90 mg dry weight per milliliter). The pellet obtained after centrifugation is washed once with 7.5% ammonium sulfate in 0.05 M Tris-chloride buffer, pH 7.4. The enzymatic activity is then eluted from the calcium phosphate pellet by suspension in 0.05 M Tris-chloride buffer containing 20% ammonium sulfate and 0.001 M dithioerythritol. The calcium phosphate is removed after standing for 15 minutes by centrifugation at 22,000 g for 10 minutes, and the procedure is repeated. The two fractions are pooled, solid ammonium sulfate (3 g/10 ml) is added, and the resulting precipitate is collected by centrifugation and dissolved (ca. 10 mg of protein per milliliter) in 0.1 M Tris-chloride buffer, pH 7.4 containing 0.001 M dithioerythritol (fraction 7). Attempts to carry the enzymatic purification further have been unsuccessful to date. The pertinent purification data are summarized in the table. As can be seen, it is difficult to determine the extent of purification obtained due to the marked activation obtained in the early fractions. The reasons for the inability to measure all enzymatic activity in crude fractions is not

PURIFICATION OF NMN PYROPHOSPHORYLASE FROM RAT LIVER

Fraction	Protein (mg/ml)	Specific activity (cpm/hr/mg protein)	Recovery (%)
1. Homogenate	47.2	4,384	100
2. Supernatant from homogenate	26.0	27,692	268
3. Supernatant from protamine sulfate precipitation	17.9	48,627	375
4. Ammonium sulfate precipitation	84.0	100,488	200
5. Dialyzed ammonium sulfate precipitate of supernatant after heat treatment (55° for 5 min)	60.0	181,933	83
6. Void volume from Sephadex G 75 column	46.5	271,739	25
7. Eluate from calcium phosphate precipitation	5.4	1,392,234	7

known. There are, however, indications that an inhibitor is removed in these early stages of purification.[1a]

Properties

ATP Requirement. One of the most interesting properties of this enzyme is that it has a requirement for ATP as well as for PRPP and nicotinamide. The requirement is highly specific; GTP, CTP, TTP, AMP, dGTP, dAMP have no effect on enzymatic activity. dATP, on the other hand, has some slight activity. (The stimulation originally observed with ADP[1a] is produced by the conversion of ADP to ATP through the action of contaminating adenylate kinase.) High levels of PRPP and Mg^{2+} do, however, activate the enzyme. The evidence to date indicates that ATP is not a substrate, but serves as an allosteric modifier of NMN pyrophosphorylase and that PRPP, at high concentrations,[2] can serve as both substrate and an inefficient modifier. Whether ATP is converted to ADP in its role as a positive modifier cannot be satisfactorily determined at the present level of purity.

Molecular Weight. Studies employing gel filtration[3] indicate that rat liver NMN pyrophosphorylase has a molecular weight of 67,000.

Kinetic Parameters. NMN pyrophosphorylase, as isolated from rat liver, undergoes an ordered sequential reaction, PRPP being the first substrate binding to the enzyme and inorganic pyrophosphate being the first product released.[2]

The apparent K_m's for nicotinamide and PRPP are 3 μM and 0.4 mM, respectively, in the presence of ATP and 3 μM and ca. 6.0 mM, respectively,

[2] M. C. Powanda, O. Muniz, and L. S. Dietrich, Biochemistry **8**, 1869 (1969).
[3] P. Andrews, *Biochem. J.* **91**, 222 (1964).

in the absence of ATP (employing PRPP as the modifier). The apparent K_m for ATP is 0.4 mM. One mole of ATP appears to add to 1 mole of enzyme. Employing PRPP as the modifier, two molecules of PRPP can be demonstrated to bind to the enzyme. It is assumed that one binds at the PRPP site and the other at the modifier site.

pH Optimum and Stability. The enzymatic activity exhibits a broad peak between pH 7.5 and 9.5 with maximal activity around pH 8.0.

Stability. Enzymatic activity is lost very rapidly below pH 7, also by dilution and by exposure to urea. Under certain conditions the loss of activity produced by dilution can be restored by concentrating the solution. Enzymatic preparations of fraction 4 and undialyzed fraction 5 are stable for months. Further purification results in a product that deteriorates rapidly.

Inhibition by Pyridine Bases, Nucleosides, and Nucleotides. NMN pyrophosphorylase activity is markedly inhibited by various nicotinamide analogs such as 6-aminonicotinamide, thionicotinamide, and 3-acetylpyridine; but not by nicotinic acid, 6-chloronicotinamide, 6-hydroxynicotinamide, aminoisonicotinamide, or ethyl nicotinate. The inhibition by nicotinamide analogs is competitive in nature.

At least two of these nicotinamide analogs serve as substrates for NMN pyrophosphorylase. Thionicotinamide is utilized as well as nicotinamide and exhibits kinetic properties very similar to those observed for the natural pyridine base. 3-Acetylpyridine, on the other hand, has an apparent K_m of 2×10^{-5} M, about ten times that observed for nicotinamide. In addition, the V_{max} observed with 3-acetylpyridine is ca. 5 times that observed with nicotinamide. 6-Aminonicotinamide, on the other hand, is an example of a pyridine analog which inhibits NMN pyrophosphorylase but does not serve as a substrate.[4] Whether the other nicotinamide analogs which inhibit NMN-pyrophosphorylase fall in the 6-aminonicotinamide group or the thionicotinamide group is not known.

Nicotinamide nucleoside is a strong competitive inhibitor of the reaction. NMN, the natural pyridine dinucleotides, including α-NAD and all the NAD analog tested, e.g., 3-acetylpyridine AD, 3-pyridine aldehyde AD, thionicotinamide AD, and hypoxanthine adenine dinucleotide, are inhibitors of the reaction. Nicotinate AD is inactive. In most instances, the inhibition is a strong "mixed type." It is thought that the level of the pyridine nucleotide in the tissue helps to regulate NAD biosynthesis from nicotinamide.[5]

Inhibition by Polyanionic Polymers. The enzymatic activity is strongly

[4] L. S. Dietrich, O. Muniz, B. Farinas, and L. Franklin, *Cancer Res.* **28,** 1652 (1968).
[5] L. S. Dietrich, O. Muniz, and M. C. Powanda, *J. Vitaminol. (Kyoto)* **14,** 123 (1968).

inhibited by polyanionic polymers such as heparin, dextran sulfate, and RNA. It is of interest that this inhibition appears to be specific, since rat liver nicotinate pyrophosphorylase, which has similar Mg^{2+}, PRPP, and ATP requirements, is unaffected by these compounds.[1]

Intracellular Location of Enzymatic Activity. With the use of the procedure of Schneider and Hogeboom,[6] 99% of the NMN pyrophosphorylase in rat liver was found associated with the supernatant fraction (i.e., that material not sedimented by centrifugation at 105,000 g for 2 hours).

Activity in Other Tissues. NMN pyrophosphorylase activity has been found in all animal tissues so far investigated.[5] The NMN pyrophosphorylase isolated from acetone powder of erythrocytes has a very high apparent K_m for nicotinamide (0.1 M).[7] The enzymatic activity from ascites cells[8] is ATP dependent and has an apparent K_m for nicotinamide of 10 μM.

Microorganisms that require nicotinamide or nicotinate appear, in general, to lack NMN pyrophosphorylase. A notable exception is *Lactobacillus fructosus* which grows on nicotinamide, but cannot utilize nicotinate. This organism has a very active NMN pyrophosphorylase.[9]

[6] W. C. Schneider and G. H. Hogeboom, *J. Biol. Chem.* **183**, 123 (1950).
[7] J. Preiss and P. Handler, *J. Biol. Chem.* **225**, 759 (1957).
[8] L. S. Dietrich, L. Fuller, I. L. Yero, and L. Martinez, *Nature* **208**, 347 (1965).
[9] F. Ohtsu, A. Ichiyama, Y. Nishizuka, and O. Hayaishi, *Biochem. Biophys. Res. Commun.* **29**, 635 (1967).

[118] NAD⁺ Kinase[1] from *Azotobacter vinelandii* and Rabbit Liver

By ALBERT E. CHUNG

$$NAD^+ + ATP \rightarrow NADP^+ + ADP$$

Assay Method

Principle. The enzyme activity is determined by reducing the NADP⁺ formed in the reaction to NADPH with NADP⁺–specific isocitrate dehydrogenase. The NADPH concentration is determined spectrophotometrically at 340 mμ.

Definition of Unit and Specific Activity. One unit of activity is defined as the amount of enzyme that synthesizes 1 micromole of NADP⁺ per hour under the specified reaction conditions. The specific activity is defined as units of activity per milligram of protein.

[1] ATP:NAD 2'-Phosphotransferase.

Reagents[1a]

ATP, 0.04 M
NAD+, 0.02 M
$MgCl_2$, 0.1 M
Phosphate buffer, 0.05 M, pH 7.0
Isocitrate, 0.04 M
Isocitrate dehydrogenase, 1 mg/ml (specific activity 5 e.u./mg)[2]

All reagents are dissolved in phosphate buffer except $MgCl_2$, which is dissolved in water.

Procedure A. This procedure is a modification of the method described by Kornberg[3] and is the method of choice for enzyme preparations with low specific activity. The reaction mixture consists of 0.2 ml of ATP, 0.1 ml of NAD+, 0.1 ml of $MgCl_2$, enzyme, and phosphate buffer to a total volume of 1 ml; it is incubated at 37° for 15–30 minutes. The reaction is initiated by the addition of enzyme. A control in which the enzyme solution is replaced by an equivalent volume of buffer is run in parallel with the experimental reaction vessel and treated in exactly the same way in the subsequent steps of the experiment. At the end of the incubation, the reaction is terminated by heating the mixture in a boiling water bath for 1 minute. To the cooled reaction mixture, 1 ml of phosphate buffer is added; the denatured protein is removed by centrifugation, and an aliquot of the supernatant solution is used for NADP+ determination. The supernatant solution (1 ml or other suitable volume) is placed in a standard glass or silica cuvette. Isocitrate dehydrogenase, 0.1 ml, and phosphate buffer to yield a final volume of 2.9 ml are then added. The absorbancy of the mixed contents of the cuvette is determined at 340 mμ in a spectrophotometer against a blank cuvette containing 3 ml of phosphate buffer. To the cuvette containing the reaction mixture, 0.1 ml of isocitrate solution is added; after mixing, the absorbancy of the contents of the cuvette at 340 mμ is recorded when it reaches a constant value (about 3 minutes). The change in absorbancy after a correction for the control has been made is used to determine the amount of NADP+ formed in the kinase reaction. The NADP+ is quantitatively converted to NADPH by the isocitrate dehydrogenase.

Procedure B. This procedure was described by Chung[4] and may be used for the purified kinase. In this method the kinase reaction is coupled directly with the isocitrate dehydrogenase reaction. The rate of reaction is

---

[1a] These reagents may be purchased from Sigma Chemical Co., St. Louis, Missouri.
[2] One enzyme unit reduces 1 micromole of NADP+ per minute at 37°.
[3] A. Kornberg, *J. Biol. Chem.* **182**, 805 (1950).
[4] A. E. Chung, *J. Biol. Chem.* **242**, 1182 (1967).

then followed in a spectrophotometer by recording the absorbancy at 340 mμ as a function of time. The reaction mixture in a standard cuvette consists of 0.2 ml ATP, 0.1 ml NAD$^+$, 0.1 ml MgCl$_2$, 0.1 ml isocitrate, 0.1 ml isocitrate dehydrogenase, and phosphate buffer to 2 ml. The reaction mixture is maintained at 37° in the thermostated cuvette holder of the spectrophotometer. After addition of the kinase, the absorbancy readings are recorded at intervals of 30 seconds. The absorbancy changes obtained for a control in which the kinase is replaced by buffer are used to correct the values obtained for the reaction cuvette.

NAD$^+$ Kinase from *Azotobacter vinelandii*

Purification Procedure

Growth of Cells. *Azotobacter vinelandii* (ATCC 9104) cells are grown aerobically at 25° for 48–72 hours on Burk's nitrogen-free medium[5] to an optical density of 200–250 Klett units (No. 66 filter). The cells are harvested with a Sorvall RC-2 refrigerated centrifuge equipped with a continuous-flow device or with a refrigerated Sharples centrifuge. Each liter of growth medium yields 4–6 g wet weight of cells. All purification steps are carried out at 0–4° unless otherwise stated.

Crude Extract. Cell paste (70 g) is resuspended in 3 volumes of 1 M glycerol in 0.02 M potassium phosphate buffer, pH 7. The cell suspension is sonically treated in a Branson sonicator (Branson Ultrasonic Corporation, Stamford, Connecticut) at 6–8 A for 10 minutes. The sonically treated suspension is centrifuged at 30,000 rpm in a Spinco No. 30 rotor for 30 minutes. The supernatant solution is used as the source of enzyme.

First DEAE-Cellulose Fractionation. Low molecular weight components in the crude extract (190 ml) are removed by gel filtration on a Sephadex G 25 column (6 × 60 cm). The enzyme is eluted from the Sephadex column in 0.05 M potassium phosphate buffer, pH 7.0, which contains 1.5 × 10^{-5} M NAD$^+$. The eluate from the Sephadex column is applied to a DEAE-cellulose column (5.5 × 13 cm) which had been previously equilibrated with 0.02 M potassium phosphate buffer, pH 7.0, containing 1.5 × 10^{-5} M NAD$^+$. The column is eluted with the buffer in which the enzyme was initially dissolved until no more protein is eluted. This fraction contains the enzyme.

First Calcium Phosphate Gel Fractionation. Calcium phosphate gel at a gel to protein ratio of 1:1 is added to the eluate from the DEAE-cellulose fractionation. The enzyme is almost completely adsorbed on

---

[5] P. W. Wilson, and S. G. Knight, "Experiments in Bacterial Physiology," p. 53. Burgess, Minneapolis, Minnesota, 1952.

the gel after an equilibration period of 10 minutes. The gel is collected by centrifugation and the supernatant solution is discarded. The gel is eluted with 30 ml of 0.10 M potassium phosphate buffer, pH 7.0, containing $1.5 \times 10^{-4}\ M$ NAD$^+$. The elution is repeated twice. These eluates contain only a small quantity of enzyme and are discarded. The gel is next eluted five times with 20-ml volumes of 0.2 M potassium phosphate buffer, pH 7.0, containing $1.5 \times 10^{-4}\ M$ NAD$^+$. These eluates which contain the enzyme are pooled.

Heat Treatment. The NAD$^+$ concentration in the pooled eluates from the gel fractionation is increased to $1.5 \times 10^{-3}\ M$. This solution is heated for 10 minutes with gentle agitation at 52–53°. The heated solution is rapidly cooled in an ice bath, and the denatured protein is removed by centrifugation at 34,000 g for 10 minutes.

Ammonium Sulfate Fractionation. Ammonium sulfate is added to the supernatant solution from the heated enzyme preparation at a concentration of 277 g of ammonium sulfate per liter of solution. The precipitate is collected by centrifugation after 15 minutes and redissolved in 80 ml of 0.05 M potassium phosphate buffer, pH 7.0, containing $1.5 \times 10^{-5}\ M$ NAD$^+$. This redissolved enzyme is filtered on Sephadex G 25 and eluted in 0.05 M potassium phosphate buffer, pH 7.0, containing $1.5 \times 10^{-5}\ M$ NAD$^+$.

Second DEAE-Cellulose Fractionation. The desalted enzyme solution is applied to a DEAE-cellulose column (2.5 $\times$ 25 cm) that had been packed under slight pressure and equilibrated with 0.02 M potassium phosphate buffer, pH 7.0, containing $1.5 \times 10^{-5}\ M$ NAD$^+$. The column is eluted successively with 150-ml volumes of 0.05, 0.10, and 0.5 M phosphate buffer, pH 7.0, containing $1.5 \times 10^{-5}\ M$ NAD$^+$. Fractions of 12 ml each, are collected, and each is assayed for enzyme and protein contents.

Second Calcium Phosphate Gel Fractionation. The active fractions from the DEAE-cellulose column are pooled, and calcium phosphate gel at a gel to protein ratio of 5:1 is added to the pooled enzyme solution. The gel adsorbs most of the enzyme and is collected by centrifugation. The gel is then eluted three times with 3-ml volumes of 0.1 M potassium phosphate buffer, pH 7.0, containing $1.5 \times 10^{-4}\ M$ NAD$^+$. There is little enzyme activity in these fractions which are discarded. The gel is next eluted five times with 3-ml volumes of 0.2 M potassium phosphate buffer, pH 7.0, containing $1.5 \times 10^{-4}\ M$ NAD$^+$. These fractions contain the enzyme.

The results of a typical purification are shown in Table I.

Properties

Stability. The purified enzyme is unstable to dialysis against 0.05 M potassium phosphate buffer, pH 7.0, at 4°. The enzyme is, however, re-

TABLE I
PURIFICATION OF NAD$^+$ KINASE FROM *Azotobacter*

Fraction	Total volume (ml)	Total protein (mg)	Total enzyme (units)	Specific activity (units/mg)	Recovery (%)
1. Desalted extract	275	3712	148.5	0.04	100
2. First DEAE-cellulose	380	2470	182.4	0.07	123
3. First calcium phosphate gel	187	455	152.0	0.33	102
4. Heated	184	266	174.1	0.66	117
5. Ammonium sulfate	100	140	126.0	0.90	90
6. Second DEAE-cellulose	120	21.6	103.2	4.78	70
7. Second calcium phosphate gel					
Fraction II	3	1.50	18.4	12.28	—
Fraction III	3	0.87	16.9	19.38	—
Fraction IV	3	1.77	33.4	18.86	44
Fraction V	3	1.02	14.3	13.98	—

markably stable in the presence of NAD$^+$. In the absence of NAD$^+$ the dialyzed enzyme when incubated at 38° loses approximately 50% of its activity in 5 minutes. However, the addition of NAD$^+$ at a concentration of $10^{-3} M$ not only stabilizes but enhances the catalytic activity after incubation at 38° for 30 minutes. ATP at a higher concentration also stabilizes the enzyme activity. At a concentration of $10^{-3} M$, 2'-AMP, 3'-AMP, 5'-AMP, ADP, NMN, and nicotinamide do not stabilize the enzyme.

Specificity. The enzyme does not exhibit a high degree of specificity for the nature of the phosphate donor. The five nucleoside triphosphates, GTP, ITP, ATP, UTP, and CTP, are effective phosphate donors. At a nucleoside triphosphate concentration of $2 \times 10^{-3} M$ the relative rates of synthesis of NADP$^+$ under the standard assay conditions are 10, 8.6, 7.8, 7.2, and 2.1 when GTP, ITP, ATP, UTP, and CTP are used as phosphate donors.

Michaelis Constants. The Michaelis constants for NAD$^+$ and ATP are $4 \times 10^{-4} M$ and $10^{-3} M$, respectively.

Metal Requirements. The enzyme requires either Mg^{2+} or Mn^{2+} ions for activity. The optimal concentration for Mg^{2+} is $5 \times 10^{-3} M$.

Inhibitors. The enzyme activity is inhibited by a number of sulfhydryl reagents such as *p*-chloromercuribenzoate, 5,5'-dithiobis(2-nitrobenzoic acid), mercuric acetate, *o*-iodosobenzoate, and methyl mercuric bromide. *p*-Chloromercuribenzoate is effective at concentrations of $5 \times 10^{-6} M$ and lower. The enzyme activity is partially protected from inactivation by *p*-chloromercuribenzoate by NAD$^+$. NADP$^+$ at a concentration of $5 \times 10^{-4} M$ decreases the enzyme activity by 76%.

Molecular Weight. The molecular weight of the enzyme estimated by sucrose density gradient centrifugation is approximately 130,000.

Reversible Inactivation. The enzyme is reversibly inactivated by exhaustive dialysis against 0.05 M potassium phosphate buffer, pH 7.0. Partial reactivation of the enzyme can be accomplished by preincubation of the dialyzed enzyme with a combination of 2-mercaptoethanol and NAD⁺. The reversible inactivation is accompanied by a change in sedimentation behavior of the enzyme. The native enzyme has a sedimentation constant of 6.1 S, the inactivated enzyme has a sedimentation constant of 4.6 S, and the reactivated enzyme has a sedimentation constant of 5.9 S.

NAD⁺ Kinase from Rabbit Liver

Purification Procedures

Crude Extract. Rabbit liver, 175 g, is cut into small pieces and homogenized in 525 ml of chilled 0.25 M sucrose solution dissolved in 0.05 M potassium phosphate buffer, pH 7.0, containing 1.5×10^{-5} M NAD⁺. The homogenization of the liver is best accomplished in small batches in a chilled Waring blendor. The homogenization is carried out for 30 seconds at maximal speed of the blender with each batch. All further manipulations are done at 0–4° except where stated. The homogenate is centrifuged at 10,000 rpm for 10 minutes in a Sorvall SS-1 rotor. The supernatant fluid is gently decanted and recentrifuged at 30,000 rpm in a Spinco No. 30 rotor for 90 minutes. The supernatant solution is used as the source of enzyme. This solution, 475 ml, is freed of low molecular weight compounds by passage through a Sephadex G 25 column (10 $\times$ 60 cm). The enzyme is eluted in 0.05 M phosphate buffer, pH 7.0, containing 1.5×10^{-5} M NAD⁺, and 0.01 M 2-mercaptoethanol.

First DEAE-Cellulose Fractionation. The enzyme solution is applied to a DEAE-cellulose column (5 $\times$ 40 cm) which had been previously equilibrated with 0.05 M phosphate buffer, pH 7.0, containing 1.5×10^{-5} M NAD⁺ and 0.01 M 2-mercaptoethanol. The column is eluted with a logarithmic gradient of phosphate. The elution is started with 2000 ml of 0.05 M phosphate buffer, pH 7.0, containing 5×10^{-5} M NAD⁺ and 0.01 M 2-mercaptoethanol in the mixing chamber, and 2000 ml of 0.5 M phosphate buffer, pH 7.0, containing 1.5×10^{-5} M NAD⁺ and 0.01 M 2-mercaptoethanol in the reservoir. Fractions, 200 drops each, are collected, and aliquots are removed for enzyme assay. The active fractions are pooled. The enzyme is eluted when the phosphate concentration is between 0.1 and 0.2 M.

Calcium Phosphate Gel Fractionation. The pooled enzyme solution from the DEAE-cellulose fractionation is applied to a Sephadex G 25 column

TABLE II
PURIFICATION OF NAD$^+$ KINASE FROM RABBIT LIVER

Fraction	Total volume (ml)	Total protein (mg)	Total enzyme (units)	Specific activity (units/mg)
1. Desalted extract	625	14062	50	0.0035
2. DEAE-cellulose	575	3680	242	0.066
3. Calcium phosphate gel	80	100	114	1.14

(10 × 60 cm), and the enzyme is eluted in 0.02 M phosphate buffer, pH 7.0, containing 0.01 M 2-mercaptoethanol and 1.5 × 10^{-5} M NAD$^+$. Calcium phosphate gel at a gel to protein ratio of 2:1 is added to the enzyme solution eluted from the Sephadex column. The enzyme is adsorbed on the gel which is collected by centrifugation for 10 minutes at 10,000 rpm in a Sorvall SS-1 rotor. The gel is eluted with 4 × 50 ml of 0.10 M phosphate buffer, pH 7.0, containing 1.5 × 10^{-4} M NAD$^+$ and 5 × 10^{-3} M 2-mercaptoethanol. These eluates contain enzyme of low specific activity and are discarded. The gel is next eluted with 5 × 40 ml of 0.25 M phosphate buffer, pH 7.0, containing 1.5 × 10^{-4} M NAD$^+$ and 5 × 10^{-3} M 2-mercaptoethanol. The bulk of the enzyme is obtained in the first two eluates. These two fractions are pooled.

A summary of the purification is shown in Table II. The increase in total enzyme activity after DEAE-cellulose fractionation is reproducible. The increased activity is probably due to removal of some inhibitory compounds by the DEAE-cellulose.

Properties

Stability. The partially purified liver kinase is labile. Attempts to concentrate the enzyme under nitrogen in the presence of NAD$^+$ and mercaptoethanol in a Diaflo (Amicon Instruments) apparatus results in about 50% loss of activity overnight. The *Azotobacter* enzyme is quite stable under these conditions. Repeated freezing and thawing also results in large losses of catalytic activity. Adequate conditions for storing the partially purified enzyme without loss of activity have not been found.

Substrate Requirements. In contrast to the *Azotobacter* enzyme, the rabbit liver enzyme is more specific for the nature of the phosphate donor in the synthesis of NADP$^+$. Under the usual assay conditions the relative rates of NADP$^+$ synthesis are 100, 10, 5, 2, and 0 for ATP, CTP, GTP, UTP, and ITP, respectively, when the nucleoside triphosphates are used at equimolar concentrations. The enzyme reaction requires a metal ion which may be Mg^{2+} or Mn^{2+}.

Kinetic Constants. The Michaelis constants for ATP and NAD^+ are $1.9 \times 10^{-3} M$ and $2.6 \times 10^{-3} M$, respectively.

Inhibitors. The enzyme is inhibited by *p*-chloromercuribenzoate, methyl mercuric bromide, mercuric acetate, and *N*-ethylmaleimide.

Molecular Weight. The molecular weight of the enzyme estimated by centrifugation in a sucrose density gradient is approximately 136,000. The enzyme does not appear to undergo reversible structural changes in contrast to the *Azotobacter* enzyme.

[119] NAD⁺ Kinase[1] in Liver Tissue

By L. S. DIETRICH and I. L. YERO

$$NAD^+ + ATP \xrightarrow{Mg^{2+}} NADP^+ + ADP$$

Assay Method

Principle. The $NADP^+$ formed in the reaction mixture is assayed by a modification of the method described by Kornberg.[1a] This method is based on the measurement of the change in optical density at 340 mμ obtained by reducing the $NADP^+$ formed with isocitrate and $NADP^+$ specific pig heart isocitrate dehydrogenase.

Reagents

ATP, 0.013 *M*
NAD^+, 0.01 *M*
$MgCl_2$, 0.01 *M*
Nicotinamide, 0.01 *M*
Sodium pyruvate, 0.01 *M*
dl-Isocitrate (allo free), 0.05 *M*
Tris-chloride, 0.35 *M*, pH 8.0
Tris-chloride, 0.05 *M*, pH 8.0
Pig heart isocitrate dehydrogenase, 10 mg/ml

Procedure. To 0.1-ml aliquots of the fractions to be assayed, or appropriate dilutions thereof, 0.4 ml of an incubation mixture containing the following is added (in micromoles): NAD^+, 10; ATP, 13; nicotinamide, 10; sodium pyruvate, 10; $MgCl_2$, 10; and Tris, 350. This high level of Tris was employed to assure that the pH remained constant throughout the

---

[1] ATP:NAD^+ 2′-Phosphotransferase, EC 2.7.1.23.
[1a] A. Kornberg, *J. Biol. Chem.* **182,** 805 (1950).

reaction. The sample tubes are incubated with shaking at 37°, and the reaction is stopped after 0, 15, and 30 minutes by heating in a boiling water bath for 1 minute. The assay is linear with time under these conditions. The boiled samples are centrifuged 2 minutes in a Beckman Microfuge, and 0.3 ml of the supernatant solution is removed and placed in a 1-ml quartz cuvette. Tris (0.05 M), 0.7 ml is added and the initial optical density reading at 340 mμ is recorded. Then 0.05 ml of isocitrate and 0.1 ml of pig heart isocitrate dehydrogenase are added and mixed, and a final optical density reading is taken after 2–3 minutes when no further increase is noted. Zero time readings are subtracted from the 15- and 30-minute readings to obtain the change in optical density per unit of time. Sodium pyruvate must be included in the reaction mixture for assay of fractions prior to fraction V to prevent the high initial absorbance (prior to addition of isocitrate and isocitrate dehydrogenase), which increases linearly with time. It is thought that contaminating enzymes and substrates could be reducing NAD⁺ and the newly synthesized NADP⁺. Greenbaum, Clark, and McLean[2] have sidestepped this obstacle by developing a completely different assay procedure. Values for NAD⁺ kinase activity in homogenates and supernatants of rat liver obtained by either method are similar. In the experiments carried out with the purified enzyme, nicotinamide and sodium pyruvate are not added; cysteine, 0.005 M, is added unless otherwise specified.

In all cases, enzymatic activity is expressed as micromoles of NADP⁺ synthesized per hour per milligram of protein.

Purification Procedure

The results of a typical purification procedure are shown in Table I. The enzyme has undergone up to a 2000-fold purification with a 4% yield. The enzyme at this stage, however, still contains contaminating proteins as measured by disc gel electrophoresis. Fraction VII, which is employed routinely, is free of detectable amounts of NAD glycohydrolase, adenylate kinase, ATPase, and the various pyrophosphatases dealing with pyridine nucleotide metabolism.

Fischer rats weighing 300–350 g are killed by decapitation. The livers are immediately removed, collected on crushed ice, and stored at $-15°$. The livers can be stored in the frozen state for several days without significant loss of enzymatic activity. Pooled frozen livers (40–80 g) are homogenized in a Teflon–glass homogenizer in 4 volumes of 0.05 M Tris-chloride buffer, pH 7.5. Rat liver, 200 g, is processed at one time. The crude homogenate (fraction I) is centrifuged at 33,000 g for 60 minutes, and the

² A. L. Greenbaum, J. B. Clark, and P. McLean, *Biochem. J.* **93**, 17c (1964).

TABLE I

PURIFICATION OF NAD$^+$ KINASE FROM FROZEN RAT LIVER[a]

Fraction	Volume (ml)	Protein (mg/ml)	Specific activity (micromoles NADP$^+$ synthesized/ hr/mg protein)	Recovery[b] (%)
I. Crude homogenate	1420	50.0	0.015	100
II. Supernatant of crude homogenate	1140	34.0	0.023	84
III. Dialyzed (NH$_4$)$_2$SO$_4$ fraction of the supernatant	330	76.0	0.047	111
IV. Heated dialyzate	230	20.8	0.191	86
V. Dialyzed (NH$_4$)$_2$SO$_4$ fraction of extract from first calcium phosphate precipitation	15	25.0	1.49	52
VI. Dialyzed (NH$_4$)$_2$SO$_4$ fraction of extract from protamine sulfate precipitation	12	5.5	6.44	40
VII. (NH$_4$)$_2$SO$_4$ fraction of extract from second calcium phosphate precipitation	2	9.8	15.7	29
VIII. (NH$_4$)$_2$SO$_4$ fraction of fraction VII after dilution with 4 M urea	0.5	0.027	30.0	4

[a] I. L. Yero, B. Farinas, and L. S. Dietrich, *J. Biol. Chem.* **243**, 4885 (1968).
[b] Based on 100% recovery for fraction I.

pellet is discarded. The addition of 35.3 g of solid ammonium sulfate per 100 ml of supernatant (fraction II) precipitates the enzymatic activity, which is collected by centrifugation at 33,000 g for 30 minutes. The pellet is then suspended in 200 ml of 0.01 M Tris-chloride buffer, pH 8.0, and this material is dialyzed overnight against 30 volumes of the same buffer. The dialyzate (fraction III) is heated for 5 minutes at 60°, cooled in an ice bath, and centrifuged at 35,000 g for 30 minutes. The pellet is discarded, and the supernatant material (fraction IV) is treated with 0.1 ml of a calcium phosphate gel (90 mg dry weight per milliliter) per milliliter of supernatant material and allowed to stand for 15 minutes. After centrifugation at 2000 g for 10 minutes, the pellet is washed once with 200 ml of 0.05 M Tris-chloride buffer, pH 8.0. The pellet is then eluted twice with 40-ml portions of 0.5 M potassium phosphate, pH 7.5, allowed to stand 15 minutes each time, and centrifuged at 2000 g for 10 minutes. Both eluates are combined, and solid ammonium sulfate (10.59 g/100 ml) is added. The material that sediments after centrifugation at 35,000 g for 15 minutes is discarded, and 14.12 g of solid ammonium sulfate per 100 ml of supernatant solution is added. The material sedimenting at 35,000 g for 15 minutes was dissolved in 10 ml of 0.05 M Tris-chloride buffer, pH 8.0, containing 0.005 M cysteine. This is then dialyzed for 3 hours against 1 liter of an identical buffer solution (fraction V). Protamine sulfate (1%), in water, is added slowly to the dialyzate until no further precipitation is detected (about 3.5 ml/10 ml of dialyzate). The pellet obtained after centrifugation at 35,000 g for 30 minutes is homogenized in 40 ml of 0.05 M Tris-chloride buffer, pH 8.0, containing 0.005 M cysteine and 5% ammonium sulfate. After standing for at least 2 hours, this material is centrifuged at 35,000 g for 30 minutes. The pellet is discarded, and 35.3 g of ammonium sulfate per 100 ml is added; the pellet obtained after centrifugation at 35,000 g for 15 minutes is dissolved in 10 ml of 0.05 M Tris-chloride buffer, pH 8.0, containing 0.005 M cysteine and is dialyzed against 1 liter of an identical buffer solution for 3 hours (fraction VI). Another calcium phosphate gel precipitation is then carried out with 0.3 ml of gel per milliliter of dialyzate. The pellet is washed twice with 0.05 M Tris-chloride buffer containing 0.005 M cysteine and 5% ammonium sulfate and is eluted twice with 10-ml portions of 0.5 M potassium phosphate, pH 7.5, containing 0.005 M cysteine; the suspended gel is allowed to stand for 15 minutes before being centrifuged. The supernatant material thus obtained is treated with 10.59 g of ammonium sulfate per 100 ml and centrifuged for 15 minutes at 35,000 g. The pellet is discarded, and 14.12 g of ammonium sulfate per 100 ml of supernatant is added. After centrifugation for 15 minutes at 35,000 g, the pellet is dissolved in 2 ml of 0.05 M Tris-chloride buffer, pH 9.3, containing 0.005 M cysteine (fraction VII). This is then diluted in 4 M urea and fractionated with ammonium sulfate (24.7–38.8 g of ammonium sulfate per 100 ml), and the pellet is dissolved in 0.5 ml of 0.05 M Tris-chloride buffer, pII 9.3, containing 0.005 M cysteine (fraction VIII). All the studies reported are carried out on fraction VII because the high specific activity of fraction VIII is not always obtained and the yield drops to around 4% in this step. In all ammonium sulfate precipitations, centrifugation is begun 15 minutes after complete solution of the salt. All operations are carried out at 4°. All the calcium phosphate gel pellets are hand-homogenized in test tubes equipped with tight-fitting pestles.

Properties

Intracellular Location of Enzymatic Activity. With the use of the procedure of Schneider and Hogeboom,[3] 98% of NAD⁺ kinase activity in rat liver was found associated with the supernatant fraction, i.e., that material not sedimented by centrifugation at 105,000 g for 2 hours. This is also true

[3] W. C. Schneider and G. H. Hogeboom, *J. Biol. Chem.* **183**, 123 (1950).

for pigeon liver,[4] yeast,[1a] and spinach.[5] NAD^+ kinase in thyroid gland tissue is associated with the nuclear fraction.[6]

Kinetic Parameters. The apparent K_m values for NAD^+ and ATP under the experimental conditions described are 1.4×10^{-3} M and 7.0×10^{-3} M, respectively. The V_{max} was calculated to be 21 micromoles of $NADP^+$ formed per milligram of protein per hour.

Substrate Specificity. The following nucleotides were tested with regard to their ability to replace ATP: GTP, CTP, TTP, dATP, ADP, UTP, and dGTP. dATP is as active as ATP and is the only compound to have any activity.

Various NAD analogs were qualitatively tested in order to determine whether or not NAD kinase could convert them to their respective NADP analogs. Thionicotinamide adenine dinucleotide, 3-acetyl pyridine adenine dinucleotide, and 6-aminonicotinamide adenine dinucleotide are so utilized. Pyridine aldehyde adenine dinucleotide, α-NAD, and hypoxanthine adenine dinucleotide do not serve as substrates for NAD kinase.

Metal Requirements. The results of substituting Mg^{2+} for other cations are shown in Table II. The optimal Mg^{2+} concentration is obtained in the presence of a $1:1$ ratio of $Mg^{2+}:ATP$. Mn^{2+} was found to be a slightly more

TABLE II

EFFECT OF VARIOUS CATIONS IN REPLACING Mg^{2+} REQUIREMENT OF NAD^+ KINASE[a,b]

Metal	Activity[c] (%)
Mg^{2+}	100
Mn^{2+}	80
Co^{2+}	6
Fe^{2+}	19
Fe^{3+}	0
Zn^{2+}	41
Cu^{2+}	0
Ni^{2+}	0
Sn^{2+}	0
Ba^{2+}	0
Ca^{2+}	10

[a] I. L. Yero, B. Farinas, and L. S. Dietrich, *J. Biol. Chem.* **243**, 4885 (1968).
[b] The concentration of all metals is 1×10^{-2} M with the exception of Co^{2+}, Fe^{2+}, and Fe^{3+}, which are 5×10^{-3} M.
[c] Based on 100% activity in the presence of Mg^{2+}.

[4] T. P. Wang and N. O. Kaplan, *J. Biol. Chem.* **206**, 311 (1954).
[5] Y. Yamamoto, *Plant Physiol.* **41**, 523 (1966).
[6] J. B. Field, S. M. Epstein, A. K. Remer, and C. Boyle, *Biochim. Biophys. Acta* **121**, 241 (1966).

powerful activator in concentrations one-fourth to one-fifth that of Mg^{2+} Both Mg^{2+} and Mn^{2+} inhibit the reaction if present in excess amounts.

pH Optimum and pH Stability. The pH optimum of the NAD⁺ kinase reaction is around 7.0. At lower pH values there is a rapid decrease in enzymatic activity, although this may be a reflection of the high instability of the enzyme at neutral and acid pH values. On the basic side of the plateau the values decrease very slowly. The enzyme is most stable at pH 9–10. Cysteine has a marked protective action and thus was included in all stages of purification after fraction IV.

NAD⁺ Kinase from Frozen Calf Liver

Purification Procedure

Fresh calf liver is obtained from a slaughterhouse. The livers are collected on crushed ice and stored at $-15°$. The livers can be stored in the frozen state for several days without significant loss of enzymatic activity.

About 200 g of calf liver is processed at one time. The frozen liver is homogenized in a Waring blendor in 9 volumes of $0.05\,M$ Tris-chloride buffer, pH 7.5.

The crude homogenate (fraction I) is centrifuged at 33,000 g for 60 minutes, and the pellet is discarded. The addition of solid ammonium sulfate to the supernatant (fraction II) precipitates the enzymatic activity in the 20–60% saturation of the salt (considering 70.6 g/100 ml to be 100% saturation).

The pellet collected by centrifugation at 33,000 g for 30 minutes is then suspended in 100 ml of $0.01\,M$ Tris-chloride buffer, pH 7.5, and this material is dialyzed overnight against 30 volumes of the same buffer.

The dialyzate (fraction III) is heated for 5 minutes at $60°$, cooled in an ice bath, and centrifuged at 35,000 g for 30 minutes. The pellet is discarded, and the supernatant material (fraction IV) is treated with 0.8 ml of a calcium phosphate gel (400 mg, dry weight, per milliliter) per 100 ml of supernatant material and allowed to stand for 15 minutes. After centrifugation at 2000 g for 10 minutes, the supernatant is discarded (fraction V), and the pellet is eluted once with 40 ml of $0.05\,M$ Tris-chloride buffer, pH 7.5, and twice with 20-ml portions of $0.5\,M$ potassium phosphate, pH 7.5, allowed to stand 15 minutes each time, and centrifuged at 2000 g for 10 minutes. All eluates are combined, and solid ammonium sulfate is added to 30% saturation (21.18 g/100 ml). The material sedimented at 35,000 g for 15 minutes is dissolved in 2 ml of $0.05\,M$ Tris-chloride buffer, pH 7.5 (fraction VI).

All operations are carried out at $4°$. All the calcium phosphate gel pellets were hand-homogenized in test tubes equipped with tight-fitting pestles.

See Table III for a summary of the purification.

TABLE III
PURIFICATION OF NAD⁺ KINASE FROM FROZEN CALF LIVER

Fraction	Volume (ml)	Protein (mg/ml)	Specific activity (micromoles NADP⁺ synthesized/ hr/mg protein)	Recovery[a] (%)
I. Crude homogenate	2000	17.25	0.0029	—
II. Supernatant of crude homogenate	1880	8.75	0.039	100
III. Dialyzed (NH₄)₂SO₄ fraction	200	32.0	0.039	46
IV. Heated dialyzate	162	10.0	0.182	39
V. (NH₄)₂SO₄ fraction of extract from calcium phosphate precipitation	2	36.75	1.89	22

[a] Based on 100% recovery for fraction I.

[120] Picolinic Carboxylase in Reference to NAD Biosynthesis

By YASUTOMI NISHIZUKA and OSAMU HAYAISHI

Tryptophan serves as a precursor to NAD in animals and certain microorganisms.[1] The liver and kidney are responsible for this conversion. The amount of NAD synthesized from tryptophan is inversely related to the activity of picolinic carboxylase.[2] This enzyme decarboxylates α-amino-β-carboxymuconic ε-semialdehyde, the primary oxidation product of 3-hydroxyanthranilic acid, to produce α-aminomuconic ε-semialdehyde.[3] The latter semialdehyde is metabolized further to acetyl-CoA, ammonia, and CO_2 by way of α-aminomuconic acid, α-ketoadipic acid, and glutaryl-

---

[1] *Neurospora crassa* and *Xanthomonas pruni* are shown to utilize tryptophan as a precursor to NAD.

[2] M. Ikeda, H. Tsuji, S. Nakamura, A. Ichiyama, Y. Nishizuka, and O. Hayaishi, *J. Biol. Chem.* **240**, 1395 (1965).

[3] A. H. Mehler, *J. Biol. Chem.* **218**, 241 (1956).

CoA (glutarate pathway).[4,5] Before being decarboxylated by picolinic carboxylase, α-amino-β-carboxymuconic ϵ-semialdehyde spontaneously cyclizes to quinolic acid.[6] No enzyme thus far has been detected to rearrange the semialdehyde to quinolinic acid. The pyridine dicarboxylic acid is converted to nicotinic acid ribonucleotide by a PRPP-dependent decarboxylation catalyzed by quinolinate phosphoribosyltransferase.[7] The ribonucleotide is further converted to NAD by way of deamido-NAD (NAD pathway).[8] The activity of picolinic carboxylase, therefore, competes with the nonenzymatic cyclization reaction for the common substrate, α-amino-β-carboxymuconic ϵ-semialdehyde, and is inversely related to the formation of quinolinic acid available for the NAD biosynthesis (Fig. 1).

The activity of picolinic carboxylase has been shown to vary with various hormonal conditions.[9] In the diabetic rat, for example, the activity increases markedly with a concomitant decrease in the formation of NAD from tryptophan, and eventually in the urinary excretion of N'-methylnicotinamide.[2,10] In this article a method is described to measure the formation of pyridine nucleotides from tryptophan and its metabolites with special reference to its relation to picolinic carboxylase or glutarate pathway.

I. Assay *in Vitro*

All measurements are performed with crude extracts of the liver and kidney. Animals are killed by exsanguination; the organs are immediately removed, chilled, and homogenized for 3 minutes with 2 volumes of 0.14 M KCl in a glass–Teflon homogenizer. The homogenates are centrifuged for 20 minutes at 20,000 g at 2°. The supernatant solution is treated with charcoal (15 g per 100 ml of the extract) to remove nucleotides.

A. Formation of Nicotinic Acid Ribonucleotide from 3-Hydroxykynurenine

Principle. Crude liver and kidney extracts contain all enzymes necessary for the conversion of 3-hydroxykynurenine to nicotinic acid ribonucleotide in the presence of PRPP. When 3-hydroxykynurenine, the benzene ring of which is uniformly labeled with ^{14}C, is employed as sub-

[4] R. K. Gholson, L. V. Hankes, and L. M. Henderson, *J. Biol. Chem.* **235**, 132 (1960).
[5] Y. Nishizuka, A. Ichiyama, R. K. Gholson, and O. Hayaishi, *J. Biol. Chem.* **240**, 733 (1965).
[6] L. M. Henderson, and H. M. Hirsch, *J. Biol. Chem.* **181**, 667 (1949).
[7] Y. Nishizuka, and O. Hayaishi, *J. Biol. Chem.* **238**, 3369 (1963).
[8] J. Preiss, and P. Handler, *J. Biol. Chem.* **233**, 488 (1958).
[9] A. H. Mehler, E. G. McDaniel, and J. M. Hundley, *J. Biol. Chem.* **232**, 331 (1958).
[10] E. G. McDaniel, J. Hundley, and W. H. Sebrell, *J. Nutr.* **59**, 407 (1956).

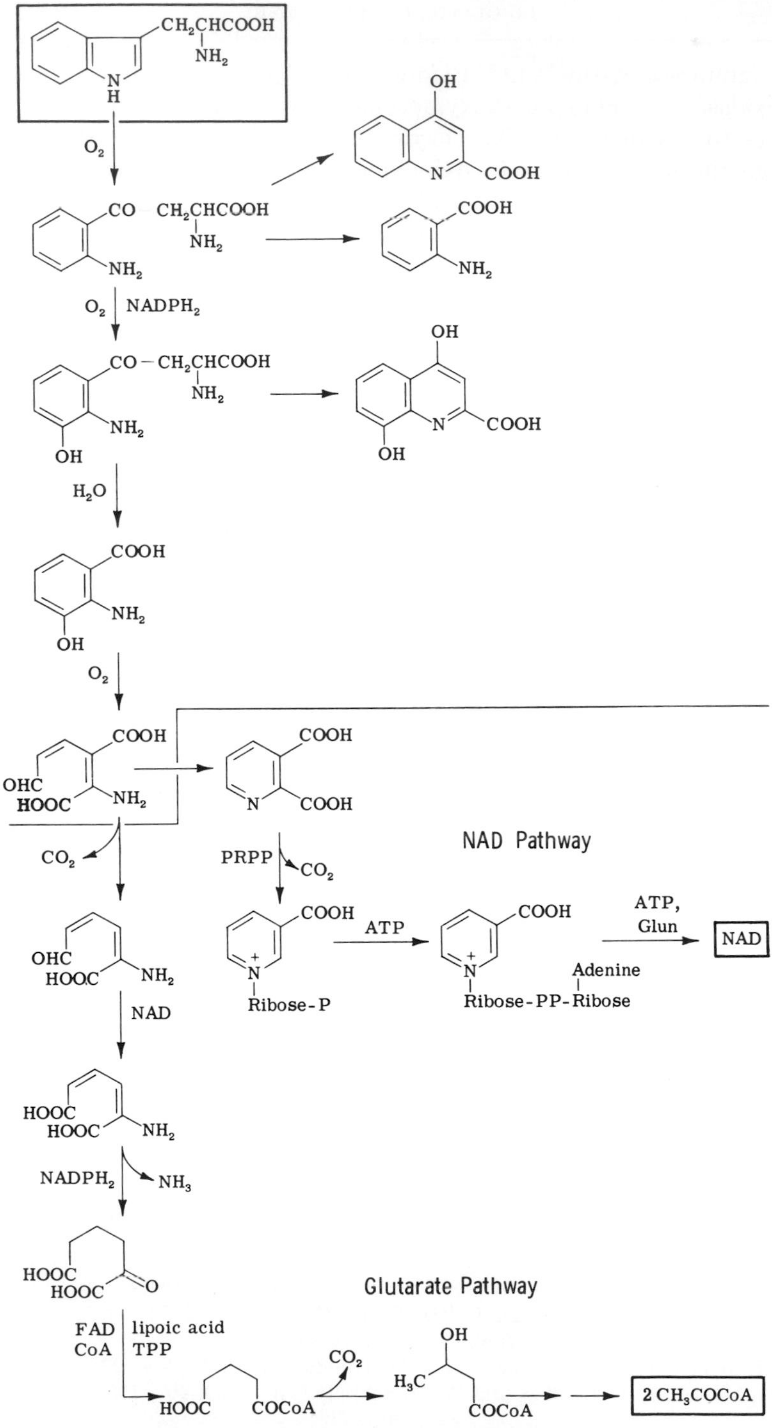

FIG. 1. Metabolic pathways of tryptophan.

strate, 3-hydroxyanthranilic acid is gradually generated from the substrate by kynureninase. 3-Hydroxyanthranilic acid is then cleaved by its oxygenase to produce α-amino-β-carboxymuconic ϵ-semialdehyde, which cyclizes spontaneously to quinolinic acid. The α-carboxyl carbon of the latter compound, which is derived from the carbon 6 of the original substrate, is radioactive and liberated as $^{14}CO_2$ by the subsequent PRPP-dependent decarboxylation catalyzed by quinolinate phosphoribosyltransferase (Fig. 1). This radioactive CO_2 is trapped in alkali and the radioactivity is determined.

Reagents

3-Hydroxy-L-kynurenine, 1.0 mM, uniformly labeled with ^{14}C in the benzene ring (0.2–0.6 μCi/micromole)[11]

PRPP, 4 mM. Dissolve about 4 mg of PRPP, magnesium salt, in 1 ml of H_2O

$MgCl_2$, 0.1 M

Potassium phosphate buffer, 1 M, pH 7.0

Crude liver or kidney extract

Procedure. The reaction mixture made to 1.0 ml with water contains 0.1 ml each of 3-hydroxykynurenine-^{14}C, PRPP, $MgCl_2$, potassium phosphate buffer, and tissue extract. This mixture is placed in a Thunberg tube. The side arm contains 0.2 ml of 0.25 N NaOH. The reaction is started by the addition of the extract. After the incubation is carried out for 30 minutes at 25°, the reaction is stopped by the addition of 0.5 ml of 1 N NaOH to the main chamber. The tube is slightly evacuated with a water aspirator. Then, the reaction mixture is acidified by introducing 0.2 ml of 5 N H_2SO_4 *via* the evacuation arm, and the tube is immediately sealed and allowed to stand at room temperature for at least 2 hours to trap $^{14}CO_2$ evolved in the side arm. A 0.1-ml aliquot of the alkali is removed and the radioactivity is determined using a gas flow Geiger-Müller counter or a liquid scintillation spectrometer.

Comments on the Procedure. The amount of $^{14}CO_2$ evolved under these conditions is entirely accounted for by the formation of nicotinic acid ribonucleotide, and equivalent quantities of $^{14}CO_2$ and the nicotinic acid derivative are produced. The ribonucleotide may be determined by Dowex 1 column chromatography as well as by microbiological assay.[12] The radio-

---

[11] The radioactive substrate may be prepared enzymatically from tryptophan whose benzene ring is uniformly labeled with ^{14}C (see Vol. XVIIA [59]).

[12] Nicotinic acid ribonucleotide does not serve as a growth factor. It may be converted quantitatively to nicotinic acid upon alkaline hydrolysis for 30 minutes at 100° in 0.1 N NaOH. The nicotinic acid is determined microbiologically with either *Leuconostoc mesenteroides* (ATCC 9135) or *Lactobacillus arabinosus* (ATCC 8014) as test organisms.

active substrate may also produce $^{14}CO_2$ when metabolized by the glutarate pathway (Fig. 1). However, the latter process absolutely requires NAD and CoA as described below. Under the conditions employed, the activity of picolinic carboxylase removes α-amino-β-carboxymuconic ϵ-semialdehyde to an extent depending upon its enzymatic activity. The product of reaction, α-aminomuconic ϵ-semialdehyde cyclizes nonenzymatically to picolinic acid, which is inert metabolically. Theoretically, radioactive 3-hydroxyanthranilic acid may be used in this assay instead of 3-hydroxykynurenine. However, 3-hydroxykynurenine may be preferentially employed as substrate for the following reasons. (1) 3-Hydroxyanthranilic acid oxygenase is the most active enzyme in the reaction sequence, and a large excess of the activity cleaves the acid immediately to produce α-amino-β-carboxymuconic ϵ-semialdehyde, which is converted predominantly to quinolinic acid before being metabolized by picolinic carboxylase. (2) 3-Hydroxyanthranilic acid is rather unstable and may not be stored in an aqueous solution.

B. Formation of Nicotinic Acid Ribonucleotide from Quinolinic Acid

Quinolinate phosphoribosyltransferase catalyzes the stoichiometric conversion of quinolinic acid and PRPP to nicotinic acid ribonucleotide and pyrophosphate with concomitant decarboxylation of the α-carboxyl group of quinolinic acid. The enzymatic activity is relatively constant under various hormonal conditions. The detailed procedure for the assay of this enzyme is described elsewhere.[13]

Principle. The activity may be assayed by measuring the PRPP-dependent formation of $^{14}CO_2$ from the α-carboxyl group of quinolinic acid that is labeled with ^{14}C.[7]

Reagents and Procedure. Reagents and procedure for this assay are exactly the same as those employed for Assay A, except that radioactive quinolinic acid is used instead of 3-hydroxykynurenine-^{14}C. A solution (1 mM) of the radioactive quinolinic acid (0.02–0.1 μCi per microatom of the α-carboxyl carbon) may be prepared as described elsewhere.[13]

Comments. This assay measured directly the formation of nicotinic acid ribonucleotide from quinolinic acid. The activity is not influenced by the picolinic carboxylase activity. The enzyme is present only in the liver and kidney, not in other tissues and organs. In addition, the enzyme is distributed in various plants and molds as well as in microorganisms.[13] Quinolinic acid may be a common intermediate in the biosynthesis of NAD from tryptophan and from glycerol and succinate.[14,15]

[13] See Vol. XVII A [59].
[14] M. V. Ortega, and G. M. Brown, *J. Biol. Chem.* **235**, 2939 (1960).
[15] A. J. Andreoli, M. Ikeda, Y. Nishizuka, and O. Hayaishi, *Biochem. Biophys. Res. Commun.* **12**, 92 (1963).

C. Formation of Glutaryl-CoA, Ammonia, and CO_2 from 3-Hydroxykynurenine

Principle. Crude liver and kidney extracts contain all enzymes responsible for the conversion of 3-hydroxykynurenine to glutaryl-CoA, ammonia, and CO_2.[16] The formation of CO_2 from the benzene ring of the substrate is absolutely dependent on the presence of NAD and CoA (Fig. 1). This assay measures the NAD and CoA-dependent formation of radioactive CO_2 which is derived from carbon 6 of 3-hydroxykynurenine.[5]

Reagents

    3-Hydroxy-L-kynurenine-^{14}C, 1.0 mM, see above
    NAD, 0.05 M
    CoA, 5 mM
    Reduced glutathione, 0.05 M
    MgCl$_2$, 0.1 M
    Potassium phosphate buffer, 1 M pH 7.0
    Crude liver or kidney extract

Procedure. The reaction mixture made to 1.0 ml with water contains 0.1 ml each of 3-hydroxykynurenine-^{14}C, NAD, CoA, reduced glutathione, MgCl$_2$, potassium phosphate buffer, and the tissue extract; and water. The mixture is placed in a Thunberg tube. The side arm contains 0.2 ml of 0.25 N NaOH. The reaction is started by the addition of the extract, and the incubation is carried out for 30 minutes at 25°. Subsequent operations are essentially the same as those of Assay A.

Comments on the Procedure. This assay does not measure the exact overall activity of the glutarate pathway, since many enzymes participate in the formation of radioactive CO_2 starting from 3-hydroxykynurenine, and an additional nonenzymatic, perhaps nonphysiological, cyclization reaction of α-aminomuconic ϵ-semialdehyde to picolinic acid is involved. However, as shown in Table I, picolinic carboxylase appears to be one of the rate-limiting enzymes in this pathway, and the results obtained by this assay may provide some information on the interrelation between the NAD and glutarate pathways.

D. Assay of Picolinic Carboxylase

Principle. The picolinic carboxylase activity may be assayed conveniently by measuring the decrease in absorbance of α-amino-β-carboxymuconic ϵ-semialdehyde, which is characterized by its intense absorption

---

[16] The enzyme responsible for the oxidative decarboxylation of α-ketoadipic acid to glutaryl-CoA and CO_2 appears to be localized in mitochondria, but the activity is partly released into the crude supernatant fraction.

of ultraviolet light with a maximum at 360 mμ (ϵ = 45,000). This decrease in absorbance is due to the formation of picolinic acid. The primary decarboxylation product, α-aminomuconic ϵ-semialdehyde, is extremely labile and cyclizes very rapidly to picolinic acid. The cyclization reaction is nonenzymatic and does not limit the overall reaction to produce picolinic acid. The validity of this assay is described elsewhere.[17]

   Reagents and Procedure. Reagents and procedure for this assay are described elsewhere.[17]

E. Application of Assay *in Vitro*

   The activity of picolinic carboxylase may be shown *in vitro* to be inversely related to the biosynthesis of NAD from tryptophan.

   Effect of Externally Added Picolinic Carboxylase on Nicotinic Acid Ribonucleotide Synthesis. The crude extract of rat liver is capable of synthesizing nicotinic acid ribonucleotide from 3-hydroxykynurenine. The amount of the ribonucleotide produced decreases sharply with the addition of increasing amounts of picolinic carboxylase as judged by Assay A (Fig. 2).[2] The picolinic carboxylase employed in this experiment is purified from cat liver by the method of Nishizuka, Ichiyama, and Hayaishi.[17] The addition of a boiled preparation of picolinic carboxylase does not affect the reaction, nor does the enzyme affect the activities of kynureninase, 3-hydroxyanthranilic acid oxygenase, and quinolinate phosphoribosyltransferase. These results may suggest that the activity of picolinic carboxylase is inversely related to the amount of nicotinic acid ribonucleotide produced from a tryptophan metabolite, 3-hydroxykynurenine.

   Effect of Hormones on Picolinic Carboxylase and Nicotinic Acid Ribonucleotide Synthesis. The activity of picolinic carboxylase is shown to be dependent on various hormonal conditions.[9,10] McDaniel *et al.*[10] have shown that urinary excretion of N'-methylnicotinamide decreases markedly in alloxan-diabetic rats. Mehler, McDaniel, and Hundley[18] subsequently showed that the liver picolinic carboxylase activity increased markedly in alloxan-diabetic rats. Hence a survey of enzymatic activities in diabetic as well as normal rats was made; the results are summarized in Table I. The picolinic carboxylase activity increases markedly in the liver of diabetic rats. The overall metabolic activity of converting 3-hydroxykynurenine to nicotinic acid ribonucleotide drops to about 40% of that in normal rats, although other related enzymes including quinolinate phosphoribosyltransferase do not show any marked change in activities. The results

---

[17] See Vol. XVII A [58].
[18] A. H. Mehler, E. G. McDaniel, and J. M. Hundley, *J. Biol. Chem.* **232**, 323 (1958).

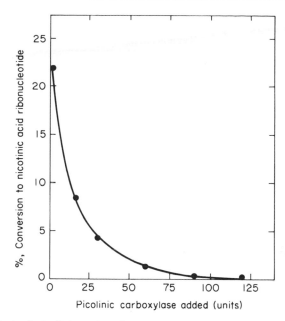

FIG. 2. Effect of picolinic carboxylase on the conversion of 3-hydroxykynurenine to nicotinic acid ribonucleotide. The incubation is carried out under the conditions of Assay A, except that a purified preparation of picolinic carboxylase is added as indicated to the reaction mixture containing the rat liver extract. The amount of nicotinic acid ribonucleotide produced is estimated from the radioactivity of $^{14}CO_2$ evolved and is expressed as percentage conversion of 3-hydroxykynurenine to the ribonucleotide. The picolinic carboxylase employed is purified from the cat liver by the method of Ichiyama *et al.* [*J. Biol. Chem.* **240**, 740 (1965)]. One unit of the enzyme is expressed as that amount of enzyme which decarboxylates 1 micromole of α-amino-β-carboxymuconic ϵ-semi-aldehyde per hour at 25°. This experiment is based on the observation described by M. Ikeda, H. Tsuji, S. Nakamura, A. Ichiyama, Y. Nishizuka, and O. Hayaishi [*J. Biol. Chem.* **240**, 1395 (1965)].

again suggest that the increase in picolinic carboxylase activity may result in the decrease in NAD synthesized from tryptophan. The administration of insulin to the diabetic rats results in a decrease in the enzymatic activity with a concomitant increase in the NAD produced. The kidney usually contains a much higher activity of picolinic carboxylase, and practically no nicotinic acid ribonucleotide is produced under the conditions. The liver, therefore, is a major organ, if not the sole one, that converts tryptophan to NAD.

Adrenal steroids have been shown to be required for an increase in picolinic carboxylase to occur. Mehler *et al.*[9] have stated that the increase in the enzyme level is not a simple response to cortisone like hormones,

TABLE I

ACTIVITIES OF ENZYMES INVOLVED IN THE NICOTINIC ACID RIBONUCLEOTIDE
FORMATION FROM 3-HYDROXYKYNURENINE IN DIABETIC RATS[a]

| Organ | Enzyme or enzyme system | Activity[b] | |
		Diabetic rats	Normal rats
Liver	3-Hydroxykynurenine to nicotinic acid ribonucleotide	0.020	0.053
	Kynureninase[c]	6.1	5.1
	3-Hydroxyanthranilic acid oxygenase[d]	400	450
	Quinolinate phosphoribosyltransferase	0.077	0.083
	Picolinic carboxylase	13.61	1.57
Kidney	3-Hydroxykynurenine to nicotinic acid ribonucleotide	0.004	0.004
	Quinolinate phosphoribosyltransferase	0.011	0.011
	Picolinic carboxylase	14.27	17.02

[a] Taken from Ikeda et al.[2]
[b] Activity is expressed as micromoles of product formed per gram wet tissue per hour. The numbers are mean values obtained from 3 to 7 animals.
[c] The activity is assayed by the method of Knox (see Vol. II [32]).
[d] The activity is assayed as described elsewhere (see text footnote 17).

however. In general, the changes of picolinic carboxylase parallel those reported for glucose-6-phosphatase.[19]

Species Difference. Nutritional studies have shown that the capacity to convert tryptophan to nicotinic acid varies with the species of animal.[20,21] Cats apparently cannot utilize tryptophan in place of nicotinic acid.[22] Suhadolnik et al.[23] have reported that picolinic carboxylase in cat liver is much higher than that in rat liver. Table II shows the comparison of enzymatic activities in the rat and cat. The quinolinate phosphoribosyl-transferase activity in cat liver is almost equal to that in rat liver. No significant difference is observed between cat and rat livers in the activity of either kynureninase or 3-hydroxyanthranilic acid oxygenase. The activities of tryptophan oxygenase, kynureninase, and kynurenine 3-hydroxylase in cat liver are also comparable with those in rat liver.[24] However,

[19] J. Ashmore, A. B. Hastings, F. B. Nesbett, and A. E. Renold, *J. Biol. Chem.* **218,** 77 (1956).
[20] J. M. Hundley, *Vitamins* **2,** 578 (1954).
[21] L. M. Henderson, G. B. Ramasarma, and B. C. Johnson, *J. Biol. Chem.* **181,** 731 (1949).
[22] A. C. Da Silva, R. Fried, and R. C. De Angelis, *J. Nutr.* **46,** 399 (1952).
[23] R. J. Suhadolnik, C. O. Stevens, R. H. Decker, L. M. Henderson, and L. V. Hankes, *J. Biol. Chem.* **228,** 973 (1957).
[24] F. T. De Castro, R. R. Brown, and J. M. Price, *J. Biol. Chem.* **228,** 777 (1957).

cat liver shows only about one-tenth of the activity for converting 3-hydroxykynurenine to nicotinic acid ribonucleotide as compared with rat liver (Table II). The picolinic carboxylase activity is 30–50 times higher in cat liver than in rat liver. In contrast, the overall activity for metabolizing 3-hydroxykynurenine to glutaryl-CoA is more active in cat liver than in rat liver. No significant difference is observed, however, between these animals in the enzymatic activities involved in the glutarate pathway except picolinic carboxylase. It may be concluded, therefore, that the cat has a potential to utilize tryptophan as a precursor to NAD, but the apparent inability of tryptophan to serve in this role is probably due to the higher level of picolinic carboxylase activity. The results may provide another piece of suggestive evidence that picolinic carboxylase is inversely related to the amount of NAD synthesized from tryptophan.

Table III shows the distribution of picolinic carboxylase in the livers of various animals. The level of the enzymatic activity varies with the source of enzyme. The cat and rat have the maximal and minimal activities, respectively. The cat is one of the animals in which tryptophan cannot replace nicotinic acid nutritionally. Calves, pigs, rabbits, mice, guinea pigs, and chicken do not require a dietary source of nicotinic acid when sufficient tryptophan is supplied, and tryptophan can substitute, at least in part, for nicotinic acid in human diets.[20]

TABLE II

COMPARISON OF ENZYME ACTIVITIES IN LIVER AND KIDNEY OF RATS AND CATS[a]

Organ	Enzyme or enzyme system	Activity[b]	
		Rat	Cat
Liver	Kynureninase[c]	5.10	4.72
	3-Hydroxyanthranilic acid oxygenase[d]	430	240
	Quinolinate phosphoribosyltransferase	0.083	0.099
	Picolinic carboxylase	1.57	50.50
	3-Hydroxykynurenine to nicotinic acid ribonucleotide	0.053	0.006
	3-Hydroxykynurenine to glutaryl-CoA	0.045	0.500
Kidney	Quinolinate phosphoribosyltransferase	0.011	0.011
	Picolinic carboxylase	16.93	68.80
	3-Hydroxykynurenine to nicotinic acid ribonucleotide	0.004	0.004

[a] Taken from Ikeda et al.[2]

[b] Activities are expressed as micromoles of products formed per gram of wet tissue per hour at 25°. Numbers are the mean values obtained from 3–7 animals.

[c] The activity is assayed by the method of Knox (see Vol. II [32]).

[d] The activity is assayed as described elsewhere (see text footnote 17).

TABLE III
LEVEL OF PICOLINIC CARBOXYLASE IN LIVER OF VARIOUS SPECIES[a]

Species	Activity[b]
Cat	50.50
Lizard	29.64
Frog	13.72
Cattle	8.30
Pig	7.12
Pigeon	6.95
Rabbit	4.27
Mouse	4.20
Guinea pig	3.97
Chicken	3.94
Man	3.18
Hamster	3.14
Toad	3.05
Rat	1.57

[a] Taken from Ikeda et al.[2]

[b] Activity is expressed as micromoles of product formed per gram wet tissue per hour at 25°.

II. Assay in Vivo

Experiments may be carried out in vivo to ascertain whether or not the activity of picolinic carboxylase is parallel with that of the glutarate pathway and is inversely related to the amount of NAD synthesized from tryptophan.

Principle.[2] Tryptophan with the benzene ring uniformly labeled with ^{14}C is injected into the peritoneal cavity of rats or mice. For various periods of time after the injection, CO_2 expired is trapped in alkali and the radioactivity is determined. The animals are killed by decapitation, and the liver and kidney are submitted to the determination of radioactive NAD as well as picolinic carboxylase activity. From the amounts of expired $^{14}CO_2$ and NAD-^{14}C in the liver, the relative amounts of tryptophan metabolized by way of the glutarate pathway and the NAD pathway may be roughly estimated. When radioactive tryptophan is converted to NAD, 1 mole of $^{14}CO_2$ is liberated from each mole of tryptophan utilized, whereas 6 moles of $^{14}CO_2$ may be produced when the tryptophan undergoes complete degradation. Thus, the amount of expired $^{14}CO_2$ may be attributable practically to the amount of tryptophan metabolized by way of the glutarate pathway. Picolinic carboxylase seems to be one of the rate-limiting steps in this metabolic pathway.

Injection of Tryptophan. Male Wistar rats, weighing about 200 g, may be used.[25] DL-Tryptophan uniformly labeled in the benzene ring with ^{14}C (20 μCi in 0.5 ml of saline, 3.95 μCi/micromole)[26] is injected intraperitoneally.

Procedure for Determination of Expired $^{14}CO_2$. Expired $^{14}CO_2$ is trapped in 1 N NaOH by the method of Weinhouse and Friedmann.[27] The rat is placed immediately in a special chamber designed for collection both of respiratory CO_2 and urine (Fig. 3). Samples of CO_2 are collected at half-

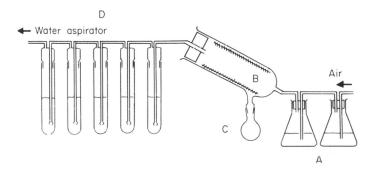

FIG. 3. Metabolic chamber for collection of respiratory CO_2. (*A*) Wash flasks containing 1 N NaOH for removal of CO_2 from incoming air, (*B*) chamber for animal with metal screen, (*C*) glass chamber to collect urine, (*D*) traps of respiratory CO_2 containing 1 N NaOH. This is essentially based on the procedure described by S. Weinhouse and B. Friedmann [*J. Biol. Chem.* **191**, 707 (1951)].

hour intervals during various periods of time (usually 2 hours).[28] To the alkaline trap containing $^{14}CO_2$, an equal volume of saturated $BaCl_2$ is added to precipitate CO_2 as barium carbonate. The heavy precipitate is collected by centrifugation and washed five times with water and subsequently with ethanol and ether. The radioactivity of the barium carbonate is determined as infinitely thick samples.[29]

Procedure for Analysis of Liver NAD. At the end of experiments to measure expired CO_2 (2 hours after the injection) the animal is immediately killed by decapitation, the liver and kidney, if necessary, are quickly removed and frozen in dry-ice–ethanol. All subsequent operations are carried out at 0–4°. The tissue is homogenized for 3 minutes in a glass–

[25] The mouse or other animals may be employed for this kind of experiment. The amount of radioactive tryptophan injected is controlled depending on the body weight.[6]

[26] The radioactive tryptophan may be obtained from the Radiochemical Centre, Amersham, Buckinghamshire, England.

[27] S. Weinhouse and B. Friedmann, *J. Biol. Chem.* **191**, 707 (1951).

[28] The radioactive CO_2 is usually expired during a period of 5–6 hours, after which CO_2 activities in all experiments are negligibly low.

[29] See Vol. IV [20].

Teflon homogenizer with 10 volumes of 2% perchloric acid. After centrifugation for 10 minutes at 20,000 g, the supernatant material is removed. The precipitate is again extracted with 10 volumes of 2% perchloric acid as above. The supernatants are combined, neutralized with 2 N KOH, and centrifuged to remove potassium perchlorate. As an optical density marker, 2 micromoles of NAD is added to the clear supernatant, which is then poured onto a Dowex 1-formate (X2, 200–400 mesh; 0.8 × 30 cm) column. The column is washed extensively with water (about 200–400 ml), then NAD is eluted from the column by an application of a formic acid gradient which is made by allowing 0.05 N formic acid in the reservoir to flow into a mixing chamber containing 300 ml of water. Fractions (10 ml each) containing NAD are collected, and the radioactivity is determined as infinitely thin samples.[29] NAD-^{14}C thus produced may be identified by paper chromatography in different solvent systems.[30]

Application of Assay in Vivo.[2] An experiment is carried out *in vivo* to obtain some information about the correlation of two metabolic pathways of tryptophan, NAD and glutarate pathways, with special reference

TABLE IV

FORMATION OF [14]CO$_2$ AND NAD-^{14}C IN NORMAL AND DIABETIC RATS FROM TRYPTOPHAN-[14]C[a]

Rat		Expired [14]CO$_2$[b]	Picolinic carboxylase[c] (liver)	NAD-[14]C[d]	
				Liver	Kidney
Normal	1	2042	3.00	11.9	0.6
	2	1534	2.40	17.5	0.4
	3	2120	2.40	14.1	0.6
	Average	1899	2.60	14.5	0.5
Diabetic	1	6409	9.60	5.9	0.6
	2	3555	4.68	7.3	0.7
	3	5316	8.52	1.8	0.5
	Average	5093	7.60	5.0	0.6

[a] Taken from Ikeda *et al.*[2]

[b] The values are calculated from the specific activity of one carbon which is expected to be one-sixth of that of the original tryptophan-[14]C.

[c] The activity is expressed as micromoles reacted per gram of wet tissue per hour at 25°.

[d] The values are calculated from the specific activity of the nicotinamide moiety of NAD, which is expected to be five-sixths that of the original tryptophan-[14]C.

[30] Solvent systems for paper chromatography, which may be recommended, are: (1) 1 M ammonium acetate, pH 5.0–ethanol (3:7), where the R_f of NAD is 0.13; (2) isobutyric acid–ammonia–water (66:1.7:33, pH 3.8), where the R_f of NAD is 0.48; and (3) pyridine–water (2:1), where the R_f of NAD is 0.62.

to the picolinic carboxylase activity. The radioactive tryptophan is injected intraperitoneally into normal and alloxan-diabetic rats, three animals each. After 2 hours, the expired $^{14}CO_2$, the radioactive NAD in the liver and kidney, and the picolinic carboxylase activity are analyzed; the results are summarized in Table IV. The increased activity of picolinic carboxylase, which is induced by alloxan treatment, is accompanied by an increased amount of expired $^{14}CO_2$. In contrast, the incorporation of radioactivity into NAD from tryptophan-^{14}C during 2 hours after the injection is approximately one-half that in the diabetic rat as compared with the normal rat liver. The radioactivity of NAD-^{14}C in the kidney is much less in both normal and diabetic rats than in the livers, and no significant difference is observed between these two groups of animals. Indeed, the activity of kidney picolinic carboxylase does not increase upon treatment with alloxan (Table I). Essentially no radioactive quinolinic acid, nicotinic acid ribonucleotide, or deamido-NAD is detected in the liver. These results may suggest that the increase of picolinic carboxylase activity is accompanied by a decrease in the NAD synthesis from tryptophan, and that picolinic carboxylase may be one of the most important factors in controlling the coenzyme biosynthesis.

[121] Picolinic Acid Carboxylase

By L. M. HENDERSON and P. B. SWAN

Picolinic acid carboxylase (2-amino-3-carboxymuconic-6-semialdehyde decarboxylase) was first detected in rat liver by Mehler.[1,2] The substrate, formed by the action of 3-hydroxyanthranilate oxygenase, undergoes decarboxylation as well as isomerization around the double bond between C-2 and C-3, followed by cyclization to form picolinic acid—hence the name, picolinic acid carboxylase. The substrate of this enzyme has been more precisely identified by hydrogenation studies.[3] Two possible routes of formation of picolinate are outlined in Fig. 1. Published reports do not permit a clear choice between these alternatives. It has been assumed that the rate of nonenzymatic formation of quinolinic acid at room temperature is determined by the rate of isomerization to give a product with the amino and carbonyl groups in juxtaposition. Ichiyama *et al.*,[4] on the basis of

[1] A. H. Mehler, *J. Biol. Chem.* **218**, 241 (1956).

[2] A. H. Mehler and E. L. May, *J. Biol. Chem.* **223**, 449 (1956).

[3] E. Kuss, *Hoppe-Seylers Z. Physiol. Chem.* **348**, 1589 (1967).

[4] A. Ichiyama, S. Nakamura, H. Kawai, T. Honjo, Y. Nishizuka, O. Hayaishi, and S. Senoh, *J. Biol. Chem.* **240**, 740 (1965).

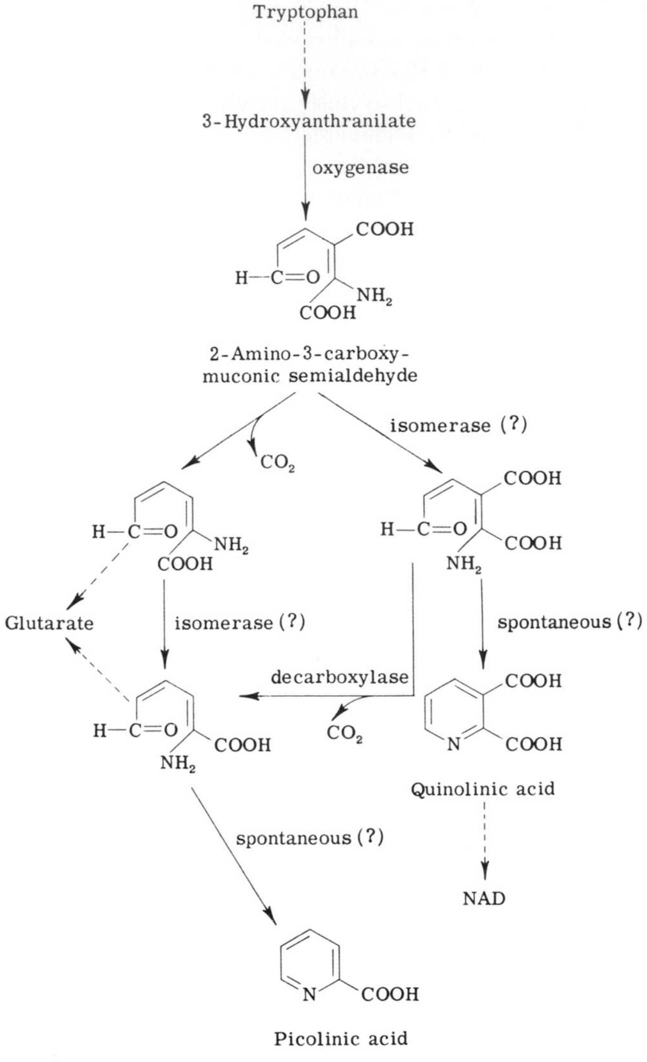

FIG. 1. Formation of picolinate.

comparative rates of $^{14}CO_2$ release and spectrophotometric changes, have suggested that decarboxylation is the rate-limiting step in picolinate formation. This would preclude the involvement of a slow isomerization step. The isomerization which would bring the polar carboxyl groups to the cis configuration might be expected to occur less readily[5] than the isomerization of the decarboxylated product. For these reasons, decar-

[5] D. Pressman, J. H. Bryden, and L. Pauling, *J. Am. Chem. Soc.* **70,** 1352 (1948).

boxylation prior to isomerization appears to be more likely. No evidence has been presented to support the involvement of enzymes in either isomerization or cyclization. Although the cyclization after decarboxylation is probably a fast reaction *in vitro*, it does not occur to an appreciable extent in the whole animal as evidenced by the failure to find significant picolinic acid excretion.[6] Instead, the decarboxylated open-chain compound is converted to 2-aminomuconic acid through the action of an NAD-requiring dehydrogenase,[4] and this compound is further metabolized to yield glutaryl coenzyme A.

Picolinic acid carboxylase catalyzes a reaction which competes for substrate with a spontaneous reaction[7] which leads to pyridine nucleotides[8,9] via quinolinate. Thus, this enzyme exerts a control on the formation of niacin coenzymes from tryptophan. The effectiveness of tryptophan as a dietary replacement of niacin in the diet can be predicted from a knowledge of the ratio of the activity of 3-hydroxyanthranilate oxygenase to that of picolinic acid carboxylase; the higher the ratio, the better is tryptophan as a source of niacin.[6,10,11]

Picolinic acid carboxylase appears to be widely distributed among vertebrates; however, it has not been found in invertebrate tissues.[11] The decarboxylase activity of liver from diabetic rats is markedly increased, resulting in a poor conversion of tryptophan or 3-hydroxyanthranilate to pyridine nucleotides.[10,12,13]

Assay Method

Principle. The enzyme activity can be determined spectrophotometrically by generating the substrate from 3-hydroxyanthranilate, then measuring the decrease in absorbance at 360 mμ.[1,4,11] Alternatively, the release of $^{14}CO_2$ from 3-hydroxyanthranilic acid-^{14}C (carboxyl labeled) can be followed.[2,4,6] The release of $^{14}CO_2$ measures picolinic acid carboxylase activity directly, whereas the decrease in absorbance results from cyclization. Therefore, the spectrophotometric measurement must be corrected for the spontaneous cyclization before decarboxylation which leads to quinolinate.

[6] R. J. Suhadolnik, C. O. Stevens, R. H. Decker, L. M. Henderson, and L. V. Hankes, *J. Biol. Chem.* **228**, 973 (1957).

[7] C. L. Long, H. N. Hill, I. M. Weinstock, and L. M. Henderson, *J. Biol. Chem.* **211**, 405 (1954).

[8] Y. Nishizuka and A. Hayaishi, *J. Biol. Chem.* **238**, 3369 (1963).

[9] R. K. Gholson, T. Ueda, N. Ogasawara, and L. M. Henderson, *J. Biol. Chem.* **239**, 1208 (1964).

[10] M. Ikeda, H. Tsuji, S. Nakamura, A. Ichiyama, Y. Nishizuka, and O. Hayaishi, *J. Biol. Chem.* **240**, 1395 (1965).

[11] S. J. Lan and R. K. Gholson, *J. Biol. Chem.* **240**, 3934 (1965).

[12] A. H. Mehler, E. G. McDaniel, and J. M. Hundley, *J. Biol. Chem.* **232**, 323 (1958).

[13] A. H. Mehler, K. Yano, and E. L. May, *Science* **145**, 817 (1964).

Apparently the cyclization after decarboxylation is faster than the decarboxylation; hence, after correction for quinolinate formation, the two methods give similar results.[4]

Reagents

3-Hydroxyanthranilic acid *or* 3-hydroxyanthranilic acid-[14]COOH
Tris-acetate buffer, pH 8.0
3-Hydroxyanthranilate oxygenase
Acetic acid, 2 N
NaOH, 0.5 N

Procedure

For the spectrophotometric assay, the incubation mixture contains 60 millimicromoles of 3-hydroxyanthranilic acid, 200 micromoles of Tris-acetate buffer, pH 8.0, and an excess of a purified preparation of 3-hydroxyanthranilate oxygenase (approximately 0.4 mg protein) in a total volume of 2.9 ml.[4] The formation of 2-amino-3-carboxymuconic-6-semialdehyde is allowed to go to completion as judged by the increase in absorbance at 360 mμ. The decarboxylase reaction is initiated by the addition of 0.1 ml of the enzyme. The effect of temperature and 3-hydroxyanthranilate concentration on the yield of product and the methods of purifying 2-amino-3-carboxymuconic-6-semialdehyde have been described by Kuss.[3] The decrease in absorbance at 360 mμ is followed at 30-second intervals against a blank containing all ingredients except the substrate. A separate blank containing no enzyme provides the correction for quinolinate formation or alternatively the rate of spontaneous decrease in absorbance at 360 mμ can be established prior to adding the decarboxylase.[11]

For the isotopic assay, the incubation is done in a vessel which has a chamber for alkali (0.2 ml of 0.5 N NaOH) to trap the radioactive carbon dioxide. The reaction is stopped by the addition of 0.5 ml of 2 N acetic acid.[4] Under these conditions, the nonenzymatic decomposition of the substrate observed in the presence of strong acid to yield 2-hydroxymuconic-6-semialdehyde does not occur.

Definition of Unit and Specific Activity. One unit of picolinic acid carboxylase activity is the amount which catalyzes the release of 1 micromole of carbon dioxide per minute or causes a decrease in absorbance at 360 mμ of 15 per minute. The molar extinction coefficient of 2-amino-3-carboxymuconic semialdehyde is reported to be 4.5 × 10^4 under these conditions.[4]

Purification Procedure

Purification of the enzyme from guinea pig liver was first described by Mehler,[1] who obtained about a 70-fold purification through ammonium sulfate precipitation and treatment with alumina. Ichiyama *et al.*[4] reported over 150-fold purification with an overall yield of 10–20% from cat liver. The procedure used is as follows (all steps at 0–4°):

1. Frozen cat liver (45 g) is thawed in 180 ml of 0.14 M KCl, homogenized in a Waring blendor for 3 minutes, and centrifuged for 15 minutes at 20,000 g.

2. To 178 ml of the supernatant solution, 36 ml of 0.4% protamine sulfate is added slowly and stirred for an additional 5 minutes. The precipitate is removed by centrifugation.

3. The supernatant solution (202 ml) is warmed rapidly to 58° with stirring, maintained at this temperature for 1 minute, and cooled rapidly to 1–2° in an ice–salt bath. The precipitate is removed by centrifugation.

4. To 192 ml of supernatant solution, 72 ml of peroxide-free acetone (−30°) is added slowly with stirring. The solution is maintained at −2° to −10° in dry-ice–methanol. The precipitate is removed by centrifugation at −10°. Cold acetone is added (final concentration of 50%) to the supernatant solution, and the solution is allowed to stand for 3 minutes. The precipitate is collected by centrifugation at −10° and put into 30 ml of 0.0075 M potassium phosphate buffer at pH 7.5. This solution is centrifuged to remove insoluble material.

5. The solution (36 ml) is placed on a DEAE-cellulose column (2.5 × 5 cm) previously equilibrated with 0.0075 M potassium phosphate buffer, pH 7.5. The column is then washed with 100 ml of the same buffer, and the enzyme is eluted with 180 ml of 0.025 M potassium phosphate buffer, pH 7.5.

6. To the eluate (180 ml), 220 ml of saturated ammonium sulfate is added with stirring. After 15 minutes the precipitate is removed by centrifugation. To the supernatant solution, 200 ml of saturated ammonium sulfate is added; after 15 minutes the precipitate is collected by centrifugation and dissolved in 13 ml of 0.0075 M potassium phosphate buffer, pH 7.5. It is then dialyzed overnight against 2 liters of the same buffer.

7. Fourteen milliliters of the enzyme solution is adsorbed on a hydroxylapatite column (1.6 × 2.5 cm) which has been equilibrated with 0.0075 M potassium phosphate buffer, pH 7.5. The column is washed with 20 ml of 0.1 M potassium phosphate buffer at the same pH and eluted with 20 ml of 0.2 M potassium phosphate buffer, pH 7.5.

Properties

Both Mehler[1] and Ichiyama *et al.*[4] have reported a broad range of maximal activity for picolinic acid carboxylase from pH 6 to 9.5 with a rather sharp drop in activity on either side of this range. The K_m for 2-amino-3-carboxymuconic-6-semialdehyde is 10^{-6} M under the assay conditions described above.[4] Very little is known about the substrate specificity of the enzyme; however, 2-hydroxymuconic-6-semialdehyde at $3.3 \times 10^{-6} M$ markedly inhibits the reaction. Hydrogen cyanide at $10^{-5} M$ inhibits the reaction[1] as does p-chloromercuribenzoate.[1,4] This latter inhibition can be prevented by the addition of cysteine.[4] There is a slight stimulation by $MgCl_2$ at $2 \times 10^{-4} M$.[4]

[122] Nicotinamide Deamidase from *Torula cremoris*[1]

By JAYANT G. JOSHI, D. F. CALBREATH, and P. HANDLER

$$\text{Nicotinamide} \rightarrow \text{nicotinic acid} + NH_4^+$$

Assay Method

Enzyme activity can be assayed either by measuring ammonia formed, colorimetrically, or by measuring nicotinic acid formed or nicotinamide hydrolyzed. The most satisfactory method is given below.

Principle. Nicotinamide deamidase activity is measured by following the formation of nicotinic acid 7-^{14}C from nicotinamide 7-^{14}C with a strip counter or liquid scintillation counter after separation of the reaction components by paper chromatography.

Reagents. Reaction mixture contains 50 millimicromoles of nicotinamide 7-[14]C (specific activity 1.4–6.0 mCi per millimole may be obtained from New England Nuclear Corporation), 10 micromoles of Tris-acetate buffer, pH 7.0, and about 20 units of enzyme in a total volume of 0.5 ml.

Procedure. After incubation of the reaction mixtures for 15 minutes at 37°, reaction is halted by either placing the tubes in a boiling water bath for 1 minute or by addition of 0.1 ml of 10 N HCl. An aliquot (25 μl) is spotted on Whatman No. 1 paper and chromatographed overnight. The solvent system consists of n-butanol saturated with 15% ammonium hydroxide. Chromatographs so obtained are analyzed for the percent conversion of nicotinamide 7-^{14}C to nicotinic acid 7-[14]C.

Definition of Units and Specific Activity. A unit of activity is defined as the amount of enzyme required for the formation of 1 millimicromole of nicotinic acid in 15 minutes. Specific activity is expressed as units per milligram of protein.

Application of Assay Method to Crude Extracts.[2] Modest requirements of nicotinamidase for optimal activity and extremely high sensitivity of the

---

[1] J. G. Joshi and P. Handler, *J. Biol. Chem.* **237,** 929 (1962).

[2] D. F. Calbreath, J. G. Joshi, and P. Handler, *Federation Proc.* **28,** 841 (1969).

assay method permits it application to crude extracts. Thus, the method has been successfully applied for measuring nicotinamidase activity in dialyzed cell-free extracts of *E. coli*, Fleischmann's yeast, *Clostridium acetobutylicum*, *Bacillus stearothermophilus*, *Aspergillus niger*, and *Micrococcus lysodeikticus*.

Purification Procedure

Growth of the Organism and Preparation of Acetone Powder. *Torula cremoris* (ATCC 2512) is grown for 24 hours at 37° with continuous vigorous aeration in batches of 10 to 12 liters of nutrient medium. One liter of medium contains glucose, 50 g; KH_2PO_4, 300 mg; $MgSO_4 \cdot 7$ H_2O, 100 mg; Bacto-peptone, 10 g; K-Citrate, 5 g; citric acid, 1 g; and yeast extract, 1 g. The cells are harvested by centrifugation in the cold, washed four times with ice-cold saline solution and stirred in a Waring blendor for 45 seconds with 5 volumes of previously chilled acetone, washed with ether and dried in a vacuum. The nicotinamidase of the powder prepared in this manner and stored at −12° remains active for several months. About 50 g of acetone powder is obtained from 12 liters of growth medium.

Autolysis. Acetone powder (36 g) is autolyzed by incubation in 200 ml of 0.1 M $KHCO_3$ for 3 hours at 37° with occasional shaking. The autolysate is centrifuged in the cold at 18,000 g for 10 minutes, and the supernatant fluid is set aside. The residue is resuspended in 200 ml of 0.1 M $KHCO_3$ and recentrifuged. The two supernatant fluids are pooled, and the residue is discarded. All subsequent steps are performed in the cold.

First Ammonium Sulfate Fractionation. To 305 ml of clear autolysate is added 143.8 g of solid ammonium sulfate (0.7 saturation) with continuous stirring. After 1 hour the mixture is centrifuged, and the supernatant fluid is discarded. The precipitate is redissolved in 100 ml of 0.001 M K phosphate, pII 7.0.

Second Ammonium Sulfate Fractionation. The above solution is brought to 0.4 saturation with solid ammonium sulfate (24.7 g) and centrifuged after 30 minutes. The residue is discarded, and the clear supernatant liquid, which contains all the activity, is again brought to 0.7 saturation with ammonium sulfate (22.5 g). After 30 minutes, the mixture is centrifuged, the supernatant liquid is discarded, and the precipitate is redissolved in 50 ml of 0.001 M K phosphate, pH 7, and dialyzed against 6 liters of this buffer for 16 hours.

Acetone Precipitation. To the dialyzed solution prepared above are slowly added 20 ml of acetone which has previously been chilled to −20°. The resultant mixture is centrifuged at 18,000 g at 4°, and the inactive precipitate is discarded. Additional acetone (40 ml) is added to the supernatant to bring the final acetone concentration to 50%, the temperature being maintained below 0° during the addition. The solution is stirred for 15 minutes and then maintained at −10° for an additional 90 minutes, after which it is centrifuged at −10° for 15 minutes at 18,000 g. The supernatant liquid is devoid of enzyme activity and is discarded. The precipitate is redissolved in 15 ml of 0.001 M K phosphate and dialyzed against 4 liters of this buffer for 8 hours.

Calcium Phosphate Gel[3] Treatment. Removal of additional inactive protein is accomplished by addition to the preceding preparation of 2 mg of calcium phosphate gel per mg of protein. The mixture is stirred for 15 minutes and centrifuged at 3,000 rpm. Approximately 75% of the activity remains in the supernatant. The gel is reextracted with 5 ml of 0.05 M phosphate, pH 7, centrifuged, and the two supernatants are pooled and dialyzed against 4 liters of 0.001 M K phosphate, pH 7, for 4 hours.

DEAE-Cellulose Chromatography. A column (16 × 120 mm) is packed with DEAE-cellulose and is equilibrated with 0.001 M phosphate, pH 7. Enzyme solution, from the earlier step, is added to the column, and elution is conducted successively with 50-ml portions of 0.001 M and 0.025 M phosphate and 150 ml of 0.05 M phosphate, pH 7, respectively. All nicotinamidase activity appears in the 0.05 M phosphate eluate.

Hydroxylapatite[4] Chromatography. The enzyme from the previous step is dialyzed against 6 liters of 0.001 M phosphate, pH 7, and applied to a column (12 × 150 mm) of hydroxylapatite which has been equilibrated with 0.001 M phosphate, pH 7. Elution is performed with 50 ml of 0.001 M phosphate followed by 50 ml of 0.025 M phosphate. Thereafter, a linear gradient is applied. The mixing chamber and reservoir contain 200 ml each of 0.05 M and 0.20 M phosphate, pH 7, respectively. Fifty 10-ml fractions are collected, and alternate tubes are assayed for protein content and nicotinamidase activity. Activity is recovered in two independent peaks (see Fig. 1). The first peak elutes at 0.025 M and the second at 0.17 M phosphate concentration, with specific activities of 5,600 and 26,000, respectively. Despite this different chromatographic behavior, both peaks have similar kinetic and inhibition profiles. Typical data is summarized in the table.

This procedure is also applicable to the isolation of nicotinamidase from commercially available Fleischmann's yeast.[5] More recently, a somewhat similar procedure has also permitted isolation of isozymes of nicotinamidase from rabbit liver.[6]

[3] S. P. Colowick, Vol. I, p. 97.
[4] A. Tiselius, S. Hjertén, and Ö. Levin, *Arch. Biochem. Biophys.* **65,** 132 (1956).
[5] D. F. Calbreath, J. G. Joshi, and P. Handler, *Am. Chem. Soc. Div. of Biochem.* 158*th ACS National Meeting New York.* Abstr. No. 267 (1969).
[6] S. Su, L. Albizati, and S. Chaykin, *J. Biol. Chem.* **244,** 2956 (1969).

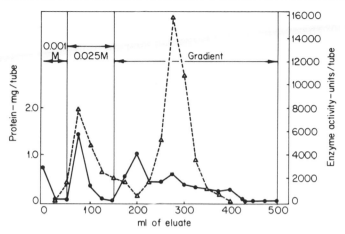

Fig. 1. Elution of nicotinamidase (Δ) and protein (○) from a hydroxylapatite column. Enzyme and protein were assayed in the standard manners.

Properties

Reversibility. The reaction is apparently irreversible.

Stability. Enzyme preparations representing the various steps of purification retain full activity for several weeks when stored at 4°. The major fraction, eluting at 0.17 *M* phosphate, is ultracentrifugally homogeneous with a molecular weight of about 100,000.

Effect of Temperature. The Q_{10} value for nicotinamidase over the range 10°–50° is approximately 2.0, with an activation energy of 10,900 cal.

PURIFICATION OF NICOTINAMIDASE FROM *Torula cremoris*[a]

Procedure	Volume (ml)	Units (thousands)	Protein (mg)	Specific activity	Yield (%)
Autolysate	305	104	3500	30	100
First ammonium sulfate fractionation	100	103	690	150	99
Second ammonium sulfate fractionation	60	97.3	338	290	93
Acetone precipitation	15	73.6	60	1,240	71
Calcium phosphate gel treatment	30	60	26.5	2,260	57
DEAE-cellulose chromatography	150	51	4.7	10,600	49
Hydroxylapatite chromatography, fraction II[b]	10	15.8	0.68	26,000	7

[a] Data taken from text reference 1.
[b] From 60 g of acetone powder.

Between 40 and 50° the substrate protects against heat denaturation. Above 50° the enzyme is rapidly inactivated.

pH Optimum. The enzyme is active over an unusually wide pH range and exhibits optimum activity between pH 6 and 7 in Tris-acetate buffer and approximately at pH 7 in K phosphate buffer.

Specificity. The enzyme appears to be highly specific for nicotinamide as a substrate. No ammonia is liberated when the enzyme is incubated with glutamine, asparagine, NAD, NADP, glycinamide, nicotinamide mononucleotide, benzamide or N^1-methylnicotinamide, nor is any amine liberated when N^2-ethylnicotinamide or N^2-diethylnicotinamide are tested as substrates. The final preparation is also free from 5'-AMP deaminase, cytosine deaminase, guanase, or any esterase activity. Thus, it appears to be a true nicotinamidase.

Kinetics. Unlike the nicotinamidase from rabbit liver,[6] the present enzyme has high affinity for nicotinamide. K_m for nicotinamide is 1.4 × 10^{-5} M. Neither nicotinic acid nor NH_4^+ at concentrations as high as 0.002 M affect the rate of enzyme activity. Likewise, a 200-fold excess of the substrate does not result in diminution of enzyme activity below V_{max}.

Activators and Inhibitors

Metals. The enzyme does not exhibit an apparent requirement for external addition of a metal. Moreover, no loss in activity is apparent after 18 to 20 hours of dialysis against metal-free buffers. The presence of metal chelating agents such as 0.002 M EDTA, 0.004 M α,α'-dipyridyl, or 8 × 10^{-6} M thyroxine, with or without preincubations, is without effect. At a concentration of 0.005 M, Mg^{2+} is without effect; however, other metals are inhibitory in varying degree. At this concentration, Mn^{2+} inhibits activity by 20%, Ca^{2+} by 26%, Cu^+ by 63%, Cu^{2+} by 83%; Hg^{2+} and Fe^{3+} inhibit by more than 90%.

At 10^{-5} M concentration, *p*CMB completely inhibits the enzyme. The inhibition is completely reversed by sulfhydryls such as cysteine, glutathione, and dithiothreitol.

Inhibition Studies. At 0.002 M N^1-methylnicotinamide, trigonelline, benzamide, nicotinic acid, N^2-ethylnicotinamide, and N^2-diethylnicotinamide are without effect on the hydrolysis of nicotinamide. 3-Acetylpyridine is a competitive inhibitor with K_i value of 3.05 × 10^{-4} M. NAD and NADP are allosteric inhibitors, with apparent K_i values of 5.3 × 10^{-4} M and 4.65 × 10^{-4} M, respectively. Reduced forms of these coenzymes are without effect. Nicotinamidase in several bacteria also exhibits similar allosteric inhibition at physiological concentrations of NAD.[2]

Urea at 4 M concentration is without effect.

Induction and Repression. Nicotinamidase in *Torula cremoris*, like in several other microorganisms, is a constitutive enzyme. However, 1 mM NAD in growth medium reduces the intracellular concentration of nicotinamidase in *Torula cremoris* by 50%.

[123] Nicotinamide Deamidase

By SHIRLEY SU and STERLING CHAYKIN

Assay Method

Principle. The deamidase has such a high K_m for its substrate that assays based on disappearance of substrate are impractical. Several assay procedures based on the determination of products have been used. In crude extracts where release of ammonia from a variety of sources is possible, estimation of nicotinic acid is the preferred method. The use of the more rapid and simpler methods for the determination of ammonia is recommended for assaying the purified enzyme.

Assay Based on the Determination of Nicotinic Acid-7-^{14}C

Reagents

Nicotinamide-7-^{14}C, 0.137 M (10^6 cpm/0.1 ml)
Potassium phosphate buffer, 0.1 M, pH 7.4
Acetic acid, 1.0 N
Sodium nicotinate, 8.1×10^{-2} M
Scintillation solvent, 4 g of 2,5-bis[2-(5-*tert*-butylbenzoxazolyl)]-thiophene per liter of reagent grade toluene

Procedure. The standard deamidase reaction mixture contains 0.1 ml of nicotinamide-7-^{14}C, 0.1 ml of phosphate buffer, and deamidase in a total volume of 0.3 ml. Incubations are carried out at 37° for 1 hour. The reaction vessels containing microsomal deamidase should be shaken during the incubation procedure. The reaction is terminated by the addition of 25 μl of 1 N acetic acid, followed by a 2-minute heat treatment in a boiling water bath.

The addition of acetic acid to the reaction mixtures, immediately before heat denaturation of the enzyme, serves to lower the pH to a point that ensures the complete coagulation of protein during the heat denaturation step. Incomplete removal of protein from reaction mixtures leads ot difficulty in the application of samples to paper chromatograms.

After completion of the heat treatment, 5 μl of unlabeled sodium nicotinate is added to each reaction mixture to serve as a chromatographic marker, and denatured protein is removed by centrifugation. A 0.2-ml portion of the deproteinized reaction mixture is subjected to paper chromatography on Whatman No. 1 filter paper strips, 1.5 inches in width and 24 inches in length. The strips are developed for 18 hours in 1-butanol saturated with 3% ammonia. Nicotinate is located on the air-dried chromatogram by its ability to quench the fluorescence of the paper which is normally visible under ultraviolet light (259 mμ). Nicotinate normally travels with an R_f of 0.11. The portion of the chromatogram that contains nicotinate is cut out, cut up, suspended in 10 ml of scintillation fluid, and counted in a scintillation counter. A control reaction mixture which contains heat-denatured enzyme is also carried through the assay procedure in order to provide a means for the correction of results for the chromatographic tailing of the fast-moving substrate into the region of the product. Since the deamidase is very sensitive to inhibition by detergents, only acid-washed glassware is used in enzyme assays. The possibility of erroneous results due to the accidental use of contaminated glassware can be further guarded against by carrying out all incubations in duplicate.

Assay Based on the Determination of NH$_3$ by Nessler's Method

Reagents
    Nicotinamide, 0.137 M
    Potassium phosphate buffer, 0.1 M, pH 7.4
    Boric acid, 2%
    Nessler's reagent (see Vol. III, p. 991)

Procedure. Incubations are carried out as described above. The action of the deamidase is terminated by the addition of 0.1-ml samples of reaction mixture to a test tube containing 0.5 ml of Nessler's reagent, 1.0 ml of 2% boric acid, and 3.4 ml of distilled water. The optical density of nesslerized samples is measured at 460 mμ, and the ammonia content is calculated from a standard curve obtained with $(NH_4)_2SO_4$. The method as described gives satisfactory results over the range 0.2–2.0 micromoles of ammonia. The presence of small quantities of purified deamidase does not interfere with the detection of ammonia. The method is not satisfactory for use with crude tissue extracts.

Assay Based on the Determination of NH_3 by the Indophenol Method[1]

Reagents

Nicotinamide, 0.41 M
Ammonium sulfate, 10^{-3} M
Potassium phosphate buffer, 0.1 M, pH 7.4
Phenol solution, 1.33%
Nitroprusside solution, 7.5%
Sodium hydroxide, 3%
Sodium hypochlorite, 6% (household bleach)
Alkaline sodium hypochlorite (prepared on the day of use): 10 ml of
    3% sodium hydroxide + 0.2 ml of sodium hypochlorite

General Comments. The presence of nicotinamide in the standard indophenol method reaction mixtures interferes with color development. It has been found that increased concentrations of nitroprusside will circumvent the problem. The method as outlined below is optimal for 0.15-ml samples containing 0.13 M nicotinamide. Adjustments of nitroprusside concentration for other levels of nicotinamide are simple, since there is a linear relation between nicotinamide concentration in the indophenol method reaction mixtures and the amount of nitroprusside required for maximum color development; 1.2 micromoles of nitroprusside is required per micromole of nicotinamide. This adjustable scale is satisfactory for concentrations of nicotinamide up to 40 micromoles per indophenol reaction mixture. In the complete absence of nicotinamide, 0.1 ml of 0.1% nitroprusside is optimal; the color produced is blue. In the presence of nicotinamide and the absence of ammonia, a yellow color results. When both nicotinamide and ammonia are present, the reaction mixtures appear green. The absorption maximum (625 mμ) for ammonia-dependent color is unaffected by the presence of nicotinamide. The method as described is most useful over the range 20–200 millimicromoles of ammonia. The range can be extended upward to at least 1.4 micromoles of ammonia simply by diluting high-absorbance samples with water. Color yield is linear over the entire range 20 millimicromoles–1.4 micromoles. Although not examined in detail, it appears possible to increase sensitivity to the 1–2 millimicromole range by reducing the size of the indophenol reaction mixtures. This assay is not recommended for use with crude tissue extracts.

Procedure. Incubations, using 0.15 ml of deamidase reaction mixtures (all reactants taken at one-half the volumes previously specified), are carried out as described above. An ammonia standard consisting of 0.05 ml

---

[1] Adapted from the method as described by H. H. Leffler, *Am. J. Clin. Pathol.* **37**, 233 (1967); see S. Chaykin, *Anal. Biochem.* **31**, 375 (1969).

of nicotinamide, 0.05 ml of phosphate buffer, and 0.05 ml of ammonium sulfate and a reagent blank, in which water is substituted for the ammonium sulfate of the ammonia standard, are also carried through the procedure. The action of the deamidase is terminated by the addition of 3.75 ml of phenol solution to the reaction vessel. Color development is initiated by the addition of 0.1 ml of nitroprusside, mixing, and the rapid addition of 1.0 ml of the alkaline sodium hypochlorite solution. It is important that the alkaline sodium hypochlorite solution be added immediately after the addition of the nitroprusside solution. Failure to do so results in high blanks and erratic results. Color development is essentially complete in 20 minutes. The absorbance of the reaction mixtures is measured at 625 mμ at that time. The colorimeter or spectrophotometer is adjusted to zero absorbance with a water blank. The absorbance of all reaction mixtures is corrected for the absorbance of a reagent blank. This blank should contain all reagents affecting color development, and in particular nicotinamide. The ammonia content is calculated from a standard curve determined in the presence of nicotinamide.

Assay Based on the Determination of NH_3 by Coupling to Glutamic Dehydrogenase[2]

Reagents

Acetic acid, 1 N
α-Ketoglutarate, 0.167 M
DPNH, 1.3 $\times$ 10^{-3} M
Potassium phosphate buffer, 0.01 M, pH 7.4
Glutamic dehydrogenase, 0.66 μM units/0.1 ml

Procedure. Incubations are carried out as described above. The deamidase is inactivated by adding 25 μl of acetic acid to the reaction mixture and immediately heating the reaction vessel in a boiling water bath for 2 minutes. Denatured protein is removed by centrifugation and 50 μl of the supernatant is added to a reaction mixture which contains 0.2 ml of α-ketoglutarate, 0.2 ml of DPNH, 2.45 ml of phosphate buffer, and 0.1 ml of glutamic dehydrogenase. Ammonia content is calculated from the decrease in absorbance at 340 mμ. Twenty minutes are required for the complete incorporation of NH_3 into glutamate.

Definition of Specific Activity. Specific activity is defined as the number of micromoles of nicotinamide hydrolyzed per milligram of protein per hour. Protein is determined by the biuret method[3] in crude fractions and a

[2] See Vol. II [27].
[3] A. G. Gornall, C. J. Bardawill, and M. M. David, *J. Biol. Chem.* **177**, 751 (1949).

spectrophotometric method based on absorbance at 215 and 225 mμ[4] is used with the more highly purified fractions.

Crude enzyme fractions contain inhibitors, presumably fatty acids, that cause variable and low specific activities. It is therefore customary to assay these fractions both in the absence and the presence of bovine serum albumin (BSA), 2.5 mg/standard assay. Since BSA is a good binder of fatty acids, it can compete with the enzyme for these inhibitors. Care should be used in the choice of BSA samples, since some commercial preparations contain enough impurities to cause the inhibition of pure deamidase fractions. The enzyme is ordinarily free of BSA effects after acetone fractionation.

Purification Procedure

Centrifugations in steps subsequent to step 1 are performed at 27,000 g for 15 minutes. In fractionation steps resulting in precipitation, solutions are stirred for 15 minutes after the last addition of precipitant before being centrifuged. All operations are carried out at 0° unless stated otherwise.

Step 1. Isolation of Microsomes. Fresh rabbit livers are cooled in an ice-water bath immediately after removal from the animals and are quickly transported to the laboratory. The livers are then homogenized in a Waring blendor with 5 volumes of cold 0.1 M potassium phosphate buffer, pH 7.4. The homogenate is filtered through cheese cloth and centrifuged at 10,000 g for 15 minutes. The supernatant solution (clarified homogenate) is brought to 0.5 saturation with solid ammonium sulfate and centrifuged at 10,000 g for 15 minutes. The precipitate obtained is suspended in a volume of 0.1 M potassium phosphate buffer, pH 7.4, equal to one-tenth that of the original homogenate, and the microsomes are collected by centrifugation at 78,000 g for 2 hours. An alternate procedure can be used to collect the microsomal fraction. The clarified homogenate is made 0.2 M in NaCl and 3% in polyethylene glycol. The microsomes can then be collected by centrifugation at 12,000 g for 15 minutes.

Step 2. Preparation of Microsomal Acetone Powder. The microsomal pellets from the preceding step are suspended in a minimal amount of 0.1 M potassium phosphate buffer, pH 7.4. The microsomal suspension is added to 10 volumes of acetone, $-10°$, and stirred on a magnetic stirring motor at $-10°$ for 15 minutes. The residue is collected by centrifugation at $-10°$. The acetone powder is stirred with a fresh portion of $-10°$ acetone, which is one-half the volume used in the initial extraction. The residue is again collected by centrifugation. The acetone powder is spread

---

[4] J. B. Murphy and M. W. Kies, *Biochim. Biophys. Acta* **45**, 382 (1960).

in a shallow dish and dried overnight, under reduced pressure, over concentrated sulfuric acid at room temperature. Acetone powders are stored over a desiccant at 0°.

Step 3. Extraction of the Deamidase. Twenty grams of microsomal acetone powder are extracted with 400 ml of 0.2 M glycine buffer, pH 8.9, at 37° for 30 minutes. Insoluble material is removed by centrifugation and the supernatant solution is subjected to fractionation with ammonium sulfate.

Step 4. Ammonium Sulfate Fractionation. The material that precipitates between 0.45 and 0.65 saturation is dissolved in water and dialyzed overnight against 0.05 M phosphate buffer, pH 7.4.

Step 5. Removal of Nucleoproteins with Protamine Sulfate and Bentonite. The protein concentration of the dialyzed solution is adjusted to 10 mg/ml with potassium phosphate buffer, 0.05 M, pH 7.4. The pH of the diluted enzyme solution is then brought to pH 6.0 with 1 N acetic acid, and 2.5 ml of 1% protamine sulfate solution per 10 ml of the enzyme solution are added. The inert precipitate is removed by centrifugation and the pH of the supernatant solution is readjusted to pH 7.4 with 1 N potassium hydroxide. Bentonite, which has been washed with 0.05 M potassium phosphate buffer, pH 7.4, is added to the supernatant solution (20 mg of bentonite per milliliter of the solution). After centrifugation the supernatant solution is subjected to acetone fractionation.

Step 6. Acetone Fractionation. Acetone at −10° is added slowly with stirring. The temperature of the protein solution is gradually lowered to −10° but the solution is kept above its freezing point throughout the process. The deamidase activity is precipitated between 40 and 55% acetone (v/v). The precipitate is collected by centrifugation at −10° and dissolved in a small volume of water. The enzyme solution is dialyzed overnight against 1 mM potassium phosphate buffer, pH 6.8.

Step 7. Hydroxylapatite Column Chromatography. Three milliliters of the dialyzed enzyme, containing 50 mg of protein, is applied to a hydroxylapatite column (2.5 × 25 cm). The column should be preequilibrated with 1 mM potassium phosphate buffer, pH 6.8. It is developed with a linear gradient prepared by using 1000 ml of 1 mM potassium phosphate, pH 6.8, as starting buffer and 1000 ml of 500 mM potassium phosphate, pH 6.8, to form the gradient. Fractions of 15 ml are collected at a flow rate of 0.5 ml/min. The protein elution profile consists of a small peak (fractions 15–22) followed by a large peak (fractions 42–54) with a front running shoulder (fractions 22–42). Both the small (F_2) and the large (F_1) peaks have deamidase activity.

A typical purification is summarized in the table.

PURIFICATION OF NICOTINAMIDE DEAMIDASE FROM RABBIT LIVER

Fraction	Specific activity		Total activity		% Recovery[a]
	−BSA	+BSA	−BSA	+BSA	
Whole liver acetone powder extract	0.047	0.064	—	—	—
Microsome acetone powder extract	0.321	0.443	474	654	100
0.45–0.65 (NH₄)₂SO₄ precipitate	1.31	1.74	404	536	82
Supernatant after protamine sulfate	1.84	2.50	368	450	69
Supernatant after bentonite	2.20	3.35	240	366	56
Acetone precipitate, 40–55%	4.37	4.30	214	228	35
Hydroxylapatite					
F₂	2.24	2.24	11	11	2
F₁	4.50	4.50	86	86	13

[a] Based on total activity in the presence of bovine serum albumin (BSA).

Properties

Nicotinamide deamidase (only the F_1 fraction has been subjected to extensive study) is stable at all stages of purification. The purified enzyme can be stored indefinitely at −15°. Under the standard assay conditions, the reaction is linear for at least 6 hours. A variety of compounds cause inhibition of the deamidase.[5] These include small amounts of surface active agents (dishwashing detergents, fatty acids, deoxycholate, etc.), thyroxine and thyroxine analogs, ammonium sulfate, diisopropylfluorophosphate, δ-hydroxyquinoline, o-phenanthroline, and carbobenzyloxyamido-2-phenylethylchloromethylketone. The pH optimum of the deamidase is 7.4. Inhibition by detergents brings about a shift in pH optimum to high values. Sulfhydryl reagents inhibit the deamidase only at high concentrations.

The deamidase catalyzes the hydrolysis of a variety of amides and esters. It does, however show specificity for the pyridine ring. The K_m of the purified enzyme is 4×10^{-2}. Its molecular weight has been variously estimated at 196,000–218,000. The enzyme has been shown to be a glycoprotein containing 2.84% mannose and 0.57% glucosamine in addition to 95.4% amino acids. Evidence has been obtained which indicates that the enzyme is made up of at least two, and probably four, subunits. Thyroxine

[5] J. Kirchner, J. G. Watson, and S. Chaykin, *J. Biol. Chem.* **241**, 953 (1966).

causes the enzyme to dissociate. An acyl enzyme intermediate appears to be an intermediate in both ester and amide hydrolysis. Two serine residues and at least two histidine residues per mole have been implicated in the catalytic process. Substrate activation of nicotinamide hydrolysis has been demonstrated.

[124] Preparation of Nicotinic Acid Mononucleotide from Nicotinamide Mononucleotide by Enzymatic Deamidation

By HERBERT C. FRIEDMANN

Preparation

Principle. An enzyme from *Propionibacterium shermanii*, partially purified in two simple steps to remove phosphatase, deamidates nicotinamide mononucleotide to nicotinic acid mononucleotide. The product is purified on a Dowex 1 column.

Reagents

> *Propionibacterium shermanii* (ATCC 9614) grown in a medium[1] as described by Friedmann[2] without aeration, iron, or yeast, but with added cobalt. The harvested bacteria, stored at $-22°$, may be used for at least 15 months.
> Acid-washed glass beads, average diameter 200 μ (see Vol. VI [22])
> Nicotinamide mononucleotide (NMN), available commercially or prepared from DPN$^+$ with snake venom phosphodiesterase (Vol. III [129]) or with potato nucleotide pyrophosphatase (Vol. II [112]). If desired, nicotinamide mononucleotide-7-^{14}C (this volume [105]) may be used.
> Tris-acetic acid buffer, 2 M, pH 8.3. Dissolve 24.2 g of Tris in 60 ml of water. Add about 17 ml of 4 N acetic acid to pH 8.3, and make up the volume to 100 ml.
> Dowex 1-X2, 200–400 mesh, cycled, acetate form
> Acetic acid, 0.1 N, 0.5 N, 1.0 N, 4.0 N
> Glycerol (12%). Dissolve 3 volumes of glycerol in 22 volumes of water.
> Protamine sulfate (salmine). Prepare a 1% solution in 12% glycerol.
> Ammonium sulfate

[1] K. Bernhauer, E. Becher, and G. Wilharm. *Arch. Biochem. Biophys.* **83**, 248 (1959).
[2] H. C. Friedmann, *J. Biol. Chem.* **243**, 2065 (1968).

Procedure

Partial Purification of the Deamidase

The purification involves three steps, namely, preparation of a glass bead extract of *P. shermanii*, removal of phosphatase and some nucleic acid by means of protamine sulfate in the presence of glycerol,[3] and fractionation and concentration of the enzyme with ammonium sulfate. All steps are carried out in the cold.

Step 1. Glass-Bead Extract.[2] A mixture of 1 part by weight of packed *P. shermanii* cells from the freezer, 3 parts of glass beads, and 1 part of water is placed in the ice-cooled chamber of a Sorvall Omni-Mixer homogenizer. It is sheared at 43% of the maximum speed for a total of 12 minutes at 2-minute intervals, with 2-minute cooling periods at 20% of the maximum setting. The mass is extracted successively with cold water corresponding to 1, $\frac{3}{4}$, and $\frac{3}{4}$ the weight of the packed bacteria. The mixture is stirred each time for 30 seconds at 25% of the maximum voltage, allowed to settle for a few minutes, and decanted from most of the glass beads. The combined extract is centrifuged for 5 minutes at 3000 *g* and decanted from any beads present. The residue and the material remaining in the mixing chamber are combined and extracted with a further $\frac{3}{4}$ volume of cold water. All extracts are combined and centrifuged for 10 minutes at 12,000 *g*. The clear yellow supernatant solution is carefully decanted, and a small volume of cloudy supernatant solution above the heavy debris is recentrifuged in clean centrifuge tubes as before. The combined clear supernatant solutions will have a volume of between 3.1 and 3.3 ml per gram of packed bacteria used. Each 22-ml aliquot of the extract now receives 3 ml of glycerol. The solution obtained contains about 14 mg of protein per milliliter, determined according to Lowry *et al.*[4] [see Vol. III (73)].

Step 2. Protamine Sulfate Treatment. The exact amount of protamine required to remove phosphatase and little if any of the enzyme is readily determined by adding increasing volumes of the protamine sulfate solution to 1-ml aliquots of the enzyme solution, followed by water to give a total volume of 1.5 ml. Increments of 50 μl protamine sulfate solution are suitable. The suspension is mixed and allowed to stand in the cold for a few minutes. The sediment is readily centrifuged down as a brownish pellet. On dilution of the supernatant solutions 120-fold with water (25 μl made up to 3 ml is convenient), it is found that the absorbance at 260 mμ and at

[3] H. C. Friedmann, *J. Biol. Chem.* **240**, 413 (1965).
[4] O. H. Lowry, N. J. Rosebrough, A. L. Farr, and R. J. Randall, *J. Biol. Chem.* **193**, 265 (1951).

280 mμ at first decreases with increasing protamine sulfate concentration, and then increases. The ratio of absorbances at 280 and at 260 mμ shows a minimum value of about 0.55 at the critical protamine sulfate concentration, but this is not very pronounced and is a far less sensitive indicator than the actual absorbance. The increase is due to a visually all but imperceptible cloudiness. The volume of protamine sulfate solution to be added to the enzyme extract should be just a trace more than that needed for minimum absorbance under the above conditions. About 0.37 ml of the protamine sulfate solution is needed per milliliter of the crude glycerol-containing bacterial extract, but on occasion less is required. No further water is added. When the protamine treatment is instituted in the absence of glycerol there is no clear-cut absorbance increase corresponding to loss of enzyme, so that a titration by means of individual enzyme assays with increasing amounts of protamine sulfate would be needed.

Step 3. Fractionation and Concentration with Ammonium Sulfate. Ammonium sulfate (30 g) is dissolved in each 100 ml of the clear supernatant solution obtained after protamine sulfate addition and centrifugation. The suspension is allowed to stand for 30 minutes, and the precipitate is discarded after the mixture has been centrifuged at 12,000 g for 10 minutes. The enzyme is salted out from the supernatant solution by dissolving 10 g of ammonium sulfate for each original portion of 30 g. The protein is centrifuged down as before after 30 minutes in the cold, and is dissolved in about 0.05 ml of cold 12% glycerol for each gram of bacteria used. The final volume is about 0.125 ml per gram of bacteria used, and the final protein concentration is between 40 and 60 mg protein per milliliter. The enzyme is stored at −22° and can be used for at least 1 month.

Incubation

A representative incubation mixture contains 70 μl of the Tris-acetic acid buffer, 50 μl of 0.25 M NMN solution (based on molar extinction coefficient[5] at 266 mμ of 4.6 × 10³) and 100 μl of enzyme containing 40 mg of protein per milliliter. The mixture is incubated at 37° for 48 hours.

Purification of the Nicotinic Acid Mononucleotide Formed

The orange mixture is diluted with about 1 ml of water and poured directly onto a column of Dowex 1, acetate form, 6 × 70 mm. A suitable container is a Pasteur pipette, with shortened tip, provided with a small pad of glass wool. Fractions are eluted at room temperature with water, followed by stepwise increased concentrations of acetic acid at a flow rate

[5] P-L Biochemicals, Inc. Circular OR-18, Milwaukee, Wisconsin, April, 1961.

of about 1.7 ml per minute. It is convenient to monitor the effluents continuously by means of a flow cell at 254 mμ. Protein and unreacted NMN are eluted in one or sometimes two sharp peaks with about 60 ml of water. When the baseline returns to the original reading, the elution is continued with 80–100 ml of 0.1 N acetic acid which removes two merging peaks, containing mainly nicotinic acid. A short treatment with 1.0 N acetic acid (about 30 ml) follows until the absorbance just begins to increase. All the nicotinic acid mononucleotide can be eluted by 1.0 N acetic acid, but this is a slow process. It is more convenient to increase the strength of the acetic acid to 4.0 N. This elutes all the product in about 50–60 ml.

The eluates obtained with 1.0 N and 4.0 N acetic acid are combined, about ⅕ volume of ethanol is added, and the mixture is evaporated *in vacuo* almost to dryness at about 35° in a rotary flash evaporator. The evaporation is repeated two or three times with small volumes of ethanol to remove the remaining acetic acid. The residue is taken to dryness to remove ethanol. The solution in 5 ml of water shows a very faint yellow color. Use of a longer column is of no advantage, since the elution takes much longer and since the product obtained is also slightly yellow. If necessary the slight color may be removed with essentially no loss of nucleotide by passage of the material through a fresh Dowex 1 column of the same dimensions as before. The column is treated with water (about 80 ml), 0.1 N acetic acid (about 20 ml), and 0.5 N acetic acid (about 80 ml). The nicotinic acid mononucleotide is then eluted in a symmetrical peak with about 110 ml of 1.0 N acetic acid. Further elution with 4.0 N acetic acid is not necessary. The eluate with 1.0 N acetic acid is freed of acetic acid *in vacuo* with the aid of ethanol as before. The solid residue is dissolved in a small volume of water and is obtained as an essentially white powder on freeze drying. The powder is stored over silica gel in the freezer.

The yield is about 70%. The procedure can be stepped up for a larger-scale preparation. The deamidase is very stable, since the amount of product is considerably greater after 48 than after 24 hours of incubation. The effect on the yield of addition of more enzyme initially or after the first 24 hours has not been investigated.

Properties

The following properties identify the product as nicotinic acid mononucleotide.

Since the compound requires much stronger acetic acid than NMN (this volume [105]) for elution from Dowex 1, it is more acidic than NMN. This is confirmed by ionophoresis in 0.5 N acetic acid on Whatman No. 1 paper at an applied voltage of 17.5 volts/cm. In 1 hour the compound and synthetic nicotinic acid mononucleotide[6,7] both move 24 mm toward the anode when applied half-way between the electrodes, while NMN migrates about 10 mm toward the cathode. Under the same conditions, free nicotinic acid and free nicotinamide migrate about 32 and 94 mm, respectively, toward the cathode. In ascending chromatography for 6 hours at room temperature on Whatman No. 1 paper in the solvent isobutyric acid–88% NH_4OH–water (66:1:33),[5] the compound and synthetic nicotinic acid mononucleotide move together with an R_f of about 0.27, while NMN is clearly separated with an R_f of about 0.41. Under the same conditions free nicotinic acid has an R_f of about 0.69, and free nicotinamide the correspondingly higher R_f of 0.86. Both on ionophoresis and on chromatography of the compound, no other UV-absorbing regions can be detected.

The absorption spectrum of the aqueous solution in the absence of cyanide has a maximum at 266 mμ, as has NMN. In the presence of 1.0 M KCN, however, a maximum, fully developed only after a few minutes,[8] appears at 315 mμ, and its height is exactly or almost exactly the same as that without cyanide at 266 mμ. This property[8,9] again distinguishes the product from NMN, which in the presence of 1.0 M KCN absorbs maximally at 325 mμ, and with an extinction 1.35 times that without cyanide at 266 mμ.[5] It also clearly distinguishes the compound from de-amido-DPN[+], which absorbs maximally in the absence of cyanide at 260 mμ, and which, in the presence of cyanide, has a maximal extinction at 315 mμ only about one-fourth that in the absence of cyanide at 260 mμ.[8]

On acid hydrolysis in 0.1 N HCl at 100° for 45 minutes, nicotinic acid and some unchanged material are found by the ionophoretic and chromatographic criteria given above. Since under these conditions of hydrolysis the glycosidic bond of nicotinic acid mononucleotide is extensively cleaved,[9] while any nicotinamide present would be only slightly deamidated,[10] it is confirmed that the compound contains nicotinic acid. In addition, the spectrum now shows an absorption maximum at about 261 mμ, expected for free nicotinic acid.[11]

The NMN deamidase can be further purified by additional protamine sulfate treatment, followed by bentonite treatment, dialysis, and fractionation on DEAE cellulose, but this is not needed to prepare the nicotinic acid mononucleotide. The existence of this enzyme, inactive toward free

[6] M. R. Atkinson and R. K. Morton, *Nature* **188,** 58 (1960).

[7] C. Wagner, *Anal. Biochem.*, **25,** 472 (1968).

[8] J. Preiss and P. Handler, *J. Biol. Chem.* **233,** 488 (1958).

[9] R. W. Wheat, *Arch. Biochem. Biophys.* **82,** 83 (1959).

[10] F. Schlenk, *Arch. Biochem.* **3,** 93 (1944).

[11] E. B. Hughes, H. H. G. Jellinek, and B. A. Ambrose, *J. Phys. Colloid Chem.* **53,** 414 (1949).

nicotinamide and DPN, makes the lengthy pyridine nucleotide cycle[12,13] seem unnecessary, and may indicate a role in the bacterial DNA ligase reaction.

[12] R. K. Gholson, *J. Vitaminol.* (*Japan*) *Suppl.* **14**, 114 (1968).
[13] A. J. Andreoli, T. Grover, R. K. Gholson, and T. S. Matney, *Biochim. Biophys. Acta* **192**, 539 (1969).

[125] Nicotinate Ribonucleotide: Benzimidazole (Adenine) Phosphoribosyltransferase (Nicotinatenucleotide: Dimethylbenzimidazole Phosphoribosyltransferase, EC 2.4.2.21)[1,2]

By JAMES A. FYFE and HERBERT C. FRIEDMANN

Nicotinate-β-R-5'-P$^+$ + base $\rightarrow$ Base-α-R-5'-P + nicotinate + H$^+$

Assay Method

Principle. The activity of the enzyme is measured by the amount of benzimidazole-2-^{14}C nucleotide formed from benzimidazole-2-^{14}C in the presence of a slight molar excess of nicotinate ribonucleotide. The product, after separation from the free base by ionophoresis on paper, is located and measured with a strip counter and attached digital integrator.

Reagents. The reaction mixture contains the following (in micromoles): nicotinate ribonucleotide, 0.5; benzimidazole-2-^{14}C, 0.32; Tris-HCl, 26, pH 8.6 (Vol. I [16], Section 15); enzyme; total volume, 50 microliters. Nicotinate ribonucleotide (Vol. VI [44]) may be made from NMN by deamidation with nitrous anhydride[3] or NOCl[4] or enzymatically (this volume [124]). It is not available commercially. Its extinction coefficient either without cyanide at 266 mμ or with cyanide at 315 mμ[5] is 4.4×10^3 M^{-1} cm^{-1}. The solution of nicotinate ribonucleotide in water is neutralized to pH 5–6 with NaOH before use. Benzimidazole-2-^{14}C may be prepared[6,7] by reaction between *o*-phenylenediamine and formic acid-^{14}C, or obtained commercially from Calbiochem or Merck Sharp and

[1] H. C. Friedmann, *J. Biol. Chem.* **240**, 413 (1965).
[2] J. A. Fyfe and H. C. Friedmann, *J. Biol. Chem.* **244**, 1659 (1969).
[3] M. R. Atkinson and R. K. Morton, *Nature* **188**, 58 (1960).
[4] C. Wagner, *Anal. Biochem.* **25**, 472 (1968).
[5] J. Preiss and P. Handler, *J. Biol. Chem.* **233**, 488 (1958).
[6] E. C. Wagner and W. H. Millett, *Org. Syn.* **19**, 12 (1939).
[7] H. C. Friedmann and D. L. Harris, *Biochem. Biophys. Res. Commun.* **8**, 164 (1962).

Dohme, Canada. Its extinction coefficient at 273 mμ in acid[8] is 7.0 $\times$ 10^3 M^{-1} cm^{-1}; for slightly different extinction values see.[9] (If the spectrophotometer is not accurately calibrated, the wavelength setting may have to be varied by about 1 mμ to obtain a maximal reading, since this peak and one at 266.5 mμ are very sharp. Hence, also, a minimum slit width should be used.) Other ^{14}C-labeled benzimidazoles such as 5,6-dichlorobenzimidazole-2-^{14}C or 5(6)-nitrobenzimidazole-2-^{14}C may be synthesized similarly[2,10,11]; 5,6-dimethylbenzimidazole-2-^{14}C may be obtained commercially from Merck Sharp and Dohme, Canada. The maximal velocity with adenine, which is also a substrate, is only about one-fifth that with benzimidazole. This is compensated by the higher specific activity of commercially available ^{14}C-labeled adenine, but one has to be aware of side reactions in relatively impure preparations.[1]

Procedure. In order to achieve efficient mixing of the small volumes employed, the solutions are deposited as far down the reagent tube (10 $\times$ 75 mm) as possible, and the tubes are centrifuged in a table top centrifuge for a few seconds before temperature equilibration and addition of enzyme. The reaction mixtures are incubated between 15 minutes and 45 minutes (depending on the number of enzyme units used, see below), usually at 37°. The tubes are sealed with parafilm before incubation. The reaction is stopped either by swirling the tubes in boiling water for 45 seconds, or by the addition of 12 μl of 2.25 N HCl, followed before ionophoresis by 12 μl of 2.25 N KOH. In the former case 25 μl is used for ionophoresis; in the latter, 50 μl. The former method is slightly less accurate because of heat activation of the enzyme just before heat inactivation; it cannot be used when adenine is the substrate, since the 7-α-D-ribofuranosyladenine 5'-phosphate formed is heat-labile[12,13] (Volume 18C [213a]). Ionophoresis is performed on Whatman No. 1 paper in Tris-borate buffer, pH 8.6[14] for 1 hour at about 18 V/cm. The sample is applied near the cathode end of the paper. After the paper is dry, the radioactivity is determined in a strip counter with attached digital integrator. Separation of free base and of nucleotide are very satisfactory, so that the maximal slit width in the particular strip counter can be used to obtain maximum counts. Whatman No. 3 MM paper offers no particular advantage over Whatman No. 1 paper for counting ^{14}C-labeled compounds. No. 3 MM paper is

[8] G. H. Beaven, E. R. Holiday, and E. A. Johnson, *Spectrochim. Acta* **4**, 338, erratum, p. 541 (1950–52).
[9] K. Hofmann, "Imidazole and Its Derivatives," Part I, p. 253. Wiley (Interscience) New York, 1953.
[10] F. Weygand, H. Simon, and H. Klebe, *Z. Naturforsch.* **9b**, 761 (1954).
[11] H. C. Friedmann, *J. Biol. Chem.* **243**, 2065 (1968).
[12] J. A. Montgomery and H. J. Thomas, *J. Am. Chem. Soc.* **87**, 5442 (1965).
[13] H. C. Friedmann and J. A. Fyfe, *J. Biol. Chem.* **244**, 1667 (1969).
[14] H. C. Friedmann and D. L. Harris, *J. Biol. Chem.* **240**, 406 (1965).

almost exactly twice as thick as No. 1 paper, but the counting efficiency on the former is only 52% of that on the latter. Thus a doubling of the amount of material is needed to give essentially the same counts.

Definition of Units and Specific Activity. A unit of enzyme is defined as that amount which catalyzes the formation of 1 micromole of 1-α-D-ribofuranosylbenzimidazole 5′-phosphate per minute at 37° under the above conditions of assay. Specific activity is expressed as units per milligram of protein. Since dithiothreitol and ethylene glycol (present in the enzyme preparation), in addition to Tris, interfere with the biuret reaction,[15] while dithiothreitol and to a very slight extent ethylene glycol interfere with the commonly used method of Lowry et al.,[16] a turbidimetric method[17] is used to measure protein concentration.

Purification Procedure

Growth of Bacteria. For preparation of the enzyme, the anaerobe *Clostridium sticklandii* (ATCC 12662) is grown on the medium, slightly modified, used by Stadtman,[18] containing per 100 ml: 2 g of Bacto-tryptone (Difco), 1 g of Bacto-yeast extract (Difco), 0.15 g of sodium formate, 0.2 g of glycerol, 0.175 g of K_2HPO_4, and 0.06 g of $Na_2S \cdot 9H_2O$. The Na_2S solution is prepared and sterilized separately just before use. Culture in 1.2 ml of medium containing the bacteria in log phase is transferred to 15 ml of sterile medium. Growth is allowed to proceed 7 hours at 30°. The bacteria are serially recultured in the same manner into 100 ml, 2250 ml, and 28 liters of medium before being harvested in a Sharples centrifuge. For the final incubation, tall flasks of 8.5-liter capacity, each containing 7 liters of sterile medium, are used. The flasks are plugged with nonabsorbing cotton. The cell paste is washed twice by suspension and recentrifugation in 0.9% NaCl and used immediately. The yield is about 2 g of packed cells per liter. All purification steps are at 0°, unless noted otherwise.

Step 1. Lysis of Bacteria and Sonication. In initial experiments bacterial extracts were obtained after grinding with acid-washed glass beads (Vol. VI [22]), but a tenaciously persistent phosphatase was present. Treatment with lysozyme, followed by very gentle sonication, was found preferable. The fresh cells from about 50 liters of culture (94 g packed cell weight) are uniformly suspended at room temperature in 940 ml of 20% sucrose, containing 0.05 M Tris, pH 8.0, and 0.001 M dithiothreitol. Addition of 18.8 ml of 0.1 M EDTA, pH 8.0, is followed by 94 mg of crystalline chicken egg white lysozyme (Worthington Biochemical Corporation) (see Vol. IX

[15] S. Zamenhof, Vol. III, p. 702.
[16] O. H. Lowry, N. J. Rosebrough, A. L. Farr, and R. J. Randall, *J. Biol. Chem.* **193,** 265 (1951).
[17] F. Heepe, H. Karte, and E. Lambrecht, *Z. Physiol. Chem.* **286,** 207 (1951).
[18] T. C. Stadtman, *J. Bacteriol.* **79,** 904 (1960).

[113]). The suspension is slowly stirred with a magnetic stirring bar. Lysis of the cells is followed by noting the decrease in the absorbance at 600 mμ of 25-μl aliquots diluted with water to 3 ml. The stirring is continued until no further decrease in this absorbance occurs (about 1 hour). The suspension is centrifuged at 3300 g for 15 minutes, and the supernatant solution is discarded after careful decantation. The "pellet" of spheroplasts is suspended in 940 ml of 20% ethylene glycol containing 0.005 M dithiothreitol at pH 7–8. The suspension is thoroughly mixed with a nonaerating stirrer (Kraft Apparatus, Inc.). The very viscous mixture is centrifuged for 2 hours at 13,200 g. The supernatant solution is carefully removed and may be stored at −15° at this stage. Sonication is necessary before the next step in order to remove viscous material. The frozen supernatant solution is rapidly melted, and sonicated at 0°. A "Sonifier" (Branson Instruments, Inc.), used for 4 seconds at a setting of 6 and tuned to 5 amperes was found suitable. This procedure is performed with 10-ml aliquots at a time in a 25-ml Nalgene bottle. The tip of the 1 cm-wide probe is immersed about 5 mm. These extracts contain no detectable phosphatase activity.

Step 2. Protamine Sulfate Treatment and pH Adjustment. To 890 ml of the sonicated solution from step 1, 89 ml of a 0.3% protamine sulfate solution in 0.01 M KH$_2$PO$_4$ are added slowly (25 minutes) with stirring. The solution is allowed to stand 10 minutes and is then adjusted to pH 5.5 by slow addition with stirring of 0.5 N H$_2$SO$_4$ in 10% ethylene glycol. The pH is lowered further to 5.05 by addition of 0.1 N H$_2$SO$_4$ in 10% ethylene glycol. The mixture is allowed to stand for 30 minutes, and centrifuged at 13,200 g for 20 minutes.

Step 3. pH Precipitation. Dithiothreitol (277 mg) is added to 950 ml of the supernatant solution from step 2 to bring its concentration to about 0.007 M. The pH is slowly adjusted to 4.45 by addition with stirring of 0.1 N H$_2$SO$_4$ in 10% ethylene glycol. After 30 minutes the solution is centrifuged at 13,200 g for 20 minutes. The supernatant solution is discarded and the precipitate is dissolved in 125 ml of 0.005 M potassium phosphate buffer at pH 6.6, containing 10% ethylene glycol and 0.005 M dithiothreitol. The solution is stored at −15°.

Step 4. Ammonium Sulfate Precipitation. The mixture from the previous step is rapidly dissolved and diluted in the same buffer to give 0.068 unit per milliliter. Dithiothreitol is added to 0.01 M, and EDTA to 0.001 M. The volume is now 202 ml. The pH is slowly adjusted to 5.5 with 0.5 N H$_2$SO$_4$ in 10% ethylene glycol, and 11.1 g ammonium sulfate is added slowly and with stirring. The pH of the solution is adjusted to 5.3, and stirring is continued for 45 minutes. The suspension is centrifuged for 20 minutes at 13,200 g, the supernatant solution is collected, and the precipitate is discarded. A further 58.2 g of ammonium sulfate is added as above. The supernatant solution is discarded after centrifugation. The precipitate is stirred for 2 hours with 60 ml of the buffer used in step 3. The remaining precipitate is removed by centrifugation and extracted with 10 ml of the same buffer for 30 minutes. The combined supernatant solutions are desalted on a column of Sephadex G 50 coarse, equilibrated with the above buffer. The column volume used is 4.3 times the volume of the applied sample, and the flow rate 9.8 ml hr^{-1} cm^{-2}. In addition to ammonium sulfate, much yellow material is removed in this step.

Step 5. Hydroxylapatite Column Chromatography. The desalted enzyme solution is poured onto a hydroxylapatite column (2.5 × 15 cm) which has been equilibrated with 0.01 M potassium phosphate buffer, pH 6.6, in 10% ethylene glycol. The column is eluted stepwise with increasing concentrations of potassium phosphate at pH 6.6. Soon after collection, each fraction in the fraction collector receives a small aliquot of a dithiothreitol solution to a final concentration of 0.005 M. Considerable protein but no enzyme is eluted by 0.01 M potassium phosphate buffer (160 ml). No protein is eluted when the strength of the buffer is increased to 0.03 M (60 ml). On increasing the buffer concentration to 0.05 M, a small amount of protein is eluted, followed after about 30 ml by a decrease in the protein peak but elution of material with enzymatic activity. Enzyme and little protein continue to be eluted by about 60 ml of the 0.05 M buffer. More enzyme is obtained when the strength of the buffer is increased to 0.07 M (about 200 ml), but the last 110 ml of this eluate contains a protein peak and relatively little of the enzyme, and is hence discarded. Very little more enzyme, but much protein, is eluted by 0.2 M buffer. The last 60 ml of the 0.05 M buffer eluate and the first 90 ml of the 0.07 M buffer eluate are combined. The hydroxylapatite was obtained from the Clarkson Chemical Company (Hypatite-C). It is not known whether other preparations of this material would show exactly comparable properties.

In order to concentrate the enzyme, ammonium sulfate is added to near saturation (90 g) at pH 6.5. After allowing 1 hour for equilibration, the suspension is centrifuged for 30 minutes at 12,100 g. The supernatant solution is discarded, and the precipitate is dissolved in 2 ml of 0.01 M potassium phosphate buffer, pH 7.6, containing 0.005 M dithiothreitol and 10% ethylene glycol.

Step 6. Sephadex Gel Filtration. To the sample from the previous step, 0.12 ml of 2 M sucrose is added. The mixture is carefully layered on a column of Sephadex, G 75 superfine (2.5 × 38 cm) which has been equilibrated with 0.01 M potassium phosphate buffer, pH 7.6 in 10% ethylene glycol, containing 0.005 M dithiothreitol. Previously, fines are removed from the G 75 by suspension and decantation, and a 1.5-cm layer of Sepha-

PURIFICATION OF N_aMN: BI (ADENINE) PHOSPHORIBOSYLTRANSFERASE

Stage of purification	Volume (ml)	Protein (mg/ml)	Total activity (units)	Specific activity (units/mg)	Purification (−fold)	Recovery (%)	NaMN/ NMN[a]
Step 1	890	8.2	25.1	0.0034	1	100	15
Step 2	950	1.25	18.0	0.015	4	71	42
Step 3	125	2.6	13.8	0.042	12	55	33
Step 4	110	1.8	6.5	0.033	10	26	43
Step 5[b]	2	2.2	2.2	0.5	140	8.6	600
Step 6	1.5	0.4	1.4	2.3	680	5.6	860

[a] These are the ratios of velocities of the enzyme preparation with N_aMN or NMN as the ribose 5-phosphate donor.

[b] During step 5, 25% of the enzyme was lost from a broken centrifuge tube. Results in steps 5 and 6 reflect this loss.

dex G 25 medium is placed at the bottom of the column to prevent clogging of the bed support. Elution is performed with the equilibrating buffer at a flow rate of 1.7 ml hour^{-1} cm^{-2}, and 2-ml fractions are collected. A pressure head of 15 cm should never be exceeded during equilibration or elution. Enzyme activity first appears after about 80 ml has been eluted. Fractions with total activity of 0.17 unit or more are combined and concentrated with ammonium sulfate at pH 6.5. The precipitate is drained well and dissolved in 1.5 ml of 0.01 M potassium phosphate buffer, pH 6.8, containing 0.005 M dithiothreitol and 10% ethylene glycol. The enzyme is stored in 0.1-ml aliquots at −15°. In this form the preparation is stable for several months.

A summary of the purification procedure is given in the table. Step 4 does not increase the specific activity, but some proteins which interfere with later steps are removed. The preparation is not pure, since disc electrophoresis on polyacrylamide gel[19] shows two heavy and six faint bands of protein.

Properties

Stability. The enzyme is relatively unstable in aqueous solution. After 32 days at −15°, a crude extract had lost 94% of its activity. Considerable protection during purification and storage is afforded by glycerol, or preferably ethylene glycol, and by dithiothreitol (see above for concentrations used).

pH Optimum. The enzyme exhibits optimum activity at pH values between 8.6 and 9.2, both with benzimidazole and with adenine as one substrate and nicotinate ribonucleotide as the other substrate.

[19] B. J. Davis, *Ann. N.Y. Acad. Sci.* **121**, 404 (1964).

Specificity. RIBOSE PHOSPHATE DONOR. The enzyme is highly specific for a ribose phosphate donor which contains nicotinic acid, not nicotinamide. Thus the activity with NMN decreases drastically during purification (see summary of purification procedure) and DPN$^+$ is not active. Both nicotinic acid ribonucleotide, and to a much lesser extent (about 10%) deamido-DPN$^+$ (nicotinic acid adenine dinucleotide) are active. Nicotinic acid ribonucleoside, nicotinamide ribonucleoside, PPRP, and ribose 5-phosphate are inactive. Most, if not all, of the activity with NMN appears to be due to generation of nicotinate ribonucleotide by a separate enzyme which causes deamidation. It is not known whether the slight residual activity with NMN in the most purified preparation available is due to traces of nicotinate ribonucleotide formed or to direct reaction of the transferase with NMN.

BASE. The specificity for the free base is much broader than that for the ribose phosphate donor. The following radioactive benzimidazoles have been found to react: benzimidazole-2-^{14}C, 5,6-dimethylbenzimidazole-2-^{14}C, 5,6-dichlorobenzimidazole-2-^{14}C, 5(6)-nitrobenzimidazole-2-^{14}C. The analogous enzyme from *Propionibacterium shermanii*[1] reacts with benzimidazole, 5(6)-methylbenzimidazole, 5,6-dimethylbenzimidazole, 5(6)-trifluoromethylbenzimidazole, and 5(6)-aminobenzimidazole. No accurate figures are available on the relative rates with the various benzimidazoles due to substrate inhibition at high base concentrations. Both 5,6-dichlorobenzimidazole-2-^{14}C and 5(6)-nitrobenzimidazole react at least twice as rapidly as benzimidazole-2-^{14}C. Adenine reacts with the enzyme from *C. sticklandii*. It appears not to react with the enzyme from *P. shermanii*, although this has to be confirmed with a more active preparation and with acid rather than heat inactivation after the reaction.

Nature of the Products. The products both with benzimidazole[14] and with adenine[13] are the α-glycosidic 5′-ribonucleotides of the respective base. In the case of adenine, the glycosidic bond is at N_7 (Volume 18C [213a]).

Inhibitors. Not many substances have been tested. The enzyme is rapidly and completely inhibited by the bifunctional mercurial 3,6-bis-(acetatomercurimethyl)dioxane[20] at 1 mM concentration. A 3-fold excess of cysteine or of dithiothreitol rapidly and almost completely reverses this inhibition. A number of bases, such as benzotriazole, benzothiazole, quinoxaline, 6,7-dimethylquinoxaline, indole, indoline, perimidine (all available from Aldrich Chemical Company) decrease the formation of the benzimidazole nucleotide. Accurate figures on a molar basis are not available for most of these, since most are rather insoluble under the conditions used,

[20] L. Eldjarn and E. Jellum, *Acta Chem. Scand.* **17**, 2610 (1963).

and a few were not completely pure by inspection under ultraviolet light after ionophoresis in 0.5 M acetic acid. In most cases this lowering of benzimidazole incorporation into nucleotide may indicate that these compounds act not as inhibitors, but as substrates. This would be in accord with the known incorporation of the first three above substances into corrinoids[21] in "guided biosynthesis." Thus both adenine and 2-methyladenine strongly lessen benzimidazole nucleotide formation, while purine in a concentration equimolar to that of benzimidazole decreases it by about 30%. On the other hand, an equimolar concentration of 2-methylbenzimidazole, which is not incorporated into corrinoid by *Escherichia coli*[22] and which is not a substrate for the *P. shermanii* enzyme,[1] lowers benzimidazole utilization about 30%, and this may represent a true inhibition of the enzyme.

Kinetic Data. The double reciprocal plots for the reaction with adenine are consistent with a single displacement[23] or sequential[24] mechanism. The K_m values for adenine and for nicotinate ribonucleotide in the presence of excess of the other are $1 \times 10^{-5} M$ and $3 \times 10^{-4} M$ respectively at 30°. The K_m values for benzimidazole and for nicotinate ribonucleotide similarly are $5 \times 10^{-4} M$ and $7 \times 10^{-4} M$, respectively. The maximal velocity for benzimidazole is approximately five times that for adenine.

[21] D. Perlman and J. M. Barrett, *Can. J Microbiol* **4**, 9 (1958).
[22] H. Dellweg, E. Becher, and K. Bernhauer, *Biochem. Z.* **327**, 422 (1956).
[23] D. E. Koshland, Jr., *in* "The Enzymes" (P. D. Boyer, H. Lardy, and K. Myrbäck, eds.), 2nd ed., Vol. 1, p. 305. Academic Press, New York, 1959.
[24] W. W. Cleland, *Biochim. Biophys. Acta* **67**, 104 (1963).

[126] Enzymatic Phosphorolysis of Pyridine Ribonucleosides by Rat Liver Nucleoside Phosphorylase

By YASUTOMI NISHIZUKA

Nicotinate ribonucleoside + P_i → nicotinate + ribose 1-phosphate
Nicotinamide ribonucleoside + P_i → nicotinamide + ribose 1-phosphate

The enzyme nucleoside phosphorylase catalyzes the phosphorolysis of nicotinate ribonucleoside as well as nicotinamide ribonucleoside. It appears to react also with ribo- and deoxyribonucleosides of adenine, guanine, and hypoxanthine, but not with pyrimidine nucleosides. The broad substrate specificity of this enzyme has been suggested by Rowen and Kornberg.[1] With purine nucleosides as substrates the equilibrium greatly favors synthesis of the nucleosides.[2] On the other hand, the phos-

[1] J. W. Rowen and A. Kornberg, *J. Biol. Chem.* **193**, 497 (1951).

phorolytic cleavage of pyridine nucleosides proceeds to near completion, with the equilibrium constant of 10^3 at pH 7.[3] This is due to the fact that the pyridine riboside linkage is a high-energy bond, and the free energy of hydrolysis has been calculated to be 8.3 kcal.[3] Methods have been described for preparing "nicotinamide nucleoside phosphorylase" from beef liver acetone powder by Kornberg[4] and "purine nucleoside phosphorylase" from beef spleen by Price, Otey, and Plesner.[5] Purine nucleoside phosphorylase has been purified also from beef liver,[6] erythrocytes,[7] fish muscle,[8] and yeast.[9] The yeast enzyme reacts with purine and pyridine nucleosides, but not with pyrimidine nucleosides.[9] Pyrimidine nucleosides are attacked by a different enzyme, pyrimidine nucleoside phosphorylase.[10–12] Although the identity of enzymes toward purine and pyridine nucleosides have not yet been fully settled, methods to assay, partial purification and substrate specificity of rat liver nucleoside phosphorylase are briefly described in this article.

Assay Method A

Principle. Enzyme activity may be assayed by measuring the formation of radioactive nicotinate or nicotinamide from the respective ribonucleoside-^{14}C as substrate in the presence of phosphate. The product is isolated by a small Dowex 1-formate column, and the radioactivity is determined by a Geiger-Müller gas-flow counter.

Reagents

Nicotinate-7-^{14}C ribonucleoside, 0.16 mM (2–10 mCi/millimole)[13]
Potassium phosphate buffer, 2 mM, pH 8.0
Enzyme solution

Procedure. The incubation mixture contains 0.05 ml of nicotinate-7-^{14}C ribonucleoside, 0.05 ml of potassium phosphate buffer, enzyme, and water in a total volume of 0.4 ml. The reaction is started by the addition of en-

---

[2] H. M. Kalckar, *J. Biol. Chem.* **167,** 477 (1947).
[3] L. J. Zatman, N. O. Kaplan, and S. P. Colowick, *J. Biol. Chem.* **200,** 197 (1953).
[4] See Vol. II [65].
[5] See Vol. II [64].
[6] E. D. Korn and J. M. Buchanan, *J. Biol. Chem.* **217,** 187 (1955).
[7] K. K. Tsuboi and P. B. Hudson, *J. Biol. Chem.* **224,** 879 (1957).
[8] H. L. A. Tarr, *Federation Proc.* **14,** 291 (1955).
[9] L. A. Heppel and R. J. Hilmoe, *J. Biol. Chem.* **198,** 683 (1952).
[10] L. A. Manson and J. O. Lampen, *J. Biol. Chem.* **193,** 539 (1951).
[11] M. Friedkin and D. Roberts, *J. Biol. Chem.* **207,** 245 (1954).
[12] E. S. Canellakis, *J. Biol. Chem.* **227,** 329 (1957).
[13] Nicotinamide-^{14}C ribonucleoside can be used as a substrate. The radioactive substrates may be prepared by the method of Ueda, Yamamura, and Nishizuka (see this volume [106]).

zyme. After incubation for 30 minutes at 37°, the reaction is stopped by heating for 1 minute in a boiling water bath. The mixture is cooled to 0–4° in an ice bath, and centrifuged to remove the denatured protein. A 0.2-ml aliquot of the supernatant is put on a Dowex 1-formate (X8, 200–400 mesh, 0.5–1.0 cm) column. After nicotinate ribonucleoside has been washed out completely with 10 ml of water, nicotinic acid produced is eluted with 4 ml of 1 N formic acid. Under these conditions more than 98% of nicotinic acid is recovered without contamination of the substrate. A 0.5-ml aliquot of the nicotinic acid fraction is determined for radioactivity.

Definition of Unit and Specific Activity. One unit of enzyme is defined as that amount which produces 1 micromole of nicotinic acid per minute under the assay conditions. Specific activity is expressed as units per milligram protein. Protein is determined by the method of Lowry et al.[14]

Application to Crude Extract. This method may be applied to crude extract. No enzyme to catalyze the hydrolytic cleavage of riboside linkage has been found thus far in animal tissues except fish muscle nucleoside hydrolase.[15]

Assay Method B

Principle. Enzyme activity may be assayed alternatively by measuring the conversion of inorganic phosphate-^{32}P to ribose 1-phosphate-^{32}P (or deoxyribose 1-phosphate-^{32}P) in the presence of nicotinate ribonucleoside or nicotinamide ribonucleoside. The radioactive organic phosphate is determined after removing a large excess of inorganic phosphate-^{32}P as a triethylamine-phosphomolybdate complex by a slight modification of the method of Sugino and Miyoshi.[16]

Reagents

Nicotinate ribonucleoside,[17] 0.16 mM
Potassium phosphate-^{32}P buffer, 2 mM, pH 8.0 (2–10 μCi/ml)

---

[14] O. H. Lowry, N. J. Rosebrough, A. L. Farr, and R. J. Randall, *J. Biol. Chem.* **193**, 265 (1951).

[15] H. L. A. Tarr, *Biochem. J.* **59**, 386 (1955).

[16] Triethylamine is shown to be an effective and selective precipitant of phosphomolybdic acid. Under optimal conditions, inorganic orthophosphate at concentrations as low as 5×10^{-6} M can be quantitatively precipitated. After the removal of inorganic orthophosphate, various phosphate esters and anhydrides, such as nucleoside mono-, di-, and triphosphates, ribose 1-phosphate, and inorganic pyrophosphate, remain in the supernatant solution. Y. Sugino and Y. Miyoshi, *J. Biol. Chem.* **239**, 2360 (1964).

[17] The substrate may be prepared by the method of Honjo and Nishizuka (see this volume [114]). Nicotinamide ribonucleoside may be prepared by the method of Kaplan and Stolzenbach (see Vol. III [129]).

Enzyme solution
Perchloric acid, 1 N
Ammonium molybdate, 10% (w/v)
Triethylamine HCl, 0.2 M
TEA mixture. Just before use, mix 4 volumes of perchloric acid, 2 volumes of ammonium molybdate, and 1 volume of triethylamine HCl.
Monobasic potassium phosphate, 0.01 M

Procedure. The incubation is carried out as Assay Method A except that nonradioactive nicotinate ribonucleoside and phosphate buffer-^{32}P are used. The reaction is stopped by the addition of 0.5 ml[18] of TEA mixture. After 3–5 minutes at room temperature 0.1 ml of monobasic potassium phosphate is added as carrier to precipitate inorganic phosphate-^{32}P completely. A yellow precipitate is immediately produced which is a complex of triethylamine and phosphomolybdate. After 5 minutes, when the formation of the complex is completed, the mixture is centrifuged for 10 minutes at 4000 g. An 0.5-ml aliquot of the supernatant is removed, placed on an aluminum planchet, dried, and counted.

Definition of Unit. One unit of enzyme is defined as that amount which produces 1 micromole of organic phosphate-^{32}P, and is identical to that described in Assay Method A.

Application. This method may be applied to an incubation mixture with a nonradioactive nucleoside, and provides a rapid and sensitive assay for the activity. However, endogeneous phosphate in enzyme solutions should be removed by dialysis or by Sephadex gel filtration before assay.

Purification Procedure

The enzyme is purified and activity followed by Assay Method A.

Step 1. Crude Extract. Albino rats are decapitated, and livers are immediately removed and chilled. Approximately 25 g of the livers are homogenized for 1 minute with 4 volumes of 0.14 M KCl with a Potter-Elvehjem type Teflon-glass homogenizer. All manipulations are carried out at 0–4° unless otherwise noted. The homogenate is centrifuged for 10 minutes at 12,000 g, and the residue is discarded.

Step 2. Protamine Sulfate. To the supernatant solution (100 ml), an equal volume of a 0.4% protamine sulfate solution, pH 7.4, is added with stirring and the resulting precipitate is removed by centrifugation.

Step 3. First Ammonium Sulfate Fractionation. To the supernatant

---

[18] A 0.5-ml aliquot of TEA mixture may precipitate approximately 5 micromoles of inorganic phosphate.

solution (190 ml), an equal volume of a saturated ammonium sulfate solution is added slowly with stirring. After 15 minutes, the mixture is centrifuged for 10 minutes at 12,000 g, and the precipitate is discarded. To the supernatant solution, an additional volume (190 ml) of the ammonium sulfate solution is added. After 15 minutes, the precipitate is collected by centrifugation and dissolved in 50 ml of 0.01 M Tris-acetate buffer, pH 7.5.

Step 4. Heat Treatment. The ammonium sulfate fraction is heated rapidly to 50° and is maintained for 3 minutes at this temperature, and then cooled to 0–4° in an ice bath. The denatured protein is removed by centrifugation.

Step 5. Second Ammonium Sulfate Fractionation. To the supernatant solution (48 ml), an equal volume of a saturated ammonium sulfate solution is added slowly. The precipitate is removed by centrifugation. To the supernatant is added an additional 40 ml of the ammonium sulfate solution, and the mixture is centrifuged. The precipitate is dissolved in 24 ml of 0.025 M Tris-acetate buffer, pH 7.5, and dialyzed for 12 hours against 5 liters of 0.01 M of the same buffer.

Step 6. Calcium Phosphate Gel Fractionation. The enzyme solution is diluted with an equal volume of water, and 20 ml of a calcium phosphate gel suspension[19] (20 mg dry weight per milliliter) are added with stirring. After 10 minutes, the gel is collected by centrifugation and washed twice, each time with 40 ml of water. Then the enzyme is eluted from the gel twice, each time with 12 ml of 0.08 M K_2HPO_4. The combined eluates are adjusted to pH 7.5 with 2 N HCl and dialyzed overnight against 5 liters of 0.01 M Tris-acetate buffer, pH 7.5. The enzyme is purified about 100-fold with an overall yield of more than 100% (Table I). A large increase of total activity is observed with the first ammonium sulfate fractionation, probably because of the removal of some unknown inhibitors. The enzyme

TABLE I
Summary of Enzyme Purification

Fraction	Volume (ml)	Units/ml $\times 10^3$	Protein (mg/ml)	Specific activity	Yield (%)
1. Crude extract	100	0.6	56.3	0.000011	100
2. Protamine sulfate	190	0.3	11.3	0.000026	95
3. First $(NH_4)_2SO_4$	50	1.9	7.75	0.00024	158
4. Heat treatment	48	1.9	5.36	0.00035	152
5. Second $(NH_4)_2SO_4$	24	3.1	6.15	0.00050	124
6. Calcium phosphate gel	24	2.7	2.31	0.00116	108

[19] See Vol. I [11].

is stable and can be stored for several weeks at $-20°$ without significant loss of activity.

Properties

The enzyme catalyzes a stoichiometric conversion of nicotinate ribonucleoside or nicotinamide ribonucleoside to the respective pyridine derivative and ribose 1-phosphate by phosphorolytic but not by hydrolytic cleavage. The reaction is absolutely dependent on the presence of phosphate. The crude as well as the purified preparations also catalyze the phosphorolysis of ribo- and deoxyribonucleosides of adenine, guanine, and hypoxanthine. Pyrimidine nucleosides are inert as substrates. Relative ratios of activities toward these substrates are given in Table II. Several

TABLE II
SUBSTRATE SPECIFICITY OF RAT LIVER NUCLEOSIDE PHOSPHORYLASE

Nucleoside	Relative activity[a] (%)
Nicotinate ribonucleoside	100
Nicotinamide ribonucleoside	114
Adenosine	105
Guanosine	130
Inosine	150
Deoxyadenosine	31[b]
Deoxyguanosine	37[b]

[a] The activities are measured by Assay Method B.
[b] The number appears to be a minimum value, since the product, deoxyribose 1-phosphate, is extremely labile and decomposed partly under the assay conditions.

experiments have been performed to examine whether or not a single enzyme is responsible for the phosphorolysis of these nucleosides, and the results are compatible with the supposition that the reactions are catalyzed by a single enzyme protein, since the ratios of activities toward several nucleosides mentioned above remain constant throughout the entire purification procedure. Ribo- and deoxyribonucleosides of 8-azaguanine[20] and of 4-amino-6-imidazolecarboxamide[21] are reported to act as substrates for the nucleoside phosphorylase.

The pH optimum of the enzyme is approximately 8. The K_m value for nicotinate ribonucleoside is $4 \times 10^{-4} M$, and is considerably lower than

[20] M. Friedkin, *J. Biol. Chem.* **209**, 295 (1954).
[21] E. D. Korn, F. C. Charalampous, and J. M. Buchanan, *J. Am. Chem. Soc.* **75**, 3611 (1953).

that for nicotinamide ribonucleoside $(1.1 \times 10^{-3} M^4)$. The constant for phosphate is $2.6 \times 10^{-5} M$, as compared with a value of $2.8 \times 10^{-4} M$ obtained by Kornberg.[4] Other properties of the enzyme obtained from beef liver have been described elsewhere.[4]

[127] The Reduction of Nicotinamide N-Oxide by Xanthine Oxidase

By KEITH MURRAY

Nicotinamide N-oxide + DPNH + H$^+$ → nicotinamide + DPN$^+$ + H$_2$O

Nicotinamide N-oxide + xanthine → nicotinamide + uric acid

The following assays and purification were developed specifically for the reduction of nicotinamide N-oxide to nicotinamide, rather than for xanthine oxidase. In Vol. XII [1], Roussos describes a method for purifying xanthine oxidase and gives references to the numerous other methods that are available for this purpose.

Assay Method

Principle. Xanthine oxidase (xanthine:oxygen reductase, EC 1.2.3.2) catalyzes the reduction of nicotinamide N-oxide by a number of different reducing agents.[1,2] The nicotinamide produced is assayed with cyanogen bromide as described by Mueller and Fox.[3] When nicotinamide reacts with cyanogen bromide in the presence of ammonia, a bright yellow color is produced. The absorbance of the color at 398 mμ is proportional to the concentration of nicotinamide. When purer enzyme fractions serve as catalyst and DPNH is used as the reducing agent, the disappearance of DPNH can be followed spectrophotometrically at 340 mμ. The formation of uric acid from xanthine cannot be conveniently monitored at 295 mμ because of the interfering absorption of nicotinamide N-oxide.

Reagents

Nicotinamide N-oxide, 15 mg/ml. The solution can be stored frozen, but the precipitate of nicotinamide N-oxide which forms on cooling should be redissolved before use. Nicotinamide N-oxide is prepared according to the method of Taylor and Crovetti.[4] A mixture of

[1] K. N. Murray and S. Chaykin, *J. Biol. Chem.* **241**, 2029 (1966).

[2] K. N. Murray and S. Chaykin, *J. Biol. Chem.* **241**, 3468 (1966).

[3] A. Mueller and S. H. Fox, *J. Am. Pharm. Assoc.* **40**, 513 (1951). Reproduced with permission of the copyright owner.

[4] E. C. Taylor and A. J. Crovetti, *J. Org. Chem.* **19**, 1633 (1954).

5.0 g (0.041 mole) of nicotinamide, 8 ml (0.096 mole) of 30% hydrogen peroxide, and 50 ml of glacial acetic acid is refluxed on a steam bath for 4 hours. The mixture is then diluted with 100 ml of water and evaporated to dryness under reduced pressure. The residue is dissolved in 20 ml of boiling water, the solution is filtered, and the filtrate is diluted with 5 ml of absolute ethanol. Cooling causes colorless crystals of nicotinamide N-oxide, m.p. 291–293°, to precipitate.

Potassium phosphate buffer, 0.4 M, pH 6.5, which is 0.14 M in 2-mercaptoethanol

DPNH, 0.013 M

Tris-HCl buffer, 0.4 M, pH 8.0, which is 0.14 M in 2-mercaptoethanol. The mercaptoethanol is added immediately before use in order to minimize its oxidation.

Xanthine, 0.013 M

Enzyme solution to be assayed

Trichloroacetic acid, 50% (w/v)

NaOH, 2 N

Cyanogen Bromide 10% (w/v).[3] To hasten solution, the mixture is stirred on a magnetic stirrer in a tightly stoppered bottle (glass stopper). The CNBr solution may be kept in the hood at room temperature when in daily use. No deterioration of this solution has been found at the end of one month. The crystalline CNBr is stored in a refrigerator. All procedures using CNBr, including the rinsing of glassware, must be performed in an efficient hood.

Buffered Ammonia Reagent.[3] Dissolve 87.0 g of dipotassium phosphate (K_2HPO_4), reagent grade, and 107.0 g of ammonium chloride (NH_4Cl), reagent grade, in about 500 ml of distilled water. Then add 6.7 ml of concentrated ammonium hydroxide (28%), reagent grade, and dilute to 1 liter with water. Filter, if a sediment forms on standing.

Procedure. The reaction mixture contains 0.25 ml of phosphate buffer, 0.05 ml of DPNH, 0.05 ml of nicotinamide N-oxide, enzyme, and water to a final volume of 0.5 ml; or the DPNH is replaced by 0.05 ml of xanthine and the phosphate buffer is replaced by 0.25 ml of Tris buffer. The size of the reaction mixture is doubled when a larger sample is needed for the nicotinamide assay. Either the enzyme or substrate is added last, and the reaction mixture is incubated, with shaking, at 50° for 10 minutes. A control tube, lacking nicotinamide N-oxide, is run with each reaction tube to correct for formation of nicotinamide other than that from nicotinamide N-oxide. The correction is particularly important when a crude homogenate is assayed, since, in addition to the endogenous material, the hydrolysis of DPN by DPN nucleosidase (EC 3.2.2.5) produces considerable amounts of nicotinamide.

The reaction is stopped with 0.05 ml of trichloroacetic acid and the precipitated protein is removed by centrifugation. Neutralize a 0.25-ml portion of the deproteinized solution with $2.0 N$ NaOH, dilute to 1 ml with deionized water, and add 1.5 ml of buffered ammonia reagent. At zero time, add 2.5 ml of cyanogen bromide solution and on the eleventh minute, read the optical density at 398 mμ. The method is equally applicable to nicotinic acid when the readings are made at 408 mμ after 2 to 2.5 minutes. The wavelength and time of maximum intensity can be used to distinguish between nicotinamide and nicotinic acid.[3]

Definition of Unit and of Specific Activity. One unit of activity is defined as that amount of enzyme which catalyzes the formation of 1.0 μg of nicotinamide in 10 minutes at 50°. An optical density increment of 0.063 at 398 mμ represents 1 μg of nicotinamide over a range of 1 to 30 μg. Specific activity is the number of units per milligram of protein under these conditions. Protein is determined by the biuret method[5] or the $A_{280} : A_{260}$ method.[6] Because 2-mercaptoethanol interferes with the biuret method, the protein is first precipitated from mercaptoethanol-containing solutions with trichloroacetic acid. The precipitated protein is recovered by centrifugation and redissolved in 3% NaOH (w/v) before being subjected to the biuret determination.

Purification Procedure

Hog liver xanthine oxidase has also been prepared by the method of Murashige, McDaniel, and Chaykin[7] and purified further by subjecting it to the diethylaminoethyl cellulose chromatography procedure described below.

Purified milk xanthine oxidase (Worthington) also reduces nicotinamide N-oxide with xanthine or DPNH as electron donors. However, it should be mentioned that the milk xanthine oxidase and the liver xanthine oxidase are not identical. They differ in both pH optima (see below) and, depending on the reductant, in the relative amounts of nicotinamide N-oxide reduced.

Enzyme Preparation. All enzyme fractions are generally kept at 0° during the purification procedure. However, chromatography on Sephadex and on DEAE-cellulose can be performed at 25° since this procedure does not appear to result in a significant loss in activity. All buffers contained 0.14 M 2-mercaptoethanol.

[5] A. G. Gornall, C. J. Bardawill, and M. M. David, *J. Biol. Chem.* **177,** 751 (1949).
[6] O. Warburg and W. Christian, *Biochem. Z.* **310,** 384 (1941).
[7] K. Murashige, D. McDaniel, and S. Chaykin, *Biochim. Biophys. Acta* **118,** 556 (1965).

Step 1. Crude Extract. Fresh hog liver is obtained at the local abattoir and stored frozen until used. The frozen hog liver (170 g) is diced and then homogenized in a Waring Blendor for 1 minute with 500 ml of 0.1 M Tris buffer, pH 8. The resulting homogenate is centrifuged for 30 minutes at 27,000 g. The supernatant solution is decanted and strained twice through a single layer of cheesecloth to remove floating fat particles.

Step 2. pH Treatment. The defatted supernatant is titrated to pH 5.0 with 1.0 N hydrochloric acid. After centrifugation at 27,000 g for 10 minutes, the precipitated protein is discarded. The supernatant solution is then titrated back to pH 8.0 with 1.0 N NaOH.

Step 3. Heat Treatment. Ten-milliliter portions of the pH-treated supernatant solution are heated to 60° in 50-ml Erlenmeyer flasks for 2.5–3 minutes in a shaking water bath. The denatured protein is removed by centrifugation, and the supernatant is subjected to ammonium sulfate fractionation.

Step 4. Ammonium Sulfate Fractionation. The protein fraction which precipitates between 0.40 and 0.55 saturation with ammonium sulfate contains most of the activity. It is collected by centrifugation and redissolved in a minimal amount of 0.01 M Tris buffer, pH 8. The latter solution is freed of ammonium sulfate by passing it through a Sephadex G-25 column which has been equilibrated with 0.01 M Tris, pH 8.

Step 5. DEAE-Cellulose Chromatography. The Sephadex eluate is applied to a DEAE-cellulose–chloride column (3.3 × 28 cm), which has been equilibrated with 0.01 M Tris, pH 8. After washing the protein onto the column with Tris buffer, the protein is eluted with a linear salt gradient, 0–0.25 M KCl in the 0.01 M Tris buffer in a total volume of 600 ml; 7-ml fractions are collected at 3-minute intervals. The activity is eluted at a KCl concentration of about 0.12 M. When the gradient is finished, the remaining protein is eluted with 1.0 M KCl in Tris buffer. The column is reequilibrated with Tris buffer and is then ready for reuse. The dilute solutions of the enzyme in the DEAE-cellulose column fractions are not as stable during storage at 4° as are concentrated solutions. Therefore, the column fractions containing the activity are combined, and the protein is precipitated by adding ammonium sulfate to 0.80 saturation. The precipitate is collected by centrifugation, redissolved in a minimal volume of 0.01 M Tris, pH 8, and freed of ammonium sulfate by chromatography on a Sephadex G 25 column equilibrated with the same buffer. This concentrated DEAE-cellulose fraction can be stored at 4°.

Step 6. CM-Cellulose Chromatography. In preparation for this step, 0.01 M citrate, pH 5.5, is substituted for the 0.01 M Tris buffer when the DEAE-cellulose fractions, which have been concentrated with ammonium sulfate, are redissolved and desalted.

The concentrated DEAE-cellulose fraction is applied to a carboxy-methylcellulose-KCl column (2.4 × 14 cm) which has been equilibrated with the citrate buffer. The pH of the buffer with which the column is equilibrated is critical because the activity is not adsorbed at pH 5.8 or higher. The activity is eluted with a linear salt gradient, 0 to 0.25 M KCl in 0.01 M citrate buffer, pH 5.5; total volume of eluent, 200 ml; 3-ml fractions are collected at the rate of 1 ml per minute. The column is washed with 1.0 M KCl, 0.01 M citrate (pH 5.5), and 0.01 M citrate buffer, pH 5.5. The active fractions, which are generally eluted at a KCl concentration of 0.1 M, are pooled and concentrated in the same manner as the DEAE-cellulose fractions. The precipitated protein is redissolved in 0.01 M Tris, pH 8, and freed of salt as described previously. The $A_{280}:A_{450}$ ratio of this fraction is approximately 12.

A typical purification is shown in the table.

PURIFICATION OF XANTHINE OXIDASE FROM HOG LIVER

Step and fraction	Volume (ml)	DPNH activity (units/ml)	DPNH total activity (units × 10⁻⁴)	DPNH specific activity (units/mg)
1. Crude extract[a]	470	107	5.02	1.8
2. pH 5 supernatant[a]	444	48	2.11	2.3
3. 60° heat-treated[a]	397	46	1.84	4.4
4. 0.40 to 0.55 (NH₄)₂SO₄	45	372	1.67	12.0
5. DEAE-cellulose eluate (after (NH₄)₂SO₄ concentration)	17.5	475	0.83	52.2
6. CM-cellulose eluate (after (NH₄)₂SO₄ concentration)	4.8	977	0.47	276.7

[a] Assay samples were chromatographed on Sephadex G 25 to remove endogenous substrates; the values were corrected for dilution.

Properties

Stability. The enzyme is not stable to freezing. Storage of the concentrated DEAE-cellulose fraction for 2 weeks at 4° results in only a 50% decrease in activity, whereas storage at −15° for the same period of time results in complete loss of activity.

Effect of pH. The pH optimum of the DPNH-dependent activity falls in the range 6.4–6.8.

When xanthine is substituted for DPNH, the pH optimum is 8.5–9.0. Nicotinamide *N*-oxide reduction by milk xanthine oxidase also has two distinct pH optima; 7.0–7.5 for DPNH and 9.0–9.5 for xanthine.[2]

Substrate Specificity and Affinity. Nicotinic acid *N*-oxide is as effective a substrate as nicotinamide *N*-oxide. Neither the *N*-oxide of nicotinamide nor that of nicotinic acid interferes with the assay of the parent compounds by the cyanogen bromide method.

DPNH, xanthine, hypoxanthine, and *N*-methylnicotinamide all effect the reduction of nicotinamide *N*-oxide. TPNH is only 40% as effective as DPNH. Hypoxanthine reduces two equivalents of nicotinamide *N*-oxide just as it reduces two equivalents of oxygen during its two-step conversion to uric acid. The carboxymethyl cellulose fraction will oxidize *N*-methylnicotinamide to its 6-pyridone at pH 10,[7] but little, if at all, at pH 8. This indicates that the enzyme is relatively free of liver aldehyde oxidase (EC 1.2.3.1).

The K_m for nicotinamide *N*-oxide in the DPNH-dependent reaction is $2 \times 10^{-3} M$ and probably less than $5 \times 10^{-3} M$ for the xanthine-dependent reaction.

Effect of Temperature. Under the given assay conditions, the enzyme is active up to at least 60°. Therefore, inactivation of the enzyme by boiling is useful when maximum conversion is desired, but not when quantitative measurements of enzyme activity are desired. As a result of the increased activity observed at temperatures above 37°, the sensitivity of the assay procedure benefits by the adoption of 50° as the incubation temperature for routine assays.

Activators and Inhibitors. Because of the presence of endogenous substrates in crude tissue fractions, only a partial dependence on DPNH can be demonstrated. However, if these substrates are removed by Sephadex chromatography, complete dependence on DPNH or on xanthine can be observed. Potassium ion stimulates the DPNH-dependent activity at least 30%. Ammonium ion appears to be nearly as effective. Sodium, chloride, and phosphate ions have no effect.

Although xanthine oxidase contains sensitive sulfhydryl groups, the presence of a sulfhydryl reagent is not normally required for activity. The relatively high levels of mercaptoethanol that are necessary for the maximal expression of the nicotinamide *N*-oxide-reducing activities are, at least partially, required in order to effect anaerobiosis. Anaerobic conditions stimulate the xanthine-dependent activity to the same extent as 2-mercaptoethanol. Anaerobiosis accounts for at least 60% of the stimulation of the DPNH activity by mercaptoethanol. Although oxygen appears to antagonize the *N*-oxide reduction, concentrations of up to 6.2 m*M* nicotinamide *N*-oxide do not inhibit the utilization of oxygen by xanthine oxidase.

Nicotinamide effects a product inhibition, achieving more than 90% inhibition at a concentration of 0.15 *M*. A 5-minute preincubation with

$5 \times 10^{-3} M$ cyanide inhibits the xanthine-dependent reduction of nicotinamide N-oxide 50%.

The following cofactors have no effect on the activity: ATP ($7.3 \times 10^{-4} M$), TPN+ ($5 \times 10^{-3} M$), DPN+ ($10^{-2} M$), FAD ($2 \times 10^{-5} M$), GSH ($10^{-2} M$), dihydrolipoate ($10^{-2} M$), and ascorbate ($2 \times 10^{-2} M$).

Stoichiometry and Product Identification. Equimolar amounts of nicotinamide and DPN+ are produced upon reaction of nicotinamide N-oxide and DPNH; when xanthine and nicotinamide N-oxide are the reactants, uric acid and nicotinamide are produced in equimolar amounts. The products and reactants present in a reaction mixture can be separated by chromatography on 1.5-inch Whatman No. 1 paper strips in a 1-butanol–acetic acid–water (4:1:1, v/v) solvent. Reaction mixtures should be freed of protein by heat denaturation and centrifugation prior to chromatography. The R_f values of uric acid, xanthine, nicotinamide N-oxide, and nicotinamide are 0.17, 0.29, 0.45, and 0.70, respectively.

Although water supplies the required oxygen atom when oxygen is the electron acceptor, more than 50% of the oxygen incorporated into uric acid under anaerobic conditions can be derived from the N-oxide oxygen atom.[8]

[8] K. N. Murray, J. G. Watson, and S. Chaykin, *J. Biol. Chem.* **241**, 4798 (1966).

[128] Rabbit Liver N^1-Methylnicotinamide Oxidase (Aldehyde : Oxygen Oxidoreductase, EC 1.2.3.1)

By RONALD L. FELSTED and STERLING CHAYKIN

$2N^1$-Methylnicotinamide $+ 2H_2O + 2O_2 \rightarrow N^1$-methyl-2-pyridone-5-carboxamide
$+ N^1$-methyl-4-pyridone-3-carboxamide $+ 2H^+ + 2H_2O_2$

Assay Method

Principle. The spectral assay is based on the increase in optical density measured at 300 mμ due to the oxidation of N^1-methylnicotinamide to N^1-methyl-2-pyridone-5-carboxamide (2-pyridone) and N^1-methyl-4-pyridone-3-carboxamide (4-pyridone). At the pH of the assay mixtures, the two pyridones have almost identical spectra.

Because of its simplicity, this assay is the one routinely used in monitoring the course of the isolation of rabbit liver N^1-methylnicotinamide oxidase. However, it has the liability that in crude homogenates it does not have the sensitivity necessary to distinguish between the oxidation of N^1-methylnicotinamide by xanthine oxidase and that by N^1-methylnicotinamide oxidase. A number of control procedures, therefore, have been adopted to assure the validity of the isolation procedure. The pH optima for the oxidation of N^1-methylnicotinamide by both N^1-methylnicotinamide oxidase and xanthine oxidase fall in the range 9.0–10.0. The activity associated with xanthine oxidase falls off rapidly when the reaction is carried out at lower pH values; at pH 7.8 it is essentially nonexistent. Considerable N^1-methylnicotinamide oxidase activity remains at that pH. In order to capitalize on this differential pH dependence, assays of N^1-methylnicotinamide oxidase activity are carried out at pH 7.8.

The xanthine oxidase catalyzed reaction produces only the 2-pyridone.[1,2] Thus an assay based on following the formation of the 4-pyridone would be an unequivocal means of following N^1-methylnicotinamide oxidase activity. Since the ratio of products, 2-pyridone to 4-pyridone, is greater than 100, the use of radioisotopically labeled substrate and the time-consuming paper chromatographic separation of the pyridones must be resorted to in order to effect the determination of the 4-pyridone. It is because of its inconvenience that this method is primarily used as a backup to the spectrophotometric method. Once xanthine oxidase has been separated out, the radioactive assay need not be used.

Two other approaches can be used to evaluate the extent of the contamination of N^1-methylnicotinamide oxidase with xanthine oxidase. These are the direct assay using xanthine as substrate[3] and the inclusion of allo purinol,[4] an inhibitor of xanthine oxidase, in spectrophotometric assays for N^1-methylnicotinamide oxidase.

Catalase, bovine serum albumin, and EDTA-Fe(III)[5] are included in the assay mixtures to destroy H_2O_2 and improve the linearity of the reaction. H_2O_2 rapidly inactivates the enzyme. Provision of means for the destruction of H_2O_2 is particularly important in the assay of purified fractions from which naturally occurring catalase has been eliminated.

Spectral Assay

Reagents

Potassium phosphate buffer, 0.1 M, pH 7.8, 0.005% EDTA sodium ferric salt[5]

Bovine serum albumin,[5] 25 mg/ml

[1] L. Greenlee and P. Handler, *J. Biol. Chem.* **239**, 1090 (1964).

[2] K. Murashige, D. McDaniel, and S. Chaykin, *Biochim. Biophys. Acta* **118**, 556 (1966).

[3] H. M. Kalckar, *J. Biol. Chem.* **167**, 429 (1947).

[4] T. L. Loo, C. Lim, and D. G. Johns, *Biochim. Biophys. Acta* **134**, 467 (1967).

[5] EDTA sodium ferric salt is available from K and K Laboratories, Inc. Crystallized and lyophilized bovine serum albumin was obtained from Sigma. Catalase was purchased from Worthington (CTR 7BC).

Catalase[5] in 0.01 M potassium phosphate buffer, 0.24 mg/ml, pH 7.0
N^1-methylnicotinamide$^+$ Cl$^-$,[6] 0.5 M

Procedure. Place 0.7 ml of the phosphate buffer, 0.05 ml of catalase, 0.05 ml of bovine serum albumin, and 0.01 ml of N^1-methylnicotinamide$^+$ Cl$^-$ in a 1-ml cuvette. Add the appropriate amount of deionized water so that the final volume after addition of enzyme is 1.0 ml. Incubate the cuvette without enzyme in a 25° water bath for at least 5 minutes. After temperature equilibrium is achieved, wipe the cuvette dry, add the appropriate amount of enzyme in a volume of 0.05 ml or less, and immediately record the increase in optical density at 300 mμ in a spectrophotometer. Activity estimations are based on the initial 2–3 minute period of linear increase in absorption at 300 mμ. The change in optical density is converted to micromoles of products formed using an extinction coefficient of 4.23 $\times$ 10^3.

Radioactive Assay

Reagents

Potassium phosphate buffer, 0.377 M, pH 7.8, 0.005% EDTA sodium ferric salt[5]
Bovine serum albumin,[5] 6.25 mg/ml
Catalase,[5] 0.01 mg/ml, in 0.01 M potassium phosphate buffer, pH 7.0
N^1-Methylnicotinamide-7-^{14}C$^+$ Cl$^-$,[7] 0.0274 M (2 μCi per micromole)

Procedure. Place 0.01 ml of the phosphate buffer, 0.01 ml of bovine serum albumin, 0.01 ml of catalase, and 0.01 ml of N^1-methylnicotinamide-7-^{14}C$^+$ Cl$^-$ in a 5-ml thin-walled microconical centrifuge tube. The tube, containing everything except the enzyme, is equilibrated for 5 minutes in a water bath at 25°. After temperature equilibrium is achieved, the reaction is started by the addition of 0.01 ml of appropriately diluted enzyme. The final volume is 0.05 ml. After incubation at 25° for 5 minutes, the enzyme is inactivated by placing the reaction tubes in a boiling water bath for 2 minutes. Insoluble protein is removed by centrifugation in a clinical centrifuge. In preparation for paper chromatography, the supernatant is applied to a 1.5-inch-wide Whatman No. 1 filter paper strip. Chromatograms must be developed in two solvent systems in order to separate the various radioactive materials present in the reaction mixtures. The strips

---

[6] N^1-Methylnicotinamide$^+$ I$^-$ was converted to the Cl$^-$ salt by exchange on a Dowex 1-X8 (50–100 mesh) column.

[7] N^1-Methylnicotinamide-7-^{14}C was synthesized from nicotinamide-7-^{14}C (New England Nuclear Corporation) by the method of W. I. M. Holman and C. Wiegand, *Biochem. J.* **43**, 423 (1948).

are developed first in 1-butanol saturated with water for 6 hours, dried, and then further developed in 1-butanol saturated with 1 N HCl for 24 hours. Nonradioactive 2- and 4-pyridones are added to all samples prior to chromatography. The two pyridones are located on chromatograms under ultraviolet light. Alternatively, the radioactive compounds can be located with a paper chromatogram counter. The appropriate areas on the chromatograms are cut out and counted in the scintillation counter. With the starting N^1-methylnicotinamide concentration known, the percentages of radioactivity converted to 2- and 4-pyridones can be converted to micromoles of pyridones formed.

Definition of Units and Specific Activity. One unit of enzyme is defined as that amount which causes the conversion of 1 micromole of N^1-methylnicotinamide to its pyridones per minute at 25°. Specific activity is expressed in units per milligram of protein. The protein contents of fractions through the DEAE-Sephadex step are determined by the biuret reaction.[8] The specific activity of the purified enzyme is based on absorption at 280 mμ and related to the dry weight through use of the extinction coefficient of the purified enzyme at 280 mμ.

Purification Procedure

The purification procedure is based on the original method of K. V. Rajagopalan *et al.*[9] Modifications of and additions to their method have resulted in a procedure that will yield a homogeneous enzyme.

Only fresh livers from New Zealand white rabbits are used. Optimal enzyme activity is dependent on the use of healthy, well-fed, and mature animals.

Step 1. Extraction and Heat Treatment. The livers are homogenized for 1.5 minutes with 5 volumes of 0.05 M potassium phosphate buffer, pH 6.8,[10] in a Waring blendor. The homogenate is heated to 55° in 300 to 350-ml portions in 1-liter flasks in a 75–80° water bath. After reaching 55°, the homogenate is placed in a 56° water bath for 11 minutes and then cooled in an ice bath to about 15°. During both the heating step and cooling procedure the homogenate is stirred. The heat-treated homogenate is then centrifuged at 30,000 g for 15 minutes, and the precipitate is discarded.

Step 2. Ammonium Sulfate Precipitation. The enzyme is precipitated by the addition of $(NH_4)_2SO_4$ to 0.50 saturation. The precipitate is collected by centrifugation and dissolved in a small volume of 0.05 M potassium phosphate buffer, pH 7.8.[10] The solution is adjusted to pH 7.8 with 2 M

---

[8] A. G. Gornall, C. J. Bardawill, and M. M. David, *J. Biol. Chem.* **177**, 751 (1949).
[9] K. V. Rajagopalan, I. Fridovich, and P. Handler, *J. Biol. Chem.* **237**, 922 (1962).
[10] Buffer includes 10^{-4} M ethylenediaminetetracetic acid.

potassium hydroxide and clarified by centrifugation. The enzyme at this stage can be stored overnight in the refrigerator without loss of activity.

Step 3. Acetone Fractionation. All operations while the enzyme is in contact with acetone are carried out as rapidly as possible, the temperature of the protein solution being maintained at $0° \pm 1°$. The solution from step 2 is diluted to a protein concentration of about 42 mg/ml. Reagent grade acetone at $-70°$ is added[11] to a final concentration of 42%. During the addition of acetone, the solution should be stirred vigorously. The 0–42% precipitate is removed by centrifugation for 5 minutes at 30,000 g. Cold acetone is then added to a concentration of 48%. The 42–48% precipitate is collected by centrifugation and dissolved in a small volume of deionized water. The enzyme is then precipitated by taking the solution to 0.60 saturation with $(NH_4)_2SO_4$; this is done to remove traces of acetone. This precipitate is collected by centrifugation, and the enzyme is extracted from a considerable quantity of insoluble material by stirring with 15 ml of deionized water. The insoluble material is removed by centrifugation, extracted twice more with 15 ml of deionized water, and the extracts are combined. The pH of the combined extracts is adjusted to 7.8 with 2 M potassium hydroxide, and the solution is clarified by centrifugation. The enzyme is stable, at this stage, when stored overnight in the refrigerator.

Step 4. Calcium Phosphate Adsorption and Elution. The protein content of the solution from step 3 is diluted to about 10 mg/ml by the addition of deionized water. Enough calcium phosphate gel[12] to remove most of the activity from the solution (determined on a small batch) is added to the diluted protein solution and the mixture is stirred in an ice bath for 15 minutes. The gel is collected by centrifugation and washed twice with deionized water by stirring for 15 minutes and collecting by centrifugation. The gel is then extracted four or five times by stirring for 15 minutes with 0.01 M potassium phosphate buffer, pH 7.8. The gel is removed after each extraction by centrifugation, and the eluates are combined. The protein is concentrated by precipitation with $(NH_4)_2SO_4$ to 0.60 saturation. The precipitate is collected by centrifugation and dissolved in a small volume of 0.05 M Tris-HCl buffer, pH 8.0[10]. The next step is carried out immediately.

Step 5. DEAE-Sephadex. The enzyme solution from step 4 is dialyzed 3–4 hours against 100 volumes 0.05 M Tris-HCl buffer, pH 8.0.[10] After dialysis the enzyme is applied to a column (2.5 × 30 cm) containing DEAE-Sephadex (A-50, coarse) which has previously been washed with 0.5 M

---

[11] The acetone is added by means of a jacketed condenser held in a vertical position and equipped with a Teflon stopcock. The acetone solution to be added is cooled by drawing a separate "dry ice–acetone" solution into the surrounding jacket by means of a vacuum.

[12] T. P. Singer and E. B. Kearney, *Arch. Biochem. Biophys.* **29,** 190 (1950).

NaOH, 0.5 M HCl, deionized water, 0.05 M Tris-HCl buffer, pH 8.0,[10] and equilibrated with the latter buffer. The protein is eluted from the column with a linear gradient of 700 ml of 0.05 M Tris-HCl buffer, pH 8.0,[10] and 700 ml 0.05 M Tris-HCl buffer, pH 8.0,[10] containing 0.2 M NaCl. Fractions of about 10 ml are collected every 12 minutes. The fractions are scanned for absorption at 280 mμ and assayed. An orange-colored protein which comes off the column immediately after the active fractions can be used as a visual landmark to locate the enzyme. The active fractions are combined, and $(NH_4)_2SO_4$ is added to 0.60 saturation. The precipitate is collected by centrifugation and dissolved in a small volume of a 1:4 dilution of the Tris-phosphate concentrating gel buffer described in the Buchler Poly-Prep Apparatus instructions (see step 6). The enzyme at this stage is stable to storage overnight in the refrigerator.

Step 6. Preparative Acrylamide Electrophoresis. Electrophoresis is carried out with a Buchler Poly-Prep apparatus (Buchler Instruments, Inc., Fort Lee, New Jersey). The method is similar to that provided in the instruction manual supplied with the apparatus. Separation into bands takes place at pH 10.3 at 0°. The upper buffer is 0.0522 M Tris and 0.0525 M glycine, pH 8.9. The concentrating gel buffer is 0.0587 M Tris and 0.032 M phosphate, pH 7.2. The resolving gel buffer is 0.36 M Tris and 0.043 M glycine, pH 10.3. The lower and elution buffer is 0.1 M Tris-HCl, pH 8.1. The concentrating gel is 2.5% acrylamide while the resolving gel is 9%. The resolving gel height is 1 cm. Photopolymerization of both gels is carried out with 0.005% riboflavin 5'-phosphate. Because of the light sensitivity of the enzyme, the whole apparatus is covered throughout the run.

Up to 100 mg of the protein from step 5 is dialyzed against a 1:4 dilution of the Tris-phosphate concentrating gel buffer and made 3% in sucrose. No tracking dye is used. The dialyzed protein (3% in sucrose) is introduced onto the top of the upper gel. The instrument is covered, then a current of 50 mA is applied until the protein is concentrated; a current of 45 mA is maintained throughout the remainder of the run (6–7 hours). The fractions containing enzyme at a constant specific activity are combined and concentrated to 10–20 mg/ml with an Amicon ultrafiltrator equipped with a UM-1 Diaflo ultrafilter. The enzyme is stored in the refrigerator and protected from exposure to light.

A typical purification is summarized in the table.

Properties

Physical Properties. The enzyme is homogeneous by sedimentation velocity, sedimentation equilibrium, and disc electrophoresis at a variety of pH values and gel concentrations. Its isoelectric point is 5.0. The $S_{20,w}$

Purification Procedure for N^1-Methylnicotinamide Oxidase from Rabbit Liver

Purification Procedure for N^1-Methylnicotinamide Oxidase from Rabbit Liver

Step	Total volume (ml)	Mg/ml	Units/ml	Total units	Specific activity (units/mg)	Recovery (%)
1. Extraction and heat treatment	1945	11.6	0.188	365	0.0162	100
2. $(NH_4)_2SO_4$, 0–0.50 M	80	80.3	4.48	358	0.0558	98
3. Acetone, .42–.48%	52.7	30.4	4.82	254	0.158	70
4. Calcium phosphate gel	13.6	19.6	11.3	153	0.577	42
5. DEAE-Sephadex	9.7	6.45	8.43	81.8	1.31	22
6. Preparative electrophoresis	3.0	9.24	18.2	51.9	1.89	15[a]

[a] Yields from electrophoresis depend upon a contaminating protein that is sometimes present and runs just ahead of the enzyme. If the contaminant is present in high amounts, overlap sometimes occurs with the enzyme and yields of pure enzyme are reduced.

of the enzyme extrapolated to zero enzyme concentration is 14.45 S. The molecular weight based on the disc electrophoretic method of Hedrick and Smith[13] is 260,000. The meniscus of depletion method of Yphantis[14] (assuming $\bar{v} = 0.725$) gives a molecular weight of 270,000. The enzyme tends to polymerize on storage, giving species with molecular weights of 460,000, 540,000, and 670,000. The polymerization can be prevented or reversed by 0.005 M cysteine.

Cofactor and Metal Content. The purified enzyme contains 2 moles of FAD, 8 atoms of iron, and 0.72–0.86 mole of molybdenum per mole.

Absorption Spectrum. The enzyme exhibits absorption maxima at 275, 340, and 450 mμ. Extinction coefficients at 280 and 450 mμ in 0.035 M potassium phosphate buffer, pH 7.8, are found to be $E^{1\%}_{280\,m\mu} = 11.2$ and $E^{1\%}_{450\,m\mu} = 1.89$. The 280 m$\mu$:450 m$\mu$ ratio varies between 5.6 and 8.4. The 280 mμ:260 mμ ratio is 1.33.

Miscellaneous Properties. The enzyme exhibits a $K_m = 6 \times 10^{-4}\,M$ and a pH optimum of 9.2. The enzyme catalyzes the oxidation of N^1-methylnicotinamide to its two products, 2- and 4-pyridones, in fixed ratio throughout the purification procedure. All the evidence presently available indicates rabbit liver aldehyde oxidase and N^1-methylnicotinamide oxidase are one and the same.

[13] J. L. Hedrick and A. J. Smith, *Arch. Biochem. Biophys.* **126**, 155 (1968).
[14] D. A. Yphantis, *Biochemistry* **3**, 297 (1964).

[129] Polymerization of the Adenosine 5'-Diphosphate-Ribose Moiety of NAD

By SHINJI FUJIMURA and TAKASHI SUGIMURA

The polymerization of the adenosine 5'-diphosphate-ribose moiety of NAD is catalyzed by certain aggregated enzyme preparations.[1-4] This enzymatic polymerization is accompanied by a release of the nicotinamide moiety of NAD. No energy source is required for the polymerization. Nicotinamide inhibits the reaction.[1-4] Under optimal conditions, about 30% of NAD added as a substrate is converted to poly (ADP-Rib).[5]

Poly (ADP-Rib) can be purified by the following series of steps: acetate buffer–ethanol precipitation, pronase digestion, phenol extraction, pan-

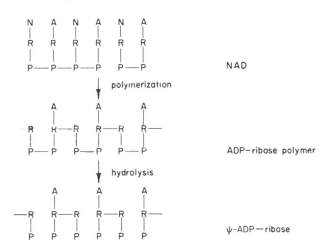

FIG. 1. The schematic representation of the formation of poly (ADP-Rib) and the breakdown of poly (ADP-Rib). Ade, adenine; Nm, nicotinamide; Rib, ribose; P, phosphate.

[1] P. Chambon, J. D. Weill, J. Doly, M. T. Strosser, and P. Mandel, *Biochem. Biophys. Res. Commun.* **25**, 638 (1966).

[2] T. Sugimura, S. Fujimura, S. Hasegawa, and Y. Kawamura, *Biochim. Biophys. Acta* **138**, 438 (1967).

[3] S. Fujimura, S. Hasegawa, Y. Shimizu, and T. Sugimura, *Biochim. Biophys. Acta* **145**, 247 (1967).

[4] Y. Nishizuka, K. Ueda, K. Nakazawa, and O. Hayaishi, *J. Biol. Chem.* **242**, 3164 (1967).

[5] T. Shima, S. Fujimura, S. Hasegawa, Y. Shimizu, and T. Sugimura, *J. Biol. Chem.* **245**, 1327 (1970).

creatic RNase and DNase digestion, and gel filtration on Sephadex. For further purification, digestion with micrococcal nuclease and spleen phosphodiesterase can be applied.[5,6]

Purified polymer behaves as a polynucleotide and can be digested to yield Ado (P)-Rib-P [2'-(5"-phosphoribosyl)-5'-AMP] and 5'-AMP by snake venom phosphodiesterase[1,7–9a] and by rat liver phosphodiesterase.[10–12a] NMN is not recovered after the hydrolysis. Recent studies suggest that ADP-ribose polymer is attached to histone at a terminus other than that with AMP.[13,14]

Enzymatic Formation of ADP-Ribose Polymer

Enzyme Preparation

Whole nuclei of rat liver prepared by the method of Chauveau *et al.*,[15] (see Vol. XII [50, 51]), the aggregated enzyme preparation obtained after disrupting the whole nuclei,[1,3,4] or the chromatin fraction obtained by the method of Marushige and Bonner,[16–18] can be used as an ADP-ribose polymerase preparation. ADP-ribose polymerase activity is recovered in soluble form after digestion of the aggregated enzyme preparation with DNase.[19] To obtain the aggregated enzyme preparation, the whole nuclear

[6] T. Sugimura, S. Fujimura, S. Hasegawa, Y. Shimizu, and H. Okuyama, *J. Vitaminol.* **14,** 135 (1968).

[7] S. Fujimura, S. Hasegawa, and T. Sugimura, *Biochim. Biophys. Acta* **134,** 496 (1967).

[8] S. Hasegawa, S. Fujimura, Y. Shimizu, and T. Sugimura, *Biochim. Biophys. Acta* **149,** 369 (1967).

[9] R. H. Reeder, K. Ueda, T. Honjo, Y. Nishizuka, and O. Hayaishi, *J. Biol. Chem.* **242,** 3172 (1967).

[9a] H. Matsubara, S. Hasegawa, S. Fujimura, T. Shima, T. Sugimura, and M. Futai, *J. Biol. Chem.,* **245,** 3606 (1970).

[10] M. Futai and D. Mizuno, *J. Biol. Chem.* **242,** 5301 (1967).

[11] M. Futai, D. Mizuno, and T. Sugimura, *Biochem. Biophys. Res. Commun.* **28,** 395 (1967).

[12] M. Futai, D. Mizuno, and T. Sugimura, *J. Biol. Chem.* **243,** 6325 (1968).

[12a] H. Matsubara, S. Hasegawa, S. Fujimura, T. Shima, T. Sugimura, and M. Futai, *J. Biol. Chem.,* **245,** 4317, 1970.

[13] Y. Nishizuka, K. Ueda, T. Honjo, and O. Hayaishi, *J. Biol. Chem.* **243,** 3765 (1968).

[14] H. Otake, M. Miwa, S. Fujimura, and T. Sugimura, *J. Biochem. (Tokyo),* **65,** 145 (1970).

[15] J. Chauveau, Y. Moulé, and C. Rouiller, *Exptl. Cell Res.* **10,** 317 (1956).

[16] K. Marushige and J. Bonner, *J. Mol. Biol.* **15,** 160 (1966).

[17] Y. Nishizuka, K. Ueda, K. Nakazawa, R. H. Reeder, T. Honjo, and O. Hayaishi, *J. Vitaminol.* **14,** 143 (1968).

[18] K. Ueda, R. H. Reeder, T. Honjo, Y. Nishizuka, and O. Hayaishi, *Biochem. Biophys. Res. Commun.* **31,** 379 (1968).

[19] Y. Shimizu, S. Hasegawa, S. Fujimura, and T. Sugimura, *Biochem. Biophys. Res. Commun.* **29,** 80 (1967).

pellet obtained from 50 g of liver is suspended in 25 ml of 0.05 M Tris-HCl buffer (pH 7.9) with a Potter-Elvehjem type homogenizer and dialyzed against the same buffer for 1.5 hours at 4°. The suspension of disrupted nuclei is centrifuged at 105,000 g for 1 hour. The pellet is resuspended in 0.25 M sucrose to give a concentration of about 10 mg protein per milliliter (roughly at the rate of 0.25 ml per gram of liver tissue). In the aggregated enzyme preparation thus obtained, the ratio of protein/DNA/RNA is 4:1:0.3.

Incubation Conditions

To follow the enzymatic formation poly (ADP-Rib), the use of NAD labeled with radioactive isotope is recommended.

Chemicals. *Escherichia coli* B is grown in a medium with added ^{32}P. The medium contained, per liter: 10 g of Bacto peptone (Difco), 5 g of NaCl, 10 ml of 50% glycerol, and 1 g of yeast extract (Difco). The cells are harvested in the late logarithmic phase, washed with saline, and disrupted by homogenization in a Virtis "45" homogenizer containing an equal volume of water and glass beads. NAD was extracted with 0.5 M $HClO_4$ and purified on a Dowex 1-X8 (200–400 mesh) formate column (1 $\times$ 20 cm) after removal of $HClO_4$ by neutralization with KOH (see Vol. III [124]).

NAD labeled with adenine-8-^{14}C or with ^{32}P at the AMP moiety is prepared by incubation of NAD pyrophosphorylase purified from hog kidney (see Vol. II [116]) with NMN and adenine-8-^{14}C labeled ATP or α-^{32}P-labeled ATP. The radioactive ATP is commercially available.

Incubation. The incubation medium contains the following in a total volume of 1.0 ml: 100 micromoles of Tris-HCl buffer (pH 8.0), 30 micromoles of $MgCl_2$, 2 micromoles of NAD (about 1 to 3 $\times$ 10^6 cpm) and 10–15 mg of protein of the aggregated enzyme preparation. The reaction product increases linearly with the amount of enzyme between 1 and 10 mg protein. The volume of the incubation medium can be proportionally increased for the purpose of the preparation of poly (ADP-Rib) on a large scale. The incubation is carried out with a gentle shaking at 37° for 10 minutes. The formation of polymer can be followed by the increase of the radioactive acid-insoluble material collected on Millipore membrane filters.

An apparent optimal pH for the formation of poly (ADP-Rib) is 8.0. The aggregated enzyme preparation is contaminated not only with phosphodiesterase which hydrolyzes the reaction product, poly (ADP-Rib), but also with NAD pyrophosphatase and NADase which degrade a precursor, NAD. Therefore, the incubation time of 10 minutes gives the maximal yield of polymer.

The reaction is effectively inhibited by nicotinamide, but not by isonicotinic acid hydrazide. p-Chlormercuribenzene sulfonate inhibits the reaction at $10^{-3} M$.

The addition of ammonium sulfate at $2 M$ concentration, which inhibits the phosphodiesterase but not polymerase action, results in the continuous formation of poly (ADP-Rib) up to 240 minutes. To obtain poly (ADP-Rib) of higher molecular weight, the inclusion of ammonium sulfate in the incubation medium is recommended.

For the preparation of polymer, the reaction is stopped by the addition of 0.12 volume of 5 M acetate buffer (pH 5.0) and 2 volumes of cold ethanol.

As a conventional way to prepare poly (ADP-Rib), NMN and labeled ATP can be incubated with the same enzyme preparation.[3] Since the enzyme preparation contains NAD pyrophosphorylase, NAD is formed and serves as a substrate for polymerase. In this case the incubation medium should contain the following in a total volume of 1.0 ml: 1 micromole of labeled ATP, 1 micromole of NMN, 100 micromoles of Tris-HCl buffer (pH 8.0), 30 micromoles of $MgCl_2$, and 10–15 mg of enzyme protein. Incubation for 30 minutes at 37° is suitable.

Isolation of Poly (ADP-Rib)

Step 1. Pronase Digestion and First Phenol Extraction. The precipitates from the incubation medium with acetate buffer–ethanol are suspended in 50 mM Tris-HCl buffer (pH 7.2) to give a concentration of 10 mg protein per milliliter. The pronase is added at a final concentration of 100 μg/ml after pH adjustment to 7.2, and the mixture is incubated for 60 minutes at 37°. The mixture is then shaken for 30 minutes at room temperature with 0.8 volume of water-saturated phenol, and the aqueous phase is separated after centrifugation. The previous digestion with pronase results in a sharp separation of two phases. The phenol phase is reextracted with 0.6 volume of water. The pooled aqueous phase is washed with ether to remove phenol. Poly (ADP-Rib) is precipitated with DNA and RNA by adding 0.12 volume of 5 M acetate buffer (pH 5.0) and 2 volumes of ethanol.

Step 2. Digestion with RNase and DNase. The precipitates are dissolved in an appropriate volume of 50 mM Tris-HCl buffer (pH 7.2) containing 5 mM $MgCl_2$, and digested by pancreatic ribonuclease (final 50 μg/ml) and pancreatic deoxyribonuclease (final 100 μg/ml) for 60 minutes at 37°. Phenol extraction is repeated as before.

Step 3. First Gel Filtration. To the aqueous phase at step 2, 0.12 volume of 5 M acetate buffer (pH 5.0) and 2 volumes of ethanol are added. The resulting precipitates are dissolved in 1.0 M NaCl and 10 mM Tris-HCl buffer (pH 8.0). This is applied onto a Sephadex G-50 column equilibrated with 1.0 M NaCl and 10 mM Tris-HCl buffer (pH 8.0). Half of the radioactivity is eluted just after the void volume followed by ultraviolet absorbing substances (oligoribonucleotides and oligodeoxyribonucleotides).

Step 4. Digestion with Micrococcal Nuclease and Spleen Phosphodiesterase. The radioactive fractions just after the void volume, are pooled, dialyzed against water, and concentrated under lyophilization. The product is incubated for 60 minutes at 37° with micrococcal nuclease (final concentration of 6 micromolar units/ml) and spleen phosphodiesterase (final 1 unit/ml) to digest the contaminated polynucleotides in 50 mM Tris-HCl buffer (pH 7.5) containing 1.5 mM CaCl$_2$. Phenol extraction is repeated as before.

Step 5. Second Gel Filtration. The aqueous layer at step 4 is adjusted to 1.0 M NaCl, pH 8.0, and applied again on Sephadex G-50 column. A single radioactive peak appears after the void volume, completely separated from the contaminating ultraviolet absorbing substances. On the radioactive peak, the ratios of radioactivity to the optical density at 260 mμ are constant in all tubes.

An example of the recovery of the poly (ADP-Rib) at each step is shown in Table I. The overall recovery is more than 60% of the initial radioactivity in the acid-insoluble form.

Properties of Poly (ADP-Rib)

Properties of Polymer. Purified polymer is precipitated by 5% trichloroacetic acid, 5% trichloroacetic acid–0.25% sodium tungstate (pH 2.0), 0.5 N perchloric acid, or 0.2 M acetate buffer (pH 5.0)–66% ethanol.

TABLE I
PURIFICATION OF POLY (ADP-RIB)

	Poly (ADP-Rib)	
Step	Prepared from NAD[a] (cpm)	Prepared from ATP and NMN (cpm)
Acetate buffer–ethanol precipitates	4.30×10^6	2.28×10^6
Pronase digestion and phenol extraction	3.02×10^6	2.12×10^6
RNase and DNase digestion, phenol extraction and gel filtration	2.74×10^6	1.78×10^6
Micrococcal nuclease and spleen phosphodiesterase digestion, phenol extraction and gel filtration	1.25×10^6	1.28×10^6

[a] Twenty-five milliliters of the incubation mixture with 50 micromoles of NAD labeled with ^{32}P (1.2×10^7 cpm) and 230 mg of protein.

[b] Ten milliliters of the incubation mixture with 10 micromoles of ATP labeled at adenine-8-^{14}C (1.50×10^7 cpm), 10 micromoles of NMN, and 200 mg of protein.

Polymer purified from the incubation without ammonium sulfate has an s value of 3–10 S on sucrose density gradient centrifugation. On cesium sulfate equilibrium density gradient centrifugation, ADP-ribose polymer has a density of 1.575. The ratio of the optical density at 280 mμ to that at 260 mμ of purified polymer in neutral solution is 0.26.[5] The digestion of polymer in the solution with snake venom phosphodiesterase or elevating temperature of solutions of the polymer increases its optical density at 260 mμ to the extent of 10–20%.[6] After incubation of poly (ADP-Rib) in 0.5 N KOH at 37° for 18 hours, most of polymer still remains in the acid-insoluble form.[8] Poly (ADP-Rib) is readily converted to the acid-soluble form by boiling in 1 N HCl for 7 minutes.[1,8,9]

Poly (ADP-Rib) is not hydrolyzed by pancreatic RNase, T1 RNase, pancreatic DNase, spleen DNase, micrococcal nuclease, or spleen phosphodiesterase.[7,8] Snake venom phosphodiesterase[1,7–9a] or rat liver phosphodiesterase,[10,11] yielding 5'-nucleotides, can hydrolyze poly (ADP-Rib) to produce Ado (P)-Rib-P (Fig. 1). The latter enzyme hydrolyzes polymer in exonucleolytic fashion.[12,12a] A small amount of 5'-AMP is recovered from a hydrolyzate of polymer with snake venom phosphodiesterase, and the molar ratio of Ado (P)-Rib-P to 5'-AMP is a function of an average chain length of the intact molecule of polymer. Polymer obtained as de-

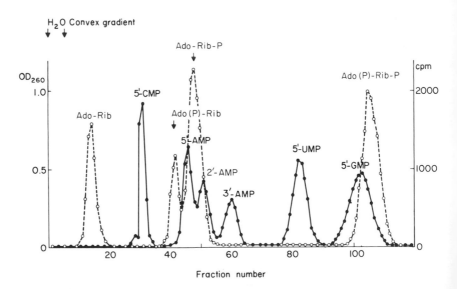

FIG. 2. The elution pattern of Ado (P)-Rib-P (labeled with adenine-8-^{14}C) and its dephosphorylated compounds from Dowex 1-X2 (Cl-form) column. Column size, 0.5 × 20 cm; fraction volume, 1.0 ml; flow rate, 12 ml per hour. 5'-GMP, 5'-UMP, 5'-CMP, 5'-AMP, and 2',3'-AMP are added as markers. ●—●, OD$_{260}$; ○—○, radioactivity.

scribed from the incubation without the addition of ammonium sulfate has a chain length of 20–30 ADP-ribose residues.

Properties of Ado (P)-Rib-P and Related Compounds. Ado (P)-Rib-P is purified from the phosphodiesterase hydrolyzate of polymer on Dowex 1-X8 (200–400 mesh) formate column chromatography.[7-9] The elution is carried out according to the procedure of Cohn and Volkin.[20,20a] Ado (P)-Rib-P is eluted with 0.45 M ammonium formate separately from other mononucleotides. Column chromatography on Dowex 1-X2 (chloride form) is also suitable for separation of Ado (P)-Rib-P and related compounds.

TABLE II

RUNNING DISTANCES OF ADO (P)-RIB-P AND RELATED COMPOUNDS

	Solvent	
	---	---
Compound	1[a] (cm)	2[b] (cm)
Ado (P)-Rib-P	16.2	43.7
Ado (P)-Rib	26.7	34.0
Ado-Rib-P	28.7	23.7
Ado-Rib	38.9	16.5
ADP-Ribose	15.2	36.8
NAD	22.5	32.2
5'-ATP	13.2	43.3
5'-ADP	19.2	36.5
3',5'-ADP	18.6	36.5
2',5'-ADP	16.8	42.3
5'-AMP	26.3	29.8
2',3'-AMP	29.9	16.2, 27.5
Adenosine	38.9	13.9

[a] Solvent 1 is the mixture of isobutyric acid–conc. NH_4OH–water (66:1:33, v/v/v), 24 hours at 25°.

[b] Solvent 2 is the mixture of 0.1 M sodium phosphate (pH 6.8)–$(NH_4)_2SO_4$–n-propanol (100:60:2, v/w/v), 36 hours at 25°.

A convex gradient of elution is achieved with 420 ml of 0.0035 N HCl in a mixing chamber and 0.25 M NaCl in a reservoir, as described by Tada et al.[21] Figure 2 illustrates a typical pattern of the elution of Ado (P)-Rib-P and its dephosphorylated derivatives, Ado (P)-Rib, Ado-Rib-P and Ado-

[20] W. Cohn and E. Volkin, *J. Biol. Chem.* **203**, 319 (1953).

[20a] T. Shima, S. Hasegawa, S. Fujimura, H. Matsubara, and T. Sugimura, *J. Biol. Chem.* **244**, 6632 (1969).

[21] M. Tada, M. Tada, and K. Yagi, *J. Biochem.* **55**, 136 (1964).

Rib. These derivatives are prepared by digestion of Ado (P)-Rib-P with alkaline phosphatase. The running distances of Ado (P)-Rib-P and related compounds on descending paper chromatography are listed in Table II for convenience.

[130] Adenosine Diphosphoribosyltransferase in Chromatin

By Yasutomi Nishizuka, Kunihiro Ueda, and Osamu Hayaishi

The adenosine diphosphoribose portion of NAD is transferred to nuclear proteins, presumably histones, with the simultaneous release of nicotinamide. A successive transfer of ADP-ribose units results in the formation and elongation of a homopolymer with repeating ADP-ribose units. The ADP-ribose units are linked together through a ribose-ribose linkage, and the site of the transfer of the ADP-ribosyl group has been established to be the 2'-hydroxyl of the AMP moiety of the adjacent ADP-ribose unit. The enzymatic synthesis of the polymer was recognized first by Chambon et al.,[1] Sugimura et al.,[2] and Nishizuka et al.,[3] and the reaction has been studied extensively by the latter two groups.[4,5] Although the precise biological significance of this reaction has yet to be elucidated, the assay as well as some properties of this enzymatic reaction are briefly described.

Assay Method

Principle. Enzymatic activity may be assayed by measuring the radioactivity incorporated into an acid-insoluble fraction with radioactive NAD as a substrate. The acid-insoluble radioactive material is collected on a Millipore filter, and the radioactivity is determined.

[1] P. Chambon, J. D. Weill, J. Doly, M. T. Strosser, and P. Mandel, *Biochem. Biophys. Res. Commun.* **25**, 638 (1966)

[2] T. Sugimura, S. Fujimura, S. Hasegawa, and Y. Kawamura, *Biochim. Biophys. Acta* **138**, 438 (1967); S. Fujimura, S. Hasegawa, Y. Shimizu, and T. Sugimura, *Biochim. Biophys. Acta* **145**, 247 (1967).

[3] Y. Nishizuka, K. Ueda, K. Nakazawa, and O. Hayaishi, *J. Biol. Chem.* **242**, 3164 (1967); R. H. Reeder, K. Ueda, T. Honjo, Y. Nishizuka, and O. Hayaishi, *J. Biol. Chem.* **242**, 3172 (1967).

[4] Y. Nishizuka, K. Ueda, T. Honjo, and O. Hayaishi, *J. Biol. Chem.* **243**, 3765 (1968).

[5] S. Fujimura, M. Miwa, H. Otake, S. Hasegawa, J. Oikawa, and T. Sugimura, *J. Japan. Biochem. Soc.* **40**, 670 (1968).

Reagents

NAD-adenine-^{14}C (5–100 mCi/millimole),[6] 5 mM
Tris-HCl buffer, 0.25 M, pH 8.0
MgCl$_2$, 0.15 M
Nuclear preparation
Trichloroacetic acid, 20%

Procedure. The incubation mixture contains 0.05 ml of Tris-HCl buffer, 0.05 ml of MgCl$_2$, 0.01 ml of radioactive NAD, nuclear preparation (either whole nuclei or chromatin preparation), and water in a total volume of 0.25 ml. The reaction is started by the addition of NAD, and the incubation is carried out for 5 minutes at 37° with shaking. The reaction is stopped by adding about 10 ml of 20% trichloroacetic acid. The mixture is then passed through a Millipore filter (pore size, 0.45 μ). The filter is washed five times with 10-ml aliquots of 20% trichloroacetic acid, placed on an aluminum planchet, and dried. The radioactivity is determined with a Geiger-Müller gas flow counter (Nuclear Chicago).

Preparation of Nuclear Fraction

The enzymatic activity is exclusively localized in nuclei from various tissues and organs, but not in cytoplasm. Nuclear preparations active for the reaction may be made as follows. All operations are carried out at 0–4°.

Preparation of Nuclei. Approximately 50 g of rat liver removed immediately after decapitation are homogenized with 10 volumes of 0.25 M sucrose containing 3.3 mM CaCl$_2$ in a Potter-Elvehjem type Teflon-glass homogenizer. The homogenate is centrifuged for 10 minutes at 2000 g. The precipitate is suspended in 200 ml of 2.2 M sucrose containing 3.3 mM CaCl$_2$ and homogenized again with the homogenizer. The suspension is centrifuged for 60 minutes at 78,000 g with a No. 30 rotor of a Spinco ultracentrifuge, Model L. Nuclei are pelleted under these conditions.[7] The nuclei thus obtained are washed twice, each time with about 50 ml of 0.25 M sucrose containing 3.3 mM CaCl$_2$, followed by centrifugation for 10 minutes at 2000 g. The nuclei thus obtained are active for the enzymatic reaction.

Preparation of Chromatin Fraction. The washed nuclei are then suspended in 20 ml of 0.01 M Tris-HCl buffer, pH 7.4, and any soluble material is extracted by stirring the suspension for 10 minutes. After centrifugation

---

[6] NAD labeled with ^{32}P or ^{14}C or ^{3}H elsewhere than at the nicotinamide moiety may be employed as substrate. These radioactive NAD's are prepared as described by K. Ueda and H. Yamamura (see [107] in this volume).
[7] J. Chauveau, Y. Moulé, and C. Rouiller, *Exptl. Cell Res.* **11**, 317 (1956).

for 10 minutes at 20,000 g, the supernatant solution is discarded. The sediment is further extracted with 20 ml of 0.05 M Tris-HCl buffer, pH 7.8, and soluble proteins are discarded after centrifugation. The precipitate is further extracted successively with 20 ml each of 0.075 M NaCl containing 0.024 M ethylenediaminetetraacetate, pH 8.0, 0.05 M Tris-HCl buffer, pH 8.0, and 0.01 M Tris-HCl buffer, pH 8.0. The final precipitate is stirred overnight with 10 ml of water at 0°, and centrifuged for 1 hour at 105,000 g. The clear viscous supernatant solution contains approximately 80% of DNA in the original nuclei; it is referred to as chromatin fraction.[8,9]

Properties

Reaction Products. Quantitative analysis of the reaction products shows a variety of size distributions from monomer up to polymer with molecular weight of several thousands. The products that are extractable from the reaction mixture with dilute HCl are mainly oligomers attached to histones with an average chain length of 1.1–1.5. These products are not associated with a particular histone species but appear to be linked covalently to various histone subfractions. The ADP-ribosyl linkage to histones is extremely labile under alkaline conditions, and approximately 50% is hydrolyzed for 20 minutes at 0° at pH 9–10. The linkage is relatively stable under acidic conditions.

The reaction products with a larger molecular weight, poly ADP-ribose, appear to be also attached to the nuclear proteins, but are not extracted by dilute HCl since the polymer itself is acid-precipitable. The polymer may be extracted from the reaction mixture by treatment with proteinases followed by phenol extraction and alcohol precipitation. The polymer is further purified by a Sephadex column after treatment with RNase and DNase. In contrast to the ribosyl linkage attached to histone, the polymer free of protein is alkali stable but is converted to adenine and ribose 5-phosphate in a ratio of 1 to 2 upon hydrolysis in 1 N HCl at 100° for 7 minutes.

Enzyme. The enzyme responsible for the reaction appears to be a kind of NAD glycohydrolase, since nicotinamide is released simultaneously. The activity is exclusively localized in nuclei. The addition of either histone, DNA or both to the cytoplasmic NAD glycohydrolase[10] does not induce the ADP-ribosyltransfer reaction. The nuclear enzyme may be distinguished clearly from the cytoplasmic enzyme, since the former does not

[8] K. Marushige and J. Bonner, *J. Mol. Biol.* **15**, 160 (1966).
[9] K. Ueda, R. H. Reeder, T. Honjo, Y. Nishizuka, and O. Hayaishi, *Biochem. Biophys. Res. Commun.* **31**, 379 (1968).
[10] See Vol. II [113].

react with NADP.[11] The activity is extremely labile upon heat treatment, and approximately 50% of the activity is lost at 45° for a minute. The transfer reaction seems to be irreversible. It is not clear as yet whether a single enzyme is responsible for both the first ADP-ribose transfer to histone and the subsequent chain elongation of the polymer.

The activity has been shown to be essentially localized in chromatin. In addition, the enzyme appears to be tightly associated with DNA.[9] DNA seems to be necessary for the chain elongation of the polymer, since a prior treatment of chromatin with pancreatic DNase makes the polymer shorter. Attempts have been thus far unsuccessful to resolve the enzyme from DNA or chromatin.

Optimal pH. The activity has a maximum at pH 8.0.

Substrate Specificity and K_m Value. The enzyme is specific for NAD. Neither NADP nor reduced pyridine nucleotides act as substrates in this reaction. Deamido-NAD is inert as a substrate. Acetylpyridine-NAD and thionicotinamide-NAD are about 4 and 9% as active as NAD, respectively. K_m value for NAD is $2.5 \times 10^{-4} M$.[11]

Inhibitor and Activator. The enzymatic activity is strongly inhibited by various pyridine derivatives such as nicotinamide and 3-acetylpyridine, which are known as inhibitors of NAD glycohydrolases. Nicotinamide inhibits the reaction nearly 100% at $10^{-2} M$. The enzymatic activity is relatively insensitive to nicotinic acid and isonicotinic acid hydrazide. $MgCl_2$ slightly stimulates the activity around a concentration of 30 mM.

Distribution. The enzymatic activity is distributed in nuclei from various tissues and organs from mammals and other vertebrates. The nuclei from liver and brain show a higher activity. No activity is found in mammalian erythrocytes, but the erythrocytes from chicken and pigeon show some activity. No activity has thus far been detected in microorganisms and molds, including *Escherichia coli*, *Pseudomonas fluorescens*, and yeast cells.

[11] K. Nakazawa, K. Ueda, T. Honjo, K. Yoshihara, Y. Nishizuka, and O. Hayaishi, *Biochem. Biophys. Res. Commun.* **32**, 143 (1968).

[131] Anaerobic Degradation of Nicotinic Acid

By L. Tsai and E. R. Stadtman

$$\text{Nicotinic acid} + 4 H_2O \rightarrow \text{propionate} + \text{acetate} + CO_2 + NH_3 \qquad (1)$$

A *Clostridium* sp. was isolated from soil enrichment cultures in which nicotinic acid was supplied as a major source of carbon, nitrogen, and

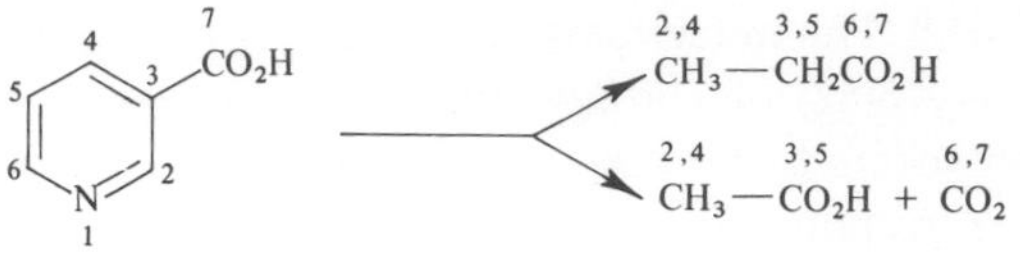

FIG. 1

energy.[1,2] Fermentation of nicotinic acid by this organism leads to the formation of 1 mole each of propionate, acetate, CO_2, and ammonia, according to reaction (1). Studies with nicotinic acid preparations that were singly labeled with ^{14}C in various carbon atoms disclosed that the methyl groups of both acetate and propionate are derived equally from carbons 2 and 4 of nicotinic acid; the 2-carbon of propionate and the carboxy group of acetate are derived equally from carbons 3 and 5 of nicotinic acid; the CO_2 and the carboxyl group of propionate are derived equally from carbons 6 and 7 of nicotinic acid.[2] This is illustrated in Fig. 1.

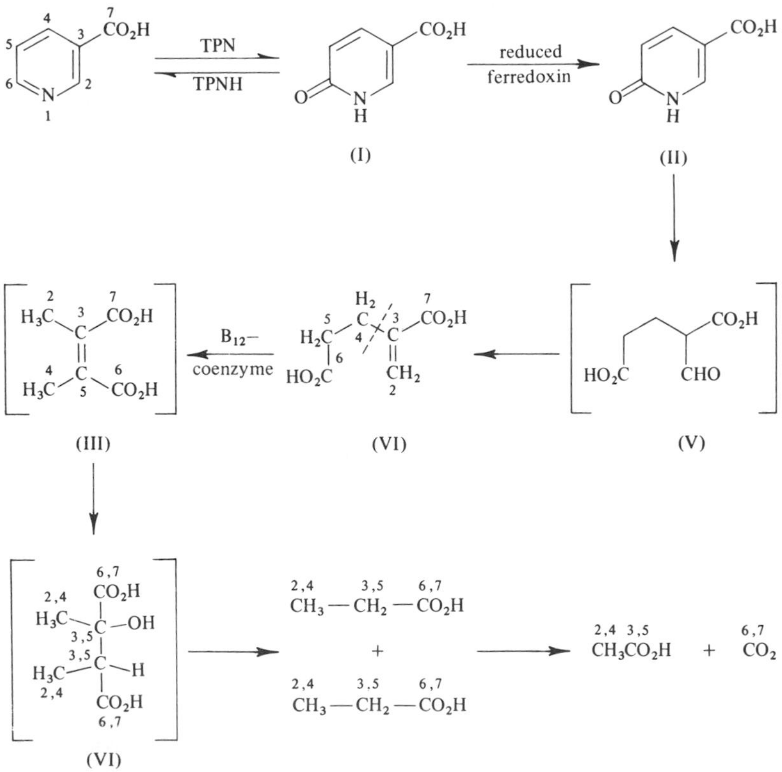

FIG. 2

[1] I. Harary, J. Biol. Chem. 227, 815, 823 (1957).
[2] I. Pastan, L. Tsai, and E. R. Stadtman, J. Biol. Chem. 239, 902 (1964).

Three compounds, 6-hydroxynicotinic acid (compound I), 1,4,5,6-tetrahydro-6-oxonicotinic acid (compound II), and α-methyleneglutaric acid (compound IV), have been isolated as intermediate breakdown products of nicotinic acid.[3] A fourth compound, tentatively identified as 2-formylglutaric acid (compound III), is produced when compound (II) is incubated with crude enzyme preparations.[4]

Finally, a fifth compound, tentatively identified as dimethylmaleic acid (compound V), is produced when α-methyleneglutaric acid is incubated with crude enzyme preparations.[5] The formation of this intermediate requires the presence of coenzyme B_{12}, and it is presumed to occur by a rearrangement of the carbon skeleton by a mechanism that involves breakage of the C-3/C-4 bond and attachment of C-3 to C-5 (see Fig. 2). Since dimethylmaleic acid is a symmetrical molecule, carbons 6 and 7, 3 and 5, and 2 and 4 are equivalent, and the subsequent decomposition of this compound to propionate, acetate, and CO_2 via dimethylmaleic acid could readily account for the observed distribution of carbon atoms in the ultimate products, as shown in Fig. 2.

I. Nicotinic Acid Hydroxylase

$$\text{Nicotinic acid} + H_2O + TPN^+ \rightarrow \text{6-hydroxynicotinic acid} + TPNH \qquad (2)$$

Assay Method

Principle. The increase in absorbancy at 340 mμ due to the reduction of TPN^+ by reaction 2 is the basis of assay. Since the hydroxylase also possesses strong TPNH oxidase activity, all traces of molecular oxygen must be removed from the reaction mixture, and it must be kept anaerobic during the time course of the assay. Anaerobiosis, even in open cuvettes, is achieved by the addition of ferrous salts, glutathione, and inorganic pyrophosphate buffer. These reagents catalyze nearly instantaneous, nonenzymatic reduction of dissolved oxygen and thereby maintain an anaerobic environment. Inorganic pyrophosphate buffer cannot be replaced by other buffers. Pyrophosphate prevents the precipitation of iron phosphate complexes, and amine buffers yield colored solutions in the presence of the iron and glutathione.

Reagents

Potassium pyrophosphate buffer, 1.0 M, pH 8.3
Potassium phosphate buffer, 1.0 M, pH 7.7

[3] L. Tsai, I. Pastan, and E. R. Stadtman, *J. Biol. Chem.* **241,** 1807 (1966).
[4] L. Tsai, J. Holcenberg, and E. R. Stadtman, unpublished results.
[5] H. F. Kung, S. Cederbaum, L. Tsai, and T. C. Stadtman, *Proc. Natl. Acad. Sci. U.S.* **65,** 978 (1970).

Potassium hydroxide, 1.0 M
Glutathionine, 0.2 M
Potassium nicotinate, 0.1 M
$FeSO_4$, 0.1 M
TPN$^+$, 0.01 M, pH 6.5
Nicotinic acid hydroxylase, dilute stock solutions to contain 0–0.1
unit per milliliter (see definition below)

In a 1.8-ml rectangular quartz cuvette (1-cm light path) are added 0.9 ml water, 0.02 ml of potassium pyrophosphate buffer, 0.05 ml of potassium phosphate buffer, 0.05 ml of glutathione, 0.01 ml NaOH, 0.05 ml of TPN$^+$, 0.03 ml of $FeSO_4$ and 0.1 ml of nicotinate hydroxylase. The final volume is 1.21 ml, pH 7.9. An identical sample containing all reagents except the hydroxylase, but with 0.1 ml water instead, serves as the reference blank. The change in absorbancy at 340 mμ during a 3- to 5-minute interval following the addition of enzyme at 23° is a measure of TPNH formation and is proportional to the enzyme concentration over the range of 0–0.02 unit.

The glutathione and $FeSO_4$ may be replaced by 50 mM dithiothreitol and 2.5×10^{-7} M cobinamide, respectively (final concentrations in the reaction mixture).

Purification Procedure

The purification procedure described below is essentially identical to that previously described by Holcenberg and Stadtman.[6]

The nicotinic acid fermenting bacterium is grown as previously described.[6] After harvesting, the cells are frozen and stored at −20°.

Step 1. To prepare cell-free extracts, 50 g of frozen cells are thawed in 50 ml of 0.05 M potassium phosphate buffer (pH 7.5) containing 4 mM dithiothreitol. The suspension is passed through a French pressure cell at 6000–9000 psi. A few crystals of DNase are added to the viscous extract, and after a few minutes the particulate material is sedimented by centrifugation at 16,000 g for 30 minutes and is discarded.

Step 2. Streptomycin Precipitation. The supernatant solution from step 1 is diluted with an equal volume of water. Then 0.1 volume each of 10% streptomycin solution and 0.2 M dithiothreitol (neutralized to pH 7.0) are added. After 10 minutes the solution is centrifuged and the precipitate is discarded. The supernatant solution contains about 20 mg of protein per milliliter and can be stored in liquid nitrogen for months without loss of activity.

Step 3. Acid Precipitation. The supernatant solution from step 2 is

---

[6] J. Holcenberg and E. R. Stadtman, *J. Biol. Chem.* **244**, 1194 (1969).

diluted with an equal volume of water. The solution is kept at 5° as the pH is adjusted to 5.7 by the careful addition of potassium acetate buffer (1.0 M, pH 4.1). After 15 minutes the solution is centrifuged and the precipitate is discarded. The supernatant solution is adjusted to pH 7.0 with careful addition of 1 M KOH.

Step 4. Acetone Precipitation. The neutralized supernatant solution (200 ml) from step 3 is chilled in a brine bath, and 44 ml of acetone (chilled to −20°) is added slowly, care being taken to keep the temperature below −5°. The precipitate is removed by centrifugation and discarded. To the supernatant solution, 23 ml of acetone is added, and the copious precipitate formed is collected by centrifugation and dissolved in 36 ml of 0.05 M potassium phosphate buffer, pH 7.5, containing 2 mM dithiothreitol.

Step 5. Ammonium Sulfate Fractionation. The dissolved enzyme from step 4 is cooled to 5°, and solid ammonium sulfate is added. The protein precipitating between 48 and 62% saturation is dissolved in 25 ml of 0.05 M potassium phosphate buffer, pH 7.5 containing 2 mM dithiothreitol. The resulting yellow solution can be stored in liquid nitrogen.

Step 6. Disc Gel Electrophoresis. Aliquots (3–10 ml) of the ammonium sulfate fraction from step 5 are taken for disc gel electrophoresis at 4°. Standard Canalco equipment and chemicals are used with the PD2-150 cm column, 3.5% acrylamide stacking gel, 10% acrylamide separating gel, and electrode buffers of 0.19 M glycine and 0.025 M Tris base. The column is eluted with 0.375 M Tris-HCl at pH 7.9. Sodium sulfide (0.03% $Na_2S \cdot 7H_2O$) is added to the eluting and upper electrode buffers. The enzyme is layered on the upper gel in 10–20% glycerol. The yellow hydroxylase is the last major protein band to be eluted from the gel. The fractions of eluate containing hydroxylase are pooled and diluted 4-fold with water. The enzyme is concentrated by passing the diluted eluates through a 1 × 5 cm column of DEAE-cellulose (previously equilibrated with Tris-HCl and washed with water; the enzyme forms a sharp brown band at the top of the DEAE column; after washing with water, it is eluted with 0.25 M potassium phosphate buffer, pH 7.5). The resulting solution is stored in liquid nitrogen. The overall purification is only about 24-fold; however, the enzyme at this state of purity appears to be nearly homogeneous as judged by sedimentation studies and disc gel electrophoresis.

A summary of a purification procedure is given in Table I.[6]

Properties

Specificity. In the standard spectrophotometric assay using the dithiothreitol-cobinamide antioxidant system, no TPN reduction occurs in the presence of the following nicotinate analogs: nicotinamide, isonicotinate,

TABLE I
PURIFICATION OF NICOTINIC ACID HYDROXYLASE

Step	Total volume (ml)	Total protein (mg)	Total units	Specific activity (units/ mg)	Yield (%)
1. Crude extract	85	2400	2900	1.2	100
2. Streptomycin precipitation	100	2000	2600	1.3	90
3. Acid precipitation	200	1800	3000	1.7	110
4. Acetone precipitation	36	520	1900	3.7	65
5. Ammonium sulfate fractionation	25	170	850	5.0	29
6. Disc gel electrophoresis	5	1.7[a]	49[a]	28.8[a]	14[a]

[a] Obtained from 3 ml of fraction 5. Data given and calculated on assumption that all of fraction 5 was used.

picolinate, nicotinic acid N-oxide, N-methylnicotinamide, trigonelline, or hypoxanthine. Pyrazine 2-carboxylate is the only analog that is oxidized; it is oxidized at about one-third the rate of nicotinate. The enzyme thus appears to require an unsubstituted pyridine or pyrazine ring nitrogen and a carboxyl group meta to the nitrogen. DPN will not replace TPN as an electron acceptor; however, artificial electron acceptors, including benzyl-viologen and 2,3,5-triphenyltetrazolium dyes will replace TPN. The apparent K_m for nicotinate is 1.1×10^{-4} M and for TPN is 2.8×10^{-5} M.

Effect of pH. In the standard phosphate–pyrophosphate buffer system the pH optimum is 8.0–8.3. The activity at pH 7.0 or 9.0 is about half that at pH 8.0.

Activation by P_i. Inorganic orthophosphate stimulates the activity about 50%. Two analogs of orthophosphate, arsenate, and methyl phosphate, are less effective activators, whereas citrate, sulfate, and inorganic pyrophosphate do not stimulate the activity. The concentration of P_i required for half-maximal activation is 50 mM.

Reversibility. The equilibrium of reaction (2) lies far in the direction of 6-hydroxynicotinate formation. Oxidation of TPNH in the presence of 6-hydroxynicotinate has not been detected directly. However reduction of 6-hydroxynicotinate to nicotinate does occur if reaction (2) is coupled with a TPNH-generating system of low potential, such as isocitrate and isocitrate dehydrogenase. The enzymatic reduction of 6-hydroxynicotinate can also be coupled with the oxidation of dithionite when either TPN or methylviologen are supplied as electron carriers.

Prosthetic Groups. The best preparations obtained to date contain about 11 moles of iron, 6 moles of labile sulfide, and 1.5 moles of FAD per mole of enzyme. These are probably minimal values.[6]

Molecular Weight. The molecular weight, estimated by sedimentation analysis, is about 300,000.

Other Activities. The purified enzyme exhibits strong TPNH oxidase and diaphorase activities.

II. 6-Hydroxynicotinic Acid Reductase

Reduced-ferredoxin + 6-hydroxynicotinate →
$$1,4,5,6\text{-tetrahydro-6-oxo-nicotinic acid} + \text{oxidized ferredoxin} \quad (3)$$

Assay Method

Principle. The conversion of 6-hydroxynicotinate to compound (II) is accompanied by a decrease in absorbancy at 310 mμ and an increase in absorbancy at 275 mμ. Therefore, the concentrations of these two compounds in a mixture can be calculated from absorbancy measurements at 275 and 310 mμ. The decrease in concentration of 6-hydroxynicotinic acid that occurs when it is incubated with the reductase in the presence of reduced ferredoxin is a measure of reductase activity. Reduced ferredoxin is generated *in situ* by the addition of an excess of sodium dithionite. This method as well as the following purification procedure were essentially the same as those previously described by Holcenberg and Tsai.[7]

Reagents

Potassium phosphate buffer, 1.0 M, pH 7.4
6-Hydroxynicotinic acid, 0.1 M
Sodium dithionite, 0.2 M. This solution is made up just before use by dissolving 48.4 mg of $Na_2S_2O_6\cdot2H_2O$ in 1.0 ml of 0.1 M NaOH which was previously gassed for 1.0 minute with an inert gas in a 12 mm $\times$ 10 cm test tube. This solution (pH $\approx$ 7.0) is stable for at least 1 hour if kept stoppered, under inert gas, at 0°.
Clostridial ferredoxin, 900 μg/ml
6-Hydroxynicotinate reductase, 0.2–1.3 units/ml (unit defined below)
Perchloric acid, 20%

Procedure. In a 10 $\times$ 75 mm test tube are added 0.5 ml water and 0.1 ml each of phosphate buffer, 6-hydroxynicotinate, and ferredoxin. The mixture is gassed for 25 seconds by passing a stream of argon or helium over the surface of the solution, then, with gassing continued, the reductase is added, and then the dithionite. The tubes are stoppered and incubated at 25°, for 30 minutes, then 0.1 ml of perchloric acid is added, and after 5 minutes, when the excess dithionite is decomposed, the samples are ad-

[7] J. Holcenberg and L. Tsai, *J. Biol. Chem.* **244**, 1204 (1969).

justed to pH 7.0 and centrifuged. The absorbancy of the supernatant solution is measured at 275 and 310 mμ. At 310 mμ, the molar absorbancies of 6-hydroxynicotinate and compound (II) are 3.54 $\times$ 10³ and 0.93 $\times$10³, respectively. At 275 mμ the molar absorbancies of 6-hydroxynicotinate and compound (II) are 3.66 $\times$ 10³ and 11.2 $\times$ 10³, respectively. Therefore the concentration of these compounds are calculated as follows:

$$\text{Compound II (m}M) = \frac{A_{275\,m\mu} - A_{310\,m\mu}}{10.3}$$

$$\text{6-Hydroxynicotinate (m}M) = \frac{A_{310\,m\mu} - (0.93 \times \text{m}M \text{ compound II})}{3.5}$$

Although theoretically the decrease in concentration of 6-hydroxynicotinate could be calculated directly from the decrease in absorbancy at 310 mμ, this procedure is unreliable, since the best reductase preparations obtained to date are contaminated with enzymes that catalyze the further partial conversion of compound II to other products that do not absorb at 310 mμ. It is therefore necessary to calculate the concentration of 6-hydroxynicotinate from the above relationships.

Units of Specific Activity. One unit of 6-hydroxynicotinate reductase is defined as the amount of enzyme that catalyzes the reduction of 1.0 micromole of 6-hydroxynicotinate per minute under the above standard conditions.

Purification Procedure

Step 1. Crude Extract. Cell-free extracts of the nicotinate-fermenting clostridium (35–40 mg of protein per milliliter) were prepared as described in Section I, step 1. The freshly prepared extracts lose appreciable reductase activity on standing several hours at 0°; therefore, the next step is carried out immediately.

Step 2. Streptomycin Treatment and Heat Step. One-tenth volume of a 10% solution of streptomycin is added to the extract from step 1. After 10 minutes dithiothreitol is added to a final concentration of 20 mM, and after an additional 15 minutes, the precipitate is removed by centrifugation. The activity of the supernatant solution is stable for several days at 5° if stored under an inert gas. For heating, the supernatant solution is placed in a stainless steel centrifuge bottle, 1.0 mM hydroxynicotinate is added, the mixture is gassed with argon for 1 minute, stoppered, and heated 30 minutes in a 62° water bath. Then, after cooling in ice, the precipitate is removed by centrifugation and discarded.

Step 3. First Ammonium Sulfate Precipitation. The highly colored supernatant solution from step 2 (8 mg of protein per milliliter) is cooled to 5°, and solid ammonium sulfate is added. The fraction precipitating between

55 and 75% ammonium sulfate saturation is collected by centrifugation and is dissolved in 0.05 M potassium phosphate buffer, pH 7.4, containing 2 mM dithiothreitol. Final protein concentration is 8–9 mg/ml.

Step 4. Second Heat Step and Ammonium Sulfate Precipitation. To the dissolved precipitate from step 3, 1.0 mM 6-hydroxynicotinate is added, the mixture is gassed with argon for 1 minute, and is heated in a water bath at 72° for 30 minutes. After cooling to 5°, the sample is centrifuged. Solid ammonium sulfate is added to the supernatant solution and the fraction precipitating between 55 and 60% saturation is collected by centrifugation and dissolved in phosphate buffer as described in step 3.

The overall purification procedure through step 4 results in a 20-fold enrichment of reductase activity and a 46% recovery of total activity. Efforts to purify the enzyme further by means of column chromatography on DEAE-cellulose, hydroxylapatite or by means of preparative disk gel electrophoresis, or by solvent and isoelectric precipitation were all unsuccessful.[7] Table II summarizes the results of a typical run.

TABLE II

PURIFICATION OF 6-HYDROXYNICOTINIC ACID REDUCTASE ACTIVITY

Step	Total volumes (ml)	Total protein (mg)	Total units	Specific activity (units/ mg)	Yield (%)
1. Crude extract	139	5400	2440	0.45	100
2. Streptomycin treatment and heat step	145	1150	2520	2.2	103
3. First ammonium sulfate precipitation, 55–75%	40	340	2120	6.2	87
4. Second heat step and ammonium sulfate precipitation, 55–60%	40	100	1120	11.0	46

Properties

Specificity. Reduced ferredoxin is the natural electron donor for 6-hydroxynicotinate reduction. Neither DPNH nor TPNH will serve as electron donors in the presence or absence of oxidized ferredoxin.[7]

In crude extracts, the CoA-dependent oxidation of pyruvate is the natural source of electrons for the generation of reduced ferredoxin; then dithionite is not needed. When dithionite is the electron donor, methylviologen will replace ferredoxin as the electron carrier; however, other dyes, including safranin O and neutral red, are inactive.

Reversibility. In the presence of appropriate electron acceptors, including benzylviologen, methylviologen, 2,3,5-triphenyl-tetrazolium dyes, and ferredoxin, the enzyme catalyzes the oxidation of compound (II) to 6-hydroxynicotinate. Slow enzymatic oxidation of compound (II) occurs also in air in the absence of added electron carriers.

The E'_0 for the reverse reaction at pH 7.4 is about -0.39 V.[7]

Stability. The reductase from step 3 is stable to freezing and thawing, storage for several months in liquid nitrogen and up to 50 hours at 5° under argon. It is very unstable in air. This instability could not be prevented by the addition of 1–10 mM FeSO$_4$, MgCl$_2$, NaMoO$_4$, 5–50 mM 6-hydroxynicotinate, compound (II), 2-mercaptoethanol, dithiothreitol, glutathione, or by 0.1 M concentrations of various inorganic salts.[7]

Inhibition. The reductase activity is inhibited by α,α'-dipyridyl or o-phenanthroline. Activity cannot be restored by the addition of FeSO$_4$.

III. Conversion of 1,4,5,6-Tetrahydro-6-oxonicotinic Acid (Compound II) to α-Formylglutaric Acid (Compound III)

An enzyme activity has been demonstrated in crude extracts of the nicotinic acid fermenting organism that catalyzes the conversion of compound (II) to a derivative that is tentatively identified as α-formylglutaric acid.[4] This product has not been isolated as a pure compound. Its tentative identification is based on the fact that it has properties of a β-keto acid and reacts with 2,4-dinitrophenylhydrazine in acid solution to yield the 2,4-dinitrophenylhydrazone derivative of glutaric semialdehyde which could have resulted from decarboxylation of α-formylglutaric acid under the conditions of hydrazone formation.

Assay Method

Principle. At pH 7.5 compound (II) has a strong absorption band at 273 mμ ($A_M = 3.67 \times 10^3$ M^{-1} cm^{-1}), whereas compound (III) does not absorb light at this wavelength (pH 7.5). Therefore under appropriate conditions [i.e., when the oxidation of compound (II) to compound (I) is prevented], the decrease in absorbancy at 273 mμ is a measure of the conversion of compound (II) to compound (III).

Preparation of the Enzyme

No serious efforts have been made to purify the enzyme that catalyzes the conversion of compound (II) to compound (III). It is present in crude extracts and in the fraction derived from step 3 in the purification of 6-hydroxynicotinate reductase described above. It is absent, however, from the fraction obtained in step 4 of that method. Since the oxidation of compound (II) to compound (I) is also associated with a decrease in absorbancy at 275 mμ, it is necessary to inhibit the reductase activity when the conversion of compound (II) to compound (III) is assayed in crude enzyme preparation. To selectively destroy the reductase activity in crude enzyme preparations, the preparations are incubated with 5 mM o-phenanthroline (pH 7.0) for 3 days at 5° and are then passed through a Sephadex G-25 column to separate the protein from reagent. The o-phenanthroline-treated preparations catalyze the conversion of compound (II) to stoichiometric amounts of NH$_3$ and compound (III) estimated as its 2,4-dinitrophenyl hydrazone derivative.[4]

IV. Conversion of (Compound III) to α-Methyleneglutarate (Compound IV)

This reaction has not been directly demonstrated. Its occurrence has been inferred from structural relationship between α-formylglutarate and α-methylene glutarate. The latter has been isolated as one of the metabolites that accumulates when nicotinate is fermented by crude extracts.[3]

V. Conversion of α-Methyleneglutarate (Compound IV) to Dimethylmaleic Acid (Compound V)

Results of studies on the conversion of α-methyleneglutarate to a compound tentatively identified as dimethylmaleic acid are too preliminary to warrant a detailed discussion. Cell-free extracts prepared as described in Section II, step 1, do catalyze the decomposition of α-methyleneglutarate.[5] A major product of the reaction reacts slowly with 2,4-dinitrophenyl hydrazine in Eq. (4) to yield a derivative that has been identified as N-2′,4′-dinitroanilino-3,4-dimethylmaleimide (Va) as represented.

(Va)

This same derivative is produced when dimethylmaleic anhydride is allowed to react with 2,4-dinitrophenylhydrazine under the assay conditions.[5] From this fact and the consideration that at neutral to acid pH dimethylmaleic acid exists in aqueous solution as the anhydride, it is assumed that dimethylmaleic acid is the product of α-methyleneglutarate metabolism and accumulates as the anhydride. This possibility is strengthened by the further consideration that the distribution of various carbon atoms of nicotinate in the ultimate fermentation product (Fig. 1) could be readily explained by the formation of a symmetrical intermediate such as dimethylmaleic acid. Its formation from α-methyleneglutarate would involve a rearrangement in which the bond between carbons 3 and 4 is broken and carbon 3 becomes attached to carbon 5. This rearrangement is very analogous to that involved in the vitamin B_{12} coenzyme-dependent conversion of glutamate to β-methylaspartate which is catalyzed by an enzyme from *Clostridium tetanomorphum*.[8] It is therefore of particular interest that the conversion α-methyleneglutaric acid to dimethylmaleic acid also requires the presence of vitamin B_{12} coenzyme.[5]

Nothing is presently known about the subsequent steps in nicotinic acid degradation; however, a plausible mechanism is illustrated in Fig. 2.

VI. Ring-Labeled Nicotinic Acids

Principle. Nicotinic acids containing ^{14}C-labeling in various positions of the pyridine ring can be obtained by oxidation of appropriately labeled quinolines. Of the many known methods for syntheses of quinolines, (a) the Skraup[9] and (b) the Friedländer[10] methods are chosen for this purpose on the basis of the commercial availability of ^{14}C-compounds required as starting materials. The overall scheme for the preparation of 2-, 5-, and 6-^{14}C-nicotinic acids is represented in Fig. 3.

Procedure a. Nicotinic Acid-2-^{14}C. Aniline hydrochloride-1-^{14}C, 9.8 mg, in a conical centrifuge tube is covered with 1 drop of water and 1 ml of ethyl ether. To this is added 25 mg of anhydrous sodium carbonate; the mixture is stirred thoroughly with a small glass rod. After the solid is settled, the ethereal solution is transferred to a 25-ml round-bottom flask with a side-arm. This process is repeated eight times. The combined ethereal

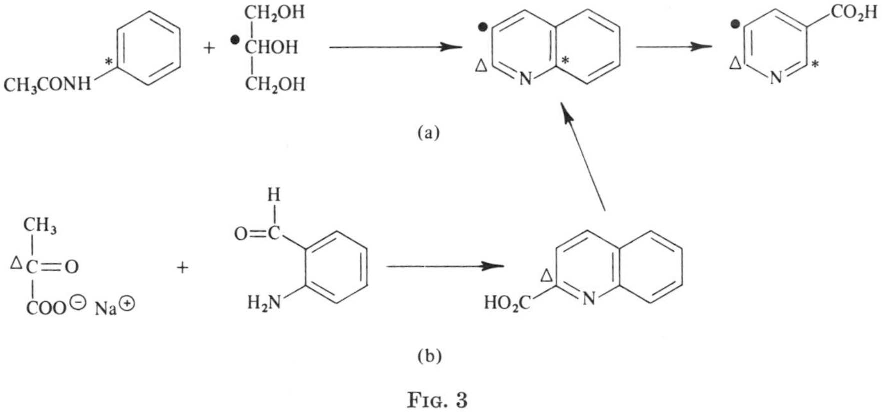

(a)

(b)

FIG. 3

[8] H. A. Barker, H. Weissbach, and R. D. Smyth, *Proc. Natl. Acad. Sci. U.S.* **44**, 1093 (1958).
[9] Z. H. Skraup, *Monatsh. Chem.* **1**, 316 (1880).
[10] P. Friedländer and H. Ostermaier, *Chem. Ber.* **14**, 1916 (1881).

solution is treated with 0.1 ml of acetic anhydride at room temperature. After 5 minutes the solution is evaporated to dryness over a steam-bath. The residue containing 1-^{14}C-acetanilide is placed in a vacuum desiccator over KOH pellets for 2 hours.

To this flask, carrier acetanilide, 0.770 g, is introduced followed by: glycerol, 2.19 g; FeSO$_4$, 0.2 g; boric acid, 0.35 g; nitrobenzene, 0.35 ml, and finally conc. H$_2$SO$_4$, 1 ml.[11] The mixture is heated in an oil-bath at 150° for 20 minutes. The temperature is then raised to 180–185° and kept there for 3.5 hours, during which the mixture turns into a very dark syrup. At the end of the reaction, it is allowed to cool to room temperature, then in an ice-water bath; 10 ml of water is added carefully. This mixture is subjected to steam distillation for 15 minutes. The distillate is discarded, and the solution in the flask is made alkaline by adding slowly, and under ice-bath cooling, 10 ml of a 35% NaOH solution. The tarry mixture resulting is steam distilled until about 40 ml of distillate is collected. The distillate is extracted four times with 5 ml of benzene each time. The combined benzene extracts are dried over anhydrous MgSO$_4$, filtered, and concentrated under reduced pressure; a pale yellow oil is obtained having a UV absorption spectrum in methanol identical with that of quinoline.

The crude ^{14}C-quinoline in a large test tube is mixed with 0.5 g of Se and 8 ml of conc. H$_2$SO$_4$.[12] The mixture is heated in a Wood's metal bath at 320–330° for 5 hours. (This stage must be conducted in a fume hood.) Upon cooling, it is mixed with 20 ml of H$_2$O; the insoluble material is filtered off and discarded. The filtrate is brought to pH 3.6 by careful addition of solid Na$_2$CO$_3$, extracted with 15 ml of n-butanol for 4 times. The combined extracts are evaporated to dryness under reduced pressure. The residue is sublimed at 0.1 mm pressure and 160° for many hours. About 0.2 g of slightly colored sublimate can be obtained. Recrystallization from methanol gives colorless crystalline nicotinic acid-2-^{14}C.

Procedure b. Nicotinic Acid-5-^{14}C. This compound can be prepared as described under procedure a, using glycerol-2-^{14}C and unlabeled acetanilide as starting materials.

Procedure c. Nicotinic Acid-6-^{14}C. A mixture of 0.3 g of aminobenzaldehyde in 10 ml of ethanol, 0.277 g of sodium pyruvate containing pyruvate-2-^{14}C in 1.0 ml of H$_2$O, and 0.2 ml of 5 N NaOH is heated under reflux and stirring for 4 hours.[13] After cooling to room temperature, it is mixed with

[11] R. H. F. Manske and M. Kulka *in* "Organic Reactions," (R. Adams, A. H. Blatt, A. C. Cope, F. C. McGrew, and C. Niemann, eds.), Vol. VII, p. 70. Wiley, New York, 1953.

[12] T. Hirakata, S. Kubota, T. Akita, and I. Aratani, *Chem. Abstr.* **50**, 999b (1956); *Ann. Rept. Fac. Pharm., Tokushima Univ.* **3**, 5 (1954).

[13] W. Borsche and W. Reed, *Ann. Chem., Liebigs,* **554**, 269 (1934).

10 ml of H_2O and extracted twice with ether to remove basic and neutral impurities. The alkaline aqueous solution is acidified with 0.5 ml of glacial acetic acid and extracted six times with 10 ml of $CHCl_3$. The $CHCl_3$ solution is dried over anhydrous $MgSO_4$ and filtered; 1 ml of benzene is added, and the solution is concentrated to about 4 ml. Upon cooling, pale yellow crystals of quinaldic acid-2-^{14}C is collected on a filter and dried over KOH pellets in a vacuum desiccator. The yield is 0.34 g, m.p. 150–153°. Further purification can be achieved by recrystallization from benzene. Quinoline-2-^{14}C can be obtained by heating a solid sample of quinaldic acid to 300° for 1 minute; it is then subjected to oxidation to nicotinic acid as described above.

VII. Preparation of 6-Hydroxynicotinic Acid-7-^{14}C

Principle. The conversion of nicotinic acid to 6-hydroxynicotinic acid is outlined in Fig. 4. The synthesis is based on two main reactions: (a) the alkaline ferricyanide oxidation of the benzylpyridinium salt (VII)[14] and (b) the removal of the N-benzyl group (VIII → IX) by treatment with a $POCl_3$–PCl_5 mixture.[15]

Procedure.[16] A solution of 1.2 mg of nicotinic acid-7-^{14}C in 4 ml of methanol is treated with excess diazomethane in ether at room temperature. After 20 minutes, 87 mg of carrier methyl nicotinate is added and the solution is evaporated to dryness under nitrogen. The residue is dried in

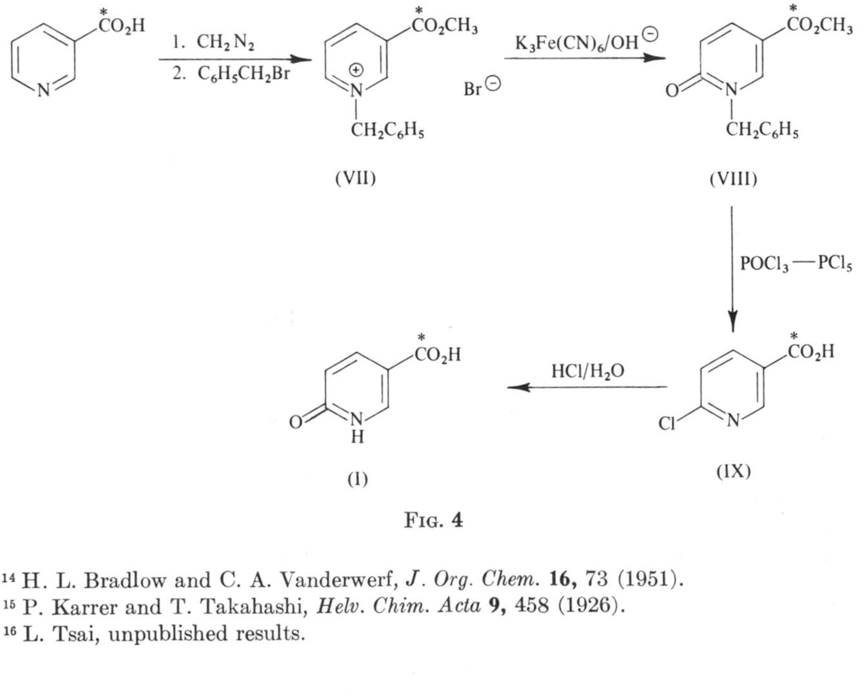

FIG. 4

[14] H. L. Bradlow and C. A. Vanderwerf, *J. Org. Chem.* **16,** 73 (1951).
[15] P. Karrer and T. Takahashi, *Helv. Chim. Acta* **9,** 458 (1926).
[16] L. Tsai, unpublished results.

vacuum, mixed with 2 ml of benzyl bromide and warmed on a steam bath for 1 hour. Upon cooling, 5 ml of water is added, and the mixture is extracted twice with 2 ml of ethyl ether to remove the excess benzyl bromide. To the ice-cooled aqueous solution, containing the pyridium salt (VII), is added under stirring alternate portions of a solution of 1.05 g of $K_3Fe(CN)_6$ and 1.2 ml of 20% NaOH. The addition requires about 10 minutes. The mixture is stirred at room temperature for 2.5 hours and then acidified with glacial acetic acid. Precipitates of the pyridone compound (VIII) are collected and dried in a vacuum overnight. The dried sample (VIII) is mixed with 2.2 ml of phosphorus oxychloride and 80 mg of phosphorus pentachloride, and heated under reflux for 2 hours. It is evaporated under reduced pressure to give 6-chloronicotinic acid (IX) as a brown residue. This is taken up in 4 ml of 6 N HCl and heated under reflux for 18 hours. After cooling to room temperature, the pH of the mixture is brought to about 1 by the careful addition of solid sodium carbonate upon which precipitates of 6-hydroxynicotinic acid-7-^{14}C (I) are obtained. These are collected and recrystallized from water. Yield: 44 mg, m.p. 280–300° (dec.).

VIII. Synthesis of 1,4,5,6-Tetrahydro-6-oxonicotinic Acid

Principle. The preparation of a closely related compound, ethyl 2-methyl-1,4,5,6-tetrahydro-6-oxonicotinate, has been reported,[17] but attempts to adapt this method for the preparation of 1,4,5,6-tetrahydro-6-oxonicotinic acid (II) have not been successful. A synthesis has been developed[18] by taking advantage of a ring-cleavage reaction of coumalic acid observed by von Pechmann.[19] Thus, the transformation of coumalic acid to 1,4,5,6-tetrahydro-6-oxonicotinic acid can be accomplished by the sequence of reactions summarized in Fig. 5.

Procedure a. Dimethyl α-Aminomethyleneglutaconate (XI).[20] A suspension of 50 g of dried coumalic acid in 500 ml of methanol, cooled in an ice-bath, is saturated with dry HCl gas. The mixture is heated under reflux for 2 hours, then evaporated under reduced pressure to dryness. Crude dimethyl α-methoxymethyleneglutaconate (X) is obtained as a semisolid mass.[21] This is dissolved in 100 ml of methanol and added slowly with stirring to 200 ml of conc. NH$_4$OH cooled in an ice-salt bath. The addition should be at such a rate as to keep the temperature of the reaction mixture below 0°. After the addition, the mixture is stirred for another 0.5 hour.

---

[17] N. F. Albertson, *J. Am. Chem. Soc.* **74**, 3816 (1952).

[18] L. Tsai and E. Caveney, unpublished results.

[19] H. von Pechmann, *Ann. Chem., Liebigs*, **273**, 164 (1893).

[20] This procedure is a modification of one reported by von Pechmann.[19] The assignment of structures (X) and (XI) is based on the spectral properties of these compounds.

[21] This compound can be purified by recrystallization, but it decomposes upon standing, hence it should be used directly for the subsequent step.

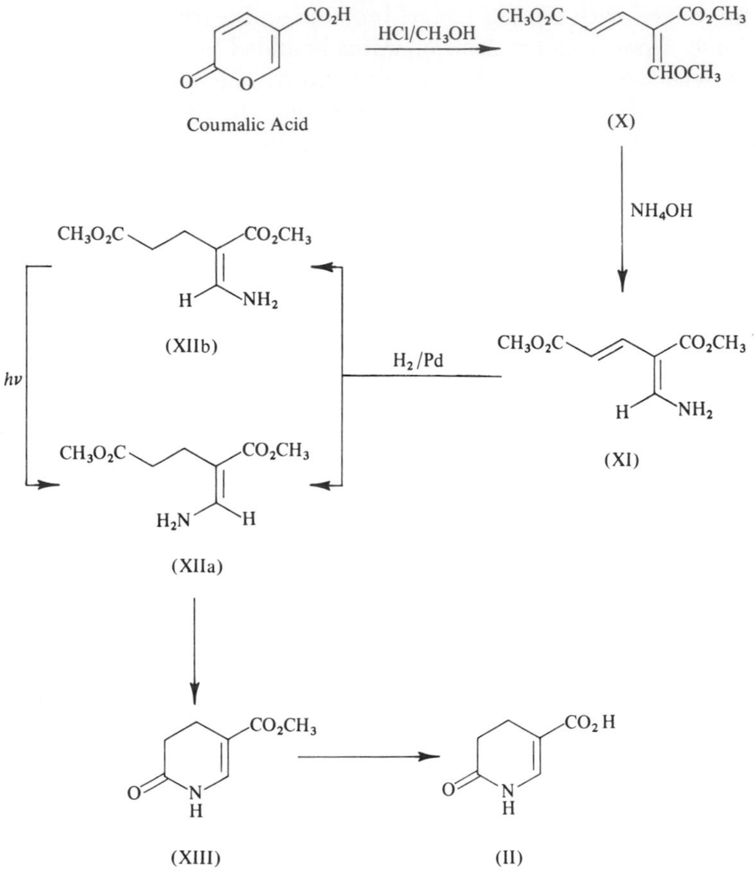

Fig. 5

The yellow precipitates are filtered and washed thoroughly with ice-water. Crystallization from dilute methanol yields 32.3 g of dimethyl α-amino-methyleneglutaconate (XI) as light yellow needles, m.p. 141–142°.

Procedure b. Dimethyl α-Aminomethyleneglutarate (XIIa). A mixture of 1.0 g of dimethyl α-aminomethyleneglutaconate (XI) and 50 mg of Pd-C (10%) catalyst in 10 ml of methanol is shaken with hydrogen at 20 psi at room temperature for 1 hour. The catalyst is filtered off, and the filtrate is evaporated under reduced pressure to give a slightly colored oil which consists of a mixture of the two geometric isomers, (XIIa) and (XIIb), of dimethyl α-aminomethyleneglutarate.[22] In general, (XIIb) is

---

[22] These two isomers are distinguishable by their NMR and UV spectra as well as their retention time on GLC. Enrichment of (XIIa) is observed after the preparation has stood in the refrigerator for many days.

the predominant component. Of these two forms, only (XIIa) possesses an arrangement of groups that is favorable for cyclization. The conversion of (XIIb) to (XIIa) is accelerated by irradiation. The mixture is dissolved in 750 ml of benzene and irradiated under a nitrogen atmosphere by a medium pressure Hg-lamp using a Pyrex filter for 5 hours. The solvent is distilled off under reduced pressure; an almost white solid residue is obtained. Recrystallization from ethyl acetate-cyclohexane gives colorless crystals of (XIIa), m.p. 91–92°, 0.8 g.

Procedure c. Methyl 1,4,5,6-Tetrahydro-6-oxonicotinate (XIII).[23] A solution of 1.09 g of dimethyl α-aminomethyleneglutarate (XIIa) in 250 ml of 0.035 M sodium methoxide in methanol is heated to reflux for 0.5 hour. The mixture is neutralized with 0.26 ml of conc. H_2SO_4, dried over anhydrous sodium sulfate, and filtered. The filtrate is evaporated to dryness under reduced pressure. Two recrystallizations of the residue from methylene chloride–cyclohexane gives 0.55 g of the methyl ester (XIII), m.p. 88–90°.

Procedure d. 1,4,5,6-Tetrahydro-6-oxonicotinic Acid (II).[23] A solution of 400 mg of the methyl ester (XIII) in 2 ml of methanol and 2 ml of 10% Na_2CO_3 is heated on a steam bath for 45 minutes. Upon cooling, the mixture is diluted with 10 ml of water, and extracted with 5 ml of chloroform to remove neutral material. The alkaline aqueous solution is acidified with 6 N HCl, and precipitates of 1,4,5,6-tetrahydro-6-oxonicotinic acid (II) are collected on a filter, washed with a small volume of water, and dried in a vacuum desiccator. Yield: 120 mg, m.p. 210–220° (dec.).

----

[23] Optimal conditions of this step have not yet been worked out.

Section VIII

Flavins and Derivatives

[132] Fluorometric Analyses of Riboflavin and Its Coenzymes

By Jacek Kozioł

Flavins are present in almost all types of biological tissues. The discovery of the vitamin activity of riboflavin initiated extensive studies of the biochemical role and functions of flavins. It is now obvious that most biological tissues show the presence of many biologically active structures in which different riboflavin derivatives are incorporated.

The most widespread flavins of biological importance are riboflavin (Rb) and the so-called nucleotides: riboflavin 5'-phosphate (flavin mononucleotide, FMN) and the intramolecular complex of FMN with adenosine 5'-monophosphate (flavin adenine dinucleotide, FAD). There is evidence that other biologically important forms of flavins[1,2] are also present in some biological tissues. These seldom occurring forms, as well as different biologically active flavin derivatives (fatty acid esters of riboflavin[3] and others[2]), artificially introduced to the tissues will be disregarded in this work.

The quantitative determination of flavins for several reasons should be done by applying methods of high sensitivity and specificity. The three most important flavins, Rb, FMN, and FAD, appear in biological tissues in the free state (mainly riboflavin) or as complexes of different types with proteins (both nucleotides) and other components of living cells. In most types of biological tissues, flavins are present in very low concentrations in the range of 1 μg/g or less. Fluorometric methods, which have been in common use for years for the analyses of riboflavin and its coenzymes, are considered to be the most sensitive physicochemical methods now available.

The fluorometric determination of flavins can be carried out in two ways; by the measurement of the intensity of the natural fluorescence of flavins and the fluorescence of lumiflavin derived from flavins. Both these methods can be equally well applied by using similar analytic procedures according to certain physicochemical properties of flavins. These properties, important as a basis for the fluorometric determination, will be described with respect to the three main forms—Rb, FMN, and FAD—

---

[1] H. Beinert, in "The Enzymes" (P. D. Boyer, H. Lardy, and K. Myrbäck, eds.), Vol. 2, p. 339. Academic Press, New York, 1960.

[2] See articles in this volume on isolation of FAD peptides [143], nekoflavin [144], and riboflavin glycosides [141] and on preparation of riboflavin analogs [145].

[3] K. Yagi, J. Okuda, A. A. Dmitrovskii, R. Honda, and T. Matsubara, J. Vitaminol. 7, 276 (1961).

and to the two biologically inactive products of their photochemical degradation: lumiflavin (Lf) and lumichrome (Lc). Lumichrome, rarely present in tissues, is easily formed when samples or flavin solutions are handled without sufficient care. The presence of lumichrome, because of its specific physicochemical properties, can cause considerable analytic errors. More exhaustive information concerning the chemical, physical, and biochemical behavior of flavins is given in some excellent reviews.[1,4,5]

Physicochemical Properties of Flavins Important for Their Fluorometric Determination

Crystalline flavins are powderlike orange-yellow (lumichrome, pale yellow) substances. FMN or its sodium salts are slightly hygroscopic; FAD and its salts are hygroscopic. Riboflavin is polymorphic; three different crystal forms of different solubility in water are well known.

Solubility

All flavins dissolve well in dilute aqueous solutions of alkalies, in pyridine, and in glacial acetic acid, and not so well in phenols. Riboflavin is poorly soluble in water (0.03–0.15 g/100 ml) with differences due to the different crystal forms. Lumiflavin and lumichrome are less soluble in water (approximately 2.6 mg and 0.8 mg per 100 ml, respectively). FMN and its sodium salt are highly soluble and FAD is freely soluble in water. All three compounds are slightly more soluble in aqueous salt solutions and dilute acids. The solubility of flavins in organic solvents depends on the polarity of compounds: FAD is only slightly soluble in benzyl alcohol; FMN dissolves a little more in benzyl alcohol and slightly in ethanol; riboflavin is much more soluble in ethyl (4.5 mg/100 ml) and benzyl alcohols, slightly soluble in propyl and amyl alcohols, cyclohexanol, and amyl acetate. Riboflavin and its nucleotides are insoluble in acetone, chloroform, ethyl ether, and hydrocarbons. Lumiflavin and lumichrome show limited solubility in alcohols (methyl, ethyl, propyl, butyl, benzyl) and in acetone, chloroform, p-dioxane, and 80% cyclohexane–dioxane mixture.

Care must be taken when preparing solutions of riboflavin, lumiflavin, and lumichrome since they show a tendency to form supersaturated solutions.

Stability of Flavins

Heat Stability. Flavins in crystalline form are stable to heating up to 100–120°. In aqueous neutral solutions they show no destruction by pro-

[4] P. Hemmerich, C. Veeger, and H. C. S. Wood, *Angew. Chem.* **77,** 699 (1965).
[5] G. R. Penzer and G. K. Radda, *Quart. Rev. (London)* **21,** 43 (1967).

longed heating up to 70°. Higher temperatures, up to 120°, have no effect on Rb, FMN, Lf, and Lc in the pH range between 2 and 5.[6] FAD in aqueous solutions heated above 70° is transformed into riboflavin 4',5'-cyclic phosphate.[7] In buffer solutions of higher concentrations, some buffer anions and cations lower the heat stability of flavins, especially at elevated pH values.

Action of Alkali and Acids. All flavins undergo gradual decomposition in the presence of alkalies. This reaction is base-catalyzed, and its effectiveness is related to alkali and flavin concentration, temperature, etc. As shown for isoalloxazines related to riboflavin in aerobic or anaerobic conditions, at different pH values different products and intermediates are formed as a result of the cleavage of the pyrimidine part of the molecule.[8] It has been claimed that ammonia has a specific destructive action on flavins. In diluted solutions of ammonia, FAD is decomposed to riboflavin 4'5'-cyclic-phosphate even in the cold.[9]

Flavins show a tendency to form complexes with silver, cuprous, and mercuric ions in water. FMN and FAD form alkaline earth metal salts (the barium salt of FAD is insoluble in water). Flavin metal chelates are generally slightly soluble in water. In nonaqueous solvents, flavins easily form weak chelates with other metal ions.[4]

Generally acids have no destructive influence on the isoalloxazine nucleus of flavins. The ester bond of flavin nucleotides, however, is readily hydrolyzed in diluted acids; FAD can be fully hydrolyzed to FMN by allowing it to stand overnight at 38° in 10% trichloroacetic acid, although no destruction occurs at 0° for 30 minutes. In solutions of diluted mineral acids (0.1 N HCl at 100°) FAD is fully hydrolyzed to FMN and partially hydrolyzed to riboflavin. Prolonged heating of flavins in more concentrated acids leads to the formation of different transient products of FAD and FMN hydrolysis, including riboflavin, and finally to its decomposition. In concentrated acids riboflavin is easily protonized to different degrees.

Action of Enzymes. Both phosphate esters of riboflavin (FMN and FAD) can be more specifically hydrolyzed by means of enzyme action. Since pure phosphatase preparations are too expensive, commercial preparations, such as clarase and Takadiastase, are in common use. Pure FAD solutions are easily hydrolyzed to FMN and partly hydrolyzed to riboflavin; FAD in extracts obtained from biological sources, especially from plant tissues, is more resistant to phosphatase action. The hydrolysis of FMN to riboflavin needs prolonged action of enzymes and is quantitative only in pure

[6] K. T. H. Farrer and J. L. MacEvan, *Australian J. Biol. Sci.* **7**, 73 (1954).

[7] K. Yagi and Y. Matsuoka, *J. Biochem.* **48**, 93 (1960).

[8] D. E. Guttman and T. E. Platek, *J. Pharm. Sci.* **56**, 1423 (1967).

[9] H. S. Forrest and A. R. Todd, *J. Chem. Soc.* p. 3295 (1950).

solutions. Phosphatases in commercial preparations can be quantitatively inhibited by means of 1 M phosphate buffer (with respect to action on FMN, they lose their activity in 0.5 M phosphate).

Oxidation–Reduction. Flavins are stable against common oxidizing agents (excepting chromic acid, persulfate, and permanganate). In slightly acidic solutions (pH 4–5) at room temperature, flavins are almost stable at low concentrations (up to 0.02 M) of potassium permanganate. Also hydrogen peroxide is without effect in low concentration and in the absence of ferrous ions. An apparent loss of flavin stability toward oxidizing agents can be observed, however, in solutions containing higher salt concentrations, depending also on the type of anions present. The use of different reducing agents (dithionite, titanous chloride, zinc, hydrochloric acid, active hydrogen, etc.) leads to different forms of flavins. The reduction is reversible, and free flavins not bound to proteins are very easily oxidized by air oxygen. Reduced forms of flavins differ in some of their properties (lower solubility, lack of fluorescence, different absorption spectra, etc.).

Photolysis. Flavins are very sensitive to light of wavelengths shorter than 600 nm. Riboflavin and FMN in neutral aqueous solutions show similar relative high photolability while FAD is photostable. This difference is due to the fact that in neutral solutions FAD exists in the form of an intramolecular complex. In conditions in which this complex is destroyed (alkaline solutions, addition of organic solvent), all three forms become almost equally sensitive to the action of light.[10]

When considering the photolytic properties of flavins, at least three different processes must be taken into account: anaerobic photoreduction, and anaerobic and aerobic photobleaching.

Anaerobic photoreduction takes place when flavins are illuminated in a solution in which external electron donors are present. Flavins are photoreduced and thereby lose their color, and characteristic light absorption. On admitting oxygen, flavins are quantitatively oxidized without decomposition.

Anaerobic photobleaching differs from photoreduction in the absence of external electron donors. Under such conditions the role of an electron donor is played by the ribityl side chain; this results in the formation of reduced flavins, lumiflavin, lumichrome, and 6,7-dimethyl-9-formylisoalloxazine (an intermediate photoproduct that is easily transformed into lumiflavin, even in the dark, when treated with alkali[11,12]). When oxygen is admitted, color is only partially restored, as lumichrome has a different absorption spectrum.

[10] G. Oster, J. S. Bellin, and B. Holmström, *Experientia* **18**, 249 (1962).
[11] E. C. Smith and D. E. Metzler, *J. Am. Chem. Soc.* **85**, 3285 (1963).
[12] P. S. Song, E. C. Smith, and D. E. Metzler, *J. Am. Chem. Soc.* **87**, 4181 (1965).

Aerobic photobleaching is a process that commonly occurs in the course of work with flavin solutions in laboratories that are not darkened. Such a process is similar to photobleaching under anaerobic conditions. The main difference lies in the facts that (1) this process is subject to general acid–base catalysis and therefore is pH dependent, (2) more intense light is required, as the presence of oxygen hinders the formation of the photolabile flavin species—the triplet. In the course of this process in aqueous solutions at pH higher than 7.0, lumiflavin is predominantly formed, whereas at pH below 7.0 the main product is lumichrome.

In nonaqueous solutions or in solutions containing the highest possible concentration of organic solvent in which flavins are still soluble, significant enhancement of photolability is observed.[13] The main product of riboflavin photolysis in ethanol, 95% acetone, 98% dioxane, glacial acetic acid, and pyridine is lumichrome. The photolability of riboflavin in such conditions is 3–5 times higher than in water. The highest observed photolability of riboflavin occurred in pyridine, whereas an almost immediate photobleaching with consequent decomposition was observed in acetic acid. Lumiflavin, which is stable in aqueous neutral solutions, becomes photolabile in organic solvents, especially in dioxane and in glacial acetic acid. Similar effects were also observed for lumichrome. The enhancement of riboflavin photolability in low polar solvents is comparable to that observed in alkaline solutions, and great care must be taken, as remarkable decomposition can be caused even by measuring the absorption or emission spectra.

Light Absorption

All flavins concerned show strong absorption in the blue and ultraviolet regions of the spectrum. The electronic spectra of riboflavin and FMN in water are apparently identical and consist of four bands centered around 223, 268, 374, and 449 nm. Absorption spectra of the two other isoalloxazine derivatives, FAD and lumiflavin, are similar to those of riboflavin and FMN.[14] The main difference lies in the shifts of maxima positions and changes in molar extinction coefficients. In the case of lumiflavin, these changes are negligible and result from the elimination of the inductive influence of the ribityl side chain. The differences in FAD spectrum are connected with the existence of the intramolecular complex between isoalloxazine and adenosine parts of the molecule.

Similar changes in the spectrum of riboflavin are caused by phenol and its derivatives—adenosine, caffeine, and other compounds which

[13] J. Kozioł, *Photochem. Photobiol.* **5,** 55 (1966).
[14] L. G. Whitby, *Biochem. J.* **54,** 437 (1953).

TABLE I
Light Absorption Characteristics of Flavins in Different Solvents

Compound	Solvent	Maxima positions (nm) (molar absorptivities $\epsilon \times 10^{-3}\ M^{-1}cm^{-1}$ in parentheses)			
Isoalloxazine derivatives					
Riboflavin	Water	223 (30.1)	268 (31.4)	374 (10.8)	448 (12.3)
	Phosphate buffer, 0.1 M, pH 7.0[a]	—	266 (32.5)	373 (10.6)	445 (12.5)
	98% dioxane in mixture with water	224 (31.0)	271 (33.1)	344 (8.4)	440 (12.1)
FMN	Water	223 (30.5)	268 (32.8)	374 (10.6)	448 (12.6)
	Phosphate buffer, 0.1 M, pH 7.0[a]	—	266 (31.8)	373 (10.4)	445 (12.5)
FAD	Phosphate buffer, 0.1 M, pH 7.0[a]	—	263 (38.0)	375 (9.3)	450 (11.3)
Lumiflavin	Water	225.5 (30.3)	265.5 (32.4)	369 (8.4)	444 (10.9)
	50% cyclohexane in dioxane	223 (27.1)	271 (32.3)	332 (6.6)	440 (10.2)
			S[b] ~ 260 (24.0)		S ~ 420 (7.9)
			S ~ 254 (16.0)		S ~ 463 (7.2)
			S ~ 248 (9.6)		
			S ~ 242 (5.6)		
Alloxazine derivative					
Lumichrome	Water	219 (30.1)	261 (30.5)		354 (8.6)
			S ~ 250 (25.4)		S ~ 384 (5.9)
	NaOH, 0.25 N, aq.	222 (30.6)	264 (43.0)	351 (6.5)	424 (5.3)
	80% cyclohexane in dioxane	221 (41.0)	258.5 (35.0)	328.5 (8.3)	379.5 (8.6)
			S ~ 248.0 (38.5)		S ~ 394.0 (7.3)

[a] L. G. Whitby, *Biochem. J.* **54**, 437 (1953).
[b] S, shoulders.

show a complexing ability toward riboflavin.[15,16] Absorption spectra of many types of flavoproteins are similar in character.[17]

Lumichrome, an alloxazine derivative, in water has a spectrum with bands around 220, 262, and 356 nm (Table I).

Absorption spectra of flavin ionic species are different as compared to the neutral forms. According to Dudley et al.,[18] flavoquinone monocation (in 6 N HCl) has its absorption maxima at 222, 264, and 390 nm; the anion (2 N NaOH) maxima are at 230, 270, 350, and 444 nm. Under the same conditions lumichrome gives quite similar spectra (in 6 N HCl its maxima are at 214, 260, and 390 nm; and in 2 N NaOH, at 224, 266, 340, and 432 nm. The solvent polarity has also a significant influence on the absorption spectra of flavins.[19,20] The solvent polarity-dependent changes in absorption spectra of isoalloxazines are most pronounced in the case of the simplest derivative—lumiflavin (Fig. 1).

Also Table I presents spectral data for riboflavin and lumiflavin in solvents of the lowest polarity in which these compounds are sufficiently soluble. The main changes are: the blue shift of 370 nm band and the decrease of absorbancy, the formation of shoulders on both sides of the longest wavelength band at 445 nm and on the short wavelength side of the 260 nm band. There is a roughly linear correlation between the shifted position of the 370 nm band and the experimentally measured polarity of organic solvents taken in order (from highest to lowest): acetic acid, ethanol, pyridine, acetone, p-dioxane, and cyclohexane–dioxane mixtures. A possible explanation for the observed solvent-sensitive changes can be connected with the tendency of the flavin to form hydrogen bonds in polar solvents and/or with the contribution of a charge-transfer structure in the part of the molecule which corresponds to the 370-nm band. The shoulders have an apparently vibrational character. All absorption bands in spectra of flavins reflect transitions of a $\pi \rightarrow \pi^*$ type. That statement is supported by high molar absorptivities of all bands, high quantum yields of fluorescence (see below), and a constant degree of fluorescence polarization across absorption bands at 370 and 445 nm indicating that there are only two independent transitions.[21] Recent optical rotatory dispersion studies on

[15] J. A. Roth and D. B. McCormick, *Photochem. Photobiol.* **6**, 657 (1967).

[16] D. B. McCormick, Heno-Chun Li, and R. E. MacKenzie, *Spectrochim. Acta* Part A, **23**, 2353 (1967).

[17] V. Massey and H. Ganther, *Biochemistry* **4**, 1161 (1965).

[18] K. H. Dudley, A. Ehrenberg, P. Hemmerich, and F. Müller, *Helv. Chim. Acta* **47**, 1354 (1964).

[19] J. Kozioł, *Photochem. Photobiol.* **5**, 41 (1966).

[20] J. Kozioł, *Photochem. Photobiol.* **9**, 45 (1969).

[21] G. Weber, *in* "Flavins and Flavoproteins" (E. C. Slater, ed.), p. 15. Elsevier, Amsterdam, 1966.

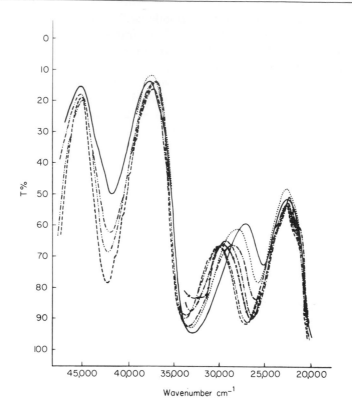

FIG. 1. Absorption spectra of lumiflavin in solvents of different polarity. Concentration of lumiflavin, $2.66 \times 10^{-5} M$; light path, 10 mm; room temperature. Water ———, acetic acid • • • • •, pyridine —•—•—, ethanol — — •• — — — ••, acetone ×—×, dioxane — •• — ••, 50% cyclohexane in dioxane — — — —.

flavins[22, 23] suggest, however, that each of the absorption bands at 220 and 268 nm corresponds to more than one $\pi \rightarrow \pi^*$ transition, and that very weak transitions of $n \rightarrow \pi^*$ type most probably take place in the 300–340 nm region of absorption spectrum. Further studies are necessary for a more complete elucidation of spectral properties of flavins.

More evident changes are caused by low polar solvents in the absorption spectrum of lumichrome (Fig. 2 and Table I). It can be supposed that in water lumichrome exists in a partly ionized form and that in low polar media the true neutral alloxazine form is present.

[22] This volume, Optical rotatory dispersion of flavins and flavoproteins.
[23] D. W. Miles and D. W. Urry, *Biochemistry* **7**, 2791 (1968).

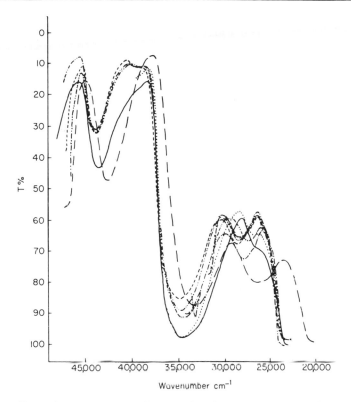

FIG. 2. Absorption spectra of lumichrome in solvents of different polarity. Concentration of lumichrome, $2.66 \times 10^{-5} M$; light path, 10 mm; room temperature. Water ———, acetic acid ·····, pyridine —·—·—, ethanol — — ·· — —, acetone ×—×, dioxane — ·· —, 80% cyclohexane in dioxane ------------, 0.25 N NaOH (aqueous) — — —.

Fluorescence

Flavin derivatives exhibit an intense greenish yellow fluorescence, and this property is widely used for their characterization and determination. The fluorescence excitation spectra of flavins within experimental error are identical with their absorption spectra; this indicates that the intensity of fluorescence is proportional to the light absorption throughout the spectral range between 210 and 550 nm. The fluorescence emission spectra of flavins in water are a broad, structureless band with maximum about 530 nm, slightly shifted for individual flavin species. In water solutions lumichrome shows a bluish green fluorescence with a broad maximum at about 485 nm. In equimolar neutral aqueous solutions, the intensity of fluorescence of Rb and FMN is practically the same whereas that of lumiflavin is slightly higher. The fluorescence intensity of FAD represents only 15% of that of riboflavin. This value apparently depends on FAD purity and other conditions, and different authors have reported values between 9 and 20%.[24,25] Higher fluorescence of FAD in aqueous solutions indicates its decomposition.

Reduced forms of flavins do not show any fluorescence. This property is widely used for preparing blanks in the course of fluorometric determinations (sodium dithionite, concentrated HCl, anaerobic photoreduction, etc.).

The fluorescence of flavins is distinctly temperature dependent and decreases with increasing temperature (approximately 1% per 1°). In rigid media at low temperature, isoalloxazines and alloxazines exhibit phosphorescence with maxima at about 605 and 545 nm, respectively.[26]

Effect of pH. The emission of fluorescence is a characteristic only of uncharged, neutral forms of isoalloxazines. Anions and cations do not fluoresce. The influence of pH on the fluorescence of flavins in aqueous solutions was carefully studied.[24,25,27] Riboflavin shows an equal fluorescence intensity in the pH range between 3.5 and 7.5. The same is true for FMN and lumiflavin. FAD reaches its maximal fluorescence intensity at pH 2.7–3.1, and at pH 1.5 it is equal to that of riboflavin. The same feature is observed also for different flavoproteins. This property is related to the fact that at pH values close to 3.0 the intramolecular complex of FAD and the flavoprotein complexes dissociate. The difference of the dependence on pH on the fluorescence intensity between FAD and other flavins can be utilized in analytical work. Changes of fluorescence intensity caused by the alteration of pH are reversible, and if no decomposition of flavins (especially in alkaline medium) occurs, the original fluorescence intensity reappears after neutralization. Changes of pH values do not influence the position of fluorescence maxima.

An alloxazine derivative such as lumichrome behaves in a different way. Lumichrome, nonfluorescent as cation or anion, shows blue fluorescence in slightly acidic solutions (at pH 3.5 maximum at about 450 nm), and with the increase of pH its fluorescence maximum shifts bathochromically. In neutral aqueous solutions the maximum fluorescence of lumichrome lies at about 485 nm. In slightly alkaline solutions or in pyridine, the shape of the lumichrome emission spectrum becomes almost identical with

[24] O. A. Bessey, O. H. Lowry, and R. H. Love, *J. Biol. Chem.* **180,** 755 (1949).
[25] P. Cerletti and N. Siliprandi, *Arch. Biochem. Biophys.* **76,** 214 (1958).
[26] J. M. Lhoste, A. Haug, and P. Hemmerich, *Biochemistry* **5,** 3290 (1966).
[27] P. Cerletti and A. Rossi-Fanelli, *J. Vitaminol.* **4,** 71 (1958).

those of isoalloxazine derivatives, reaching comparable intensity in pyridine (see the section on influence of solvent polarity, below).

The riboflavin fluorescence intensity dependence on pH in 1:1 (by volume) mixtures of ethanol, acetone, or dioxane with 0.04 M Britton-Robinson buffer is practically the same as in buffer alone.

Quenching. The fluorescence intensity of flavins is restricted by many factors. The analyst must take into account at least three different processes known as "quenching"; these result from too high concentration of solutions, collisions between molecules, and formation of complexes.

Concentration quenching[28] (self-quenching, inner filter effect) can easily be observed as lack of linearity of correlation between fluorescence intensity and concentration of fluorescent species. It takes place when not all the molecules are excited or when part of the emitted fluorescence is reabsorbed by molecules in the solution. In the first case the intensity of exciting light, and in both cases the concentration of fluorescent molecules and of other absorbing species is important. As the presence of absorbing contaminants is to be avoided in all cases, the practical consequence is the necessity of using sufficiently dilute solutions of the fluorescent compounds under examination. The limits of flavin concentration must be defined for each apparatus, no concentration quenching usually occurs in solutions containing no more than 1 μg of flavin per milliliter.

Collisional quenching is due to collisions between molecules of fluorescent species and molecules of the quencher, in which the excited molecules lose their energy and return to the ground state without the emission of fluorescence. This process is diffusion-controlled and takes place only when the lifetime of the excited state involved is longer than 10^{-9} second. The fluorescence of flavins is quenched by most electrolytes, including different metal ions. Specially efficient are halide salts, even of simple alkali metals.

The formation of nonfluorescent complexes between flavins and phenols, purines, pyrimidines, and other organic species including solvents is a more specific way of quenching. Such complexes are formed between molecules in the ground state and generally have no influence on the lifetime of the excited state. Quenching of this type takes place within the molecule of FAD.[29] The internal complex between isoalloxazine and adenine moieties is nonfluorescent, and there is an equilibrium between undissociated and dissociated (fluorescent) forms. This equilibrium depends on pH, salt concentration, solvent polarity, etc. Generally, complexes between flavins and organic molecules are of the donor–acceptor type with

---

[28] D. J. Laurence, Vol. IV, p. 174.
[29] G. Weber, *Biochem. J.* **47**, 114 (1950).

partial charge-transfer effects. Many flavoproteins also belong to such complexes. The quenching of flavin fluorescence by heavy metal ions is caused by formation of chelate-type complexes.

In the course of sample preparation for fluorometric analyses, all types of flavin complexes with other molecules (except in special cases, such as when FAD is to be determined) are to be destroyed and the complexing agents should be eliminated.

There are also other possibilities for fluorescence quenching, such as noncollisional energy transfer and specific and nonspecific dipole-dipole interactions. Uncontrolled quenching effects are to be eliminated if possible, but if controlled they could be a useful tool for an analyst.

Influence of Solvent Polarity. The fluorescence of flavins is to a great extent influenced by the solvent polarity. Generally, the intensity of fluorescence increases with decrease of solvent polarity, though with different effectivity for individual flavins. Bessey et al.[24] observed that fluorescence intensity of FAD in mixtures of ethanol and benzyl alcohol is considerably higher than in water. The author, studying the desorption of flavins from phenolic resins, observed significant changes of quantum yield of riboflavin and lumiflavin fluorescence in organic solvent–water mixtures (Table II) and of Rb, Lf, and Lc in pure organic solvents (Table III). These changes differ in character in the case of isoalloxazine and alloxazine derivatives. Parallelly, considerable changes were observed in the shape of emission spectra. For lumiflavin, because of the lack of the hydrophilic ribityl chain, which promotes nonradiative energy dissipation of the excited molecule, the changes are more evident than in the case of riboflavin.

It can be seen from Fig. 3 that with the decrease of the solvent polarity the main changes in lumiflavin emission spectra are increase of intensity, a

TABLE II

QUANTUM YIELDS ($\phi \times 10^2$) OF FLAVIN FLUORESCENCE IN SOME ORGANIC
SOLVENT–WATER MIXTURES

Organic solvent concentration % (v/v)	Riboflavin				Lumiflavin		
	Ethanol	Acetone	Dioxane	Pyridine	Ethanol	Acetone	Dioxane
0 (water)	25.2	—	—	—	28.2	—	—
10	25.6	25.6	25.8	10.4	29.0	29.5	29.6
25	26.2	26.5	27.0	8.8	30.1	32.0	32.5
50	27.4	27.7	29.0	12.6	32.4	36.0	35.6
80	28.4	29.0	32.4	24.4	36.5	43.0	44.6
90	29.7	32.0	35.0	28.0	39.0	40.0	51.9
95	31.0	35.0	38.5	—	—	—	—

TABLE III

QUANTUM YIELDS ($\phi \times 10^2$) OF FLAVIN FLUORESCENCE IN SOME
ORGANIC SOLVENTS

Solvent	Riboflavin	Lumiflavin	Lumichrome
Water[a]	25.2	28.2	8.8
Acetic acid	18.4	23.5	7.0
Pyridine	32.6	52.0	38.0
Ethanol	32.0	42.0	7.6
Acetone[b]	35.0	56.7	3.5
Dioxane[c]	41.0	63.6	3.0
50% Cyclohexane in dioxane	—	65.2	2.8

[a] Recently A. Bowd, P. Byrom, J. B. Hudson, and J. H. Turnbull, *Photochem. Photobiol.* **8**, 1 (1968) reported quantum yields of some flavins in water and ethanol. Assuming quantum yield of riboflavin in water at room temperature to be equal to 0.26, they found the following values: in water 29 (Lf), 25 (FMN), 38 (FAD); in ethanol 45 (Lf), 27 (FMN), 20 (FAD), and 32 (Rb).

[b] For riboflavin 95%.

[c] For riboflavin 98%.

slight blue shift of the maximum about 535 nm, and significant broadening toward short wavelengths with the formation of a new maximum at about 500 nm (similar changes were recently reported for FMN in ethanolic solution at 77°K[30]). The overall shape of spectra indicates that there exists a mirror symmetry to the absorption maximum at longest wavelengths. Roughly linear correlation exists between changes of quantum yields of fluorescence, positions of the solvent-sensitive absorption maximum at 370 nm, and the experimentally measured polarity of solvents applied. A plausible explanation of the effect of the solvent may be as follows. Flavin molecules in water and other solvents of relatively high polarity tend to form hydrogen bonds with solvent molecules. Also nonspecific dipole–dipole interactions between solute and solvent molecules may be taken into consideration here. In the case of FAD an intramolecular complex between isoalloxazine and adenine moieties exists in solvents of high polarity. In solvents of low polarity, the solute-solvent interactions become weaker or display a character that does not hinder the process of radiative emission. In such conditions the intramolecular complex in FAD is decomposed. It was already shown that complexes of flavins with different organic molecules are formed in the presence of water molecules, and it is very possible that there exists a specific interaction with water. For instance, riboflavin in water–pyridine mixtures, when the concentration of pyridine

[30] A. Bowd, P. Byrom, J. B. Hudson, and J. H. Turnbull, *Photochem. Photobiol.* **8**, 1 (1968).

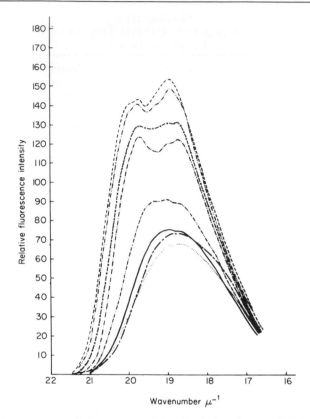

FIG. 3. Fluorescence emission spectra of lumiflavin in solvents of different polarity (corrected). Concentration of lumiflavin $1.33 \times 10^{-6} M$, exciting light 436 nm, room temperature. Water ———, acetic acid ······, pyridine —·—·—, ethanol — — ·· — —, acetone ×—×, dioxane — ·· —, 50% cyclohexane in dioxane - - - - - - -. For comparison, riboflavin in water $(1.33 \times 10^{-6} M)$ — · — · -.

is increased, shows changes in the absorption spectra and the quenching of fluorescence, which can be regarded as evidence of the formation of a ternary complex between the molecules of flavin, pyridine, and water.[21,31]

In the case of lumichrome, an alloxazine derivative, the changes caused by solvent polarity are of a different type (Fig. 4).

The emission spectra in acetic acid and pyridine, both acting probably as electron donors for lumichrome, in shape, maximum position, and intensity are close to those of isoalloxazine derivatives. Since absorption spectra in these solvents change as for other low polar solvents, it can be concluded that excited molecules of lumichrome are transformed into

[31] J. Kozioł, *Zeszyty Nauk. Wyzszej Szkoły Ekonomicanej w Poznaniu*, Ser. I. Nr. 26, p. 91 (1966); cf. *Chem. Abstr.* **66**, 68960 (1967).

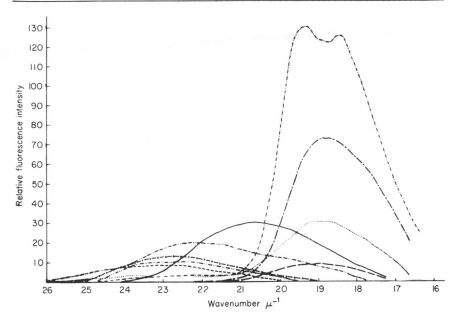

Fig. 4. Fluorescence emission spectra of lumichrome in solvents of different polarity (corrected). Concentration of lumichrome $2.66 \times 10^{-6} M$; exciting light, 365 nm; room temperature. Water ————, acetic acid ••••••, pyridine —•—•—, ethanol —— •• —— ••, acetone ✕—✕, dioxane — •• —, 80% cyclohexane in dioxane - - - - - - -, 0.25 N NaOH — — —. For comparison, riboflavin in water ($1.33 \times 10^{6} M$) —•—.

excited isoalloxazine species (abstraction or delocalization of H⁺ at N-1). The emission spectrum in water seems to result from the equilibrium of excited neutral molecules and of excited molecules of the isoalloxazine type. In such solvents as ethanol, acetone, dioxane, and dioxane–cyclohexane mixtures, the neutral form prevails, and the decrease of solvent polarity causes the decrease of quantum yield of fluorescence and a significant blue shift of emission maximum. In solvents of lowest polarity the emission maximum is very broad, flat, and structureless. It is reasonable to assume that in such solvents excited molecules of lumichrome tend to form weak fluorescent dimers or complexes of excimer or exciplet type.[32]

The Use of Fluorescence in Analyses of Flavins. When using fluorescence for qualitative and quantitative analyses, the properties of flavins discussed above must be taken into consideration. Each wavelength within the main absorption maxima bands can be used for fluorescence excitation. For quantitative work it is most convenient, however, to use light of wavelengths within the 430–460 nm region because: (1) only a few fluorescent

32 J. B. Birks, *Nature* **214**, 1187 (1967).

compounds present in extracts from biological tissues besides flavins (iso-alloxazines) are excited with this light; (2) the absorption maximum of flavins (isoalloxazines) at 440–450 nm is almost insensitive to changes in solvent polarity and pH values between 3.5 and 7.5; and (3) light of these wavelengths is not absorbed by sodium dithionite used for flavin reduction in blank preparation (solutions of sodium dithionite show very strong absorption below 390 nm, causing a strong inner filter effect). For qualitative purposes, such as identification of flavins on chromatograms (paper, thin layer, etc.) the 360–370 nm band can be recommended (visible light dazzles the eyes of the observer), as most flavin decomposition products and fluorescent contaminants fluoresce under such conditions. The same wavelength should be used for the quantitative determination of lumichrome. For the identification of flavins on chromatograms, besides R_f values, the following specific tests are also useful: (1) reduction with a drop of sodium dithionite solution to a nonfluorescent form and then reoxidation with air oxygen to a fluorescent form within a few seconds; (2) conversion to lumiflavin (photolysis of Rb, FMN, and FAD in alkaline solution on the chromatogram or after elution) and development in another direction; (3) for lumichrome, change of spot color from blue to greenish yellow with a drop of acetic acid solution (in solvent systems containing acetic acid, lumichrome often exhibits greenish yellow fluorescence which changes to blue after some time when acetic acid evaporates).

Different types of the quenching of the fluorescence of flavins occur in most cases in crude extracts from biological tissues. Effects caused by concentration or collisional quenching can be avoided to some extent, without the use of special purification procedures, by extensive dilution, but only when concentration of flavins is sufficiently high. It must, however, be remembered that in dilute solutions all decomposition processes of flavins are more rapid than in concentrated solutions.

The analyst working with flavins should remember also that physicochemical properties of synthetic flavin preparations in pure solutions are to some extent different from native flavins in multicomponent extracts. Generally, native flavins show slightly higher resistance toward destructive agents; this resistance is due most probably to the protective action of different compounds in extracts from biological tissues.

All operations with flavins should be carried out in laboratories protected from direct sunlight, illuminated only with yellowish red light (wavelengths longer than 550 nm).

Determination of Riboflavin, FMN, and FAD

The assay procedures described below are worked out in the author's laboratory on the basis of a typical scheme for the fluorometric determination of flavins. It should be stressed, that in our opinion no fully universal procedure of equal applicability for any type of biological tissue can be proposed. In the procedures described the author suggests some alternative methods according to their applicability. The analyst must, however, be aware of the necessity of choosing an optimal combination of procedures to be applied if more specific types of tissues are to be studied.

The procedures described could be used when determining (1) the so-called "total flavins," in course of which the sum of flavin forms (Rb, FMN, and FAD) is estimated as riboflavin; and (2) "individual flavins," in which the three flavin forms are estimated separately. Each procedure destined for individual flavin determination can be applied also for the total flavins determination, but the reverse is not applicable.

Principle. Flavins are extracted from the tissues by using different extraction procedures which release individual flavin forms from different complexes without the hydrolysis of FMN and FAD or including the hydrolysis of these forms. According to the quantity of compounds which influence the fluorescence of flavins, their fluorescence can be measured without further purification or after different purification procedures. The purification step can include the isolation of all flavins as well as the separation of individual flavin forms. Fluorescence of flavins is then measured by using internal or external standard techniques. In all cases, instead of measuring the natural flavin fluorescence, flavins can be photolyzed in alkaline solution to lumiflavin and its fluorescence measured.

Extraction

The great variability of chemical and structural composition of biological tissues and different stability of individual flavin forms and their complexes with proteins are the most important reasons for the use of different extraction procedures. No acid extraction (excepting cold trichloroacetic acid) can be used for individual flavin determination, as FAD very quickly splits to FMN and Rb under such conditions. In all cases, in the course of extraction, heating is applied for denaturation of proteins and inactivation of different enzymes. Most flavin–protein complexes are readily destroyed by any change of conditions (pH, temperature, etc.), but some types of flavoproteins are extraordinarily resistant. Only prolonged extraction in hot mineral acids ($1 N$ or more) or trypsin digestion are able to release the flavin moiety from such complexes. As the use of hot, concentrated acids results in the formation of many intermediate products of unknown structure in the course of hydrolysis of flavin coenzymes and also of partial decomposition of riboflavin, only trypsin digestion can be recommended. In addition, trypsin digestion used in the case of protein-rich samples tends to counteract the occlusion of flavins in precipitated proteins.

For each type of tissue analyzed the effectiveness of such processing should be examined by using extraction procedures with and without trypsin digestion. If results are closely similar, trypsin digestion can be omitted. In some specific tissues (as liver, heart muscle, kidney, etc.), trypsin digestion releases the flavin component in the form of flavopeptides.[33,34] These flavopeptides show maximum fluorescence intensity at pH 3.1; their lumiflavin-type derivatives are not soluble in chloroform, and therefore for their determination a specific method of Cerletti et al.[35] should be used. Recently, for extraction of flavins from plant tissues rich in cellulose, thermal hydrolysis of the sample in concentrated solutions of lithium chloride was proposed.[36]

To obtain quantitative extraction, samples should be prepared by cutting, crushing, milling, or, best of all, homogenizing to destroy, if possible, the cell structure of tissues. Before extraction, samples containing significant amounts of fat should be defatted by extraction with ethyl ether or another solvent not miscible with water and subsequent evaporation of the organic solvent.

According to the flavin concentration expected, an amount of tissue should be taken which contains 5–20 μg of flavins (usually 1–10 g). Effective extraction is possible only when the extractant is applied in sufficient excess; this must be greater when the sample contains a great amount of insoluble substances such as cellulose derivatives or starch. It is usually sufficient, for most animal tissues, to use 10 ml of extractant for 1 g of tissue; for plant tissues 10–50 ml/g is needed. Since readsorption of flavins on different insoluble structures or precipitates might occur, before separation of the clear filtrate or centrifugate, all operations in the extraction procedure should be standardized in time.

Procedure for the Determination of Total Flavins

Acid Extraction. The sample is mixed with 75 ml of 0.1 N sulfuric acid aqueous solution (to liquid samples an appropriate amount of more concentrated acid is added) until full dispersion of the sample is obtained. The mixture is heated on a boiling water bath for 45 minutes or autoclaved at 1 atm for 30 minutes. The mixture is then cooled to room temperature, neutralized with 5 ml of 2.5 M sodium acetate solution, and made up to

[33] Y. L. Wang, *Rev. Roumaine Biochim.* **3**, 149 (1966).
[34] This volume [143].
[35] P. Cerletti, R. Strom, and M. G. Giordano, *Arch. Biochem. Biophys.* **101**, 423 (1963).
[36] A. Lempka and H. Andrzejewski, *Przemysl. Spozywczy.* **10**, 41 (1966).

100 ml with water. Separation of insoluble components of the mixture is obtained by filtration through medium fast paper (Whatman 2 or 4), or by centrifugation. In the case of filtration, the first 15–20 ml is discarded.

Procedure for the Determination of Individual Flavin

Extraction with Hot Water or Buffers. The sample is mixed with 20 ml or more of hot (according to the sample volume, the temperature can be 80–100° to raise the temperature of the mixture immediately to about 80°) water or phosphate buffer (0.01 M KH_2PO_4).[37] The temperature is kept at 80° for 3–5 minutes and then the sample is carefully disperged (homogenized) and the temperature is maintained at 80° for 3–5 minutes. During this operation phosphatases and other enzymes present in the sample are inactivated. Then another 20 ml (or greater) portion of hot water or buffer is added, and the mixture is maintained for 15 minutes at 80° with occasional stirring. After cooling to room temperature the mixture is separated in a centrifuge. The residue (precipitate and other insoluble particles) should be further extracted with small portions of hot water or buffer with energetic stirring. The washings are centrifuged, and the supernatant solutions are collected and added to the main extract. This washing procedure should be repeated at least twice.

Enzyme Treatment

In all cases where samples contain measurable amounts of flavins tightly bound to proteins, it is necessary to release flavins by using trypsin digestion after acid, hot water, or buffer extraction (before filtration or centrifugation). After trypsin digestion in the course of the determination of total flavins (acid extracts), the released coenzymes should be hydrolyzed by the use of phosphatases. Commercial diastase preparations usually contain sufficient amounts of phosphatases to obtain full hydrolysis of FAD to FMN and partial hydrolysis of FMN to Rb in the course of a short digestion.

In hot water or buffer extracts from samples that contain starch, the starch must be hydrolyzed by using phosphatase-free amylase preparations. Commercial diastase preparations can be used also under conditions that ensure full inactivity of phosphatases (high concentration of phosphates). If starchy samples contain flavins tightly bound to proteins, starch must be hydrolyzed prior to trypsin digestion.

After each enzyme digestion the enzymes in the mixture should be inactivated by heating at 80° for 5 minutes at pH 4.5–5.0.

---

[37] In some cases (see amylase digestion) 1 M KH_2PO_4 solution is used.

TRYPSIN DIGESTION

Procedure for the Determination of Total Flavins

The pH of the cooled acid extract is adjusted to 7.5–8.0 by using 1 N NaOH solution, crystalline trypsin (100 mg per gram of protein present in the sample); a few drops of toluene are added, and the mixture is placed in a thermostatted bath at 45°. After each 30-minute period, the pH is readjusted until it remains constant (usually 3–4 hours). Then the mixture is cooled to room temperature, the pH is adjusted to 4.5–5.0 with glacial acetic acid, and the mixture is heated on a boiling water bath to 80°. This temperature is maintained for 5 minutes (inactivation of trypsin and possible bacterial enzymes), then the mixture is cooled to room temperature.

For the hydrolysis of FAD to FMN and for partial splitting of FMN to Rb, pure phosphatase or a commercial diastase preparation is used. Flavin-free enzyme preparation is added to the mixture (1 mg of acid phosphatase or 20 mg of diastase per gram of sample). The mixture is then placed in a thermostatted bath at 45° for 3 hours (at 37° overnight). After incubation, the pH is adjusted and the mixture is heated on a boiling water bath to 80° and maintained at this temperature for 5 minutes, then cooled to room temperature, diluted with water to a desired volume, and filtered or centrifuged.

Procedure for the Determination of Individual Flavins

Hot water or buffer extracts from starch-free samples are cooled after the first 3–5 minutes heating period. The pH value is adjusted to 7.5–8.0 by using 0.05 M K_2HPO_4 for water extracts or 1 N NaOH for buffer extracts; trypsin is added (100 mg per gram of protein in the sample). After the addition of a few drops of toluene, the mixtures are placed in a thermostatted bath at 45° for 4 hours, then cooled to room temperature; the pH is adjusted to 4.5–5.0 by using 1 N sulfuric acid. Acidified extracts are heated at 80° for enzyme inactivation (5 minutes), cooled, diluted to a desired volume, and filtered or centrifuged.

AMYLASE DIGESTION

Procedure for the Determination of Individual Flavins

This step is prior to trypsin digestion only when starch is present. Hot-water or buffer extracts after the first heating period are cooled, and the pH of the hot-water extracts is adjusted to 4.5–5.0 by using 0.05 M KH_2PO_4. Then pure (phosphatase-free) amylase is added (10–20 mg per gram of starch in the sample) followed by a few drops of toluene, and the mixture is kept in a thermostatted bath at 45° until all starch is liquefied (usually 2–3 hours). If necessary more amylase may be added. After incubation the mixture is heated at 80° for 5 minutes and then cooled.

If pure phosphatase-free amylases are not available, commercial diastase preparations can be utilized according to the following procedure: for extraction 1 M KH_2PO_4 is used; and after cooling, a commercial, flavin-free diastase preparation is added (20–50 mg per gram of starch). Further operations are the same as above, the only difference being that for acidification or alkalization (trypsin digestion) more concentrated acid or alkali must be used.

If after the hydrolysis of starch trypsin digestion is not needed, the mixtures are diluted to a desired volume with water and filtered or centrifuged.

Deproteinization

In the course of determining individual flavins in biological liquids or animal tissues very rich in proteins, special treatment is sometimes necessary for deproteinization. Such treatment is needed mainly when hot water or buffer extraction is performed without enzyme treatment and the proteins remain in a colloid phase and filtration or centrifugation is ineffective. In such cases it is desirable to limit the volume of the extracts as much as possible. The two following treatments can be recommended.

Procedures for the Determination of Individual Flavins

Trichloroacetic Acid Treatment. This method is used only when direct fluorescence measurement without any further purification can be applied. The extract, cooled to 0°, is immediately mixed with an equal volume of ice-cold 20% trichloroacetic acid aqueous solution. The mixture is kept at 0° (to prevent hydrolysis of FAD) for 10 minutes and rapidly centrifuged; the supernatant is neutralized at once by the addition of one-fourth of its volume of 4 M K_2HPO_4. Instead of neutralization, the trichloroacetic acid can also be removed using several extractions with ethyl ether at a temperature close to 0°. From the neutral extract, ether must be evaporated under reduced pressure until no odor of ether is detectable.

Acetone Treatment. This method should be used when further purification is needed. To the extract, 3–5 (or more) volumes of acetone are added and the mixture is cooled in a refrigerator for better precipitation of proteins and other impurities. Flavins remain in the water–acetone mixture, since acetone prevents adsorption on the precipitate. After filtration or centrifugation, acetone must be evaporated from the clear extract under reduced pressure at a temperature not exceeding 50° until no odor of acetone is detectable.

Removal of Interfering Substances

Most extracts, especially those obtained from plant tissues, contain significant amounts of organic and inorganic compounds sufficient to quench the fluorescence of flavins, which are usually present in very low concentration. These impurities may also exhibit similar fluorescence. Several different methods can be used to achieve the removal of all, or of a significant part, of the interfering substances or to isolate and concentrate the flavins.

OXIDATION

Because of the relatively high resistance of flavins toward oxidizing agents, extracts are treated with $KMnO_4$, excess of which is then destroyed by hydrogen peroxide. In this process most of the pigments and other organic compounds in extracts besides flavins are irreversibly oxidized. However, although under suitable conditions flavins are not destroyed during the oxidation, their lability is considerably increased, and if oxidized solutions are allowed to stand for some time, considerable flavin losses are observed. Therefore, purification with this method should preferably not be used, but, if necessary, oxidation should be performed just before fluorescence measurements.

Procedure for the Determination of Total and Individual Flavins

Appropriate equal portions of filtered or centrifuged extracts are prepared. For 10-ml portions the following amounts of reagents are added: 0.1 to 1.0 ml of glacial acetic acid (depending on the desired pH of solution at the moment of fluorescence measurement—from about 6 to about 3—respectively), and, after mixing, 1 ml of 3% $KMnO_4$. The mixture is thoroughly stirred; after exactly 2 minutes the excess of $KMnO_4$ is discharged by adding of 1 ml of 3% H_2O_2 (if portions of extract are of other volumes, amounts of reagents should be proportionally changed). The color of $KMnO_4$ should dissappear within a few seconds, and no precipitate should be formed. All the portions are then centrifuged to eliminate bubbles of oxygen and immediately used for further operations.

USE OF ORGANIC PHASE

Some organic solvents can be used for the isolation of flavins or for removal of interfering substances present in the extracts.

The different solubility of riboflavin and its nucleotides in benzyl alcohol can be used for their separation. The method of fluorometric determination of flavins based on such a procedure can be applied.[38]

---

[38] H. B. Burch, Vol. III, p. 960.

All flavins can be extracted from aqueous solutions by using phenol or its derivatives.[39] To the phenol layer containing flavins, a small volume of water is added and then an excess of ethyl ether. Under these conditions, most of the flavins are displaced to the water layer, and in this way multifold concentration of flavins can be achieved. This procedure can be used for flavin concentration for qualitative paper chromatographic methods. If quantitative results are needed, each step must be repeated many times. This method is rather inconvenient for fluorometric analyses, since it is very hard to discharge traces of phenol that effectively quench fluorescence of flavins.

Chloroform is commonly used in two different ways: for the removal of contaminants soluble in chloroform before photolysis (preextraction) and for the isolation of lumiflavin from photolyzed extracts (procedures are given in the description of photolysis for the lumiflavin method).

Resin Chromatography

Phenol-formaldehyde type resins were found to be very convenient for the isolation, concentration, and partial separation of flavins present in extracts. Best results were obtained with the resorcinol-formaldehyde resin R-15.[40] This resin retains quantitatively riboflavin, lumiflavin, and lumichrome from aqueous solutions and FMN and FAD from aqueous solutions containing well dissociated salts in a concentration higher than $0.05 M$. Salt concentrations higher than $1 M$ can cause, in the course of adsorption, slight splitting of FMN to Rb and are to be avoided. The character of flavin adsorption on phenol-formaldehyde type resins is not as yet quite clear. It seems very probable that formation of charge-transfer type complexes between phenols and flavins take place there. The role of dissociable salts in the adsorption process of FMN and FAD probably lies in the limitation of dissociation of their phosphate groups. Quantitative elution of all flavins can be achieved by using mixtures of some organic solvents (propanol, acetone, p-dioxane) with water (1:1, by volume) or a saturated aqueous solution of urea with ethanol (1:1, by volume). Flavin nucleotides can be separated quantitatively from riboflavin by a two-step elution—using a mixture of lower concentration of organic solvent (10–15%) for the elution of the sum of FMN and FAD, and then a mixture of higher organic solvent concentration (40–50%) for riboflavin elution. To avoid uncontrolled changes of fluorescence intensity due to organic solvent concentration, the internal standard technique or evaporation of the organic solvent under reduced pressure should be applied. This method

[39] K. Yagi, in "Methods of Biochemical Analysis" (D. Glick, ed.), Vol. 10, p. 349. Wiley (Interscience), New York, 1962.
[40] A. Koziołowa and J Koziol, J. Chromatog. 34, 216 (1968).

of purification, after the evaporation of organic solvent, yields a small volume of aqueous solution containing highly concentrated flavins and sometimes (only in case of extracts of highest degree of contamination, as from wheat bran) negligible traces of other pigments.

Procedure for the Determination of Total and Individual Flavins

Columns 1 cm in diameter are charged to a height of 5 cm with R-15 resin previously suspended in water. Small plugs of glass wool are placed on top of the resin. First, 25-ml portions of aqueous salt solutions at a concentration similar to the mean salt concentration in analyzed extracts (0.05 M or 1 M KH$_2$PO$_4$, 0.2 M Na$_2$SO$_4$, etc.) are passed through the columns. If hot-water extracts without enzyme treatment and/or the oxidation step are being analyzed, 0.05 M KH$_2$PO$_4$ is passed through the columns and a suitable amount of solid KH$_2$PO$_4$ is added to the extract to make the solution about 0.05 M. Then carefully measured volumes of extracts are placed on the columns (usually 20–50 ml containing no more than 20 μg of flavins) and sorption is performed at a flow rate not higher than 2 ml per minute. The columns are then washed with two 20-ml portions of 0.05 M KH$_2$PO$_4$ and with a 5-ml portion of water. Elution is then performed at the same flow rate as follows: For the determination of individual flavins, small portions of 10% acetone in mixture with water are passed through columns, and 50 ml of eluate is collected which contains FMN and FAD; then small portions of 50% acetone–water mixture are passed and 50 ml of eluate containing riboflavin is collected. For the determination of total flavins, only 50% acetone–water mixture is used and 50 ml is collected.

Acetone-containing eluates can be used directly for fluorescence measurements for total flavins determined by using internal standard techniques (also for riboflavin determination in the course of determining individual flavins). From eluates containing FMN and FAD, acetone is evaporated under reduced pressure at temperatures not exceeding 40–50° (this can be applied also for eluates containing riboflavin or total flavins). Evaporation of acetone should be performed by avoiding foaming, evaporation to dryness, and any imperfections that could cause losses of flavins. Concentrated solutions of flavins should be diluted with water to a desired volume and analyzed for fluorescence intensity.

It is convenient to use part of these concentrated solutions for paper or thin-layer chromatography to test the effectiveness of removal of interfering substances and to identify the flavins present.

Photolysis for the Method Involving Lumiflavin Fluorescence

The lumiflavin fluorescence method is based on the fact that three main flavin forms that are insoluble in chloroform (Rb, FMN, and FAD) when irradiated in alkaline solution are readily converted to lumiflavin that is soluble in chloroform. This reaction is specific for flavins, as only very few other compounds become fluorescent under such conditions. The most serious disadvantage of the method is the fact that the transformation of Rb, FMN, and FAD into lumiflavin is not quite quantitative. The velocity and yield of this reaction is different for each form and depends on spectral behavior and intensity of the light used, the pH, the flavin concentration, the time of exposure, the presence of different contaminants, the temperature, etc. Photolysis must be precisely standardized in these aspects to achieve constant yields of lumiflavin from all flavin forms and to avoid the formation of lumichrome or the photolytic degradation of lumiflavin. It was claimed,[41] that under correctly selected conditions flavins at concentrations lower than 2.4 $\mu g/ml$ are quantitatively converted into lumiflavin. If in the course of photolysis any other compound (nonflavin) becomes fluorescent in the spectral region close to that of lumiflavin and the compound is soluble in chloroform, it is necessary to apply a more extensive purification of the extract including the use of R-15 resin. Usually oxidation with $KMnO_4$ and preextraction with chloroform are sufficiently effective.

In the lumiflavin method it is most convenient to use the internal standard technique, since in this way most of the possible differences in processing of sample and standard solutions due to unequal absorbancy, influence of catalytically active contaminants, nonuniform separation of lumiflavin between water and chloroform, etc., can be avoided.

Procedure for the Determination of Total Flavins

Preextraction. To appropriate (usually 5–25 ml) portions of sample solutions, equal volumes of chloroform are added, the mixtures are shaken vigorously for 1 minute, and layers are separated by centrifugation. The chloroform layer (bottom) is tested fluorometrically; if measurable fluorescence in the 525–535-nm region is present, this procedure should be repeated until no fluorescence of the chloroform layer can be detected.

Photolysis. Appropriate volumes of sample solutions preextracted with chloroform and of parallel standard flavin solutions (if external standard is applied) are placed in a series of glass-stoppered centrifuge tubes (or other glassware of uniform transparency, wall thickness, and diameter). If the internal standard technique is applied, appropriate volumes of standard flavin solution are added to one series, and equal volumes of water to the other. The amount of flavin added should be almost equal to that in the sample solution. A volume of 10 N NaOH is added to each

---

[41] K. Yagi and S. Mitsuhashi, *Japan. J. Exptl. Med.* **21**, 353 (1951).

tube (with careful mixing) in order to obtain a pH of about 12 (for 10-ml portions, 5 ml of sample solution plus 5 ml of standard flavin solution, 0.5 ml of 10 N NaOH is needed). The tubes are then irradiated for an appropriate period (usually 20–30 minutes) making sure that conditions are equal for all the tubes. After irradiation, glacial acetic acid is added to each tube (volume equal to that of 10 N NaOH) and the contents are mixed. If necessary, as for extracts from plant tissues not purified with R-15 resin, oxidation with $KMnO_4$ can be performed according to the described procedure. An appropriate volume of chloroform is now added to each tube (equal to the volume of photolyzed solution), and the tubes are stoppered, shaken for 30 seconds, and centrifuged; the aqueous (upper) layer is removed. Blanks are prepared by an identical procedure without an irradiation step.

Estimation of Fluorescence

All fluorometric methods are based on a comparison of the fluorescence of the sample solutions to the fluorescence intensity of standard solutions of known flavin concentration. Standard solutions of flavins could be applied by using the so-called "external standard" or "internal standard" techniques. External standard can be applied as a standard flavin solution of known concentration and fluorescence intensity or as a standard solution of known concentration of flavin or flavins run through all operations in parallel with the sample. Internal standards are used in the form of standard solution of flavin or flavins of a known concentration added to the sample before processing or at any step of the procedure.

Both kinds of standard application have advantages and disadvantages. External standards can be used in all cases where extracts do not contain significant amounts of interfering substances or where the applied purification procedures are highly effective, as is the case when the R-15 resin step is used. Internal standard should be used when extracts are not precisely purified and when in the course of analysis any uncontrolled losses occur, such as flavin reactions that are not fully quantitative or changes in the composition of extracts. Therefore internal standard techniques should always be applied before oxidation with $KMnO_4$, before photolysis in lumiflavin method, and before measurement of flavin fluorescence intensity of eluates from which acetone was not evaporated. The use of an internal standard, however, does not assure "absolute" correctness of the determinations, since synthetic flavins added to samples or to extracts may behave in a slightly different manner than native flavins. The error caused by this phenomenon is the greater the higher the concentration of interfering substances in the solution.

In the course of determination of the total flavins, fluorescence is measured for sample solutions containing only Rb and FMN—and in the lumiflavin method, only lumiflavin—and compared to fluorescence intensity of suitable standard solutions such as riboflavin in water or lumiflavin in chloroform, respectively. Results are calculated as riboflavin content in the analyzed sample.

When determining individual flavins, the fluorescence intensity of the eluates containing only riboflavin is measured as for total flavins. For the eluates containing FAD and FMN, fluorescence is measured twice: directly and after hydrolysis of FAD to FMN. From the difference between fluorescence intensity before and after hydrolysis, and taking into account differences between fluorescence intensities of these two flavin coenzymes, the concentration of each of them can readily be calculated, as internal standards solutions of riboflavin are commonly used in all cases.

To avoid any possible error due to the presence even of traces of nonflavin fluorescent compounds in a sample solution, the fluorescence also of blanks is tested in the course of fluorometric measurements. Commonly sodium dithionite in solid form is used to reduce flavins to nonfluorescent forms. It is, however, a method that is not specific for flavins, since some fluorescent compounds may behave exactly like flavins, whereas other compounds may become fluorescent after reduction. Such phenomena were observed for highly contaminated extracts from plant tissues to which the R-15 resin step was not applied. Although photoreduction of flavins in the presence of EDTA[42] is more specific, we recommend the use of the sodium dithionite method for its simplicity.

In the lumiflavin method, chloroform extracts from nonirradiated sample solutions, processed in parallel, are used as blanks. This method is correct only when no other fluorescent compound, or compound soluble in chloroform, other than lumiflavin, is formed in the course of photolysis. It should be tested chromatographically, and, if necessary, additional purification must be applied or conditions of photolysis must be changed.

RIBOFLAVIN FLUORESCENCE METHOD

Procedure for the Determination of Total Flavins

Sample solutions, purified or not purified, acetone-free eluates from R-15 resin, and external standards processed in parallel with sample solutions are placed in cuvettes of the fluorometer, and their fluorescence is measured several times for each solution to obtain mean values. If fluorescence of any sample solution is higher than the fluorescence of the standard solutions used for fluorometer calibration, this sample solution must be

[42] E. Knobloch, *Intern. Z. Vitaminforsch.* **37,** 38 (1967).

diluted with water. When the internal standard technique is applied, fluorescence intensities are measured separately for sample solutions with and without added riboflavin. If internal standard is to be applied (as in the case of acetone-containing eluates from the R-15 resin), two series of equal volumes of sample solutions are prepared, an appropriate volume of standard riboflavin solution is added to one (the amount of riboflavin should be close to that expected in this volume of sample solution), and an equal volume of water to the other, and fluorescence is then measured.

Then for all the solutions, blanks are prepared. After measurement of the fluorescence intensity, a minute amount of solid sodium dithionite is added directly to the cuvette, the content is mixed with a glass rod or by shaking, and fluorescence is measured within a few seconds. A new portion of sodium dithionite is then added, and fluorescence is measured again. If a change in fluorescence intensity for these two measurements is observed, this operation should be repeated once more, but the total amount of sodium dithionite should not exceed 20 mg per 5 ml of analyzed solution. Sodium dithionite is poorly soluble in solutions containing about 50% acetone (unevaporated eluates from the R-15 resin), and it is recommended that sodium dithionite be added directly to cuvettes placed in the fluorometer, illuminated with exciting light; the solution should be mixed until rapid decrease of fluorescence begins. Immediately afterward, the fluorescence of the blank should be measured. One must remember, however, that an excess of sodium dithionite can lead to erroneous results (reduction of nonflavin compounds, formation of colloidal sulfur, etc.). Sometimes it may be more convenient to use sodium dithionite in solution, and if so a correction must be made for any change in volume.

Calculations with External Standard

$$\text{Riboflavin } (\mu g/g) = A - B \times \frac{c}{a - b} \times \frac{V}{d} \times \frac{V_2}{V_1} \cdots$$

where A is fluorescence of the sample solution; B, fluorescence of sample solution after the addition of sodium dithionite (blank); a, fluorescence of riboflavin standard solution; b, fluorescence of riboflavin standard solution after addition of sodium dithionite (blank); c, concentration of riboflavin standard solution in $\mu g/ml$; V, volume to which sample is diluted after extraction and enzyme treatment; V_1, volume of extract taken for oxidation; V_2, volume of extract after oxidation; d, sample weight in grams.

When chromatographic purification on R-15 resin is included instead of oxidation, V_1 equals volume of extract taken for purification, and V_2 equals the volume to which eluate is diluted after evaporation of acetone (or volume of acetone-containing eluate). When the oxidation step too is involved, or any other dilutions must be made, suitable correction factors should be applied in the same way.

Calculations with Internal Standard

$$\text{Riboflavin } (\mu g/g) = \frac{(A - B) \times R}{C - A} \times \frac{V}{V_1 \times d}$$

where A equals fluorescence of the sample solution without standard added; B, blank of A; C, fluorescence of the sample solution with standard riboflavin solution added; R, riboflavin added in micrograms per V_1 volume of sample solution; V_1, volume of extract taken for oxidation; for the R-15 resin step, or for measurement of fluorescence before addition of standard riboflavin solution (if standard riboflavin solution is added to the sample before or in the course of extraction, this factor must be omitted); V, volume to which sample is diluted after extraction and enzyme treatment; d, sample weight in grams. If any dilutions are made before the addition of standard riboflavin solution, appropriate corrections must be applied (as V_1 and V_2 in the calculation with external standard).

Procedure for the Determination of Individual Flavins

Fluorescence intensity of the eluates containing only free riboflavin is measured and calculated in the same manner as for total flavin determination. For determination of flavin coenzymes, FMN and FAD, to one part of eluates, completely freed from acetone, containing these forms, 3 N HCl is added in an amount necessary to obtain 0.1 N HCl concentration. Then this solution is heated on a boiling water bath for 15 minutes, cooled, neutralized with 2.5 M sodium acetate solution, and diluted with water to a desired volume; the fluorescence intensity is measured as for free riboflavin (including blanks). To the other part is added an equivalent volume of a mixture of 3 N HCl and 2.5 M acetate. Then this solution is diluted with water and its fluorescence intensity is measured as for free riboflavin (including blanks).

Calculations

$$\text{FAD } (\mu g/g) = \frac{(C - D) \times 100}{82} \times \frac{c}{a - b} \times \frac{V}{d} \times \frac{V_2}{V_1} \times \frac{V_4}{V_3} \times \frac{785}{376}$$

where C is fluorescence of hydrolyzed eluate; D, fluorescence of nonhydrolyzed eluate; 82, percentage increase of fluorescence after hydrolysis of FAD to FMN and Rb; c, concentration ($\mu g/ml$) of riboflavin in standard solution used for fluorometer calibration; a, fluorescence of standard riboflavin solution; b, blank of a; V, volume to which sample was diluted after extraction and enzyme treatment; d, sample weight in grams; V_1, volume of extract taken for the R-15 resin chromatography; V_2, volume of eluate containing FMN and FAD after evaporation of acetone and eventual dilution; V_3, volume of eluate taken for hydrolysis with HCl; V_4, volume to which the hydrolyzed eluate is diluted after hydrolysis; 785 and 376 are the molecular weights of FAD and riboflavin, respectively.

$$\text{FMN } (\mu g/g) = E - F \times \frac{c}{a - b} \times \frac{V}{d} \times \frac{V_2}{V_1} \times \frac{V_4}{V_3} \times G - \frac{456}{376}$$

where E and F are fluorescence intensities of the hydrolyzed eluate before and after the addition of sodium dithionite, respectively;

$$G = \frac{(C - D) \times 100}{82} \times \frac{c}{a - b} \times \frac{V}{d} \times \frac{V_2}{V_1} \times \frac{V_4}{V_3}$$

i.e., contents of riboflavin in form of FAD; 456 and 376 are the molecular weights of FMN and Rb, respectively; other symbols are the same as for the FAD calculation.

LUMIFLAVIN FLUORESCENCE METHOD

Procedure for the Determination of Total Flavins

Separated chloroform layers containing lumiflavin from photolyzed sample solutions with and without added standard riboflavin solution and/or from standard riboflavin solution processed in parallel with sample solutions are placed in the cuvette of the fluorometer, and the fluorescence intensity is measured. Then the fluorescence of the blanks (nonphotolyzed sample solutions) is measured, and the flavin concentration in the sample is calculated as riboflavin, using the same equations as for the riboflavin fluorescence method.

Reproducibility

The reproducibility of the described methods is very high, and for most tissues it is better than 5% for separate analyses. For some types of plant tissues containing very low amounts of flavins and large amounts of other extractable interfering substances, the reproducibility reaches values about 5%. These data were proved using the proposed extraction and purification steps in the course of determination of flavins according to riboflavin and lumiflavin fluorescence methods as well.

Appendix

Reagents

All reagents must be of the highest purity and should not exhibit any fluorescence in the spectral region 510–550 nm. Water and other solvents should be redistilled from all-glass apparatus. In the course of all operations any contact of solutions with rubber, plastics, stopcock greases, and other possible sources of fluorescent substances should be avoided.

Riboflavin stock solution. Crystalline riboflavin is dried over phosphorus pentoxide in a desiccator for 72 hours. Twenty milligrams of riboflavin, carefully weighed, is placed in a 2-liter volumetric flask; 300 ml of water is added, and the mixture is heated to 70° and shaken to aid solution. If a less soluble form of riboflavin is used, 2.4 ml of glacial acetic acid can be added. The preparation is cooled and made up to volume, then a few drops of toluene are added and the solution is stored in an amber bottle in the refrigerator. It is stable for months.

Riboflavin standard solution. Stock solution of riboflavin is diluted with water to obtain a riboflavin concentration of exactly 1 μg/ml. This solution is used as external or internal standard. Stored in the dark, it is stable for days.

FMN and FAD standard solutions. These solutions should be prepared and used only when preparations of very high purity are obtainable. Solutions of these flavins are of limited stability because of bacterial action, catalytic action of many substances, and other factors. Therefore amounts as small as possible are weighed and dissolved in small volumes of freshly distilled water (1 mg to 100 ml); a drop of toluene is added, and the solutions are stored in the refrigerator. Before use, the absence of degradation products should be controlled by using paper or thin-layer chromatography. If free riboflavin is present, solid KH_2PO_4 is added to make the solution 0.1 M, then it is passed through the R-15 resin, the column is washed with a very small volume of water (3–5 ml), and elution is performed by using 10% acetone in water. After evaporation of acetone under reduced pressure, the solution is diluted with water to obtain the desired concentration of flavin. If FAD contains FMN, another chromatographic method should be used for separation.[43,44]

Solvents. Water, acetone, chloroform, and toluene are redistilled; chloroform and toluene are saturated with water. Acetone–water mixtures should be cooled to room temperature before use.

Buffers, salt solutions, acids and alkali should be prepared by using reagents of high analytical purity and redistilled water.

Sodium dithionite of high purity should be tested for reducing ability. If solution is to be used, it is prepared as follows: 2 g of sodium bicarbonate is dissolved in 100 ml of water; the solution is cooled in an ice bath or refrigerator, and 5 g of sodium dithionite is dissolved in it. When kept in an ice bath or refrigerator, it is stable for about 3 hours.

---

[43] For the preparation of flavin coenzymes, see this volume [148].
[44] F. M. Huennekens and S. P. Felton, Vol. III, p. 950.

R-15 resin for chromatography. The resin is prepared as follows: 15 g of
solid NaOH and 33 g of freshly sublimed resorcinol are dissolved in
60 ml of water and cooled on an ice bath. Carefully, with continuous
stirring, 45 ml of 40% aqueous formaldehyde solution is added (the
temperature should not rise above 50°) followed by 30 ml of water. The
reaction mixture is placed on a water bath, warmed to 50°, and kept at
this temperature for the next 25–30 minutes. After this time, the mixture
begins to form long threads when a glass rod is drawn through it (beginning of gelatinization), and from this moment the mixture is kept
at 50° for another 15 minutes. The formed gelatinous, reddish brown
resin is poured into excess of cold tap water and pressed through a sieve
(1 mm mesh diameter) under a stream of tap water. Then the resin is
placed in a great excess of tap water and washed several times with
fresh water portions until the water becomes almost colorless. After
1 hour, the resin is pressed through another sieve (0.5 mm mesh diameter) and washed in a stream of tap water overnight. The resin is then
placed in a large chromatographic column and washed successively
with water, 3 *N* HCl, water, 0.5 *N* NaOH, water, 3 *N* HCl, water, 50%
acetone in water, 0.5 *N* NaOH, water, 3 *N* HCl, and finally with water
(redistilled). Washing with the acetone–water mixture and NaOH solution must be repeated until the effluents become nonfluorescent and
colorless, respectively. After each use of the resin, regeneration should
be performed directly in the analytical columns by using 0.5 *N* NaOH,
water, 3 *N* HCl, and water. The resin should be stored in water or acidified with HCl–water in the refrigerator. It is stable for months. Before
stored resin is used, it must be regenerated. The resin should not be
dried, because under such conditions further condensation occurs that
results in the decrease of resin capacity toward flavins. The mean capacity of the R-15 resin in relation to riboflavin is in the range of 23×10^{-3}
meq of riboflavin per gram of air-dry resin.

Enzyme preparations. Crystalline trypsin, crystalline or high-purity (phosphatase-free) amylases, pure acid phosphatase, commercial diastase
preparations (Takadiastase, diastase, clarase, etc.). Commercial complex
enzyme preparations usually contain different amounts of riboflavin,
which should be removed. For this purpose 1 g of the preparation is
dissolved in 10–20 ml of water (if a suspension forms, it should be filtered
or centrifuged) and passed through an analytical column of the R-15
resin (flow rate 1–2 ml/min). The water effluent is used.

Apparatus

Device for photolysis. Any type of lamp can be used that gives light of a
spectral range within the absorption spectrum of flavins (200–500 nm)
and allows for the illumination of an appreciable area with equal in-

tensity. Fluorescent daylight lamps are to be preferred. Transparent glass tubes containing solutions to be photolyzed are placed in an appropriate holder (with glass stoppers removed) vertically or at any other angle if the tubes are of equal size and wall thickness. Each place under the lamp should be checked so that it is illuminated with equal intensity. If another type of lamp is used, uniformity of temperature under the lamp should be proved (when not uniform, cooling must be applied). Before work is started, the following should be fixed: the optimal lamp distance, tube position, time of exposure, etc.; photolysis of a series of flavin solutions of different concentration should be done and the solutions be tested fluorometrically and chromatographically for yield of lumiflavin and of other fluorescent species (lumichrome, products of lumiflavin photolysis, etc.).

Fluorometer. Any type of selective and sensitive fluorometer can be used. For the excitation of fluorescence, light of wavelengths within the range of 430–460 nm should be used and fluorescence should be observed within the range of 520–550 nm (for lumiflavin in chloroform, 510–540 nm). It is highly preferable to use monochromatic or almost monochromatic excitation and emission light beams; exciting light of lowest possible intensity; highest possible sensitivity of detector; rectangular, all walls polished, cells; and a constant temperature cell compartment. It must be made certain that all cells (especially cylindrical ones) in each position give equal values of fluorescence intensity with standard solutions. If necessary, the correct position must be marked for each cuvette, or only one cuvette in constant position for all measurements should be used. The instrument should be adjusted using a freshly prepared standard riboflavin solution of concentration not exceeding 1 μg/ml.

[133] Determination of Flavin Compounds in Tissues[1]

By Paolo Cerletti and Maria Grazia Giordano

The majority of flavin compounds in vegetal or animal tissues—namely, flavine adenine dinucleotide (FAD), flavin mononucleotide (FMN), and riboflavin—are easily extracted by dilute saline or acid solutions. Some, however, are covalently bound to proteins[2-7] and can be solubilized as

[1] In partial fulfillment of a grant by the National Research Council of Italy (CNR).
[2] E. B. Kearney, *J. Biol. Chem.* **235**, 865 (1960).
[3] T. F. Chi, Y. L. Wang, C. L. Tsou, Y. C. Fang, and C. H. Yu, *Scientia Sinica (Peking)* **14**, 1193 (1965).
[4] V. N. Boukine, *Proc Intern. Congr. Biochem. 3rd, Brussels* 1955, p. 61.
[5] W. R. Frisell and C. G. Mackenzie, *J. Biol. Chem.* **237**, 94 (1962).

flavin peptides only after proteolysis. In this group as yet only the flavin prosthetic group of mammalian succinate dehydrogenase has been thoroughly documented.[2,3]

Principle

Water-soluble flavins are first extracted. The residue is digested with proteolytic enzymes so as to solubilize peptide-bound flavins. The extracted compounds are individually determined without requiring a complete separation of the mixture, taking advantage of the different behavior of their fluorescence at acid and neutral pH. Indeed, between pH 5 and 8 the fluorescence of FAD in the experimental conditions used is 18% of that of an equimolar amount of FMN or of riboflavin.[8] Natural flavin peptides, either at the dinucleotide or at the mononucleotide level, emit fluorescence only in the acidic range, with a maximum at pH 3.1 (see footnotes 2, 6, and 9).

When organisms like flavinogenic ascomycetes are examined, which produce overwhelming amounts of one or more flavins, the preliminary separation of the compound in excess is required.

Reagents

Potassium phosphate buffer 0.03 M, pH 6.8

KCl-phosphate buffer containing 96 volumes of 0.01 M KCl and 4 volumes of 0.03 M potassium phosphate buffer, pH 6.8

Riboflavin, 10 μg/ml, in 0.01 M buffer

Dithionite solution, freshly prepared 10% $Na_2S_2O_4$ in 5% $NaHCO_3$

Trichloroacetic acid (TCA) in water, 10% and 1%

Procedure

Determination of Water-Soluble Flavins.[10] The tissue is weighed; to it is added (7 ml/g wet wt.) to ice-cold KCl-phosphate and it is homogenized at 0°, in the dark, with a homogenizer with a Teflon pestle. The homogenate is heated at 80° for 12 minutes in the dark with stirring. It is then rapidly cooled and centrifuged at 24,000 g for 15 minutes. The supernatant is collected, and the precipitate is homogenized in the same volume as that previously used of 0.013 M buffer (4 parts of water and 3 parts of 0.03 M

[6] T. P. Singer, *in* "Oxidases and Related Redox Systems" (T. E. King, H. S. Mason, and M. Morrison, eds.), p. 448. Wiley, New York, 1965.

[7] C. Bhuyaneshwaran and T. E. King, *Biochim. Biophys. Acta* **132**, 282 (1967).

[8] P. Cerletti and N. Siliprandi, *Arch. Biochem. Biophys.* **76**, 214 (1958).

[9] P. Cerletti, R. Strom, and M. G. Giordano, *Arch. Biochem. Biophys.* **101**, 423 (1963).

[10] P. Cerletti and P. L. Ipata, *Biochem. J.* **75**, 119 (1960).

buffer) and extracted as above. After centrifugation, a third extraction in diluted buffer is carried out on the precipitate.

The supernatants from the three extractions are combined. One portion of 4 ml (A + B) and two of 2 ml (C and D) are taken. Portion A + B is passed through a 0.8 × 14 cm column of Amberlite IRC-50 (H$^+$), 150–200 mesh, packed to a height of about 7 cm with 1.01 g (dry wt) of ion exchanger; it is eluted with water at a flow rate of 0.5–1 ml/min. The first 10 ml of the eluate is collected and divided into 5-ml fractions in graduated tubes (A and B). Samples C and D are brought to 5 ml with water.

Samples A and C receive 1 ml of 0.6 N HCl and are hydrolyzed for 15 minutes at 100°. They are then cooled and neutralized with 1 ml of 0.595 N KOH and are made to a convenient volume with 0.03 M buffer, the final pH being between 4.5 and 6.8. Samples B and D receive the same amount of HCl, KOH, and buffer, mixed together before addition.

The distribution of flavins in the samples is for B: FMN + FAD; D: FMN + FAD + riboflavin; A and C: before hydrolysis as B and D, respectively; FAD disappears after hydrolysis, being quantitatively converted into FMN or riboflavin.

The fluorescence of each sample is measured with a spectrofluorometer, excitation wavelength 445 nm, emission wavelength 520 nm. Readings are made (1) after addition of buffer (F_1), (2) after addition of an internal standard of riboflavin (0.1–1 μg) (F_2), and (3) after addition of 0.01 volume of dithionite solution.[11] F_2 and F_3 are corrected for dilution (F'_2 and F'_3), and the apparent riboflavin content of each sample is given by:

$$\text{Riboflavin (μg) added} \times \frac{F_1 - F'_3}{F'_2 - F_1} = F_A, F_B, F_C, F_D, \text{respectively}$$

The real amounts of each flavin in the tissue are calculated as follows:

$$\text{Riboflavin (μg/g tissue wet wt)} = \frac{\text{total volume of extract}}{2} \times (F_D - F_B)$$

$$= \frac{\text{total volume of extract}}{2} \times (F_C - F_A)$$

$$\text{FAD (μg/g tissue wet wt)} = \frac{F_A - F_B}{0.82} \times \frac{\text{total volume of extract}}{2} \times \frac{785}{376}$$

FMN (μg/g tissue wet wt)

$$= \left(F_B - \frac{F_A - F_B}{0.82}\right) \times \frac{\text{total volume of extract}}{2} \times \frac{456}{376}$$

[11] A solution of fluorescein is used to check the stability of the spectrofluorometer during each group of readings. Flavin concentrations in samples and the amount of riboflavin added should be such as to have all readings made on the same scale on the microammeter. The fluorescence can be measured also with a filter fluorometer, lamp filter Corning glass 5543, and photocell filter Corning 3385.

When separate values for FMN and riboflavin content are not required, treatment with ion exchanger is omitted and determinations are run only on samples C and D. The amount of FAD is calculated as above, but where

FMN + riboflavin (μmoles/g tissue wet wt)

$$= \left(F_C - \frac{F_C - F_D}{0.82} \right) \times \frac{\text{total volume of extract}}{2 \times 376}$$

A less precise determination of total flavin in the extract is obtained by reading absorbancies at 450 nm before and after addition of 0.1 volume of dithionite solution. Molar extinction coefficients of oxidized minus reduced flavin are $11.5 \times 10^3 \times 0.91$ for FAD, and $12.2 \times 10^3 \times 0.93$ for FMN and for riboflavin.[8]

Flavins can be extracted also by dilute acid. The tissue is homogenized with 7 ml/g wet weight of 10% ice-cold TCA in the dark. The homogenate is centrifuged at 24,000 g for 10 minutes. The supernatant solution is separated, and the precipitate is extracted twice with the same volume as above of 1% TCA. The supernatant solutions from the three extractions are mixed. All operations are performed at 0–2°. The presence of TCA in the extract, however, makes the treatment with ion exchanger and spectrophotometric measurements difficult.

RECOVERY. When known amounts of flavins were added to the homogenate, total recoveries were obtained.[10]

Determination of Peptide-Bound Flavins.[9] The residue from the extraction of water-soluble flavins is homogenized in 1% TCA. The homogenate is centrifuged for 15 minutes at 24,000 g, and the supernatant is discarded. This treatment does not extract flavins, but it improves the efficiency of the subsequent tryptic digestion. It is omitted when previous extractions are made by diluted acid.

The precipitate is suspended in phosphate buffer, pH 7.8, and 20 mg of trypsin and 20 mg of chymotrypsin per gram of tissue, wet weight (approximately 0.1 g of each enzyme per gram of protein) are added. The digestion is carried out at 38°, the pH being adjusted with KOH at intervals of about 30 minutes. After about 3 hours of digestion, it does not change any more. The proteolytic enzymes are then denatured by heating for 3 minutes at 100°, and the suspension is centrifuged.[12]

An aliquot (2–3 ml) is brought to pH 1 with 1 N HCl and hydrolyzed 15 minutes at 100° so as to degrade to the mononucleotide level the flavin peptide freed by proteolysis.

The pH of the hydrolyzate is adjusted with KOH, and the fluorescence is measured first at pH 3.1 (F_E) and then at pH 6.8 (F_F). An internal

---

[12] With some tissues (e.g., liver, brain), a few drops of 100% TCA is added before heating at 100°, and freezing and thawing after hydrolysis helps obtain clear solutions.

standard of riboflavin (0.1–1 μg) is then added,[11] and the fluorescence is measured again (F_S). The values are corrected for dilution (F'_F, F'_S). The amount of flavin peptide expressed as micrograms of riboflavin per gram of tissue is calculated as follows:

$$\text{Riboflavin (μg) added} \times \frac{F_E - F'_F}{F'_S - F'_F} \times \frac{\text{ml supernatant from proteolysis}}{\text{ml aliquot taken} \times \text{g tissue}}$$

Flavin peptides showing the emission characteristics of that deriving from succinate dehydrogenase are a minor part of the total flavins in tissues. For this reason, unless FAD, FMN, and riboflavin are previously extracted, the fluorometric determination of bound flavin is impracticable.

Special Cases. When broths or mycelial extracts from cultures of flavinogenic ascomycetes are examined, or other materials in which one or some flavins are in great excess over the others, the component in excess is separated as follows: If riboflavin is in excess, the mixture is passed through a column of Amberlite IRC 50 (Na+). Flavin coenzymes and 6-methyl-7-hydroxy-8-ribityllumazine are eluted from the resin with water in a first peak, 6,7-dimethylribolumazine in a second. Riboflavin is eluted only by 1 N HCl. Recoveries are 95% for flavin coenzymes and lumazines and about 85% for riboflavin.[13] By this procedure, extracts from *Ashbya gossypii* containing a 2000-fold excess of riboflavin could be resolved and analyzed.[13] If FAD or FMN is the component in excess, the mixture is passed through a column of Dowex 1-X2 ion exchanger. Riboflavin is eluted by water; and FMN, by 0.05 M NH₄Cl, pH 6. FAD is then eluted with 0.2 M NaCl in 10⁻³ M HCl.

The elution can be followed spectrophotometrically by recording the absorption at 350 nm, 400 nm, and 450 nm, corresponding, respectively, to 6-methyl-7-hydroxy-8-ribityllumazine, 6,7-dimethyl-8-ribityllumazine, and the flavins; or by measuring the fluorescence at the following wavelengths: 6,7-dimethyl-8-ribityllumazine, 415 nm excitation, 475 nm emission; 6-methyl-7-hydroxy ribolumazine, 340 nm excitation, 425 nm emission; flavins, 445 nm excitation, 520 nm emission.

Analysis

Complex mixtures containing flavins (and lumazines from flavinogenic ascomycetes) in microgram amounts are conveniently resolved by paper electrophoresis[14] in sodium acetate buffer, pH 5.1, $I = 0.05$, on Munktell 20 or Whatman No. 3 MM paper, at a voltage of 15 V/cm.

A quantity containing not more than 7 μg of each flavin is applied to the paper. A moderately yellow spot is obtained. The paper is then

[13] P. Cerletti, R. Strom, S. Giovenco, D. Barra, and M. A. Giovenco, *J. Chromatog.* **29**, 182 (1967).

[14] P. Cerletti and N. Siliprandi, *Biochem. J.* **61**, 324 (1955).

ELECTROPHORETIC MOBILITIES OF FLAVINS AT pH 5.1, $I = 0.05$, R_f VALUES
IN 5% (w/v) Na_2HPO_4, AND FLUORESCENCE OF THE COMPOUNDS ON PAPER

Compound	Mobility relative to riboflavin[a]	R_f	Fluorescence
Riboflavin	−1	0.30	Bright yellow
FMN	+3.9	0.54	Bright yellow
FAD	+5.7	0.40	Pale yellow
Flavin peptides			Pale yellow
Lumiflavin	−0.3	0.18	Greenish yellow
Lumichrome	−0.6	0.07	Bright sky blue

[a] Compounds marked + move to the anode; those marked −, to the cathode.

moistened with a buffer solution, except where the mixture has been applied, and lightly pressed between two sheets of filter paper to remove excess liquid. An adequate separation of the components is obtained after a 4-hour run (see the table). After drying, the flavin zones are located by visual inspection or by contact printing under an ultraviolet lamp emitting in the 366 nm region (Mineralight long wave)[15] (see the table).

If necessary, the identification is confirmed by reference to the spots obtained from a known mixture in a parallel run. The fluorescence of FAD and of flavin peptides is enhanced when the paper is acidified by fumes of concentrated HCl. The spot of flavin peptides gives a weak ninhydrin reaction 1 or 2 days after spraying the reagent.

Mixtures such as those mentioned above can also be resolved by paper chromatography. A convenient solvent (see the table) is 5% (w/v) Na_2HPO_4 (footnote 16). The sensitivity of the method, however, is less than that using electrophoresis.

[15] A high intensity ultraviolet source, e.g., the Chromatovue cabinet with Transilluminator (U.V. Products, San Gabriel, Calif.), increases severalfold the limit of detection
[16] E. Dimant, D. R. Sanadi, and F. M. Huennekens, *J. Am. Chem. Soc.* **75**, 3611 (1953).

[134] Simultaneous Microdetermination of Riboflavin, FMN, and FAD in Animal Tissues

By KUNIO YAGI

Principle

Flavin compounds contained in animal tissues are generally restricted to riboflavin, FMN, and FAD. In the simultaneous determination of each of these flavin compounds, care should be taken to prevent the decomposition of FAD and FMN during the procedure of their extraction from animal tissues. To extract flavins quantitatively from animal tissues without decomposition of flavin nucleotides, the "warm-water extraction method"[1] is recommended when the lumiflavin fluorescence method is applied to the determination of flavins. The extraction method consists of the heating of animal tissues in water at 80°, the homogenization and the subsequent heating of the homogenate at the same temperature. This procedure was established on the basis of the following experimental results: (1) FAD in the tissue or the extract is stable to heating at 80°, though pure FAD is unstable and is converted into riboflavin 4′,5′-cyclic phosphate upon heating above 70°.[2] (2) Many kinds of phosphatases, which may attack flavin nucleotides, can be denatured by heating above 60°.[3] (3) Generally, extraction of tissue flavins with water can be completed by heating at 80° for 15 minutes.[1] (4) During the heating at 80°, most of the protein is denatured and precipitated.

For the microdetermination of the total amount of flavins in the above-mentioned aqueous extracts from animal tissues, the lumiflavin fluorescence method is suitable. The principle of this method is based on the fluorescence measurement of lumiflavin which is derived from riboflavin, FMN, or FAD by photolysis in alkaline medium and is extracted with $CHCl_3$ from acidified medium. This confers high specificity on this method. Using fluorometry, low concentrations of flavin (0.001–1 μg/ml) can be determined. In this method, the test solution need not be clear, because slight turbidity of the solution does not interfere with the photolysis of flavins and the $CHCl_3$ extraction of lumiflavin.

Since riboflavin, FMN, and FAD are equally converted into lumiflavin by photolysis in alkaline medium and the total flavin is calculated as riboflavin,[4] the amount of each flavin can be calculated if the molar ratio of the three flavins is measured.[1] The determination of the ratio can be made upon separation of these flavins by paper chromatography.[5] The detection of the spot of each flavin separated on paper is carried out under ultraviolet light, and each flavin is eluted from the filter paper and estimated by the lumiflavin fluorescence method. From the estimated relative values, the molar ratio of these flavins can be determined.

[1] K. Yagi, *J. Biochem.* **38,** 161 (1951).

[2] K. Yagi and J. Okuda, *J. Biochem.* **47,** 77 (1960).

[3] J. Okuda and K. Yagi, *Chem. Pharm. Bull.* (*Tokyo*) **7,** 456 (1959).

[4] K. Yagi, *J. Biochem.* **43,** 635 (1956).

[5] Instead of paper chromatography, paper electrophoresis is also applicable.

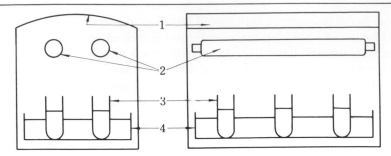

Fig. 1. Apparatus for photolysis to convert flavins into lumiflavin: *1*, mirror; *2*, fluorescent lamp (e.g., Mazda FL 20D); *3*, glass-stoppered centrifuge tube without stopper; *4*, holder for centrifuge tube.

Apparatus

Apparatus for photolysis. Visible light of wavelengths shorter than 530 mμ, in addition to ultraviolet, is effective for photolysis of these flavins to lumiflavin,[6] and a fluorescent daylight lamp (e.g., Mazda FL 20D) is to be preferred to ordinary tungsten lamps or a high pressure mercury lamp as a practical light source for photolysis.[7] The apparatus shown in Fig. 1 is an example suitable for the photolysis.[4]

Glass-stoppered centrifuge tube. For the photolysis of the flavins and CHCl$_3$ extraction of lumiflavin, a glass-stoppered centrifuge tube

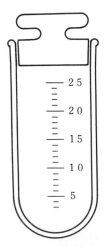

Fig. 2. Glass-stoppered centrifuge tube for photolysis and CHCl$_3$ extraction procedure.

[6] K. Yagi, *Med. Biol.* (*Japan*) **28**, 54 (1953).
[7] K. Yagi, *Vitamins* **7**, 493 (1954).

(Fig. 2) is used.[4] It can also be used for the extraction of flavins from animal tissues, as well as for the pretreatment procedure for paper chromatography.

Reagents

Reagents for the Lumiflavin Fluorescence Method

Chloroform. Before use, it must be tested for fluorescence. If it has fluorescence, one must distill it carefully in an all-glass apparatus.

Riboflavin standard solution. Five milligrams of pure riboflavin crystals (recrystallized from dilute acetic acid) is dissolved in 1.0% acetic acid (about 70 ml) at 60°, cooled, and made up to 100 ml with 1.0% acetic acid. This stock solution should be put into a brown bottle and stored in an icebox. It may be used for several months. From this stock solution, 1.0 µg/ml standard solution is prepared by dilution with distilled water, and the concentration is checked by its optical density at 450 mµ.[8] Other standard solutions (0.1 µg/ml or other appropriate concentrations) are then prepared. The dilute standard solutions should be prepared for each use.

NaOH, 1 N

Acetic acid, glacial

Reagents for Pretreatment for Paper Chromatography

$(NH_4)_2SO_4$

Phenol

Ethyl ether

Solvent for Paper Chromatography

n-Butanol–acetic acid–water (4:1:5, v/v/v), upper layer

Procedures

Extraction of Flavins from Animal Tissues

The tissue is excised fresh from animals, immediately weighed and cut into small pieces.[9] It is put into a few milliliters of water previously heated

---

[8] The $E_{1\,cm}^{1.0\,\mu g/ml}$ at 450 mµ is 0.324.

[9] Ordinarily, 0.1–1.0 g of tissue is used for the determination of total flavin. For the simultaneous determination of the three flavins, however, several grams of tissue are needed because the net amount of total flavin in an extract should be more than 20–30 µg to obtain reliable data on free riboflavin, the amount of which is usually small.

at 80°, and kept at 80° for 3–5 minutes. The tissue is ground in a glass homogenizer, then the resulting suspension is transferred quantitatively to a graduated tube (Fig. 2), diluted with water to a volume of over 20 ml per gram of tissue, and heated at 80° for 15 minutes with occasional stirring. After cooling at room temperature, the suspension is made up with water exactly to the volume before heating, then is stirred and centrifuged. An aliquot of the supernatant is used for the estimation of the total amount of flavins.

Determination of Total Amount of Flavins

Photolysis. In tube A are placed 1.0 ml of test solution[10] and 1.0 ml of distilled water. In tube B are placed 1.0 ml of the test solution and 1.0 ml of standard riboflavin solution.[11] Then 2.0 ml of 1 N NaOH is added to each tube. The tubes are irradiated for 30 minutes in the apparatus for photolysis (Fig. 1), and then 0.2 ml of acetic acid is added to each tube.

Extraction of Lumiflavin with Chloroform. In tube C, 1.0 ml of the test solution, 1.0 ml of water and 0.2 ml of acetic acid are placed. Then 2.0 ml of 1 N NaOH is added and mixed. To each A, B, and C tube, is added 6.0 ml of $CHCl_3$; the tubes are stoppered, cooled with tap water, shaken vigorously, and then centrifuged upon removal of the stopper.

Estimation of Fluorescence of Lumiflavin. From tubes A, B, and C, 5.0-ml aliquots of the $CHCl_3$ layers are taken for the measurement of fluorescence. The reading of B is designated as f_1, and those of A and C as f_2 and f_3, respectively.

Calculation. Here $(f_2 - f_3)$ corresponds to the flavin content of the test solution, and $(f_1 - f_2)$ to that of riboflavin added. Therefore, the total amount of flavins calculated as riboflavin in the sample when 0.1 µg of riboflavin is added is

$$0.1 \times [(f_2 - f_3)/(f_1 - f_2)](b/a)\mu g/g$$

where a is the amount of sample in grams and b is the volume in milliliters after dilution of the homogenate.

[10] It must be tested whether a sample solution contains fluorescent substances soluble in $CHCl_3$, although the usual extract from animal tissues does not contain such substances. If the extract contains $CHCl_3$-soluble fluorescent substances, the following preextraction procedure is needed. In a glass-stoppered centrifuge tube [A], 5.0 ml of test solution and 5.0 ml of distilled water are placed. In another centrifuge tube [B], 5.0 ml of test solution and 5.0 ml of standard riboflavin solution are placed. To each tube, 5.0 ml of $CHCl_3$ is added. The tubes are cooled with tap water, shaken, and centrifuged. The fluorescence of the $CHCl_3$ layer is tested with a microphotofluorometer. The $CHCl_3$ extraction must be repeated until all the $CHCl_3$-soluble fluorescent substances have been removed. After the preextraction, 2.0-ml aliquots of the aqueous layers of [A] and [B] are placed into tubes A and B, respectively.

[11] To eliminate experimental errors, the concentration of the standard solution should be close to that of the test solution. For this, a preliminary experiment is required.

Determination of the Molar Ratio of the Separated Flavins

Paper Chromatography. After an aliquot of the tissue extract is taken for the measurement of the total amount of flavins, the remaining extract is used for paper chromatography. It is saturated with $(NH_4)_2SO_4$ with stirring. After several minutes, the mixture is centrifuged or filtered to remove the precipitated protein, and the supernatant or filtrate is collected in a glass-stoppered centrifuge tube (Fig. 2).[12] To this tube is added 2 ml of phenol. The tube is stoppered, shaken vigorously, and centrifuged. The upper phenol layer is transferred to another centrifuge tube using a pipette, and then 2 ml of phenol is added to the residual water layer to repeat the extraction in the same manner.

To the combined phenol layers is added 0.1 ml of water (when the quantity of flavin is small, less water is used). After gentle shaking, 10–20 ml of ethyl ether is added. The tube is stoppered, cooled with tap water, shaken vigorously, and centrifuged. Most of the flavins are concentrated in the water layer at the bottom of the tube. After removal of the phenol–ether layer, the water layer[13] is collected with a micropipette and spotted on the paper in the usual fashion. The paper is dried in the darkroom, and developed using the upper layer of *n*-butanol–acetic acid–water mixture (4:1:5, v/v/v) as solvent.[15]

Estimation of Individual Flavin after the Paper Chromatography. After the development, the spots of FAD, FMN, and riboflavin on filter paper are located by ultraviolet light. Each spot area containing flavin is cut out, washed with ether twice, cut into small pieces, and put into a centrifuge tube (Fig. 2). To this tube, are added 2.0 ml of water and 2.0 ml of 1 N NaOH. Tubes are then placed in the photolysis apparatus for 30 minutes.

After the photolysis is over, each solution is acidified with 0.2 ml of glacial acetic acid, and 6.0 ml of chloroform is added. The tube is cooled with tap water, shaken vigorously, and centrifuged. Then, 5.0 ml of the chloroform layer is transferred into a cuvette. The relative fluorescence intensity for each sample of FAD, FMN, and riboflavin is recorded as f_{FAD}, f_{FMN}, and f_{FR}, respectively.

---

[12] When adsorption of flavin to the protein precipitated by $(NH_4)_2SO_4$ is observed, it can be avoided by acidifying the extract with acetic acid to bring the solution to a pH of about 5.0 before the addition of $(NH_4)_2SO_4$, followed by centrifugation. Then the supernatant is saturated with $(NH_4)_2SO_4$.

[13] This aqueous solution can be submitted to paper electrophoresis to separate flavins from each other without their decomposition. As an electrolyte, 0.05 M phosphate buffer, pH 8.0, is recommended. A current of 2.44 mA per centimeter of paper width is required for complete separation of flavins within an hour.[14]

[14] K. Yagi, *in* "Methods of Biochemical Analysis" (D. Glick, ed.), Vol. 10, p. 319. Wiley (Interscience), New York, 1962.

[15] When the upper layer of *n*-butanol–acetic acid–water (4:1:5, v/v/v) is used as solvent, R_f values of FAD, FMN, and riboflavin are 0.03, 0.1, and 0.3, respectively.

As a blank test, a part of the filter paper that does not contain flavin is treated in the same way as above. Then, the blank value is subtracted from each value to give f'_{FAD}, f'_{FMN}, and f'_{FR}, respectively.

Calculation

If total flavin in the sample is estimated as e $\mu g/g$ and the relative values of the three types of flavin are estimated as f'_{FAD}, f'_{FMN}, and f'_{FR} respectively, the amount of each flavin, expressed as riboflavin, is calculated as follows:

$$FAD = e[f'_{FAD}/(f'_{FAD} + f'_{FMN} + f'_{FR})] \,\mu g/g$$
$$FMN = e[f'_{FMN}/(f'_{FAD} + f'_{FMN} + f'_{FR})] \,\mu g/g$$
$$Riboflavin = e[f'_{FR}/(f'_{FAD} + f'_{FMN} + f'_{FR})] \,\mu g/g$$

To obtain the absolute amount of each flavin, the following equations may be used:

$$FAD \,(abs.) = e[f'_{FAD}/(f'_{FAD} + f'_{FMN} + f'_{FR})] \times 2.09 \,\mu g/g$$
$$FMN \,(abs.) = e[f'_{FMN}/(f'_{FAD} + f'_{FMN} + f'_{FR})] \times 1.21 \,\mu g/g$$
$$Riboflavin \,(abs.) = e[f'_{FR}/(f'_{FAD} + f'_{FMN} + f'_{FR})] \times 1.0 \,\mu g/g$$

Examples of the Measurement

Flavins in the tissues of normal rats measured by this method are listed in the table.[14]

AMOUNTS OF FLAVINS IN TISSUES OF NORMAL RATS

Tissue	Total flavin ($\mu g/g$)	FAD		FMN		Riboflavin	
		$\mu g/g$	%	$\mu g/g$	%	$\mu g/g$	%
Liver	29.5	22.1	75.0	6.7	22.7	0.8	2.3
Kidney	34.9	24.1	69.0	9.9	28.4	0.9	2.6
Heart	15.9	14.3	90.0	1.5	9.7	0.1	0.3
Intestine	4.3	3.2	74.5	1.0	23.0	0.1	2.5

[135] Identification of Crystals of Isoalloxazines by Microscopy in Polarized Light

By E. C. OWEN

An invaluable aid in the identification of isoalloxazines during their isolation from natural materials is the mineralogical microscope.[1,2] This has a rotatable stage graduated in degrees and is provided with a plate of polaroid as the polarizer, below the condenser, and with a second polaroid plate, the analyzer, between the objective and the eyepiece. With polarizer and analyzer out, crystals mounted on microscope slides can be examined for color and shape in ordinary light with or without color filters in front of the light source. Isoalloxazines are faint yellow to brown according to their thickness. With the polarizer in place, the crystals can be examined in plane-polarized light; and with the analyzer also in place, examinations can be made with the polars either crossed or parallel. When they are crossed, the light from the polarizer cannot pass through the analyzer unless the crystals being examined are birefringent. In plane-polarized light or with the polars parallel, pleochroism, a phenomenon that is not uncommon in certain minerals, was found[1] to be a general property of isoalloxazines, though it had already been reported for riboflavin by Sakate.[3] The optical properties of isoalloxazines are discussed in the following sections.

Riboflavin

Ordinary Light. Riboflavin crystallized from water appears as long needles (2.6 × 155 mm) with parallel sides, but not ending in well-defined crystal faces.

Plane-Polarized Light. The crystals show pleochroism, being brown or yellow, according to the depth of the crystal, when they are oriented perpendicular to the plane of polarization and being yellow or colorless when they are lying parallel to the polarization plane. Crystals large enough for the more detailed microscopic measurements of riboflavin and arabo-flavin made by Sakate[3] were not encountered, but it has been found that there are marked changes of refractive index (R.I.) of riboflavin and other isoalloxazines as the stage is rotated in plane-polarized light.[2] With ribo-flavin in the parallel position, the R.I. is almost equal to that of the mountant (R.I. = 1.54). The mountant is polystyrene in xylol (Kirk-patrick & Lendrum, 1941) and is obtainable from G. T. Gurr Ltd., London. By contrast, in the perpendicular position, the R.I. of riboflavin is much greater than that of the mountant. The Becke test, which is described in detail in any book on optical mineralogy, is used to judge the differences of R.I. When riboflavin is crystallized on a slide from glacial acetic acid, shorter wider rods are formed, but still the ends are not eumorphic.

---

[1] E. C. Owen, *Proc. 1st Intern. Congr. Food Sci. Technol.* (*London*) **III**, 669–678 (1962).
[2] E. C. Owen, *Biochem. J.* **84**, 96P (1962).
[3] H. Sakate, *Nagoya J. Med. Sci.* **18**, 203–214 (1956).

Crossed Polars.[4] The rods of riboflavin, whether from water or glacial acetic acid, have straight extinction as would be expected from Sakate's measurements[3] which describe riboflavin as orthorhombic. As would be expected from the R.I. differences with orientation in polarized light, the birefringence gives rise to very bright polarization colors, which are also seen in all other isoalloxazines so far examined, a finding which helps in the isolation of isoalloxazines from milk and urine. Many birefringent and some isotropic colorless substances are sorted out with isoalloxazines on paper or thin-layer chromatograms. All of these birefringent substances exhibit only first-order polarization colors and so are easily seen when they contaminate isoalloxazine concentrates. One such substance which contaminates 7,8-dimethyl-10-(2-hydroxyethyl)isoalloxazine has been described.[1]

Hydroxyethylflavin [7,8-dimethyl-10-(2-hydroxyethyl)isoalloxazine]

Ordinary Light. Crystals of hydroxyethylflavin from acetic acid, like those of other isoalloxazines are honey-yellow to amber to brown according to their depth as they lie on the slide. Their size and variety of shape can be seen from Fig. 1.

Plane-Polarized Light. The crystals exhibit (see Figs. 1a and 1b) pleochroism of the same type as that shown by riboflavin. They also show the same change of relief when the stage is rotated, except that when crystals are perpendicular to the plane of polarization they are both colorless and have no outline, and the Becke test shows their R.I. in this position to be equal to that of the medium (R.I. = 1.54). When lying parallel to the plane of polarization, the R.I. of crystals is much greater that 1.54.

Crossed Polarizers. As with riboflavin, the polarization colors are very

[4] N. H. Hartshorne and A. Stuart, "Crystals and the Polarising Microscope. A Handbook for Chemists and Others," 3rd ed. Edward Arnold, London, 1960.

FIG. 1. Additional blue filter. All these show the same field of crystals of 7,8-dimethyl-10-(2-hydroxyethyl-)isoalloxazine. (a) Parallel polars to enhance pleochroism. The variation of crystals from black to gray is due to pleochroism. The arrow indicates the plane of polarization of light. (b) Stage rotated 90° to position in 1a. Crystals lying parallel to the plane of polarization are black. Those lying perpendicular to it are gray. The arrow indicates the plane of polarization of light. (c) Polars crossed. Most crystals are now brightly colored (white in the picture) against a black background. Crystals that have completely disappeared are in the extinct position. Angle of extinction, 17°. (d) As in 1c, but stage is rotated 45°. Comparison with 1a, 1b, and 1c shows that crystals extinct in 1c are different from those extinct in 1d.

Each photomicrograph is at the same magnification (×193). The microscope used was a Swift Model PN (Hilger & Watts, London); camera, a Land Polaroid. Illumination was from a Pointolite platinum arc, with a Calorex (Chance Bros., London) filter in front of the condenser. The actual diameter of every field was 0.465 mm.

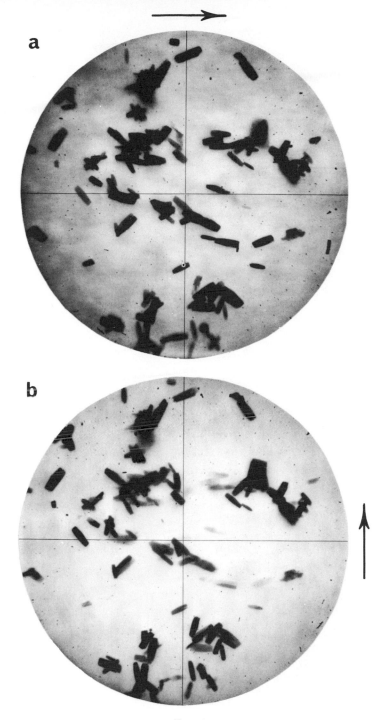

Fɪɢ. 1.
299

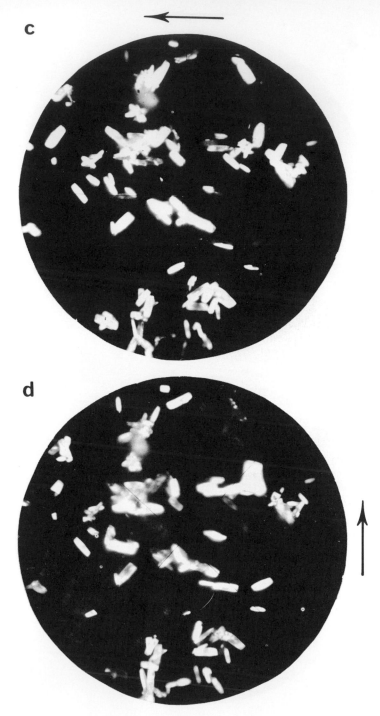

Fig. 1.

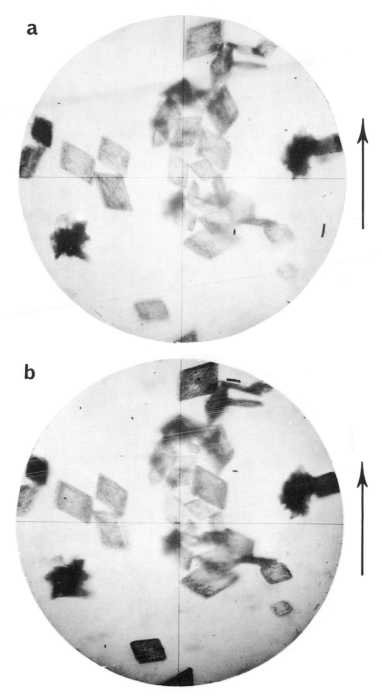

a

b

Fig. 2.

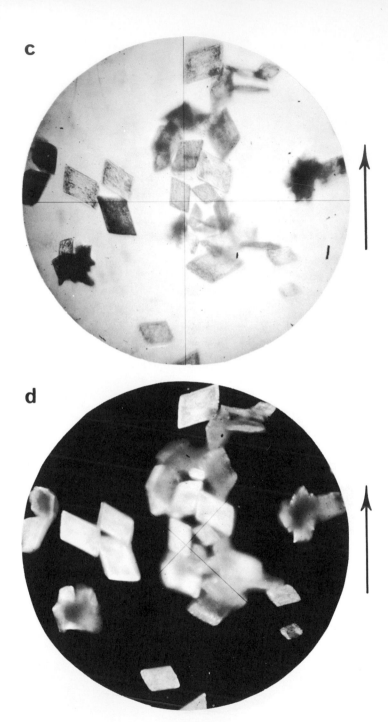

c

d

Fig. 2.

302

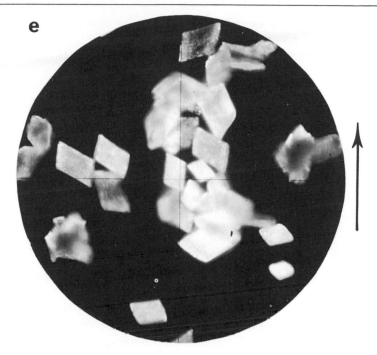

FIG. 2. Same field in all pictures of 7,8 dimethyl-10-(2-acetoxyethyl)isoalloxazine. No blue filter. (2a) Ordinary light. The different shades of gray of various crystals are due to variations of tint from yellow to amber due to crystal depths. Some of the crystals have domed tops. (2b) and (2c) Plane-polarized light, but with stage in 2c at 90° to position in 2b. Variation of grayness between crystals in 2b and the same crystal in 2a are due to pleochroism. (2d) and (2e) Polars crossed. Extinction is parallel to the shorter side of the rhomboid crystals. Stage in 2e is at 45° from 2d.

Same magnification and instrumentation as in Fig. 1.

bright. The extinction angle is 17° from the axis of elongation of the crystals. Some of the crystals exhibit geniculate, and others simple, twinning (Figs. 1c and 1d).

7,8-Dimethyl-10-(2-acetoxyethyl)isoalloxazine from Chloroform

Ordinary Light (Fig. 2a). This acetoxy compound crystallizes in rhombic forms that vary from light yellow to amber according to their depth. Some crystals exhibit interpenetrative twinning that produces stellate clusters.

Plane Polarized Light. Pleochroism with the same type of color changes is very evident, as can be seen from the apparently random changes of shade of almost any crystal from Figs. 2b and 2c, the difference between these two photographs being produced by a rotation of the stage through

90°. The deeper hue is seen when the crystals lie with their axes of elongation perpendicular to the plane of polarization.

When the Becke test is applied, the crystals have R.I. greater than 1.54 in the perpendicular position. In the parallel position their R.I. is less than 1.54.

Crossed Polars. Polarization colors are bright and the extinction is parallel to the shorter side of each rhomboid.

7,8-Dimethyl-10-formylmethylisoalloxazine from Acetonitrile

Ordinary Light. The compound crystallizes in yellow to amber blades and stellate clusters. Ends of the blades are not eumorphic. The color varies with depth from yellow to amber.

Plane-Polarized Light. By contrast with each of the three preceding compounds, this aldehyde shows an entirely different positioning of the maximum and minimum of pleochroism. The maximum and minimum are at right angles to one another as in the other isoalloxazines, but they occur when the plane of polarization is at 45° to the axis of elongation of each crystal.

Crossed Polars. The frequency of twinning makes it difficult to find crystals which are lying on a face, so that measurements of the extinction do not give a result which is repeatable from one crystal to another. Polarization colors are very bright.

Other Isoalloxazines

Microscopic examinations were also made of 7,8-dimethyl isoalloxazines with ω-hydroxyalkyl side chains. These side chains are normal and contain

EXTINCTION ANGLES OF 7,8-DIMETHYL-10-(ω-HYDROXYALKYL)ISOALLOXAZINES
MEASURED FROM THE AXIS OF ELONGATION OF THEIR CRYSTALS

Value of ω	Nature of n-alkyl side chain at N-10	Relief[a] and pleochroism. Maximum color and relief when axis is parallel ‖, or when perpendicular ⊥, to polarization plane	Extinction angle[b] (degrees)
2	—$CH_2 \cdot CH_2OH$	‖	17
3	—$(CH_2)_2 \cdot CH_2OH$	‖	15.6
4	—$(CH_2)_3 \cdot CH_2OH$	‖	14.8
5	—$(CH_2)_4 \cdot CH_2OH$	‖	0
6	—$(CH_2)_5 \cdot CH_2OH$	‖	0
Riboflavin	—$CH_2(CHOH)_3 \cdot CH_2OH$	⊥	0

[a] The greater the difference between 1.54 (the R.I. of the mountant) and the crystal, the greater is the relief.

[b] Each figure is the mean of at least 10 measurements.

The author thanks Dr. D. B. McCormick for samples of isoalloxazines including those of ω = 3, 4, and 5 of the above table.

3, 4, 5, or 6 carbon atoms (see table). All these showed the same color variation from crystal to crystal and the same type and orientation of pleochroism as the 2-OH compound described above, and these changes of color are accompanied by analogous changes of R.I. in plane-polarized light. Polarization colors are very bright.

[136] Polarography of Flavins

By EDUARD KNOBLOCH

Introduction

Polarography has opened up new possibilities for electrochemical and analytical research into biochemically significant redox systems. These possibilities are comparatively well demonstrated in the polarographic research into flavins. The use of the dropping mercury electrode as a source of electrons has provided new experimental conditions for studying the transport mechanisms of electrons. With the help of the expanded theory of polarography, it is possible to classify and explain the observed phenomena. At present, polarography is considered to be a biochemical method on the level, for example, of molecular spectroscopy, nuclear magnetic resonance, and electron paramagnetic resonance. The application of physicochemical methods to fundamental enzymological research is typical of the present period. From polarography and its branches it may be expected that the methods will further contribute to the study of the reversibility and reaction mechanisms of enzyme-catalyzed reactions. Biochemically significant redox systems are mostly electroactive, and they are capable of electron exchange with the electrode, so that it is possible to measure their oxidation-reduction potentials, which are a direct indicator of the electron activity of the system. The application of polarography to electrochemical research into flavins is especially suitable, as these substances reduce in a perfectly reversible manner on a mercury electrode. The work of Professor Heyrovský's collaborators led by Brdička, which was concerned with the study of flavins and related redox systems, has provided new theoretical information for polarography itself, especially for the investigation of the formation of intermediate reduction products and for the application of adsorption during the electrode process.

In an introduction to the polarography of flavins, it is necessary to mention the results of the polarographic studies of redox systems that have a general validity and that are important for understanding the sections that follow. We shall limit ourselves to that knowledge which is directly related to flavins. The main attention is devoted to the analysis of the polarographic wave with a view to the formation of semiquinone, as this intermediate reduction product plays an ever more recognized role in the enzymatic transport of hydrogen. Also the dependence of the reduction on pH has been studied, which also relates to the application of dissociation constants of oxidized and reduced forms of flavins, etc. A review is given here of the polarographic behavior of the most important flavins that occur as coenzymes. The review includes results of the polarographic study of photolysis and photoreduction of riboflavin which have provided new data for the research into the reaction kinetics and photodynamic functions of flavins. Also described is the catalytic separation of hydrogen on the dropping mercury electrode, which occurs in the presence of isoalloxazine derivatives and some other nitrogen substances. Catalytic currents belong to the characteristic polarographic phenomena. An example of the use of polarography for studying the reaction kinetics of reversible reduction in the model flavin–DPNH system is also discussed. Finally, an example is given of the application of polarographic methods to the research into catalytic functions of the Schardinger enzyme. In the theoretical treatment some fundamental information on the electrochemical characterization of redox systems will be mentioned in order to avoid the necessity of having to return to it in the text.

Theory and Methods

Polarography is an electrochemical method, employing mercury, regenerating droplets as electrodes. During the polarographic study of substances in solutions, the transit of current is recorded on the polarograph indicator, under continuously increasing voltage, and characteristic curves of the current intensity and voltage are obtained; from these it is possible to draw qualitative and quantitative conclusions about the depolarizer in the solution. The intensity of the current being passed through is the indicator of concentration, and the potential at which reduction occurs is the qualitative indicator. If the system in question is reversible, frequently the value $E_{1/2}$, (the half-wave potential), corresponds to the redox potential of the system. The electrochemical characteristics of redox systems and the utilization of polarography for the study of reversibility, especially in regard to flavins are mentioned below.

It is possible to express the dependence of the potential on the ratio of the oxidized and reduced forms of the system electrochemically by the Peters formula:

$$E = E_0 - \frac{RT}{nF} \times \ln \frac{[\text{RED}]}{[\text{OX}]} \tag{1}$$

where E_0 is the normal redox potential of the system when the ratio of the reduced and oxidized forms is 1:1. The potential is most frequently expressed relative to a normal hydrogen electrode with a unit concentration of [hydrogen ions, or as E'_0, which represents the redox potential of the system at pH 7. In experiments, measurements are usually made relative to a saturated calomel electrode (SCE), and the results are then reduced to normal potentials; R is the gas constant, 8.314 joules; T is the absolute temperature; F is 1 Faraday = 96.495 coulombs; n is the number of electrons taking part in the electrochemical reaction. After transformation to decadic logarithms and summing of the constants that occur in the formula, we arrive at a simplified expression for the temperature of 30°:

$$E = E_0 - 0.06 \log \frac{[\text{RED}]}{[\text{OX}]} \tag{2}$$

For $n = 1$, 0.06 is valid; for $n = 2$, 0.03 is valid. The free energy related to the change of potential is given by the expression

$$\Delta G_0 = -nF \times \Delta E \tag{3}$$

ΔE is the potential difference after the redox reaction. If the usual symbols are used, the energy comes out in joules. To transform a value to calories, it is necessary to divide this value by 4.184.

For the extension of the redox-system theory, the papers published by Michaelis and Elema[1,2] are the most important. Michaelis has proved that on the basis of the analysis of the potentiometric curve of various organic redox systems it is possible to follow the formation of intermediate reduction products, which have the nature of free radicals. The number of electrons used up during a given redox reaction can be determined from the tangent to the potentiometric curve, or from the so-called index potentials, E_i, which may be derived from the system potentials, where the ratio of the oxidized and reduced forms is 1:3 and 3:1. The values of these

[1] A. H. Friedheim and L. Michaelis, *J. Biol. Chem.* **91**, 355 (1931).
[2] B. Elema, *Rec. Trav. Chim.* **50**, 807 (1931); **54**, 76 (1935).

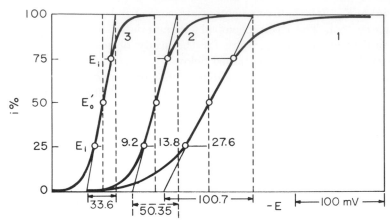

Fig. 1. Potentiometric titration curves for the one-, two-, and three-electron process in the arrangement currently used in polarography. (According to R. Brdička.)

potentials for normal redox systems, in which the exchange of one, two, or three electrons takes place, are 27.6, 13.8, and 9.2 mV at 20°. With many of the systems Michaelis found values that did not correspond to whole numbers. These anomalies were later explained as the effect of the formation of intermediate reduction products. Michaelis derived, from thermodynamic equilibria, equations for the shape of the potentiometric curves of oxidation-reduction processes during which semiquinone is formed. The basis of the Michaelis hypothesis is the conception that during a two-electron reversible process the reception of electrons takes place gradually under the formation of the intermediate stage (semiquinone, S),[3-5] which is in reversible equilibrium with the fully oxidized and reduced forms. This radical can dimerize. The formation of the dimer can also be followed by means of the potentiometric curve (see Fig. 1).

$$OX + e \rightleftharpoons S \tag{4}$$

$$S + e \rightleftharpoons RED \tag{5}$$

S is in mobile equilibrium with the oxidized and the reduced forms:

$$2S \rightleftharpoons OX + RED \tag{6}$$

Semiquinone may exist as a free radical or it can dimerize:

$$s^2 = K_s \times rt \tag{7}$$

[3] L. Michaelis, *Chem. Rev.* **16**, 243 (1935).
[4] L. Michaelis and M. P. Schebert, *Chem. Rev.* **22**, 437 (1938).
[5] L. Michaelis and C. V. Smith, *J. Biol. Chem.* **116**, 587 (1936).

where K_s is the semiquinone formation constant expressing its stability.

$$K_s = s^2/rt \tag{8}$$

The formation of the dimer is expressed by the dimer formation constant K_d:

$$K_d = d/s^2 \tag{9}$$

The concentration of semiquinone $= s$, dimer $= d$, locally reduced form $= r$ and oxidized $= t$.

The computation of the semiquinone formation constant can be most simply carried out according to Elema from the index potentials:

$$K_s = \left(\alpha - \frac{3}{\alpha} \right)^2 \tag{10}$$

For 30°

$$\alpha = 10^{E_i/0.0601}$$

$$E_i = E_{\frac{1}{4}} - E_{\frac{1}{2}}, \; E_{\frac{3}{4}} - E_{\frac{1}{2}}$$

$$E^2 - E^1 = 0.0601 \times \log K_0 \tag{11}$$

Table I shows the index potentials and the corresponding values of the constant, K_s.

On the basis of a mathematical analysis of polarographic curves of some redox systems (riboflavin, methylene blue), under the assumptions derived by Michaelis, Brdička[6-8] carried out the general solution of the course of the polarographic curves for the cases of formation of intermediate reduction products. The solution is based on the assumption that at a given value of the potential the ratio of the oxidized and reduced forms on the surface of the electrode corresponds to the relation given by the Peters formula. For applications in polarography, the Peters formula can be suitably arranged by substituting the concentrations of the oxidized and reduced forms by the appropriate current values.

$$E = E'_0 - \frac{RT}{nF} \ln \frac{i}{i_d - i} \tag{12}$$

where

$$\frac{[\text{RED}]}{[\text{OX}]} = \frac{i}{i_d - i}$$

The equation for the current generated during the reduction of the de-

[6] L. Michaelis and G. Schwarzenbach, *J. Biol. Chem.* **123,** 527 (1938).
[7] R. Brdička, *Z. Elektrochem.* **47,** 314 (1941).
[8] R. Brdička, *Z. Elektrochem.* **48,** 686 (1942).

TABLE I
INTERPOLATION TABLE OF INDEX POTENTIALS (MICHAELIS)[a,b]

Log K	0.0601 Log $K = E_2 - E_1$	E_i (V)
-3	-0.1803	0.0146
-2	-0.1202	0.0151
-1.5	-0.0901	0.0157
-1.3	-0.0781	0.0160
-1.0	-0.0601	0.0167
-0.0	-0.0481	0.0174
-0.6	-0.0361	0.0181
-0.4	-0.0240	0.0191
-0.2	-0.0120	0.0203
-0.0	0.0000	0.0218
$+0.2$	$+0.0120$	0.0237
$+0.4$	$+0.0240$	0.0258
$+0.6$	$+0.0361$	0.0286
$+0.8$	$+0.0481$	0.0310
$+1.0$	$+0.0601$	0.0362
1.2	$+0.0721$	0.0404
1.4	$+0.0841$	0.0448
1.6	$+0.0962$	0.0494
1.8	$+0.1082$	0.0546
2.0	$+0.1202$	0.0601

[a] Temperature $= 30°C$.
[b] Whenever $\log K > 2$, then $E_i = 0.03005 \log K$.

polarizer on the dropping mercury electrode, was derived by Ilkovič[9,10] using Fick's laws and by solving the concentration gradient at the surface of the growing dropping electrode. For i, Ilkovič obtained the expression (so-called immediate current)

$$i = 0.732nF(c - c_0) \cdot D^{1/2}m^{2/3}t^{1/6} \tag{13}$$

where c_0 is the concentration in the electrode intermediate phase; c is the concentration in the solution, expressed in mole cm^{-3}; D is the diffusion coefficient having the dimension cm^2 s^{-1}; m is the flow velocity of mercury in grams per second; and t is time in seconds. When these dimensions are used, the current comes out in amperes. The so-called diffusion current is generated in polarography when the depolarizer is exhausted in the intermediate phase, so that further transport of the depolarizer is limited by the velocity of the diffusion. This limiting, or diffusion, current no longer depends on the increase of the potential.

$$i_d = 0.627nFcD^{1/2}m^{2/3}t^{1/6} \tag{14}$$

[9] D. Ilkovič, *Collect. Czeck. Chem. Commun.* **6**, 498 (1934).
[10] D. Ilkovič, *J. Chem. Phys.* **35**, 129 (1938).

Equation (14) corresponds to the so-called mean limiting current. For a given capillary the following expression from the Ilkovič equation is a constant:

$$k = 0.627Fm^{2/3}t^{1/6} \tag{15}$$

From the Ilkovič equation it follows that the current used for the reduction of the depolarizer is directly proportional to its analytical concentration. This is the basis for the utilization of polarography as a very accurate electrochemical analytical method.

$$i_\mathrm{d} = kc \tag{16}$$

In this equation k is the Ilkovič constant.

By solving the diffusion of the increasing dropping electrode, Ilkovič contributed to building up the theoretical foundations of polarography. An approximation of the Ilkovič solution is given by the assumption of linear diffusion to the spherical surface of the electrode. The exact equation of the diffusion current, based on considering spherical diffusion to an increasing spherical electrode, was solved by Koutecký[11,12]; by the method of dimensionless parameters:

$$i_\mathrm{d} = 0.627nFcmz_1(1 + 3.4z_1 + z_1^2) \tag{17}$$

The expressions z and z_1 are defined as follows:

$$z = \frac{D^{1/2}t^{1/6}}{m^{1/3}}$$

$$z_1 = \frac{D^{1/2}t^{1/6}}{m^{1/3}}$$

Brdička applied the simple relations, valid according to Michaelis for potentiometric curves, for polarographic curves. The deriving is based on the use of the Ilkovič equation under certain assumptions. Brdička assumes that the diffusion coefficients of semiquinone and the reduced forms are approximately identical. A further assumption is that the concentration of semiquinone and the reduced forms in the electron intermediate phase are equal to the concentration of the oxidized form in the solution.

$$i = 2k(r_0 - r) + k(s_0 - s) \tag{18}$$

where s_0 is the concentration of semiquinone in the electrode intermediate phase and s the concentration in the solution.

$$r_0 + s_0 + t_0 = a = i_\mathrm{d}/2k \tag{19}$$

[11] J. Koutecký, *Czech. J. Phys.* **2**, 50 (1953).
[12] J. Heyrovský and J. Kůta, Základy polarografie, nakl. ČSAV-Praha, 1962.

where t_0 is the concentration of the oxidized form in the intermediate phase, and a is the analytical concentration in the solution.

The electrode potential is defined by the ratio of the concentrations of the oxidized and reduced forms in the electrode intermediate phase.

$$\frac{t_0}{r_0} = e^{2E(-E_{1/2})F/RT} = P \tag{20}$$

With the help of Eqs. (18), (19), and (20) we can express the concentrations of the individual forms in the electrode intermediate phase as a function of the current and of the potential.

$$t_0 = \frac{P(i_d - 2i)}{2(P - 1)k} \tag{21}$$

$$s_0 = \frac{i(P + 1) - i_d}{(P - 1)k} \tag{22}$$

$$r_0 = \frac{i_d - 2i}{2(P - 1)k} \tag{23}$$

The dependence of concentration of the individual components on the current intensity and the potential sought was derived after substitution into the equation for the formation of semiquinone.

$$i = \frac{i_d}{2} \frac{\sqrt{K_s P} + 2}{P + \sqrt{K_s P} + 1} \tag{24}$$

For the potential

$$E - E_{1/2} = -\frac{RT}{2F} \ln \frac{i}{i_d - i}$$
$$- \frac{RT}{2F} \ln \frac{\sqrt{(2i - i_d)^2(K_s - 4) + 4i_d{}^2} + (2i - i_d)\sqrt{K_s}}{(2i - i_d)^2(K_s - 4) + 4i_d{}^2 - (2i - i_d)\sqrt{K_s}} \tag{25}$$

The shape of the curves for various constants of semiquinone formation is shown in Fig. 2. The values of the index potentials are also shown with the curves. The curves are practically identical with those derived by Michaelis for potentiometry, except that the percentage of oxidation has in this case been substituted by the current intensity. In this way the identity of redox-system polarographic curves and of potentiometric curves was derived and proved mathematically. The difference is only that, with the potentiometric measurements, reduction occurs in the whole solution after the reduction agent is added, whereas in polarography, reduction (oxidation) of the substance under investigation occurs only in the electron intermediate phase due to the effect of the electrons, the donor of which is the electrode. This was well described in his time by

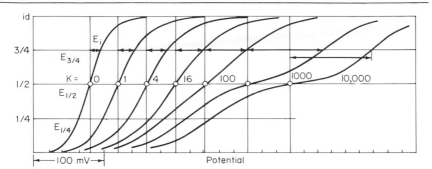

FIG. 2. Polarographic curves derived for the two-step reduction of the redox system in the formation of semiquinone. The curves express the course for different constants of semiquinone formation from 0 to 10,000. On the curves the values of $E_{\frac{1}{2}}$ and those of index potentials are indicated. (According to R. Brdička.)

Müller[13] as electron titration in the intermediate phase. During polarographic analysis, due to the effect of the external inserted voltage on the electrode, the ratio of the oxidized and reduced forms is set, which corresponds to a given potential. In recording the polarographic curve, the current flowing through is recorded under continuously increasing voltage. The polarographic curve obtained expresses the current as a function of the potential. The potential corresponding to half the diffusion current is called, according to Heyrovský,[14] the half-wave potential; this value corresponds to the redox potential of the system. From the polarographic curves, as from the potentiometric ones, it is possible to read off the index potentials.

It is also necessary to mention the dependence of the half-wave potentials, $E_{\frac{1}{2}}$, on pH. With systems in which hydrogen ions take part in the electrochemical reaction, a shift of half-wave potentials occurs. If the generated products of the electrode reaction have the nature of acids, as is the case, e.g., with quinone, the dissociation equation is valid for the appropriate equilibria:

$$H_2 \, RED \rightleftharpoons H \, RED^- + H^+ \qquad (26)$$

$$H \, RED^- \rightleftharpoons RED^{2-} + H^+ \qquad (27)$$

Both dissociation stages are defined by the appropriate dissociation constants, K_{a_1} and K_{a_2}; by substituting into the Peters formula, using the Ilkovič equation, we arrive at the relation expressing the dependence of the redox-system half-wave potential on pH. The assumption here is that the analytical concentration of the oxidized form is identical to the

13 O. H. Müller, *Ann. N.Y., Acad. Sci.* **40**, 91 (1940).

14 J. Heyrovský and D. Ilkovič, *Collect. Czech. Chem. Commun.* **7**, 198 (1935).

analytical concentration in the solution, and that the diffusion coefficients of the oxidized and reduced forms are identical.

$$E_{1/2} = (E_0) + \frac{RT}{2F} \ln (H^+)^2 + K_{a_1}(H^+) + K_{a_1}K_{a_2} \tag{28}$$

where (E_0) is the normal redox potential of the system for unit concentration of the hydrogen ions (pH = 0). The dependence of the half-wave potentials on pH according to Eq. (28) is expressed by a curve, composed of three practically linear sections, of which the section in the strongly acid region, where the expression $(H^+)^2$ is much larger than the expression $K_{a_1}(H^+) + K_{a_1}K_{a_2}$, has a gradient of 58 mV at 20°. The second section, dominated by the term $K_{a_1}(H^+)$, has a gradient of 29 mV, and the third section in the alkaline section has a zero gradient. The discontinuities along the curve correspond to the appropriate pK of the acid studied. In case the dissociation constant of the oxidized form is also effective, the change is displayed by an increase in the potential gradient on pH. The theoretical derivation by Clark and Cohen[15] was proved by several of Michaelis'[3] examples. For the potential dependence of a redox system on pH, it is generally valid, according to Clark, that, in case the acid function of the reduced forms prevails, a decrease occurs in the gradient of the dependence, whereas if this is the case with the acid dissociation constant of the oxidized form, an increase of the gradient occurs. With isoalloxazine derivatives, as will be seen later on, the application of both kinds of dissociation is observed. In this way it is possible to determine with comparative accuracy the appropriate dissociation constants of redox systems, which are especially difficult to measure directly for the reduced forms. The assumption of the validity of these conclusions is the preservation of the concentration of hydrogen ions in the electrode intermediate phase, i.e., work in a buffer medium of sufficient capacity. Otherwise an anomalous course of the polarographic reduction may occur.

Among the advantages of the polarographic method with automatic recording is speed, so that it is possible in a comparatively short time to obtain an idea of a whole series of experiments (e.g., in studying the dependence of the polarographic reduction on pH). Polarography may be used also as a very sensitive and specific analytical method. The method makes it possible, for example, to determine several substances of varying oxidation-reduction potentials next to one another and to determine the oxidized and reduced forms of a redox system in an admixture by a simple experiment without any external intervention; this has considerable significance in studying reaction kinetics. As far as the electrochemical act itself

[15] W. M. Clark and B. Cohen, *U.S. Publ. Health Rept.* **38,** 666 (1923).

is concerned, polarography may be characterized as a dynamic process taking place in the electrode intermediate phase under a continuously increasing voltage. The transit currents are only a quantitative indication of the process. In potentiometry, the whole process is considered statically, and the measurements are made only after the potential in the solution is stabilized.

The polarographic method was discovered by Heyrovský[16,17] in 1925, and it was further developed and supplemented by further branches, which have introduced new experimental possibilities. An example is oscillographic polarography. Matheson and Nichols[18] were the first to use an oscillograph for studying the polarization of a droping mercury electrode. The use of alternating current in oscillographic polarography was introduced in 1941 by Heyrovský.[19] In this branch of polarography the dropping electrode is polarized by alternating current of a high frequency. As an indicator, an oscillograph is included instead of the galvanometer; its electron ray reacts practically without inertia to the changes of voltage on the electrode. Instruments recording the derivative of the potential with respect to time[20,21] have so far found the most widespread application. With this arrangement it is possible to follow oxidation and reduction processes on the electrode in very short time intervals, the electrode working in the alternating current interval as a cathode or an anode. This method led to the discovery of new possibilities for the study of reversibility and for the investigation of unstable intermediate products of electrode reactions. More detailed information about the use of oscillographic polarography may be found in the cited monographs.[22-24]

For purposes of studying the differential capacity of an electrode double layer, which varies very characteristically during adsorption and desorption of the depolarizer, it is convenient to use the Breyer[25] method. With this method the droplet is polarized by an increasing direct current voltage, increased by an alternating current voltage with an amplitude of 15 mV. The so-called tensametric curve obtained provides informa-

[16] J. Heyrovský and M. Shikata, *Rec. Trav. Chim.* **44**, 496 (1925).

[17] J. Heyrovský, *Chem. Listy* **16**, 256 (1922).

[18] L. A. Matheson and N. Nichols, *Trans. Electrochem. Soc.* **73**, 193 (1938).

[19] J. Heyrovský, *Chem. Listy* **35**, 155 (1941).

[20] J. Heyrovský and J. Forejt, *Z. Physik. Chem.* **193**, 77 (1943).

[21] J. Heyrovský, *Anal. Chem. Acta* **2**, 533 (1948).

[22] J. Heyrovský and R. Kalvoda, *Oscil. Polarographie mit Wechselstrom*, Academie-Verlag, Berlin, 1960.

[23] L. Meites, "Polarographic Techniques." Wiley (Interscience) New York, 1955.

[24] I. M. Kolthoff and J. J. Lingane, "Polarography." Wiley (Interscience), New York, 1952.

[25] B. Breyer and F. Guttmann, *Australian J. Sci.* **8**, 21, 163 (1945).

tion about the qualitative changes of the differential capacity during electrolysis.[26]

In applying polarography as an electrochemical method for studying biochemical and biophysical problems, it is necessary to get acquainted at least briefly with the contemporary state of the theoretical development of this discipline, with the characteristics of various currents encountered in polarography, and with the effect of the medium on the polarographic wave. Ignorance of these basic principles may lead to erroneous explanations and incorrect conclusions. In this work it is not possible to discuss the methods in greater detail. The author therefore recommends a number of monographs that discuss these problems.[12,27-31] We shall show in greater detail in the following sections the applicability of polarography to the electrochemical and biochemical research into flavins. The polarographic method was used to discover a number of interesting properties of flavins, for which as yet no connections have been found with their biochemical function of electron carrier. This includes the catalytic evolution of hydrogen, derivatives of isoalloxazine on a dropping mercury electrode, or strong adsorptive properties of the reduced forms of riboflavin.

Polarography of Riboflavin, FMN, and FAD

Riboflavin used in this work was manufactured by Farmakon, Olomouc (ČSSR) and was recrystallized from 1 N acetic acid. For studying the properties of FMN, a substance produced by Hoffmann-La Roche was used, and for studying FAD, a preparation from British Drug Houses, England. The lumiflavin and lumichrome used were prepared synthetically in our institute, the 6,7-dimethyl-9-formylmethylisoalloxazine was prepared according to available literature by the oxidation of riboflavin using periodic acid and by recrystallization from aqueous dioxane. Other flavins were from the author's own experimental material. The purity of the substances obtained was checked spectroscopically, polarographically, and by the paper chromatographic method. The amino acids were of high degree of purity. The properties of DPN were studied using preparations from Sigma, USA.

The polarographic curves were recorded by the Heyrovský polarograph

[26] B. Breyer, F. Guttmann, and F. Bauer, Öster. Chem. Z. **57**, 67 (1956).

[27] J. Heyrovský, "Polarographie." Springer, Vienna, 1941.

[28] R. Brdička, in "Polarographie (Die Methoden der Fermentforschung)" (E. Bamann and K. Myrbäck, eds.) Thieme, Leipzig, 1941.

[29] M. von Stackelberg, "Polarographische Arbeitsmethoden." Gruyter, Berlin, 1950.

[30] G. V. Milner, "Principles and Applications of Polarography." Longmans, Green, London, 1957.

[31] M. Březina and P. Zuman, "Die Polarographie in der Medizin, Biochem, und Pharmazie." Academie-Verlag, Leipzig, 1956.

LP 55 with a Jena glass capillary, internal diameter 0.06 mm, $t = 4.5$ seconds, $m = 1.48$ mg/sec in a 0.1 N KCl medium, and a reservoir height of 60 cm. The dV/dt curves, recorded by a P-524 (Tesla) polaroscope, and either normal dropping or streaming electrodes were used. A saturated calomel electrode was used as a reference in all cases. The it curves were recorded by means of a Kipp and Zonen string galvanometer in the J. Heyrovský Polarographic Institute. For work in an inert atmosphere electrolytic hydrogen or bulb nitrogen was used. The ultraviolet spectra were recorded by a Unicam SP 700 spectrophotometer, the infrared spectra by a Unicam 200 G.

Riboflavin (I) is a derivative of isoalloxazine, and its corresponding structure is 6,7-dimethyl-9-(D-1'-ribityl)isoalloxazine; molecular weight, 376.4. A further derivative of riboflavin, occurring in the form of a coenzyme, is riboflavin 5-phosphate (II), denoted in the literature as flavin adenine mononucleotide (FMN). The most significant coenzyme of this group is flavin adenine dinucleotide (III, FAD); molecular weight, 785.6 (see Fig. 3).

Polarographic research into riboflavin was first taken up by Brdička

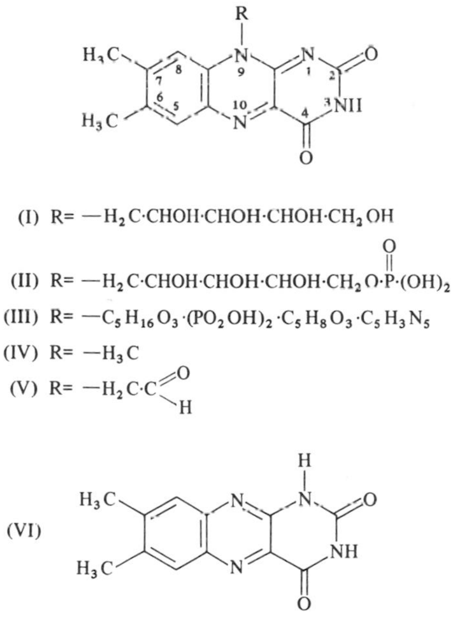

(I) R= —H$_2$C·CHOH·CHOH·CHOH·CH$_2$OH

(II) R= —H$_2$C·CHOH·CHOH·CHOH·CH$_2$O·P·(OH)$_2$

(III) R= —C$_5$H$_{16}$O$_3$·(PO$_2$OH)$_2$·C$_5$H$_8$O$_3$·C$_5$H$_3$N$_5$

(IV) R= —H$_3$C

(V) R= —H$_2$C·C$\lesssim$ ^O_H

(VI)

FIG. 3. Structural formulas of isoalloxazine derivatives. I = riboflavin, II = FMN, III = FAD, IV = lumiflavin, V = 6,7-dimethyl-9-formylmethylisoalloxazine, VI = lumichrome.

and Knobloch[32] in 1941. Simultaneously the application of polarography as an analytical method for determining riboflavin was described by Lingane and Davis.[33] In the paper of Brdička and Knobloch, the polarographic behavior of riboflavin was investigated in the whole pH range by using a universal buffer according to Britton and Robinson; the half-wave potentials were measured accurately and compared with results obtained by applying potentiometric methods of Kuhn and Boulangere,[34] and Michaelis and Smith.[5,6] A comparison showed that the polarographic reduction potentials are practically identical with the values obtained by potentiometric methods, and that they correspond to real redox-potentials of this substance. In applying various reducing agents, such as dithionite and $TiCl_3$, perfect reversibility was observed in the riboflavin–dihydroriboflavin system (Fig. 4).

In the graphic illustration (Figs. 5 and 6) of the dependence of the half-wave potentials on pH, it may be seen that the curve shows two charac-

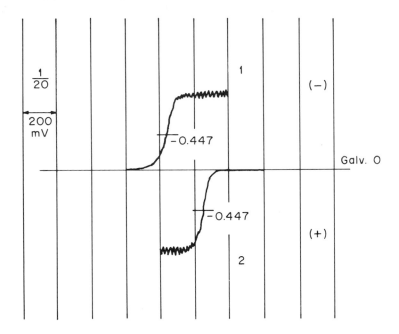

FIG. 4. Polarographic waves expressing the reduction of riboflavin (*1*) and the oxidation of the leuco form (*2*) in phosphate buffer, pH 7; $c = 2.5 \times 10^{-4} M$; 200 mV/absc.; potential vs. saturated calomel electrode.

[32] R. Brdička and E. Knobloch, *Z. Elektrochem.* **47**, 721 (1941).
[33] J. J. Lingane and O. L. Davis, *J. Biol. Chem.* **137**, 567 (1941).
[34] R. Kuhn and P. Boulangere, *Ber.* 69, 1557 (1936).

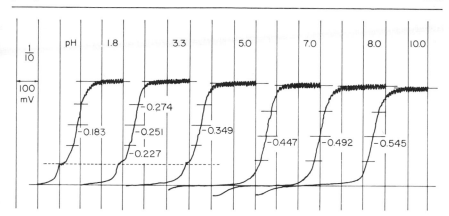

FIG. 5. Polarographic waves of riboflavin ($c = 2.5 \times 10^{-4}\,M$) dependent on pH, in the medium of the universal Britton-Robinson buffer. 100 mV/absc.; the potential vs. saturated calomel electrode.

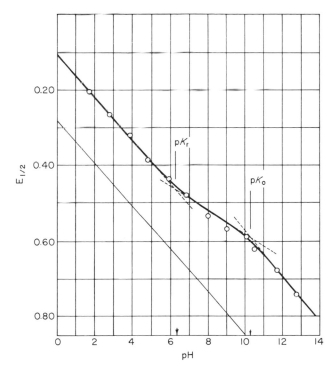

FIG. 6. Graphic illustration of the dependence of the $E_{\frac{1}{2}}$ value on the pH for riboflavin at 20°, the potential vs. NKE; —, values according to Knobloch and Brdička; ○, values according to Michaelis.

teristic jumps at pH 6.3 and 10.2 (Fig. 6) which correspond, according to the criteria derived by Michaelis,[6] to the action of the acid dissociation constants of dihydroriboflavin and of riboflavin ($K_r = 5.01 \times 10^{-7}$ and $K_0 = 6.3 \times 10^{-11}$). Both values are in good agreement with values that were recorded by potentiometric methods. Brdička and Knobloch[32] derived for the dependence of half-wave potentials on pH at 20° a relation dependent on the solution of the equation

$$E_{1/2} = E_0 + \frac{RT}{2F} \ln \frac{(K_r + [\mathrm{H}^+])[\mathrm{H}^+]^2}{(K_0 + [\mathrm{H}^+])} \qquad (29)$$

The value for E_0, determined by extrapolation at pH = 0, and the potential gradient of the function, is 0.058 V. For the dependence of $E_{1/2}$ on the pH of riboflavin, it holds that

$$E_{1/2} = +0.188 + 0.029 \times \frac{5.01 \times 10^{-2} + [\mathrm{H}^+]}{6.31 \times 10^{-10} + [\mathrm{H}^+]} - 0.058\,\mathrm{pH} \qquad (30)$$

For $E_{1/2}$ 20° in a medium of phosphate buffer pH 7, the value -0.447 V relative to the SCE was observed. Reduced to the E'_0 value, we obtained -0.197 V relative to the N hydrogen electrode, which is in good agreement with values observed by potentiometric and polarographic methods by a number of authors, as mentioned in Beinert's comprehensive work.[35]

Polarographic curves differ from the potentiometric ones in that they are not centrosymmetric, their course being perturbed by a sharply marked adsorption fore-wave in the more positive potential part of the curve. This fore-wave is most clearly developed in the acid region of the buffer, but it is still clearly to be seen around pH 6. The fore-wave has a characteristic shape, illustrated by a sharp increase in current. In the concentration region under 4×10^{-5} M of riboflavin only this fore-wave occurs; during the increase of concentration, it rises to a certain value and then it remains constant. It was not possible to remove it by surface-active substances, but it disappears with the increase of temperature. The character of the fore-wave and its origin were explained later, after more experimental material had been assembled by Brdička,[8] as the adsorption effect, by measuring the electrocapillary curves and by studying it curves, recorded oscillographically. On the it curves, recorded in a pH 1.8 buffer medium at various values of inserted voltage, two maxima may be observed, corresponding probably to two adsorbable forms of riboflavin, an observation in good agreement with the formation of two products of riboflavin reduction (semiquinone and dihydroriboflavin). In this manner the formation of semiquinone during the reduction of riboflavin on a mercury electrode

[35] H. Beinert, in "The Enzymes" (P. Boyer, H. Lardy, and K. Myrbäck, eds.), Vol. 2, p. 370. Academic Press, New York, 1960.

was verified thermodynamically by a kinetic experiment. On the basis of a theoretical analysis of the polarographic wave, considering the adsorption of semiquinone and leuco forms, Brdička[8] found that after the adsorption current has been achieved the polarographic curve attains a centrosymmetric shape. It is thus possible to read off from this part the half-wave reduction potential and the index potentials, which correspond to the constants obtained from the curve of undeformed adsorptions (Fig. 7).

To conclude, it can be said that the fore-wave is an adsorption stage and that it corresponds to the adsorption of semiquinone. A similar prestage was observed on the polarographic wave of methylene blue.[36] From the dependence of the diffusion current on the concentration, Brdička calculated that the value of the diffusion coefficient for riboflavin for the concentration range $5 \times 10^{-5} M$ to $3.8 \times 10^{-4} M$ is in the range 7.4×10^{-6} to 5.6×10^{-6} cm^2/sec.

During further electrochemical research into riboflavin, this substance was subjected to orientational oscillopolarographic study where the image of the dV/dt-curves was investigated as a function of pH using the stream-

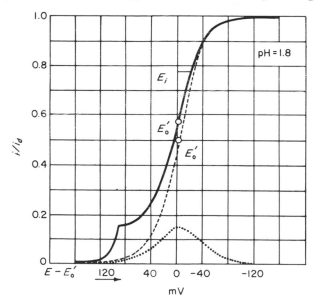

FIG. 7. Analysis of the polarographic curve of riboflavin in the medium of the universal buffer, pH 1.8; $c = 3.2 \times 10^{-4} M$; constants of the capillary $m = 0.00125$ g/sec, $t = 2.76$ sec; —, the experimental curve; — — —, the calculated curve without absorption; ...., the differentiation of the current corresponding to the formation of semiquinone (i_s/i_d). (According to R. Brdička.)

[36] R. Brdička, Z. Elektrochem. **48**, 278 (1942).

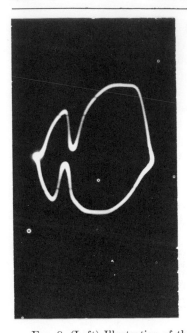 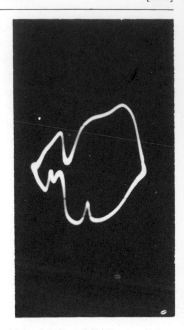

FIG. 8. (Left) Illustration of the oscillopolarographic reduction of FAD; $c = 2.5 \times 10^{-4}\ M$ in acetate buffer, pH 5.4; $\mu = 0.5$; the direct current component = 0.2 mA; the streaming mercury electrode, sensitivity 1/20, and the Polaroscope 524-Tesla were used.

FIG. 9. (Right) Illustration of the oscillopolarographic reduction of FAD in acetate buffer, pH 5.4, under the same conditions as in the preceding experiment, but with the use of the dropping mercury electrode.

ing and dropping electrodes (see Figs. 8 and 9). Interesting facts were discovered that supplement well the study of *it* curves carried out by Brdička. Oscillopolarograms were recorded by a P-524 Tesla polaroscope using a pH 5.0 acetate buffer medium and a pH 8.6 Britton buffer medium. It was found that in an acid as well as alkaline buffer only the reversible character of the riboflavin reduction, displayed by symmetric notches in the cathode and anode branches, was observable on a streaming electrode; the notches are only slightly mutually shifted as a result of the infrared value. On the other hand, using a dropping electrode, one notch is developed on the cathode branch in each of the buffers, whereas on the anode branch both notches, probably appropriate to the reduced adsorbable forms of riboflavin, may be clearly seen in an acid medium. In a pH 8.6 buffer medium, where the adsorption fore-wave is no longer detectable, the image on the streaming electrodes, as well as on the dropping electrodes, is similar. The two notches on the anode branch can no longer be detected. The reason why this phenomenon cannot be observed on a streaming electrode, is that adsorption equilibrium is not established because of the velocity of the current and the variability of the surface. This is also supported by the fact that on a small drop even the reduction image changes during the development, and that the two notches appear on the anode branch just before dropping. An image similar to that of riboflavin is obtained also for FMN and FAD. Figures 8 and 9 show the behavior of FAD.

A suitable medium was sought for the polarographic determination of riboflavin because of the low solubility of this substance in the common aqueous buffers. The most suitable solvent medium was found to be 5% sodium salicylate, in which the solubility of riboflavin is 10 times that in water. A review of the prescriptions for the analytical determination of riboflavin in pharmaceutical and other materials, together with examples, may be found in the monograph on vitamin methods compiled by Knobloch.[37]

Flavin mononucleotide (FMN) and flavin adenine dinucleotide (FAD) were studied polarographically for the first time by Kay and Stonehill,[38] who investigated the semiquinone formation with various respiration bacteriological catalyzers and inhibitors. They computed the semiquinone formation constants from the index-potential values, determined polarographically. For riboflavin, they found the semiquinone formation constant to be $K_s = 1.4$, and they proceeded similarly with the various derivatives of acridine and DPN. However, these values are not accurate because the application of index potentials is connected with a strictly reversible polarographic wave, which is not valid for acridine derivatives, nor for the DPN wave. Polarographic DPN waves and acridine derivatives are not considered to be reversible.

Ke[39] studied the polarography of FMN and FAD in greater detail in 1957. He investigated the dependence on pH of the polarographic reduction and found that the reduction potentials of both substances are practically identical with that of riboflavin. In a medium of pH 8.0 phosphate buffer, he observed with FMN a further polarographic wave, which he ascribed to the reduction of the adsorbed oxidized form of riboflavin (according to the criteria derived by Brdička). This wave appears at a potential value of about -1.6 V relative to NCE. It will further be shown that this is not an adsorption wave, but that it relates to catalytic evolution of hydrogen, which still occurs in this pH range. This conclusion is in agreement with another observation of the author, who mentions that this wave merges with the hydrogen-evolution wave during the decrease of pH, and

[37] E. Knobloch, "Phys. Chem. Vitaminbestimungs-methoden," Academie-Verlag, Berlin, 1963.
[38] R. C. Kay and H. J. Stonehill, *J. Chem. Soc.* p. 3240 (1952).
[39] B. Ke, *Biochim. Biophys. Acta* **68**, 330 (1957).

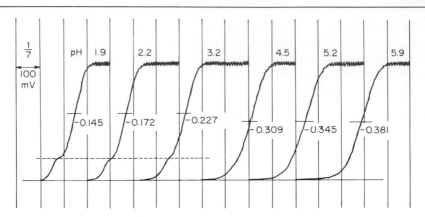

Fig. 10. Dependence of polarographic reduction of FAD on the pH value; $c = 2.5 \times 10^{-4}\,M$, the Britton-Robinson buffer was used; 100 mV/absc., potential vs. saturated calomel electrode.

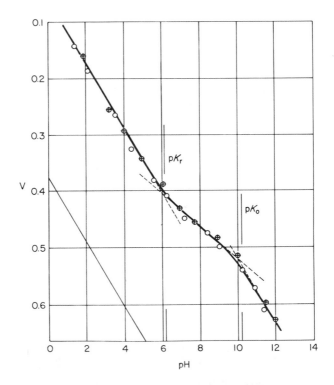

Fig. 11. Graphic illustration of the dependence of $E_{\frac{1}{2}}$ values of the substances FAD (O) and FMN (+) on pH at 20°; potential vs. saturated calomel electrode.

in the potential region of its generation strong oscillations of the galvanometer may be observed. Exceedingly large potential differences between the two waves (about 1 V) are also in contradiction to the polarographic reduction of the adsorbed form. The Japanese scientist Asahi[40] also discusses FAD polarography in his review paper. He compares the polarographic behavior of both substances and finds, in agreement with us, that the potential difference of FAD reduction and of riboflavin reduction is so small that it cannot be detected by the polarographic method. He also discusses the adsorption fore-wave, which we observed with riboflavin, and he finds that this fore-wave also occurs with FAD.

For purposes of verifying the results of the previous papers, we measured the dependence of the half-wave potentials of FMN and FAD on pH. From Figs. 10 and 11 it may be seen that both substances reduce identically and that the potential dependence curves with respect to pH show two characteristic jumps, corresponding to the constants of both the substances. The values of both dissociation constants are very close to that of riboflavin. For K_r at 20° the value 4.0×10^{-7} and for K_0 4.5×10^{-11} was determined; Lowe and Clark,[41] using the potentiometric method, found for FAD at 30° the value $K_r = 2 \times 10^{-7}$ and for $K_0 = 4 \times 10^{-11}$. For FAD at 20° we found the value of E'_0 to be $E_{\frac{1}{2}} = -0.440$ V relative to SCE, which corresponds to the value $E'_0 = -0.195$ V; Lowe and Clark[41] found for the same substance on the basis of potentiometric measurements at 30° the value $E'_0 = -0.187$ V. On the polarographic waves of both substances in the acid region, one may observe a characteristic adsorption fore-wave, similar to that with riboflavin. This adsorption fore-wave with FAD has a slightly different character than with riboflavin—the slope is less.

In the following chapters the differing behavior of FMN and FAD will be shown, e.g., during the catalytic separation of hydrogen during photolysis and photoreduction. We also attempted a direct polarographic study of the yellow enzyme isolated from yeast. This experiment, however, was unsuccessful, because even at high concentrations of the enzyme we obtained only a negligible diffusion current in the potential region of riboflavin reduction. This phenomenon may be explained by the fact that the limiting polarographic current is determined by the diffusion rate toward the surface of the mercury electrode; in the case of the diffusion of large protein molecules of the enzyme, only a small number of reducible prosthetic groups arrives at the electrode as a result of the shielding and slow diffusion, so that the resulting diffusion current is negligible. At these small values of the current it was not certain whether or not it was a question of the part of the dissociated prosthetic group in the given medium.

[40] Y. Asahi, *J. Pharm. Soc. Japan* **76**, 365 (1956).
[41] H. J. Lowe and W. M. Clark, *J. Biol. Chem.* **221**, 983 (1956).

Decomposition Products of Riboflavin

The polarographic characterization of the products of the photolytic decomposition of riboflavin was carried out by Brdička and Knobloch[32] in their fundamental paper devoted largely to riboflavin. It was found that lumichrome (VI) 6,7-dimethylalloxazine, as the main product of riboflavin photolysis, is reduced more negatively than riboflavin, and that it is very easy to determine lumichrome polarographically. In order to supplement this fundamental research and to verify the results, we studied the dependence of the polarographic reduction on pH using pure lumichrome prepared synthetically (Fig. 12). Lumichrome, being a derivative of alloxazine, differs substantially in its polarographic behavior from the derivatives of isoalloxazine. The half-wave potentials are shifted to more negative values. The total reduction corresponds to the consumption of two electrons, and the curve no longer has an expressively reversible character as with riboflavin. On the potential pH curve a z-shaped discontinuity appears. The pK of the corresponding dissociation constant of the reduced form has a value in the vicinity of 6.4; and the pK of the dissociation constant of the oxidized form is around pH 7.2. On the lumichrome curves the characteristic adsorption fore-wave was not observed in an acid medium, this being typical of isoalloxazine derivatives. In studying the photolysis of riboflavin, Moore et al.[42] use the polarographic method for proving the formation of lumichrome.

Lumiflavin 6,7,9-trimethylisoalloxazine (IV) is generated, according to Kuhn,[43] during the alkaline photolysis of riboflavin and by the decomposition of Warburg's yellow enzyme.[44] Lumiflavin behaves polarographically in the same way as riboflavin. Polarographically, these two substances cannot be distinguished in a mixture.

In the next part we shall consider the polarography of the recently isolated and identified key product of riboflavin photolysis, 6,7-dimethyl-9-formylmethylisoalloxazine (V). The photoproduct mentioned, was isolated by Smith and Metzler,[45] using thin-layer chromatographic methods, from irradiated riboflavin solutions. The structure was proved by synthesis carried out by Fall and Petering,[46] based on the oxidation of riboflavin by means of periodic acid. This substance has not yet been polarographically studied in a pure state.

The image of polarographic reduction of this substance is considerably

[42] W. M. Moore, J. T. Spence, and F. A. Raymond, *J. Am. Chem. Soc.* **85,** 3367 (1963).
[43] R. Kuhn and T. Wagner-Jauregg, *Ber.* 66, 1577 (1931).
[44] O. Warburg and W. Christian, *Biochem. Z.* **258,** 496 (1933).
[45] E. C. Smith and D. E. Metzler, *J. Am. Chem. Soc.* **85,** 3285 (1963).
[46] H. H. Fall and H. G. Petering, *J. Am. Chem. Soc.* **78,** 377 (1956).

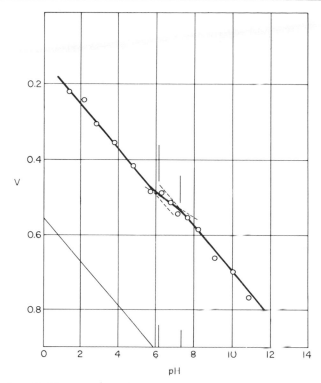

Fig. 12. Graphic illustration of the dependence of the $E_{\frac{1}{2}}$ values on pH at 25° for umichrome; potential vs. saturated calomel electrode.

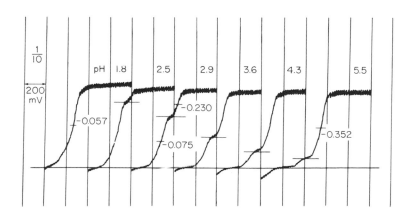

Fig. 13. Dependence of the polarographic reduction of 6,7–dimethyl–9–formylmethyl-isoalloxazine on the pH value; $c = 2.10^{-4}\,M$, 200 mV/absc.; potential vs. saturated calomel electrode.

different from that of the other isoalloxazine derivatives. The reduction takes place in the acid pH region, as can be seen in Fig. 13, in two stages. The more positive stage decreases with increasing pH and vanishes in the region around pH 7. The more negative stage appears at pH 2 and predominates in the neutral region. If the amplitude of the wave of the more positive stage is plotted against pII, we obtain a curve corresponding in shape to the dissociation curve of weak acids, with a pK value of about 3. During potentiometric titration it was found that the substance behaves like an acid and that it probably forms a protonized form, similarly to riboflavin, only with the difference that riboflavin is a much weaker base, as will be seen later on. According to the potentiometric determination, the acid constant of the protonized form has a pK value of 2.9. A further proof of protonization was obtained after studying the dependence of UV spectra

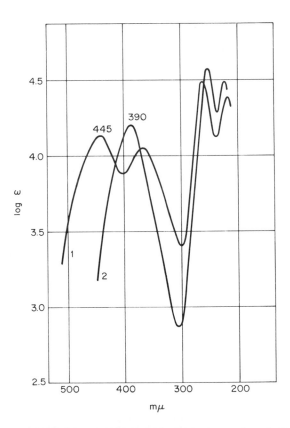

FIG. 14. Ultraviolet absorption spectra of riboflavin (curve *1*) and of 6,7-dimethyl-9-formylmethylisoalloxazine (curve *2*) in the Britton-Robinson buffer medium at pH 1.8; $c = 2.5 \times 10^{-5} M$, $d = 1$ cm; Unicam SP 700 was used; the absorption maxima of substance (1): 444, 370, 267, 222 mμ, of substance (2): 388, 262, 219 mμ.

of the isoalloxazine form on pH. It was found that the UV spectrum of this substance differs from that of riboflavin in a weakly acid medium, but is practically identical with the protonized form of riboflavin, measured in a medium of perchloric or sulfuric acid. An absorption belt at a wavelength of 390 mμ (Fig. 14) is characteristic for the spectrum of the protonized form. According to the spectrum, the protonized form predominates in the region around pH 1. From these experimental results we have deduced that the two-stage reduction of 6,7-dimethyl-9-formylmethylisoalloxazine corresponds to the equilibrium state of the protonized form with a free base, the protonized form being associated with a more positive reduction stage. In illustrating the dependence of half-wave potentials on pH (Fig. 15), we obtain a representation similar to that of riboflavin when investigating the more negative branch. The curve displays two discontinuities, corresponding to the dissociation constants of the reduced (pK 6.1) and the oxidized (pK = 9.6) forms. The potential gradient of the more positive branch is 36 mV and that of the more negative branch 56 mV per pH unit. On the whole, the reduction corresponds to a two-electron process, similar to that of riboflavin. In reducing the substance by means of dithionite or active hydrogen in the presence of palladium, we encountered another interesting phenomenon. The substance, after reduction, produced an anode wave at a potential value corresponding to the protonized form. This more positive wave is in evidence also in the range where the concentration of the protonized form is already negligible, in the region of pH 7–8 (Fig. 16).

In this way the polarographic reduction of this substance receives an irreversible character due to the fact that the cathode reduction takes place at a different potential value than the anode oxidation. After reoxidation by oxygen, we again obtain the original image, which has also been proved spectrophotometrically. This interesting behavior of the substance (V) will be explained on the basis of the hypothesis of the application of tautomeric equilibria, assumed in solutions of varying pH. The reduction of the aldehydic group is not displayed in the whole of the pH range, as it is probably covered by catalytic currents of hydrogen, which accompany the reduction of the substance (V) in the strongly hydrated acid and the neutral regions.

An orientational study of the polarographic behavior of another isoalloxazine derivative, obtained from 6,7-dimethyl-9-formylmethylisoalloxazine after reduction by lithium aluminum hydride, was then carried out. Through reduction, the appropriate alcohol with the structure 6,7-dimethyl-9-(2'-hydroxyethyl)isoalloxazine was obtained. Polarographically, this substance is similar to riboflavin; it does not display two-stage reductions in the acid region, and the dependence of the half-wave potentials on pH is practically identical with riboflavin (cf. Fig. 15).

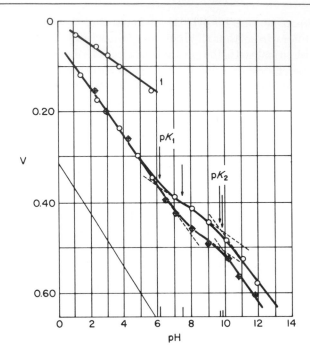

Fig. 15

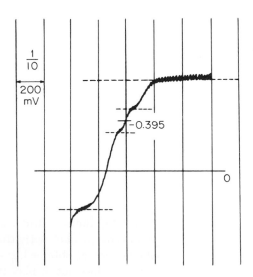

Fig. 16

Discussion

For purposes of explaining the different image of the polarographic reduction of 6,7-dimethyl-9-formylmethylisoalloxazine, the possibility was considered of the application of tautomeric equilibria of isoalloxazine derivatives, discussed frequently in the literature (Beinert). For the purpose of elaborating the hypothesis, the latest research of isoalloxazine derivatives by the EPR method in an acid medium was used. Guzzo and Tollin[47] studied the EPR spectra of various isoalloxazine derivatives in HCl and DCl media, and they compared the experimental results with theoretical computations of the electron densities in a molecule according to the Hückel theory of molecular orbits. They reached the conclusion that the protonization of an isoalloxazine nucleus takes place on oxygen in position 2, not on nitrogen atoms in a heterocycle. From our previous investigations it follows that 6,7-dimethyl-9-formylmethylisoalloxazine is more easily protonized than riboflavin, perhaps due to the effect of the aldehydic group in position 2′, which may influence significantly the electron density on the nitrogen in position 1, and indirectly also the neighboring enolizable carbon in position 2. We may assume that equilibrium sets in between the protonized and free forms, determined by the concentration of hydrogen ions. Because the protonized form and the free base reduce separately, we must assume that the establishment of this equilibrium is slow in respect to the electron process, and we think that the main reason lies in tautomerism; in other words, that protonization is connected with a deeper change of the electron structure of the substance. The assumed equilibrium may be illustrated as follows.

In the case where perfect mobile equilibrium is established, the well-known phenomenon of recombination would occur; this has been frequently described in polarography, for example, in connection with the polarographic reduction of keto acids, in which case the undissociated acid and the anion reduce separately.

In the case of the irreversible shift of the potential, observed during the anode oxidation of substance (V) after reduction, we explain this shift as due to the establishment of mobile equilibrium. We think that in this case this mobile equilibrium is not disturbed, so that the equilibrium

[47] A. V. Guzzo and G. Tollin, *Arch. Biochem. Biophys.* **103**, 231 (1963); **103**, 244 (1963).

FIG. 15. Dependence of the $E_{\frac{1}{2}}$ values of 6,7-dimethyl-9-formylmethylisoalloxazine (+) and 6,7-dimethyl-9-(2′-hydroxyethyl)isoalloxazine (○) on the pH values; potential vs. saturated calomel electrode; temperature, 25°.

FIG. 16. 6,7-Dimethyl-9-formylmethylisoalloxazine in the buffer, pH 3; $c = 2.5 \times 10^{-4}$ M after 30 minutes of irradiation, potential vs. SCE, 200 mV/absc. (anodic-cathodic polarization).

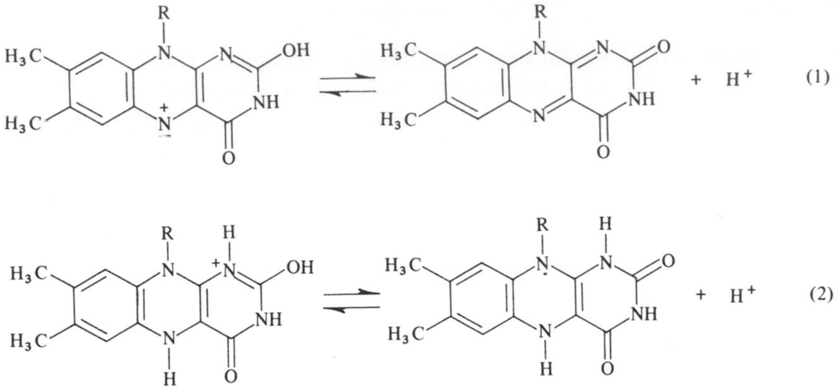

FIG. 17. Equilibria of the protonized forms of riboflavin. (1) Oxidized form. (2) Reduced forms.

between the protonized form of the reduced substance and its free base is established immediately. If this equilibrium is disturbed (by removing the more positively oxidizing protonized form), recombination occurs, so that the observed anode positive wave is actually a recombined wave of the protonized form. The equilibrium may be represented as shown in Fig. 17.

A further possible hypothetical explanation of the observed effect could be based on the assumption of changes of the dissociation constant of the reduced form. We have not encountered a description of the phenomenon with any of the isoalloxazine derivatives so far. We know that with riboflavin protonization also occurs, but only in a medium of diluted mineral acids. In this medium the observed distribution of the polarographic wave is probably disturbed by the generation of a very stable radical of ionic character, which strongly influences the shape of the polarographic waves. The contingent consequences of these facts for enzymatic kinetics will be discussed further on.

Electrochemical Study of Semiquinone Formation in Flavins

Considerable attention has lately been devoted to the formation of the semiquinone radical with flavin because of the significant function of this intermediate reduction product for the transport of electrons by flavin enzymes. The main attention was centered on the hydrogen-transport mechanism from the systems containing diphosphopyridine nucleotides in the form of a coenzyme to the cytochrome series. As early as 1938, Haas[48] drew attention to the orange-red color, appearing during the reduc-

[48] E. Haas, *Biochem. J.* **290,** 291 (1937).

tion of the yellow enzyme, and he voiced the hypothesis that this is a question of the formation of a radical. Also Kuhn[49,50] drew attention to the differently colored products, generated during the reduction of riboflavin in a mineral acid medium. During the last few years a number of papers have appeared, devoted to the study of a semiquinone during enzymatic oxidoreduction. These papers were monographically treated by Beinert.[51]

For purposes of identifying semiquinone and for investigating its formation, a series of methods has been elaborated, most of them based on physicochemical principles. In previous sections, papers of Michaelis[6] and Elema[2] have already been mentioned that are based on the analysis of the potentiometric curves of isoalloxazine derivatives. The stability of semiquinone is characterized here by the appropriate index potentials. Lowe and Clark[41] studied FAD and FMN with respect to the oxidation-reduction potentials. Also polarography has been used in a large number of cases for studying the formation of semiquinone. Theoretical conditions for the utilization of polarographic curves, in this connection, were derived by Brdička.[7] Merkel and Nickerson[52] used polarographic methods for determining the index potentials of isoalloxazine derivatives. Similar studies were done by Kay and Stonehill[38] and Ke.[39] Beinert[51] considered the spectroscopic characterization of semiquinone and derived the criteria for assessing the specificity of the absorption bands ascribed to semiquinone. For identifying the presence of semiquinone he used difference spectra. In this investigation he also considered the application of various isomeric or dissociation forms, the existence of which can be assumed in the solutions at various pH values. The generation of characteristic absorption bands, differing from the spectra of the oxidized and reduced forms of isoalloxazine derivatives, was observed after reduction also with various flavin enzymes; for example, Massey and Veeger[53] noticed them during the reduction of cytochrome reductase, Beinert[54] during the study of mitochondrial enzymes, and Wellner[55] during the study of amino acid oxidase. New possibilities were introduced by electron paramagnetic resonance (EPR), based on the determination of radical forms containing odd electrons. These new methods were applied to free isoalloxazine derivatives

[49] R. Kuhn and R. Stübel, Ber. **70**, 753 (1937).
[50] R. Kuhn and T. Wagner-Jauregg, Ber. **67**, 361 (1943).
[51] H. Beinert, J. Biol. Chem. **225**, 465 (1957).
[52] J. R. Merkel and W. J. Nickerson, Biochim. Biophys. Acta **14**, 303 (1954).
[53] V. Massey and C. Veeger, Biochim. Biophys. Acta **40**, 184 (1960).
[54] H. Beinert and R. H. Sands, Symp. Biolog. Free Radicals, Academic Press, New York, 1960.
[55] D. Wellner and A. Meister, J. Biol. Chem. **235**, PC-12 (1960).

and to enzymes.[47,56] With respect to spectroscopic characterization, Massey[57] divides flavin enzymes into two groups. The yellow enzyme, the oxidase of glucose, etc., belong to the first group. The spectra of these enzymes show expressive bands at 570 to 610 mμ in a partially reduced state. The second group, represented mainly by oxidases of amino acids, provides spectra with characteristic bands at 490, 400, and 370 mμ. During the study of these spectra, Massey very conveniently applied the photoreduction of flavinenzymes in the presence of EDTA. In both cases the EPR signal was recorded parallel to the spectrum. With all the enzymes studied, the presence of a radical structure was detected. To the third group belongs diaphorase (formerly called the oxidase of lipoic acid),[58] the enzyme catalyzing the oxidation of α-keto acid. With this enzyme, using the stop-flow technique,[59] it was shown that while it functions alteration occurs between the fully oxidized and semiquinone forms. If full reduction of the enzyme occurs, the system is inactivated. A further condition for the function of the enzyme is the presence of DPN in an oxidized form. During the reduction of this enzyme, no EPR signal is observed, although it displays a characteristic red coloring with expressive bands at 490–510 mμ. This anomalous phenomenon was explained by Massey and Gibson[60] by the presence of a second prosthetic group besides FAD (SH), which they assume also goes over to the radical during the functioning of the enzyme. The mutual effect of both radicals (coupling effect) results in compensation, so that no EPR signal can be generated. The authors assume that flavin enzymes with a neutral nonionized prosthetic group correspond to group I, whereas with group II an ionic nature is assumed. The spectroscopic investigation of semiquinone formation becomes slightly complicated when it is realized that isoalloxazine derivatives also form charge-transfer complexes,[61] with various substrates and color complexes, e.g., the red complex with tryptophan.[62] The formation of semiquinone, with respect to kinetics, was also proved during photoreduction of FMN in the presence of EDTA by Hölmström[63] using flash photometry. By the use of trace amounts of β-alanine during the enzymatic oxidation by aminooxidase, the generation of trace enzyme was proved.[64]

[56] A. Ehrenberg and G. D. Ludwig, *Science* **127**, 1177 (1958).
[57] V. Massey and G. Palmer, *Biochemistry* **5**, (10), 3181 (1966).
[58] V. Massey, *Biochim. Biophys. Acta* **30**, 205 (1958); **37**, 314 (1960).
[59] Q. H. Gibson, *Discussions Faraday Soc.* **17**, 137 (1954).
[60] V. Massey and Q. H. Gibson, *Federation Proc.* **23**, 18 (1964).
[61] H. A. Harbury and K. A. Foley, *Proc. Natl. Acad. Sci.* **44**, 662 (1958).
[62] I. Isenberg and A. Szent-György, *Proc. Natl. Acad. Sci.* **44**, 857 (1958).
[63] B. Holmström, *Arkiv Kemi* **22**, 281 (1964).

The mechanism of hydrogen transport by means of a semiquinone radical as yet has not been solved uniquely, and various alternatives have been suggested, considering the presence of one or two prosthetic groups in the enzyme molecule. In recent years, Massey and Gibson[60] studied this complex problem, and Wellner[65] systematically treated the hitherto available results in this area.

Polarographically it is possible to investigate very simply the formation of semiquinone according to the shape of the intensity–voltage curve and according to the index potentials to determine the amount of semiquinone generated in the electrode intermediate phase.[66] The analogy between polarographic and potentiometric curves was reliably proved by Brdička,[36] as mentioned in the review contained in the general part of this paper. By its nature, the polarographic method provides a new electrochemical criterion for the formation of semiquinone. As a result of inserted voltage, equilibrium is established on the electrode, given by the ratio of semiquinone of the oxidized and reduced forms. This mobile equilibrium is a function of the potentials and the medium. We carried out measurements of the index potentials for the most important isoalloxazine derivatives considered in enzymology, and we calculated the appropriate values of the semiquinone-formation constants for various pH values.

For computing the semiquinone-formation constant Table I, after Michaelis with corrections adopted from the Elema equation,[3] was used. In the table we also included data on semiquinone concentration at the maximum of formation, i.e., at a 1:1 ratio of the oxidized and reduced forms. These values were computed using an approximate formula derived for the conditions of maximum semiquinone concentration, from the equation expressing semiquinone formation.

$$S = \sqrt{K}/(2 + \sqrt{K}) \tag{31}$$

From Table II it may be seen that the values of the semiquinone-formation constant in neutral and weakly acid media are around 1; this corresponds to a semiquinone concentration of 15–25%. Slightly higher values were observed for FAD. In a strongly acid medium a large increase in semiquinone concentration may be observed. We compared the values obtained with those of Michaelis, Lowe, and Clark, and we found satisfactory agreement on the whole, considering the error connected with the reading of index potential values off the polarographic curves. Michaelis

[64] D. S. Cofey and L. Hellerman, *J. Biol. Chem.* **240**, 4058 (1965).

[65] D. Wellner, *Ann. Rev. Biochem.* **36**, 655 (1967).

[66] E. Knobloch, *Abhandl. Deut. Akad. Wiss. Berlin*, Klasse Chemie **1**, A-2, (1964).

TABLE II
INDEX POTENTIALS AND FORMATION CONSTANTS OF SEMIQUINONE FOR ISOALLOXAZINE DERIVATIVES (SCE)[a]

pH	$E_{1/2}$	$E_{3/4}$	E_i (V)	K	C_s (%)	$E_{1/2}$	$E_{3/4}$	E_i (V)	K	C_s (%)
	Riboflavin					FMN				
1.8	0.150	0.175	0.025	1.70	40	0.145	0.169	0.024	1.60	38
3.2	0.251	0.274	0.023	1.47	36	0.236	0.259	0.023	1.47	36
5.0	0.348	0.370	0.022	1.10	34	0.330	0.350	0.020	0.63	28
7.0	0.445	0.462	0.022	1.10	34	0.455	0.473	0.018	0.25	20
8.0	0.492	0.515	0.023	1.40	36	0.485	0.503	0.18	0.25	20
1 N HClO$_4$	0.072	0.106	0.034	6.3	55	0.485	0.109	0.034	6.3	55
40% H$_2$SO$_4$	0.358[b]	0.416[b]	0.058	71.0	80	—	—	—	—	—
	FAD					6,7-Dimethyl-9-formylmethylisoalloxazine				
1.8	0.145	0.171	0.026	2.5	44	0.057	0.090	0.033	6.4	56
3.2	0.230	0.254	0.024	1.6	38	0.081	0.108	0.027	3.1	45
5.0	0.330	0.352	0.022	1.1	34	—	—	—	—	—
7.0	0.454	0.472	0.088	0.26	21	—	—	—	—	—
8.0	0.494	0.512	0.018	0.26	21	—	—	—	—	—
1 N HClO$_4$	0.068	0.109	0.041	13.2	64	0.018	0.036	0.054	50	70

[a] $E_{1/2}$ = Half-wave potential, $E_{3/4}$ = potential of 3/4 wave, E_i = index potential, K = formation constant of semiquinone, C_s = maximal concentration of semiquinone. SCE = saturated calomel electrode.

[b] Standard electrode, 80% sulfuric acid + Na$_2$SO$_4$ + Hg.

found a value of 17 mV in a neutral medium for the index potentials, which corresponds to 15% of semiquinone. For FAD, Lowe and Clark found values practically identical to ours. Also the polarographic determination of index potentials, carried out by Ke, are in good agreement with our results. In computing the index potentials we used the corrections, elaborated by Brdička for purposes of eliminating the effect of the adsorption fore-wave.

The next part of the work was devoted to the study of semiquinone formation in a strongly acid medium, where a large stability of the radical could be anticipated. We worked in a medium of diluted perchloric and sulfuric acid. It is known that riboflavin is comparatively stable in this strongly acid medium. The character of the polarographic wave in a strongly acid medium changes considerably, the adsorption fore-wave vanishes, and the gradient changes. In a medium of 40% sulfuric acid, a characteristic division of the polarographic wave occurs, which corresponds to theoretical curves, derived by Michaelis and Brdička, for a semiquinone-formation constant approximating 100.

From the curve in Fig. 18 it is clear that the index potentials have reached a value of 0.058 V, which corresponds to the semiquinone-formation

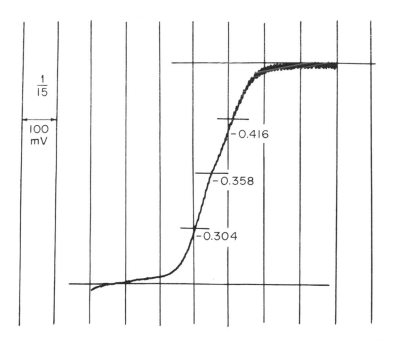

FIG. 18. Polarogram of riboflavin in a medium of 40% sulfuric acid; $c = 5 \times 10^{-4} M$, 100 mV/absc., potential vs. electrode (60% H_2SO_4 + Na_2SO_4 + Hg).

constant of $K = 64$ and a semiquinone concentration maximally around 70%. The whole curve is symmetrical. It is assumed that in a strongly acid medium protonization of riboflavin occurs. With a view to determining the effect of the protonization on the shape of the polarographic wave in a strongly acid medium, we attempted a more accurate determination of the dissociation constant of the protonized form. The first to draw attention to the formation of the protonized form with riboflavin were Kuhn and Moruzzi,[67] who estimated the pK value of this constant at 1.7 on the basis of a study of the fluorescence dependence on pH. Weber[68] noted that fluorescence cannot be used for the accurate determination of the dissociation constant relative to the lifetime of the excited state of riboflavin. Michaelis[5] made a more accurate determination of the constant, using spectrometry. We repeated this procedure according to the method of Flexser and Hammett,[69] based on investigating the extinction at a wavelength of 395 mμ depending on pH. We used diluted sulfuric acid and a buffer medium according to Clark and Lubs. According to the computation, as well as to the graphic representation, a dissociation constant of 0.26 was found for pK. This value was verified, taking errors into consideration, by the dependence of the half-wave potentials on pH and H_0. We obtained a dependence with a gradient of around 36 mV, from which it is possible to detect clearly the presence of a discontinuity in the region of pH $= 0.4$. According to the criteria derived by Michaelis, the course of this dependence corresponds to the application of the dissociation constant of the oxidized form (the gradient of the curve increases after the discontinuity). The value found spectroscopically we consider to be more accurate. From this it follows that the formation of semiquinone and the increase in its stability occur in a region where the protonized form is effective.

To verify the high semiquinone concentration, computed according to the index potentials in a 40% medium of sulfuric acid, we carried out an investigation of the UV spectrum under optimal conditions for the formation of semiquinone.

In order to obtain optimum conditions reliably, the experiment was arranged as follows: the solutions of the oxidized and the reduced forms were prepared separately in a hydrogen atmosphere medium. The reduction was carried out by active hydrogen in the presence of a small amount of Pd, then the hydrogen was extruded from the apparatus by pure nitrogen and the reduced solution was filtered through asbestos, then was combined with the oxidized part and transferred to a quartz ground-glass cell; the spectrum was recorded. In Fig. 19 are compared the spectra of the oxidized

[67] R. Kuhn and G. Moruzzi, *Ber.* **67**, 888 (1934).

[68] G. Weber, *Biochem. J.* **47**, 114 (1950).

[69] L. A. Flexser and L. P. Hammett, *J. Chem. Soc.* **57**, 2103 (1935).

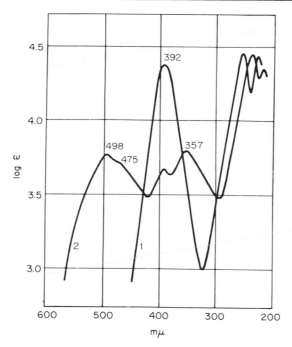

Fig. 19. Ultraviolet absorption spectra of riboflavin in a 40% sulfuric acid medium; $c = 2.5 \times 10^{-5} M$, $d = 1$ cm; the oxidized form (curve *1*); mixture of oxidized and reduced form in the ratio of 1:1 (curve *2*); a Unicam SP 700 spectrophotometer was used.

form of riboflavin and the mixture with the reduced form at a ratio of 1:1 in a 40% medium of sulfuric acid. The comparison shows at first glance the substantial difference in both the spectra. The spectrum displays several difference bands, which, according to literature, may be ascribed to semiquinone. The spectrum is, on the whole, in good agreement with the calculation of semiquinone concentration, based on the analysis of the polarographic wave. With a view to this calculation, the sample measured is about 70% semiquinone and 30% a mixture of the oxidized and reduced forms of riboflavin. The spectrum belongs to the protonized ionogene form of the radical, which is apparently very stable. According to Guzzo and Tollin,[47] the protonization occurs on oxygen in position 2. FMN behaved similarly to riboflavin, and FAD could not be studied in this medium because of its instability.

Discussion

The polarographic method may be used for the electrochemical characterization of semiquinone formation, and it is possible to determine the dependence of the semiquinone-formation constant on the medium, pH, etc., by means of a very simple experiment. A comparatively accurate reading of the index-potential values makes it possible to determine the semiquinone-formation constants and to calculate the maximum concentration of the radical. On the whole it may be said that polarography provides a new electrochemical criterion for the study of semiquinone.

Polarography provides new knowledge of the exceptionally high adsorptive properties of semiquinone, which are demonstrated on the polarographic waves of riboflavin by a characteristic prestage which disturbs the symmetry of the wave. This interesting property was not discovered by potentiometric research conducted by Michaelis. This wave depends on concentration only up to an order of about 5×10^{-5}, when it stops being dependent on concentration due to its nature (the intensity of the current is determined by the surface of the electrode). In the paper of Merkel and Nickerson,[51] the occurrence of this adsorption wave is erroneously considered to be a two-stage reduction of riboflavin with a separate stage for semiquinone. The authors worked in a pH 2.2 medium with a concentration of $10^{-4} M$. According to our paper[37] it was found that the division of the polarographic wave into the stage appropriate to semiquinone and into the stage appropriate to two electrons occurs only in a medium of 40% of sulfuric acid. Under the conditions presented by the authors, the state in which the adsorption wave approximates half the total diffusion current may occur, so that the overall image may imitate the two-stage reduction. If the concentration of riboflavin were decreased, this ratio would change until only one wave would occur, as was shown experimentally by Brdička and Knobloch.[32] Measuring the index potentials under the assumptions given by Merkel and Nickerson, we would arrive at false results.

According to the classification of semiquinone spectra, published by Massey, it would be possible to include the riboflavin spectrum in a medium of 40% sulfuric acid, which we observed in the second category. According to Massey, the bands at 490, 400, and 350 mμ are characteristic of this category. We found the bands to be at 498, 392, and 350 mμ, which is in very good agreement, considering the different media. According to Massey, this group has an ionogene character, the same as with the protonized form we studied. From this one could conclude that the structure of semiquinone generated, for example, during the reduction of the amino acid oxidase, is structurally very similar to semiquinone derived from the protonized form. In other words, the protein bond of the carrier affects the electron structure of semiquinone the same as with the protonized form. So far no suitable model has been found for polarographic research to prove this hypothesis, because unfortunately it is not possible to experiment with enzyme directly in polarography for the reasons given in the preceding paragraph.

In a medium of sulfuric acid, where there is the possibility of investigation with respect to solubility in a sufficiently wide concentration range, we attempted to investigate the dimerization of semiquinone in the concentration range 2×10^{-3} to $2 \times 10^{-4} M$. The half-wave potentials did not differ in this concentration range by more than 5–7 mV; from this we conclude that dimerization is negligible in solutions of the order of $10^{-4} M$. As for structural dependence, it may be said that FMN and riboflavin behave similarly, and that FAD displays a slightly higher stability of semiquinone. The highest semiquinone-formation constants in a medium of acid buffers were found for 6,7-dimethyl-9-formylmethylisoalloxazine, probably because they form the protonized form more easily than riboflavin and FMN. The complex of various metals with riboflavin and their influence on semiquinone formation was studied in an orientational experiment with a red complex of silver. After adding one equivalent of an Ag ion to the solution of riboflavin in a medium of sodium acetate, we obtained a polarographic record composed of three stages. The first, the most positive at -0.180 V, corresponds to the extruding of silver; the second, at -0.480 V, corresponds to free riboflavin; and the third, at -0.572 V, corresponds to the riboflavin complex with silver. After further addition of silver, a red sediment falls out. On the basis of this experiment it is not possible to prove the existence of a complex with a semiquinone structure because on no account did the reduction of riboflavin occur. Among the other polarographic phenomena contributing to the characterization of semiquinone, there is the adsorption on a mercury electrode, demonstrated by the known adsorption fore-wave, that takes effect in an acid medium during flavin polarography. The adsorption effect was proved oscillographically, on the basis of the deformation of the electrocapillary curve and by means of the changes of the differential capacity of an electric double-layer. The presence of semiquinone in the electrode intermediate phase was thermodynamically explained by the adsorption effect.

Catalytic Evolution of Hydrogen on a Mercury Electrode through the Effect of Flavins

The catalytic current is one of the significant polarographic phenomena, and it is explained by the decreased bias of the hydrogen due to the catalyzer. The catalytic current is displayed by a characteristic wave at more positive potential than in the normal process of extruding hydrogen from buffer or acid solutions.

The study of these catalytic currents has contributed to polarography, and thus also to electrochemistry, a series of new discoveries; for example, it has contributed material to the study of adsorption effects on electrodes and of the effect of a reaction double-layer on electrode processes, to the application of free radicals in the mechanism of the electrode process, and generally to the study of the mechanism of electron transport. A further contribution is in the study of the dependence of the catalyzer structure on the catalytic current. A very extensive part of the literature is devoted to catalytic currents caused by proteins and amino acids. This part of the research brought practical applications in the form of the Brdička reaction as an auxiliary clinical diagnostic method.

A further practical application of catalytic currents is in the analytical region. Recently there have been published several experiments on utilizing this effect to determine the dissociation rate of weak acids. The paper presented is a contribution to the electrochemical study of riboflavin. From the point of view of historical development of catalytic currents two trends can be recognized. In the first, catalytic currents caused by proteins are investigated; the second trend is devoted to low-molecular nitrogen substances. Attention was drawn to the catalytic wave caused by proteins first in 1932,[70] when it was named the prenatrium wave. Babička and Heyrovský[71] described this wave in greater detail in a medium of ammoniacal buffer and explained it as a wave caused by the catalysis of hydrogen ions. Brdička[72] shortly afterward proved that this wave appears in all other buffers and that hence it is not subject to the presence of the ammonium ion. Pech[73] noticed that catalytic currents appear in buffer solutions of some of the alkaloids of the quinoline series, and Kirkpatrick[74] noticed the same thing with a series of other alkaloids. Brdička[75-78] described the catalytic currents caused by proteins and some amino acids in the presence of cobalt salts. The catalytic effect is ascribed in this case to SH-groups of proteins; this has been proved by investigation of numerous model substances, including cysteine. Catalytic currents of protein fission products were used as auxiliary diagnostic methods in clinical practice in connection with malignant diseases under the name of the Brdička filtrate reaction. A large amount of material was devoted to verifying the specificity of the reaction and determining its applicability as a diagnostic method. Knobloch,[79] in his paper of 1945, drew attention to catalytic currents caused by pyridine and associated substances. These

[70] F. Herles and A. Vančura, *Rozpravy Česk. Akad.* **42**, 21 (1931).

[71] J. Babička and J. Heyrovský, *Collect. Czech. Chem. Commun.* **2**, 270 (1930).

[72] R. Brdička, *Collect. Czech. Chem. Commun.* **8**, 366 (1936).

[73] J. Pech, *Collect. Czech. Chem. Commun.* **6**, 190 (1934).

[74] H. F. Kirkpatrick, *Quart. J. Pharm.* **18**, 245, 338 (1945); *ibid.*, **19**, 8, 127, 526 (1946).

[75] R. Brdička, *Rozpravy Česk. Akad.* 1 (1936).

[76] R. Brdička, *Collect. Czech. Chem. Commun.* **5**, 112 (1933); *ibid.*, **5**, 148 (1933); *ibid.*, **8**, 366 (1936).

[77] R. Brdička, *Nature* **139**, 1020, 1937.

[78] R. Brdička, *Acta Radiol. Canc. Boh. Mor.* **2**, 7 (1938).

[79] E. Knobloch, *Collect. Czech. Chem. Commun.* **12**, 407 (1947); *Chem. Listy* **39**, 54 (1945).

low-molecular, heterocyclic compounds behave differently from proteins. The paper was devoted to the study of the mechanism of the electrode process during the generation of the catalytic current. From the dependence of the effect on pH it was proved that the catalyzer is the cation of the reduced form base, and that the catalytic effect is qualified by the adsorption of the catalyzer on the electrode. The catalytic separation of hydrogen by nitrogen compounds was later studied in detail by von Stackelberg and his co-workers.[80] This group checked 230 nitrogen compounds and devoted its attention to the dependence of the catalytic effect on the chemical structure. They assume, in accordance with Knobloch, that nitrogen compounds, capable of protonization, are effective, and that the condition of their effectiveness is the adsorption of the reduced form on the electrode. A significant contribution of the von Stackelberg school to the research into catalytic currents is the suggestion concerning the mechanism of the catalytic process. They suggest the following schematic description of the electrochemical process accompanying the catalytic evolution of hydrogen:

$$B_{ads} + H_3O^+ \rightarrow BH^+_{ads} + H_2O \tag{32}$$

$$BH^+_{ads} + e \rightarrow \dot{B}H_{ads} \rightarrow B + \tfrac{1}{2}H_2 \tag{33}$$

The donor of the proton is the acid. Catalytic evolution is determined by the protonization of the catalyzer in an adsorbed state. The protonized form accepts the electron from the electrode, changes into a radical that disintegrates into atomic hydrogen and the regenerated catalyzer. A further significant step in the study of catalytic currents is due to the work of Majranovskij.[81–83] He connects up his work with that of Knobloch and mainly goes into the theory of the electrode mechanism. He starts with the same concept as the previous authors, founded on the function of the protonized form of the catalyzer. The basis of the mechanism, published by Majranovskij, is a protolytic reaction taking place during the catalytic separation of hydrogen as the first stage.

$$B + AH^+ \rightleftharpoons BH^+ + A^- \tag{34}$$

AH is the proton donor, A^- is the conjugate base. A reaction is taking place on the electrode, similar to the second stage assumed by von Stackelberg, i.e., the acceptance of an electron from the electrode.

$$BH^+ + e \rightarrow \dot{B}H \tag{35}$$

[80] M. von Stackelberg and H. Fassbender, *Z. Elektrochem.* **62**, 834 (1958); *ibid.*, **62**, 839 (1958).

[81] S. G. Majranovskij, *Izv. Akad. Nauk USSR*, OXCH 615 (1953); *ibid.*, 805 (1953).

[82] S. G. Majranovskij, *Dokl. Akad. Nauk* 114, 1272 (1957); *ibid.*, 142, 1327 (1962); *ibid.*, 132, 1352 (1960).

[83] S. G. Majranovskij, J. Koutecký, and V. Hanus, *Zh. Fiz. Khim.* **37**, 18 (1963).

A particle is created on the electrode, which has the character of a free radical; Majranovskij thinks that it dimerizes under the evolution of molecular hydrogen.

$$2\dot{B}H \overset{k_d}{\rightleftharpoons} 2B + H_2 \tag{36}$$

Majranovskij assumes that the evolution of hydrogen according to the reaction given, is acceptable from the energetic point of view, as it causes the creation of molecular hydrogen simultaneously with the regeneration of the catalyzer. The defining steps of the catalytic process are, according to Majranovskij, the protonization and the dimerization of the free radical.

In studying heterocyclic compounds as catalyzers for the evolution of hydrogen, a characterization of the catalyzer was derived on the basis of plentiful experimental material. The capability of the catalyzer to accept the proton was set as the fundamental condition. von Stackelberg and his co-workers verified this fact on a large number of compounds. A further condition was set in the adsorbability on a dropping mercury electrode. Among nitrogen compounds found to be ineffective were the quaternary ammonium bases, which are not capable of protonization, and the lower aliphatic amines. Only arginine and histidine provide a catalytic effect of the amino acids. Very efficient catalyzers are heterocyclic nitrogen bases, such as the amide of nicotinic acid, pyridine, quinoline, piperidine, etc.

Compounds found to have a catalytic effect include the chromone derivatives, described by Knobloch,[84] and the catalytic currents of iso-alloxazine derivatives, of a recent date, described also by Knobloch in 1966.[85]

The catalytic currents of hydrogen, caused by the presence of iso-alloxazine derivatives in buffer solutions, were noticed by Knobloch[85] and Jambor[86] as late as in 1966. In the preceding literature only Kočent's[87] paper mentions the fact that in buffer solutions there sometimes appears with riboflavin, under a more negative potential, a wave that resembles by its character a catalytic current.

Catalytic evolution of hydrogen was observed in the presence of riboflavin FMN and FAD, and an insignificant wave was also observed, in relation to riboflavin, with lumiflavin. No occurrence of catalytic current was observed with lumichrome.

In Fig. 20 the dependence of the catalytic current on the concentration

[84] E. Knobloch, *Collect. Czech. Chem. Commun.* **25**, 3330 (1960).
[85] E. Knobloch, *Collect. Czech. Chem. Commun.* **31**, 4503 (1966).
[86] E. Jambor, *Acta Chim. Acad. Sci. Hung.* **48**, 89 (1966).
[87] A. Kočent, *Chem. Listy* **97**, 195, 1953.

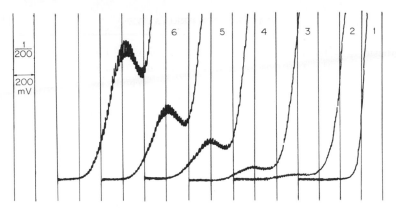

FIG. 20. Catalytic current of hydrogen evoked by the substance FMN in the concentration $10^{-4}\ M$ in a medium of $0.2\ N$ acetate after the addition of acetic acid. Curve *1*, beginning at -1.4 V, pH 7.1; curve *2*, at -1.0 V, pH 6.1; curve *3*, at -1.0 V, pH 5, 8; curve *4*; pH 5.6; curve *5*, pH 5.3; curve *6*, pH 5.0.

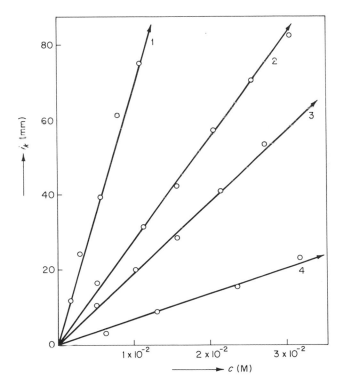

FIG. 21. Dependence of the catalytic current of hydrogen on the concentration of different acids as proton donors. Riboflavin $c = 2.5 \times 10^{-5}\ M$. The acids are: curve *1*, formic acid; curve *2*, acetic acid; curve *3*, benzoic acid; curve *4*, phosphoric acid.

of acetic acid in the presence of FMN is shown. The catalytic current observed has the features of a maximum and depends significantly on the height of the reservoir. In intensity it exceeds by several orders of magnitude the current that would correspond to the reduction of the catalyzer. The current can be suppressed by surface-active substances by gelatin, tested on isoalloxazine derivatives, and by polysorbate (Tween 80). The current depends not only on the concentration of the catalyzer, but also on the concentration of the buffer, as in the case of pyridine derivatives. It depends also on the properties of the acid, its dissociation constant, etc. This dependence may be seen well in Fig. 21, which illustrates the dependence of the intensity of the catalytic current on the concentration of various acids of the same molarity.

The experimental results of the arrangement were such that an acid 0.2 N was added to a solution of a 0.1 N acid salt. The concentration of the catalyzer was $2.5 \times 10^{-5} M$; the ion strength was preserved by adding KCl at a constant value. The experimental results were plotted on a graph, the acid concentration in moles along the x-axis, the intensity of the catalytic current in millimeters along the y-axis; the sensitivity of the galvanometer was 1/100. The dependence was checked with the following acids: formic acid, acetic acid, benzoic acid, and phosphoric acid. A solution of primary phosphate was added to the solution of the secondary phosphate with the phosphoric acid. The dependence displays considerable differences between the individual acids. The highest current was obtained in a medium of formic acid, and the lowest in a medium of phosphoric acid. The dependence of the intensity of the catalytic current on the concentration is nearly linear in the range investigated and the concentration given. The different behavior of the acids with a different dissociation constant can probably be explained by the rate of dissociation. A nearly linear dependence of the catalytic current on the concentration of weak acids proves that in this case the proton donors are mainly the undissociated acid molecules. The shape of the catalytic wave is, therefore, different with the different donors.

The catalytic effect, caused by riboflavin, may also be investigated in a medium of mineral acids. The catalytic evolution of hydrogen occurs in a medium of mineral acids in the presence of riboflavin at a potential which is more positive by 240 mV than in the case of the normal hydrogen wave in this medium.

Figure 22 shows the effect of the concentration of riboflavin on the character of the hydrogen wave in a medium of $10^{-3} M$ HCl. The polarogram shows that this is an unusually sensitive reaction. Already at a concentration of riboflavin of the order of $10^{-8} M$, a clear effect may be observed. At a concentration of $10^{-6} M$, only one wave is created, which

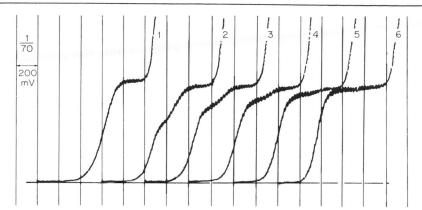

FIG. 22. Influence of riboflavin on the evolution of hydrogen from a strong acid medium, 0.001 M HCl, 0.1 M KCl; riboflavin concentration: curve 1, 0; curve 2, 4.9 $\times$ $10^{-8} M$; curve 3, 9.6 $\times$ $10^{-8} M$; curve 4, 1.8 $\times$ $10^{-7} M$; curve 5, 3.4 $\times$ $10^{-7} M$; curve 6, 4 $\times$ $10^{-6} M$. The waves are from -1.0 V; 200 mV/absc.; sensitivity, 1:70; potential vs. saturated calomel electrode.

differs from the original not only in potential, but also in shape. At a concentration of the order of $10^{-4} M$ of the catalyzer, a maximum appears in the diffusion current. The assumption expressed by Knobloch in studying pyridine derivatives by, that the cause of the catalytic current is in the reduced form, was proved. It was found that the leuco form of riboflavin produces catalytic current in the same way as the oxidized form, the only difference being that the current of the reduced form is slightly lower. This

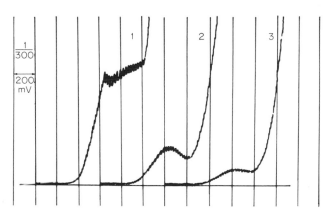

FIG. 23. Comparison of catalytic current evoked by the presence of FAD (curve 1), riboflavin (curve 2), and FMN (curve 3); the concentration of the catalyst is 5 $\times$ $10^{-5} M$ in an acetate buffer medium, pH 5.4; waves from -1.0 V; 200 mV/absc.; saturated calomel electrode.

is explained by the effect of the strong adsorption of the leuco form and by its negligible solubility.

The dependence of the catalytic current on structure was studied in a number of isoalloxazine derivatives, FAD, FMN, riboflavin, and lumiflavin. In Fig. 23 are compared the intensities of catalytic currents, caused by the substances mentioned in an identical medium and at an identical catalyzer concentration. The comparison indicates that the highest intensity of catalytic current is observed with the FAD substance, the next with riboflavin, and the lowest with the FMN substance. The lowest current of the group mentioned is given by lumiflavin. With lumichrome, which is a derivative of alloxazine, no characteristic catalytic current was observed. Between the currents due to FAD and riboflavin, there is a difference of more than an order of magnitude. There are also differences in the shape of the catalytic currents; the characteristic current, accompanied by a maximum, is developed only with riboflavin and with FMN.

The dependence of the catalytic current on the catalyzer concentration was studied with riboflavin, FAD, and FMN. This dependence formally satisfies the Langmuir isotherm, as is the case with the pyridine derivatives. The increase in current is limited here not only by the catalyzer concentration, but also by the buffer capacity. After the capacity of the buffer has been exhausted, no further increase in current with increasing catalyzer concentration can be observed.

The effect of the reaction layer on the catalytic current may be investigated with respect to dependence on the height of the reservoir and to the shape of the it curves. If the reaction layer is decreased due to the effect of the anion concentration, kinetic currents are obtained. The current depends on the rate of regeneration of the catalyzer in the electrode intermediate phase. In a reaction layer which has been realized in the presence of a small anion concentration with predominance of an undissociated acid, the current has a diffuse character. These conclusions were proved by studying the dependence of the current on the height of the reservoir and on it curves (Figs. 24 and 25).

The study of the dependence of the current on temperature substantiates the opinion about the adsorption of the catalyzer on the electrode. Due to the temperature a decrease in the adsorption occurs, as well as a decrease in the catalytic current. Also the deformations of the electrocapilliary curves speak in favor of the application of adsorption.

Discussion

The dependence of the catalytic effect on the structure of the catalyzer is an interesting contribution to the electrochemical study of isoalloxazine derivatives appearing in the function of coenzymes. The oxidation-reduc-

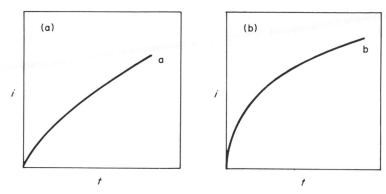

FIG. 24. It curves of catalytic currents of riboflavin; $c = 5.10 \times 10^{-5} M$. (a) 0.2 M acetate buffer, pH 5.3, at -1.6 V. (b) 0.02 M acetate buffer, pH 4.7, at -1.55 V. The exponent in the equation $i = kt^x$ is: a, 0.66; b, 0.33.

tion potentials of riboflavin, FAD, and FMN are practically identical, but if they are compared with respect to the catalytic effect, we find remarkable differences. FAD appears to be the most effective catalyzer; it is the most common coenzyme with the enzyme containing the isoalloxazine nucleus. So far, we have explained the mentioned differences in catalytic effectiveness by the assumption of varying degrees of adsorption on the mercury electrode. Another possible explanation could lie in the intensity of the catalytic effect itself, expressed as the rate of proton exchange. With both

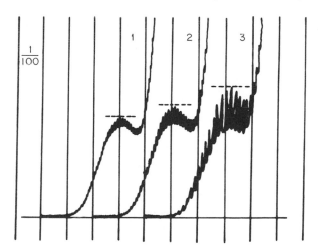

FIG. 25. Dependence of catalytic hydrogen waves of riboflavin ($c = 5 \times 10^{-5} M$) on the height of the reservoir in an acetate buffer, pH 5, 2; h for wave $1 = 70$ cm; for wave $2 = 50$ cm; for wave $3 = 30$ cm; 200 mV/absc., sensitivity $= 1:200$.

the systems it is a question of a certain, although very removed, parallelism determined by the catalytic transport of hydrogen, in one case onto the substrate, in the other case onto the electrode. It is interesting to note that the catalytic effect is most perfectly developed with nitrogen heterocyclic compounds, from which the most significant coenzymes of electron transport are also derived (diphosphopyridine nucleotides and isoalloxazine derivatives). DPN and the amide of nicotinic acid catalyze the evolution of hydrogen on the mercury electrode significantly (Fig. 26).

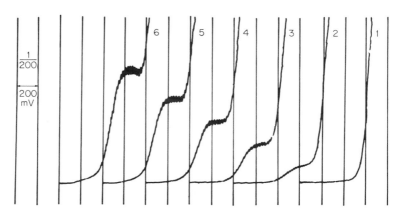

Fig. 26. Polarogram expressing the dependence of the catalytic wave of hydrogen evoked by the presence of DPN ($c = 10^{-4} M$) on the concentration of acetic acid in the acetate, 0.2 M. Acetic acid (0.2 M) added; the acetic acid concentrations: wave *1*, 0; wave *2*, 0.1 ml; wave *3*, 0.2 ml; wave *4*, 0.3 ml; wave *5*, 0.4 ml; wave *6*, 0.5 ml. Beginning of the waves: *1*, −1.6 V, *2*, etc. −1.0 V; potential vs. saturated calomel electrode; 200 mV/absc.

Only a hypothesis is available for explaining the mechanism of catalytic evolution of hydrogen by isoalloxazine derivatives. It is necessary to mention that the behavior of this group disturbed, to a certain extent, the concepts created over a number of years by Knobloch, von Stackelberg, Majranovskij and others. In all hitherto known cases, the protonization of the catalyzer is assumed to be the basic process of the catalysis. In the case of isoalloxazine derivatives, protonization is hardly acceptable because the catalytic effect occurs in the pH 7.5 region, which is in contradiction with the value of the dissociation constant of the protonized form. It is known that the pK value of this constant is near 0.26. We expressed a hypothesis concerning the generation of this current with the isoalloxazine derivatives, based on the concept of the dissociation of the reduced form in an adsorbed state and on the recombination of the catalyzer. We assume that the principle of the catalytic reaction is the dissociation of the reduced form in the electrode intermediate phase into the first degree, the value of which is $pK = 6.2$.

$$R + 2e + 2H \rightleftharpoons RH_2 \tag{37}$$

$$2RH_2 + 2e \rightarrow 2RH^- + H_2 \tag{38}$$

$$RH^- + AH \rightleftharpoons RH_2 + A^- \tag{39}$$

Reaction (37) represents the creation of the reduced form of the catalyzer; the second reaction stage corresponds to the catalytic process, and the last represents the regeneration of the catalyzer in the electrode intermediate phase. The condition for the whole reaction is the presence of a suitable proton donor. This diagram does not consider the formation of an intermediate reaction product of a radical nature. The catalytic function of isoalloxazine derivatives could be explained with a view to the original hypothesis, founded on the protonization and the formation of the radical under the assumption that the reduced form, adsorbed in the electrode intermediate phase, is more easily protonized than the oxidized form. This concept, however, has no experimental backing. For purposes of comparison of the catalytic effect of isoalloxazine derivatives on a mercury electrode with the function of the coenzyme, no further conclusions can be derived because neither of the mechanisms has been solved to such a degree as to enable a comparison.

Photolysis and Photoreduction of Flavins

The sensitivity of flavin to light has been known from the very beginning of the research on this group of substances. The UV absorption spectra of riboflavin and all isoalloxazine derivatives are very characteristic, and they differ from the spectra of the derivatives of alloxazine.[35] Absorption spectra are currently used as characteristic constants of these derivatives, and they were used for proving the structure of coenzymes containing riboflavin. Pullman[88] studied an electron model of riboflavin with reference to light excitation. A characteristic property of riboflavin and of related substances is their yellow-green fluorescence, which has often been used for purposes of analytical determination of riboflavin in natural materials. The effect of solvents on the intensity of fluorescence and on the form of UV absorption spectra of flavin has recently been studied by Kozioł and Knobloch.[89,90] It was observed that the intensity of the fluorescence increases with the decrease of water content in a mixture with organic solvents (dioxane, acetone, ethanol). The increase in fluorescence was accompanied by a decrease in extinction at 365 mμ. This effect was compared with

[88] B. Pullman and A. Pullman, *Proc. Natl. Acad. Sci. U.S.* **45**, 139 (1959).
[89] J. Kozioł and E. Knobloch, *Biochim. Biophys. Acta* **102**, 289 (1965).
[90] J. Kozioł, *Experientia* **21**, 189 (1965).

the dielectric constant and with the polarity of the solvent system, expressed, according to Kosower,[91] by the value Z. The effect observed was explained by the solvation of riboflavin molecules. Theorell and Halwer[92,93] studied the photolysis of riboflavin in detail. Photolysis is a complex process, which depends on the pH of the solution, on temperature, etc. Of the known disintegration products, we have lumiflavin, generated mostly in an alkaline medium, and lumichrome, generated in an acid medium. 6,7-Dimethyl-9-formylmethylisoalloxazine has lately been isolated as a new significant intermediate product of photolysis.[45] The photolytic disintegration of riboflavin as yet has not been definitely solved, although since the time of Karrer a large number of papers have been published on the problem. Especially complicated and hard to follow is the disintegration of the side chain. The products of the disintegration were determined to be formaldehyde, acetaldehyde, glyceraldehyde, methanol, acetic acid, formic acid, etc. Paper chromatography of the products of riboflavin photolysis was carried out by Hais and co-workers[94] and by Metzler and co-workers.[45,95] It has been known for a long time that riboflavin functions as a photodynamically effective substance. With a view to its great biological significance, considerable attention has been devoted to the study of this property. The basis of the photodynamic effect is the transport of energy, obtained by light activation, to other substances in the solution. The light energy, absorbed by riboflavin, can be emitted in the form of fluorescence, or it can be displayed by photolysis, or utilized intermolecularly, i.e., by passing on to other substances. The photosensitivity effect of riboflavin was observed with a large number of biologically interesting substances. One might mention the photooxidation of indoleacetic acid,[96,97] which may be the basis of phototropism of plants. Also the catalysis of the oxidation of various amino acids,[98] amines,[99] ascorbic acid,[100] and pyridoxine[101] has been studied. A negative photodynamic effect was observed also in the effect on the functions of some enzymes.[102] The assumed function of riboflavin in the sight mechanism has not yet been explained, although

[91] E. Kosower, *J. Am. Chem. Soc.* **80**, 3253 (1958); *ibid.*, **80**, 3261 (1958).

[92] H. Theorell, *Biochem. Z.* **279**, 186 (1935).

[93] M. Halwer, *J. Am. Chem. Soc.* **73**, 4870 (1951).

[94] I. Hais, S. Svobodová, and J. Košťíř, *Chem. Listy* **47**, 205 (1953).

[95] P. S. Song and D. E. Metzler, *Photochim. Photobiol.* **6**, 691 (1967); **6**, 113 (1967).

[96] A. M. Galston, *Science* **111**, 619 (1950).

[97] L. Brauner and M. Brauner, *Z. Botan.* **42**, 83 (1954).

[98] K. V. Giri and G. D. Kalyankar, *Naturwissenschaften* **41**, 88 (1954).

[99] W. R. Frisell, C. W. Chung, and C. G. Mackenzie, *J. Biol. Chem.* **234**, 1297 (1959).

[100] D. B. Hand and E. S. Gutrie, *Science* **87**, 439 (1938).

[101] M. N. Maisel and E. M. Dikanskaja, *Dokl. Akad. Nauk USSR*, **85**, 1317 (1952).

[102] D. Shugar, *Bull. Soc. Chim. Biol.* **33**, 710 (1951).

free riboflavin was found in increased concentration in the retina of sea fish and mammals.[103] Oster and co-workers[104] studied the biochemistry of these problems in greater detail.

During photoreduction riboflavin functions as the acceptor of electrons. The contemporary literature describes a number of systems in which, in the presence of certain substances, the reduction of riboflavin occurs when it is exposed to light. For example, Merkel and Nickerson[51] used EDTA as the proton donor in the photoreduction of riboflavin. These authors, however, assumed that the electron donor in this case is water, and they explained the effect of the EDTA acid as the influence of a chelate-forming agent. In our experiments, mentioned further on, it was found, however, that the presence of trace metals is not necessary and that reduction takes place even without them. A valuable contribution to the study of riboflavin photoreduction is the work of Frisell and co-workers,[99] who investigated the photoreduction of riboflavin in the presence of sarcosine. They used a labeled substance, and they designated the following as products of the photoreduction: dihydroriboflavin, methylamine, CO_2, and formaldehyde. They observed the formation of formaldehyde with all amino acids investigated. For purposes of identification, however, they employed a comparatively inaccurate colorimetric reaction with chromotropic acid. These authors were also the first to publish a diagram of the mechanism of riboflavin photoreduction in the presence of amino acids. The basis is deamination and decarboxylation of the amino acids:

$$R + H_3C \cdot NH \cdot CH_2 \cdot COOH + H_2O = RH_2 + H_3C \cdot NH_2 + CH_2O + CO_2 \quad (40)$$

Considerable discussion was devoted to resolving the function of the electron donor during the photolysis of riboflavin. Merkel and Strauss with co-workers[51,105,106] assume that disintegration of water occurs during the photolysis of riboflavin, so that the source of the electron is water. They believe that the activator of this process may be one of the amino acids, such as methionone, or even trace metal elements. The second group, represented by Holmström, Oster, and Smith,[104,107,108] defend the opinion that if there is no suitable electron donor in the reaction, the reduction takes place to the detriment of the side chain of riboflavin. The paper of Moore and co-workers[108] is a contribution to the solution of this problem; the authors investigated the photolysis of riboflavin in a medium of heavy

[103] A. Pirie, *Nature* **186,** 352 (1960).
[104] G. Oster, J. S. Bellin, and B. Holmström, *Experientia* **18,** 249 (1962).
[105] J. Nickerson and G. Strauss, *J. Am. Chem. Soc.* **82,** 5007 (1960).
[106] G. Strauss and J. Nickerson, *J. Am. Chem. Soc.* **83,** 3187 (1961).
[107] E. C. Smith and D. E. Mekler, *J. Am. Chem. Soc.* **85,** 3285 (1963).
[108] W. M. Moore, J. T. Spence, and E. J. Raymond, *J. Am. Chem. Soc.* **85,** 337 (1960).

water. With regard to the isotope effect they found that water does not take part in the photolytic reaction. They also investigated the photolysis spectrophotometrically and polarographically. On the basis of the polarographic image, they assume the generation of alloxazine and acetaldehyde. These authors have suggested the modification of the original diagram of riboflavin photolysis, worked out in 1935 by Karrer and Meerwein.[109] The diagram is based on the concept of the function of active carbon in position 2'. According to this diagram, during light activation the proton is accepted on nitrogen in position 1. Protonization results in the rearrangement of the electrons simultaneously with the generation of a reaction diradical, which is subject to further disintegration. The reaction diagram, based on this diradical form was suggested as early as in 1943 by Brdička.[110] Brdička also studied the photolysis image by using the polarographic method, and he found in a strongly alkaline medium at a potential of -1.65 and -1.85 V two waves; these he ascribes to the disintegration products of a sugar chain, and he assumes that they belong to formaldehyde and erythrose. The problem of disintegration of the side sugar chain of riboflavin was also studied by Kočent,[87] who stated that the -1.65 V wave corresponds to glyceraldehyde, not to formaldehyde; he explains this by the dependence of this wave on pH.

Photolysis of Riboflavin

Knobloch connected up his work with his papers from 1948 and 1967[32,111] and carried on in the study of flavins. He worked in a medium of inert atmosphere in order to eliminate contingent effects of air oxygen. The main attention was concentrated on the investigation of the photolysis of the recently identified fundamental product of riboflavin photolysis, 6,7-dimethyl-9-formylmethylisoalloxazine (V). He used a point-source lamp of 100 W with a condensor as a source of radiation, and in some cases he employed a filter to eliminate the radiation wavelengths under 400 mμ. He carried out the irradiation in a normal polarographic vessel, according to Kalousek, fitted with a cover to preserve constant temperature.

The polarograms in Fig. 27 show the time distribution of the photolysis of riboflavin and of substance (V) in a medium of phosphate buffer, pH 8. The photolysis with both the isoalloxazine derivatives is displayed by a decrease in the wave in the region of anode oxidation, which indicates reduction. In case of riboflavin it was possible to determine by means of the polarographic record the presence of three substances.

The substance, displaying a cathode wave at a potential of -0.580 V

[109] P. Karrer and H. F. Meerwein, *Helv. Chim. Acta* **18,** 1126 (1935).
[110] R. Brdička, *Chem. Listy* **36,** 3286 (1942); **36,** 299 (1942).
[111] E. Knobloch, *Biochim. Biophys. Acta* **141,** 19 (1967).

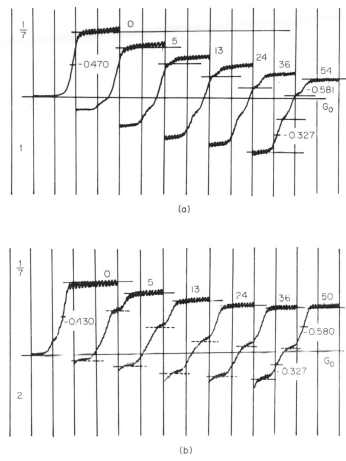

FIG. 27. Time distribution of the photolysis of riboflavin (panel 1) and 6,7-dimethyl-9-formylmethylisoalloxazine (panel 2) in a phosphate buffer, pH 8.0; the concentration of both the isoalloxazine derivatives $= 1.5 \times 10^{-4} M$. The beginning of the waves is at -0.1 V; 200 mV/absc.; potential vs. SCE. Waves 0, 5, 13, 24, . . . 54 correspond to the intervals of light exposure in minutes. In panel 1, the distance of the source of light is 10 cm; in panel 2, 15 cm.

is lumichrome. This was substantiated by the polarographic half-wave potential, by paper chromatography, and spectrally after isolation on Zerrolith 225 in the ammonium cycle as an ion exchanger. A further stage, which is displayed by a wave at $E_{\frac{1}{2}} - 0.470$ V is riboflavin; from the polarogram it may be seen that it has partly deviated to the leuco form. The third stage can be determined in the region of anode oxidation at a potential of -0.327 V; it was identified by paper chromatography, polarographically, and, after isolation on Zerrolith, also spectroscopically as 6,7-dimethyl-9-formylmethylisoalloxazine.

The photolysis of substance (V) takes place faster than with riboflavin. (The radiation source with experiment No. 2 was at double the distance in comparison to riboflavin.) The image of polarographic reduction of the photolysis product can be solved very easily. There are basically two products, one of which is lumichrome at a potential of -0.580 V, and the other, a substance displayed by an oxidation wave at a potential of -0.327 V, identically to riboflavin. This product is the leuco form of the initial substance. This was proved after isolation on an ion exchanger by the polarographic behavior, which is very characteristic and differs from that of other isoalloxazine derivatives, as described on page 326. It

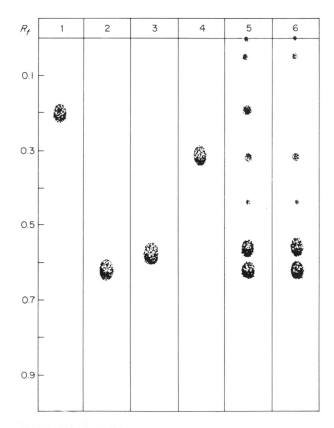

FIG. 28. Chromatogram of photolysis products of riboflavin and 6,7-dimethyl-9-formylmethylisoalloxazine in a phosphate buffer, pH 8.0. Riboflavin (1), lumichrome (2), substance V (3), 6,7-dimethyl-9-(2'-hydroxyethyl)isoalloxazine (4), products of riboflavin photolysis (5), products of photolysis of substance V (6). In a butanol–acetic acid–water system.

has been shown experimentally that, after repeated irradiation and reoxidation, all of substance (V) may be transformed to lumichrome. Comparing both experiments, it is clear that the main products of the photolysis, i.e., lumichrome and the leuco form of substance (V), are common for both isoalloxazines.

To obtain a comprehensive picture of the photolysis of both substances, paper chromatography was carried out in the arrangement recommended by Hais and Svobodová[94] in a butanol–acetic acid–water system.

According to the chromatogram shown in Fig. 28, it is clear that the main products of riboflavin photolysis are lumichrome and substance (V). In the R_f region of lumiflavin a small spot appears with yellow-green fluorescence. It was proved that it belongs to mixtures of small amounts of 6,7-dimethyl-9-(2'-hydroxyethyl)isoalloxazine and lumiflavin. The identity of the former isoalloxazine derivative was determined after isolation on Zerrolith and after investigating the infrared spectrum. Bands were found at 464, 608, and 1270 cm⁻¹, corresponding to the hydroxyl group. Lumiflavin was determined by column chromatography. The total amount of both the substances, judging by the magnitude of the spots, was around 5–8%. The chromatogram also displayed a small spot of a substance with a low R_f. We think that it represents a substance denoted by Hais as 27 CX, and later identified by the Japanese scientists Fukumachi and Sakurai[112] as 6,7-dimethyl-9-(2'-carboxymethyl)isoalloxazine, which has also been confirmed by Metzler and his co-workers.[95]

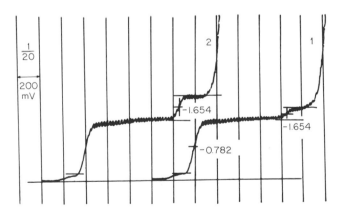

FIG. 29. Polarographic evidence of formaldehyde formation during the photolysis of 6,7-dimethyl-9-formylmethylisoalloxazine in a phosphate buffer, pH 7.0, after conversion into an alkaline medium, pH 11.5. Wave 2 after the addition of formaldehyde. The beginning of waves is at −0.1 V; potential vs. saturated calomel electrode; $c = 2.5 \times 10^{-4}\ M$ riboflavin.

112 C. Fukamachi and Y. Sakurai, *J. Vitaminol.* (*Japan*) **1,** 217 (1955).

The next step was to attempt to identify the product of disintegration of the side chain of riboflavin and of substance (V). It was proved in both case by irradiation in a medium of pH 7–8 formaldehyde. The presence of formaldehyde was proved polarographically after transformation of the reaction mixture into an alkaline medium and adding formaldehyde (glycolaldehyde and glyceraldehyde behave differently in these media in comparison to formaldehyde) (see Fig. 29).

Further proofs of the formation of formaldehyde were presented by transformation to semicarbazone and finally by gas chromatography. Further study was devoted to determining the product created from the second carbon of the side chain. The large volume of the solution of substance (V) in distilled water was subject, after photolysis, to chromatography on Dowex, and an acid substance was isolated with a pK value of 3.6, corresponding to the dissociation constant of formic acid. For further proof, qualitative colorometric determination was carried out after transformation to hydroxamic acid. A distinct maximum, corresponding to the hydroxamic acid complex with Fe at 508 mμ, was found in the difference spectra. The formation of formaldehyde was investigated quantitatively, and it was found that 1 mole of substance (V) corresponds to 0.9 mole of formaldehyde.

With riboflavin, a product obtained by the irradiation of saturated solutions of riboflavin in methanol and water was subjected to analytical investigation at temperatures of $-70°$, atm. CO_2; and water $0°$, atm. N_2. The solutions were evaporated to a smaller volume, and by filtration through a column of Florisil the photolysis product of the isoalloxazine nucleus was eliminated; the evaporation residue of the nearly pure solution was subjected to chromatographic analysis on Whatman No. 2 paper in a pyridine–butanol–water (6:4:3) system. The chromatogram was detected after oxidation by KIO_4 and benzidine; with both samples the presence of glycerin was proved.

Photooxidation of Some Amino Acids and Peptides in the Presence of Isoalloxazine Derivatives

Knobloch[111,113,114] applied the polarographic method to the study of the photoreduction of isoalloxazine derivatives in the presence of amino acids. The photoreduction of flavins can be characterized as a reversible process taking place during the irradiation of solutions of these substances in the presence of a suitable electron donor. In order to eliminate the participation of air oxygen in this reaction, work was done under anaerobic conditions.

[113] E. Knobloch, *Abhandl. Deut. Acad. Wissensch., Berlin, Klasse Chem. Geol. Biol.* p. 565 (1966).

[114] E. Knobloch, *Proc. IV Intern. Polarog. Congr. Prague* p. 44 (1966).

Various amino acids and nitrogen substances were tested as electron donors. It was found that by the polarographic method it is possible to investigate the reversible reduction of flavins without difficulty, and also, in some cases, to investigate the formation of products of the photooxidation of amino acids. Polarography contributed to the progress in the study of this significant reaction, which can be employed as a model for studying the photodynamic effect of flavins. Simultaneously, it is possible to determine polarographically the extent to which photoreduction is disturbed by the photolysis of flavins. For this reason, it is possible to divide the photooxidation of amino acids into two groups according to the degree of photolysis. Group 1 contains the amino acids where no photolysis takes place under the conditions mentioned. The measure for investigating the degree of photolysis is the formation of lumichrome, which, next to riboflavin may be most easily determined polarographically. The following belong to the first group of amino acids: proline, histidine, methionine, iminodiacetic acid, nitrilotriacetic acid, EDTA, cysteins. With Group 2, under the same experimental conditions, i.e., under a surplus of amino acids with respect to riboflavin, it was possible to find, apart from reversible reduction, also photolytic disintegration. The following amino acids belong to group 2: glycocol, alanine, lysine, valine, serine. The main reason for the difference in behavior is probably in the rates of photolysis and photoreduction under the conditions given. The products of the photooxidation of amino acids were determined polarographically according to the half-wave potentials, according to the characteristic absorption spectra, and mainly by chromatographic methods. For irradiation the same experimental arrangement was used as with the photolysis.

We give below the experimental results obtained with some of the amino acids. A typical representative of group 1 is EDTA. The polarogram in Fig. 30 shows the time investigation of photoreduction of riboflavin in the presence of EDTA. The polarogram shows that with time the reduction of riboflavin takes place with perfect reversibility (through the reversible decrease of the polarographic wave of riboflavin into the region of anode oxidation). After transformation into an alkaline medium, a wave appears from the oxidation product of the EDTA acid at -1.362 V. This is not due to an aldehyde, as Frisell[99] thought, because as can be seen from the polarogram, formaldehyde reduces at a much more negative potential (curve 2, Fig. 31). Merkel and Nickerson[51] were the first to describe the reversible photoreduction of riboflavin in the presence of EDTA. Massey and his co-workers made use of this reaction to study the formation of semiquinone during the reduction of various flavin enzymes.[60]

From the polarogram illustrating the photoreduction of riboflavin in the presence of glycocol, it can be seen that besides photoreduction photol-

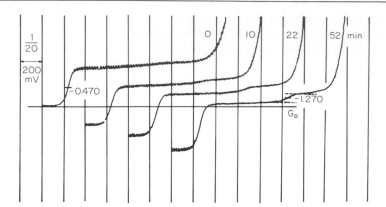

FIG. 30. Photoreduction of riboflavin in the Britton-Robinson buffer, pH 8.5 ($c = 2 \times 10^{-4} M$) in the presence of EDTA ($2 \times 10^{-3} M$); potential vs. saturated calomel electrode; 200 mV/absc. The intervals of light exposure are expressed in minutes.

ysis also is taking place simultaneously; this can be substantiated by the formation of lumichrome. Formaldehyde and ammonia were determined as products of the photooxidation of this acid. Also the behavior of nitrilotriacetic and iminodiacetic acids was studied, and it was found with both these substances that reversible photoreduction takes place. In the first case, iminodiacetic acid and formaldehyde were proved as products of the photooxidation of the corresponding amino acids; and in the second case, glycocol and formaldehyde. With sarcosine the formation of formaldehyde, already earlier observed by Friesell, was proved.

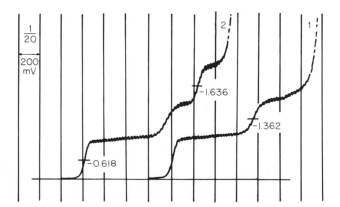

FIG. 31. Polarographic recording of the photooxidation product of the EDTA acid in a phosphate buffer, pH 8.0; riboflavin concentration, $2 \times 10^{-4} M$; EDTA, $2 \times 10^{-3} M$. Wave 1, after conversion into the alkaline medium (pH 11.5); wave 2, after the addition of formaldehyde; potential vs. saturated calomel electrode.

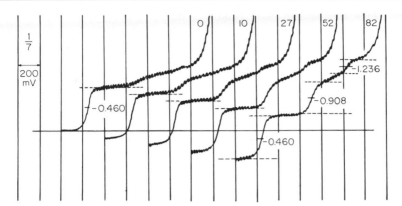

Fig. 32. Study of riboflavin photoreduction in the presence of tripetide (II) in the Tris buffer, pH 8.1; 0, 10, 27, 52, and 82, time in minutes; the beginning of the wave is at −0.2 V; 200 mV/absc.; potential vs. saturated calomel electrode.

Very interesting results emerged from the study of photoreduction in the presence of peptides. The behavior was studied of two peptides that represent building stones of the hormone oxytocin: L-glutaminyl-L-asparaginyl-L-cysteinemethyl ether (I) and L-prolyl-L-leucyl-glycinamide (II) (Figs. 32 and 33).

Peptide (II) provides a very characteristic picture of polarographic reduction, different from the individual amino acids which form this chain, and basically different from the polarographic picture of the second pep-

$$\begin{array}{c}
CH_2-CONH_2 \\
| \\
CH_2 \\
| \\
CH-CO \\
| \\
CH_3
\end{array}
\;\;\Big|\;\; NH-
\begin{array}{c}
CH_2-CONH_2 \\
| \\
CH-CO \\
\end{array}
\;\;\Big|\;\; NH-
\begin{array}{c}
CH_2-S-H_2C- \\
| \\
CH-COOCH_3
\end{array}
$$

(1)

$$\begin{array}{c}
H_2C-CH_2 \\
| \quad\quad | \\
H_2C\diagdown_{\underset{H}{N}}\diagup CH-CO-NH-
\begin{array}{c}
CH \\
| \\
CH_2 \\
| \\
CH \\
\diagup\;\diagdown \\
H_3C \quad CH_3
\end{array}
-CO-NH-CH_2-CONH_2
\end{array}
$$

(2)

Fig. 33. Structural formulas of peptides: L-glutaminyl-L-asparaginyl-S-benzyl-L-cysteinemethyl ether (I); L-prolyl-L-leucyl-glycinamide (II).

tide. This means that the products of the photochemically catalyzed oxidation of peptides can have different chemical properties, and especially structure, from the products derived from the amino acids forming this chain.

Very interesting behavior, different from that of the other amino acids, was found with cysteine. The polarogram in Fig. 34 shows the photoreduction of riboflavin in the presence of cysteine. From the time dependence it can be seen that reversible reduction of riboflavin takes place simultaneously with the oxidation of cysteine to cystine. Both components of this system can be followed polarographically. The mechanism of this significant reaction will be the subject of another paper. Unknown, polarographically determinable, products are formed with methionine, proline, and histidine. $E_{1/2}$ of these products corresponds to the values -1.31, -1.30, -1.10 V relative to SCE in a medium of pH 8 buffer. With tryptophan no photoreduction was observed. Of other biochemically interesting substances, the behavior of adenine, 3-indolylacetic acid, uracil, and ATP was tested. With none of these substances was photoreduction found. With compounds other than nitrogen compounds, photoreduction was found only in a single case with thiomalic acid.

In the next part of the paper attention is centered on the dependence of the rate of photoreduction on the structure of isoalloxazine derivatives and on the structure of the electron donor.

The dependence of the rate of photoreduction on the structure of the catalyzer was tested using EDTA acid as the electron donor in a phosphate buffer, pH 8, medium. The photoreduction was investigated polarographi-

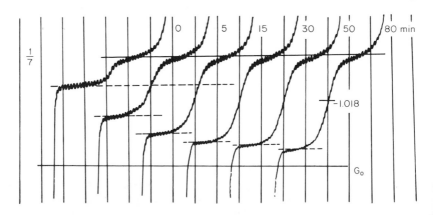

FIG. 34. Polarogram of riboflavin photoreduction in phosphate buffer, pH 8.0, in the presence of cysteine ($2 \times 10^{-2} M$) and riboflavin ($2 \times 10^{-4} M$); potential vs. saturated calomel electrode; 200 mV/absc. The waves indicate time of exposure to light in minutes.

TABLE III
Comparison of the Velocity of Photoreduction of Isoalloxazine Derivatives under Identical Experimental Conditions

Substance	Relative reaction velocity[a] (%)
FMN	120
Riboflavin	100
Lumiflavin	100
6,7-Dimethyl-9-formylmethylisoalloxazine	95
Isoriboflavin	53
FAD	43
Lumichrome	30

[a] Reaction velocity of photoreduction of riboflavin = 100%.

cally, and the results are expressed graphically. The reaction rates are expressed as the tangent to the time-dependence curve. Table III gives the results obtained for several isoalloxazine derivatives; it shows that the first four substances have rates close to that of riboflavin, which was chosen as reference standard (100%). A lower rate is displayed by isoriboflavin (antivitamin B_2), and a very low value is displayed by FAD, which is remarkable. The least effective is the alloxazine derivative lumichrome.

It was found earlier, in studying the photodynamic effect of riboflavin, that the condition for photoreduction is the presence of a nitrogen substance. Friesell and co-workers are of the opinion that the most effective are substances containing secondary or tertiary amino groups. We studied this with a number of amino acids, and we found that photoreduction is more effective with amino acids than with amines. From a systematic study of the photoreduction of amino acids certain examples were obtained of the dependence of the electron donor on structure. The dependence is clearly apparent in the series glycine, iminodiacetic acid, nitrilotriacetic acid, and EDTA. From a graphic representation of the rate of photoreduction with the group mentioned, it can be seen that the rate increases with the amount of carboxyl in the molecule. We were also able to prove that with α-amino acids (α-alanine) the photolysis takes place faster than with β-amino acids (β-alanine). It was also found that with phenylalanine the photoreduction is faster than with alanine, so that not only carboxyl substitutes, but also phenyl speeds up electron transport. The fastest, and fully reversible, rate of photoreduction was found with EDTA, histidine, proline, sarcosine, arginine, methionine, cysteine. With these amino acids a secondary or tertiary amino group or sulfur usually occurs. Aromatic amino acids do not serve as electron donors in photoreduction; on the contrary, they retard the reaction, as was found from a consideration of the effect of PAB on the photooxidation of EDTA. The inhibition effect, in this case, cannot be explained by the action of these substances as filters because the work was carried out in glass vessels, which allowed the passage of light above 320 mμ.

Discussion

From the comparison of the polarographic image of the photolysis of riboflavin and of 6,7-dimethyl-9-formylmethylisoalloxazine it follows that the products are identical. It is thus possible to conclude that substance (V) is created immediately as a product of the photolysis of riboflavin. The main products of the photolysis are lumichrome and the leuco form of substance (V). A detailed study of the photolytic disintegration of substance (V) showed that the side chain is subject to oxidative disintegration accompanied by the generation of formic acid and formaldehyde. With riboflavin it was possible to prove glycerin quantitatively. In studying the stability of substance (V) which is dependent on pH, we found that in a medium of pH 9 and 10 alkaline hydrolysis becomes effective, and that it takes place without lighting; its main products are lumichrome and lumiflavin. With a view to orientation experiments, formic and acetic acids are probably created by the disintegration of the side chain.

The polarogram in Fig. 35 demonstrates the disintegration of 6,7-dimethyl-9-formylmethylisoalloxazine in a pH 12 buffer medium. The identity of lumichrome was proved polarographically, and the generation

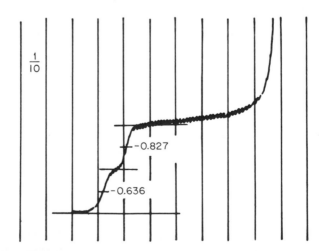

Fig. 35. Polarogram demonstrating the hydrolysis of 6,7-dimethyl-9-formylmethyl-isoalloxazine in a buffer medium, pH 12.0; $c = 2.5 \times 10^{-4} M$; the beginning of the waves is at -0.4 V; potential vs. saturated calomel electrode; the wave was recorded during 1 hour in the absence of light.

of lumiflavin was proved in parallel by paper chromatography. On the basis of these experimental results, it is possible to form a concept as to the basic degree of photolytic disintegration of riboflavin.

Figure 36 shows the three basic stages of the photolysis of riboflavin; the first and second have the character of a photochemical reaction, whereas the third is alkaline hydrolysis, which takes place in an alkaline medium at a high rate and is only very slightly effective in a weakly alkaline medium. Apart from these main stages, a subsidiary disintegration is effective of which it cannot as yet be said whether or not it is a purely photochemical process. The generation of side products cannot yet be proved polarographically, but their presence has been substantiated by paper chromatography. One spot with an R_f value near that of lumiflavin probably corresponds to a mixture of lumiflavin and 6,7-dimethyl-9-(2'-hydroxyethyl)isoalloxazine. Another substance with a low R_f and a spot near the start corresponds probably to 6,7-dimethyl-9-(2'-carboxylmethyl)isoalloxazine.

The generation of this substance can be explained chemically by dismutation of aldehyde into an oxidized and reduced form, or by a photochemical reaction based on the oxidation of a hydrated aldehydic group. So far, we have no reliable explanation for the formation of the leuco form of riboflavin during photolysis. Perhaps this effect can be explained by a redox reaction between the reduced form of substance (V) and riboflavin. (According to the $E_{1/2}$-values of both substances, they are different redox systems.) A different explanation could be based on a mechanism that assumes the formation of glyceraldehyde. It cannot be reliably proved by polarographic methods that small amounts of glyceraldehyde exist next to formaldehyde because the $E_{1/2}$ values of both aldehydes are very close. By using paper chromatography and suitable detection, we proved the existence of glyceraldehyde, next to glycerin, in the products of riboflavin photolysis. The system isopropanol–ammonia–water was used (9:1:2), and detection was carried out by periodate-benzidine or 2,4-dinitrophenylhydrazine. It was found that the spot with $R_f = 0.38$ corresponds to glycerin, and the one with $R_f = 0.33$ corresponds to glyceraldehyde. The generation of glyceraldehyde is connected with a release of two electrons and two protons, which may be used in the reduction of riboflavin. Kočent[87] assumes that glyceraldehyde is the main product of the disintegration of the side chain of riboflavin.

The suggested general diagram does not consider as yet the radical stages that accompany photolysis, and these are discussed in contemporary literature.[63,93,104] From the comparison of the rate of photolysis of riboflavin and 6,7-dimethyl-9-formylmethylisoalloxazine, it follows that the disintegration of substance (V) is faster. For this reason we assume that the

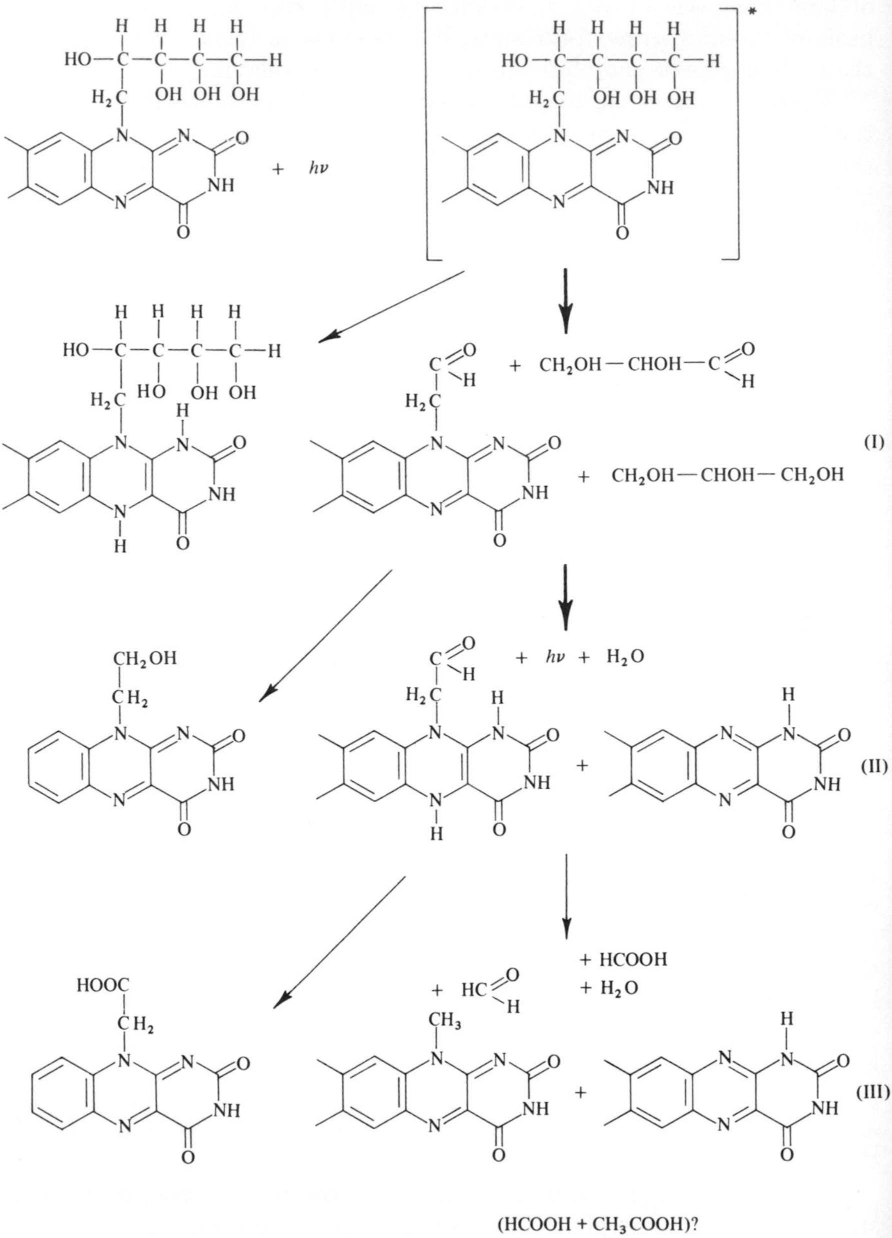

Fig. 36. Diagram of the photolysis of riboflavin.

defining step in the photolysis rate is the disintegration of the side chain of riboflavin. According to a private communication from I. Hais (Faculty of Medicine, Charles University, Hradec Králové) during the photolysis of riboflavin in the presence of acetic acid a hitherto unknown photoproduct is created, the UV spectrum of which is similar to that of the flavins. The author assumes that acetic acid takes part in its generation. This assumption is substantiated by using labeled acetic acid. The substance was proved by paper chromatography and denoted by 69 CX.

The mechanism of photoreduction is based on the disintegration of the CN bond with simple amino acids such as glycine, sarcosine, alanine, and iminodiacetic acid. A similar situation exists with oxidative deamination due to aminooxidase.[65,115] The assumed intermediate stage of this reaction is the formation of imino acids. Frisell[99] published a diagram based on the radical reaction and on the transport of one electron:

$$
\begin{array}{c}
CH_3 \quad H \quad O \quad H^+ + e \quad CH_3 \qquad CO_2 \\
\searrow \quad | \quad \nearrow \qquad \nearrow \qquad \searrow \qquad \nearrow \\
N\!-\!C\!-\!C \qquad \rightarrow \qquad N\!-\!\dot{C}\!-\!COO^- + H_2O \\
\nearrow \quad | \quad \searrow \qquad \nearrow \quad | \\
CH_3 \quad H \quad O \qquad CH_3 \quad H
\end{array}
$$

$$
\begin{array}{c}
CH_3 \quad OH \quad CH_3 \qquad O \\
\searrow \quad | \qquad \searrow \qquad \nearrow \\
N\!-\!CH \rightarrow \qquad NH + HC \\
\nearrow \quad | \qquad \nearrow \qquad \searrow \\
CH_3 \quad H \quad CH_3 \qquad H
\end{array}
$$

We studied the course of photoreduction quantitatively with iminoacetic acid. The reaction can be expressed by the following equation:

$$
\begin{array}{c}
CH_2COOH \\
\diagup \\
HN \qquad\qquad + H_2O = H_2 + NH_2CH_2COOH + HCOH + CO_2 \qquad (41) \\
\diagdown \\
CH_2COOH
\end{array}
$$

We do not as yet know the products of the photooxidation of histidine, proline, methionine, EDTA, etc., so that we cannot form an idea as to a generally valid photoreduction diagram. A significant feature of these photooxidations is the disintegration of the bond between C and N and the formation of aldehydes. The photooxidation in the cysteine–riboflavin system is quite different, and the whole reaction has a reversible character. We shall discuss the mechanism of this reaction due to its significance for the transport of flavin electrons. The discovery that the presence of cysteine and other SH-substances increases the stability of riboflavin with respect to light, is also interesting, so that photolysis is practically eliminated. This is probably a question of influencing the radical intermediate products of the reaction.

[115] A. Cantarow and D. Schepartz, "Biochemistry," Saunders, London, 1962.

With respect to the discussion concerning the function of water during photolysis and photoreduction, in which one group claims that water is the source of electrons during photoreduction catalyzed by flavins, we should like to note that our own experimental results rather support the other group, represented by Oster and his co-workers.[83,104,108] We have proved experimentally that during the photolysis of riboflavin a reduced form of substance (V) is created, which can be explained by the oxidative disintegration of the side chain. Using an amino acid as an electron donor, we proved that these amino acids oxidize and provide products that can be characterized polarographically without difficulty. It is also necessary to note that in no case were we able to prove the formation of hydrogen peroxide, which would necessarily accompany the cleavage of water, although hydrogen peroxide is otherwise easily proved polarographically, e.g., during the reoxidation of dihydroriboflavin by air oxygen.

We must also mention in the discussion the assumed explanation of phototropism on the basis of the photodynamic effect of riboflavin on 3-indolylacetic acid.[96] We tried tryptophan and this acid as electron donors during photoreduction, but in neither case did we observe the formation of the leuco form. The participation of riboflavin in the mechanism of sight is frequently discussed, due to the discovery of Euler and Adler,[116] who found high concentrations of free riboflavin in the retinas of sea fish. Our experiments prove that free riboflavin is much more effective as a photodynamic catalyzer than FAD.

On the basis of the study of the photoreduction of flavins, a new method was worked out for determining riboflavin in biological or other material, subject to certain limitations. Among such materials are urine, yeast, and synthetic fodder mixtures. The new method, worked out by Knobloch,[117,118] is based on the measurement of the intensity differences of fluorescence after the selective photoreduction of riboflavin. It is known that the reduced form of riboflavin does not fluoresce. The removal of the interfering materials from the extract is carried out by Florisil-column chromatography. The trapped riboflavin is eluted by a suitable buffer. The fluorescence is measured, surplus EDTA is added, then photoreduction is carried out; fluorescence is again measured at a wavelength of 518 mμ. The concentration of riboflavin is then calculated from the difference of both measurements. The quantitative evaluation is carried out by the method of standard increment. The bound forms of riboflavin have to be enzymatically released before measurement. The specificity of this method, in comparison to currently used methods, was proved on various materials.

[116] H. v. Euler and E. Adler, *Z. Physiol. Chem.* **228**, 1 (1934).
[117] E. Knobloch, *Intern. Z. Vitaminvorsch.* **37**, (1), 38 (1967).
[118] E. Knobloch, *Československ. Farm.* **16**, 64 (1967).

Kinetic Study of the Redox System Riboflavin–DPNH

A large part of the cell oxidation-reduction processes is controlled by the pyridine nucleotides, flavin enzymes, and the cytochrome group system. Frequent experiments have been conducted to study this system on a suitable model. Singer and Kearny[119] studied the enzymatic reduction of cytochrome c, using pyridine nucleotides in the presence of flavins. They found that the reduction of the cytochrome takes place very quickly in this system, even in the absence of the cytochrome reductase enzyme, and that flavin is sufficient as the catalyzer. The highest catalytic effect was observed with riboflavin, isoriboflavin, and FMN, but FAD had a much lower effect. In their paper they discussed the differences observed in the effectiveness of the individual flavins, and they reached the conclusion that it cannot be explained by the action of oxidation-reduction potentials, which are very close with the group studied. Our experiment is devoted to investigating the reaction kinetics of a simple model of electron transport in a riboflavin–DPNH system, using the polarographic method. The purpose of the work was to determine the equilibrium constant of this reaction and the appropriate E'_0-values. The oxidation-reduction potential of the DPN–DPNH system has already been measured by a number of authors. The direct potentiometric method[120–122] was used, and the values observed in a pH 7 medium at 25° were between -0.260 and -0.310 V. A number of authors employed indirect measurements, based on determining the equilibrium constants of systems in the presence of a suitable dehydrogenase as a catalyzer. For example, Burton and Wilson[123] used the dehydrogenase of alcohol in an ethanol–acetaldehyde medium. Olsen and Anfinsen[124] used the dehydrogenase of glutamic acid; Kaplan and Colowick[125] used the transdehydrogenase. All values of E'_0 obtained were more negative than the potentiometric measurements, being around -0.320 V. Theorell and Bonnichsen[126] studied the dependence of the E_0 values on experimental conditions. In using the dehydrogenase from horse liver they found that the formation of a complex between the reduced DPNH form and the enzyme takes place. They found that in comparing the measured potentials of a free system and of a system with an enzyme, the difference was in the region of 60 mV, which is quite considerable. The effect

[119] T. P. Singer and E. B. Kearny, *J. Biol. Chem.* **183**, 409 (1950).
[120] F. Schlenk, H. Hellström, and H. v. Euler, *Ber.* **71**, 1471 (1938).
[121] H. Borsook, *J. Biol. Chem.* **133**, 629 (1940).
[122] W. M. Clark, *J. Appl. Chem.* **9**, 99 (1938).
[123] K. Burton and F. H. Wilson, *Biochem. J.* **54**, 84 (1953).
[124] J. H. Olsen and C. B. Anfinsen, *J. Biol. Chem.* **202**, 841 (1953).
[125] N. O. Kaplan, S. P. Colowick, and E. F. Neufeld, *J. Biol. Chem.* **205**, 1 (1953).
[126] H. Theorell and R. Bonnichsen, *Acta Chem. Scand.* **5**, 1105 (1951).

TABLE IV

VALUES CALCULATED FROM EQUILIBRIUM CONSTANTS FOR THE REDOX POTENTIALS
OF THE ADH-BOUND DPN–DPNH SYSTEM[a]

	E'_0 (V)		
pH	DPN, DPNH free	DPN, DPNH bound	Ethanol-acetaldehyde
6.4	—0.258	−0.196	−0.121
7.0	−0.275	−0.208	−0.156
8.0	−0.304	−0.244	−0.214
9.0	−0.333	−0.302	−0.272
10.0	−0.361	−0.351	−0.330

[a] Theorell and Bonnichsen.[126]

of the enzyme is displayed in that the potential value is shifted in the positive direction. For the sake of comparison, Table IV shows some of the results in buffer media of pH 7 and 8.

The formation of the assumed complex was proved spectroscopically in another paper by Kaplan and co-workers.[127] Ball and Ransdell[128] used the DPNH–DPN–flavin enzyme system isolated from milk, and as the substrate dichlorophenolindophenol, safranine, methylene blue, hypoxanthine,

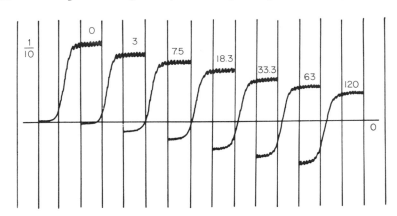

FIG. 37. Polarogram demonstrating the time dependence of DPNH–riboflavin reduction in a phosphate buffer medium, pH 7.8; c of riboflavin = $2.5 \times 10^{-4} M$; c of DPNH = $5 \times 10^{-4} M$; time is indicated in minutes; the beginning of the curves is at −0.2 V; 200 mV/absc.; potential vs. saturated calomel electrode.

[127] N. O. Kaplan and M. M. Ciotti, J. Biol. Chem. 211, 431 (1954).
[128] E. G. Ball and P. A. Ransdell, J. Biol. Chem. 131, 767 (1934).

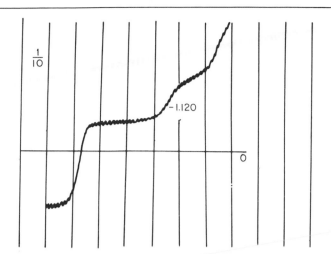

FIG. 38. Polarogram of the same dependence as in the Fig. 37; time 120 minutes. The polarogram demonstrates DPN formation at a potential of -1.120 V.

etc., to determine the E'_0 values. With all these systems, after establishing equilibrium, they calculated the value $E_0 = -0.260$ V in a pH 7.2 buffer medium at 25°. Comparison of the results mentioned shows that there is a considerable difference between the values observed. The whole problem is discussed by Kaplan,[129] who expressed the opinion that with the calculation based on equilibrium states the results will be affected by complex-forming properties of the pyridine nucleotides, and as a result also by the dependence on the medium.

We attempted the direct investigation of riboflavin reduction in the presence of DPNH in a phosphate buffer, pH 7.8, medium under anaerobic conditions. The reduced form of DPNH was prepared according to Lehninger[130] from a pure DPN (Sigma, USA) substance. The reduction of riboflavin was investigated polarographically, direct light being excluded. The polarogram in Fig. 37 shows the time distribution of the reaction. From the course of the reaction it is clear that it approaches the equilibrium state as a limit. It is assumed that the reaction takes place as follows:

$$DPNH_2 + R \rightleftharpoons DPN + RH_2 \qquad (42)$$

The mathematical model of this reaction is represented by a system of nonlinear differential equations:

[129] N. O. Kaplan, *in* "The Enzymes" (P. Boyer, H. Lardy, and K. Myrbäck, eds.), Vol. 3, p. 152. Academic Press, New York, 1960.
[130] A. L. Lehninger, Vol. III, p. 885.

$$-\frac{d(\text{DPNH}_2)}{dt} = k_1(\text{DPNH}_2) \cdot (\text{R}) - k_2(\text{DPN}) \cdot (\text{RH}_2) \tag{43}$$

$$-\frac{d(R)}{dt} = k_1(\text{DPNH}_2) \cdot (\text{R}) - k_2(\text{DPN}) \cdot (\text{RH}_2) \tag{44}$$

$$\frac{d(\text{DPN})}{dt} = k_1(\text{DPNH}_2) \cdot (\text{R}) - k_2(\text{DPN}) \cdot (\text{RH}_2) \tag{45}$$

$$\frac{d(\text{RH}_2)}{dt} = k_1(\text{DPNH}_2) \cdot (\text{R}) - k_2(\text{DPN}) \cdot (\text{RH}_2) \tag{46}$$

From these equations it can be seen that the right-hand sides are identical, so that the left-hand sides are also equal. It is, therefore, sufficient to solve any one of them and to attach on stoichiometric computation of the concentration for the remaining reaction components. If the reaction kinetics really correspond to this selected mathematical model, there must exist a pair of positive real numbers, which, when substituted into this system for k_1 and k_2, will provide a solution agreeing with the experimental results. With the following initial concentrations:

$$[\text{DPNH}_2] = 5 \times 10^{-4} M$$
$$[\text{R}] = 2.5 \times 10^{-4} M$$

the following values were obtained:

Time (min)	0	3.0	7.5	18.3	33.3	63.0	120.0
[R] $(2.5 \times 10^{-4} M)$	1.0	0.96	0.87	0.735	0.635	0.52	0.43

The problem is to find a pair of real numbers, k_1 and k_2, that correspond to the conditions given above. This solution is conveniently carried out under our conditions by an analog computer. In this case other methods would be applicable, because the solution of the nonlinear differential equation used is known, but the computation is tedious. A program was written for the computation, and this was carried out by means of a computing network. Because the data obtained are subject to some experimental error, it is necessary to carry out several computations and take the average of the individual results. After averaging, the following values were obtained:

Time (min)	0	3.0	7.5	18.3	33.3	63.0	120.0
[R] $(2.5 \times 10^{-4} M)$	1.0	0.953	0.885	0.752	0.628	0.502	0.435

Differences	0.0	−0.007	+0.015	+0.017	−0.007	−0.018	+0.005
Sum of deviations	+0.005						

Computed rate constants $\quad k_1 = 6.48$ liter/mole-sec
$$k_2 = 11.35 \text{ liter/mole-sec}$$
Equilibrium reaction constant $\quad k_1/k_2 = K = 0.571$

For a graphic illustration, see Fig. 39.

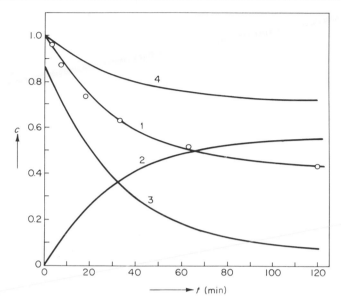

FIG. 39. Graphic illustration of the solution of the mathematical model of the reduction of riboflavin in the presence of DPNH; curve *1*, riboflavin ($1.25 \times 10^{-4} M$); curve *2*, dihydroriboflavin ($1.5 \times 10^{-4} M$). Axis x = time in minutes, axis Y = concentration. Analog computer Tesla AP 4; graphic record; ordinate plotter ARITHMA BAK.

A further possible alternative of the solution of the reaction kinetics given is provided by the assumption that, apart from the said reversible reaction of the second order, riboflavin photolysis takes place which results in a decrease by one reaction component. This disintegration cannot be seen in the polarogram, but it may, according to our experience, reach 3%. In solving this alternative it was found that the results differ only slightly and that they are within the limits of the error with which the height of the polarographic wave is recorded ($\pm 2\%$).

Computation proved the assumed hypothesis about the effect of the reversible reaction of the second order. Rate constants of both reactions, as well as the equilibrium constant, were determined.

The equilibrium constant K was also used for calculating the equilibrium potential E_e for the whole system under the assumption that the following relation holds:

$$E_D{}^0 - 0.03 \log \frac{[DPNH_2]}{[DPN]} = E_R{}^0 - 0.03 \log \frac{[RH_2]}{[RH]} \qquad (47)$$

$E_D{}^0$ and $E_R{}^0$ are redox-potentials of both systems in a buffer, pH 7.8, medium.

The concentration of the reduced form of riboflavin in an equilibrium state can be computed using the following equation:

$$0.571 = x^2/[(5 \times 10^{-4} - x)(2.5 \times 10^{-4} - x)] \tag{48}$$

By solving this quadratic equation we arrive at the real term $x = 2.05 \times 10^{-4}$. By substituting into the Peters formula, we arrive at the value of the appropriate equilibrium potential E_e.

$$E_e = -0.485 - 0.03 \log (2.05 \times 10^{-4}/0.45 \times 10^{-4}) \tag{49}$$

$E_e = -0.504$ V relative to the SCE.

From the value of the equilibrium potential we can compute E'_0 for the DPN–DPNH system under the assumption of the reversibility of this system.

$$-0.504 = E_D^0 - 0.030 \log \frac{[DPNH_2]}{[DPN]} = \frac{2.95 \times 10^{-4}}{2.05 \times 10^{-4}} \tag{50}$$

$E_D^0 = -0.499$ V relative to the SCE; after reduction to the potential relative to a normal hydrogen electrode, we arrive at the value -0.249 V. For the DPN–DPNH system at 20° in a pH 7.8 buffer medium $E'_0 = -0.249$ V.

Discussion

The foregoing experiment and computation proved that the reduction of riboflavin in a system with DPNH was a reversible reaction of the second order. Rate constants of both reactions, as well as the equilibrium constant, from which the E'_0 values were derived, were determined. It may thus be said that the system behaves reversibly. The E'_0 value that we obtained for the DPNH–DPN system, -0.249 V (pH 7.8, +20°), is close to the value obtained for this system by Ball in a pH 7.6 buffer medium at a temperature of 25°. Ball obtained a value of -0.260 V in a system with the xanthineoxidase enzyme. Theorell and his co-workers, using the dehydrogenase from horse liver, obtained the value -0.208 V for pH 7 and -0.244 V for pH 8. Values close to these were also obtained by direct potentiometry. More negative values were obtained using other enzymes, as has been mentioned earlier. The considerable differences in the observed values of the redox potential are explained by the effect of the complexes of the reduced form with enzymes. In the reduction of DPN on the mercury electrode, the substance behaves differently from riboflavin, the reduction of which takes place completely reversibly. DPNH gives a polarographic wave of irreversible character at a much more negative potential than would correspond to the data on this redox system. The value obtained for pH 7 at 20° relative to the SCE was $E_{\frac{1}{2}} = -0.910$ V. It was observed that, during the reduction of DPN by dithionite, an intermediate product of the reduction is created; this was explained by Colowick and his co-workers[131] as a complex of dithionite and the reduced form. During polarographic reduction in the presence of dithionite, we observed the creation of an intermediate product also; this did not belong to DPN or dithionite, and it was displayed by a reversible polarographic wave at a much more positive potential than that at which the reduction of DPN took place; for this stage we found a value of -0.630 V. Hanschmann[132] observed a similar product at an even more positive potential during the photoreduction of DPN. On the basis of these experiments it is possible to allow for the creation of an intermediate reversible product of the DPN reduction that would be electroactive and would be effective in the establishment of redox equilibrium, which would then probably be more complicated than it is usually considered to be.

Polarographic Study of the Schardinger Enzyme

This chapter reviews the experimental work of E. Knobloch,[133-138] which was designed to study the function of the Schardinger enzyme, using the polarographic method. The work was conducted within the framework of the polarographic research into riboflavin being done by Brdička and Knobloch. During an orientational polarographic study of the preparations of the Schardinger enzyme and of the yellow enzyme isolated from yeast, it was found that both enzymes contain riboflavin as a prosthetic group, and that neither of them, under normal conditions, yields measurable polarographic waves. The probable cause of this effect is the size of the molecules and diffusion and steric conditions in the electrode intermediate phase. Thus, direct polarographic research into flavin enzymes was at that point impossible; therefore, the work was delineated in another direction – the utilization of polarography as a sensitive analytic method for studying enzymatic functions, e.g., for investigating the kinetics and the reaction mechanism. The Schardinger enzyme was chosen for the experiments as the most suitable. Basically, it was a question of checking the possibilities of polarography as a new method for research into biochemical redox systems.

131 M. Yarmolinsky and S. P. Colowick, *Federation Proc.* **13**, 327 (1954).
132 H. Hanschmann, *Proc. IV Intern. Polarog. Congr. Prague*, p. 32 (1966).
133 E. Knobloch, *Chem. Listy* **37**, 10 (1943).
134 E. Knobloch, *Naturwissenschaften* 1/4, 43 (1944).
135 E. Knobloch, *Collect. Czech. Chem. Commun.* **12**, 407 (1947). (English Transl.)
136 E. Knobloch, *Z. Vitamin, Hormon. Fermentforsch.* **1**, 3/4, 358 (1947).
137 E. Knobloch, *Collect. Czech. Chem. Commun.* **12**, 281 (1947).
138 E. Knobloch, Thesis, Vysoká škola chem. techn. inženýrstvi, Praha (1967).

By way of introduction, it is necessary to say a few words about the contemporary state of research on this enzyme. Schardinger, as early as in 1902,[139] observed that fresh milk discolors methylene blue in the presence of formaldehyde. Deeper research into this enzymatic function was started by the papers of Wieland,[140] who found that various aldehydes could serve as substrates for this enzyme. In 1922, Hopkins[141] proved that the preparation of the Schardinger enzyme can catalyze the oxidation of some purines, such as hypoxanthine and xanthine to uric acid. The enzyme occurs in animal tissues, and it has been isolated from the liver, as well as other organs. Comparatively pure preparations were made from milk.[142] Ball[143] proved by studying spectra that this was a problem of flavoprotein; this was substantiated by Corran.[144] Green and Beinert[145] proved that the prosthetic group of the enzyme contains molybdenum, and Richert[146] also found iron. The crystalline form was prepared in 1954.[147]

The problem of the specific activity of the enzyme was discussed immediately after the discovery. Wieland[148] assumed that there are two enzymes, whereas the English school, represented by Dixon,[149] was prone to the opinion that both functions are due to one enzyme. At present this problem is the subject of a large amount of experimental work, connected with the study of the inhibition and the reaction kinetics using various substrates. It has not yet been proved conclusively whether there are two enzymes or only one. Fridowitsch and Handler[150,151] assume, on the basis of the characteristic substrate inhibition, that the enzyme contains two active centers. In studying the mechanism by the ESR method, it was found that the function of enzymes results in the formation of radicals, perhaps of semiquinone nature.[152]

Little is known of the biological function of this enzyme, but the knowledge so far obtained is very important. Bauer[153] found that the activity

[139] K. Schardinger, *Z. Unter. Nahr. Genuss.* p. 5 (1902).

[140] H. Wieland, *Ber.* p. 47 (1914).

[141] T. E. Morgan, F. G. Hopkins, and C. T. Stewart, *Proc. Roy. Soc. (London) Ser. B.* **94**, 109 (1922).

[142] C. Dixon and J. Thurlow, *Biochem. J.* **18**, 51 (1939).

[143] E. G. Ball, *J. Biol. Chem.* **128**, 51 (1939).

[144] H. S. Corran, A. H. Gordon, and D. E. Green, *Biochem. J.* **33**, 1694 (1939).

[145] D. E. Green and H. Beinert, *Biochim. Biophys. Acta* **11**, 599 (1953).

[146] D. A. Richert and W. W. Westerfield, *J. Biol. Chem.* **209**, 179 (1954).

[147] P. G. Avis, F. Bergel, and C. F. Brag, *Nature* **173**, 1230 (1954).

[148] H. Wicland, *Liebigs Anal. Chem. B* p. 483 (1930).

[149] M. Dixon, *Enzymologia* **5**, 198 (1938).

[150] I. Fridovich and P. Handler, *J. Biol. Chem.* **231**, 899 (1958).

[151] B. H. Hofstee, *J. Biol. Chem.* **216**, 235 (1955).

[152] R. C. Brayer and R. Peterson, *Biochem. J.* **81**, 178 (1961).

[153] D. J. Bauer and P. C. Bradley, *Proc. Intern. Congr. Biochemie, 4th, Vienna* 1958, *Summaries* **7**, 142 (1959).

of xanthine oxidase increases during the multiplication of viruses. In this connection, it was hypothesized that the enzyme has a regulatory function in mitosis and in the creation of nucleic acids. It was also found that the level of the enzyme is lower in tumors than in normal tissue.[154,155] Positive results were obtained in the treatment of mice with malignant tumors.[156] These results were extraordinarily stimulating for research into the function and mechanism of enzyme effects. Further details and critical evaluations of the state of contemporary research into this enzyme are contained in a monograph published by Bray.[157]

Experimental Part

Fresh milk and the isolated enzyme, prepared by the Corran and Green method,[158] were used as the enzymatic preparation. The rate of the enzymatic reaction was investigated directly in a polarographic vessel, according to Kalousek, connected with a second vessel for purpose of separating the substrate solution from the enzyme. This was done in an inert atmosphere. Hypoxanthine and salicylaldehyde were used as substrates. As oxidizing agents, oxygen, benzoquinone, methylene blue, safranine, dichlorophenolindophenol, etc., were used. The rate of the enzymatic reaction was investigated, e.g., in a salicyl–aldehyde–oxygen system according to the decrease of the polarographic wave of oxygen. In studying the reaction of salicylaldehyde and hypoxanthine, in the presence of oxygen, it was found that oxygen reacts with hypoxanthine, forming hydrogen peroxide, whereas if salicylaldehyde is used, hydrogen peroxide is not formed.

Further work was devoted to the study of inhibitory effects. The effect of cyanide ions was studied, as well as of uric acid, salicylic acid, hypoxanthine, heat, and radiation. If the enzymatic preparation was heated to 80°, inhibition of both the enzyme functions occurs. In following the rate of reaction in the presence of cyanide ions, it was found that the oxidation of hypoxanthine is delayed to a higher degree than the oxidation of salicylaldehyde (Figs. 40 and 41).

An inhibitory effect was also observed with an increased amount of hypoxanthine; this is in agreement with the general theory of the effect of the substrate surplus with high affinity on the enzyme function. Dixon and Lemberg[159] observed that purine derivatives also, such as adenosine

[154] J. S. Colter and H. H. Birt, *Cancer Res.* **17**, 815 (1957).

[155] I. Lewin and R. Lewin, *Nature* **180**, 763 (1957).

[156] P. Feigelson, J. Ultmann, and S. Avis, *Cancer Res.* **19**, 1230 (1959).

[157] C. R. Bray, in "The Enzymes" (P. Boyer, H. Lardy, and K. Myrbäck, eds.), Vol. 7, p. 555. Academic Press, New York, 1963.

[158] F. Lynen, in "Polarographie (Die Methoden der Fermentforschung)" (E. Baumann and K. Myrbäck, eds.), pp. 2352. Thieme, Leipzig, 1941.

[159] M. Dixon and L. Lemberg, *Biochem. J.* **18**, 2065 (1934).

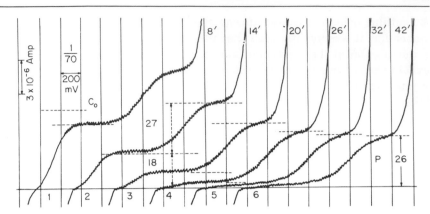

FIG. 40. Polarogram expressing the time dependence of hypoxanthine oxidation with air oxygen in the presence of xanthineoxidase in a pH 6.2 buffer medium; hypoxanthine, $c = 3 \times 10^{-4} M$. On the polarogram, it is possible to see the decrease of oxygen dependence on time and the formation of hydrogen peroxide; 200 mV/absc.

and adenylic acid, the oxidation of which is not catalyzed by the enzyme, have the effect of inhibitors.

A very interesting contribution was the study of the inhibitory effect of uric acid. In a parallel experiment with both substrates, it was found that in enzymatic oxidation of hypoxanthine by oxygen, the presence of uric acid has no effect on the rate of the reaction, whereas if salicylaldehyde is used for the substrate a total inhibition of the reaction occurs at a uric acid concentration of $10^{-3} M$. Salicylic acid inhibits both processes intensively to about the same degree. Interesting results were also provided

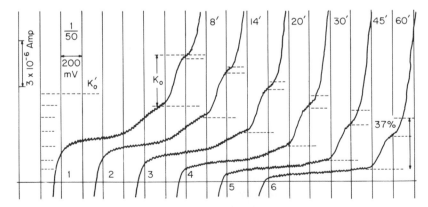

FIG. 41. Polarogram demonstrating the time-dependent action of xanthine oxidase on a salicylaldehyde–oxygen system under the same experimental conditions as indicated for Fig. 40.

by inhibition by light. The preparation was irradiated by sunlight, filtered through ordinary glass, so that only the region above 360 mμ was effective. In testing the preparations irradiated in this way, it was found in case of salicylaldehyde that the rate of the enzymatic oxidation is not affected, whereas in the case of hypoxanthine the reaction rate is decreased by 42% relative to the nonirradiated preparation.

During inhibition experiments with uric acid in the presence of oxygen, we observed that if hypoxanthine is used as the substrate, hydrogen peroxide is not formed. A more detailed study of this phenomenon indicated that uric acid is further oxidized in a milk medium. It was proved that this oxidation takes place optimally at pH 5 and that it is inhibited by a very slight concentration of cyanide ions. These indications would suggest that this is a question of peroxidase.

Discussion

The experimental results given indicate that milk xanthine oxidase does not appear to be an enzyme with two functions, which are inhibited to a different degree, as was shown during the testing of the inhibitory effect of uric acid and light. It is also interesting to note that, in using oxygen as an oxidizing agent, hydrogen peroxide is formed during the enzymatic oxidation of hypoxanthine, whereas in the presence of salicyl-aldehyde this does not occur. This fact would point to the varying reaction mechanisms of both catalytic processes. The experimental results mentioned correspond either to two enzymatic functions, or, as has been forecast by Fridowitsch, to the function of one enzyme with two active centers.

As a result of these experiments, conducted in comparatively primitive conditions, one may conclude that the polarographic method is suitable as an analytical method for the study of oxidation-reduction enzymatic systems, due to its sensitivity and to the possibility of investigating several component systems in one experiment. It may be anticipated that increasing the research in this direction, considering the contemporary state of knowledge and using pure enzymatic preparations, would probably bring new information, in the study of enzymatic oxidation–reduction systems bound to mitochondria and ribosomes.

General Discussion

Polarography has shown new routes in the research into flavins as coenzymes, and it may be expected that a similar development will be repeated with the other vitamins, occurring as prosthetic groups of enzymes. Similar possibilities are given by the amide of nicotinic acid and DPN; both these substances also reduce on the mercury electrode. Polarography has brought new information of two kinds concerning flavins. First, it has been effective as a specific and very sensitive analytical method, enabling the quantitative determination of flavins in solutions and their reliable differentiation from alloxazine derivatives. The analytical application of polarography is suitable in studying reaction kinetics in various systems and under different conditions (anaerobic, aerobic, etc.). Other information derived from behavior during electroreduction are a contribution to the electrochemical characterization of flavins. From this point of view, polarography has brought new criteria of reversibility and has provided conditions for studying the electrochemistry of semiquinone, its stability, etc. An original contribution to the study of the electrochemical properties of flavins is the catalytic separation of hydrogen and the adsorption phenomena accompanying the reduction of flavin on a mercury electrode. In the area of electrochemistry one may expect further contributions from the study of fast electrode processes, using the oscillopolarographic method, or other related methods employing current or voltage impulses. Also for the photochemistry of flavins, this method has made some contributions. The quantitative investigation of the photolysis of the isoalloxazine nucleus has contributed to the solution of the mechanism of this process, which has already been the subject of research by different methods over the past 30 years. A further contribution to photochemical study is the reduction of riboflavin in the presence of certain amino acids and peptides. Some products of the photooxidation of amino acids were identified polarographically, by paper chromatography, and spectroscopically. The photolysis of riboflavin and its photoreduction in the presence of suitable electron donors are separable processes. These studies will be carried on because of their significance for the photodynamic function of riboflavin and for the study of the mechanism of electron transport through the riboflavin–cysteine system. The transport of hydrogen has been demonstrated on another suitable model, DPNH–riboflavin, in which, on the basis of mathematical treatment of the time record of the riboflavin reduction in the presence of DPNH, the equilibrium constant of this system was determined and the E'_0 value for the DPN–DPNH system was computed. The determination of the E'_0 value for the DPNH–DPN system has already been the subject of a number of papers, providing results of considerable scatter. Using as an example the Schardinger enzyme, the possibility of investigating the function of the enzyme in different systems and the effect of the inhibitory influences was shown. New knowledge may be expected from the application of polarography and derived electrochemical methods to the study of reaction kinetics of pure flavin enzymes in defined conditions and to research into complicated enzymatic systems contained in mitochondria.

Acknowledgments

The author wishes to thank Academician R. Brdička (Institute of Physical Chemistry, ČSAS, Prague) for numerous discussions and for reading the manuscript; Professor K. Wiesner, Ph.D. (Fredericton University, Canada) for his help with the theoretical treatment of the experimental results of the photolysis of riboflavin; Ing. Z. Peřina for the mathematical treatment of the experimental material obtained in the study of the reduction of riboflavin in the presence of DPNH; Dr. Ing. Buděšinský, DrSc for preparing some isoalloxazine derivatives used in the work; Mrs. M. Dočekalová for typing the manuscript and Ing. L. Křenová (all from the Institute of Pharmacy and Biochemistry, Prague) for her help in drawing the figures; and RNDr. J. Tauer CSc for translating the manuscript into English.

[137] Assay of Flavin Mononucleotide (FMN) and Flavin Adenine Dinucleotide (FAD) Using the Bacterial Luciferase Reaction

By EMMETT W. CHAPPELLE and GRACE LEE PICCIOLO

$$\text{FMNH}_2 + \text{aldehyde} + \text{O}_2 \xrightarrow{\text{Luciferase}} \text{light} + \text{products}$$

Principles

The reaction that produces an *in vitro* light emission from extracts of photobacteria involves an enzyme (luciferase), a long-chain saturated aldehyde, reduced flavin mononucleotide (FMNH$_2$), and oxygen.[1]

The role of FMNH$_2$ in the reaction and the nature of the reaction products have not been established, although several theories have been presented.[2] Nevertheless, the fact that the system requires FMNH$_2$ for the light-emitting reaction provides the basis for a sensitive assay for this compound[3] as well as for FAD, under the following conditions:

From an acetone-powder extract of *Photobacterium fischeri*, a luciferase preparation can be obtained which (after partial purification by Sephadex-gel chromatography) will give rise to light emission with a maximum intensity proportional to the quantity of FMNH$_2$ added. The relationship between FMNH$_2$ and maximum light intensity is linear between 10^{-8} and 10^{-4} micromole of FMNH$_2$. FMN and FAD are extracted from cellular material by cold perchloric acid containing lumiflavin to protect FMN and FAD against photodegradation. FMN, after reduction by sodium boro-

[1] B. L. Strehler and M. J. Cormier, *Arch. Biochem. Biophys.* **53**, 138 (1954).
[2] M. J. Cormier and J. R. Totter, *in* "Photophysiology" (A. C. Giese, ed.), Vol. 4, p. 315. Academic Press, New York, 1968.
[3] E. W. Chappelle, G. L. Picciolo, and R. H. Altland, *Biochem. Med.* **1**, 252 (1967).

hydride with palladium chloride as a catalyst, is determined by measuring the maximum intensity of the light emitted when an aliquot of the extract is added to a mixture of luciferase and dodecyl aldehyde.

The total phosphoflavin (FMN plus FAD) content of the sample is measured after boiling the perchloric acid extract for 30 minutes to convert FAD to FMN. After reduction and assay, FAD is calculated as the difference between total phosphoflavin and FMN.

Preparation of Luciferase

Culture and Harvest of Photobacteria. The large-scale culture and harvest of photobacteria is carried out using a method described by Picciolo et al.[4] *Photobacterium fischeri* ATCC 7744 is grown in the dark with shaking at ambient temperatures in Difco photobacterium broth. When bright luminescence is obtained, after approximately 15 hours, the broth culture is transferred to Difco photobacterium agar contained in a large shallow tray at an approximate thickness of 5 mm. The culture is spread in an even layer on the agar surface in a ratio of 1 ml per 100 cm². The inoculated trays are covered with Kraft paper to prevent excessively rapid evaporation and kept in the dark at ambient temperatures. After 24 hours, at which time the culture should be luminescing brightly, the cells are removed from the agar surface with a rubber spatula.

Preparation of Acetone Powder. The cell paste is suspended in 20 volumes of acetone at 3°. After blending at approximately 15,000 rpm for 1 minute, the suspension is filtered through a Büchner funnel under reduced pressure; the resulting filter cake is air-dried. The yield of acetone powder is about 1 g per 1000 cm² of agar surface. The powder is stable for at least one year when stored in the presence of desiccant at −80°.

Extraction and Purification of Luciferase. A crude luciferase preparation is obtained by extracting the acetone powder with 0.05 M tris(hydroxymethyl)aminomethane-sulfonic acid (TES), pH 7.0 (10 mg acetone powder per milliliter of TES) for 10 minutes at room temperature with occasional stirring; the suspension is then centrifuged at 10,000 g for 20 minutes at 5°. The luciferase activity is present in the supernatant solution.[5]

To reduce the amount of light emitted because of endogenous FMN in the enzyme preparation, the crude enzyme is partially purified by chromatography on Sephadex G-100. The Sephadex G-100 column before use is equilibrated with 0.05 M TES, pH 7.0, and also eluted with the same TES. Active fractions are determined by measuring the intensity of light

[4] G. L. Picciolo, E. W. Chappelle, and E. Rich, *Appl. Microbiol.* **16,** 954 (1968).
[5] Bacterial luciferase of a purity corresponding to that of the acetone powder extract may be obtained from Sigma Chemical Co., St. Louis, Missouri; Worthington Biochemical Corp., Freehold, New Jersey; and Nutritional Biochemical Corp., Cleveland, Ohio.

emission when 0.1 ml of $FMNH_2$ (10^{-6} micromole) is injected into a small test tube containing 0.2 ml of the eluted fraction and 0.1 ml of 0.05 M TES, pH 7.0, saturated with dodecyl aldehyde. The most active fractions are pooled. The enzyme solution, frozen and stored at $-80°$, is stable for several weeks; if longer stability is desired, it may be lyophilized and stored with desiccant at $-80°$.

Preparation of Lumiflavin

One gram of riboflavin is dissolved in 1 liter of 1 N KOH; the solution is exposed to fluorescent illumination for 15 hours. Lumiflavin is obtained from the solution by chloroform extraction. The lumiflavin solution is then taken to dryness by flash evaporation. The final yield is approximately 250 mg.

Assay Method

Instrumentation. Light-measuring instrumentation required for this assay includes:

High-gain, low-noise photomultiplier tube (S-20 response) with a well-stabilized high-voltage power supply

Light-tight reaction chamber in which a small test tube containing the luciferase solution can be positioned in front of the cathode surface of the photomultiplier tube without allowing the entry of extraneous light. The top surface of the chamber should include a small, rubber-sealed orifice to permit injection of the FMN solution into the luciferase without affecting the light-tight integrity of the chamber

High-gain, high-impedance direct current amplifier with a wide range of sensitivity

Means of measuring and recording the output signal: e.g., a recorder, oscilloscope, etc.

A system similar to that used here is described in detail elsewhere.[6]

Commercial sources of instrumentation specifically designed for bioluminescent assays, or which may be modified for such use, include E. I. duPont, Wilmington, Delaware; Photovolt Corp., New York, New York; American Instrument Co., Silver Spring, Maryland; and Gamma Scientific, San Diego, California.

Reagents

Sodium borohydride ($NaBH_4$), 0.2 M, in 0.2% KOH (prepare fresh immediately before use); used as reducing agent

---

[6] E. W. Chappelle and G. V. Levin, *Biochem. Med.* **2,** 49 (1968).

Palladium chloride (PdCl$_2$), 0.002 M, in 1 N KOH; used as reduction catalyst

Perchloric acid (HClO$_4$), 1 N. This solution, used at 2°, effectively extracts FMN and FAD from cellular material; used at 100°, it also hydrolyzes FAD to FMN.

Lumiflavin: immediately before use, 10 mg is dissolved in 250 ml of distilled H$_2$O; 1 ml of this solution is added to 100 ml of the HClO$_4$ extracting solution. This photolysis product of riboflavin retards the photodegradation of FMN and FAD but does not inhibit the luciferase reaction.

Tris(hydroxymethyl)aminomethane-sulfonic acid (TES), 0.2 M, pH 7.0; used as a buffer

Dodecyl aldehyde (prepare immediately before assay); 0.1 ml of concentrated aldehyde is suspended and dispersed with shaking in 5 ml of luciferase solution

Luciferase solution, pooled active fractions from Sephadex G-100 column in 0.05 M TES, pH 7.0

Flavin mononucleotide (FMN); used for preparation of standard curves

Extraction of Flavins from Cellular Material. The extraction and subsequent assay steps should be carried out with minimal exposure to light. In order to minimize surface adsorption and degradation of the flavins, it is advisable to use polypropylene containers, pipettes, and syringes as much as possible.

One milliliter of an aqueous cellular suspension is added to 10 ml of the HClO$_4$-lumiflavin solution and allowed to stand for 15 minutes at 2°. At this point, the cells have ruptured and FMN and FAD are free of protein.

Assay for FMN. A 1-ml aliquot of the above suspension is added to each 1-ml volume of the PdCl$_2$ solution. This step and all subsequent ones take place at room temperature. The mixture is allowed to stand for 5 minutes for complete precipitation of the perchlorate ion, after which 7 ml of 0.2 M TES buffer is added. The FMN contained in the solution is then reduced by adding 1 ml of the NaBH$_4$ solution. After 15 minutes, not more than 30 minutes, 0.1 ml of the solution is injected by means of needle and syringe into 0.3 ml of the luciferase solution contained in a 6 × 50-mm test tube mounted in front of the cathode surface of the photomultiplier tube, and the emitted light is measured. The FMN content is calculated by reference to a standard curve prepared from measurements of the light intensity obtained with known concentrations of FMN subjected to the same conditions as the unknowns. The concentrations prepared should be such that the injected 0.1 ml volumes contain quantities of FMN covering the range of 10^{-8} to 10^{-4} micromole. Values from blank assays using water in place of the FMN solution should be subtracted both from the standard curve and from the unknown values. Light emission as a function of FMN content becomes nonlinear above 10^{-4} micromole; therefore, the quantity of sample used must be adjusted so that its flavin content falls on the linear portion of the standard curve.

Assay for Total Phosphoflavin (FMN plus FAD). The procedure for total phosphoflavin assay is the same as that described for FMN except that, before adding $PdCl_2$, the $HClO_4$-treated samples are boiled in a covered water bath for 30 minutes to convert FAD to FMN. The samples are then assayed for FMN, and the FAD content is calculated as the difference between total phosphoflavin and FMN.

Efficiency of Procedures

Use of purified preparations of three FAD flavoproteins (diaphorase, glucose oxidase, and L-amino acid oxidase) showed less than 1% conversion of FAD to FMN during treatment with $HClO_4$ at 2°, whereas approximately 100% conversion resulted from boiling with $HClO_4$ for 30 minutes. There was no measurable destruction of FMN during this period. The ability of cold $HClO_4$ to remove FMN from its protein moiety was shown by the detection of 100% of the calculated flavin when the FMN flavoprotein (cytochrome *c* reductase) was exposed to the cold acid for 15 minutes.

The assay is highly specific for FMN; FAD produces less than 5% of the light intensity obtained with an equivalent amount of FMN, much of this being due to the spontaneous hydrolysis of FAD to FMN. Other flavin compounds (such as riboflavin and lumiflavin) give less than 1% of the response of an equivalent quantity of FMN.

As examples of the small quantities of samples required in the assay of total phosphoflavin, the following minimal quantities of certain materials are given: 100,000 *Bacillus subtilis* cells; 0.1 ml of whole blood (human); 10 mg of beef liver (dry).

[138] Extraction, Purification, and Separation of Tissue Flavins for Spectrophotometric Determination

By Arpad G. Fazekas and Karoly Kokai

Among the enzymes connected with oxidoreductive processes those utilizing flavin adenine dinucleotide (FAD) or flavin mononucleotide

(FMN) as coenzymes are especially important. Biochemical experiments concerned with the study of flavin enzymes and the biosynthesis of flavin coenzymes all require pure preparations and reliable quantitative measurement of flavin compounds. The most important representatives of the isoalloxazine-containing flavin compounds in the living organism are riboflavin (RF, vitamin B₂), FMN, and FAD. Their physicochemical properties, biochemical functions, and methods of determination were summarized in the excellent papers of Yagi[1] and Fragner.[2]

In this paper, we describe briefly the most important characteristics of tissue flavins. Crystalline FAD (mol. wt. 758.58) is an orange, hygroscopic substance. It dissolves well in water, pyridine, phenol, and p-cresol. It is insoluble in ethyl ether, chloroform, ethanol, and acetone. FMN (mol. wt. 456.36) is a greenish yellow substance, soluble in water and in glacial acetic acid, phenol, and p-cresol. It is not soluble in acetone or ether. RF (mol. wt. 376.36) is a greenish yellow substance, slightly soluble in water but quite soluble in glacial acetic acid, pyridine, phenol, and p-cresol. In solution, RF and FMN show a greenish yellow fluorescence, and FAD an orange fluorescence. In the solubilized state, FMN and FAD are destroyed by light. In alkaline medium, riboflavin is converted to lumiflavin and lumichrome by light irradiation (photolysis).[1] The presence of phenol[3] or fructose[4] derivatives strongly inhibit the photochemical degradation process. The flavins have characteristic absorption maxima in the yellow, blue, and ultraviolet regions of the light spectrum. Absorption maxima of FAD are at 263, 375, and 450 mμ according to Yagi[1] or at 265, 375, and 450 mμ according to Kokai et al.[5] The spectrum of FMN, however, is slightly different, having peaks at 266, 373, and 445 mμ.

Warburg and Christian[6] were the first to isolate FAD from horse liver by phenolic extraction. Since then numerous authors have published various procedures for the preparation of FAD from animal tissues. Some methods[7,8] are modified versions of the original one by Warburg and Christian.[6] Others introduced new procedures: Whitby,[9] for example, used cellulose columns for the purification of FAD from bakers' yeast extracts. Siliprandi and Bianchi[10] described the application of ion-exchange chroma-

[1] K. Yagi, *Methods Biochem. Anal.* **10**, 319 (1962).
[2] J. Fragner, *in* "Vitamine: Chemie und Biochemie," Vol. II, p. 1962. VEB Gustav Fisher Verlag, Jena, 1965.
[3] K. Yagi and I. Ishibashi, *Vitamins (Kyoto)*, **7**, 935 (1954).
[4] Y. Sakurai and H. Hioki, *Vitamins (Kyoto)*, **7**, 913 (1954).
[5] K. Kokai, A. G. Fazekas, and G. Domjan, *Kiserl. Orvostud.* **19**, 241 (1967).
[6] O. Warburg and W. Christian, *Biochem. Z.* **150**, 298 (1938).
[7] A. Hellerman, A. Lindsay, and M. R. Bovarnich, *J. Biol. Chem.* **163**, 553 (1946).
[8] A. W. Schrecker and A. Kornberg, *J. Biol. Chem.* **182**, 795 (1950).
[9] L. G. Whitby, *Biochem. J.* **54**, 437 (1953).
[10] N. Siliprandi and P. Bianchi, *Biochim. Biophys. Acta* **16**, 425 (1955).

tography and column electrophoresis to prepare FAD from bakers' yeast. Huennekens and co-workers[11] purified the extract by paper chromatography.

The separation of FAD, FMN, and RF and their derivatives was performed in most cases by ascending[12,13] or circular[14] paper chromatography.

Ochoa and Rossiter[15] determined the FAD content of various rat tissues with a manometric method by measuring the increase of O_2 consumption in an alanine-oxidase–alanine system on the effect of FAD isolated from the tissues. Fluorometric methods were also elaborated by Bessey et al.[16] and Cerletti and co-workers.[17] Yagi[18] applied paper chromatography followed by fluorometric titration for quantitative FAD determination. DeLuca and Kaplan[19,20] used spectrophotometry for the quantitative measurement of FAD obtained by synthesis and isolated by column chromatography.

Purity is the most important requirement for the exact and reproducible spectrophotometric determination of tissue flavins. Previous methods fulfilled this requirement only partly, and complicated and lengthy procedures were used. Under such conditions a significant part of the flavins is lost or destroyed. Methods based on derivative formation are quick, but give only total flavin contents.

It was our aim to develop quick and reliable methods for the quantitative spectrophotometric measurement of tissue flavins. Column chromatography followed by thin-layer chromatography proved to be the best tools for the purification and separation of FAD, FMN, and RF isolated from animal tissues or bakers' yeast.

Extraction of Tissue Flavins

Since FAD and FMN are structurally bound to tissue proteins, first they must be liberated and solubilized. This can be achieved by hydrolysis or by boiling the tissue.[18] Boiling was found to give completely satisfactory results. The tissue sample (4–6 g) is chopped with scissors into cubes of 1–2 mm diameter, then homogenized in a Potter-Elvehjem type all-glass homogenizer in distilled water for 5 minutes on an ice bath (5 ml of distilled water is added per gram of wet tissue). The homogenate is boiled for 3

[11] F. M. Huennekens, S. P. Felton, S. P. Colowick, and N. O. Kaplan, Vol. III, p. 950.

[12] J. L. Crammer, Nature 161, 349 (1948).

[13] L. G. Whitby, Nature 166, 479 (1950).

[14] K. V. Giri and P. R. Krishnaswamy, J. Indian Inst. Sci. 38, 232 (1956).

[15] S. Ochoa and R. J. Rossiter, Biochem. J. 33, 2008 (1939).

[16] O. A. Bessey, O. H. Lowry, and R. H. Love, J. Biol. Chem. 180, 755 (1949).

[17] P. Cerletti, R. Strom, and M. G. Giordano, Arch. Biochem. Biophys. 101, 423 (1963).

[18] K. Yagi, J. Biochem. 38, 161 (1951).

[19] C. DeLuca and N. O. Kaplan, J. Biol. Chem. 223, 569 (1956).

[20] C. DeLuca, M. M. Weber, and N. O. Kaplan, J. Biol. Chem. 223, 559 (1956).

minutes, cooled, and centrifuged at 9000 g for 5 minutes. The sediment is resuspended in distilled water (5 ml/g), homogenized again, then centrifuged. The procedure is similar when bakers' yeast is used as raw material. The yeast is suspended in distilled water (5 ml/g), boiled, centrifuged, then the sediment is washed again. The supernatant solutions are combined, and ammonium sulfate (analytical grade) is added (0.4 g/ml). Further extraction is performed in a separating funnel with 2 × 10 ml of phenolic solution (500 g phenol + 125 ml distilled water). The use of the phenol–water solution is convenient because it is not necessary to liquefy the solvent by heating it before use.

The phenolic (upper) phases are combined and centrifuged in glass centrifuge tubes at 1500 g for 5 minutes in order to break the emulsion. After centrifugation the sharply separated upper phenolic phase is siphoned off with Pasteur pipette. The phenolic extract is mixed with an equal volume of ethyl ether (v/v) and extracted with 4 × 5 ml of distilled water. The flavin-containing water fractions are combined and filtered through paper. The filter paper is washed with 2 ml of distilled water, and the greenish-yellow clear filtrate is concentrated to 1.5–2 ml volume under vacuum in a 60° water bath. This concentrated extract is subjected to further purification by column chromatography.

Column Chromatography

Purification of the Crude Extract and Separation of FAD and FMN from RF by Partition Chromatography on a Cellulose Column. To prepare the cellulose column, Whatman cellulose powder is used. To ensure adequate purity, prior to use, the cellulose powder has to be washed three times with hot distilled water (5 liters/kg) and dried over anhydrous $CaCl_2$ in an oven at 60°. The column is prepared by mixing 7.5 g of washed cellulose powder with 50 ml of water saturated with i-amyl alcohol, and the suspension is poured into a 10-mm wide all-glass chromatography column. After the cellulose has settled, quartz sand is carefully layered onto the top in a 2–3 mm layer. In this way a 35-cm long column can be obtained with a convenient flow rate. The crude tissue extract is chromatographed on this column, water saturated with i-amyl alcohol being used as solvent. The effluent is collected in 1-ml fractions.

Separation of a mixture of authentic FAD (Light), FMN (Sigma), and RF (Sigma) is shown in Fig. 1. FAD and FMN are eluted in fractions 21–28; and RF, in fractions 31–41.

By the chromatography of the crude tissue extract flavins are eluted in fractions identical to those obtained with authentic preparations. The unspecific contaminations (brownish red substances) precede FAD during chromatography. Elution of flavins can be controlled by examining the

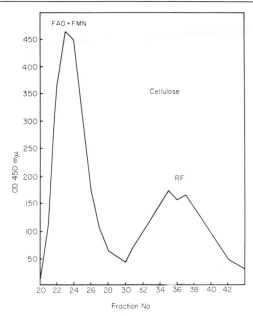

FIG. 1. Separation of FAD and FMN from riboflavin on a cellulose column.

fractions in normal and/or UV light (Wood's light). Fractions containing FAD + FMN and RF are combined and evaporated to dryness under vacuum on a 60° water bath. The dry materials are dissolved in 1 ml of 50% methanol in water and subjected to further purification and separation by thin-layer chromatography.

Purification and Separation of Flavins on a Carboxymethyl Cellulose Column. The quick separation of FAD + FMN from RF can be best achieved by chromatography on carboxymethyl cellulose (Whatman CM-70) ion-exchange column. The column is prepared by suspending 7 g of cellulose powder in about 50 ml of 0.003 M ammonium carbonate solution, and the slurry is poured onto a column with a 10-mm diameter. The height of the column will be 31 cm, and 1–2 ml of the tissue extract is chromatographed using the 0.003 M ammonium carbonate solvent and collecting 1-ml fractions. FAD and FMN are eluted together (Fig. 2). The main advantage of this column is its quickness, but purification is not so good as in the previous system.

Purification of the Extract and Separation of FAD + FMN from RF by Gel Filtration on Sephadex Columns. Molecular filtration on Sephadex gel columns can be used with advantage for the purification of the crude extract and separation of FAD from FMN + RF. The Sephadex columns are prepared as follows: the Sephadex powder (granulation: coarse) is

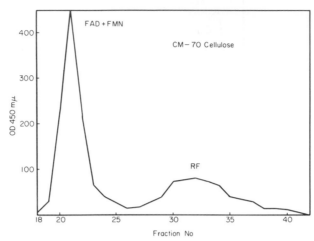

Fig. 2. Separation of FAD + FMN from riboflavin on Whatman CM-70 ion exchange cellulose column.

suspended in an appropriate volume of 0.025 M, pH 7.0, phosphate buffer and then left to swell for the necessary time, prescribed by the manufacturer. The swollen gel is then poured onto a glass column with a 10-mm diameter. The most convenient column heights are: G-10, G-15: 20 cm; G-25: 30 cm. Care should be taken to eliminate air bubbles trapped in the column to ensure uniform density and flow rate. One to two milliliters of the tissue extract is put on the column and chromatographed with 0.025 M phosphate buffer (pH 7.0) as solvent. The sequence of elution is FAD + FMN close together then RF. The nonspecific chromogenic substances in the extract are eluted preceding FAD (brownish red chromogens) and the contaminating free nucleotides (NAD, ATP, etc.) between FAD and FMN; 1-ml fractions are collected. The approximate times necessary to run the columns are: G-10: 85 minutes; G-15: 50 minutes; and G-25: 135 minutes. The best separations can be obtained on G-15 columns (Figs. 3–5). Column fractions containing FAD + FMN and RF are combined, evaporated to dryness, and either subjected to thin-layer chromatography or the flavin content measured directly by spectrophotometry at 450 mμ. The complete separation of FAD from FMN can be achieved on a 35-cm long Sephadex G-15 column.

Thin-Layer Chromatographic Purification and Separation of Tissue Flavins

The flavin-containing fractions obtained by the various column chromatographic methods can be further purified and separated by thin-layer chromatography. Quantitative measurements in the ultraviolet at 265 mμ require highly purified preparations. Thin-layer chromatography is a very useful tool for this purpose if it is carried out with certain precautions. An important point is the analytical purity of the silica gel powder. Commercially available silica gel generally contains a significant amount of impurities that seriously interfere with the spectrophotometric measurements. Prior to use, the silica gel powder (Silica gel-G, Merck) must be purified by washing (500 g gel with 2 liters of analytical grade hot methanol on three subsequent days (totally 6 liters). After the washing is completed, the methanol is decanted and the gel spread on flat glass trays and dried in an oven at 45°. The use of plastic polythene or other synthetic trays should be avoided, since this would interfere with the spectrophotometry. The chromatoplates are prepared by mixing the dry gel with distilled water (18 g/40 ml) and spreading the slurry on glass plates at 250 μ thickness. The ready chromatoplates are dried and stored over $CaCl_2$ in an oven at 45°. Activation by heating is not necessary. The materials to be separated are spotted on to the plates, 2 cm from one end. Volumes ranging from 0.05 to 0.1 ml are applied on one spot. For better identification the parallel running of authentic FAD, FMN, and RF markers is recommended (5 μg of each). The length of running is 15 cm from the start.

The composition of the solvent systems and R_f values of authentic flavin compounds are summarized in Tables I and II. Numerous running

TABLE I
R_f VALUES OF FLAVINS ON KIESELGEL-G LAYERS

Flavins	Solvent systems[a] and $R_f \times 100$					
	TLC-I	TLC-II	TLC-III	TLC-IV	TLC-V	TLC-VI
RF	81	58	61	12	56	8
FMN	65	38	61	27	19	22
FAD	49	16	91	41	26	41

[a] Composition of solvent systems: TLC-I: n-butanol–water–glacial acetic acid–methanol (7:7:0.5:3); TLC-II: n-butanol–glacial acetic acid–water (12:3:5); TLC-III: water saturated with i-amyl alcohol; TLC-IV: 5% Na_2HPO_4·12 H_2O in distilled water; TLC-V: 160 g phenol–30 ml n-butanol–100 ml distilled water (lower phase); TLC-VI: 3% $Na_2B_4O_7$·10 H_2O in distilled water.

tests, using authentic flavin preparations and tissue extracts, proved the system TLC-IV to be the most suitable for the quick separation of these compounds. Running time is comparatively short (40 minutes at 15 cm front distance). Flavins are perfectly separated and appear as sharp, easily distinguishable fluorescent spots. After chromatography, the plates

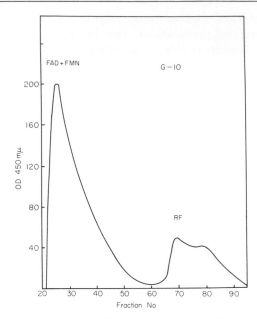

Fig. 3. Separation of FAD and FMN from riboflavin on Sephadex G-10 column.

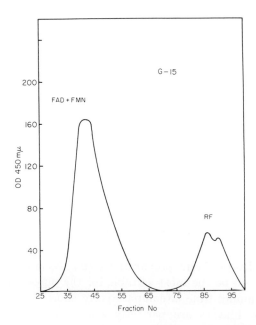

Fig. 4. Separation of FAD and FMN from riboflavin on Sephadex G-15 column.

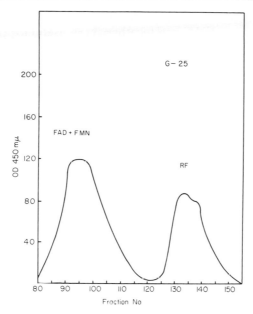

FIG. 5. Separation of FAD and FMN from riboflavin on Sephadex G-25 column.

are dried by means of a hair drier and examined under UV light. (Use a filter with a maximal transmission at 365 mμ.)

Characteristics of the other TLC-systems are as follows:

TLC-I. Components are very well separated, appear as sharp spots; runs relatively slowly.

TABLE II

R_f VALUES ($\times$ 100) OF FLAVINS AND SOME FREE NUCLEOTIDES ON KIESELGEL-HF$_{254}$ LAYERS

Compounds	System[a]		
	TLC-IV	TLC-VII	TLC-VIII
FAD	60	35	49
FMN	46	17	39
RF	30	8	22
NAD	76	69	73
ATP	90	71	83

[a] Composition of solvent systems: TLC-IV: 5% $Na_2HPO_4 \cdot 12\ H_2O$ in distilled water; TLC-VII: 160 ml saturated $(NH_4)_2SO_4$ solution–36 ml 8.2% Na acetate solution–4 ml i-propanol; TLC-VIII: 10% $Na_2HPO_4 \cdot 12\ H_2O$ in distilled water.

TLC-II. Spots are not very sharp; tailing occurs.

TLC-III. RF and FMN run together. Very useful for the separation of FAD.

TLC-V. Gives very good separations; very slow runs; complete evaporation of phenol takes a long time. In case of incomplete drying, flavins do not fluoresce.

TLC-VI. Good separations, but spots are not sharp.

The fluorescent spots are marked under UV light, then flavins are extracted from the TLC plates. Extraction can be easily performed by means of a simple zone extractor (Fig. 6). The zone extractor can be made

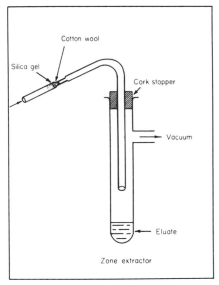

FIG. 6. Zone extractor.

in the laboratory from a glass tube of 7–8 mm diameter. The cotton wool is previously washed with water and methanol and dried in an oven. To reuse the extractor, the cotton wool is removed with a hypodermic needle. Elution of flavins is quantitative in 3–4 ml of 50% methanol in water.

Quantitative Measurement of Tissue Flavins by Ultraviolet or Visible Spectrophotometry after Thin-Layer or Column Chromatography

After thin-layer chromatography, the quantity of flavins can be determined by ultraviolet spectrophotometry based on the optical density measured at 265 mμ. Readings are taken against the 50% methanol extract of a blank spot of the same size of the same chromatoplate. The purity of the sample is checked by readings taken at 250, 260, and 270 mμ. The absorption maximum should be at 265 mμ. Quantities of flavins can be simply calculated from optical density values based on a previously prepared calibration curve. This is obtained by diluting stock solutions of FAD, FMN, and RF with 50% methanol and measuring optical densities at 265 or/and 450 mμ against the solvent. Practical experience shows, that at 265 mμ: 1 μg/ml and at 450 mμ: 3–4 μg/ml concentrations of FAD can still be satisfactorily measured. Flavin values obtained this way are calculated per gram wet tissue weight or 100 mg of protein nitrogen.

By examining the spectrum of FAD isolated from rat liver or kidney, it can be seen that after phenol-ether extraction and chromatography on cellulose columns using water saturated with i-amyl alcohol, the isolated FAD shows an absorption maximum at 450 mμ, measurable at this wavelength. However, this preparation has no absorption maximum at 265 mμ, and its absorbance is increasing toward shorter wavelength. If the extract is purified on Sephadex columns, the absorption maximum of the FAD-containing fraction is at 260 mμ indicating the presence of impurities consisting mainly of free nucleotides (ATP, NAD, NADP). The character of the contaminating substances can be best studied by thin-layer chromatography on a fluorescent layer (Kieselgel HF$_{254}$) where the impurities appear as dark spots in UV light, separated from the fluorescing flavins (Table II). Yeast extracts contain mainly ATP contamination after column chromatography. Most of the free nucleotide contamination is eluted after FAD and FMN from Sephadex columns. The FAD-containing fractions obtained by the different column chromatographic methods and further purified by thin-layer chromatography have a spectrum identical to that of the authentic FAD preparate with absorption maxima at 265 and 450 mμ (Fig. 7).

The FAD content of rat liver after phenol-ether extraction and chromatography on cellulose column using water saturated with i-amyl alcohol and spectrophotometry at 450 mμ has been found to range from 25.1 to 36.4 μg/g wet wt. These data can be compared to the values of 634 to 831 and 1483 to 2070 μg/100 mg protein nitrogen for the FAD content of rat liver and kidney, respectively, measured at 265 mμ after cellulose column and thin-layer chromatography. In the case of liver, the 729.4 μg/100 mg protein nitrogen corresponds to 33 μg/g wet tissue. It can be seen that this is close to the 30.3 μg/g wet tissue obtained after column chromatography, e.g., values are somewhat (10%) higher after thin-layer chromatography.

Values for the FAD content of rat liver obtained by the present method are compared to those published by others (Table III).

To determine FAD + FMN in animal tissues, cellulose column chro-

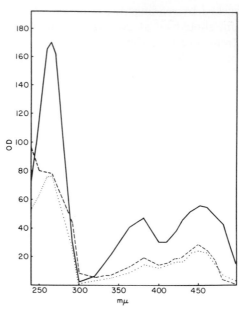

FIG. 7. Spectra of authentic and tissue FAD samples between 240 and 500 mμ. (—) Authentic FAD; (- - -) tissue FAD after cellulose column; (......) tissue FAD purified on TLC.

TABLE III

FAD CONTENT OF RAT LIVER AS DETERMINED BY VARIOUS METHODS

Authors[a]	Purification[a]	Determination[a]	FAD (μg/g wet tissue)
Ochoa and Rossiter[15]	Phenolic extraction (Warburg and Christian[6])	Manometry	77.3
Bessey et al.[16]	Extraction by trichloroacetic acid	Fluorometry	30.5
Yagi[18]	Paper chromatography	Fluorometric titration	36.6
Navazio and Siliprandi[21]	According to Bessey et al.[16]	Enzymatic assay	29.3
DeLuca et al.[20]	Extraction by water	Spectrophotometry	54.8
Domjan and Kokai[22]	Paper chromatography	Fluorometric titration[18]	18.2
Present work	Column chromatography, TLC	Spectrophotometry	33.0

[a] Superscript numbers refer to text footnotes.

[21] F. Navazio and N. Siliprandi, *Experientia* **11**, 280 (1955).
[22] G. Domjan and K. Kokai, *Acta Biol. Hung.* **16**, 237 (1966).

matography and measurement at 450 mμ is generally satisfactory. This simple procedure is especially convenient in cases where plenty of starting material is available (e.g., liver). Since, on the cellulose column, FAD is completely separated from RF, there is no contamination by the latter compound. The FAD, purified in this way has the characteristic absorption maximum at 450 mμ with an optical density high enough for reliable measurements. In those cases when the amount of material at disposal is limited or the FAD content is low (examination of FAD synthesis *in vitro*, determination of FAD in brain tissue), the material should be further purified by thin-layer chromatography and the FAD measured at 265 mμ. (Molar extinction coefficients for FAD are at 265 mμ: 46.2 $\times$ 10^3; at 450 mμ: 11.3 $\times$ 10^3 cm^2/mole.) Thin-layer chromatography on silica gel-G adsorbent results in spectrophotometrically pure FAD (Fig. 7). The sensitivity of quantitative measurements can be increased many times by using fluorometry instead of spectrophotometry.

The methods described can be successfully used also for the determination of flavins in bakers' yeast. The FAD and FMN content of different yeast populations has been determined as 7.7–23.7 μg/g and 30–45 μg/g, respectively. For the determination of yeast flavins the most useful combination is purification on a cellulose column or Sephadex G-15 column followed by thin-layer chromatography.

General Aspects

By the determination of the flavin content of animal tissues, certain variations can be experienced due to the age, sex, and environmental conditions of the experimental animals. If the results are expressed on a tissue wet weight basis, variations can be caused by the tissue fat content or in the case of the liver the time elapsed since the last feeding. It is advisable to sacrifice the experimental animals after 12 hours fasting and to give the results on a protein nitrogen basis.

The flavins are light-sensitive substances; therefore the experiments should be performed in a dark room illuminated by red light. In solution, 15–20% of the FAD is transformed in 24 hours to FMN and RF even under such conditions. Flavin-containing solutions should be carefully protected from direct light. One of the advantages of the present method is that tissue flavins can be rather quickly extracted, purified, and determined when their decomposition is still negligible. In those cases when the FAD, eluted from the column, was subjected to thin-layer chromatography the following day, the presence of FMN and RF was observed on the chromatoplates.

It is a well known fact that exact reproducibility of R_f values is rather difficult by thin-layer chromatography. For this reason the use of standards in parallel runs is advisable. According to practical experience, 4–6 g of liver and 2–3 g of kidney are the optimal quantities of tissue necessary for the determination. Analytical grade purity of the solvents for extraction and chromatography are essentially important requirements. Silica gel and cellulose powder should be washed prior to use as described before.

Parallel determinations of the same tissue give reproducible results. Recovery values are generally about 80% based on determinations using authentic FAD, FMN, and RF. Special care should be taken to avoid the use of polyethylene tubes or trays, since this would seriously interfere with spectrophotometry.

Useful Combinations

The different column and thin-layer chromatographic separation and purification methods described here can be used in various combinations, to obtain the best results, depending on the type of flavins to be determined or the tissue source. The most useful combinations are given below.

Determination of FAD in Animal Tissues. For this purpose cellulose column chromatography and subsequent thin-layer chromatography is generally satisfactory. If FMN and RF are also to be determined, further purification of the column eluate by thin-layer chromatography is advisable.

Determination of FAD in Bakers' Yeast. The crude extract should be purified and FAD separated on a Sephadex column followed by spectro-photometry at 450 mμ. For the determination of FMN and RF, the appropriate column fraction should be subjected to thin-layer chromatography.

Determination of FAD and FMN in Tissues or Bakers' Yeast. These two compounds can be quickly isolated from the crude extract on a carboxy-methyl-cellulose column. If total FAD + FMN values are required, the flavin-containing eluate can be directly measured at 450 mμ. For the separation of these compounds, subsequent thin-layer chromatography is necessary.

Determination of Riboflavin Alone. Riboflavin can easily be freed of contaminations and other flavins by chromatography on a carboxymethyl-cellulose column. After concentration, the RF-containing eluate can be directly measured by the spectrophotometer at 450 mμ.

[139] Preparation of Pure Flavin Adenine Dinucleotide[1]

By Paolo Cerletti and Paola Caiafa

Preparation Method

Principle. Flavin adenine dinucleotide (FAD) is extracted from bakers' yeast or from cultures of the flavinogenic ascomycete *Eremothecium ashbyi*. It is then purified by column electrophoresis and/or ion exchange chromatography. Chemical synthesis of FAD has also been described in a final yield of 40%, from adenosine 5'-phosphoramidate and riboflavin 5'-phosphate (FMN) using a mixture of pyridine and *o*-chlorophenol as solvent.[2]

Reagents

Phenol, freshly distilled under reduced pressure on Zn powder, stored in the dark
Diethyl ether
Methanol
Acetone
Dithionite solution: freshly prepared 10% $Na_2S_2O_4$ in 5% $NaHCO_3$
Acetate buffer, pH 5.1, $I = 0.05$: 75.8 ml of 1 M acetic acid + 50 ml of 1 M NaOH, diluted to 1 liter
$(NH_4)_2SO_4$, finely ground
$AgNO_3$, 30% in water
KCl, saturated solution in water
LiCl, 0.03 M in water, 0.015 M and 0.035 M in 0.003 M HCl
HNO_3, 1 N
$NaHCO_3$

Extraction from Bakers' Yeast.[3,4] Five kilograms of bakers' yeast is suspended in 6 liters of water, heated 10 minutes at 80° with stirring, then cooled and centrifuged 30 minutes at 12,000 g. The supernatant solution is collected, solid ammonium sulfate is added to 66% saturation, and the solution is shaken with 0.08 volume of phenol. After several hours, the aqueous and the phenol phases are separated and the intermediate emulsion is resolved by centrifuging. The combined aqueous phases are extracted twice again as above.

The phenol layers are combined, an equal volume of diethyl ether is

[1] In partial fulfillment of a grant by the National Research Council of Italy (C.N.R.).
[2] J. G. Moffatt and H. G. Khorana, *J. Am. Chem. Soc.* **80**, 3756 (1958).
[3] O. Warburg and W. Christian, *Biochem. Z.* **298**, 377 (1938).
[4] P. Cerletti and N. Siliprandi, *Arch. Biochem. Biophys.* **76**, 214 (1958).

added, and FAD is extracted with water (10 subsequent aliquots each 0.06 volume of the phenol–ether layer). The combined aqueous extracts are brought to pH 3 with 1 N HNO_3, and 0.1 volume 30% $AgNO_3$ is added. The resulting Ag salt of FAD is separated by centrifugation.[3]

The crude Ag salt of FAD is washed twice with 10 ml of water, resuspended in 100 ml of water, and dissolved by adding a slight excess of saturated KCl; the solution is brought to pH 3 by addition of 1 N HNO_3, and centrifuged. The precipitate is washed three times with 5 ml of water, the washings are combined to the clear brownish supernatant (final solution = 115 ml) and brought to pH 4.9 by addition of solid $NaHCO_3$ (ref. 4).

Extraction from Eremothecium ashbyi. Strains from collections do not in general produce significant amounts of FAD. Strains producing high amounts of FAD should be selected from them, if necessary, after ultraviolet irradiation.

The ascomycete is grown in submerged culture.[5] Three-day-old cultures are filtered, and the mycelial cake is extracted by hot diluted saline solution.[6] FAD is determined in the extract and in the broth[6]: if rewarding amounts are found, the two solutions are combined; otherwise the purification is carried on in the solution with highest FAD content.

The solution adjusted to pH 6 is concentrated under reduced pressure at 30° until a dense brown syrup is obtained. This is frozen, thawed, and centrifuged 15 minutes at 25,000 g in a refrigerated centrifuge. The supernatant solution is applied to a column of Amberlite IRC 50, Na form, and eluted with water.[7]

FAD appears in the first peak with yellow fluorescence.[6] Tubes with R 260/450 values less than 9 are collected. Average R 260/450 value of the combined tubes is 7. Recovery at this stage is 70%.

Purification.[4] The solutions obtained from bakers' yeast or from *E. ashbyi* are concentrated under reduced pressure at 30°–40° to about 35 ml and are submitted to electrophoresis in acetate buffer pH 5.1, $I = 0.05$,

---

[5] K. Yagi, *in* "Biochemical Preparations" (H. A. Lardy, ed.), Vol. 7, p. 51. Wiley, New York, 1960.

[6] P. Cerletti and M. G. Giordano, this volume [133].

[7] The elution of flavins is monitored by reading absorbancies at 260 and at 450 nm. The ratio between the two values (R 260/450) can be taken as a measure of the purity of the preparation of FAD. The value R 260/450 = 3.28 has generally been considered to correspond to pure FAD. Higher values indicate contamination by nucleotides and other impurities that absorb in the ultraviolet. For a more selective analysis of the eluate the absorption can be recorded also at 350 nm and at 400 nm, corresponding respectively to 6-methyl-7-hydroxy-8-ribityllumazine and to 6,7-dimethyl-8-ribityllumazine. Fluorescence is measured at the following wavelengths: flavins, 445 nm excitation, 520 nm emission; 6,7-dimethyl-8-ribityllumazine, 415 nm excitation, 475 nm emission; 6-methyl-7-hydroxy ribolumazine, 340 nm excitation, 425 nm emission.

on a column (4 × 40 cm) packed with cellulose powder. The mixture is displaced 10–15 cm down the column by allowing a sufficient amount of buffer to flow; the original brown-yellow zone splits into a faster moving brownish red zone and a slower yellow zone containing FAD and FMN. A potential of 12 V/cm is then applied, and the electrophoresis is run for 14 hours in the dark. The temperature of the column should not exceed 30°. Under the influence of the electric field, FAD is separated from FMN while the brownish red zone moves fastest and is eluted first.

The elution is performed with the same buffer as above.[7] Recovery in this step is about 50%.

The combined FAD fractions are concentrated under reduced pressure at 32° to a volume of about 10 ml and are then applied to a column of Amberlite IRC-50, Na form, and eluted with water.[7] A single yellow fluorescent peak is obtained, corresponding to FAD pure with respect to other flavins or other nucleotides. Two minor peaks corresponding to ultraviolet-absorbing compounds are eluted immediately before and after the FAD fraction and partially overlap it. The contaminated tubes of FAD can be completely purified, however, by submitting them again to ion exchange chromatography.

The total yield from 5 kg of bakers' yeast is 12.8 mg of pure FAD.[4] From *E. ashbyi* it depends on the producing capacity of the strain used.

An alternative to electrophoresis for the preparative separation of FAD from FMN is chromatography on DEAE-cellulose.[2] The concentrated solutions mentioned above are adjusted to pH 7 and applied to a column packed with DEAE-cellulose, chloride form. Riboflavin is eluted with water, FMN with aqueous 0.03 M LiCl or with 0.015 M LiCl in 3 mM HCl, and FAD by 0.035 M LiCl in 3 mM HCl. The FAD peak, adjusted to pH 5.8 with diluted LiOH, is evaporated to a gum under reduced pressure at 30°. Methanol (20 ml) is added, and the mixture is well swirled. Addition of 200 ml of acetone and 20 ml of ether results in the precipitation of the orange lithium salt of FAD; this is collected by centrifugation and washed three times by stirring in 4 ml of methanol and adding 40 ml of acetone and 4 ml of ether. The final precipitate is dissolved in water and lyophilized, giving Li_2-FAD as a fluffy yellow solid. This can be converted to the free acid by passage through a small column of Dowex 50 (H+) resin.

Recovery in the chromatographic procedure is 60%.

Analysis

FAD is quantified spectrophotometrically (see below for extinction coefficients, and footnote 7) or by differential fluorometry.[6] The qualitative assay is done by paper electrophoresis.[6]

The biological activity of FAD is assayed as coenzyme of D-amino acid oxidase from pig kidney.[8] The apparent dissociation constant for FAD-protein has a value of 2.2×10^{-7} mole/liter (ref. 8).

Properties

Spectra. The spectra are almost identical in neutral or acid solution, with maxima at 264, 375, and 450 nm. At alkaline pH the maximum in the ultraviolet region is found at 270 nm, and the peak in the region of 360–380 nm is displaced toward shorter wavelengths, having its maximum at 357 nm (ref. 4 and 9). The molar absorbancy at 450 nm at neutral pH is 11.3×10^3 cm^{-1} mole^{-1}. At the same wavelength the value for oxidized minus reduced form is 10.2×10^3 (ref. 4).

Fluorescence. The excitation spectrum (uncorrected) has maxima at 375 and 450 nm. Emission maximum (corrected) is at 525 nm at 25° in 0.05 M phosphate buffer pH 7, the quantum yield being 0.038 and the relative intensity of emission (photomultiplier anode current/correction factor) 3.26×10^{-5} (ref. 10). The intensity of fluorescence varies with pH. The maximum is at pH 2.9, the fluorescent efficiency falls between pH 3 and 5.5, then has a constant value up to pH 8.5 and almost disappears at higher pH (ref. 4 and 11). Reduction by 0.1 volume of dithionite solution completely destroys the fluorescence throughout the pH range mentioned.[4]

The fluorescence lifetime of FAD at 25°, pH 7, is 3.30 ± 0.05 nsec (ref. 12) and phosphorescence lifetime at 77°K, 0.137 second (ref. 10). Fluorescence polarization of an aqueous solution of FAD at 20° is 3.12%.[13]

ORD and Circular Dichroism. FAD has a negative Cotton effect centered at 375 nm, a complex region of negative rotation between 280 and 330 nm, and a large positive peak in the 260 nm region. The rotary dispersion of FAD is markedly diminished below pH 4 and by increasing urea concentration from 0 to 8 M, or dioxane concentration from 0 to 67% (ref. 14), or dimethyl sulfoxide from 0 to 50% (ref. 15). In strong acidic solutions (pH 1) the ORD of FAD approximates quite closely the sum of rotation of FMN and AMP.[14] The absorption band at 450 remains optically inactive.[14] Also the circular dichroism of FAD is influenced by the solvent

[8] P. E. Brumby and V. Massey, *in* "Biochemical Preparations" (W. E. M. Lands, ed.) Vol. 12, p. 29, Wiley, New York, 1968.

[9] G. R. Penzer and G. K. Radda, *Quart. Rev. (London)* p. 43 (1967).

[10] A. Bowd, P. Byrom, J. B. Hudson, and J. H. Turnbull, *Photochem. Photobiol.* **8**, 1 (1968).

[11] J. C. M. Tsibris, D. B. McCormick, and L. D. Wright, *Biochemistry* **4**, 504 (1965).

[12] A. de Kok, R. D. Spencer, and G. Weber, *Federation Proc.* **27**, 298 (1968).

[13] G. Weber, *in* "Flavins and Flavoproteins" (E. C. Slater, ed.), B. B. A. Library, Vol. 8, p. 15. Elsevier, Amsterdam, 1966.

[14] R. T. Simpson and B. L. Vallee, *Biochem. Biophys. Res. Commun.* **22**, 712 (1966).

[15] I. M. Gascoigne and G. K. Radda, *Chem. Commun.* p. 533 (1965).

TABLE I

PURITY DEGREE OF FAD AT DIFFERENT STAGES OF PREPARATION CALCULATED
FROM THE RATIOS OF LIGHT ABSORPTION AT 260 AND 450 NM (R 260/450)

Stage	R 260/450 found	$\dfrac{R\ 260/450 \text{ of pure FAD}}{R\ 260/450 \text{ found}}$
Extract from bakers' yeast (Y)	41	0.08
Extract from *E. ashbyi* (*E.A.*)	7	0.47
Y after electrophoresis	6.2	0.53
E.A. after electrophoresis	5	0.66
Y and *E.A.* after ion exchanger	3.276	1

and by pH.[16] FAD has a magnetic circular dichroism spectrum similar to FMN in the visible region; in the ultraviolet it is approximately the sum of the FMN and AMP spectra.[17]

Redox Potential, Ionization, Electron Spin Resonance (ESR). The midpoint potential is a function of pH. At pH 7.0 and 30° $E_M = -0.219$ V.[18] The pK for ionization of the 3-amino residue of FAD is 10.4 (ref. 19); for the 6-amino group in the adenine moiety it is 3.5 (ref. 20).

The ESR spectrum of the FAD radical is pH dependent and is discussed in ref. 21. The metal affinity of FAD is similar to that of other flavins.[22]

The nature of intramolecular complexes between the isoalloxazine and the adenine moieties of the dinucleotide and of intermolecular associations, and their influence on the physical properties of the compound are discussed in ref. 9, 11, 16, 23, and 24.

TABLE II

RATIOS OF ABSORPTION OF LIGHT BY PURE FAD AT
DIFFERENT WAVELENGTHS AND pH

pH	R 260/450	R 264/450	R 270/450	R 357/450	R 375/450	R 360/375
2.01	5.49	3.67	3.38	0.70	0.81	4.41
6.98	3.276	3.34	3.09	0.65	0.75	4.24
11.29	3.55	3.89	3.92	1.01	0.80	4.32

[16] D. W. Miles and D. W. Urry, *Biochemistry* **7**, 2791 (1968).
[17] G. Tollin, *Biochemistry* **7**, 1720 (1968).
[18] H. J. Lowe and W. M. Clarck, *J. Biol. Chem.* **221**, 983 (1956).
[19] V. Massey and H. Ganther, *Biochem.* **4**, 1161 (1965).
[20] G. Weber, *Biochem. J.* **47**, 114 (1950).
[21] A. Ehrenberg, L. E. Göran Eriksson, and F. Müller *in* "Flavins and Flavoproteins" (E. C. Slater, ed.) B.B.A. Library, Vol. 8, p. 37. Elsevier, Amsterdam, 1966.
[22] P. Hemmerich, *in* "Mechanismen Enzymatischer Reaktionen," p. 183. Springer, Berlin, 1964.
[23] P. S. Song, *J. Am. Chem. Soc.* **51**, 1850 (1969).
[24] R. M. Sarma, P. Daniels, and N. O. Kaplan, *Biochemistry* **7**, 4359 (1968).

[140] Glycosides of Riboflavin

By L. G. WHITBY

A number of glycosidic derivatives of riboflavin and other isoalloxazines have been described. The first of these compounds, identified as 5'-D-riboflavin-D-glucopyranoside (riboflavinyl glucoside), was detected when an enzyme in rat liver was incubated with riboflavin.[1] The compound was later prepared by enzymatic synthesis from riboflavin followed by purification.[2] The enzyme was shown to be a transglucosidase, and the specificity requirements were investigated.[3] The formation of a much wider range of glycosidic and oligosaccharide derivatives of riboflavin by a number of bacteria and molds has been reported, but so far no physiological function or role in intermediary metabolism has been clearly attributed to any of this group of flavin derivatives, either in animals or microorganisms, nor has their natural occurrence been demonstrated.

Isolation of Riboflavinyl Glucoside (RFG) from Animal Sources[2]

Acetone powder of rat liver is extracted in small quantities (5 g) by grinding in a mortar with distilled water (250 ml); the coarse particles are removed by filtration through gauze. Six lots of enzyme (1.5 liters approximately; 30 g of acetone powder) are combined, saturated with riboflavin (300 mg), and incubated in darkness at 37° for 6 hours. The reaction is then stopped by the addition of 100 ml of trichloroacetic acid (50 g dissolved in 50 ml of water), and the suspension is cooled rapidly to 0°. Ammonium sulfate (800 g) is added, the precipitate is removed by centrifugation, and the supernatant is shaken with 300 ml of aqueous phenol (prepared by warming crystalline phenol to 45° and adding small volumes of water, with stirring, until a solution is obtained).

In the original preparation, this procedure was carried out six times (total weight of acetone powder, 180 g) and 280 mg of RFG, in the presence of a considerable excess of riboflavin (about 1.6 g) was collected in the phenol extract. The following description assumes that similar quantities are being processed. The phenolic solution is shaken with 200 ml of water and 2 liters of ether, and the aqueous phase is collected (240 mg of RFG and 500 mg of riboflavin). Further reduction in the relative preponderance of riboflavin is effected by adding 20 ml of trichloroacetic acid (10 g in 10 ml of water) and 100 g of $(NH_4)_2SO_4$ and extracting into 200 ml of

[1] L. G. Whitby, *Nature* **166**, 479 (1950).
[2] L. G. Whitby, *Biochem. J.* **50**, 433–438 (1952).
[3] L. G. Whitby, *Biochem. J.* **57**, 390–396 (1954).

aqueous phenol. Water (20 ml) and ether (200 ml) are added to the phenolic layer, and the aqueous phase is collected (230 mg of RFG and 300 mg of riboflavin). Subsequent purification is effected by chromatography on columns of powdered cellulose. The material first used, Brown's Solka Floc (B. W. Grade, 200 mesh), requires extensive pretreatment,[2] but other forms of powdered cellulose (e.g., Whatman ashless powdered cellulose) can now be used, and a pilot experiment will show whether the preparation selected for use requires any pretreatment.

To prepare a column, 600 g of powdered cellulose is suspended in acetone and poured into a long glass tube (5 cm in diameter) tapered at one end and plugged with glass wool. The suspension is allowed to settle under gravity (no pressure should be applied), and the column is washed in turn with 50% ethanol, water, 5% glycine, and water, followed by equilibration with water saturated with 3-methyl-1-butanol. The rate of flow during preparation of the column may be 100–150 ml/hour, but this must be reduced to 20–25 ml/hour before the mixture of flavins is added; the capacity of a 600-g column is capable of separating 15 mg of RFG from 20 mg of riboflavin. The columns are run in a dark room, and the position of the bands is observed occasionally by brief examination under ultraviolet light. Several columns may be run in parallel, and fractions corresponding to the band of RFG are collected hourly about 80 hours after the start of each separation. The fractions containing RFG are pooled, extracted with ether (to remove the 3 methyl-1-butanol), and evaporated to dryness in the dark at 45° under reduced pressure with a slow stream of nitrogen entering the flask. (Yield 200 mg of RFG, contaminated with traces of riboflavin and a number of impurities that absorb light below 300 nm.)

Further purification is effected on a second set of columns of powdered cellulose. These are prepared similarly up to the stage of removing glycine by washing with water; thereafter, they are washed with 50% followed by 85% ethanol before being equilibrated with 1-butanol, 77; formic acid, 10; water, 13 (solvents at this stage have all been redistilled). The capacity of these columns for RFG is about 10 times greater than the first series of columns, and 100–150 mg of RFG can be applied after solution in the minimum amount of solvent; the impure RFG from the first set of columns is dissolved in warm water, and the corresponding proportions of formic acid and of 1-butanol are then added, this last addition being made slowly and with careful mixing to avoid precipitation of RFG. The columns are developed slowly, at a flow rate of 20–25 ml/hour, and the band of RFG is collected in fractions at hourly intervals; these are individually taken to dryness under reduced pressure (in darkness at 45° and in an atmosphere of nitrogen), and then dissolved in water for examination in a spectrophotometer. All fractions with a ratio of light absorption at 260:450 nm greater than 2.40 are pooled, evaporated to dryness, and subjected to further chromatographic purification on another column of powdered cellulose in the 1-butanol–formic acid–water system. All fractions with a ratio of light absorption of 260:450 nm of 2.40 and below are pooled and concentrated under reduced pressure as before. (Yield 160 mg of RFG, chromatographically free from fluorescent contaminants.) Final purification (with reduction of the 260:450 nm ratio to 2.25) is achieved by slowly evaporating the concentrated aqueous solution under reduced pressure at 20°, with a current of nitrogen impinging on the surface, until crystals of RFG appear. Recrystallization is performed from 80% ethanol.

Isolation of Riboflavinyl Glucoside from Cultures of Leuconostoc mesenteroides

The synthesis of RFG by bacteria and molds has been investigated mostly on a very small scale, but one method capable of producing RFG in large yields has been described.[4] This utilizes dextran-producing cultures of L. mesenteroides, strains I.F.O. 3426 or L.20, growing in the presence of riboflavin and sucrose. Other strains have been investigated and appear to have even higher productivity, at least on a small scale.[5]

Stock cultures of L. mesenteroides I.F.O. 3426 or L.20, maintained on bouillon–sucrose agar slants and successively transferred every 2 months, are used to prepare inocula by suspending a loopful of stock culture in 10 ml of inoculum medium, containing 4% sucrose, 0.5% K_2HPO_4, 0.1% NaCl, 0.02% $MgSO_4 \cdot 7H_2O$, 0.002% $FeSO_4 \cdot 7H_2O$, 0.002% $MnSO_4$, 0.06% $(NH_4)_2SO_4$, 0.2% "yeast extract" (no further description), and 0.05% polypeptone. After incubation at 25°–28° for 24 hours, 2 ml of culture is inoculated into 50 ml of sterilized fermentation medium, which contains 10% sucrose, 0.5% KH_2PO_4, 0.1% NaCl, 0.02% $MgSO_4 \cdot 7H_2O$, 0.03% $(NH_4)_2SO_4$, 0.05% polypeptone, 0.15% "yeast extract," and 50 mg of riboflavin. This medium has an initial pH of 7.0, being adjusted to this value with NaOH prior to sterilization, but the pH falls to about 5.0 during the fermentation.[5]

The best yield of RFG reported was obtained by supersaturating the fermentation medium with riboflavin and incubating at 27°–30° for 2 days in a shaking culture; up to 34.7 mg of R.F.G. can be produced from 50 mg of riboflavin.[4] At the end of the fermentation, an equal volume (52 ml) of ethanol is added slowly with stirring, and the mixture is left to stand overnight. A gummy precipitate is removed by decantation, and the

[4] Y. Suzuki and H. Katagiri, J. Vitaminol. (Kyoto) 9, 285–292 (1963).
[5] Y. Suzuki and K. Uchida, J. Vitaminol. (Kyoto) 11, 313–319 (1965).

supernatant solution is concentrated to a small volume under reduced pressure and filtered if necessary; $(NH_4)_2SO_4$ is added, and the flavins are extracted into phenol. The phenolic extract is shaken with 1 volume of water and 10 volumes of ether, and the whole process of extraction into phenol and return of flavins to water repeated twice.

Further purification is effected by multiple two-dimensional ascending chromatography using as the first solvent 1-butanol–pyridine–water (6:4:3, v/v). The band of RFG is cut out, eluted with distilled water, concentrated, and reapplied to paper; the second chromatogram is developed with water saturated with 3-methyl-1-butanol. After elution with water, the RFG solution is evaporated to dryness at 40° under reduced pressure, dissolved in a minimal amount of water, and centrifuged to remove insoluble impurities; then sodium hydrosulfite is added. The precipitate of reduced RFG is collected, dissolved in a small volume of water, and adsorbed on a column of Florisil. The column is washed with 5% acetic acid, to remove sugars and other impurities, and RFG is eluted with 5% pyridine. Pyridine is removed by shaking the preparation with chloroform, and the aqueous solution of RFG is evaporated to dryness under reduced pressure. The residue was recrystallized several times from 80% ethanol.

Properties of Pure Riboflavinyl Glucoside Prepared Enzymatically[2,3]

The reddish yellow crystalline powder is readily soluble in water; solubilities at 20° and 37° are 2.2 and 3.5 mg/ml, respectively (to be compared with riboflavin, which has a solubility of 0.1 and 0.2 mg/ml at these temperatures). It is sparingly soluble in ethanol and insoluble in ether. It melts with decomposition at 247°–248°, and has empirical and molecular formulas of $C_{23}H_{30}N_4O_{11}$ (mol. wt. 538.5). The absorption spectrum shows maxima at 445, 375, 266, and 225 nm and minima at 400, 305, and 240 nm; the molecular extinction coefficient at 450 nm (ϵ) = 12.2 × 10^3 liter mole^{-1} cm^{-1}, is identical with the value for riboflavin.[6] Minor differences between the absorption spectra of riboflavin and RFG are found in the region 440–310 nm, where RFG shows slightly less strong absorption (maximum difference 3% at 400 nm) than riboflavin.

Further proof[2] that RFG is the 5'-glucoside of riboflavin may be summarized as follows: (1) formation of lumiflavin under alkaline conditions[7]; (2) formation of riboflavin and glucose by acid hydrolysis, glucose being identified both by chromatography and by quantitative specific enzymatic assay using glucose oxidase; (3) oxidation with sodium metaperiodate[8]

[6] T. P. Singer and E. B. Kearney, Arch. Biochem. **27**, 348–363 (1950).

[7] O. Warburg and W. Christian, Biochem. Z. **298**, 150–168 (1938).

[8] H. S. Forrest and A. R. Todd, J. Chem. Soc. pp. 3295–3299 (1950).

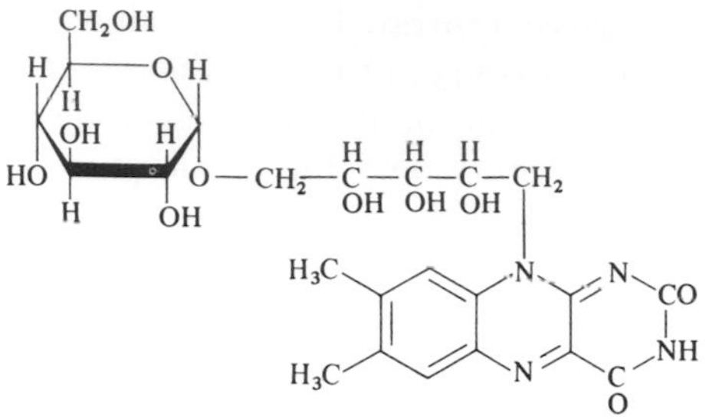

Fig. 1. Riboflavinyl glucoside: 6,7-dimethyl-9-(5'-[α-D-glucopyranosyl]-D-ribityl)-isoalloxazine.

shows an uptake of 4 moles of periodate per mole of RFG, and the small amount of formaldehyde detected among the oxidation products is attributed to overoxidation of one of the end products—it is less than 25% of the amount that would have been formed if RFG had been the 2'-glucoside; (4) the involvement of the reducing group of glucose in the linkage to riboflavin, indicated by the instability to acid hydrolysis, was further supported by the failure of RFG to react with dinitrosalicylic acid.[9] Figure 1 shows the formula of RFG.

Configuration of the Glucosidic Link in RFG

On the basis of enzymatic evidence, RFG is thought to be an α-glucoside. It can be formed by transglucosidation from maltose and from glycogen, and almost all glycoside transosylases hydrolyze glycosidic bonds with retention of the configuration at the glycosidic link.[10] Furthermore, the active nonspecific β-glycosidase in sweet almond emulsin does not hydrolyze RFG, whereas a preparation from bakers' yeast containing both α- and β-glycosidases breaks it down slowly with the production of riboflavin,[2] and rat liver glucosidase hydrolyzes enzymatically prepared RFG[3] but has no action on synthetic RFG[11] (described below), which is thought to be the β-glucoside. More recently, RFG prepared by dextran sucrase action in cultures of four strains of *L. mesenteroides* has been shown to be hydrolyzed by an α-glucosidase present in *L. mesenteroides* whereas the β-glucosidase in emulsin had no effect.[12]

An attempt was made to prove the structure of RFG unambiguously

---

[9] J. B. Sumner and E. B. Sisler, *Arch. Biochem.* **4**, 333–336 (1944).

[10] R. A. Dedonder, *Ann. Rev. Biochem.* **30**, 347–382 (1961).

[11] D. Plant and B. Lythgoe, unpublished, *in* "Some Carbohydrate Derivatives of Biological Interest." Ph.D. Thesis of D. Plant, available for consultation in the University Library, Cambridge, England, 1952.

[12] Y. Suzuki and K. Uchida, *J. Agr. Chem. Soc. Japan* **41**, 125–129 (1967).

by chemical synthesis,[11] starting with the preparation of the disaccharide 5-(D-glucopyranosyl)-D-ribofuranose by the Koenigs-Knorr reaction, and then completing the synthesis of RFG by condensing this disaccharide instead of ribose with o-4-xylidine[13] followed by catalytic reduction of the product and completion of a standard isoalloxazine ring synthesis.[14] This indirect route was chosen so as to avoid problems arising from the small solubility of riboflavin in most solvents, and the tendency of the isoalloxazine nucleus to interfere in the methods usually adopted for the synthesis of glucosides. Since no general method was available for the synthesis of disaccharides having an α-configuration, the synthesis of riboflavinyl-β-D-glucoside was performed, the object being to establish the configuration of the glycosidic link in enzymatically prepared RFG by comparison with the synthetic specimen. The melting point and chromatographic properties of the synthetic material were closely similar to those of the product obtained enzymatically, but the elementary analysis corresponded to $C_{23}H_{30}O_{11}N_4$, $2H_2O$, and the synthetic substance was found to be hygroscopic.[11] There was insufficient enzymatically prepared and synthetic RFG available for comparison of physical properties such as specific optical rotation, and the stability of synthetic RFG in the presence of emulsin or other β-glucosidases was not investigated. However, the use of 5-β-(D-glucopyranosyl)-D-ribofuranose in the synthesis of RFG, and the enzymatic experiments described above, all provide evidence consistent with the view that RFG prepared enzymatically is the α-D-glucoside of riboflavin and that synthetic RFG is the β-D-glucoside.

Characteristics of the Transglucosidases that form RFG

pH and Temperature. The enzymes from rat liver[3] and *Escherichia coli*[15] have a pH of optimal activity in the 6.7–6.9 region, depending slightly on the temperature of incubation, and these enzymes retain activity above 30°. The *E. coli* enzyme shows little loss of activity at 50° compared with the maximum (at 30°). The enzyme in *Aspergillus oryzae* shows optimal activity about pH 5.0, and RFG formation increases with rise in temperature at least up to 50°.[16] By contrast, the enzyme in *L. mesenteroides* has an optimal pH of 5.3 with activity falling off sharply above 30°,[17] and the enzyme in *Ashbya gossypii* shows maximal activity at pH 4.0 and 20°.[18]

Specificity of RFG-Forming Enzymes for Glucose Donor. The specificity

[13] W. A. Wisansky and S. Ansbacher, *J. Am. Chem. Soc.* **63**, 2532 (1941).

[14] P. Von Karrer and H. F. Meerwein, *Helv. Chim. Acta* **18**, 1130–1134 (1935).

[15] H. Katagiri, H. Yamada, and K. Imai, *J. Vitaminol. (Kyoto)* **3**, 264–273 (1957).

[16] S. Tachibana and H. Katagiri, *Vitamins (Kyoto)* **8**, 304–308 (1955).

[17] Y. Suzuki, *J. Vitaminol. (Kyoto)* **11**, 95–101 (1965).

[18] H. Onozaki and T. Takakuwa, *Vitamins (Kyoto)* **28**, 45–48 (1963).

of these transglycosidases for their glucose donors varies with the source of the enzyme and the degree of purification. Taking the activity with maltose as 100, the relative activities of glucose donors with a partially purified preparation of the enzyme from rat liver are glycogen, 100; maltulose, 90; and turanose, 60.[2,3] With crude *E. coli* enzyme, activities relative to maltose (100) were reported as trehalose, 47; glycogen, 43; sucrose, 29; dextrin, 22; glucose, 7; and salicin, 7.[15] In a later paper,[19] using partially purified enzyme, the same authors found that neither sucrose nor glucose acts as a glucose donor (salicin was not reinvestigated) and trehalose shows only a trace of activity. Sucrose, salicin, lactose, melibiose, and glucose also have been reported as potential glucose donors for RFG synthesis in cultures of *A. oryzae*.[16]

Taking account of the large number of carbohydrates and their derivatives that have been found incapable of acting as glucose donors, it may be concluded that the enzymes in rat liver and *E. coli* require the disaccharide maltose or one of its higher homologs to act as glucose donor, or a limited number of oligosaccharides containing an α-glucopyranosyl residue and possessing a free reducing group. Similar enzymes from other sources may be correspondingly specific after partial purification. By contrast, the dextran sucrase from *L. mesenteroides* shows a high specificity for sucrose as glucose donor in the synthesis of RFG.[17]

Some of the inconsistent results in the earlier experiments with microorganisms on the glucose-donor specificity may have been due to failure to exclude light during the prolonged incubations, especially at acid pH, since the photochemical formation of flavin glycosides in aqueous solutions of glucose, fructose, and galactose has been reported.[20] The extent of photoglycosidation is pH-dependent, with an optimum at pH 4.2, and the relative extent to which different wavelengths of light can effect photoglycosidation is closely similar to the absorption spectrum of riboflavin. In the presence of light the formation of RFG from maltose and riboflavin by a purified enzyme preparation from *E. coli* is enhanced, and its formation from glucose and riboflavin occurs, whereas in darkness no formation of RFG is observed when glucose is provided as glucose donor.[20]

Specificity for the Flavin Acceptor. With maltose as glucose donor, purified rat liver transglucosidase yields flavin derivatives provisionally identified as the corresponding glucosides (mainly on chromatographic behavior) from a number of isoalloxazine derivatives.[8] These include L- and D-araboflavin, L-lyxoflavin, and isoriboflavin, and two compounds lacking nuclear methyl groups, 9-L-arabityl- and 9-dihydroxypropylisoalloxazine. Very

[19] H. Katagiri, H. Yamada, and K. Imai, *J. Vitaminol (Kyoto)* **5**, 1–7 (1959).
[20] H. Katagiri, H. Yamada, and K. Imai, *J. Vitaminol. (Kyoto)* **6**, 98–102 (1960).

little glucoside formation occurs with D-galactoflavin, and none with 9-oxyethylisoalloxazine, indicating that the length of the side chain has some importance. Somewhat similar results are obtained with purified *E. coli* enzyme.[19]

Inhibitors of RFG Formation. The formation of RFG by the rat liver and *E. coli* enzymes from riboflavin and maltose is inhibited competitively by α-D-glucose 1-phosphate[3,15]; this compound has no effect on the enzyme from *L. mesenteroides*[17] which is, however, inhibited by maltose and to a lesser extent by fructose and by α-methylglucoside. Inhibition of the *E. coli* enzyme by α-D-glucose 1-phosphate, α-methylglucoside, isomaltose, inorganic phosphate and arsenate, EDTA, *p*-chloromercuribenzoate (reversed by cysteine or glutathione), iodine, and a number of heavy metals (Ag$^+$, Cu^{2+}, Hg$^+$), and of the *L. mesenteroides* enzyme by heavy metals (Cu^{2+}, Fe^{2+}, and Fe^{3+} especially)[17] have also been described.

The inhibition of RFG formation by isoalloxazine derivatives and related compounds has been investigated to a limited extent. In *E. coli*, the extent of the inhibition by lumiflavin, ribityl lumazine, and "B$_2$ keto acid" (in which the pyrimidine portion of the isoalloxazine ring has been removed) varies with the time and temperature of the experiments.[15,19] Lumiflavin acts as a competitive inhibitor.[15] It is not clear whether light was excluded during the conduct of these experiments with heterocyclic inhibitors.

Formation of Glucosyl Oligosaccharides of Riboflavin

The formation of several derivatives of riboflavin by prolonged incubation of *E. coli* with riboflavin and maltose has been described.[21] The synthesis of RFG reaches a maximum after 20 hours in these experiments, and curves showing the formation of lesser amounts of compounds described as B$_2$-isomaltoside, B$_2$-dextrantrioside, and B$_2$-dextrantetraoside with maxima about 40, 50, and 70 hours are reported; the yield of RFG falls off after 20 hours, with the appearance of the oligosaccharides. With *L. mesenteroides*, the formation of two bioside, two trioside, a tetraoside, and a pentaoside derivative of riboflavin has been described,[22] and the products have been partially characterized.[23] For the enzymatic formation of RFG by microorganisms, the optimal conditions of time, pH, and temperature have all to be established, or the yield of RFG falls, and the formation of oligosaccharide derivatives of riboflavin may increase.

[21] H. Katagiri, H. Yamada, and K. Imai, *J. Vitaminol. (Kyoto)* **4**, 126–131 (1958).
[22] Y. Suzuki and K. Uchida, *Vitamins (Kyoto)* **35**, 27–30 (1967).
[23] Y. Suzuki and K. Uchida, *Vitamins (Kyoto)* **35**, 31–37 (1967).

Formation of Other Flavin Glycosides

The formation of 5′-D-riboflavin-D-galactoside by *Clostridium aceto-butyricum* and by *A. oryzae* when incubated with lactose and riboflavin was reported,[24,25] and the glycosidic linkage was suggested to have the β-configuration. The enzyme was also found in *E. coli*, and prolonged incubation of *E. coli* with lactose and riboflavin produced compounds identified as riboflavinyl galactobioside and galactotrioside.[26] The enzyme was partially purified; it shows properties very similar to the transgluco-sidase in *E. coli* and is highly specific for lactose as the galactose donor.[27]

Transfructosidations from sucrose (and to a lesser extent raffinose) by enzymes detected in *E. coli*, *A. oryzae*, and *C. acetobutyricum* have similarly been described.[24,28] When incubated with a cell-free extract of *E. coli*, several glucose-containing and fructose-containing derivatives were obtained,[28] but a partially purified enzyme preparation gave only riboflavinyl fructoside.[29] The properties of the enzyme appear to be similar to those transglucosidases and transgalactosidases which react with riboflavin and related compounds.

Conclusion

The transglycosidases that catalyze the formation of RFG have been partially purified.[3,19] They, and the related enzymes that form riboflavinyl galactoside and riboflavinyl fructoside, and the derived oligosaccharides, all belong to the group of transosylase-glycosidases.[10] These enzymes have been found in rat liver[3] and in mice[30] and are widespread in micro-organisms.[31] Despite the widespread occurrence of enzymes capable of forming RFG and related compounds, and despite the interest shown in flavin glycosides, particularly by the Japanese workers, none of these compounds has been shown to occur naturally,[3,30,32] although it is true that RFG has been reported in homogenates of mouse tissue after loading the animals with riboflavin and ATP.[30]

So far it has not been possible to define a physiological role for RFG and the other flavin glycosides. It has been suggested that the greater solubility of RFG in water, compared with riboflavin, may mean that it is

[24] S. Tachibana, *Vitamins* (*Kyoto*) **8**, 363–365 (1955).
[25] S. Tachibana, *Vitamins* (*Kyoto*) **9**, 119–124 (1955).
[26] H. Katagiri, H. Yamada, and K. Imai, *J. Vitaminol.* (*Kyoto*) **5**, 8–12 (1959).
[27] H. Katagiri, H. Yamada, and K. Imai, *J. Vitaminol.* (*Kyoto*) **5**, 13–18 (1959).
[28] H. Katagiri, H. Yamada, and K. Imai, *J. Vitaminol.* (*Kyoto*) **5**, 298–303 (1959).
[29] H. Katagiri, H. Yamada, and K. Imai, *J. Vitaminol.* (*Kyoto*) **6**, 94–97 (1960).
[30] S. Watanabe, *J. Vitaminol.* (*Kyoto*) **5**, 254–260 (1959).
[31] H. Katagiri, H. Yamada, and K. Imai, *J. Vitaminol.* (*Kyoto*) **6**, 139–144 (1960).
[32] J. L. Peel, *Biochem. J.* **69**, 403–416 (1958).

important in the transport of this relatively insoluble compound.[3] A possible role for RFG as an intermediate in the synthesis of FAD has been proposed.[30] Another suggestion is that flavin glycosides play a part in directing hexose metabolism either into the glycolytic or oxidative pathways,[33] and an ill-defined mononucleotide of RFG has also been reported.[34] None of these suggestions has been properly substantiated and, until proved otherwise, RFG and its related compounds can only be regarded as rather exotic products of transglycosidase reactions. They attracted attention when transosylation reactions were relatively little understood, partly because of their novelty and partly because the flavin glycosides are easy to detect and measure in small amounts, but it must be stressed that the conditions under which their formation has mostly been studied, i.e., in the presence of saturated or supersaturated solutions of the parent flavins, are grossly unphysiological.

[33] S. Tachibana, *Vitamins (Kyoto)* **9**, 125–129 (1955).
[34] S. Tachibana, *Vitamins (Kyoto)* **22**, 291–294 (1961).

[141] Isolation of Riboflavin Glycosides

By SEI TACHIBANA

Preparation Method

Principle. Not only the riboflavinyl glucoside (5'-D-riboflavin-α-D-glucopyranoside), which was first obtained by Whitby[1] with the acetone-dried powder of rat liver, but also various kinds of riboflavin glycosides: riboflavinyl galactoside, riboflavinyl fructoside, and riboflavin oligosaccharides, are easily formed from riboflavin and saccharide with *Aspergillus oryzae* or Taka-Diastase.[2-4] In addition to these enzyme sources, *Escherichia coli*,[5] *Clostridium acetobutyricum*,[6] *Leuconostoc mesenteroides*,[7] and cotyledons of pumpkin *Cucurbita pepo*, and of sugar beet *Beta vulgaris*[8] are available for the biosynthesis of riboflavinyl glucoside and other riboflavin oligosaccharides. Maltose, dextrin, starch, glycogen, and salicin serve as glu-

[1] L. G. Whitby, *Biochem. J.* **50**, 433 (1952).
[2] S. Tachibana and H. Katagiri, *Vitamins (Kyoto)* **8**, 304 (1955).
[3] S. Tachibana, *Vitamins (Kyoto)* **9**, 119 (1955).
[4] S. Tachibana, H. Katagiri, and H. Yamada, *Proc. Intern. Symp. Enzyme Chem. Kyoto Tokyo* p. 154 (1958).
[5] H. Katagiri, H. Yamada, and K. Imai, *J. Vitaminol. (Kyoto)* **3**, 264 (1957).
[6] S. Tachibana, *Vitamins (Kyoto)* **8**, 363 (1955).
[7] Y. Suzuki and H. Katagiri, *J. Vitaminol. (Kyoto)* **9**, 285 (1963).
[8] S. Tachibana, *Vitamins (Kyoto)* **16**, 459 (1959).

cosyl[2,4,6,8] donors for riboflavinyl glucosides. Lactose and melibiose serve as galactosyl[3] (β and α configuration) donors for riboflavinyl galactosides. Sucrose serves both as fructosyl[4] and glucosyl[7] donor for riboflavinyl fructoside and riboflavinyl glucoside.

Growth of Aspergillus oryzae. Subcultures of *A. oryzae* are kept on Czapek-Dox agar slants by standard methods (for full details, the "Manual of the Aspergilli"[9] should be consulted). A liquid medium, containing 2% corn meal, 0.12% $(NH_4)_2SO_4$, 0.1% KH_2PO_4, 0.05% $MgSO_4 \cdot 7 H_2O$, per 200 ml tap water in a 500-ml Fernbach flask is sterilized; previously sterilized 0.5% $CaCO_3$ is then added to the medium. After inoculation by transferring three loops of spores from an agar slant subculture, the flask is incubated for 5 days at 30° under stationary conditions; during this time a surface mycelium develops. At the end of the growth period, the broth is filtered and used as an enzyme solution.

Taka-Diastase Powder. Taka-Diastase powder, which is available from Sankyo Company, Ltd., Ginza, Tokyo, is used as an enzyme preparation. The activity of 1 mg of the powder corresponds to about 0.2 ml of the filtrate described above.

Enzyme Solution from Cotyledons. The sprouting of pumpkin or sugar beet is carried out on sand culture for 6 days at 25° in light. The cotyledons detached from the stems are ground and extracted with a 2-fold volume of cold water in a chilled mortar. The mixture is centrifuged at 15,000 g for 15 minutes. The resulting supernatant solution is used as an enzyme source.

Enzymatic Synthesis. Thirty milliliters of the reaction mixture, containing 0.15 M saccharide, 2.1 × 10^{-4} M riboflavin, 2 ml of 0.1 M acetate buffer (pH 4.5), and 10 ml of the filtrate (or 50 mg of Taka-Diastase powder) is incubated with 1 ml of added toluene for 24 hours at 45°. In the case of the plant enzyme, 30 ml of the reaction mixture, containing 0.1 M maltose, 3 × 10^{-4} M riboflavin, 1.8 ml of 0.1 M acetate buffer (pH 6.2), 9.0 ml of enzyme solution, and 1 ml of toluene is incubated for 5 days at 25°.

Purification Procedure

The method of Crammer[10] is applied with some modification. To 10 volumes of reaction mixture, $(NH_4)_2SO_4$ is added up to saturation. Five volumes of phenol are added to the supernatant solution obtained by centrifugation to extract flavin compounds. After shaking, the mixture is centrifuged, and the clear phenol layer is separated. The phenol extract

[9] C. Thom and K. B. Raper, "A Manual of the Aspergilli," Williams & Wilkins, Baltimore, 1945.

[10] J. L. Crammer, *Nature* 161, 349 (1948).

is shaken with 1 volume of water and 10 volumes of ether in order to transfer water-soluble riboflavin compounds into the aqueous layer (lower layer). The last procedure is repeated 2 or 3 times, and the combined aqueous layer is subjected to filter paper chromatography on thick sheets with n-butanol–80% formic acid–water (4:1:1) as the running fluid. The yellow band on the paper corresponding to riboflavinyl glucoside and other riboflavin glycosides are cut out, and each of them is eluted with water by the descending method. The eluates are subjected again to paper chromatography with water saturated with isoamyl alcohol to separate completely the contaminants of phosphorus compounds of riboflavin, which show significantly higher R_f values near the front of solvent than those of sugar compounds (usually lower than 0.6). The material freed from phosphorus compounds is eluted from the paper with water and evaporated to dryness under reduced pressure at 50°; the residue of riboflavin glycoside is dissolved in a minimal amount of water, followed by removal of the insoluble impurities by centrifugation. To the clear supernatant, an adequate amount of sodium hydrosulfite is added; the resulting precipitate of reduced riboflavin glycoside is collected and dissolved in a minimal amount of water. The flavin solution is subjected to column chromatography on Florisil. After charging, the column is washed with 2% acetic acid, and then the flavin compound is eluted with 5% pyridine. The eluate is washed with chloroform to remove pyridine. The aqueous solution is evaporated under reduced pressure at lower temperature, and the resulting flavin powder is recrystallized 2 or 3 times from 80% ethanol. The yellow crystalline powder thus obtained is dried *in vacuo* in a $CaCl_2$ desiccator.

Properties

With the ascending method with Toyo No. 50 filter paper, R_f values of riboflavinyl glucoside, riboflavinyl galactoside, riboflavinyl fructoside,

R_f VALUES OF RIBOFLAVIN GLYCOSIDE

Compound	B:Ac:W[a]	P:W[b]	W:isoA[c]	B:F:W[d]	B:P:W[e]
Riboflavinyl glucoside	0.21	0.40	0.48	0.05	0.32
Riboflavinyl galactoside	0.18	0.43	0.49	0.04	0.24
Riboflavinyl fructoside	0.25	0.39	0.47	0.06	0.38
Riboflavinyl isomaltoside	0.10	0.47	0.53	—	0.23
Riboflavinyl dextrantrioside	0.05	—	0.57	—	0.14

[a] B:Ac:W = n-butanol–acetic acid–water, 4:1:5, v/v.
[b] P:W = $Na_2HPO_4 \cdot 12H_2O$–water, 5:100.
[c] W:isoA = water saturated with isoamyl alcohol.
[d] B:F:W = n-butanol–formic acid (80%)–water, 77:10:13, v/v.
[e] B:P:W = n-butanol–pyridine–water, 6:4:3, v/v.

riboflavinyl isomaltoside, and riboflavinyl dextrantrioside are obtained as given in the table.

All of the glycosides are readily soluble in cold water, sparingly soluble in ethanol, but insoluble in ether.

The melting point is 248°–249° for riboflavinyl glucoside and 249°–250° for riboflavinyl galactoside. Crystals of both turn brown and decompose.

Acid hydrolysis with 1 N hydrochloric acid at 100° for 2.5 hours gives glucose, galactose, fructose, and riboflavin, respectively. By partial hydrolysis with 0.2–0.5 N hydrochloric acid for 1 hour at 100°, isomaltose and dextrantrioside are shown in hydrolyzates from riboflavinyl isomaltoside and riboflavinyl dextrantrioside, respectively.

The ratio of absorbancy at 375 to 450 nm is 0.85 for riboflavinyl glucoside, 0.86 for riboflavinyl galactoside, and 0.87 for riboflavinyl fructoside. These three compounds exhibit distinct ultraviolet light absorption maxima at 266, 373, and 446 nm, which are almost the same as for riboflavin. Molar extinction coefficients of these compounds at 450 nm are almost the same as that of riboflavin.

[142] Flavin Peptides

By T. P. Singer, J. Salach, P. Hemmerich, and A. Ehrenberg

While in most flavoproteins FAD and FMN are held by noncovalent linkages and are thus liberated as free flavin nucleotides on denaturation of the protein, the flavin component of succinate dehydrogenases from aerobic cells is covalently linked to the protein, so that even drastic denaturation fails to release the flavin. When succinate dehydrogenase is subjected to extensive proteolytic digestion, the flavin is released in the form of FAD peptides of varying chain lengths,[1,2] depending upon the conditions of proteolysis. Several of these peptides have been purified, and a hexapeptide has been isolated in pure form.[3] Most of our knowledge of the chemistry of covalently bound flavins has come from studies on flavin peptides, although more recent work on the probable localization of the site of attachment of the peptide to the isoalloxazine ring has utilized flavin peptides subjected to hydrolysis in strong acid which hydrolyzes the peptide chain and yields a simpler derivative of riboflavin.

[1] E. B. Kearney and T. P. Singer, *Biochim. Biophys. Acta* **17**, 596 (1955).
[2] T. Y. Wang, C. L. Tsou, and Y. L. Wang, *Scientia Sinica* **5**, 73 (1956).
[3] E. B. Kearney, *J. Biol. Chem.* **235**, 865 (1960).

Discovery and Occurrence

Green and colleagues[4] first noted that fragments of the mitochondrial respiratory chain liberate appreciable amounts of flavin-type material that is not extractable by acidification or boiling. The source of this tightly bound flavin turned out to be succinate dehydrogenase, since purified, soluble preparations of the enzyme liberate their flavin only on proteolytic extraction, although not in the form of FAD or FMN.[1] At about the same time Boukine[5] noted that a variety of plant and animal tissues, particularly those rich in succinate dehydrogenase, liberate their total flavin content only on proteolytic digestion. These studies laid the groundwork for the widely used fluorometric determination of flavin peptides[6] to be described below.

The first pure flavin peptide was isolated by Kearney, who was also responsible for the most extensive description of the chemical properties of flavin peptides to date.[3,7,8] Purification and partial characterization of flavin peptides has also been reported by Wang et al.[2,9]

Chemical Properties

The flavin component of succinate dehydrogenase from mammalian or aerobic yeast[10] mitochondria contains FAD covalently linked to the peptide chain. The absorption spectrum (Fig. 1) shows the normal 3-banded spectrum of free flavins but the 375-mμ band is shifted to 345–350 mμ, the exact position depending on the particular peptide. (Digestion of succinate dehydrogenase with trypsin and chymotrypsin yields several flavin peptides varying in chain length.)

Flavin peptides and their acid-hydrolyzed products (6 N HCl, N$_2$, 95° for 12 hours) also show an anomalous pH-fluorescence emission curve (Fig. 2). Like the anomalous absorption spectrum, this unusual pH dependence of the fluorescence is seen at all stages of degradation from FAD-peptide to riboflavin-peptide.[3] Although at pH 3.2–3.4, flavin peptides at the monophosphate or riboflavin level show the same molar fluorescence as FMN or riboflavin, at pH 7 their fluorescence is insignificant (Fig. 2).[3] The

[4] D. E. Green, S. Mii, and P. M. Kohout, J. Biol. Chem. 217, 551 (1955).

[5] V. N. Boukine, Congr. Intern. Biochim. 2e Résumés des Communications, p. 61. Societé Belge de Biochemie, Liege, Belgium, 1955.

[6] T. P. Singer, J. Hauber, and E. B. Kearney, Biochem. Biophys. Res. Commun. 9, 146 (1962).

[7] T. P. Singer, E. B. Kearney, and V. Massey, Arch. Biochem. Biophys. 60, 255 (1956).

[8] T. P. Singer and E. B. Kearney, in "Vitamin Metabolism" (W. Umbreit and H. Molitor, eds.), p. 209. Macmillan (Pergamon), London, 1959.

[9] T. Y. Wang, C. L. Tsou, and Y. L. Wang, Scientia Sinica 7, 65 (1958).

[10] T. P. Singer, V. Massey, and E. B. Kearney, Arch. Biochem. Biophys. 69, 403 (1957).

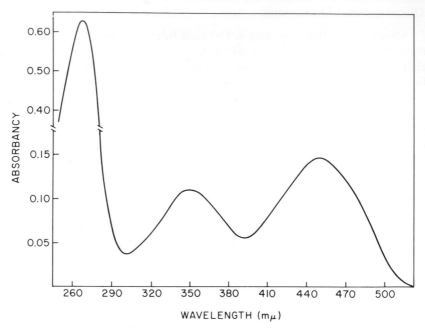

FIG. 1. Absorption spectrum of FAD-peptide in water. From Kearney,[3] reproduced by permission of the *Journal of Biological Chemistry.*

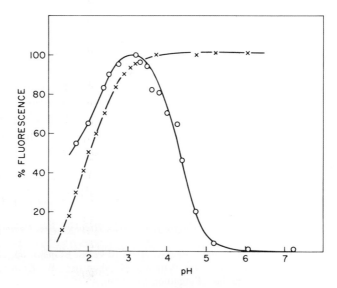

FIG. 2. Comparison of pH–fluorescence emission curves of FMN (×—×) and of pure flavin hexapeptide at the mononucleotide level (○—○). From Kearney,[3] reproduced by permission of the *Journal of Biological Chemistry.*

pK of the fluorescence quenching in the FMN-hexapeptide from heart muscle is 4.3 at 25°. It remains uncertain whether this quenching is due to the glutamic acid present in the peptide or to another, as yet unidentified group. The unusual dependence of fluorescence on pH is the basis of the method for the estimation of flavin peptides (and of succinate dehydrogenase) in tissues and is described below.

Flavin peptides exhibit a strongly cationic character at the FAD, FMN, or riboflavin level, so that they may be readily chromatographed on cation exchangers (e.g., Dowex 50).[3] This positive charge is not accounted for by the amino acid composition of the pure hexapeptide isolated by Kearney,[3] which contains 1 mole each of glutamate, alanine, valine, and threonine and 2 moles of serine, one of which is the N-terminal group. Thus an as yet unidentified positively charged group may be present in proximity to the isoalloxazine nucleus. Whether or not this group and the one responsible for the fluorescence quenching are the same remains to be established.

As a result of the attachment of hydrophilic groups to riboflavin, flavin peptides are considerably more water soluble than free flavins (FAD, FMN, or riboflavin) in the corresponding form. This property is evident even after photodegradation in alkali (lumiflavin degradation), which yields products insoluble in $CHCl_3$, whereas lumiflavin itself is $CHCl_3$ soluble.[3] Interestingly, while free flavins yield only lumiflavin on alkaline photolysis, flavin peptides yield a mixture of lumiflavin peptides and of lumichrome peptides on photolysis at alkaline or neutral pH.[3,11]

Another noteworthy difference between flavin peptides and the corresponding free flavins is that while the latter are stable in concentrated formic acid, the former appear to be destroyed by dehydration of HCOOH.[11]

Point of Attachment of Peptide to Riboflavin

It is known from the studies of Kearney and Singer that the peptide must be attached to the isoalloxazine ring system (rather than to the ribityl, ribose, or adenine moieties), since alkaline photolysis, resulting in the removal of the 10-side chain, does not liberate the peptide.[3,8] Since alkaline hydrolysis removes free urea, rather than a ureido peptide (Wang et al.[12]), the substitution cannot be in the 1, 2, or 3 positions. Positions 4 and 5 can be excluded on the basis of the fact that all possible substitutions in these positions have been examined. Thus substituents OR, NR_2, and SR are split easily from position 4 by acid as well as alkaline hydrolysis, yielding "normal" (i.e., not covalently substituted) flavin.[13] Hydrogen

[11] P. Hemmerich, A. Ehrenberg, W. H. Walker, L. E. Eriksson, J. Salach, P. Bader, and T. P. Singer, *Federation European Biochem. Soc., Ltrs.* **3**, 37 (1969).

[12] T. Y. Wang, C. L. Tsou, and Y. L. Wang, *Scientia Sinica* **14**, 1193 (1965).

[13] K. H. Dudley and P. Hemmerich, *Helv. Chim. Acta* **50**, 355 (1967).

as well as alkyl[14,15] and even aryl[16] residues can be removed from position 4 by photooxidation, yielding again "normal" flavin. Removal of alkyl substituents from N(5), on the other hand, occurs easily in acid medium.[17] No transformation of succinate dehydrogenase flavin into "normal" flavin occurs under similar hydrolytic and/or oxidative conditions.

Comparison of the hyperfine structures of the electron spin resonance (ESR) spectra of the semiquinone of acid-hydrolyzed succinate dehydrogenase flavin and of its alkaline photoderivative with that of lumiflavin, both measured in 6 N HCl, gives evidence for a flavin nucleus exhibiting at least one strongly ESR-active proton less than "normal" flavin.[11] ESR-active nonlabile protons in flavin radical cations are found in positions 6, 8α, and 10α. Therefore, peptide connection to flavin must be sought in one of these positions. The 10α position may be excluded because of the photolysis data mentioned above, and position 6 need not be considered because of its inertness. Furthermore, substitution in position 6 would alter the flavin absorption spectrum to an extent not shown by succinate dehydrogenase flavin. On the other hand, "bi-flavin" condensation[18] and selective deuteration of the 8-CH$_3$ group[19] show that the 8α protons are not quite inert.

Apart from evidence by elimination, there are several lines of independent evidence suggesting that position 8α may be the locus of the flavin-peptide interconnection. Recently,[20] 8α-hydroxylumiflavin and 8α-hydroxylumichrome have been synthesized, and comparison of their ESR spectra in the radical cation form with those of succinate dehydrogenase flavin at the lumiflavin and lumichrome level showed gratifying agreement, while being quite different from the spectra of free lumiflavin and lumichrome, respectively.[11]

While the light absorption spectra of succinate dehydrogenase flavin and of "normal" flavin neutral molecules (cf. above) do not differ significantly in the location of the first band, appreciable differences are observed in the protonated state: the first absorption maximum of protonated riboflavin is situated at 392 mμ while the succinate dehydrogenase analog shows a hypsochromic shift to 375 mμ. Exactly the same shift is found for 8α-hydroxylumiflavin compared to the parent compound.

---

[14] F. Muller and P. Hemmerich, *Helv. Chim. Acta* **49**, 2353 (1966).
[15] F. Muller, V. Massey, C. Heizmann, P. Hemmerich, J. M. Lhoste, and D. C. Gould, *Europ. J. Biochem.* **9**, 392 (1969).
[16] P. Bamberg, P. Hemmerich, and H. Erlenmeyer, *Helv. Chim. Acta* **43**, 395 (1960).
[17] W. H. Walker, P. Hemmerich, and V. Massey, *Helv. Chim. Acta* **50**, 2269 (1967).
[18] P. Hemmerich, B. Prijs, and H. Erlenmeyer, *Helv. Chim. Acta* **42**, 2164 (1959).
[19] F. Y. Bullock and O. Jardetzky, *J. Org. Chem.* **30**, 2056 (1965).
[20] S. Ghisla and P. Hemmerich, in press.

The 8α-substituted model compounds were found to be readily converted to the parent compound (lumiflavin or lumichrome) by reductive removal of the 8α-substituent. This behavior may explain why preparations of succinate dehydrogenase flavin are converted to "normal" flavin under certain strongly reducing conditions (e.g., heating with tryptophan in acid solution).[11]

The proposed site of attachment is shown in Fig. 3. The exact structure

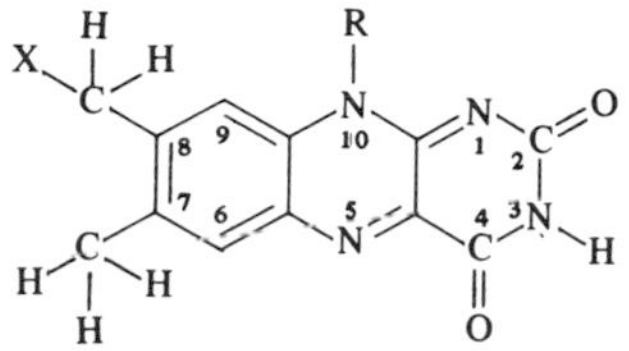

FIG. 3. Proposed structure of flavin peptides: R denotes the ribityl chain to which the nucleotide is attached; X is the substituent to which the peptide is bonded.

of the compound has been established in recent investigations.[20a-20d] The assignment of substituent was fully confirmed by ENDOR spectra.[20a,20b] Comparison of the pH-fluorescence curves of succinate dehydrogenase flavin with a series of 8α-substituted model compounds suggested that the 8α-CH$_3$ group is bound to a secondary or tertiary nitrogen.[20b] High resolution NMR measurements showed that the 7-CH$_3$ group is free[20c] and, hence, that speculations[20e] concerning its attachment to the peptide are mistaken. On drastic acid hydrolysis (125°, 6 N HCl, 16 hours) one mole of histidine was liberated from the acid hydrolyzed flavin peptide: Thus, X was identified as histidine and the bond to the 8α-CH$_3$ group of FAD as a C-N linkage to a ring nitrogen of the imidazole group of histidine.[20c,20d] This assignment was verified by catalytic hydrogenation of the acid hydrolyzed flavin peptide, which liberated histidine under the same conditions as hydrogenation of benzyl histidine and by liberation of free histidine on neutral photolysis.[20c] The complete structure of succinate dehydrogenase flavin is published elsewhere.[20c]

[20a] W. H. Walker, J. Salach, M. Gutman, T. P. Singer, J. S. Hyde, and A. Ehrenberg, *Federation European Biochem. Soc., Ltrs.* **5**, 237 (1969).

[20b] T. P. Singer, J. Salach, W. H. Walker, M. Gutman, P. Hemmerich, and A. Ehrenberg, *in* "Flavins and Flavoproteins" (H. Kamin, ed.), in press.

[20c] W. H. Walker and T. P. Singer, *J. Biol. Chem.*, submitted.

[20d] T. P. Singer, W. H. Walker, E. B. Kearney, and J. I. Salach, *Abstracts*, "Structure and Function of Redox Enzymes" (A. Ehrenberg, ed.), in press.

[20e] P. Nánási, P. Cerletti, G. Magni, and E. Nemes-Nánási, *Abstracts*, FEBS meeting, Madrid, April, 1969, p. 73.

Determination of Flavin Peptides[6]

Reagents

Trichloroacetic acid, 55% (w/v)
Trichloroacetic acid, 1% (w/v)
Trichloroacetic acid, 0.1% (w/v)
HCl, 6 N
KHCO$_3$ or NaOH, 2 M
Tris base, 0.1 M
Acetone, reagent grade
Trypsin, crystallized
Chymotrypsin, crystallized
Phosphate buffer, 0.1 M, pH 7.0
Citrate-phosphate buffer, pH 3.4. This is prepared by mixing 14.3 ml
of 0.05 M citric acid with 5.7 ml of 0.1 M Na$_2$HPO$_4$
Riboflavin standard: 1 $\times$ 10^{-4} M in 0.01 M HCl
Diluted riboflavin standard: prepared daily by diluting the above
standard to 0.4 $\times$ 10^{-6} M or 0.8 $\times$ 10^{-6} M concentration. Both
riboflavin standards are protected from light.
Fluorescein: 0.0004% (w/v) sodium fluorescein in water ($E_{494} \cong$
0.320 in 1-cm light path) is stored in a dark bottle in the
cold.
Dilute fluorescein standard: 0.02 ml of the stock fluorescein per 50 ml,
prepared weekly and protected from light.
Sodium hydrosulfite

Preparation of Samples. Throughout the procedure the flavin samples are protected from light. An aliquot containing 5–15 mg of protein (in the case of heart mitochondria or submitochondrial particles; correspondingly less in the case of purified succinate dehydrogenase) is placed in a chilled 7-ml Pyrex centrifuge tube which fits the multispeed attachment of International Equipment Co. PR-2 centrifuge (or similar glass centrifuge tube), and sufficient cold 55% trichloroacetic acid is added to give a final concentration of 5% (w/v). After brief sedimentation the clear supernatant solution, containing acid-extractable flavins, is discarded. The sediment is resuspended with a glass stirring rod in 2 ml of cold acetone and 0.016 ml of 6 N HCl. The centrifugation is repeated and the supernatant solution is again discarded. The sediment is resuspended in 0.2 to 0.3 ml of 1% trichloroacetic acid, centrifuged, and then resuspended and washed 5 times by centrifugation with 0.2–0.3 ml of 0.1% trichloroacetic acid, the supernatant solution being discarded in each case. The pellet tends to be loose at these low acidities, and care must be exercised to avoid losing the precipitate on decantation. The last supernatant solution must be completely free from flavin as judged by fluorescence (cf. below).

To the precipitate 0.1 mg each of trypsin and chymotrypsin (dissolved in 0.1 M Tris base at a concentration of 10 mg/ml) are added. The resulting pH should be 8.0 at 38°; if not, it is adjusted to this value with Tris or HCl. The suspension is incubated for 4 hours at 38° with occasional agitation. It is then cooled in an ice bath, and 0.1 volume of 55% trichloroacetic acid is added (pH = 0.3–0.4 at 0°). The sample is then incubated again at 38° overnight in a water bath in order to hydrolyze the flavin dinucleotide to the mononucleotide level.

Up to this point photolysis can be prevented by covering the samples with aluminum foil and by working in diffuse light. From this point on it is advisable to use a red photographic light (such as Kodak Safelight, filter series 1A) in handling the sample.

Depending on the flavin content, the sample is quantitatively transferred to a 1-, 2-, or 5-ml volumetric flask, using glass-distilled water for rinses. Sufficient 2 M $KHCO_3$ or NaOH is added to half neutralize the trichloroacetic acid added in the last step and thus yield a pH of 3.5 ± 0.1 at room temperature. (This is determined by titrating an aliquot of the 55% trichloroacetic acid used.) The sample is made to volume with glass-distilled water, mixed, and briefly centrifuged in 7-ml glass or polycarbonate centrifuge tubes. The clear supernatant solution is used for fluorometry.

Fluorometry. The procedure below is given for use with the Farrand fluorometer. Triplicate aliquots of the flavin sample, containing in the order of 0.01–0.02 millimicromole of flavin, are placed in scrupulously clean fluorometer tubes, each containing 2 ml of 0.1 M phosphate, pH 7.0, and diluted with water to 2.5 ml. A similar setup of tubes is prepared containing 2 ml of citrate–phosphate buffer, pH 3.4, instead of neutral phosphate. Similar sets of blanks (no flavin) and riboflavin standards are prepared, both at pH 7.0 and 3.4, each containing 0.04 ml of diluted riboflavin standard in 2.5 ml total volume.

The fluorometer, equipped with suitable filters for riboflavin determination, is standardized with the dilute fluorescein standard so that with the smallest aperture (No. 6 on the Farrand fluorometer) a galvanometer deflection of 50 is obtained. The riboflavin standards are then read, releasing the shutter only momentarily and noting the *maximum* fluorometer deflection. One tube in each set of unknowns is then read, noting the galvanometer deflection, which should be between 25 and 60 units on a scale of 100. If the reading is outside this range, the aliquot is adjusted accordingly. (If the sample is insufficient for larger aliquots, the sensitivity of the instrument may be increased by using the next larger aperture. Alternatively, the total volume may be decreased to 1.0 ml.)

All unknown samples are then read, noting in each case the maximum galvanometer deflection, and checking frequently the standardization with fluorescein. If the instrument response has varied over 0.5 galvanometer unit, it is restandardized with riboflavin, as above.

After all samples are read, 0.02 ml of dilute riboflavin standard is added to each sample and the readings are repeated, as a check on internal quenching. Subsequently, a few milligrams of sodium hydrosulfite is added with a microspatula to each tube, including blanks, riboflavin standards, and unknown, one at a time, and the residual fluorescence is immediately read.

Calculation of Results. Let ΔHS = difference in fluorometer readings before and after hydrosulfite addition, and let ΔIS = difference in fluorometer readings before and after addition of the internal riboflavin standard. The value for ΔHS is determined for blank, riboflavin standard, and each unknown; replicate samples are averaged. If the fluorometer reading for the blank is reduced by more than 0.5 galvanometer units by hydrosulfite (i.e., if the buffer contains measurable fluorescent material which is quenched by hydrosulfite), then the ΔHS value given by the blank must be subtracted from the ΔHS of the standard and of each unknown. The ΔHS value for the riboflavin standard at pH 7, divided by the millimicromoles of riboflavin present, equals fluorometer units per millimicromole of riboflavin. The same calculation for the standard measured at pH 3.4 yields the conversion factor at pH 3.4. The factor at pH 3.4 is 90–95% of that measured at pH 7.

Once this factor is determined at both pH values, it is easy to ascertain whether a significant amount of material is present in the unknown which quenches riboflavin fluorescence. This is done by comparing the experimental value of ΔIS for each unknown with the value calculated from the amount of internal standard added, corrected for dilution by the volume of the internal standard. If the observed and calculated values agree within 1 galvanometer unit, no correction is applied. If significant quenching is evident, the corrected value of the fluorescence is calculated from the relation:

$$\Delta HS_{obs} \times \frac{\Delta IS \text{ (theoretical)}}{\Delta IS \text{ (observed)}} \times \frac{2.5}{2.52} = \Delta HS_{corrected}$$

Taking the value of $\Delta HS_{corrected}$ and dividing it by the value for fluorometer units corresponding to 1 millimicromole of riboflavin gives the apparent riboflavin content of the sample at the particular pH and, with suitable correction for aliquots, the content per milliliter of original enzyme sample. The value at pH 3.4 less that observed at pH 7.0 is the millimicromoles of covalently bound flavin per milliliter of enzyme solution, since flavin peptides have negligible fluorescence at pH 7.0.[21] It should be mentioned that the pH of maximal fluorescence may vary slightly from tissue to tissue (e.g., heart vs. liver or yeast) and should be determined for each tissue.

Isolation of Covalently Bound Flavin

The procedures for the isolation of the flavin hexapeptide and of the acid-hydrolyzed bound flavin are briefly summarized below. Details of the methods are given in the relevant publications.[3,11]

The starting material for the isolation of the peptide in Kearney's studies[3,8] was succinate dehydrogenase, purified from 400 beef hearts by the procedure of Singer et al.[23] Acid-extractable flavin was removed by repeated trichloroacetic acid extractions, and the covalently bound flavin was solubilized by digestion for 4 hours at 38° at pH 8 with 0.1 mg each of trypsin and chymotrypsin per mg of protein. Undigested protein was coagulated by heating for 3 minutes at 95°–100°, sedimented, and discarded. The flavin dinucleotide peptide in the extract was chromatographed on Dowex 50-X4 in the NH_4^+ cycle: the sample was applied in 0.2 M ammonium acetate–acetic acid, pH 5.0, and the main flavin dinucleotide band collected with the same buffer was hydrolyzed (100°, 10 minutes, pH 0) to the mononucleotide form and rechromatographed on Dowex-50, NH_4^+ cycle. The sample was applied in 0.2 M ammonium acetate–0.2 M acetic acid, pH 4.0, and the column was successively eluted with ammonium acetate buffers at pH 5, 6, and 9.5. The fraction collected at pH 6.0 was freed from ammonium acetate and applied to DEAE-cellulose in the acetate cycle. After the elution of impurities with ethanol–1 M acetic acid (70:30, v/v), the flavin was eluted with ethanol–1 M acetic acid (30:70, v/v), dried, and rechromatographed on DEAE-cellulose, acetate cycle, in 2% (w/v) pyridine. The resulting material was chromatographed on filter paper in butanol–acetic acid–water (4:1:5, v/v, upper phase).

Owing to the presence of multiple forms of flavin peptide and unavoidable losses, the yield of pure flavin hexapeptide was 0.18 micromole from 4.4 micromoles of total covalently bound flavin in the original enzyme digest.

Since Kearney[3] has shown that hydrolysis in 6 N HCl at 100° does not break the linkage uniting riboflavin to the peptide chain and thus yields, instead of free flavin, a covalently bound derivative (presumably the component linked to the 8α group), in more recent studies[11] the acid-

---

[21] Yeast preparations show appreciable residual fluorescence at pH 7 (footnote 22).

[22] T. P. Singer, E. Rocca, and E. B. Kearney, in "Flavins and Flavoproteins" (E. C. Slater, ed.), p. 391. Elsevier, Amsterdam, 1966.

[23] T. P. Singer, E. B. Kearney, and P. Bernath, J. Biol. Chem. 223, 599 (1956).

hydrolyzed product has been used. This simplifies the procedure and increases the yield substantially, since the formation of several flavin peptides of varying chain lengths and the consequent need for laborious chromatographic steps is avoided. A further simplification has been the use of ETP$_H$,[24] a particulate preparation of the inner mitochondrial membrane rich in succinate dehydrogenase, in lieu of the highly purified enzyme, since ETP$_H$ may be prepared in much better yield and in far less time than the latter.

Acid-extractable flavin and hemes are removed from beef heart ETP$_H$ (25–50 g of starting material) by extraction with trichloroacetic acid and acid–acetone, by slight modifications of the analytical method described. The extracted residue is then digested with trypsin and chymotrypsin as in Kearney's procedure; residual proteins are removed by precipitation with 5% (w/v) trichloroacetic acid in the cold, and the flavin dinucleotides are hydrolyzed to the FMN level by incubation for 18–20 hours at 38°. The material is then applied to a Florisil column exhaustively washed with water, 5% (w/v) acetic acid, water, and 0.5% (v/v) pyridine and eluted with 5% (v/v) pyridine. Pyridine is removed by extraction with CHCl$_3$, and the aqueous phase is lyophilized. The flavin peptides are then chromatographed on Sephadex G-25, equilibrated with 0.01 M NaCl. Two flavin bands are obtained: the flavin peptides in the major band are concentrated and subjected to acid hydrolysis (6 N HCl under N$_2$ for 12 hours at 95°). If the chromatography on Sephadex is omitted, extensive conversion of flavin peptides to free flavin occurs at this stage due to the presence of tryptophan in peptide impurities, which reduces succinate dehydrogenase flavin to riboflavin. This appears to be avoided by prior removal on Sephadex of tryptophan containing peptides. The hydrolyzate is evaporated to dryness, taken up in water, and chromatographed on a Sephadex G-10 column.

Applications

The determination of flavin peptide content in tissues such as heart, where all covalently bound flavin belongs to succinate dehydrogenase, provides an unambiguous means of determining the true succinate dehydrogenase content and the stoichiometry of succinate dehydrogenase to other members of the respiratory chain. This is of considerable importance since the dehydrogenase is notoriously unstable in the soluble state and since it occurs[25] in both activated (fully active) and unactivated forms in all cells examined; hence reliance on catalytic activity alone may lead

[24] M. Hansen and A. L. Smith, *Biochim. Biophys. Acta* **81,** 214 (1964).
[25] E. B. Kearney, *J. Biol. Chem.* **229,** 363 (1957).

to erroneous results.[26] Succinate dehydrogenase also appears to be the main or sole source of covalently bound flavin in aerobically grown yeast cells, since anaerobically grown cells and "petite" mutants, which are devoid of the enzyme, contain only traces of covalently bound flavin.[22] On the other hand, there is some evidence that in mammalian liver mitochondria part of the covalently bound flavin may originate from sarcosine and dimethylglycine dehydrogenases[27] and amine oxidase.[27a] The presence of covalently-linked FAD in rat and beef liver (and kidney) monoamine oxidase has been confirmed in the San Francisco laboratory[11d] and, on the basis of a greatly improved procedure for the isolation of the enzyme, flavin peptides originating from this oxidase have been obtained which show significantly different properties than those obtained from succinate dehydrogenase.

Since the chemical determination of flavin peptides includes all forms of the enzyme: fully activated, *un*activated, and *in*activated, it has provided an insight into the mechanism of the reconstitution of the respiratory chain from soluble dehydrogenase and alkali-inactivated respiratory chain, and has permitted the determination of the turnover number of the enzyme even in crude preparations.[28-30] An interesting application of the method has been the demonstration that the flavin peptide core of succinate dehydrogenase is formed during the biogenesis of mitochondria considerably earlier than the catalytically active enzyme.[22]

[26] T. P. Singer, *in* "Comprehensive Biochemistry," Vol. 14 (M. Florkin and E. Stotz, eds.), p. 127. Elsevier, Amsterdam, 1966.

[27] W. R. Frisell and C. G. Mackenzie, *J. Biol. Chem.* **237**, 95 (1962).

[27a] I. Igaue, B. Gomes, and K. T. Yasunobu, *Biochem. Biophys. Res. Communs.* **29**, 562 (1967).

[28] T. P. Singer, J. Hauber, and O. Arrigoni, *Biochem. Biophys. Res. Commun.* **9**, 150 (1962).

[29] T. Kimura, J. Hauber, and T. P. Singer, *Nature* **198**, 362 (1963).

[30] P. Cerletti, R. Strom, M. G. Giordano, F. Balestrero, and M. A. Giovenco, *Biochem. Biophys. Res. Commun.* **14**, 408 (1964).

[143] Isolation and Purification of Flavin Peptides

By PAOLO CERLETTI,[1] GIULIO MAGNI, and GIULIO TESTOLIN

The flavin prosthetic group of mammalian succinate dehydrogenase (EC 1.3.99.1) is covalently bound to the protein. Only proteolytic digestion separates the flavin from the protein core and solubilizes a dinucleotide

[1] In partial fulfillment of a grant by the National Research Council of Italy (CNR).

bound to peptides of various dimensions.[2-4] Peptide-bound flavins from other sources are mentioned in the literature.[5-8] None of them, however, is sufficiently documented.

Preparation Method

Principle. Purified beef heart succinate dehydrogenase is digested with proteolytic enzymes. The soluble peptides thus produced are fractioned by solvent extraction and gel filtration until the peptide bound flavin is separated from all other nonflavin peptides.[9]

Reagents

> Phenol and *p*-cresol, freshly distilled under reduced pressure over Zn powder, stored in the dark
> Benzene
> Caprylic acid
> Diethyl ether, distilled
> Buffer: 50 mM K phosphate–50 mM K borate, pH 7.6
> Tris buffer: 0.1 M, pH 8.5, containing 10 mM CaCl$_2$
> Acid ammonium sulfate: 2.6 M (NH$_4$)$_2$SO$_4$ solution, adjusted to pH 3 with H$_2$SO$_4$
> (NH$_4$)$_2$SO$_4$, finely ground
> Trichloroacetic acid (TCA), 100% and 1%
> HCl, 1 N and 1 mM
> KOH, 6 N and 1 N

Step 1. Proteolysis. Succinate dehydrogenase from beef heart is purified.[10] The second ammonium sulfate precipitate from several preparations (about 2 g protein) is collected in 50 mM phosphate–50 mM borate buffer, pH 7.6; TCA is added (final concentration 10%), and the suspension is stirred 1–2 minutes at 2° and then centrifuged 15 minutes at 20,000 g. The sediment washed with 1% TCA is suspended in 10 mM Tris buffer containing 10 mM CaCl$_2$, pH 8.5, and pronase (10–15 mg per gram of protein)

[2] D. R. Green, S. Mii, and P. M. Kohout, *J. Biol. Chem.* **217,** 551 (1955).

[3] E. B. Kearney, *J. Biol. Chem.* **235,** 865 (1960).

[4] T. F. Chi, Y. L. Wang, C. L. Tsou, Y. C. Fang, and C. H. Yu, *Sci. Sinica (Peking)* **14,** 1193 (1965).

[5] V. N. Boukine, *Proc. Intern. Congr. Biochem., 3rd. Brussels,* 1955 p. 61 (1956).

[6] W. R. Frisell and C. G. Mackenzie, *J. Biol. Chem.* **237,** 94 (1962).

[7] C. Bhuyaneshwaran and T. E. King, *Biochim. Biophys. Acta* **132,** 282 (1967).

[8] T. P. Singer, *in* "Biological Oxidations" (T. P. Singer, ed.), p. 339. Wiley (Interscience), New York, 1968.

[9] P. Cerletti, G. Magni, C. Rossi, and G. Testolin, *6th FEBS Meeting, Madrid 1969, Abstracts of Communs.,* p. 150.

[10] D. V. Dervartanian and C. Veeger, *Biochim. Biophys. Acta* **92,** 233 (1964).

is added. Digestion is done in a stoppered flask under N_2 at 40° under continuous shaking. A separate tube containing a few drops of benzene and caprylic acid is inserted in the flask to prevent microbial growth.

The pH is checked at intervals (1 hour at first, then longer) and is adjusted with 6 N KOH. When it does not change any more (about 2 days), some more pronase is added (1–2 mg/g protein), and the digestion is continued until the pH remains constant (about 1 day).

About 2.5–3.0 micromoles of soluble flavin peptide per gram of protein digested are obtained.

Step 2. Chromatography on Sephadex. The clear mixture is centrifuged 15 minutes at 100,000 g. The supernatant is separated, concentrated under reduced pressure to 6 ml and made 2.6 M with solid $(NH_4)_2SO_4$. The precipitate is washed with acid ammonium sulfate. The combined washings and supernatant are applied to a column (2.7 × 150 cm) of Sephadex G-15 (particle size 40–120 μ) equilibrated with acid ammonium sulfate. Elution is done with the same solution at a rate of 30 ml/hour.[11]

Flavin peptides appear in two separate peaks (Table I). Recovery in this step is 58%, of which about 20% is in the first peak. The peaks are separately treated in the subsequent purification steps.

Step 3. Extraction with p-Cresol and Hydrolysis. The tubes of each peak are combined, solid ammonium sulfate is added to 0.7 saturation, and flavin peptides are extracted with p-cresol (10 subsequent aliquots, each

TABLE I

PURIFICATION OF FLAVIN PEPTIDES FROM BEEF HEART SUCCINATE DEHYDROGENASE

Fraction	Total amount (μ moles)	Yield (%)	R 280/445	Purity[a]	Purification at each step
Pronase digest (from 1840 mg protein)	4.75	100	25.3	0.06	1
Sephadex G-15: Peak I	0.85	18	2.42	0.62	10
Peak II	1.83	38	2.81	0.53	8.8
After acid hydrolysis (combined Sephadex peaks)	2.07	44	—	—	—
Biogel eluate	1.26	26	1.5	1	1.7

[a] Purity with respect to nonflavin peptides and other compounds that absorb in the ultraviolet is given as the reciprocal ratio between values of R 280/245 at each step and the value in the Biogel eluate, in which the absence of contamination was shown by chromatography and electrophoresis. This is an approximate estimate.

[11] The elution of flavin peptides is followed by measuring fluorescence excited at 466 nm and emitted at 528 nm and by reading absorbancies at 280 nm and 450 nm.

0.2 the volume of the aqueous phase[12]). An equal volume of ether is added to the combined cresol layers, and flavin peptides are extracted with water (10 subsequent aliquots, each 0.1 the volume of the cresol-ether layer). Recovery in this step is 80%. R 280/450 cannot be determined because of the presence of traces of p-cresol. Paper chromatography shows, however, that a great amount of contaminating peptides and amino acids are eliminated at this step.

The combined aqueous phases are concentrated under reduced pressure to 10 ml, acidified to pH 1 with HCl, and hydrolyzed 18 minutes at 100°. No loss of flavin is observed at this step.

Step 4. Chromatography on Biogel. The hydrolyzate is adjusted to pH 3 with 1 N KOH and is applied to a column (3.2 × 170 cm) of Biogel P2 (200–400 mesh) equilibrated with 1 mM HCl. Elution is performed with 1 mM HCl.[12]

Flavin peptides appear in one peak which in its descending part may show a broad shoulder and tail. This reflects incomplete digestion by pronase. Paper and thin-layer chromatography and the R 280/445 value show that at this stage flavin peptides are free from other peptides and from amino acids. The terminal fractions of the peak, however, may be still contaminated. These fractions are purified by submitting them again to chromatography. In this step recovery is 50% in the uncontaminated fractions, 10% in the contaminated ones.

Step 5. Chromatography on Silica Gel. The uncontaminated fractions eluted from Biogel are concentrated under reduced pressure to 0.5 ml and are applied to a column (1 × 10 cm) of silica gel G equilibrated with phenol–water (75:25, by volume). Flavin peptides are eluted with the same solvent in two separate peaks. The tubes in each peak are combined, an equal volume of ether is added, and flavin peptides are extracted with water (5 subsequent aliquots each 0.2 the volume the phenol ether layer). Recovery in the combined peaks is 95%.

Table I summarizes a typical run.

Determination

Flavin peptides are quantified spectrophotometrically or fluorometrically. In the first case, the spectrum is recorded and the tracing from 600 to 530 nm is extrapolated to 445 nm. The value obtained is subtracted from the absorbancy at 445 nm experimentally measured, and the difference is used for calculating the concentration. The fluorometric determination is based on the fact that flavin peptides emit fluorescence only

---

[12] If necessary the extractions are continued until no more fluorescence appears in the aqueous phase at visual inspection with an ultraviolet lamp.

in weakly acidic medium. The procedure is described elsewhere in this volume [133].

The purity of a preparation may be estimated by calculating the ratios of absorbancies at wavelengths corresponding to maxima and to minima in the pure compound and comparing them to the proper values of the pure flavin peptide (Table II). The ratio between absorbancies at 280 and

TABLE II

Maxima, Minima, and Ratios of Absorbancies of Flavin Peptides in 30 mM Phosphate Buffer, pH 6.85

Maxima (nm)			Minima (nm)		
442	350	266	388	298	237
Ratios:	$R\dfrac{350}{442} = 0.82$	$R\dfrac{266}{442} = 3.30$	$R\dfrac{388}{442} = 0.37$	$R\dfrac{298}{442} = 0.20$	$R\dfrac{237}{442} = 0.60$

at 445 nm, R 280/445, can be taken as a measure of the purity of the preparation with respect to nonflavin peptides. The value R 280/445 = 1.5 based on the extinction coefficients of the purest extractive preparations can be assumed to correspond to pure flavin peptide. Higher values indicate contamination by nonflavin peptides and other impurities that absorb in the ultraviolet.

Paper electrophoresis is used to assay purity with respect to free flavins. The presence of nonflavin peptides and of amino acids is assayed by paper or thin-layer chromatography. The following solvents are used in a single run or in combination: (1) n-butanol–pyridine–acetic acid–water (90:60:18:72, by volume); (2) n-butanol–acetic acid–water (60:20:20 by volume); (3) phenol–water (75:25, by volume). Solvent 1 followed in a second dimension by solvent 2 is routinely used to check for contaminating peptides and amino acids during purification. Preparations of flavin peptide with different peptide length are resolved in solvents 2 and 3.

The flavin peptides are located by visual inspection with an ultraviolet lamp. If necessary the fluorescence is enhanced by exposing the chromatogram to fumes of concentrated HCl. Ninhydrin is then sprayed to locate peptides or amino acids.

Properties

Spectra. Absorption spectra of flavin peptides are similar to those of free flavins. The maxima and minima in the visible, however, are shifted to shorter wavelength (Table II). The peak at 445 nm is bleached by dithionite. The molar absorbancy at 445 nm of pure flavin peptide is assumed to be as for free flavins 11.3×10^3 cm^2 mole^{-1} at the dinucleotide level[3,13] and 12.2×10^3 cm^2 mole^{-1} at the mononucleotide level.[3,14]

At mononucleotide level the spectra are not much influenced by the pH. In alkaline medium there is minor shift to longer wavelengths in the ultraviolet.[15,16,16a]

Fluorescence. The excitation spectrum (uncorrected) shows maxima at 286, 366, and 466 nm. The fluorescence emitted (uncorrected) has a maximum at 528 nm (excitation 366 and 466 nm).[15,17]

Flavin peptides fluoresce only in acid medium. The quenching of fluorescence varies in peptides obtained by digesting succinate dehydrogenase with different proteolytic enzymes. Peptides prepared with trypsin and chymotrypsin have pH–fluorescence curves with a sharp maximum at pH 2.7, and the midpoint in the slope of quenching is about pH 4.6. Peptides prepared with pronase (a mixture of endo- and exopeptidases[18]) have a broad maximum from pH 3 to pH 4, and the midpoint in the slope of quenching is about pH 5.5. In both cases, at neutral pH practically no fluorescence is measured.[15,16]

The behavior of fluorescence is the same at the dinucleotide and at the mononucleotide level except for minor variations in the position of the slope of quenching.[19] The length of the peptide chain has an influence on the quenching of fluorescence in neutral medium. Flavin peptides at the mononucleotide level, treated 24 hours at 105° with 30 mM HCl or with leucine aminopeptidase undergo a partial hydrolysis of the peptide chain and are moderately fluorescent also at neutral pH.[9,16]

Changes in the polarity of the medium (e.g., obtained by adding alcohols to a low ionic strength buffer) greatly reduce the internal quenching of flavin adenine dinucleotide[20] but scarcely influence the quenching of fluorescence at neutral pH in flavin peptides at the mononucleotide level.[9]

Structure. The native flavin peptide is a dinucleotide containing 5′-adenylic acid joined by a pyrophosphate bond to the flavin phosphate.[3,4] Positions 7 and 8 in the isoalloxazine nucleus are substituted.[15] The peptide

[13] O. Warburg and W. Christian, *Biochem. Z.* **298**, 150, 377 (1938).

[14] L. G. Whitby, *Biochem. J.* **54**, 437 (1953).

[15] P. Nànàsi, P. Cerletti, G. Magni and E. Nemes Nànàsi, *6th FEBS Meeting, Madrid 1969, Abstracts of Communs.*, p. 73.

[16] P. Cerletti, *in* "Flavins and Flavoproteins" (H. Kamin, ed.), University Park Press, in press.

[16a] P. Cerletti and G. Magni, submitted to *FEBS letters.*

[17] D. F. Wilson and T. E. King, *J. Biol. Chem.* **239**, 2683 (1964).

[18] Y. Narahashi, K. Shibuya, and M. Yanagita, *J. Biochem.* (*Tokyo*) **64**, 427 (1968).

[19] P. Cerletti and R. Strom, *Nature* (*London*) **198**, 1094 (1963).

[20] O. A. Bessey, O. M. Lowry, and R. M. Love, *J. Biol. Chem.* **180**, 755 (1949).

is covalently bound to the isoalloxazine nucleus.[3,4] There is evidence indicating that the site of binding is at position 8 (benzenoid ring).[4,15,21]

The number of amino acids in the peptide depends on the enzymes used in digesting succinate dehydrogenase, and even purified preparations may be inhomogeneous with respect to the length of the peptide bound to the flavin. Amino acids separated and identified with an automatic analyzer in peptides prepared by pronase digestion from beef heart succinate dehydrogenase are, per mole of flavin: Ser, Thr, Ala, Gly, Leu, Val, Asp or Asp-NH$_2$, Pro, and 2 Glu or Glu-NH$_2$.[9] In tryptic chymotryptic digests, Ser, Thr, Ala, Gly, Glu (or Asp), and Val were identified by paper chromatography.[3] The partial sequence: flavin—Ser [Gly Glu (or Glu-NH$_2$) Asp (or Asp-NH$_2$)]—[Gly Glu (or Glu-NH$_2$) Asp (or Asp-NH$_2$) Ala Thr CySO$_3$H]—was suggested on the basis of separations by electrophoresis–chromatography for flavin peptides in the pronase digest of the pig heart enzyme.[4]

[21] P. Hemmerich, A. Ehrenberg, W. M. Walker, L. E. G. Eriksson, J. Salach, P. Bader, and T. P. Singer, *FEBS Letters* **3**, 37 (1969).

[144] Isolation of Nekoflavin

By Kunio Matsui

Nekoflavin is a flavin compound that is found in cat eyes.[1] Its structure has not yet been determined, but it is thought that the N^9-side chain of nekoflavin differs from that of riboflavin. A flavin compound (or compounds) which has the same R_f values in chromatography as nekoflavin is also found in cat liver, frog liver, and the eyes of the dog and the frog. It is not known yet whether or not it is really nekoflavin. The content of nekoflavin in the cat eye is very small (about 8% of the total flavins in absorbance), and pure crystals of nekoflavin have not yet been obtained. However, it is not so difficult to obtain a preparation from cat eyes which does not contain any other fluorescent impurities. Cat liver is an unsuitable source of nekoflavin, because it contains many fluorescent substances and the procedures of purification are tedious.

Identification

Thin-layer chromatography on silica gel is suitable for the identification of nekoflavin. Paper chromatography is also useful. Not only the chromatography of nekoflavin itself, but also that of the oxidation product of

[1] K. Matsui, *J. Biochem.* (*Tokyo*) **57**, 201 (1965).

TABLE I
R_f VALUES OF NEKOFLAVIN AND RELATED COMPOUNDS

Solvent	Thin-layer chromatography[a]			Paper chromatography[b]			
	BAW[c]	AMAW	APAW	SP	BA	SPP	BAP
Nekoflavin	0.37	0.29	0.42	0.43	0.53	—	—
Riboflavin	0.51	0.46	0.60	0.33	0.44	—	—
Riboflavinyl glucoside	0.30	0.21	0.30	0.46	0.58	—	—
FMN	0.07	0.04	0.09	0.58	0.87	—	—
FAD	—	—	—	0.47	0.89	—	—
Oxidation product of neko-flavin with periodic acid	0.74	0.70	0.68	0.18[d]	—	0.48	0.50
Oxidation product of ribo-flavin with periodic acid	0.76	0.80	0.76	0.14[d]	—	0.41	0.39

[a] At 30°. Wakogel B-5 (silica gel for thin-layer chromatography containing 5% gypsum) was used.

[b] Ascending method at 27°. Toyoroshi No. 51A, which is similar to Whatman No. 1, was used. Data (except those of oxidation products) from K. Matsui, *J. Biochem. (Tokyo)* **57**, 204 (1965).

[c] Abbreviation: BAW = n-butanol–acetic acid–water, 20:1:4 by volume. AMAW = isoamyl alcohol–methyl ethyl ketone–acetic acid–water, 16:16:2:5 by volume; APAW = isoamyl alcohol–pyridine–acetic acid–water, 30:8:5:6 by volume; SP = 5% $Na_2HPO_4 \cdot 12\ H_2O$; BA = 5% H_3BO_3; SPP = 5% $Na_2HPO_4 \cdot 12\ H_2O$–90% phenol, 15:1 by volume; BAP = 5% H_3BO_3–90% phenol, 15:1 by volume.

[d] Spot is highly tailing.

nekoflavin with periodate is important, because the reaction product of nekoflavin is different from that of ordinary riboflavin derivatives.

The oxidation reaction of flavin with periodate is performed in a capillary or on a glass plate. Several microliters of an aqueous solution of the sample are mixed with about one-third the volume of a 20% (w/v) aqueous solution of periodic acid. After 2 hours the reaction mixture is spotted on a silica gel plate or paper and developed. Spots are detected by their greenish yellow fluorescence under an ultraviolet lamp. R_f values are shown in Table I.

Isolation

Principle. Flavins are extracted from the choroids of cat eyes with dilute perchloric acid and are adsorbed on Florisil. They are eluted with dilute pyridine. Nekoflavin is separated from riboflavin and purified by column chromatography on powdered cellulose. The first solvent system is n-butanol–formic acid–water, the second is n-butanol–formic acid–water–ether, and the third is isoamyl alcohol–methyl ethyl ketone–acetic acid–water.

Procedure. Cat eyes, which have been pooled in a frozen state (about $-5°$) for a week, can be used. The connective tissues, conjunctiva, and muscles are removed from the eyeballs. The sclerotic and choroid are cut along the equator with scissors. The vitreous body is removed, and the inner surface of the choroid is wiped lightly with absorbent cotton to remove the retina. The choroid is detached and taken off from the sclerotic with a small pincette. The choroids are pooled in a frozen state at $-20°$.

It is preferable to perform the following procedures in a dim place or under the light of a sodium lamp. Seventy-five grams of the pooled choroid (from about 1000 eyeballs) are suspended in 400 ml of water. The suspension is heated at $80°$ for 5 minutes under gentle stirring, and homogenized in a Waring blendor for 10 minutes. (Thorough homogenization is essential for good extraction of flavins.) The homogenate is heated again at $80°$ for 30 minutes and is cooled with water to room temperature. Ten and a half milliliters of 60% perchloric acid are added to the homogenate, which is centrifuged at 6500 g for 5 minutes. The black precipitate is suspended in 400 ml of water. The suspension is neutralized with 1 M sodium hydroxide to pH 5.0–7.0 (about 15 ml) and heated at $80°$ for 30 minutes. It is then cooled to room temperature with water. Ten and a half milliliters of 60% perchloric acid are added to the suspension, which is then centrifuged. The precipitate is further extracted six times as described above. The supernatants are combined.

The following procedures should be performed under the light of a sodium lamp. The extract is passed through a column of Florisil (2.5 × 15 cm), which is washed successively with 400 ml of 0.2 M perchloric acid, 700 ml of water, and 300 ml of 0.5% pyridine. Then flavins are eluted with 5% pyridine (3 ml/minute). The greenish yellow fluorescent fractions are combined (about 350 ml). Pyridine in the eluate is extracted four times with 100 ml of chloroform. The aqueous layer is evaporated to dryness in a rotary evaporator at a temperature lower than $60°$. The residue is dissolved in 20 ml of water. The turbid solution is saturated with ammonium sulfate. The flavins are extracted into four 5-ml aliquots of liquid phenol. The extracts are combined and mixed with 10 ml of water and 130 ml of ether. The layer of ether is washed four times with 5 ml of water. The water layer and washings are combined and washed with 100 ml of ether, and dried *in vacuo*.

The residue is extracted repeatedly with 2-ml aliquots of *n*-butanol–80% formic acid–water (77:12:11 by volume) until the greenish yellow fluorescence of the extract becomes faint. (The aqueous phase is separated from the organic phase, which is passed through a column, and repeated extraction with solvent diminishes the aqueous phase.) The extracts are passed successively through a column of powdered cellulose (1.5 × 45 cm). Flavins are eluted with the same solvent (about 2 ml/minute and 8 ml/fraction) and are analyzed by absorption at 450 mμ and fluorescence under an ultraviolet lamp. The first large, greenish yellow fluorescent peak contains riboflavin. (The fore part of the peak contains a blue fluorescent substance.) The second peak, which is smaller than the first and exhibits greenish yellow fluorescence, contains nekoflavin. The latter peak (about 90 ml) is mixed with 5 volumes of petroleum ether. The organic phase is washed three times with 10 ml of water. The aqueous phase and washings are combined and dried *in vacuo* over potassium hydroxide and concentrated sulfuric acid. The residue is dissolved in about 1 ml of water and 1 drop of concentrated aqueous ammonia, and dried again *in vacuo*. (The residue contains a small amount of flavin compounds which are thought to be formyl esters of nekoflavin and are hydrolyzed by ammonia.) This preparation weighs 35–40 mg and contains 12–15 absorbance units of flavins at 445 mμ.

The residue, which is pooled and contains about 23 absorbance units of flavins at 445 mμ, is extracted repeatedly with 2-ml aliquots of *n*-butanol–80% formic acid–water–ether (77:12:11:38 by volume), and is added successively to a column of powdered cellulose (1.3 × 45 cm). Flavins are eluted with the same solvent. The main peak (about 140 ml) is eluted after two small ones. It is treated with petroleum ether (700 ml) and water (10 ml × 3), dried, and treated with dilute ammonia, and dried as described above. The yield is about 18 absorbance units at 445 mμ and weighs about 21 mg.

The orange-yellow residue is dissolved in isoamyl alcohol–methyl ethyl ketone–acetic acid–water (8:8:1:2.5 by volume) as described above, and is added to a column of powdered cellulose (1.3 × 45 cm). The flavins are eluted with the same solvent. The main peak (about 100 ml) is treated with petroleum ether (500 ml) and with water (10 ml × 3), dried, and treated with dilute ammonia and dried again as described above. The residue contains about 14 absorbance units of nekoflavin at 445 mμ and weighs about 1.6 mg. The preparation is orange yellow and contains scarcely any fluorescent impurities. It contains fine crystalline material, but it is not perfectly pure.

Properties

Nekoflavin has the general properties of a flavin. It is reduced by sodium dithionite losing fluorescence and is reoxidized by oxygen restoring fluorescence. The absorption spectrum is very similar to that of riboflavin (Table II). It is stable in 0.05 M sodium hydroxide at room temperature for 18 hours and also in 1 M hydrochloric acid at 100° for 2 hours. The photodegradation reaction of nekoflavin in alkaline solution produces similar products as riboflavin. The characteristic reaction that differentiates nekoflavin and riboflavin is the oxidation with periodate (see Table I).

TABLE II
ABSORPTION MAXIMA (mμ) OF NEKOFLAVIN AND RIBOFLAVIN

	In 0.1 M HCl	In water	In 0.1 M NaOH
Nekoflavin	222, 267, 370, 440	222, 267, 368, 442	217, 271, 354, 445
Riboflavin	223, 267, 376, 446	223, 268, 375, 447	222, 271, 358, 452

[145] Analogs of Riboflavin

By JOHN P. LAMBOOY

The following description of the preparation of analogs[1] of riboflavin has been limited to three examples. These examples utilize procedures of considerable general usefulness, and if they are employed a number of analogs can be prepared. Two of the examples make use of an aldose for the incorporation of the side chain at position 10; the other makes use of a glycamine for this purpose. The selection of the three examples was also influenced by other considerations. Interest in the use of radioactive riboflavin justifies a description of its preparation. The preparation of diethylriboflavin and 7-methyl-8-chloroflavin is justified on the following grounds. The former is, in the classical sense, the most potent competitive inhibitor of riboflavin in the rat, and it is equivalent to riboflavin for *Lactobacillus casei* throughout limiting concentrations. The latter is the most potent competitive inhibitor of riboflavin for *L. casei*, but it is almost devoid of any form of activity in the rat.[2]

Figure 1 illustrates the two numbering systems used for the flavins. Formula A shows the system employed by *Chemical Abstracts;* it will be used in this presentation. A large part of the literature, especially the older references, made use of the system shown by Formula B. The numbering system used will be immediately apparent in any case because the designation of the position of the R groups as being at 9 or 10 differentiates the two.

[1] The term analog is used to include homologs of riboflavin as well.
[2] A comprehensive coverage of the synthesis of isoalloxazines (flavins) has been presented by Lambooy, The Alloxazines and Isoalloxazines, *in* "Heterocyclic Compounds" (R. C. Elderfield, ed.), Vol. 9, Chap. 2. Wiley, New York, 1967.

(A)   (B)

FIG. 1. The two numbering systems used for flavins. (A) *Chemical Abstracts* system. (B) Other systems used especially in the older literature.

I. Analogs with N^{10}-Side Chain Derived from Aldoses

A. The Use of Anilines. Preparation of Radioactive Riboflavin[3]

1. *Preparation of 3,4-Dimethylaniline-N-D-ribopyranoside[4]*

A 500-ml flask is fitted with a sintered-glass filter in an adapter so that a solution can be drawn into the flask by suction. The assembly is carefully freed of foreign material and thoroughly dried. 3,4-Dimethylaniline is purified before use by distillation under diminished pressure (b.p. 106–107°/11 mm). The distillate, which should be a solid at room temperature, is then recrystallized from *n*-hexane to produce material that melts at 46°–48°. The solid aniline, 25 g, is added to the flask and the assembly is arranged for suction filtration. D-Ribose, 30 g, 400 ml of absolute alcohol,[5] and 5 drops of 12 N sulfuric acid are placed in another flask; the flask is stoppered and the contents heated on a hot plate until the ribose is in solution. The sugar solution is added to the aniline by filtering it through the sintered filter. The solution is permitted to cool to room temperature, and then refrigerated overnight. The white riboside is filtered, washed with 10–15 ml of cold absolute ethyl alcohol, and dried *in vacuo*. The yield is 28.0 g (56%) of material melting at 115°–116° (dec).[6]

2. *Preparation of N-D-Ribityl-3,4-dimethylaniline*

Six grams of 3,4-dimethyl-*N*-D-ribopyranoside, 9 g of palladium on calcium carbonate,[7] and 300 ml of absolute ethyl alcohol are placed in a hydrogenator bottle. The mixture is hydrogenated at 60 psi and 70° for 3–4 hours.[8] The catalyst is removed by filtration, the filtrate passing di-

---

[3] E. E. Haley and J. P. Lambooy, *J. Am. Chem. Soc.* **76**, 2926 (1954).

[4] L. Berger and J. Lee, *J. Org. Chem.* **11**, 84 (1946).

[5] The yield is sensitive to the degree of dryness of the alcohol.

[6] Prepared in this way the product is uncontaminated and suitable for analysis. All melting points are corrected.

[7] R. Kuhn and R. Strobele, *Chem. Ber.* **70**, 773 (1937).

[8] Some ribosides of different structure require 24 hours.

rectly into a 500-ml flask as in 1, above. The flask is flushed with nitrogen, sealed, and refrigerated overnight. The product is filtered and recrystallized from 50% ethyl alcohol to yield 3.0 g (50%) of material melting at 141°–142°.

3. Preparation of 1-(D-Ribitylamino)-2-p-tolylazo-4,5-dimethylbenzene[9]

p-Toluidine, 5.9 g, is dissolved in a mixture of 100 ml of glacial acetic acid, 12.5 ml of water, and 11.2 ml of concentrated hydrochloric acid; the solution is cooled in a salt–ice bath. Solid sodium nitrite, 3.5 g, is added slowly with stirring. After all the sodium nitrite is dissolved, 8.4 g of N-D-ribityl-3,4-dimethylbenzene is added with stirring, to the diazonium salt solution (dark green color formed). When all the compound has been added, a solution of 2.6 g of sodium hydroxide in 15 ml of water is added dropwise with stirring over a period of 30 minutes (brown color formed). The stirred mixture is kept at 8°–10° for 2 hours. Water, 100 ml, is added and the solution is extracted 4 times with 100-ml portions of ether. The combined ether extracts are washed with 10% sodium bicarbonate solution until neutral and then washed 4 times with water. As the acetic acid is removed from the extract, the azo compound precipitates. The precipitate is collected on a filter, and the ether is evaporated to produce the remainder of the product. The combined product is recrystallized from 95% alcohol to yield 7.2 g (59%) of material melting at 162°–165°.

4. Preparation of Radioactive Riboflavin

This procedure makes use of barbituric acid-2-^{14}C to incorporate the radioactive carbon atom at the 2 position of the riboflavin. The level of radioactivity desired in the final product must be considered in the selection of barbituric acid-2-^{14}C of sufficient specific activity.

A 100-ml three-necked flask is equipped with a stirrer, a reflux condenser, and a stopper in the center and side necks, respectively. The equipment is assembled so that the contents can be heated to refluxing temperature while being stirred; the reaction is carried out in dim light. The desired mixture of radioactive and inactive barbituric acid, 2.20 g, is added to 50 ml of n-butyl alcohol[10] which is being stirred in the flask. The tolylazo compound, 6.30 g, is next added slowly so that all is thoroughly suspended. Glacial acetic acid, 8 ml, is then added, the side arm is stoppered, and the mixture is heated to the refluxing temperature while stirring is continued. At first, the reaction mixture is a solution, but after a variable

[9] M. Tishler, K. Pfister, R. D. Babson, K. Ladenburg, and A. J. Fleming, J. Am. Chem. Soc. 69, 1487 (1947).
[10] C. H. Shunk, J. P. Lavigne, and K. Folkers, J. Am. Chem. Soc. 77, 2210 (1955).

length of time the product begins to precipitate. Refluxing and stirring is continued for $2\frac{1}{4}$ hours. The reaction mixture is refrigerated overnight. The product is collected on a filter, the transfer being accomplished by the use of the filtrate. The precipitate is washed with 30 ml of cold n-butyl alcohol and dried in air to yield 4.9 g of crude flavin. The 4.9 g of crude flavin is triturated with 50 ml of water on the steam bath for 30 minutes. While still hot, the product is collected on a filter, washed twice with 10-ml portions of methyl alcohol, and dried in air to yield 4.4 g. The 4.4 g of flavin is dissolved by heating to boiling with 20 ml of 18% hydrochloric acid to which a little decolorizing charcoal has been added. While still hot, the mixture is carefully filtered by suction through a sintered filter. The accumulated precipitate is washed on the filter with 12 ml of 18% hydrochloric acid. To the combined filtrate is added 64 ml of hot water, and the mixture is cooled in the dark to room temperature and then refrigerated overnight. The product is collected on a filter, washed with 10 ml of water and then 10 ml of methyl alcohol, and dried in a desiccator to yield 3.88 g (61%) of riboflavin-2-^{14}C melting at 296°–298° (dec, uncorr, rapid heating to 250°).

B. The Use of Nitroanilines. Preparation of 6,7-Diethylriboflavin[11]

1. *Preparation of 3,4-Diethylbromobenzene*

A 1000-ml three-necked flask is equipped with a stirrer in the center neck and a reflux condenser in one side neck and an addition funnel in the other. The flask is surrounded by a water bath. o-Diethylbenzene, 200 g, 4.8 g of No. 40 mesh iron filings, 1.2 g of *ferrum reductum*, and a crystal of iodine are added to the flask. The water bath is heated to 30°–35°, and 210 g of bromine are added with stirring at such a rate as to maintain a gentle but continuous evolution of hydrobromic acid. Stirring is continued for 30 minutes after the last of the bromine has been added. The reaction product is poured into 1 liter of water and filtered to remove the unreacted iron. The oil layer is extracted into 400 ml of ether; the ether solution is washed successively with two 200-ml portions of 3% sodium bicarbonate solution and once with 200 ml of water.

The product is steam distilled until no more oil is found in the condensate. The water phase of the distillate is extracted with ether, and the oil phase of the distillate is added to the ether extract. The ether solution is dried over anhydrous sodium sulfate and filtered; the ether is removed by distillation. The residue is distilled through a 30-cm fractionating column. Only that portion boiling at 122°–125°/20 mm is collected as product to yield 204 g (65%).

[11] J. P. Lambooy, *J. Am. Chem. Soc.* **72**, 5225 (1950).

2. 3,4-Diethylaniline

A stainless steel hydrogenation bomb is used for this reaction. An extremely minute amount of copper is plated out on the walls of the bomb, but this is easily removed with a brush and dilute nitric acid. The bomb can then be used for precious metal reductions. To the bomb are added 127 g of 3,4-diethylbromobenzene, 657 ml of 28% ammonium hydroxide containing 12 g of cuprous chloride,[12] and 14 g of flattened and cleaned copper wire. The sealed bomb is heated to 195°–200° with continuous rocking for 18 hours. The product is removed from the bomb, and an ether wash is used to remove the last traces of product. The reaction mixture is steam distilled. The distillate is made acid by the addition of 100 ml of concentrated hydrochloric acid. The acid solution is extracted with ether to remove a small amount of unreacted bromo compound. The acid solution is made alkaline by the addition of sodium hydroxide, and 1000-ml portions are extracted three times with 100 ml of ether. The combined ether extracts are dried over anhydrous sodium sulfate and after filtration, the solvent is removed by distillation. The residue is distilled under reduced pressure, and that portion which distills at 129°–131° at 20 mm yields 72 g (81%).

If a high pressure hydrogenation bomb is not available, 3,4-diethylaniline can be prepared from 3,4-diethylnitrobenzene,[13] but in this case the 3,4-diethylnitrobenzene requires careful fractionation.

3. 3,4-Diethylacetanilide

The 72 g of 3,4-diethylaniline from the above reaction is treated cautiously with 72 ml of acetic anhydride; the mixture is refluxed for 2 minutes and permitted to cool to room temperature. If the product has not solidified, it is poured into a liter of cold water; if it has solidified, some water is added, and the mass broken is up and transferred to 1 liter of water. The solid product is thoroughly subdivided in the water and left overnight. The product is collected on a filter, sucked as dry as convenient, dissolved in 475 ml of 50% alcohol, and refrigerated overnight. The precipitated product is collected on a filter and sucked until all of a small amount of oil has passed through the filter. The product is dissolved in 450 ml of 50% alcohol and refrigerated overnight. The product resulting from this recrystallization is of sufficient purity to use. The yield is 64 g (70%) of material melting at 117°–118°.

4. Preparation of 4,5-Diethyl-2-nitroacetanilide

A 500-ml three-necked flask is equipped with an efficient stirrer in the center neck, a thermometer, the bulb of which reaches below the surface

---

[12] This procedure was devised by W. A. Wisansky and S. Ansbacher, *Org. Syn. Coll. III*, p. 307, for the preparation of 3,4-dimethylaniline.

[13] J. P. Lambooy, *J. Am. Chem. Soc.* **71**, 3756 (1949).

of the contents in one side arm, and a stopper in the other. The flask is surrounded by a salt–ice bath. A mixture of concentrated nitric acid, 185 ml and 70 ml of concentrated sulfuric acid is added to the flask, and the contents are cooled to $-2°$ to $-4°$. The 3,4-diethylacetanilide, 40 g, is ground to a fine powder and 4 g of the powder is added to the rapidly stirred acid mixture. The nitration mixture becomes brownish-orange immediately, but after some minutes the color changes to bright yellow orange. The second addition of 4 g is not made until this color change has taken place (usually 5–10 minutes). The additions are made (approximately 1/10 of total each time) until the 40 g has been added. Stirring is continued for 30 minutes after the last of the acetanilide has been added. The temperature does not exceed $0°$.

When the nitration is finished, the product is poured over 2 liters of crushed ice in a 2-liter beaker. The product, which solidifies, is left in water overnight and is then collected on a filter. The precipitate is dissolved in 500 ml of ether and washed in succession with two 100-ml portions of water, once with a 100-ml portion of 0.5 N sodium hydroxide,[14] and once with a 100-ml portion of water. The well-settled ether layer is evaporated on a steam bath, the residue is dissolved in 75 ml of ethanol, and the solution is refrigerated overnight. The product which melts at $74°–75°$ is collected on a filter, sucked as dry as convenient, and immediately recrystallized from 70 ml of ethanol as above. The precipitated product weighs 27 g[15] (55%) and melts at $75°–77°$.

5. *Preparation of 4,5-Diethyl-2-nitroaniline*

The above nitroacetanilide, 30 g, in 500 ml of 50% alcohol is heated to $80°$ on a steam bath. A warm solution of 56 g of sodium hydroxide in 112 ml of water is added with stirring. Heating is continued for 15 minutes, and the reaction mixture is refrigerated overnight. The product is collected on a filter, where it is washed with water until free of alkali and then spread to dry in air. The nitroaniline weighs 23.6 g (96%) and melts at $65–66°$. If the melting point is as low as $63°–64°$, the material should be recrystallized from the minimum amount of 50% ethyl alcohol.

6. *Preparation of 4,5-Diethyl-2-nitroaniline-N-D-ribopyranoside*[4]

D-Ribose, 4.5 g, and 50 ml of absolute ethyl alcohol are warmed in a stoppered bottle until the sugar is in solution. The anhydrous solution is filtered into a second flask with as little atmospheric exposure as possible. An additional 10 ml of absolute alcohol is used to wash the first flask and

---

[14] If the first alkali wash is not alkaline, washing should be repeated until it is.
[15] An additional 2 g is available by careful workup of the mother liquors for an overall yield of 59%.

the filter. The second flask is capped and the contents cooled to room temperature. One drop of 6 N sulfuric acid is added and quickly followed by 6.0 g of the 4,5-diethyl-2-nitroaniline. The nitroaniline is dissolved by gently swirling, and the flask is left at room temperature for 3 hours and then refrigerated for 48 hours. The product is collected on a filter, pressed quickly, washed with 10 ml of cold absolute alcohol, and dried *in vacuo*, to yield 6.46–7.35 g (67–75%) of material melting at 171°–173° (dec).

7. *Preparation of 7,8-Diethylriboflavin*

a. *Preparation of Primary Sodium Borate Solution.* Boric acid, 4.28 g, is dissolved in 200 ml of water. Three grams of sodium hydroxide is dissolved in 25 ml of water. The sodium hydroxide solution is added to the boric acid solution until the pH is 11.1. The resultant solution is diluted to 400 ml.

b. *Preparation of 4,5-Diethyl-2-amino-(1'-ribityl)aniline.* The following materials are added to a stainless steel hydrogenation bomb: 57 ml of the primary sodium borate solution prepared above, 15 ml of water, 300 ml of ethanol, 5.00 g of the ribopyranoside prepared above, and 5.00 g of palladium on calcium carbonate. The reduction is carried out for 6 hours at 70°–80° and 600 psi.

The hydrogenation bomb is opened, and 200 mg of ascorbic acid added immediately. The contents are transferred to a flask, and the flask is flushed with nitrogen and stoppered. The product is filtered from the catalyst directly into a 1-liter flask to be used to evaporate the solvents at 35° under nitrogen. During the evaporation, periodic additions of absolute alcohol are made in order to reduce foaming. Finally, three successive additions of 50 ml of absolute alcohol are made and evaporated, the product thus being brought to dryness.

c. *Preparation of 7,8-Diethylriboflavin.* To the flask containing the dry phenylenediamine is added 60 ml of glacial acetic acid; the flask is swirled in warm water to dissolve the material. Boric acid, 2.9 g, and alloxan, 2.9 g, are suspended in 60 ml of warm glacial acetic acid and added to the flask. The flask is shaken for 30 minutes at 40°–50° and then placed in the dark for 3 days.

The contents of the flask are evaporated to dryness; three successive additions of 70 ml of absolute alcohol are made, and each is evaporated in turn. To the product is added 500 ml of 5% acetic acid, and the contents are heated to boiling for 2–3 minutes. The hot solution is filtered, and the filtrate is refrigerated overnight. The crystalline product is collected on a filter and then immediately dissolved with boiling in 600 ml of water, filtered, and the filtrate refrigerated. The product is collected on a filter to yield 1.64 g (26%) of flavin which melts at 275°–280° (dec, uncorr, rapid heating to 250°).

II. Analogs with N^{10}-Side Chain Derived from Glycamines

A. Preparation of 7-Methyl-8-chlororiboflavin[16]

1. *Preparation of 3-Chloro-4-methylacetanilide*

3-Chloro-4-methylaniline, 72 g, is treated exactly as described above for 3,4-diethylaniline (I, B, 3). The crude product is left in water overnight, then collected on a filter and dried in air to constant weight. The dried material is dissolved by boiling in 360 ml of benzene, 360 ml of *n*-hexane is added, and the solution is left at room temperature overnight. The product is collected on a filter and dried in air to yield 78 g (83%) of material melting at 101°–106°.[17]

2. *Preparation of 3-Chloro-4-methyl-2-nitroacetanilide*

3-Chloro-4-methylacetanilide, 40 g, is treated exactly as described for 3,4-diethylacetanilide above (I, B, 4). The product is suspended in water, left overnight, then collected on a filter; it is resuspended in a liter of 5% sodium bicarbonate solution, and stirred occasionally for 1 hour. If the water gives an acid reaction, the material is resuspended in a fresh bicarbonate solution; if the water gives an alkaline reaction, the material is collected on a filter, washed with water, and sucked as dry as convenient. The product is dissolved by boiling with 330 ml of 95% alcohol, treated with decolorizing charcoal, filtered, and refrigerated overnight. The product is collected on a filter to yield 44–45 g (65%) of material melting at 116°.

3. *Preparation of 3-Chloro-4-methyl-2-nitroaniline*

The above nitroacetanilide, 30 g, is treated exactly as described for 4,5-diethyl-2-nitroaniline above (I, B, 5). After being refrigerated overnight the material is filtered and washed by resuspension in water until neutral. It is then collected on a filter and air dried to yield 22 g (90%) of product melting at 167°–169°. This material does not need to be recrystallized, but if this is desired it can be dissolved in 200 ml of hot benzene, and 200 ml of *n*-hexane added and the solution refrigerated.

---

[16] E. E. Haley and J. P. Lambooy, *J. Am. Chem. Soc.* **76**, 5093 (1954).

[17] This material shows dimorphism. If the above material had been recrystallized from benzene or alcohol, for example, the m.p. would have been 83°; if from pet ether or ligroin its m.p. would have been 105°–106°. The use of the above mixed solvents is convenient.

4. *Preparation of 2,4-Dichloro-5-nitrotoluene*

A 1000-ml three-necked flask is equipped as described above (I, B, 4). Concentrated sulfuric acid, 225 ml, is added to the flask and 20.5 g of solid sodium nitrite is added in small portions with stirring until all is in solution. The temperature may go as high as 70° during this time. The contents of the flask are cooled to 30°.

The above nitroaniline, 50 g, is dissolved in 550 ml of glacial acetic acid on the hot plate. This solution is poured slowly into the nitrite solution while the latter is surrounded by an ice bath. The temperature may go as high as 40° during the addition. Stirring is continued for 30 minutes while the following solution is prepared.

Cuprous chloride, 59 g, is dissolved in 550 ml of concentrated hydrochloric acid in a 4-liter beaker. The beaker is surrounded by a cold water bath and, while the contents are stirred vigorously, the diazonium salt solution is poured into it through a small-bore side-arm adapter. The mixture is stirred while the temperature of the contents is slowly raised to 60° on a steam bath. When nitrogen gas is no longer evolved, 1.5 liters of water is added and the mixture is refrigerated overnight. The product is collected on a filter, the solid lumps are mashed in a mortar, and the precipitate is washed by suspension in water until neutral, collected on a filter, and dried in air. The dry product is dissolved in 300 ml of *n*-hexane and refrigerated. The precipitate is collected on a filter and dried to yield 36–37 g (68%) of material melting at 52°–54°.

5. *Preparation of* D-*Ribamine*[18]

D-Ribose, 25.0 g, 165 ml of methyl alcohol, and 17.9 g (18.2 ml) of benzylamine are added to a hydrogenator bottle; after the ribose has dissolved, the contents are left at room temperature for 24 hours. Platinum oxide, 400 mg is added to the flask and the contents are hydrogenated at 60 psi and room temperature for 24 hours. At the end of this time 3.5–3.7 liters of hydrogen will have been consumed. If a Parr hydrogenation apparatus is used, this corresponds to a pressure fall of from 13.3 to 13.7 psi. The solution is filtered from the catalyst, concentrated at reduced pressure at 50° to approximately 80 ml, and 250 ml of ethyl acetate is added. The reaction product is refrigerated for 24–48 hours. The crystalline *N*-benzyl-D-ribamine is collected on the filter, washed with a cold mixture of 3 ml of methyl alcohol and 9 ml of ethyl acetate, and dried in the air to yield 30 g (75%) of material melting at 99°–101°.

---

[18] This procedure is essentially that described by J. Davoll and D. D. Evans, *J. Chem. Soc.*, p. 5041 (1960). The procedure is especially well suited to small-scale preparations, and the product requires no purification.

The 30 g of *N*-benzyl-D-ribamine, 300 ml of methyl alcohol, and 10 g of 5% palladized charcoal are added to a hydrogenator bottle and hydrogenated at 60 psi, at room temperature for 24 hours. At the end of this time, approximately 3.0 liters of hydrogen will have been consumed. If a Parr hydrogenation apparatus is used, this corresponds to a pressure drop of 11.0–11.3 psi.[19] The solution is filtered from the catalyst, the catalyst is washed with a little methyl alcohol, and the solvent is removed under reduced pressure at 50°. Water, 50 ml, is added to the flask and evaporated as above; the addition and evaporation of water is repeated to ensure the removal of the toluene formed during the reduction. Absolute ethyl alcohol, 100 ml, is added and evaporated as above. The product is obtained as a clear, water white, viscous syrup and may be assumed to be the theoretical amount (19 g) of D-ribamine. This material can be used in the next step without purification.[20]

6. *Preparation of 2-Nitro-4-methyl-5-chloro-N-D-ribitylaniline*

D-Ribamine, 24 g, 12 g of the 2,4-dichloro-5-nitrotoluene, and 300 ml of pyridine are heated under reflux in a nitrogen atmosphere for 10 hours. The reaction mixture is refrigerated overnight, and then the supernatant is poured from the unreacted D-ribamine. The solution is evaporated under vacuum to remove as much pyridine as possible. The residue is extracted 4 successive times by warming to boiling with 100 ml of *n*-hexane to remove unreacted 2,4-dichloro-5-nitrotoluene. The residue is dissolved in 225 ml of hot methyl alcohol, filtered, and refrigerated overnight. The product is collected on a filter, pressed, washed with 10 ml of cold methanol, and air dried. The product is obtained in a yield of 6.2 g (33%) and melts at 169°–170°. This material is dissolved in 60 ml of hot methanol, and after refrigeration a yield of 5.3 g (28%) of material melting at 170°–171° is obtained.

7. *Preparation of 7-Methyl-8-chlororiboflavin*

To a hydrogenator bottle are added 5.12 g of the above 2-nitroribitylaniline, 200 mg of platinum oxide, 88 ml of glacial acetic acid, and 20 ml of water; the mixture is hydrogenated at room temperature and 60 psi for 2 hours or until the solution has become colorless. The reaction mixture is filtered from the catalyst in such a way that the filtrate is collected directly into a solution of 2.9 g of alloxan and 5.92 g of boric acid in 275 ml

---

[19] Usually more hydrogen is consumed. When shaking is continued over a period of 24 hours care must be exercised to correct for temperature changes, small leaks, and hydrogen adsorption by the catalyst. The reaction is successful if approximately the correct amount of hydrogen is consumed.

[20] If other glycamines are desired, the procedure described by C. H. Winestock and G. W. E. Plaut, *J. Org. Chem.* **26**, 4456 (1961) should be consulted.

of glacial acetic acid. The catalyst is washed with 10 ml of acetic acid. The mixture is heated under reflux for 5 minutes and placed in the dark for 2 days. The contents of the flask are evaporated to dryness under reduced pressure, and the residue is dissolved by boiling with 1700 ml of 2% acetic acid, filtered, and refrigerated for 48 hours. The precipitate is collected on a filter and dried to yield 4.75 g (74%) of the flavin melting at 280° (dec). The above flavin is dissolved in 1700 ml of 2% acetic acid and refrigerated as above. The precipitate is collected on a filter and dried to yield 4.18 g (66%) of flavin melting at 290° (dec, uncorr, fast heating to 250°).

[146] Fat-Soluble Riboflavin Derivatives

By KUNIO YAGI

Fat-soluble riboflavin derivatives were synthesized by esterifying alcoholic hydroxyl groups in the ribitol moiety of riboflavin with some carboxylic acid.[1-3] The basic idea of the synthesis is to modify the properties of riboflavin without losing the intactness of the structure of the isoalloxazine nucleus of riboflavin. As expected, the solubility of riboflavin is markedly changed by this esterification; the esters obtained are soluble in organic solvents. A noticeable solvent effect is found in the absorption and fluorescence emission spectra of these riboflavin derivatives.[4,5] The observed results may be useful in interpreting the environmental change surrounding the isoalloxazine nucleus of flavin coenzyme upon complex formation with apoprotein and further with substrate or substrate-substitute.[6]

In addition, nutritional experiments revealed that some of these riboflavin derivatives act as a so-called "deposit" type of the vitamin.[7-9] This may be attributed chiefly to the fact that the ester absorbed without hydrolysis is hydrolyzed slowly by some esterase(s) in mammals. In this connec-

[1] K. Yagi, J. Okuda, and A. A. Dmitrovskii, *J. Biochem.* **48**, 621 (1960).
[2] K. Yagi, J. Okuda, A. A. Dmitrovskii, R. Honda, and T. Matsubara, *J. Vitaminol.* (*Kyoto*) **7**, 276 (1961).
[3] A. Kotaki and K. Yagi, *J. Vitaminol.* (*Kyoto*) **14**, 247 (1968).
[4] A. Kotaki, M. Naoi, J. Okuda, and K. Yagi, *J. Biochem.* **61**, 404 (1967).
[5] K. Yagi, N. Ohishi, M. Naoi, and A. Kotaki, *Arch. Biochem. Biophys.* **134**, 500 (1969).
[6] A. Kotaki, M. Naoi, and K. Yagi, *J. Biochem.* **59**, 625 (1966).
[7] K. Yagi, J. Okuda, and M. Kobayashi, *J. Vitaminol.* (*Kyoto*) **9**, 168 (1963).
[8] K. Yagi, J. Okuda, and T. Matsubara, *J. Vitaminol.* (*Kyoto*) **10**, 275 (1964).
[9] A. Kotaki, K. Kato, M. Okumura, T. Sakurai, and K. Yagi, *J. Vitaminol.* (*Kyoto*) **14**, 253 (1968).

tion, the ester is considered to be one of synthetic substrates of lipase,[10,11] and the characteristic properties of the riboflavin moiety may render some advantages in analyzing the enzymatic hydrolysis of this synthetic substrate.[12] Thus, the ester may be useful also for the study on the mode of action of lipase.

Chemical Synthesis of Fat-Soluble Riboflavin Derivatives

For synthesizing acyl derivatives of riboflavin, reagents such as an acid chloride or acid anhydride can be effectively used.[1-3] Three examples of the synthetic procedures of fat-soluble riboflavin derivatives are described below.

Synthesis of Riboflavin Tetrapalmitate[1,2]

In a 500-ml round-bottomed three-necked flask fitted with an air-tight stirrer, a dropping funnel, and a reflux condenser closed with a calcium chloride tube, is placed riboflavin (1 g, 2.7 millimoles) suspended in 300 ml of dried chloroform–pyridine mixture (1:1, v/v); a mixture of palmitoyl chloride (15 g, 55 millimoles) and 50 ml of dried chloroform is added dropwise at 0° over the course of an hour with vigorous stirring. After gentle stirring at 33° for 14 hours, a transparent yellowish orange solution thus obtained is evaporated *in vacuo*. To the residue is added 100 ml of 90% aqueous pyridine solution; the mixture is heated at 70° for 15 minutes to decompose the remaining palmitoyl chloride to palmitic acid. The solution is evaporated *in vacuo*. The residue is dissolved in 100 ml of methanol at room temperature, and the solution is stored in a refrigerator. The precipitate formed is then dissolved by warming in a minimal volume of absolute ethanol, and is stored in a refrigerator. About 500 mg of chromatographically pure riboflavin tetrapalmitate is obtained as yellow powder after this procedure has been repeated three times (m.p. 78.5°).

Synthesis of Riboflavin Tetrabutyrate[2]

Riboflavin tetrabutyrate can be synthesized using butyric anhydride.[13] Dried riboflavin (1 g, 2.7 millimoles) is suspended in butyric anhydride

---

[10] K. Yagi, Y. Yamamoto, and J. Okuda, *Nature* **191**, 174 (1961).

[11] K. Yagi, J. Okuda, S. Niwa, and Y. Yamamoto, *J. Vitaminol. (Kyoto)* **7**, 281 (1961).

[12] K. Yagi, N. Ohishi, and M. Osamura, *Compt. rend. Soc. Biol.*, in press.

[13] Without the use of acid chloride or acid anhydride, the following method is also applicable and is especially recommended for large-scale synthesis. Butyric acid (10 ml, 0.109 mole), dried pyridine (10 ml, 0.125 mole), and *p*-toluene sulfonyl chloride (10 g, 0.054 mole) are mixed in a round-bottomed flask provided with a reflux condenser closed with a calcium chloride tube. To the mixture, riboflavin (2 g, 5.4 millimoles) is added with vigorous stirring. After heating at 85° for 3 hours, the reaction mixture is poured into 700 ml of ice-water. The solid material is collected and dried. About 3 g of riboflavin tetrabutyrate is obtained as yellow powder.

(5 ml, 30.6 millimoles) in a round-bottomed flask equipped with an air-tight stirrer and a reflux condenser with a calcium chloride tube, and then heated in an oil bath at 130°–140° for 6 hours with stirring. After heating, the solution is evaporated *in vacuo*. The residue is dissolved in a minimal volume of methanol, which is poured into an excess volume of ice-water. Crude riboflavin tetrabutyrate precipitated is collected and dried. For crystallization,[14] 1 g of the crude preparation is suspended in 8 ml of aqueous methanol (methanol–water, 19:5, v/v) and heated at 80° to dissolve (in a water bath). The dark brown solution thus obtained is filtered through a glass filter (G-3) and placed in an incubator at 25° overnight. Reddish yellow, rhombic crystals appear. The crystallization is repeated three times (m.p. 145°–147°).

Synthesis of Riboflavin Tetranicotinate[3]

Nicotinyl chloride monohydrochloride (17.8 g, 0.1 mole) is mixed with a fine powder of dried riboflavin (9.4 g, 0.025 mole) and transferred into a 200-ml round-bottomed, three-necked flask equipped with an air-tight stirrer, a dropping funnel, and a reflux condenser connected with a calcium chloride tube. The mixture is heated to 120° in an oil bath, and 16 g (0.203 mole) of dried pyridine is added dropwise with stirring in a period of 10 minutes. The mixture is further heated for 90 minutes with vigorous stirring. After cooling, the reaction mixture is dissolved cautiously in 1 N HCl and made pH 5 with a saturated solution of Na_2HPO_4. The yellow precipitate formed is collected, washed with cold water and dried. The brownish yellow powder thus obtained is dissolved in chloroform, and the insoluble material is discarded. The yellow precipitate (15 g) is obtained by adding a large volume of ethyl ether to the chloroform solution. This precipitate is crystallized from benzene (m.p. 147°–150°).

Physicochemical Properties of Fat-Soluble Riboflavin Derivatives

Solubility. Riboflavin tetrapalmitate is freely soluble in various organic solvents such as ethanol, pyridine, benzene, ethyl ether, chloroform, and neutral fat, but is insoluble in water.[2]

Riboflavin tetrabutyrate is easily soluble in organic solvents such as chloroform, acetone, ethanol, dioxane, carbon tetrachloride, ethyl ether, and benzene.[2] However, its solubility in water is very low: 0.0024% at 30°.[14]

On the other hand, riboflavin tetranicotinate is fairly soluble in water and is easily soluble in various organic solvents, such as chloroform, pyridine, methanol, and dioxane, but it is scarcely soluble in carbon tetrachloride and in ethyl ether. It is practically insoluble in saturated hydro-

---

[14] K. Yagi, H. Ōhama, Y. Takahashi, and J. Okuda, *J. Vitaminol. (Kyoto)* **13**, 191 (1967).

TABLE I
SOLUBILITY OF RIBOFLAVIN TETRANICOTINATE IN VARIOUS SOLVENTS

Solvent	Solubility (%, w/v)
Water	0.0204 (20°); 0.24 (85°)
Benzene	0.046 (23°)
Acetone	Easily soluble
Ethanol	1.95 (23°)
Isopropanol	0.32 (23°)
Methanol	Easily soluble
Chloroform	Easily soluble
Dioxane	Easily soluble
Soybean oil	0.0083 (25°)

carbons, but is soluble in aromatic hydrocarbons, such as benzene and toluene (see Table I).[3]

Visible and Near-Ultraviolet Absorption Spectra.[3,14] By changing the solvent from water to a nonpolar one, the absorption spectrum of riboflavin tetranicotinate is modified as indicated in Fig. 1A,[3] which is essentially similar to that observed for 3-methyllumiflavin by Harbury et al.[15] In water, the absorption maxima of riboflavin tetranicotinate are, in accord with riboflavin, found at 267, 374, and 445 mμ. With the decrease in the polarity of the solvent, fine structure appears in the two peaks of longer wavelengths (Fig. 1A). The absorption band around 370 mμ shifts to the blue and is split into 2, and that around 450 mμ into 3 (compare curve I with curve III in Fig. 1A). These effects can also be observed in the case of riboflavin tetrabutyrate.[14,16] Such observations are seen to be useful to elucidate the environment surrounding the isoalloxazine chromophore and the mode of binding in flavoproteins. In the case of D-amino acid oxidase, the appearance of fine structure is restricted to only one side of the peaks, around 370 mμ.[6] It would be interpreted that the transition around 370 mμ occurs in the hydrophobic region of the protein. Upon complex formation with benzoate, a substrate-substitute, the splitting in the peak around 370 mμ disappears accompanying the appearance of fine structure around 450 mμ peak. It would be interpreted, therefore, that the environment that affects the transition around 370 mμ changes from hydrophobic to hydrophilic and the environment that affects the transition around 450 mμ changes from hydrophilic to hydrophobic. As to the fine structure

[15] H. A. Harbury, K. F. LaNoue, P. A. Loach, and R. M. Amick, *Proc. Natl. Acad. Sci. U.S.* **45**, 1708 (1959).
[16] J. Kozioł, *Photochem. Photobiol.* **5**, 41 (1966).

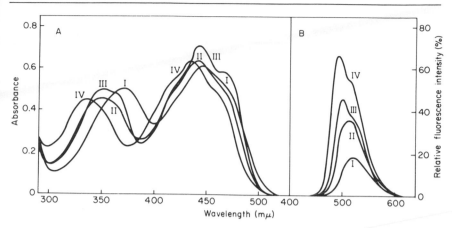

FIG. 1. Absorption and fluorescence emission spectra of riboflavin tetranicotinate in various solvents. (A) Absorption spectra (concentration of riboflavin tetranicotinate, $5.3 \times 10^{-5} M$). (B) Fluorescence emission spectra [recorded with a Hitachi MPF-2A fluorescence spectrometer (uncorrected). Excitation wavelength: 340 mμ]. Curves: I, in water; II, in methanol; III, in chloroform; IV, in dioxane.

around the 450-mμ peak, Massey and Ganther[17] had reached the same interpretation as above.

On the other hand, it may be expected that in the case of lipoyl dehydrogenase both the transitions are in the hydrophobic region.[18]

Fluorescence Spectra.[5] When the solvent is changed from water to organic solvent of low polarity, the fluorescence emission spectra of these esters shift to the blue accompanying the appearance of fine structure and the increase in relative fluorescence intensity. As shown in Fig. 1B, riboflavin tetranicotinate in water shows a fluorescence emission peak at 522 mμ,[18a] and the shape of the spectrum is similar to those of free riboflavin, FMN, and FAD. In chloroform, however, the relative fluorescence intensity increases, and the peak of the spectrum shifts to the blue accompanying the appearance of fine structure. With riboflavin tetrabutyrate, similar effects are also observed.[5,16]

Lipoyl dehydrogenase exhibits greater fluorescence than does free FAD; the emission maximum shifts slightly to the blue with the appearance of a shoulder.[19] These phenomena can also be interpreted as indicating that the flavin chromophore resides in a hydrophobic region of the protein.

Hydrolysis by Lipase. The esters are hydrolyzed by pancreatic lipase.

[17] V. Massey and H. Ganther, *Biochemistry* **4**, 1161 (1965).
[18] V. Massey, T. Hofmann, and G. Palmer, *J. Biol. Chem.* **237**, 3820 (1962).
[18a] Uncorrected value.
[19] L. Casola, P. E. Brumby, and V. Massey, *J. Biol. Chem.* **241**, 4977 (1966).

TABLE II

HYDROLYSIS OF FATTY ACID ESTERS OF RIBOFLAVIN BY PANCREATIC LIPASE[a]

Substrate	Rate of hydrolysis (%)
Riboflavin tetrapalmitate	1.2
Riboflavin tetracaprate	3.4
Riboflavin tetrabutyrate	45.4
Riboflavin tetrapropionate	4.3
Riboflavin tetraacetate	2.0

[a] The reaction mixture consisted of 0.5 ml of the substrate [(final concentration, $1 \times 10^{-3} M$) suspended in 1% polyvinyl alcohol (polymerization degree, 500)], 0.5 ml of enzyme solution (20 mg/ml), and 1.0 ml of McIlvaine buffer. It was incubated at pH 7.3 and 37° for 2 hours. The rate of hydrolysis was expressed as percent of riboflavin liberated from substrates per 2 hours.

The rate of hydrolysis is dependent on the structure of the carboxylic acid moiety. Among the esters of riboflavin with straight-chain fatty acids, riboflavin tetrabutyrate is most easily hydrolyzed, as shown in Table II.[10,11]

When riboflavin tetrabutyrate is incubated with pancreatic lipase, riboflavin mono-, di-, and tributyrate are found besides free riboflavin, indicating that the cleavage of these four ester bonds occurs stepwise.[12]

The rate of hydrolysis of riboflavin tetranicotinate is much less than that of riboflavin tetrabutyrate.[20]

Vitamin Effect. Among the fat-soluble riboflavin derivatives synthesized, riboflavin tetrabutyrate and riboflavin tetranicotinate were found to possess vitamin action,[7,9] though other esters scarcely showed it; riboflavin tetrapalmitate had no vitamin action.[7] Riboflavin tetrabutyrate and tetranicotinate may be hydrolyzed by enzymes, such as pancreatic lipase, or in a nonenzymatic manner so that some part of them, when they are administered orally, may be absorbed from the intestinal tract after hydrolysis to free riboflavin. However, a fairly large part of the administered esters could be absorbed without hydrolysis.[21,22]

Since these esters cannot always easily be excreted into urine, the time requirement for complete hydrolysis by some esterase(s) within the body of mammals may confer the "deposit" nature on these esters. When riboflavin tetrabutyrate[8] or riboflavin tetranicotinate[9] is administered orally or by injection, the blood flavin level increases and is maintained

[20] T. Tanaka, H. Tanaka, and T. Tsubaki, *Vitamins* **38**, 229 (1968).
[21] H. Ohkawa, A. Kotaki, and K. Yagi, *J. Vitaminol.* (*Kyoto*) **15**, 185 (1969).
[22] K. Yagi, Y. Yamamoto, I. Nishigaki, and H. Ohkawa, *J. Vitaminol.* (*Kyoto*), **16**, 247 (1970).

for a fairly long time, and a considerable delay in the excretion of flavin is observed.

Utilization of the riboflavin moiety of riboflavin tetrabutyrate was confirmed by showing its incorporation into both FMN and FAD of rat tissues by use of riboflavin-2-^{14}C-tetrabutyrate.[23]

[23] K. Yagi, M. Yamada, and J. Okuda, *J. Vitaminol. (Kyoto)* **15**, 155 (1969).

[147] Synthesis of 2-Substituted Riboflavin Analogs

By FRANZ MÜLLER

In the article by J. P. Lambooy[1] the three basic procedures for the synthesis of riboflavin are described. These procedures allow chemical variation of the isoalloxazine ring in positions 3, 6, 7, 8, 9, and 10, respectively. Riboflavin derivatives with substituents other than oxygen or sulfur in the 2-position cannot be obtained by these procedures. There is also a need for compounds modified in the pyrimidine part of the isoalloxazine ring in order to obtain a more specific insight into the structural requirement of flavins participating as coenzymes in biological reactions. This article describes the procedure for the preparation of such riboflavin analogs (see Scheme 1) which serve as starting material for the synthesis of the corresponding FAD and FMN analogs.[2]

Synthesis of Riboflavin Analogs[3]

2-Thioriboflavin (Ia) ($C_{17}H_{20}O_5N_4S \cdot \frac{1}{2} H_2O$, *mol. wt. 401.5*). 2-Thioriboflavin can be synthesized by the following two methods:

METHOD 1. Four grams (0.01 mole) of 2-phenylazo-4,5-dimethyl-*N*-ribitylaniline (II)[1] is suspended in 150 ml of glacial acetic acid in a round-bottomed flask with three necks. The reaction vessel is equipped with a condenser, a mechanical stirrer, and an N_2-inlet tube. The suspension is warmed to 53°, 3 g (0.021 mole) of 2-thiobarbituric acid (III) is added, and the reaction mixture is kept under nitrogen and with stirring for 8 hours at 53°. The reaction mixture is then filtered while warm, and the purple residue is washed successively with glacial acetic acid, ethanol, and ether. The product is suspended in 50 ml of 1 *N* NaHCO₃ solution and stirred at room temperature for 20 minutes to remove unreacted 2-thiobarbituric

---

[1] J. P. Lambooy, this volume [145].
[2] W. Föry and D. B. McCormick, this volume [148].
[3] F. Müller and P. Hemmerich, *Helv. Chim. Acta* **49**, 2352 (1966).

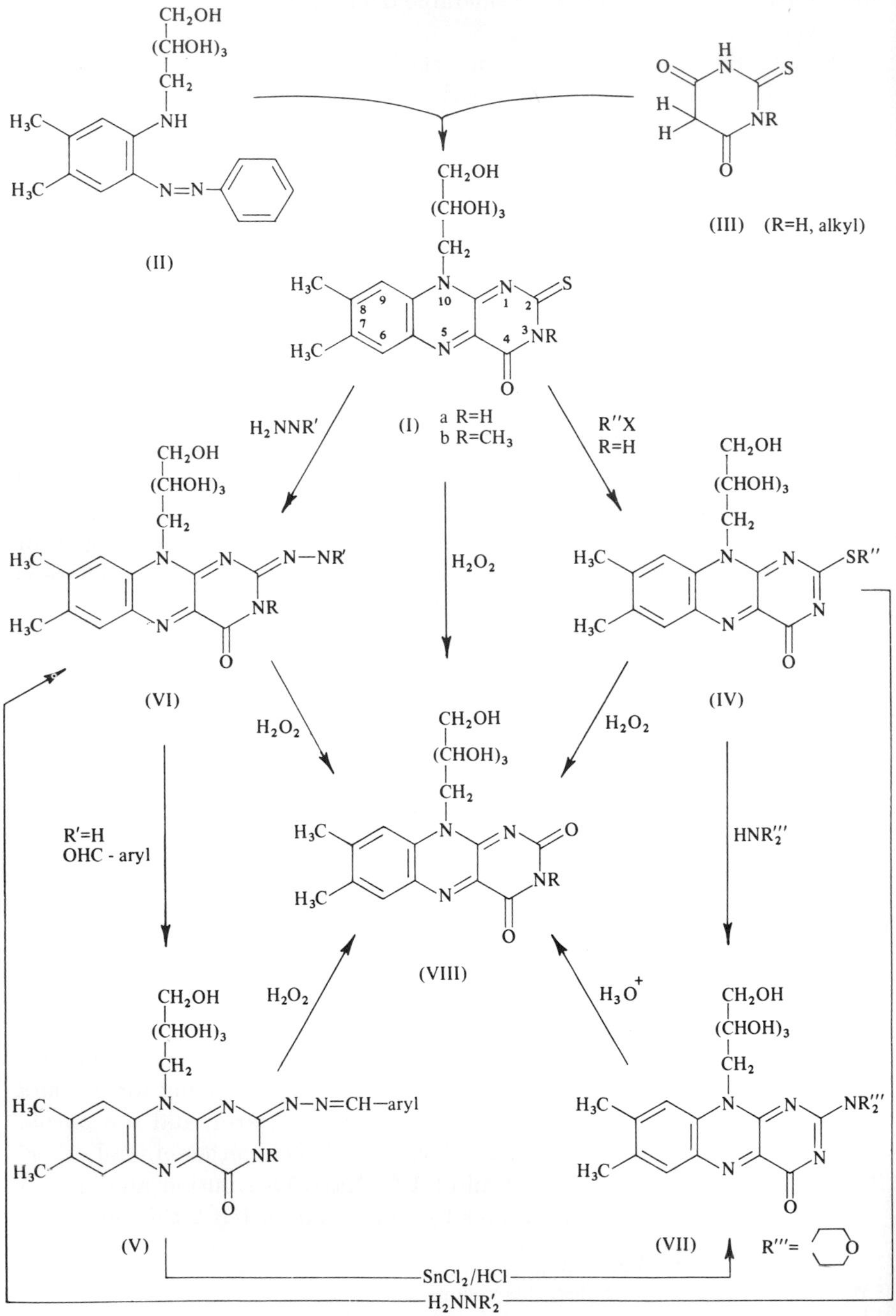

SCHEME 1

acid. The suspension is then filtered and the red product is washed successively with water, ethanol, and ether. The yield of compound (Ia) is 64%. Unreacted (II) can be recovered by a 3-fold dilution of the glacial acetic acid mother liquor with water. For recrystallization, if necessary, (Ia) is dissolved in warm 6 N HCl, treated with activated charcoal, and filtered. 2-Thioriboflavin crystallizes upon dilution with water. Compound (Ia) is very soluble in strong acids and bases, dimethyl formamide and dimethyl sulfoxide, but only slightly soluble in other organic solvents.

METHOD 2. Berezovskij et al.[4] synthesized (Ia) by suspending (II) and (III) in n-butyl acetate in the presence of a small volume of glacial acetic acid as a catalyst and refluxing the reaction mixture for 5 hours. The product is crystallized from conc. HCl and water. The yield reported by Berezovskij et al. is about the same as that obtained by method 1. 2-Thioriboflavin is sensitive to oxygen, hence the product obtained by this procedure is always contaminated with its oxidation product riboflavin.

N^3-Methyl-2-thioriboflavin (Ib) ($C_{18}H_{22}O_5N_4S \cdot H_2O$, mol. wt. 424.5). The procedure is the same as that described for 2-thioriboflavin (Ia) in method 1 except that 2-thiobarbituric acid is replaced by 1-methyl-2-thiobarbituric acid.[5] (Ib) shows the same chemical and physical properties as (Ia) except that (Ib) is insoluble in bases.

S-Methyl-2-thioriboflavin (IV, $R'' = CH_3$) ($C_{18}H_{22}O_5N_4S \cdot H_2O$, mol. wt. 424.5). Three-tenths gram (0.75 millimole) of 2-thioriboflavin (Ia) is suspended in 50 ml of absolute methanol, and 10 ml of iodomethane is added. The mixture is refluxed for 5 hours and then filtered while hot. The residue is washed with a small volume of warm 2 N perchloric acid; the filtrates are combined, treated with activated charcoal and filtered again. The filtrate is concentrated to 10 ml under reduced pressure at 30°. The resulting brown-red solution is brought carefully to pH 7 by adding small volumes of a saturated solution of sodium bicarbonate. The solution is allowed to stand at 0° for 12 hours; during this time orange-yellow crystals form. These are collected and washed successively with acetone and ether. The yield of (IV) ($R'' = CH_3$) is 52%. Compound (IV) is slightly soluble in water and alcohols, very soluble in diluted acids, and insoluble in ether. Many other derivatives (R'' being an alkyl or aryl group) can be synthesized by this procedure by replacing iodomethane by any iodo-or bromoaliphatic or bromoaromatic compounds.

2-Benzylazinoriboflavin (V) ($C_{24}H_{26}O_5N_6 \cdot HClO_4$, mol. wt. 578.5). Two-tenths gram (0.5 millimole) of 2-thioriboflavin (Ia) is suspended in 10 ml of

---

[4] V. M. Berezovskij and L. M. Melnikova, Zh. Obshch. Khim., 31, 3827 (1961).

[5] A. H. Cook, I. Heilbron, S. F. MacDonald, and A. P. Mahadevan, J. Chem. Soc. p. 1064 (1949).

hydrazine hydrate and allowed to react for 2 hours at room temperature with frequent stirring. The purple precipitate [2-riboflavinhydrazone (VI, R' = H$_2$)] is filtered off, washed with large volumes of ethanol, and dried with ether. The purple product is suspended in hot ethanol, and 6 N perchloric acid is added slowly until a clear yellow-orange solution is obtained. This is treated with charcoal and filtered. One drop of freshly distilled benzaldehyde is added to the filtrate. The solution is allowed to stand at room temperature for 6 hours and for a further 6 hours at 0°. The yield of (V) is 25.7% [brown needles of (V) as perchlorate].

2-Morpholyl-2-deoxyriboflavin (*VII*) (*C$_{21}$H$_{27}$O$_6$N$_5$·$\frac{1}{2}$ H$_2$O, mol. wt. 454.5*). Two-tenths gram (0.5 millimole) of S-methyl-2-thioriboflavin (IV) is suspended in 10 ml of morpholine, and the suspension is maintained at 50° for 2 hours with frequent stirring. The reaction mixture is diluted with an equal volume of acetone and then allowed to stand for 12 hours at 0°. The yellow-brown crystals obtained are dissolved in a small volume of 2 N perchloric acid, treated with activated charcoal, and filtered. The filtrate is neutralized with a saturated sodium bicarbonate solution. The solution is kept for 6 hours at 0°; during this time yellow-orange crystals of (VII) form. The yield of (VII) is 58.5%. Compound (VII) is the most water-soluble of the 2-substituted riboflavin derivatives.

To obtain imino analogs other than that described, morpholine is replaced by primary aliphatic or aromatic amines. If the amine is a solid, the reaction is carried out in dimethylformamide as a solvent, and the imino analog is precipitated by ether and crystallized as described above. If the amine is used in the salt form, it is necessary to add an equivalent amount of water-free triethylamine in order to liberate the base.

Peracetylation of the Riboflavin Analogs. One-tenth gram of the riboflavin analog is suspended in 10 ml of a mixture of glacial acetic acid and acetic anhydride (1:1, v/v); 0.5 ml of 6 N perchloric acid is added carefully during 10 minutes with stirring. The temperature should not exceed 40° throughout this procedure. A 4-fold volume of water is added to the clear solution, and the reaction mixture is extracted with chloroform. The separated organic phase is washed three times with water and then dried with anhydrous sodium sulfate. A small amount of activated charcoal is added, and the suspension is filtered. The filtrate is evaporated to dryness under reduced pressure at 30°. The crystals that form are collected. Yields are 70–80%.

The purity of the riboflavin analogs described is easily checked by thin-layer chromatography using solvent mixtures such as n-butanol–glacial acetic acid–water (3:1:1, v:v:v); n-butanol–ethanol–water (3:1:1, v:v:v); or n-butanol–2 N ammonia–water (3:1:1, v:v:v). The procedures described are very reproducible.

Discussion

2-Thioriboflavin (I) serves as starting material for the synthesis of various riboflavin analogs modified in position 2 of the isoalloxazine ring (Scheme 1). 2-Thioriboflavin in solution is easily oxidized to riboflavin by oxygen or peroxides. Hence the modified Tishler procedure working under N_2 and lower temperature yields a purer product than the procedure described by Berezovskij et al.[4]

Hydroxylamine and hydrazines $H_2NNR'_2$ (R' = H, alkyl, aryl), known as carbonyl reagents, react readily with 2-thioriboflavin and its S-methylated derivative (IV) forming flavoxime and flavohydrazones (VI). These derivatives are blue in their neutral form and deteriorate slowly. However, they are stable as salts. The hydrazones VI (R' = H) react with aldehydes to form the flavazines (V). The procedure described for their synthesis can also be applied to the preparation of flavazines involving aldehydes other than benzaldehyde. The flavazine (V) and hydrazone (VI) derivatives can be converted into the imino analog (VII) (R''' = H) by reduction with stannous chloride in 2 N hydrochloric acid at 60° and followed by neutralization of the reaction mixture.

2-Thioriboflavin does not react with amines to form the imino analogs (VII). However, its thioester (IV), obtained by alkylation of 2-thioriboflavin (Ia) in a nonaqueous solvent, reacts readily with amines to form the corresponding imine derivative (VII). The procedure given earlier for the conversion of the thioester (IV) into imino analogs (VII) can also be used for the preparation of compounds such as 2-hydroxyethyl- or phenylimino analogs employing hydroxyethylamine or aniline, respectively.

N^3-alkylated 2-thioriboflavin can be obtained only by condensation of the azo compound (II) with N^1-alkylated 2-thiobarbituric acid, since alkylation of 2-thioriboflavin leads to S-alkyl-2-thioriboflavin (IV).

PHYSICAL DATA OF MONO- AND DIVALENT SUBSTITUTED RIBOFLAVINS

Substituent in position 2	Number of substance	Absorption maxima (280–500 mμ) λ (mμ) ϵ (M^{-1}cm^{-1})		pK$_{LH+}^L$
		μ = 0 3, phosphate buffer, pH 7.0	μ = 0.3, sulfate buffer, pH 2.0	
=O	(VIII)	446 (12,000), 370 (10,000)	446, 370	0.2
=S	(Ia)	498 (13,000), 390 (500), 312 (23,000)	498, 390, 312	0.2
—SCH$_3$	(IV)	454 (16,000), 392 (15,000)	426, 286	4.4
—N⟨ ⟩O	(VII)	462 (9200), 371 (5900)	466, 396	4.2

The riboflavin analogs described are converted under mild conditions by acid hydrolysis or air oxidation to their parent compound riboflavin (VIII).

The derivatives with a univalent functional group in position 2 of the isoalloxazine ring are much more basic and therefore protonated at a higher pH than for example 2-thioriboflavin (see the table).[6,7] The effect of pH on these derivatives thus distinguishes them from other derivatives with similar spectra (see the table).

[6] K. H. Dudley, A. Ehrenberg, P. Hemmerich, and F. Müller, *Helv. Chim. Acta* **47**, 1354 (1964).
[7] F. Müller, W. Walker, and P. Hemmerich, *Helv. Chim. Acta* **49**, 2365 (1966).

[148] Chemical Synthesis of Flavin Coenzymes

By WERNER FÖRY and DONALD B. McCORMICK

General Techniques and Materials

All liquid reagents should be distilled before use. Pyridine is refluxed over sodium hydroxide for 4 hours and stored over sodium hydroxide. Aqueous solutions are concentrated under reduced pressure at low temperature (30° or less). For separation of the flavin mononucleotides and flavin adenine dinucleotides, the anion exchanger DEAE-cellulose is used in its chloride form. The columns are prepared by packing a thin slurry of the anion-exchange cellulose. Thin-layer chromatograms are run on MN Silica Gel S as stationary phase and developed in suitable solvents (noted at appropriate places in the text). Spots are detected by visual examination under ultraviolet light. All syntheses involving flavin compounds are carried out in a darkened room.

2-Thioriboflavin 5'-Phosphate[1]

Principle. 1-D-1'-Ribitylamino-3,4-dimethyl-6-phenylazobenzene 5'-phosphate[2] (I) is condensed with 2-thiobarbituric acid (II) to give 2-thioriboflavin 5'-phosphate (III) isolated as the monosodium salt.

2-Thioriboflavin 5'-Phosphate. 2-Thiobarbituric acid (II) (3.8 g, 27 millimoles) is suspended in 250 ml of a mixture of acetic acid–*n*-butyl alcohol (1:4). The monosodium salt of the monophosphoric acid ester of 1-D-1'-ribitylamino-3,4-dimethyl-6-phenylazobenzene[2] (I) (6.17 g, 13.4 millimoles) is added, and the reaction mixture is stirred under nitrogen

[1] W. Föry and P. Hemmerich, *Helv. Chim. Acta* **50**, 1766 (1967).
[2] L. A. Flexser, U. Montclair, and W. G. Farkas, U.S. Patent 2,610,176 (1952).

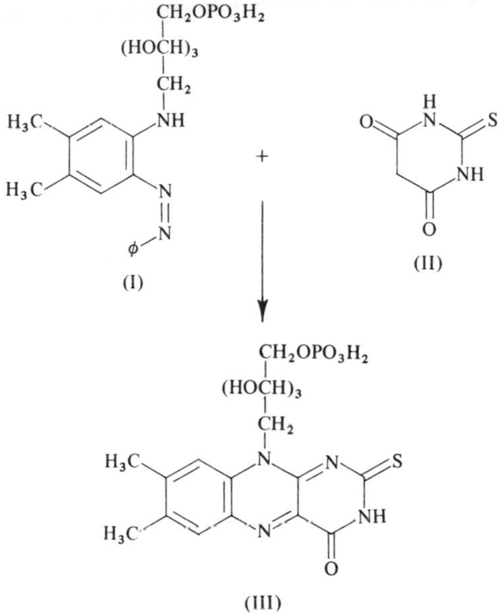

CH₂OPO₃H₂ structures...

$$\text{CH}_2\text{OPO}_3\text{H}_2$$
$$(\text{HOCH})_3$$
$$\text{CH}_2$$

(I) + (II)

↓

(III)

at 53°–55° for 36 hours. The mixture is kept at room temperature over-night; the dark red precipitate formed is collected on a filter and washed with a mixture of n-butyl alcohol–acetic acid (99:1), acetone, and ether. The crude product is suspended in dimethyl sulfoxide (150 ml) at 50°–60° and filtered; the filtrate is poured into vigorously stirred acetone (600 ml). The precipitate is collected on a filter, washed with acetone, and dried (4.4 g). This crude product is suspended in water (50 ml) and adjusted to pH 7.0 with 0.1 M sodium hydroxide; the slightly turbid solution is treated with decolorizing carbon and filtered. The filtrate is adjusted to pH 4.5 with 2 M acetic acid, treated again with decolorizing carbon, and filtered into ethyl alcohol (250 ml). The precipitate is kept for 1 hour at 0° and then collected on a filter, washed with 80% aqueous ethyl alcohol, anhydrous ethyl alcohol, and ether to give 2.7 g (31%) of (III). The product has one λ_{max} at 494 mμ (pH 7) with an ϵ of 20,800 and is chromatographically homogeneous in acetonitrile–water (7:3) and n-butyl alcohol–acetic acid–water (3:1:1).

2-β-Hydroxyethyliminoriboflavin 5′-Phosphate[1]

Principle. S-Methyl-2-thioriboflavin[3] (IV) is phosphorylated by mono-chlorophosphoric acid. Without further purification, the intermediate

[3] F. Müller and P. Hemmerich, *Helv. Chim. Acta* **49**, 2352 (1966).

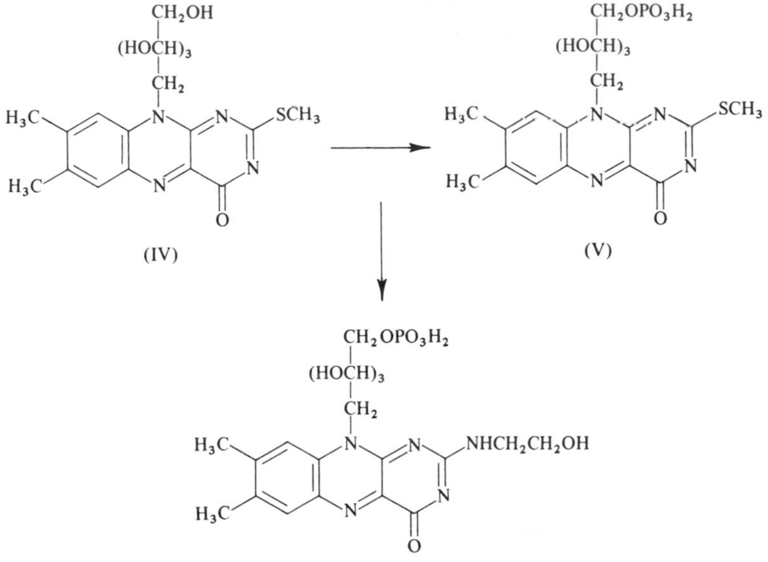

(IV) → (V)

(VI)

S-methyl-2-thioriboflavin 5'-phosphate (V) is reacted with β-hydroxyethyl-amine to give 2-β-hydroxyethyliminoriboflavin 5'-phosphate (VI).

S-Methyl-2-Thioriboflavin 5'-Phosphate.[1] Water (6.9 g, 37 millimoles) is added dropwise during 2–3 hours at 0° under nitrogen to phosphorus oxychloride (28.9 g, 18 millimoles) in a three-necked flask provided with condenser and drying tube. The reaction mixture is stirred for 2 hours at room temperature and then allowed to stand overnight. The mixture is stirred until all gas evolution has ceased. S-Methylriboflavin[3] (IV) (2 g, 4.7 millimoles) is added, and the mixture is stirred under nitrogen for 8–9 hours at room temperature. The solution is added dropwise to stirred anhydrous ether (500 ml) within 2–3 minutes. The temperature is main-tained at 5°–10° by external cooling. The suspension formed is stirred for 15 minutes at room temperature; the crude material is collected on a fritted Büchner funnel and washed with anhydrous ether (100 ml). The hygroscopic precipitate is dried *in vacuo* at room temperature and treated for 15–20 minutes in 30 ml of a mixture of dioxane–water (6:1). The solu-tion is treated with decolorizing carbon and filtered into ether (700 ml). The oily precipitate formed is dissolved in minimal amounts of methyl alcohol and precipitated with ether. This procedure (dissolving in methyl alcohol and precipitating with ether) is repeated twice more. The crude (V) is dissolved in methyl alcohol and evaporated several times from a mixture of benzene–ethyl alcohol. The residue is used in the following synthesis without further purification.

2-β-Hydroxyethyliminoriboflavin 5'-Phosphate.[1] The crude *S*-methyl-2-thioriboflavin 5'-phosphate (V) is dissolved in anhydrous methyl alcohol (30 ml) and reacted with β-hydroxyethylamine (30 ml) for 90 minutes at 50° under nitrogen in a three-necked flask provided with condenser and drying tube. Acetone (250 ml) is added to the reaction mixture, which is kept 1 hour at 0°. The crude material (1.8 g) is filtered and washed with acetone. The filtered residue is suspended in water (100 ml), and the suspension is adjusted to pH 8 with 1 *M* sodium hydroxide, treated with decolorizing carbon, and filtered. The filtrate is applied to a column (80 × 2.5 cm) of DEAE-cellulose (chloride). The nonphosphorylated flavins (riboflavin, β-hydroxyethyliminoriboflavin) are eluted with water (2 liters). The column is then eluted by using a linear gradient with water (2 liters) in the mixing chamber and 0.8 *M* lithium chloride (2 liters) in the reservoir. The flow rate should be 2–2.5 ml/min, 20-ml fractions being collected. The elution of flavin mononucleotide is followed spectrophotometrically at 451 mμ. The fractions that contain 2-β-hydroxyethylimino-riboflavin 5'-phosphate (Fig. 1) are combined and concentrated at 30° *in vacuo* until a small precipitate is formed. The suspension is adjusted to pH 5 with 0.1 *M* perchloric acid, and the solution is treated with decolorizing

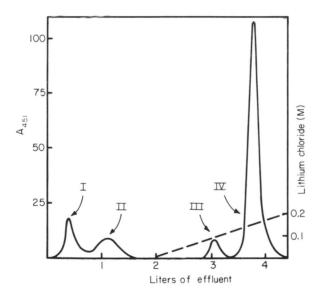

Fig. 1. Elution pattern from DEAE-cellulose of a reaction mixture containing: (I) riboflavin; (II) β-hydroxyethyliminoriboflavin; (III) riboflavin 5'-phosphate; (IV) β-hydroxyethylriboflavin 5'-phosphate. Compounds are eluted from an 80 × 2.5 cm column of DEAE-cellulose (chloride) with a linear gradient from 2 liters of water to 2 liters of 0.8 *M* LiCl. Fractions of 20 ml are collected, and aliquots are diluted appropriately with water for measurements of A_{451}.

carbon and filtered into acetone (250 ml). A liquid- liquid phase separation is removed by dropwise addition of methyl alcohol. The precipitate obtained is kept for 1 hour at 0°, collected on a filter, washed with acetone and ether, and dried *in vacuo* to give 0.6 g (20%) of (VI). The product has one λ_{max} at 451 mμ (pH 7) with an ϵ of 17,300 and is chromatographically homogeneous in acetonitrile–water (7:3) and *n*-butyl alcohol–acetic acid–water (3:1:1).

The phosphorylation procedure for the preparation of N-3-methylriboflavin 5'-phosphate[1] from N-3-methylriboflavin[1,4] is the same but for the following modifications: The phosphorylation reaction requires 36 hours. The hygroscopic precipitate (mixture of mono- and polyphosphate, see under *S*-methyl-2-thioriboflavin 5'-phosphate, p. 460) is treated with 68 ml of a mixture of dioxane–water (10.3:1) at room temperature overnight.

Flavin 8-Bromoadenine Dinucleotide[5]

Principle. 8-Bromoadenosine 5'-phosphoromorpholidate is reacted with tri-*n*-octylammonium FMN to give flavin 8-bromoadenine dinucleotide (VII).

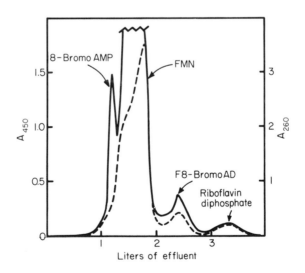

Fig. 2. Elution pattern from DEAE-cellulose of a reaction mixture containing F8-bromoAD. Compounds are eluted from a 40 × 2.5 cm column of DEAE-cellulose (chloride) with a linear gradient from 2 liters of 0.003 *N* HCl to 2 liters of the same plus 0.1 *M* LiCl. Fractions of 100 ml are collected, and aliquots are diluted 10-fold with water for measurements of A_{450} (dashed line) and A_{260} (solid line).

[4] P. Hemmerich, *Helv. Chim. Acta* **47**, 464 (1964).
[5] D. B. McCormick and G. E. Opar, *J. Med. Chem.* **12**, 333 (1969).

Flavin 8-Bromoadenine Dinucleotide. 8-Bromoadenosine 5'-phosphoro-morpholidate (0.39 g, 0.5 millimole) is condensed with tri-*n*-octylammonium FMN in anhydrous pyridine (100 ml) plus dimethyl formamide (5 ml) by a modification of the method for FAD synthesis by Moffatt and Khor-ana.[6,7] After 1 week at room temperature, the pyridine is evaporated off, the residue is dissolved in water (25 ml), and the solution is extracted twice with ether (25 ml). The aqueous phase is carefully neutralized with 1 *M* ammonium hydroxide and the solution is poured over a column (40 × 2.5 cm) of DEAE-cellulose (chloride). The small amount of riboflavin is washed through with water (2 liters), and the column is then eluted by using a linear gradient with 0.003 *M* hydrochloric acid in the mixing chamber and 0.1 *M* lithium chloride in 0.003 *M* hydrochloric acid (2 liters) in

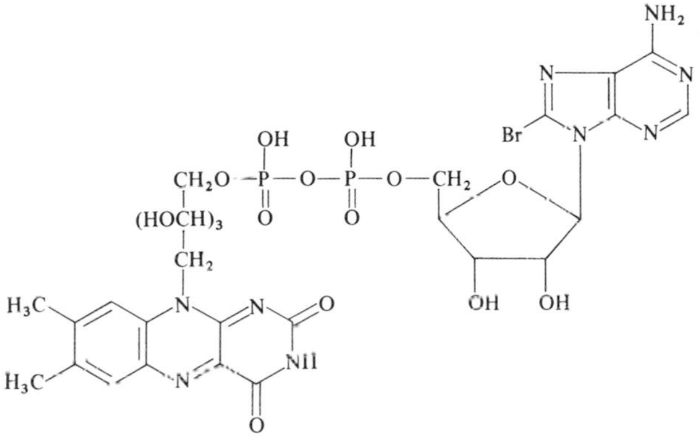

(VII)

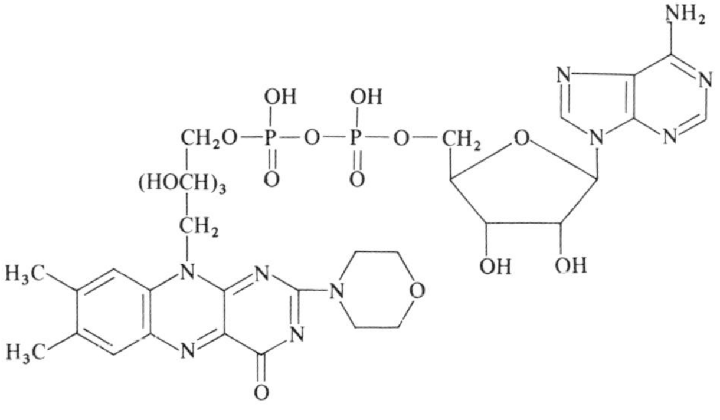

(VIII)

[6] J. G. Moffatt and H. G. Khorana, *J. Am. Chem. Soc.* **83**, 649 (1961).
[7] J. G. Moffatt and H. G. Khorana, *J. Am. Chem. Soc.* **80**, 3756 (1958).

the reservoir. The elution is followed spectrophotometrically at 260 and 450 mμ. The F8-bromoAD, which exhibits a 260/450 ratio near 3.6, is eluted after 8-bromo-5'-AMP and FMN (Fig. 2), the pH adjusted to 6 with 0.1 M lithium hydroxide, and the solution lyophilized. The residue is stirred in methyl alcohol (20 ml) and the dilithium salt of F8-bromoAD precipitated with acetone (200 ml) plus ether (20 ml). The precipitate is collected by centrifugation and washed twice more to remove all lithium chloride by resuspending in methyl alcohol (3 ml) and reprecipitating with acetone (30 ml) plus ether (3 ml). This material is dissolved in water (10 ml), filtered, and lyophilized to give 0.13 g (28%) of (VII).

The syntheses of 2-morpholino-2-deoxy FAD[1] (VIII) from 2-morpholino-2-deoxyriboflavin 5'-phosphate and adenosine 5'-phosphoromorpholidate[6] is the same but for the following modifications: The mixture of 2-morpholino-2-deoxy FMN and adenosine 5'-phosphoromorpholidate in the pyridine-dimethyl formamide solvent is kept 1 week at 40°. The FAD derivative is eluted by using a linear gradient with 0.01 M lithium chloride in 0.01 M lithium acetate (pH 5.5, 2 liters) in the reservoir. The 2-morpholino-2-deoxy FAD exhibits a 260/464 ratio near 1.2 and an R_f ratio of FMN/FAD near 3.2 on thin-layer chromatography with n-butyl alcohol–acetic acid–water (2:1:1).

Comments

The above syntheses exemplify methods useful for obtaining derivatives of flavin coenzymes in which either isoalloxazine or adenyl moieties can be variously substituted. In particular, the 2-thio and S-methyl-2-thio analogs are useful as intermediates for the formation of other 2-substituted flavins, as these functions can be displaced with appropriate nucleophiles.[1,5]

[149] Riboflavin 5'-Monosulfate

By KUNIO YAGI

Riboflavin 5'-monosulfate is a sulfate analog of FMN. It inhibits D-amino acid oxidase in competition with the FMN part of FAD.[1] This analog shows an antivitamin action for riboflavin in bacteria such as *Lactobacillus casei* and *Streptococcus faecalis*[2] as well as in mammals,[3] though the action in mammals is very weak. The method of preparation and some properties of riboflavin 5'-monosulfate are described below.

Chemical Synthesis and Purification[4]

In a 300-ml round-bottomed three-necked flask fitted with an air-tight stirrer, a dropping funnel, and a reflux condenser closed with a calcium chloride tube, dried riboflavin (200 mg, 0.54 millimole) is dissolved in 200 ml of dried pyridine by heating at 80°. The solution is cooled to 30°, then a mixture composed of 1 ml of chlorosulfonic acid (15.3 millimoles) and 3 ml of dried chloroform is added dropwise with vigorous stirring, and heated at 40° for 30 minutes.[5] The solution is evaporated *in vacuo* at 40°. The residue is dissolved in a minimal volume of water. Unreacted riboflavin, which is less soluble, is filtered off and washed with cold water. Filtrate and washings are combined and neutralized with $CaCO_3$. Calcium sulfate

[1] F. Egami and K. Yagi, *J. Biochem.* **43**, 153 (1956).

[2] F. Egami, M. Naoi, M. Tada, and K. Yagi, *J. Biochem.* **43**, 669 (1956).

[3] K. Yagi and S. Yamada, *Acta Biochim. Polon.* **11**, 315 (1964).

[4] N. Takahashi, K. Yagi, and F. Egami, *J. Chem. Soc. Japan, Pure Chem. Sect.* **78**, 1287 (1957).

[5] By heating at 60°–70° for 1 hour after the addition of chlorosulfonic acid, all the alcoholic hydroxyl groups in the ribitol chain of riboflavin are esterified to produce riboflavin 2',3',4',5'-tetrasulfate.

formed is removed by filtration. A further small amount of CaCO₃ is added to the filtrate, and the filtrate is concentrated *in vacuo* until the odor of pyridine is eliminated. The residue is dissolved in water and made up to 50 ml, and insoluble materials are filtered off. Barium chloride (0.13 g) is added to the filtrate, and BaSO₄ formed is removed by centrifugation. The supernatant is condensed *in vacuo*, and the crude calcium salt of riboflavin 5′-monosulfate is precipitated by adding excess ethanol. The precipitate is collected by centrifugation and washed with ethanol.

Since the crude preparation is contaminated with a small amount of unreacted riboflavin and other sulfuric acid esters of riboflavin,[6] further purification is needed to obtain pure riboflavin 5′-monosulfate. For this purpose, cellulose column chromatography using a mixture of benzyl alcohol–ethanol–water (3:2:1, v/v/v) as a mobile phase is recommended. After riboflavin 5′-monosulfate is clearly separated from the above-mentioned contaminating materials on the column,[7] the fraction corresponding to riboflavin 5′-monosulfate is taken out, and eluted with water. The eluate is washed with ethyl ether to eliminate benzyl alcohol and condensed under reduced pressure. From the condensate, the calcium salt of riboflavin 5′-monosulfate is precipitated by adding excess ethanol. It is crystallized from water.

Properties

Solubilities[4] of riboflavin 5′-monosulfate in various solvents are similar to those of FMN: insoluble in benzene, chloroform, or ethyl ether; slightly soluble in methanol or ethanol; but easily soluble in water. The visible absorption spectrum[4] of riboflavin 5′-monosulfate in water is identical with that of FMN, having absorption peaks at 375 and 445 mμ. In paper chromatography[2] using solvents such as the upper layer of *n*-butanol–acetic acid–water (4:1:5, v/v/v), benzyl alcohol–ethanol–water (3:2:1, v/v/v), or *n*-propanol–pyridine–water (5:3:2, v/v/v), riboflavin 5′-monosulfate can be separated from riboflavin, FMN, and FAD and migrates between FMN and riboflavin.

By kinetic analysis of the enzymatic reaction of D-amino acid oxidase,

---

[6] Usually a small amount of riboflavin 2′,3′,4′,5′-tetrasulfate and a trace amount of riboflavin tri- and/or disulfate are found. To check these substances, silica-gel G thin-layer chromatography using the upper layer of *n*-butanol–acetic acid–water (4:1:5, v/v/v) as a mobile phase is useful. In this case, R_f values of riboflavin 2′,3′,4′,5′-tetrasulfate, riboflavin 5′-monosulfate, and riboflavin are 0.03, 0.23, and 0.47, respectively. Riboflavin tri- and disulfate migrate between tetra- and monosulfate.

[7] The mobilities of these substances can be judged by paper chromatography using the same solvent; R_f values of riboflavin 2′,3′,4′,5′-tetrasulfate, riboflavin 5′-monosulfate, and riboflavin are 0.03, 0.30, and 0.46, respectively. Also in this case, riboflavin tri- and disulfate migrate between tetra- and monosulfate.

riboflavin 5'-monosulfate has been shown to compete with the FMN part of FAD, resulting in inhibition of this enzyme.[1] The dissociation constant between riboflavin 5'-monosulfate and the D-amino acid oxidase apoprotein is evaluated to be $3.2 \times 10^{-5} M$.

Using this substance as a specific indicator, the mode of action of an inhibitor which inhibits a flavin enzyme in competition with FAD can be analyzed.[8] In the presence of this indicator, it can be elucidated which moiety of FAD actually competes with the inhibitor to be tested.

In contrast with the case of D-amino acid oxidase, however, riboflavin 5'-monosulfate cannot combine with the apoprotein of the old yellow enzyme, showing no inhibitory action.[9]

As expected from the chemical structure, riboflavin 5'-monosulfate can act as an antivitamin B_2 when examined with *Lactobacillus casei* and *Streptococcus faecalis*.[2] This sulfate inhibits the growth of these bacteria, which is a bacteriostatic action.

In nutritional experiments with mammals,[3] riboflavin 5'-monosulfate, when injected, shows an antivitamin B_2 effect; this cannot, however, be shown when it is administered orally. Riboflavin 5'-monosulfate, when injected, is detected in the liver, kidney, heart, and small intestine at 30 minutes after the administration and almost disappears after 12 hours. Most of the injected riboflavin 5'-monosulfate is excreted in urine so rapidly that it shows no serious toxicity. After injection of riboflavin 5'-monosulfate, the amounts of FAD, FMN, and riboflavin in the liver, kidney, heart, and small intestine are not changed significantly, indicating that the physiologically existing flavin in these organs might not be exchanged easily for the injected riboflavin 5'-monosulfate.

Determination of Riboflavin 5'-Monosulfate in Animal Tissues[10]

Riboflavin 5'-monosulfate can be analyzed using both the lumiflavin fluorescence method and paper chromatography.[11] Analytical procedures, e.g., warm-water extraction, photodecomposition, fluorometric assay of lumiflavin formed, and paper chromatography, are essentially the same as those for the separate determination of the physiologically existing flavins, except that in this case the use of the mixture of n-butanol–acetone–acetic acid–water (5:2:1:3, v/v/v/v) is recommended as a suitable mobile phase for paper chromatography. R_f values of FAD, FMN, riboflavin 5'-monosulfate, and riboflavin are 0.08, 0.15, 0.25, and 0.42, respectively.

[8] K. Yagi and T. Ozawa, *Biochim. Biophys. Acta* **42**, 381 (1960).
[9] H. Theorell, K. Yagi, G. D. Ludwig, and F. Egami, *Nature* **180**, 922 (1957).
[10] K. Yagi and S. Yamada, *Nagoya J. Med. Sci.* **25**, 228 (1963).
[11] K. Yagi, this volume, [134].

[150] Sulfite Interaction with Free and Protein-Bound Flavin

Flavin + $SO_3^{2-} \rightleftharpoons$ flavin-SO_3^- complex

By FRANZ MÜLLER and VINCENT MASSEY

It is well documented that free and protein-bound flavins form complexes with a variety of organic compounds. The formation of these complexes causes either slight bathochromic or hypsochromic shifts of the long wavelength absorption maximum of the flavin spectrum and often a somewhat better resolution of the 450 mμ absorption band. Occasionally the appearance of a long wavelength absorption induced by the formation of charge transfer complexes is observed. The formation of such complexes does not change formally the redox state of the flavin involved.[1]

More recently, a different type of flavin complex formation was observed that caused a dramatic change on the visible part of the flavin absorption spectrum. The complex formation in consideration was detected by the experimental observation that flavoproteins reduced by dithionite very often did not yield the expected absorbancy upon air oxidation. It was found that sulfite, the oxidation product of $S_2O_4^{2-}$, interacted with flavoproteins, forming an enzyme–sulfite complex.[2,3]

Procedures

Preparation of Sulfite Solutions

Sulfite solutions are prepared in the buffer to be used for the experiments and adjusted to the desired pH with 2 N NaOH or 2 N HCl. For sulfite solutions of pH > 7, preferably Na_2SO_3 is used; and for solutions of pH < 7, $NaHSO_3$. Furthermore, to minimize loss of sulfite by air oxidation, molar solutions are prepared daily. If desired, the sulfite solution is diluted prior to the experiment.

Preparation of Flavin and Flavoprotein–Sulfite Complexes for Spectrophotometric Studies

The *purified* enzymes are dissolved in the buffer (0.1–0.3 M, pH range 6.5–8.5) to be used in the experiment, and the desired volumes of the sulfite solution are added and mixed. The free flavin–sulfite complexes are

[1] G. Palmer and V. Massey, in "Biological Oxidation" (T. P. Singer, ed.), p. 263. Wiley (Interscience), New York, 1968.

[2] B. E. P. Swoboda and V. Massey, J. Biol. Chem. **241**, 3409 (1966).

[3] V. Massey, F. Müller, R. Feldberg, M. Schuman, P. A. Sullivan, L. G. Howell, S. G. Mayhew, R. G. Matthews, and G. P. Foust, J. Biol. Chem. **244**, 3999 (1969).

prepared in a similar way. If a series of experiments at a constant pH are planned, then it is advantageous to prepare a flavin stock solution. Take the same aliquot of the flavin solution for each experiment, add the desired sulfite concentrations and make up to a total volume kept constant throughout the experiments. By this procedure samples with either a constant concentration of flavin and various concentrations of sulfite or samples with various concentrations of flavin and constant concentration of sulfite are easily and accurately prepared. The light absorption spectrum is recorded after each addition of sulfite to the enzyme or free flavin solution when no further change of the spectrum is observed, i.e., when the solution is equilibrated. This may take from a few seconds to 30 minutes, depending on the concentration of the reactants, pH, temperature, and whether or not enzymes or free flavins are investigated. Furthermore, to the reference cell the same amount of sulfite is added as to the sample cell.

Preparation of Crystalline Flavin–Sulfite Complexes[4]

1,10-Ethylene-7,8-dimethyl-5-sulfonate-1,5-dihydroisoalloxazine ($C_{14}H_{13}O_5N_4SNa$, mol. wt.: *372.34*). Three-tenths gram of 1,10-ethylene-7,8-dimethylisoalloxazinium perchlorate[4] is suspended in 10 ml of 3 M NaHSO$_3$ solution at pH 5.3 and warmed up to 60° for 5 minutes in a water bath with stirring. The reaction mixture is then removed from the water bath and stirred at room temperature for a further 30 minutes; during this time the yellow color of the reaction mixture is gradually bleached and pale yellow crystals are formed. The precipitate is collected, washed twice with small volumes of cold water, and dried with ethanol and ether, respectively. The reaction yields 0.22 g (73%) of the sulfite complex of decomposition point 250°. For recrystallization, 0.1 g of the complex is dissolved in 5 ml 0.3 M NaHSO$_3$ solution with gentle warming, the clear solution is brought to room temperature, and an equal volume of a 2 M NaCl solution is added. The mixture is allowed to stand overnight at room temperature; during this time nearly colorless microcrystals are formed. The collected crystals are washed successively with 0.5 ml of cold water, several times with ethanol and ether. The purity of the flavin–sulfite complex is checked by thin-layer chromatography employing a solvent mixture of *n*-butanol–2 N ammonia–ethanol (3:1:1). The chromatogram is developed over I$_2$-vapor showing a single spot.

5,7-Disulfonate-10-methyl-1,5-dihydroisoalloxazine ($C_{11}H_8O_8N_4S_2Na_2\cdot3$ $H_2O\cdot\frac{1}{2}$ NaHSO$_3$, mol. wt., *547.60*). Fifty milligrams of 7-sulfonate-10-methylisoalloxazine is suspended in 3 ml of NaHSO$_3$ solution at pH 5.3 and stirred mechanically for 3 days at room temperature. The pale yellow precipitate formed is collected and recrystallized twice from 3 M NaHSO$_3$

[4] F. Müller and V. Massey, *J. Biol. Chem.* **244**, 4007 (1969).

solution. The microcrystals thus obtained are collected and washed successively with a very small volume of cold water, ethanol, and ether to give 35 mg (53%) of the sulfite complex. This complex in solution is very unstable and dissociates into its elements, flavoquinone and sulfite; therefore the complex has to be recrystallized from a concentrated sulfite solution. Hence the crystallizate always contains some uncomplexed $NaHSO_3$. The purity of the isolated sulfite complex is checked in the same manner as outlined above.

Discussion

The free flavocoenzymes and some of their simple derivatives, i.e., riboflavin and lumiflavin, show a relatively low affinity for sulfite compared to flavoproteins and certain lumiflavin derivatives. Only a few models show a high reactivity toward sulfite and are therefore useful models for the preparation of crystalline flavin–sulfite complexes. Such useful models are flavinium salts obtained by alkylation of positions 1, or 2 and 4, or 2 and 3 of lumiflavin (I, Scheme 1).[5]

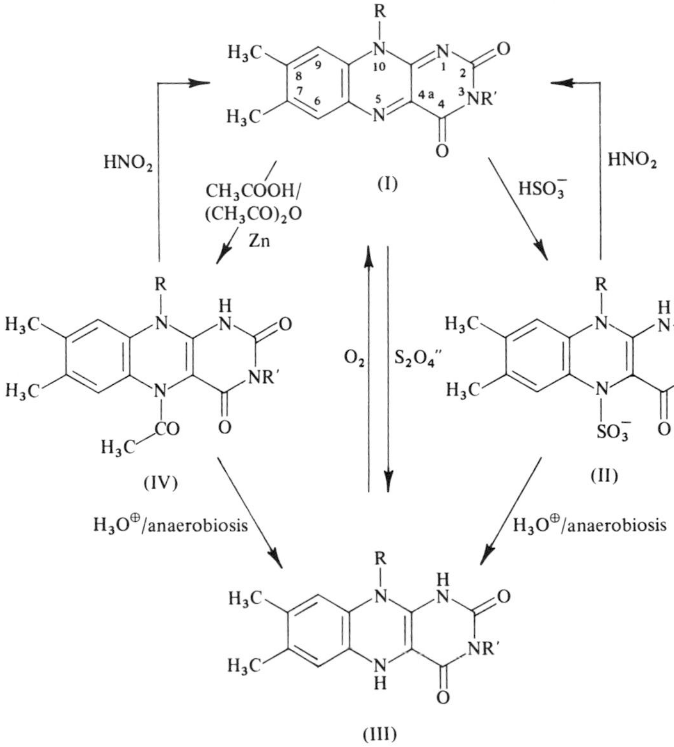

SCHEME 1

Sulfite interacts with the N-5 atom of the oxidized flavin yielding N^5-sulfite-1,5-dihydroflavin (II, Scheme 1). Evidence for the structure is obtained by elemental analysis giving a flavin-sulfite ratio of 1. Furthermore, the infrared spectrum shows besides the SO_2 stretchings an additional absorption at about 900 cm^{-1} representing the N—S-stretching. Both carbonyl stretchings of the oxidized flavin remain practically unchanged by the flavin–sulfite complex formation.

Acid hydrolysis of (II) under anaerobic conditions yields 1,5-dihydroflavin (III) and sulfate. At pH 2, the flavin–sulfite complex dissociates into its elements, oxidized flavin and sulfite. Furthermore, the light absorption spectra of N^5-acylated flavin derivatives (IV) are practically identical with their analogous sulfite complexes (II), both with respect to the spectral forms and the molar extinction coefficient (7000). Moreover, the acid hydrolysis product of N^5-acyl flavins (IV) under anaerobic conditions is identical in light and infrared absorption to that obtained from flavin–sulfite complexes under the same conditions, namely 1,5-dihydroflavin (III).

Sulfite induces the same spectral changes on the light absorption spectrum of flavoproteins and free flavins. With small amounts of sulfite the enzyme or free flavins are only partially converted into the complex, as demonstrated in Table I for tetraacetylriboflavin. The observed spectral changes produced by sulfite bleaching are similar but not identical with those observed by anaerobic reduction of flavins. Thus the spectrum of the sulfite complex shows only small absorption in the visible region and an absorption maximum in the region of 310–340 mμ whereas the spectrum

TABLE I

VARIATION OF K OF TETRAACETYLRIBOFLAVIN–SULFITE COMPLEX
WITH SULFITE CONCENTRATION[a]

Sulfite conc. (M)	Absorbancy		K^b (M)
	370 mμ	445 mμ	
0	0.343	0.403	
0.12	0.295	0.340	0.65
0.24	0.265	0.305	0.75
0.36	0.244	0.275	0.77
0.60	0.212	0.240	0.88
0.90	0.180	0.205	0.94
1.20	0.155	0.140	0.64

[a] $3.55 \times 10^{-5} M$ Tetraacetylriboflavin in 0.3 M phosphate buffer at pH 7.6 and 25°.
[b] In the calculation of K, the total concentration of sulfite and bisulfite has been used. The pH was checked after each experiment.

[5] K. H. Dudley and P. Hemmerich, *Helv. Chim. Acta* **50**, 355 (1967).

of the reduced form of flavins shows a higher absorption at wavelength $>$ 380 mμ and only marked shoulders in the visible region. Furthermore, the isosbestic point obtained by the sulfite titration of a flavin solution indicates that only two flavin species are present during the sulfite bleaching, i.e., oxidized flavin and the sulfite complex. From these results it is possible to calculate a dissociation constant for the flavin sulfite complexes:

$$K = \frac{[\text{flavin}][\text{sulfite}]}{[\text{flavin–sulfite complex}]} \tag{1}$$

By analysis of spectral results at any one wavelength (usually at about 450 mμ), the equilibrium concentration of flavin, sulfite, and complex can be determined. Thus values of K, the dissociation constants, of the free flavins and enzymes studied are summarized in Tables II and III, respectively. It is interesting to note that the flavoprotein oxidases react with sulfite, but not the flavoprotein dehydrogenases (Table III).

The flavin–sulfite interaction is reversible, and dependent on the absolute concentrations of the reactants. Thus dilution (or dialysis in the case of a flavoprotein) of a solution of flavin–sulfite complex yields the oxidized form of flavin and sulfite.

TABLE II

Dissociation Constants, K, of Various Substituted Flavin–Sulfite Complexes[a]

Substituent on the isoalloxazine ring in position (cf. Scheme 1)	K^b (M)
CH$_2$COOH(3) CH$_3$(7,8,10)	18.2
H(3) CH$_3$(7,8,10)	4.37
FAD	2.52
FMN	1.91
H(3,8) CH$_3$(6,7,10)	1.60
H(3) CH$_3$(7,8) CH$_2$(CHOCOCH$_3$)$_3$CH$_2$OCOCH$_3$(10)	1.21
H(3) CH$_3$(7,8) CH$_2$(CHOH)$_3$CH$_2$OH(10)	1.16
H(3,7,8) CH$_3$(10)	1.52×10^{-1}
CH$_3$(3,7,10) COOH(8)	1.14×10^{-1}
H(3) Cl(7) CH$_3$(8,10)	6.38×10^{-1}
S(2) H(3) CH$_3$(7,8,10)	3.85×10^{-2}
H(3) Cl(8) CH$_3$(7,10)	5.80×10^{-2}
CH$_3$(3,10) Cl (7,8)	1.45×10^{-2}
H(3,8) SO$_3^-$(7) CH$_3$(10)	1.10×10^{-2}
SCH$_3$(2) CH$_3$(7,8,10)	2.10×10^{-2}
S(2) H(3) CH$_3$(7,8)CH$_2$(CHOH)$_3$CH$_2$OH(10)	3.95×10^{-3}
CH$_2$(1,10) H(3) CH$_3$(7,8)	1.32×10^{-4}

[a] Determined in 0.3 M phosphate buffer at pH 7.0 and 25°.[4]

[b] In the calculation of K, the total concentration of sulfite and bisulfite has been used rather than the sulfite or bisulfite concentration.

TABLE III
DISSOCIATION CONSTANTS, K, OF VARIOUS FLAVOPROTEIN–SULFITE COMPLEXES[a]

Enzyme	pH	K^b (M)	Effect of temperature on K
Glucose oxidase	7.0	7.2×10^{-4}	Large
D-Amino acid oxidase	7.0	2.1×10^{-3}	Small
L-Amino acid oxidase	7.5	8.5×10^{-5}	Small
Oxynitrilase	7.0	2.9×10^{-3}	Large
Lactate oxidase (decarboxylating)	7.0	3.7×10^{-6}	Small
Glylolate oxidase	7.0	8.6×10^{-7}	Large
Flavoproteins showing no observable sulfite complex			
Flavodoxin (*P. elsdenii*)			
Lipoyl dehydroglucose			
Glutathione reductase (yeast)			
Thiorexdoxin reductase (*E. coli*)			
Old yellow enzyme			
Butyryl-CoA dehydrogenase (pig liver)			
Long-chain acyl CoA dehydrogenase (pig liver)			
Azotobacter flavoprotein			
Glyoxylate carboligase (*E. coli*)			
Ferredoxin TPN reductase (spinach)			

[a] In phosphate buffer, at 25°.[3]

[b] In the calculation of K, the total concentration of sulfite plus bisulfite has been used rather than the sulfite or bisulfite concentration.

That the flavin–sulfite interaction does not cause a simple reduction of the flavin is proved by the fact that the sulfite complex is unaffected by electron acceptors such as oxygen or ferricyanide. Furthermore, the formation of the sulfite complex is unaffected by oxygen.

The formation of the sulfite adduct is also dependent on the temperature and the pH. Thus lowering the temperature results in a decrease, and raising the temperature results in increase, of K [Eq. (1)]. The pH maximum was found to be in the pH range of 6.5–8.5 indicating that SO_3^{2-} rather than HSO_3^- is the reacting nucleophilic species.

Enzyme solutions in the presence of high concentration of urea or inhibitors are much less bleached by sulfite or the bleaching is completely depressed. Enzyme–sulfite adducts are catalytically inactive. Furthermore, substrates and inhibitors slowly displace the equilibrium of the adduct (Eq. 1) yielding the reduced enzyme in the former case under anaerobic conditions and oxidized enzyme in the latter case.

[151] On the Reduction of D- and L-Amino Acid Oxidases and Free Flavins by Borohydride Yielding 3,4-Dihydroflavin Analogs

By Franz Müller, Vincent Massey, and Peter Hemmerich

Chemical or substrate reduction of flavoproteins and free flavins yields 1,5-dihydroflavin compounds that are sensitive to oxygen. A quite different two-electron reduction was found recently[1] by the treatment of amino acid oxidases with borohydride. The light absorption spectrum of the product thus obtained is different from that of the oxidized and substrate reduced enzymes showing absorption maxima at about 410 and 330 mμ. The BH_4^--reduced enzymes are stable toward oxygen and are catalytically active. The structure of the modified prosthetic group was elucidated by model studies. The synthesis and the properties of such modified flavins are reported in this article.

Procedures

Reduction of L- *and* D-*Amino Acid Oxidases by* BH_4^-.[1] To a solution of enzyme, 4×10^{-5} M with respect to FAD content, in 0.1 M phosphate buffer of pH 7.0–8.5 at 5°–10° is added about 5 mg of $NaBH_4$ and the reaction followed by spectrophotometric means. With L-amino acid oxidase, the reaction is completed within a few minutes. With D-amino acid oxidase, 90% of the reaction occurs within 10 minutes, but it takes about one to a few hours to complete the reaction, which may, therefore, require an additional 1–2 mg of $NaBH_4$. The BH_4-reduction may be conducted under aerobic or anaerobic conditions without alteration of the product. The modified enzymes thus obtained are purified either by $(NH_4)_2SO_4$ precipitation or by dialysis and can be stored for several days without deterioration at 0°–5° if protected from light. The effect of borohydride of modifying the flavin structure is so far unique to the amino acid oxidases. The effects of borohydride on all the flavoproteins tested are given in Table I.

N^3-*Methyl-3,4-dihydrolumiflavin* (*Ia*) ($C_{14}H_{16}O_2N_4$, mol. wt. *272.30*).[2] In a 1000-ml Erlenmeyer flask is placed 300 ml of anhydrous ether cooled in an ice bath to 3°–6°, and 350 mg of $LiBH_4$ is added. The mixture is illuminated by two 100-W Tungsten lamps placed about 30 cm from the reaction vessel. To the illuminated, practically clear ethereal solution is

---

[1] V. Massey, B. Curti, F. Müller, and S. G. Mayhew, *J. Biol. Chem.* **243**, 1329 (1968).
[2] F. Mülier, V. Massey, C. Heizman, P. Hemmerich, J. M. Lhoste, and D. C. Gould, *European J. Biochem.* **9**, 392 (1969).

TABLE I
EFFECT OF BOROHYDRIDE ON FLAVOPROTEINS

Enzyme	Product	Comments
D-Amino acid oxidase	3,4-Dihydroflavin	Somewhat slow reaction
L-Amino acid oxidase	3,4-Dihydroflavin	Fast reaction
Glucose oxidase	1,5-Dihydroflavin	Very rapid
Old yellow enzyme	1,5-Dihydroflavin	Very rapid
Glycolate oxidase	1,5-Dihydroflavin	Very rapid
Oxynitrilase[a]	1,5-Dihydroflavin	Very rapid
Lactate oxidase (decarboxylating)[b]	1,5-Dihydroflavin	Very rapid
Flavodoxin	Almost quantitative production neutral flavin radical, followed by much slower conversion to 1,5-dihydroflavin	Probably radical formation due to slow production of 1,5-dihydroflavin, which then reacts with remaining oxidized enzyme (FH$_2$ + F $\rightleftharpoons$ 2FH·)
Shethna flavoprotein[c]	Same as for flavodoxin	Same as for flavodoxin
Ferredoxin TPN reductase	No reaction, even with multiple additions of NaBH$_4$	
Lipoyl dehydrogenase	Very rapid production of red intermediate, identical with that seen in substrate reduction; followed by much slower conversion to 1,5-dihydroflavin	Intermediate probably due to rapid reduction of active center disulfide, which then reacts with flavin
Glutathione reductase	Same as for lipoyl dehydrogenase	Same as for lipoyl dehydrogenase
Thioredoxin reductase[d]	1,5-Dihydroflavin	Very rapid

[a] Gift of Dr. E. Pfeil.
[b] P. A. Sullivan, unpublished.
[c] Gift of Dr. H. Beinert.
[d] Results obtained with Dr. C. H. Williams.

slowly added, during the course of 1 hour, 100 mg of N^3-methyllumiflavin (II) dissolved in 300 ml of absolute methanol. The reaction mixture is illuminated throughout the procedure, and air is admitted frequently until no further change of the light absorption spectrum of a diluted sample is observed, i.e., loss of the 445 mμ absorption peak to about 90%. If at this stage a colorless precipitate is formed (due to the formation of boric acid or other boron compounds), 50–100 ml of absolute methanol is added in order to obtain a clear solution. The ice bath is then removed, and the reaction mixture is illuminated for a further 8 hours with gentle stirring at room temperature. The ether is removed from the reaction mixture under reduced pressure at room temperature. To the residual methanolic solution is added 400 ml of H_2O, and the mixture is extracted 6 times with 150-ml volumes of $CHCl_3$. The $CHCl_3$ phase is washed 3 times with 150-ml volumes of 0.3 M phosphate buffer, pH 7.0; the organic phase is dried with anhydrous Na_2SO_4 and evaporated to dryness under reduced pressure at room temperature. The yellow brown residue thus obtained is dissolved in 10 ml of warm 0.1 N $HClO_4$ and filtered; the filtrate is neutralized by the addition of a $NaHCO_3$ solution, whereupon bright yellow crystals of decomposition 280° form. The yield is 40 mg (51%). The purity of compound (I) is checked by thin-layer chromatography using solvent mixtures such as $CHCl_3$–ethanol (47:3), n-butanol–ethanol–water (7:2:1) or benzene–isopropylether–ethanol (2:2:1).

Total Synthesis of 3,4-Dihydrolumiflavin (Ib) ($C_{13}H_{15}O_2N_4 \cdot C_2H_5OH$, *mol. wt. 293.79*).[2] One-tenth gram of $N^{3,4}$-trimethyl-6-(p-carboxyphenyl-azo)aniline[3] is dissolved in 150 ml of glacial acetic acid and hydrogenated with palladium on carbon as catalyst at room temperature and atmospheric pressure, yielding (III). The suspension is filtered, and to the filtrate 150 mg of isodialuric acid (IV) and 150 ml of absolute methanol are added. The reaction mixture was stirred mechanically at room temperature for 30 minutes and then allowed to stand at 6° for 6 hours. To the brown-red reaction mixture is added 700 ml of water, and the mixture is extracted twice with 300-ml volumes of $CHCl_3$. The combined organic phase is washed 3 times with equal volumes of 0.3 M phosphate buffer, pH 7.0, in order to remove the acetic acid. The organic phase is dried with anhydrous Na_2SO_4 and evaporated to dryness under reduced pressure at room temperature. The residue is dissolved in 7 ml of $CHCl_3$, filtered, and allowed to stand at −6° for 10 hours; during this time yellow crystals form. The collected crystals are dried with ether. The filtrate is concentrated to about 3 ml and allowed to stand at −6° for a further 10 hours. The total yield is 40 mg (42%) of (Ib) of decomposition ∼150°. The product is recrystallized several times from a small volume of $CHCl_3$ until only one spot is observed on the thin-layer chromatogram (solvent systems as described above). $CHCl_3$ contains ethanol as stabilizer, hence the ethanol content of (Ib). Compound (Ib) has enhanced solubility in all organic solvents compared to (Ia). This indicates that the 4-hydroxyl group is replaced by an ethoxy function.

Compound (Ib) can also be synthesized by BH_4^- reduction of lumiflavin by the procedure given above for (Ia), but the yield is only about 20%

---

[3] J. P. Lambooy, this volume [145].

of that of (Ia), and the product thus obtained is somewhat difficult to purify.

Discussion

As a model flavin for the synthesis of 3,4-dihydroflavin (I), N^3-methyl-lumiflavin (IIa) (Scheme 1) was found most suitable because of its photo-stability and adequate solubility. Anaerobic photocatalyzed reduction of (II) by BH$_4^-$ yields mainly 1,5-dihydroflavin (V), which is stable toward prolonged treatment with BH$_4^-$ and is rapidly reoxidized by air to give (II). However, flavins reduced by BH$_4^-$ in the presence of light are not quantitatively reoxidized. The irreversible slow reaction becomes dominant under aerobic conditions yielding (I). These facts indicate that the formation of (I) does not occur via 1,5-dihydroflavin (V), but via the excited

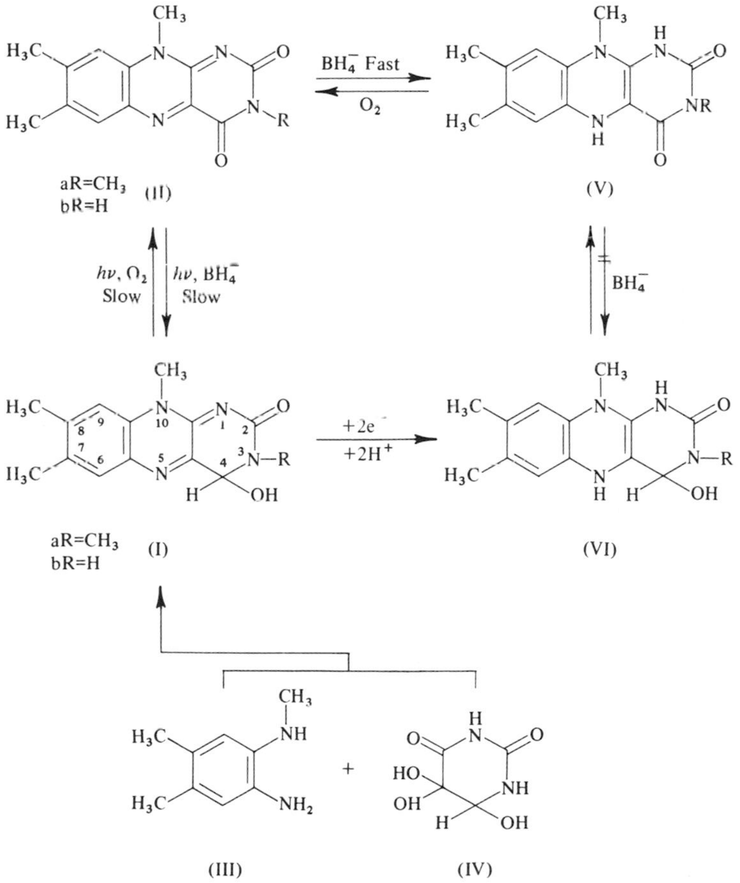

SCHEME 1

states of the oxidized flavin. The flavoprotein amino acid oxidases, on the other hand, react with BH_4^- under aerobic and anaerobic conditions in the absence of light to form the enzymes with the modified coenzyme. This suggests that the 4-CO group of the flavocoenzyme of the amino acid oxidases is not polarized by the apoprotein, i.e., by a hydrogen bond. No other flavoprotein so far tested (Table I) gave the same product as that obtained by BH_4^- treatment of amino acid oxidases.

The structure of (I) was ascertained by unequivocal synthesis by condensation of (III) and (IV), yielding (Ib). The structure was furthermore confirmed by nuclear magnetic resonance (NMR) and infrared (IR) spectroscopy. Thus the IR spectra of (I) showed lack of the 4-CO stretching and an additional OH stretching compared to the parent compounds (II) (Table I), whereas the NMR spectra exhibited an additional CH absorption and the exchangeable proton of the OH group (Ia) or a quartet and a triplet representing 4-OCH$_2$CH$_3$ (Ib).

The light absorption spectra of (Ia), (Ib), and the product of BH_4^--treated amino acid oxidases are practically identical apart from a small shift of the maxima and a somewhat better resolution of the enzyme spectra (Table II). Protonation of (I), which occurs at a higher pH than for the corresponding flavoquinones (II) (Table I), induces a hypochromic shift

TABLE II

COMPARISON OF PHYSICAL DATA OF L- AND D-AMINO ACID OXIDASE AND
FREE FLAVINS AND THE CORRESPONDING BH_4^- PRODUCTS[a]

Compound	Light absorption maxima $\lambda_{max}(m\mu)/\epsilon(M^{-1}cm^{-1})$	Infrared absorption[e]	pK_a
Ia	405/13,900; 331/6,100[b]	CO(2); 1625 cm^{-1}, OH(4); 3130, 2880 cm^{-1}	2.5
Ib	400/14,000; 330[c]/6,100[d]	CO(2); 1655 cm^{-1}	2.5
IIa	445/12,500; 373/10,600[d]	CO(2); 1662 cm^{-1}, CO(4); 1695 cm^{-1}	0.18
IIb	445/12,500; 373/10,600[d]	CO(2); 1665 cm^{-1}, CO(4); 1718 cm^{-1}	0.18
Borohydride-treated L-amino acid oxidase	409/13,000; 334/6,400[d]		
L-Amino acid oxidase-BH_4^- product after release with trichloroacetic acid	405/12,400; 334[c]/6,000[d]		

[a] Measured at 25°.
[b] In 0.3 M phosphate, pH 7.8.
[c] Shoulder.
[d] In 0.3 M phosphate, pH 7.0.
[e] KBr disk.

of about 30 mμ on the light absorption spectrum, as to be expected in analogy to the protonation of (II).

Compound (I) is slowly photooxidized in an aqueous solution in the presence of oxygen to give the starting material (II). In strong acid solutions, the photooxidation is much faster. However, methanolic solutions of (I) are indefinitely stable to oxygen in the absence of light.

3,4-Dihydroflavins (I) are easily reduced by EDTA and light under anaerobic conditions yielding (VI), but reduction with $S_2O_4{}^{2-}$ is rather slow. Air reoxidation of (VI) yields a mixture of (II) and (I), and after three cycles of reduction and reoxidation of (I) the final product obtained is flavoquinone (II).

$BH_4{}^-$-treated L- and D-amino acid oxidase are catalytically fully active. When substrate was added anaerobically to the modified enzymes, the spectra of the $FADH_2$ form were obtained. On admitting air, the spectra of the native oxidized enzymes were regained.

Unlike the $BH_4{}^-$-modified L-amino acid oxidase, which has no detectable fluorescence, the $BH_4{}^-$-modified D-amino acid oxidase has an intense blue fluorescence, characterized by an emission maximum at 475 mμ and an excitation spectrum which is similar to that of the absorption spectrum. The fluorescence spectrum of the modified D-amino acid oxidase is identical with that of (Ib), except for higher quantum yields in the case of the model compound.

The $BH_4{}^-$-modified FAD is still firmly bound to the protein. This can be demonstrated by the fact that the modified FAD can be removed only in the same ways that FAD is removed, e.g., by dialysis against 1 M KBr, by heat denaturation or by trichloroacetic acid precipitation. The released modified flavin is still in the adenine dinucleotide state, since treatment with snake venom phosphodiesterase results in an approximately 10-fold increase in fluorescence intensity without affecting the spectral characteristics of the fluorescence. Furthermore, the resulting $BH_4{}^-$-modified FMN can be bound stoichiometrically to the apoenzyme of the FMN-enzyme flavodoxin, with complete fluorescence quenching.[1]

[152] Photochemistry of Flavins

By G. R. Penzer and G. K. Radda

General Considerations

Under anaerobic conditions flavins are photoreduced by various substrates including amines, amino acids, alcohols, and reduced pyridine nucleotides.[1]

The efficiency of the reducing agents is generally in the order: tertiary amines > secondary amines > primary amines > alcohols. Since it is the unprotonated form of the amine that reacts, the reactions are quickest between pH 7 and 10 (above which the flavin ionizes). In addition, the rates of photoreductions will vary with ionic strength and the nature of the buffer. The use of amine buffers should be avoided, since they react; 0.1 M phosphate, however, is convenient.

These reactions can be used to study properties of the flavin nucleotides that are not otherwise easily observed; for instance, the disproportion rate of flavin semiquinones has been measured by flash photolysis.[2] Photochemical methods have been used to generate flavin semiquinones and dihydroquinones both in the free state[3] and in flavoproteins,[4] to study the triplet states of flavins[5] and their interactions with a variety of aromatic compounds,[6] to measure the dissociation of flavin coenzymes from the apoprotein,[7] and to investigate reactions that may serve as models for flavoprotein catalysis.[8,9]

Photoreductions

Apparatus

Light source: A 500-W tungsten projection lamp connected with a potentiometer to allow easy adjustments of the intensity is convenient. Less powerful (100 W) filament lamps are adequate for most reactions, and white light from fluorescent tubes has sometimes been used.

Filters: To avoid exciting simple aromatic substrates as well as the the coenzyme, it is best to illuminate with visible light. The radiation from a tungsten filament must thus be passed through a water filter to remove infrared wavelengths and through a suitable glass cut-off filter (e.g., Chance Brothers, Ltd., OY 18 or Corning 3-73) to remove wavelengths shorter than 400 mμ.

A set of neutral density filters with transmissions from 5–80%

[1] G. R. Penzer and G. K. Radda, Quart. Rev. (London) 21, 43 (1967).
[2] L. Tegnér and B. Holmström, Photochem. Photobiol. 5, 223 (1966).
[3] G. K. Radda and M. Calvin, Biochemistry 3, 384 (1964).
[4] V. Massey and G. Palmer, Biochemistry 5, 3181 (1966).
[5] J. M. Lhoste, A. Haug, and P. Hemmerich, Biochemistry 5, 3290 (1966).
[6] (a) T. Shiga and L. H. Piette, Photochem. Photobiol. 3, 213 (1964). (b) G. K. Radda, Biochim. Biophys. Acta 112, 448 (1966).
[7] J. F. Koster, C. Veeger, and D. B. McCormick, Biochim. Biophys. Acta 153, 724 (1968).
[8] P. Hemmerich, V. Massey, and G. Weber, Nature 213, 728 (1967).
[9] G. R. Penzer and G. K. Radda, Biochem. J. 109, 259 (1968).

are required for quantitative measurement of the effects of varying light intensity.

Glass lens: Typical dimensions are diameter 5 cm, focal length 12 cm.

Glass water bath with thermostat

Black cloth and a frame to support it

Stoppered spectroscopic cuvettes, and some means to support them in the water bath

The arrangement of the apparatus is shown in Fig. 1; the black cloth is used to make a lightproof enclosure for it. Focusing the light on the cuvette gives maximum efficiency of sensitization, but if only part of the solution is illuminated the frequency with which the mixture is shaken affects the rate of reaction measured. The same effect is unavoidable, even under homogeneous illumination with a parallel beam, for solutions with

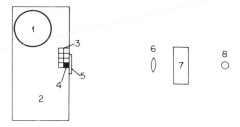

FIG. 1. Experimental arrangement for the illumination of a reaction mixture. (1) Thermostat heater and stirrer; (2) water bath; (3) cuvette basket; (4) cuvette; (5) filter(s) [cut-off, (neutral density)]; (6) lens; (7) infrared filter (water); (8) lamp.

optical densities greater than 0.1 because of attenuation of the light as it passes through the cell. Constant stirring solves this problem, but it is not easily achieved except in vessels larger than a cuvette. Thus to obtain accurately comparable results using the arrangement described, it is necessary to use a standard method, taking readings at fixed time intervals.

Method

All photoreductions must be performed under anaerobic conditions because reduced flavins react rapidly with oxygen. A variety of methods can be used to remove oxygen from solutions. An effective and simple procedure involves bubbling an inert gas (oxygen-free nitrogen, argon, or helium) through the reaction mixture for 20–30 minutes. It is advisable to remove all traces of oxygen from the inert gas by passing it through a solution (alkaline pyrogallol, or Fieser's solution,[10] or preferably vanadyl

[10] A. I. Vogel, "A Textbook of Practical Organic Chemistry," p. 186. Longmans, London, 1956.

sulfate in 2 N sulfuric acid over zinc amalgam[11]) and then through distilled water to wash it.

The simplest way to achieve the deoxygenation of a solution in a cuvette is to fit it with a rubber bung or a Teflon stopper with two fine holes in it, through which two syringe needles are pushed. One of these is connected to a supply of the inert gas, and the other is left open as a gas outlet (Fig. 2). After deoxygenation, the syringe needles are withdrawn and, in the case of the Teflon stopper, the small apertures are sealed with grease.

When the reaction mixture contains a protein, this deoxygenation technique may cause an unacceptable amount of frothing. This can be overcome by adding an inert antifoaming agent (e.g., caprylic alcohol) to the solution or by applying a thin ring of silicone grease (e.g., Hopkin Williams Silicone MS Antifoam A) to the inside of the cuvette just above the surface of the solution. If these methods interfere with the protein, the inert gas supply needle can be lifted to just above the surface of the solution. Occasional shaking of the solution during deoxygenation is now necessary, and the time allowed should be increased to 1–2 hours.

This method can be used for the deoxygenation of larger and more complicated vessels than a cuvette, under conditions where it may be possible to monitor oxygen concentration electrochemically (by polarography or an oxygen electrode).

Another common method is repeated degassing of frozen solutions under vacuum with equilibration of the solutions with an inert gas on

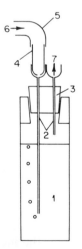

Fig. 2. Arrangement for deoxygenation of a reaction mixture in a spectroscopic cuvette. (1) Solution; (2) syringe needles; (3) rubber bung or Teflon stopper; (4) glass joint; (5) rubber tube; (6) nitrogen inlet; (7) nitrogen outlet.

[11] L. Meites and T. Meites, *Anal. Chem.* **20**, 984 (1948).

warming to room temperature in between the evacuations. This method may provide a slightly better deoxygenation, but it is more time consuming and costly (a mercury vapor pump is usually required in addition to a normal vacuum pump) and is generally unnecessary.

Reaction rates are measured by illuminating the anaerobic solution and observing the change in concentration of oxidized flavin with time. The light intensity required to promote a reasonable rate of reaction (optical density falling to one half during 5 minutes of illumination) varies with the concentration of reactants and the nature of the reducing substrate, but it is typically in the range 1000–4000 lumens per square foot.

There are three main ways to follow flavin concentrations, spectrophotometry, spectrofluorometry, and polarography.

Spectrophotometry. The long wavelength part of the absorption spectrum of riboflavin is shown in Fig. 3. The long wavelength absorption maximum is at 447 mμ, and most simple derivatives have maxima within 5 mμ of this. The long wavelength maximum of reduced flavin is in the near ultraviolet, and it has a negligible component at 447 mμ.[12] Thus, in simple photoreductions the change in optical density at this wavelength is a direct measure of the change in concentration of oxidized flavin.

Spectrofluorometry. Riboflavin has a fluorescence maximum at 530 mμ, but the reduced form has no visible fluorescence.[13] Changes in fluorescence at this wavelength (exciting at 440 mμ) are thus a measure of changes in the concentration of oxidized flavin, but the relationship is not linear above $\sim$2 × 10^{-5} M flavin because of the effects of self-quenching and reabsorption. This does not affect the usefulness of the method for comparing relative reaction rates, but it does for absolute measurements. Overall spectrofluorometry is less convenient than spectrophotometry except for very dilute flavin solutions ($\sim$10^{-6} M), where the extra sensitivity of fluorometry is an advantage, and for opaque samples, where front face fluorescence measurements alleviate some of the problems of light scattering.

Polarography. Riboflavin has a polarographic half-wave potential of -0.48 V, and its derivatives fall mainly in the range -0.2 to -0.5 V (Table I). The polarographic behavior of a reacted solution depends on the precise nature of the reducing agent, but reduced flavins have an anodic wave with a half-wave potential in the region of -0.2 V; lumichrome, another common reaction product, has its potential at -0.58 V. Thus, by following the changes in overall shape of the polarographic wave during reaction, it is possible to observe the loss of oxidized flavin and the build-up of products.[14] Application of this technique produces useful results, but

[12] H. Beinert, *in* "The Enzymes" (P. D. Boyer, H. Lardy, and K. Myrbäck, eds.), Vol. 2, p. 340. Academic Press, New York, 1960.

[13] G. Weber, *Biochem. J.* **47**, 114 (1950).

[14] M. M. McBride and W. M. Moore, *Photochem. Photobiol.* **6**, 103 (1967).

TABLE I
HALF-WAVE REDUCTION POTENTIALS OF ISOALLOXAZINES

Compound	Temperature (°C)	pH	Half-wave reduction potential (volts)	Reference
Riboflavin	20	3.0	+0.013	m
	20	4.0	−0.045	m
	20	5.0	−0.109	m
	20	6.0	−0.161	m
	20	7.0	−0.200	m
	20	7.0	−0.186	a
	20	7.0	−0.195	e
	20	8.0	−0.233	m
	20	9.0	−0.260	m
	20	10.0	−0.301	m
	20	11.0	−0.350	m
	20	12.0	−0.410	m
	25	8.0	−0.215	j
	30	5.0	−0.117	b
	30	6.0	−0.170	b
	30	7.0	−0.208	b
	30	7.0	−0.220	f
	30	7.0	−0.185	g
	30	8.0	−0.238	b
FMN	20	3.0	−0.013	m
	20	4.0	−0.045	m
	20	5.0	−0.102	m
	20	6.0	−0.161	m
	20	7.0	−0.206	m
	20	7.0	−0.187	a
	20	8.0	−0.242	m
	20	9.0	−0.271	m
	20	10.0	−0.303	m
	20	11.0	−0.352	m
	20	12.0	−0.405	m
	22	8.0	−0.248	n
	22	8.0	−0.228	j
	30	7.0	−0.219	i
	30	7.0	−0.190	g
FAD	22	8.0	−0.230	j
	25	5.0	−0.100	l
	25	6.0	−0.170	l
	25	7.0	−0.220	l
	25	8.0	−0.250	l
	30	7.0	−0.219	i

TABLE I (*Continued*)

Compound	Temperature (°C)	pH	Half-wave reduction potential (volts)	Reference
6,7,9-Trimethylisoalloxazine (lumiflavin)	0	6.9	−0.181	a
	18	6.9	−0.201	a
	20	4.7	−0.089	a
	20	7.0	−0.207	a
	22	8.0	−0.215	j
	30	5.0	−0.132	b
	30	6.0	−0.185	b
	30	7.0	−0.223	b
	30	7.0	−0.227	c
	30	8.0	−0.258	b
	38	6.9	−0.216	a
6,8,9-Trimethylisoalloxazine	20	4.7	+0.024	a
	20	7.0	−0.109	a
5,6-Dimethyl-9-D-ribitylisoalloxazine (isoriboflavin)	22	8.0	−0.208	n
6,7-Dimethyl-9-D-xyloisoalloxazine phosphate	20	7.0	−0.179	a
6,7-Dimethyl-9-D-glucoisoalloxazine	30	7.0	−0.208	b
6,7-Dimethyl-9-L-araboisoalloxazine	30	7.0	0.208	b
6,7-Dimethyl-9-hydroxyethylisoalloxazine	22	8.0	−0.205	j
		6.5–7.0	−0.25	k
6,7-Dimethyl-9-formylmethylisoalloxazine		6.5–7.0	−0.21	k
6,9-Dimethylisoalloxazine	20	4.7	−0.059	a
	20	7.0	−0.176	a
3,9-Dimethylisoalloxazine	20	4.7	−0.055	a
	20	7.0	−0.175	a
9-Methylisoalloxazine	20	4.7	−0.048	a
	20	7.0	−0.167	a
	22	8.0	−0.190	j
	30	5.0	−0.092	b
	30	6.0	−0.145	b
	30	7.0	−0.183	b
	30	8.0	−0.213	b
9-Phenylisoalloxazine	20	4.7	0.000	a
	20	7.0	−0.126	a
9-Cylohexylisoalloxazine	20	4.7	−0.012	a
	20	7.0	−0.134	a
9-Dioxypropylisoalloxazine	20	4.7	−0.034	a
9-Oxyethylisoalloxazine	20	4.7	−0.036	a
	20	7.0	−0.156	a
9-Acetoisalloxazine	20	4.7	−0.036	a
9-Dimethylaminoethylisoalloxazine	22	8.0	−0.140	j
6-Methyl-9-phenylisoalloxazine	20	4.7	−0.002	a
6-Methyl-9-dimethylaminoethylisoalloxazine	22	8.0	−0.105	j
Tetra acetylriboflavin	20	7.0	−0.19	d

(*Continued*)

TABLE I (*Continued*)

Compound	Temperature (°C)	pH	Half-wave reduction potential (volts)	Reference
1,6,7,9-Tetramethylisoalloxazine	20	7.0	−0.215	d
1,3,6,7,9-Pentanethylisoalloxazine	20	7.0	−0.20	d
5,6-Benzo-9-methylisoalloxazine	20	12.8	−0.50	a
6,7-Dichloro-9-D-riboisoalloxazine		7.0	−0.095	h
6,7-Dichloro-9-methylisoalloxazine	22	8.0	−0.088	j
6-Chloro-9-diethylaminopropylisoalloxazine	22	8.0	−0.056	j
7-Chloro-9-diethylaminopropylisoalloxazine	22	8.0	−0.070	j
6-Chloro-7-methoxy-9-(ethyl-2'-N-pyrrolidino) isoalloxazine	22	8.0	−0.140	j

[a] R. Kuhn and P. Boulanger, *Ber. Deut. Chem. Ges.* **69B,** 1557 (1936).
[b] L. Michaelis, M. P. Schubert, and C. V. Smythe, *J. Biol. Chem.* **123,** 587 (1936).
[c] K. G. Stern, *Nature* **133,** 178 (1934).
[d] R. Kuhn and G. Moruzzi, *Ber. Deut. Chem. Ges.* **67,** 1220 (1934).
[e] R. Brdička and E. Knobloch, *Z. Elektrochem.* **47,** 721 (1941).
[f] F. J. Stare, *J. Biol. Chem.* **112,** 223 (1935).
[g] C. S. Vestling, *Acta Chem. Scand.* **9,** 1600 (1955).
[h] R. Kuhn, F. Weygand, and E. F. Moller, *Ber. Deut. Chem. Ges.* **76B,** 1044 (1943).
[i] H. J. Lowe and W. M. Clarke, *J. Biol. Chem.* **221,** 983 (1956).
[j] I. M. Gascoigne and G. K. Radda, *Biochim. Biophys. Acta* **131,** 498 (1967).
[k] M. M. McBride and W. M. Moore, *Photochem. Photobiol.* **6,** 103 (1967).
[l] B. Ke, *Arch. Biochem. Biophys.* **68,** 330 (1957).
[m] R. D. Draper and L. L. Ingram, *Arch. Biochem. Biophys.* **125,** 802 (1968).
[n] G. R. Penzer and G. K. Radda, unpublished observations.

neither the experimental measurements nor the analysis of the results is as straightforward as for the spectrophotometric method.

The observed reaction time curves are most easily interpreted from the gradient at $t = 0$ because photochemical reactions are subject to many complex quenching processes and product inhibition. The shape of the kinetic curves provides a check that deoxygenation has been complete, because in the presence of oxygen the reaction shows an induction period.

Using the simple spectrophotometric technique, it is possible to achieve better than 5% reproducibility for the initial rates.

Characterization of the Nature of the Reaction

Reactive Species

In a photochemical reaction, the reactive species may be either an excited singlet or a triplet of the chromophore. These may be distinguished by one or more of the following methods.

Quenching Experiments. The extent to which quenchers deactivate the first excited singlet is directly measured by the reduction in fluorescence, and so inhibition of the reactions of the first excited singlet parallel fluorescence quenching. If a photochemical reaction is inhibited by compounds (like some paramagnetic ions) that have no effect on fluorescence, or at concentrations where fluorescence quenching is negligible, a triplet mechanism is implicated. For instance, 10^{-5} M potassium iodide will inhibit the reactions of triplet flavins, but fluorescence quenching becomes important only in solutions stronger than 10^{-3} M.

Lifetime Calculations. The lifetime of the first excited singlet of riboflavin is 4.2×10^{-9} seconds,[15] while that of the triplet in a rigid matrix at room temperature is 37×10^{-3} seconds.[16] By making simplifying assumptions, it is possible to calculate a minimum lifetime for the reactive species in a particular photochemical reaction and to compare this with the lifetimes expected for the various reactive species.

The simplest form of rate equation for a photochemical reaction is

$$\phi = \frac{k_E k_D (X):}{Q + k_D(X)}$$

where ϕ is the quantum yield of reaction; k_E is the efficiency with which the reactive species is formed from the product of the primary light absorption and must be between 0 and 1; k_D is the second-order rate constant for a diffusion-controlled reaction; (X) is the concentration of the compound that reacts with the excited state; and Q is a term to cover all deactivating processes for the reactive species other than reaction.

ϕ can be measured (see below), a likely value for k_E can be assumed (0.5 is reasonable for riboflavin), (X) is known, k_D can be calculated to a first approximation from the expression $k_D = (SRT)/(3 \times 10^3 \eta) M^{-1} sec^{-1}$ (where η is the viscosity of the solution).[17] Q can then be calculated. $1/Q$ now gives the minimum lifetime the reactive species must have to react with the measured quantum yield, as it is not possible to exceed the diffusion-controlled rate. When $1/Q$ is longer than the fluorescence lifetime (10^{-8} to 10^{-9} seconds), the triplet is the likely reactive species.

Light Intensity Dependence. Measurements of light intensity dependence can give useful information about the nature of the reaction. It is important that in such experiments the spectral composition of the incident light is constant. This can be achieved by placing neutral density filters in the light path. Changing the lamp operating current can only give a qualitative measure, since the color temperature of the light source alters.

Generally the rates are proportional to light intensity, but for biphotonic

[15] R. F. Chen, G. G. Vurek, and N. Alexander, *Science* **156**, 949 (1967).
[16] G. K. Radda and A. R. Watson, *Chem. Commun.* p. 469 (1968).
[17] J. Umberger and V. LaMer, *J. Am. Chem. Soc.* **67**, 1099 (1945).

processes they depend on its square. Alternatively, at high light intensities the rate of a triplet reaction may cease to be proportional to the incident light intensity because of second-order deactivation, e.g., $T + T \to S^* + S$. This has been observed for the anaerobic photoreduction of FMN by ethylenediaminetetraacetic acid.[9] Second-order excited singlet deactivation is unlikely, but saturation of the ground state by exciting light of very high intensity is possible, and would produce similar experimental results to triplet–triplet deactivation. Saturation can be detected by the effect of varying incident light intensity on fluorescence intensity, and can thus be distinguished from the triplet–triplet effect.

Direct Methods. Flash photolysis[18] and delayed light emission[16] have both been used to identify flavin-excited states. Both are potentially powerful methods, but they require elaborate instrumentation.

Intermediates

Reaction intermediates can be characterized by the usual physical techniques in flash experiments,[19] and may be indirectly inferred from kinetics and from the effects of structural changes in both flavin and substrate.[9]

Quantum Yield

The simplest way to measure the quantum yield of a flavin photoreaction is to compare the rate of the unknown reaction with that of a reaction of known quantum yield under identical conditions. Known flavin reactions which span a reasonable range of rates are tabulated.

Reaction	Quantum yield	Reference
Anerobic photobleaching of FMN alone	0.006	20
Anerobic photoreduction of FMN by EDTA ($10^{-3} M$) at $20°$	0.06	3
Anaerobic photoreduction of FMN by NADH ($10^{-3} M$) at $20°$	0.25	3

The comparison method is simple to apply, but it is not as accurate as methods that involve direct measurement of the output of the light source. The radiation is first monochromated, and its intensity (I_0) is measured either by using a standard actinometer (e.g., ferrioxalate[21]), or a thermopile (which must itself be calibrated against a standard light

[18] A. Knowles and E. M. F. Roe, *Photochem. Photobiol.* **7**, 421 (1968).
[19] B. Holmström, *Arkiv Kemi* **22**, 329 (1964).
[20] B. Holmström and G. Oster, *J. Am. Chem. Soc.* **83**, 1867 (1961).
[21] C. G. Hatchard and C. A. Parker, *Proc. Roy. Soc. (London) Ser. A* **235**, 518 (1956).

source). It is then possible to calculate the total number of quanta absorbed by the chromophore at the excitation wavelength from I_0, the optical density, and the period of the illumination.

Action Spectrum

The action spectrum of photoreaction of a mixture of chromophores supplies information about the primary light acceptor. This may provide the only way to identify the relevant chromophore in a complex biological system.[22] When measuring action spectra, it is necessary to use monochromatic incident light of known energy distribution with wavelength. A very simple way to achieve these conditions is to follow flavin concentration spectrofluorimetrically and to use the light source of the instrument to induce photochemical reaction. To avoid diffusion problems the flavin may be held in the solid matrix of a methyl cellulose film, but the same method applies for solutions, providing the volume of solution illuminated is kept constant.

Film Preparation. Powdered methyl cellulose, 2.25 g, is added to 100 ml of water at 60° with constant stirring. The suspension is cooled to 0° on ice, when the methyl cellulose dissolves to give a transparent and viscous solution. The reactants (flavin, reducing substrate, etc.) are stirred into this solution carefully to avoid forming bubbles. Thin layers of this mixture are poured onto glass microscope slides and allowed to dry (about 24 hours). The films produced in this way are about 0.0025 cm thick, and the included molecules retain some microscopic freedom of motion but little macroscopic freedom.[23] Diffusion of oxygen into the films is slow, and the photochemical behavior of flavins in this situation closely resembles their anaerobic solution photochemistry.

Lamp Calibration. The quantum yield of fluorescence of FMN at 530 mμ is independent of the wavelength of excitation between 350 and 500 mμ. Thus by measuring the excitation spectrum of a solution of FMN and comparing it with the absorption spectrum it is possible to calculate the spectral distribution of the light emitted by the lamp. In Fig. 3 the variation of b/a with wavelength gives this distribution.

Measurement of the Action Spectrum. The fluorescence of a chromophore held in a film must be observed from the face of the sample. By varying the excitation wavelength and the slitwidths of the instrument it is usually possible to find conditions under which plenty of fluorescence is observed while there is no bleaching of the emitter. All concentrations are measured under these conditions throughout the entire observation of the action

[22] A. D. McLaren and D. Shugar, "Photochemistry of Proteins and Nucleic Acids," p. 320. Macmillan (Pergamon), New York, 1964.
[23] G. R. Penzer and G. K. Radda, *Nature* **213**, 251 (1967).

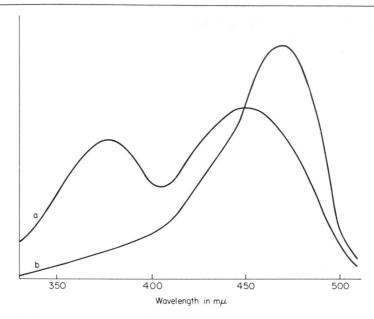

Wavelength in mμ

FIG. 3. Absorption spectrum (a) and fluorescence excitation spectrum (b) for the calibration of a xenon lamp using flavin mononucleotide solution.

spectrum. To cause reaction, the desired wavelength is selected on the excitation monochromator, its slits are opened, and the sample is illuminated for a measured time. Now the instrument is reset to the optimal observation conditions and the change in fluorescence of the flavin is noted. In this way, kinetic runs can be performed at a series of excitation wavelengths, and by correcting the initial rates thus measured for the variation in lamp intensity, a true action spectrum is obtained.

This technique is not applicable for those spectral regions where methyl cellulose absorbs, which limits it to wavelengths greater than 350 mμ.

Other methods for both exciting and observing the flavins can readily be adapted for the measurement of action spectra.

Reaction Products

When reduced flavin is a reaction product, it can be identified by its spectrum and by the fact that it is quantitatively reoxidized to the oxidized form by oxygen. Polarography is more discerning because it distinguishes some of the derivatives of reduced flavin (in which the N_9 side chain has been partially degraded) by their different half-wave potentials (Table I). The most thorough method of all is to identify the reoxidized reaction products by thin-layer chromatography.[24] A suitable support is silica gel

[24] M. M. McBride and D. E. Metzler, *Photochem. Photobiol.* **6,** 113 (1967).

G (thickness 250 μ), with a solvent system of butanol–ethanol–water, 7:2:1. The spots are most easily detected on the chromatogram by their fluorescence under ultraviolet light.

Some photochemical reactions of flavins produce air-stable adducts between substrate and chromophore.[8] These can be characterized by their

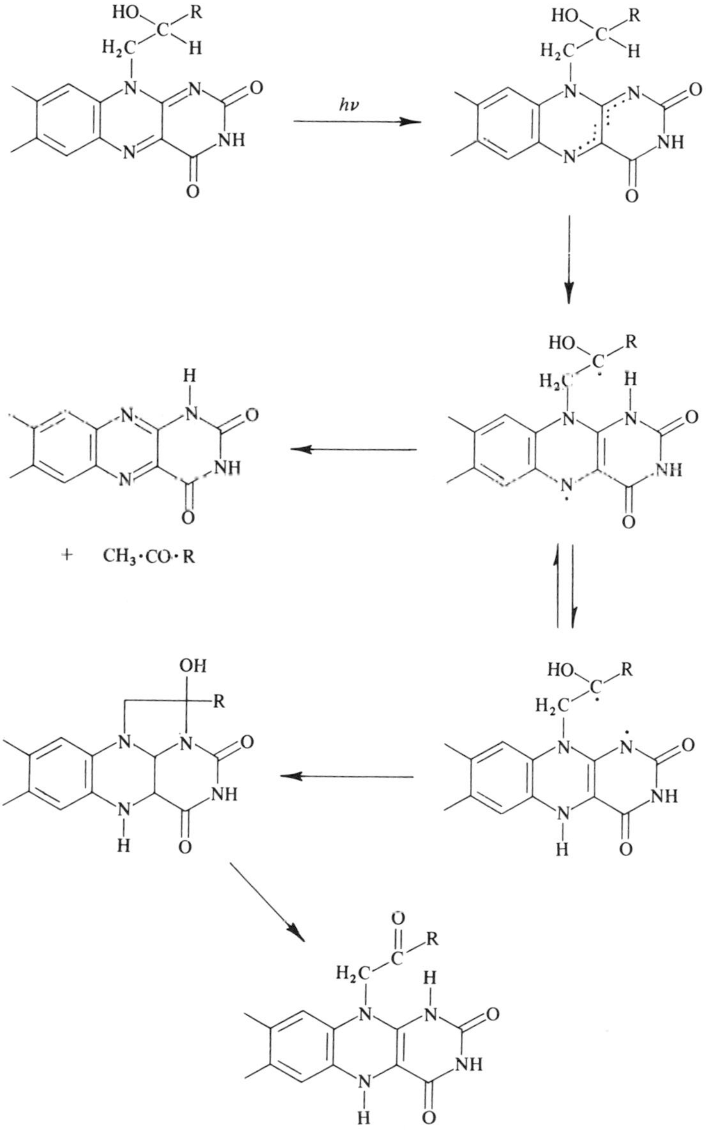

FIG. 4. The mechanism of anaerobic photobleaching.

TABLE II

INHIBITORS OF THE ANAEROBIC PHOTOREDUCTION OF FLAVINS

Flavin	Added reducing agent	Inhibitor	Reaction conditions	% Inhibition	Reference
FMN	None	DL-Histidine (10^{-3} M)	pH 7.0, 30°	0	a
	None	DL-Phenylalanine (10^{-3} M)	pH 7.0, 30°	0	a
	None	DL-Tryptophan (10^{-3} M)	pH 7.0, 30°	89	a
	None	DL-Tryptophan (4×10^{-4} M)	pH 7.0, 17°	96	b
	None	DL-Tryptophan (4×10^{-4} M)	pH 7.0, 38°	92	b
	None	L-Tyrosine (10^{-3} M)	pH 7.0, 30°	90	a
	None	Caffeine (10^{-3} M)	pH 7.0, 30°	65	a
	None	Muscle adenylic acid (10^{-3} M)	pH 7.0, 30°	41	a
	None	Phenol (10^{-3} M)	pH 7.0, 30°	97	a
	None	Serotonin creatinine sulfate (5×10^{-5} M)	pH 7.0, 30°	2.5	d
	None	Serotonin creatinine sulfate (10^{-4} M)	pH 7.0, 30°	57	d
	None	Serotonin creatinine sulfate (2×10^{-4} M)	pH 7.0, 30°	86	d
	None	Serotonin creatinine sulfate (10^{-3} M)	pH 7.0, 30°	94	a,d
	None	Serotonin creatinine sulfate (10^{-4} M)	pH 7.0, 17°	96	d
	None	Serotonin creatinine sulfate (10^{-4} M)	pH 7.0, 43°	43	d
	None	Creatinine (10^{-3} M)	pH 7.0, 30°	0	a
	None	Methyl viologen (10^{-3} M)	pH 7.0, 17°	0	a
	None	CMU[e] (10^{-3} M)	pH 7.0, 35°	86	a
	None	CMU (10^{-3} M)	pH 7.0, 19°	95	a
	None	CMU (4×10^{-4} M)	pH 7.0, 38°	76	b
	None	CMU (4×10^{-4} M)	pH 7.0, 17°	82	b
	None	2,4-Dinitrophenol (10^{-3} M)	pH 7.0, 30°	100	a
	None	Urea (6.5 M)	pH 7.0, 25°	16	b
	None	Barbiturate (3×10^{-3} M)	pH 9.0, 20°	100	c
FMN	EDTA ($\sim 10^{-3}$ M)	Serotonin creatinine sulfate (10^{-3} M)	pH 7.0, 30°	80	a
	NADH ($\sim 10^{-3}$ M)	Serotonin creatinine sulfate (10^{-4} M)	pH 7.0, 20°	78	b
	NADH ($\sim 10^{-3}$ M)	Serotonin creatinine sulfate (10^{-4} M)	pH 7.0, 40°	71	b
	NADH ($\sim 10^{-3}$ M)	CMU (10^{-3} M)	pH 7.0, 30°	65	a
	NADH ($\sim 10^{-3}$ M)	CMU (5×10^{-4} M)	pH 7.0, 10°	63	b

			pH, temp		
	NADH (~10⁻³ M)	CMU (2 × 10⁻⁴ M)	pH 6.6, 30°	50	f
	NADH (~10⁻³ M)	DCMU (2 × 10⁻⁶ M)	pH 6.6, 30°	50	f
	NADH (~10⁻³ M)	DL-Tryptophan (5 × 0⁻⁴ M)	pH 7.0, 10°	63	b
	NADH (~10⁻³ M)	DL-Tryptophan (5 × 0⁻⁴ M)	pH 7.0, 40°	68	b
	NADH (~10⁻³ M)	DL-Tryptophan (2 × 10⁻⁴ M)	pH 6.6, 30°	50	f
	NADH (~10⁻³ M)	Phenol (1.25 × 10⁻⁴ M)	pH 6.6, 30°	50	f
	NADH (~10⁻³ M)	L-Tyrosine (2.9 × 10⁻⁴ M)	pH 6.6, 30°	50	f
FAD	NADH (~10⁻³ M)	Urea (6.5 M)	pH 7.0, 25°	0	b
FMN	NADH (5 × 10⁻⁴ M)	Adenine in FAD relative to FMN	pH 7.0, 30°	92	a
	NADH (5 × 10⁻⁴ M)	Phenol (10⁻³ M)	pH 7.0, 30°	68	a
	NADH (5 × 10⁻⁴ M)	Serotonin creatinine sulfate (10⁻³ M)	pH 7.0, 30°	69	a
	NADH (5 × 10⁻⁴ M)	NAD (10⁻³ M)	pH 7.0, 30°	0	a
	NADH (5 × 10⁻⁴ M)	DCMU (10⁻³ M)	pH 7.0, 17°	43	a
FAD	NADH (5 × 10⁻⁴ M)	DL-Lipoic acid (10⁻³ M)	pH 7.0, 17°	39	a
	NADH (5 × 10⁻⁴ M)	Adenine in FAD relative to FMN	pH 7.0, 25°	87	a
Lumiflavin	EDTA (10⁻³ M)	CMU (2 × 10⁻⁴ M)	pH 6.6, 30°	50	f
	EDTA (10⁻³ M)	DCMU (2 × 10⁻⁴ M)	pH 6.6, 30°	50	f
	EDTA (10⁻³ M)	DL-Tryptophan (2 × 10⁻⁴ M)	pH 6.6, 30°	50	f
	EDTA (10⁻³ M)	Phenol (1.25 × 10⁻⁴ M)	pH 6.6, 30°	50	f
Lumiflavin	EDTA (10⁻³ M)	L-Tyrosine (2.9 × 10⁻⁴ M)	pH 6.6, 30°	50	f
	EDTA (10⁻³ M)	DL-Phenylglycine (5 × 10⁻³ M)	pH 7.0, 30°	0	b
	EDTA (10⁻³ M)	DL-m-Fluorophenylglycine (5 × 10⁻³ M)	pH 7.0, 30°	0	b
	EDTA (10⁻³ M)	DL-m-Chlorophenylglycine (5 × 10⁻³ M)	pH 7.0, 30°	0	b
	EDTA (10⁻³ M)	DL-p-Methylphenylglycine (5 × 10⁻³ M)	pH 7.0, 30°	0	b
	EDTA (10⁻³ M)	DL-p-Fluorophenylglycine (5 × 10⁻³ M)	pH 7.0, 30°	0	b
	EDTA (10⁻³ M)	DL-p-Methoxyphenylglycine (5 × 10⁻³ M)	pH 7.0, 30°	0	b
	EDTA (10⁻³ M)	DL-m-Nitrophenylglycine (5 × 10⁻³ M)	pH 7.0, 30°	94	g

[a] G. K. Radda, Biochim. Biophys. Acta 112, 448 (1966).
[b] G. R. Penzer and G. K. Radda, Biochem. J. 109, 259 (1968).
[c] A. Giuditta and A. Vitale-Neugebauer, Biochim. Biophys. Acta 110, 32 (1965).
[d] G. K. Radda and M. Calvin, Nature 200, 464 (1963).
[e] CMU, 3-(p-chlorophenyl)1,1-dimethylurea, DCMU, 3-(2,4-dichlorophenyl)-1,1-dimethylurea.
[f] P. Homann and H. Gaffron, Photochem. Photobiol. 3, 499 (1964).
[g] G. R. Penzer and G. K. Radda, unpublished observations.

optical and magnetic resonance spectra, and their atomic constitution. A simple test for the formation of an adduct of this type is the effect of oxygen on the photolyzed reaction mixture. The rate at which the optical density at 447 mμ increases is slow and may be enhanced by illumination of a mixture containing the adducts, but it will be very rapid and unaffected by light for a solution of reduced flavin.

Mechanisms of Photoreduction

The following mechanisms for flavin photoreduction by various substrates have been suggested. In all of them the reactive species has been shown to be the flavin triplet.

For riboflavin, FMN, and other derivatives with hydroxyl or amino side chains in the absence of added substrates, an intramolecular reaction is postulated on the basis of isotope effects and of other kinetic studies (Fig. 4).[3,25]

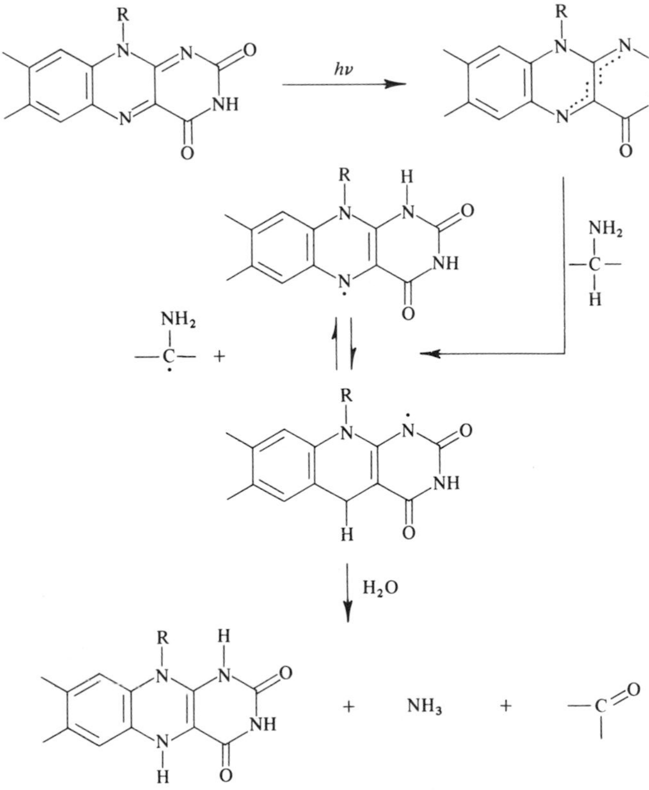

FIG. 5. The mechanism of anaerobic photoreduction.

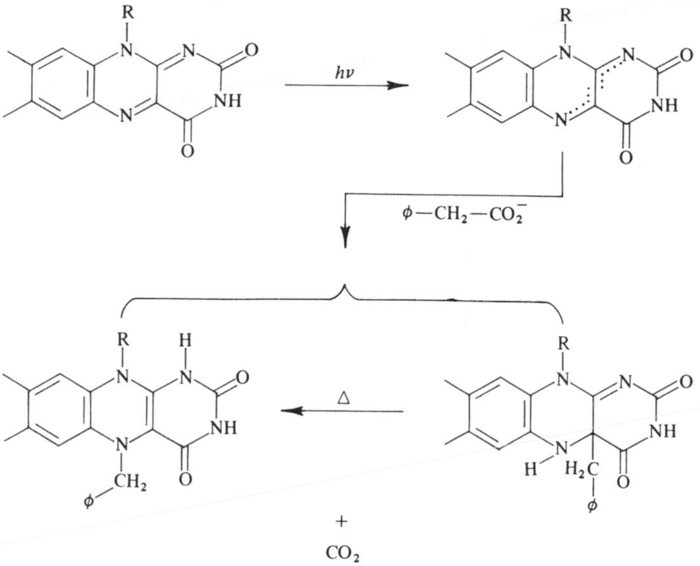

FIG. 6. The mechanism of anaerobic photoaddition.

Reduction of flavins by alcohols, amino acids, and amines (only in the unprotonated form) is generally quicker than the intramolecular reaction, and probably occurs by the two-step mechanism shown in Fig. 5.[9] Photoreaction between flavins and phenylacetic acid leads to the formation of adducts, as shown in Fig. 6.[8]

Inhibition of Photoreduction

All these reactions are inhibited by many aromatic compounds (see Table II) as well as by inorganic ions [e.g., Mn (II) and KI]. The aromatic quenching probably involves complex formation between the quencher and triplet flavin.[6] This places a limitation on the use of quantitative flavin photochemistry in the presence of the inhibitors, but it does not necessarily restrict the use of photosensitized reactions to produce particular chemical modifications in, for example, a protein (see this volume [153]).

Acknowledgment

Part of this work was supported by the Medical and Science Research Councils of Great Britain.

---

[25] W. M. Moore, J. T. Spence, F. A. Raymond, and S. D. Colson, *J. Am. Chem. Soc.* **85,** 3367 (1963).

[153] Flavins As Photosensitizers

By M. B. Taylor and G. K. Radda

General Considerations

When light energy is absorbed by a chromophore and this energy is passed on to other molecules which themselves cannot absorb at the excitation wavelength, we speak of photosensitization. The excitation energy may be directly transferred to the acceptor, bringing it to its excited state, or it may be utilized in some reaction between the sensitizer and acceptor molecules (e.g., in a redox reaction) with subsequent regeneration of the sensitizer by a secondary process. The latter process is not a true photosensitization, but mechanistically it is not always possible to distinguish the two processes on available data, so that we shall discuss both in this article.

Flavins, like many other highly absorbing dyes, such as fluoresceins, methylene blue, and acridine orange, can sensitize the photooxidation of a large variety of substrates[1,2] (e.g., amines, amino acids, nucleotides, and metal ions). They can also bring about isomerization[3,4] and polymerization[5] of olefins and deiodination of aromatic iodo compounds.[6] Thus photosensitization by flavins may be used as a method for introducing a given amount of energy into a system and producing reasonably selective reactions. For instance, the sensitized photooxidation of enzymes leads to destruction of some of the amino acids.[7] DNA can also be selectively modified by this method.[8] Since flavins are present in many parts of living cells, they may possibly be used as a method of introducing changes in the cell.

Photooxidations

Apparatus

The components of the photochemical set up are similar to those described in the companion paper,[9] except that the reaction vessel is designed

[1] G. R. Penzer and G. K. Radda, *Quart. Rev. (London)* **21**, 43 (1967).

[2] H. Beinert, *in* "The Enzymes" (P. D. Boyer, H. Lardy, and K. Myrbäck, eds.), Vol. 2, p. 339. Academic Press, New York, 1960.

[3] J. Posthuma and W. Berends, *Biochim. Biophys. Acta* **112**, 422 (1966).

[4] A. Gordon Walker and G. K. Radda, *Nature* **215**, 1483 (1967).

[5] G. Oster, *Nature* **173**, 300 (1954).

[6] S. Lissitzky, M. T. Benevent, and M. Rogues, *Biochim. Biophys. Acta* **51**, 407 (1961).

[7] A. D. McLaren and D. Shugar, *in* "Photochemistry of Proteins and Nucleic Acids," Macmillan (Pergamon), New York, 1964.

to suit the particular method chosen to follow the reaction. The most general method is to follow oxygen uptake by manometry or by an oxygen electrode. We shall describe the use of the oxygen electrode only as it is the more accurate method of the two and is more convenient to use.

Procedure

A thermostatted reaction vessel (approximate size 10–15 ml) with clear glass sides is blacked out on three sides and placed in the light beam. Lenses and filters are necessary for quantitative work.[9] The entrance of the vessel should be tapered and ground to fit an oxygen electrode assembly (E.I.L. Model SOH33). We found that a Perspex stopper of the design shown in Fig. 1 is efficient in excluding air. The reaction vessel is slightly

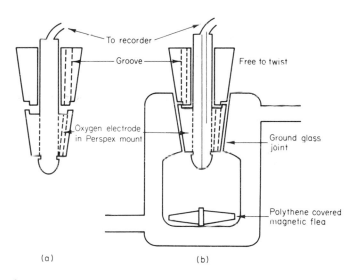

Fig. 1. Oxygen electrode in thermostatically controlled reaction vessel. (a) Electrode ready for insertion with grooves aligned to remove displaced reactants. (b) Final position, showing how the grooves are rotated to provide an efficient water seal.

overfilled with the reactants, and the electrode assembly is placed in the vessel. Care must be taken to see that all air bubbles are forced out of the side grooves. The temperature and pressure of the solution in contact with the electrode membrane are extremely critical, so that temperature equilibration of the trapped solution should be allowed before the electrodes are sealed. Care should also be taken to avoid illumination of the electrode membrane, which could lead to anomalous results. Stirring in the vessel

[8] J. S. Sussenbach and W. Berends, *Biochim. Biophys. Acta* **76,** 154 (1963).
[9] G. R. Penzer and G. K. Radda, this volume [152].

TABLE I
FLAVIN PHOTOSENSITIZED REACTIONS

Compound photooxidized	Method of study	Other details	Reference[a]
Various amino acids	O_2 Electrode	—	1
Phenylglycine, and substituted derivatives	O_2 Electrode	—	2
Cysteine and derivatives			3
Dihydroxyphenylalanine	Manometry; colorimetric analysis	Inhibition by Cu^{2+} investigated	4
Histidine, tyrosine, bovine serum albumin	Colorimetric tests after paper chromatography	—	5
Tyrosine, p-cresol, ascorbic acid, dihydroxymaleic acid, phloroglucinol	Manometry	H_2O_2 formed as by-product	6
Sarcosine, N,N-dimethylglycine	Manometry	—	7
Methionine, EDTA	EMF changes	—	8
Urease, tyrosinase, α-amylase	Change in activity	—	9
Bacterial α-amylase	Inactivation	—	10
Trypsin	Inactivation	Quantum yield stated in (13)	11, 12, 13
Taka-amylase A, tryptophan, tyrosine, histidine	Inactivation, manometry	—	14 14
Tyrosinase	Inactivation	—	15
Clostridium perfringens toxin	Inactivation	—	16
Snake venom	Inactivation	—	17
Guanine	Spectral changes	Lumichrome-sensitized oxidation	18
Purines	Flash photolysis and spectrophotometry	Mainly anaerobic	19
Purines and pyrimidines	Manometry	Comparison with other dyes	20
NAD	Spectrophotometry	—	21
NAD and NADP	Enzymatic assay of NAD, NADP	pH max = 4.0	22
Adenine	Spectra	Product is hypoxanthine	23
DNA	Changes in melting curves	—	24
DNA, p-toluenediamine	Inactivation of transforming principle in mice tumor cells	—	25
Soluble RNA from E. coli and TMV	Inactivation and loss of amino acid acceptor properties	—	26

TABLE I (*Continued*)

Compound photooxidized	Method of study	Other details	Reference[a]
RNA in *Saccharomyces* cells	Change in aggregation	—	27
DNA	Spectrophotometric method	—	28
Ascorbic acid	Manometry; pH-stat titrator	—	29
Ascorbic acid	—	Studied in rabbit eye humor	30
Indoleacetic acid	Colorimetric analysis	—	31, 32
Indoleacetic acid	—	—	33
Indoleacetic acid	Spectra	—	34
Indoleacetic acid	Inactivation	Riboflavin was found to be more efficient than other dyes examined	35
Monuron	Loss of herbicidal activity	—	36
EDTA, dimedone, oxalic acid, NaNO₂, phenyl-hydrazine, hydrazine, hydroxylamine, semicarbazone	O_2 uptake	—	— 37
Coenzyme A, ATP	Enzymatic assay	More rapid at pH 8.0	38
Deiodination of thyroxine	$^{131}I^-$ release	Comparison with enzymatic methods	39, 40
Thyroxine	Paper chromatographic separation of products		
Activation of flavin-sensitized thyroxine diodination by added proteins	Radioactive iodide release	—	41, 42
Deiodination of thyroxine	As above	—	43
Deiodination of thyroxine	—	Reaction fails in red light or if no O_2 is present. Catalase slows reaction.	44
		Serotonin prevents reaction	45
Photopolymerization of acrylamide, methacrylic acid	Adiabatic temperature rise	—	46, 47

[a] Reviews on photosensitization:
J. D. Spikes and C. A. Ghiron, *Phys. Proc. Rad. Biol. Proc. Intern. Symp. Mich. State Univ.* p. 309 (1963).
A. A. Krasnovskii, *J. Chim. Phys.* **55**, 968 (1958).
G. Oster, *J. Chim. Phys.* **55**, 899 (1958).

A. D. McLaren and D. Shugar *in* "Photochemistry of Proteins and Nucleic Acids," pp. 133–156, 313–319. Macmillan (Pergamon), New York, 1964.

[1] J. S. Bellin and C. A. Yankus, *Arch. Biochem. Biophys.* **123**, 18 (1968).

[2] G. R. Penzer, *Biochem. J.* **116**, 733 (1970).

[3] Y. Obata and H. Tanaka, *Agr. Biol. Chem. (Tokyo)* **29**, 196 (1965); *Chem. Abstr.* **63**, 2305h (1965).

[4] S. Isaka, *Chiba Daigaku Bunri Gakuba Kiyô. Shizen Kagaku* **1**, No. 4, 263 (1955); *Chem. Abstr.* **52**, 16046g (1958).

[5] Z. Vodrazka and J. Sponar, *Chem. Listy* **51**, 1649 (1957); *Chem. Abstr.* **52**, 1311i (1958).

[6] P. A. Kolesnikov, *Biokhimiya* **23**, 434 (1958).

[7] W. R. Frisell, C. W. Chung, and C. G. Mackenzie, *J. Biol. Chem.* **234**, 1297 (1959).

[8] G. Strauss and W. J. Nickerson, *J. Am. Chem. Soc.* **83**, 3187 (1961).

[9] A. W. Galston and R. S. Baker, *Science* **109**, 485 (1949).

[10] K. Sone, *Nippon Nogeikagaku Kaishi* **36**, 7 (1962); *Chem. Abstr.* **61** 4647f (1964).

[11] C. A. Ghiron and J. D. Spikes, *U.S. At. Energy Comm.* TID-20001, 164 pp. (1963).

[12] C. A. Ghiron and J. D. Spikes, *Photochem. Photobiol.* **4**, 13 (1965).

[13] B. W. Glad and J. D. Spikes, *U.S. At. Energy Comm.* C00-875-27, 106 pp. (1964).

[14] S. S. Kim, *Nippon Nogei Kagaku Kaishi* **40**, 73 (1966).

[15] K. Sone, *Bitamin* **27**, 13 (1963); *Chem. Abstr.* **60**, 8288g (1964).

[16] R. M. Bunkus, *Tr. Irkutsk. Nauchn.-Issled. Inst. Epidemiol. i Gigieny* No. 5, 181 (1960); *Chem. Abstr.* **61** 16675b (1964).

[17] R. Guidolin and R. G. Ferri, *O. Hospital (Brazil)* **38**, 737 (1950).

[18] W. Berends, J. Posthuma, J. S. Sussenbach, and H. I. X. Mager *in* "Flavins and Flavoproteins" (E. C. Slater, ed.), p. 22. Biochim. Biophys. Acta Library No. 8, 1966.

[19] A. Knowles and E. M. F. Roe, *Photochem. Photobiol.* **7**, 421 (1968).

[20] M. I. Simon and H. Van Vunakis, *Arch. Biochem. Biophys.* **105**, 197 (1964).

[21] A. A. Krasnovskii and G. P. Brin, *Dokl. Akad. Nauk SSSR* **153**, 212 (1963); *Chem. Abstr.* **60**, 7045g (1964).

[22] K. Uehara, T. Mizoguchi, Y. Okada, and J. Kuwashima, *J. Biochem. (Tokyo)* **59**, 433 (1966).

[23] K. Uehara, T. Mizoguchi, and S. Hosomi, *J. Biochem. (Tokyo)* **59**, 550 (1966).

[24] J. S. Bellin and L. I. Grossman, *Photochem. Photobiol.* **4**, 45 (1965).

[25] J. S. Bellin and G. Oster, *Biochem. Biophys. Acta* **42**, 533 (1960).

[26] A. Tsugita, Y. Okada, and K. Uehara, *Biochim. Biophys. Acta* **103**, 360 (1965).

[27] E. R. Lochmann, W. Stein, and C. Umlauf, *Z. Naturforsch.* **20b**, 778 (1965).

[28] J. S. Sussenbach and W. Berends, *Biochim. Biophys. Acta* **76**, 154 (1963).

[29] P. Homann and H. Gaffron, *Photochem. Photobiol.* **3**, 499 (1964).

[30] A. Pirie, *Nature* **204**, 500 (1965).

[31] T. Goto and D. Yamamoto, *Meiji Daigaku Nôkakubu Kenkyû Hômkoku*, No. 7, 35 (1958).

[32] A. W. Galston, *Science* **111**, 619 (1950).

[33] E. R. Waygood and G. A. Maclachlan, *Physiol. Plantarum*, **9**, 607 (1956).

[34] B. Nathanson, M. Brody, S. Brody, and S. B. Broyde, *Photochem. Photobiol.* **6**, 177 (1967).

[35] L. Brauner, *Naturwissenschaften* **39**, 282 (1952).

[36] P. B. Sweetser, *Biochim. Biophys. Acta* **66**, 78 (1963).

[37] P. A. Kolesnikov, *Dokl. Akad. Nauk SSSR* **133**, 1462 (1960); *Chem. Abstr.* **54** 22763a (1960).

[38] K. Uehara, T. Mizoguchi, Y. Okada, and J. Unenot, *J. Biochem. (Tokyo)* **59**, 556 (1966).

[39] C. Jacquemin, J. Nunez, L. Rappaport, D. Brun, and J. Roche, *Prepn. Bio-Med. Appl. Labelled Mol. Proc. Symp. Venice*, 1964.

[40] C. Jacquemin, J. Nunez, and J. Roche, *Gen. Comp. Endocrinol.* **3**, 226 (1963).

[41] S. Lissitzky, M. T. Benevent, and M. Rogues, *Biochim. Biophys. Acta* **51**, 407 (1961).

[42] M. Suzuki, I. Ishikawa, S. Shinizu, and K. Yamamoto, *Biochim. Biophys. Acta* **51**, 403 (1961).

[43] G. Escobar, R. L. Rodriguez, T. John, and F. E. del Rey, *J. Biol. Chem.* **238**, 3508 (1963).

[44] V. A. Galton and S. H. Ingbar, *Endocrinology* **70**, 210 (1962).

[45] M. T. Benevent, M. Rogues, J. Torresani, and S. Lissitzky, *Bull. Soc. Chim. Biol.* **45**, 969 (1963).

[46] G. Delzenne, W. Dewinter, S. Toppet, and G. Smets, *J. Polymer Sci. Pt A* **2**, 1069–83 (1964).

[47] G. Oster, *Nature* **173**, 300 (1954).

(with a magnetic stirrer) is essential and should be as fast as is compatible with nonturbulent motion.

If a constant voltage of -0.6 V is applied across the electrodes, the current flowing in the cell varies with oxygen concentration because it is controlled by diffusion of oxygen. The detector should be capable of recording the current in microamperes continuously. A polarograph (e.g., Radiometer PO 4b) combines both these functions, but less elaborate equipment can be used instead. The membrane sensitivity is not constant over long periods, and it is advisable to carry out calibration daily as follows. Two solutions with known oxygen concentrations should be used. These should be chosen to include the range being studied. Air-saturated water is generally chosen as one. It is prepared by sucking air through a thermostatted Dreschel bottle for about 1 hour. A solution containing zero concentration of oxygen is easily prepared photochemically as follows. A solution of riboflavin (10^{-4} M) and EDTA (10^{-3} M) at pH 9 is illuminated in the reaction vessel. Oxygen uptake can be followed on the recorder. When all the oxygen is used up, the current levels out. Deoxygenation using an inert gas usually leads to a sluggish response of the electrode. Further checks on the calibration may be made by using solutions equilibrated with oxygen at different partial pressures.

Other Methods for Following the Reaction

Optical density changes in the solution, monitored continuously or by sampling techniques, have been used to follow photosensitized reactions. Examples are shown in Table I.

The spectrophotometric cell may be coupled to the electrode vessel in an arrangement (Fig. 2) in which the rate of oxygen uptake (or changes

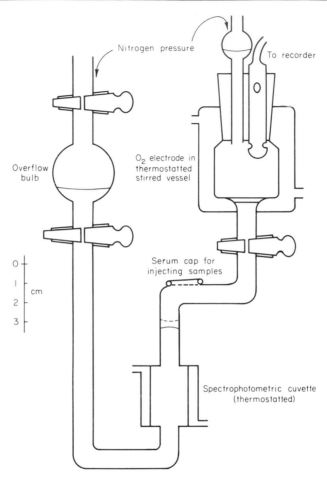

Nitrogen pressure

To recorder

Overflow bulb

O_2 electrode in thermostatted stirred vessel

cm

0
1
2
3

Serum cap for injecting samples

Spectrophotometric cuvette (thermostatted)

FIG. 2. Spectrophotometric cuvette coupled to oxygen electrode reaction vessel so that readings can be made on the same sample. The pressure of nitrogen moves the sample between the positions shown. Liquid levels marked by solid or dashed curve.

in redox potentials) and absorbancy changes can be recorded as a function of time. An inert gas (argon interferes least with the oxygen electrode) is used to move the solution between the observation chambers. Small electrodes may be directly fused onto the spectroscopic cuvette.

Flavins sensitize the photoinactivation of many biological substances so that changes in activity can be used to follow the reaction.[7] For proteins the results are difficult to interpret because several amino acids are affected. Relative rates of photooxidation of amino acids sensitized by riboflavin are listed in Table II. The rates vary with pH (examples are

TABLE II

COMPARATIVE INITIAL RATES OF OXYGEN UPTAKE BY DIFFERENT AMINO ACIDS
AT 25° IN PHOSPHATE BUFFER CONTAINING 0.15 g/100 ml FMN
AND $10^{-3} M$ AMINO ACID

Addition	pH 6.6	pH 7.9
Phenylalanine	0.25	0.40
Cysteine	0.70	1.70
Histidine	3.2	3.1
Tryptophan	4.7	7.2
Methionine	1.50	1.35
Tyrosine	2.20	3.70
No addition	0.20	0.20

shown in Fig. 3), so that some selectivity can be achieved by a suitable choice of pH. Many enzymes are photoinactivated by this method (Table I). The question of specificity of destruction using different dyes has not been fully explored.[10] In principle, further selective chemical modifications may be produced by using different dyes.[10]

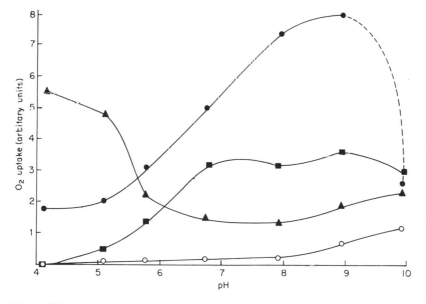

FIG. 3. Plot of oxygen uptake from a sealed vessel when equimolar concentrations of different amino acids are illuminated in the presence of flavin mononucleotide at different pH's. (■) Histidine, $\sim 10^{-3} M$; (▲) methionine, $\sim 10^{-3} M$; (●) tryptophan, $\sim 10^{-3} M$; ○, no amino acid. [FMN] 150 mg/100 ml.

[10] W. J. Ray, Jr., Vol. XI [57].

Nucleotides have also been photodecomposed by this treatment, and guanine is oxidized much faster than the other bases.[11] The treatment leads to the deactivation of certain phages[12] and to reduction in transforming ability of DNA.[13]

Recent studies have shown that some substances, including adenine, considerably increase the rates of photosensitized destructions.[14]

Reaction Conditions

The optimal conditions for the reaction depend on the nature of the substrate to be oxidized (for example, see Fig. 3 for pH effect). The use of amine buffers should be avoided because these may react under certain conditions. Phosphate buffers are suitable, but the rates often depend on ionic strength. In quantitative experiments care should be taken to exclude quenchers, such as paramagnetic metal ions, or compounds containing highly polarizable groups (e.g., KI, KBr, sulfides). The useful concentration of the sensitizer is determined by two considerations. Because of the "inner filter effect," at high optical densities, the exciting light is all absorbed in the first part of the vessel so that reaction is limited to a small volume. In addition, at high dye concentrations, dimerization of the dyes or self-quenching of triplets may become important. Similarly, high substrate concentrations may lead to quenching processes. Light intensities of the same magnitude as in photoreduction experiments[9] are usually adequate. If the method of following the reaction requires intermittent illumination, tests should be made to show that no reversal occurs in the dark period. This can easily be done by varying the relative lengths of light and dark periods.

Products and Mechanism

The products of riboflavin sensitized photooxidations have rarely been clearly identified, and in most cases complex mixtures are obtained. Thus reaction with tryptophan gives about fourteen products.[15] ^{14}C-labeling of aliphatic amino acids has been used to analyze the nature of the decomposition[16]:

$$\overset{0}{R}NH_2 - \overset{+}{C}H_2 - \overset{*}{C}O_2H \rightarrow \overset{0}{R}NH_2 + \overset{+}{C}H_2O + \overset{*}{C}O_2$$

[11] W. Berends, J. Posthuma, J. S. Sussenbach, and H. I. X. Mager, *in* "Flavins and Flavoproteins," (E. C. Slater, ed.), p. 22. Biochim. Biophys. Acta Library No. 8, 1966.

[12] A. W. Galston and R. S. Baker, *Science* **109**, 485 (1949).

[13] J. S. Bellin and G. Oster, *Biochim. Biophys. Acta* **42**, 533 (1960).

[14] K. Uehara, T. Mizoguchi, and S. Hosani, *J. Biochem. (Tokyo)* **62**, 507 (1967).

[15] P.-S. Song, personal communication.

[16] W. R. Frisell, C. W. Chung, and C. G. Mackenzie, *J. Biol. Chem.* **234**, 1297 (1959).

In the oxidation of guanine several products are formed, but ^{14}C labeling shows that about 50% of the observed degradation can be expressed by[17] The riboflavin-sensitized photooxidation of cholest-4-en-3β-ol (in pyridine–methanol, 4:1) gives rise to two products: cholest-4-en-3-one and 4α,5-epoxy-5α-cholestan-3-one in a 30:1 ratio.[18] This product ratio decreases to 1:5 with decreasing substrate concentration.[19]

The various arguments put forward in favor of one or another mechanism in photooxidations will not be discussed here. Nevertheless a knowledge of these is important from the practical point of view because the specificity of the reaction, the nature of the products, and the variation of rate with reaction conditions depend on the detailed pathway. Most of the dye-sensitized photooxidations, including those of flavins, involve the dye triplet as the initial reactive species, although cases where the singlet excited state is involved have been documented.[20,21] The triplet flavin may then react directly with the substrate by hydrogen atom abstraction, or with oxygen to give either a reactive peroxide intermediate[22] or singlet oxygen.[19] Hydrogen abstraction, followed by a fast reoxidation by oxygen of reduced or semiquinoid flavin is undoubtedly the predominant pathway in the oxidation of aliphatic amines and amino acids.[23] For these substances, the reaction rates decrease slightly with increasing oxygen concentration. In contrast, the rates of reaction of aromatic amino acids such as tyrosine and tryptophan increase with oxygen concentration.

In the reaction of riboflavin with cholest-4-en-3β-ol, the high enone: epoxy ketone product ratio has been attributed to a competition between direct reaction of riboflavin with the steroid and energy transfer to oxygen

[17] W. Berends and J. Posthuma, *J. Phys. Chem.* **66**, 2547 (1966).
[18] A. Nickon and W. L. Mendelsohn, *J. Am. Chem. Soc.* **87**, 3921 (1965).
[19] D. R. Kearns, R. A. Hollins, A. U. Khan, and P. Radlick, *J. Am. Chem. Soc.* **89**, 5456 (1967).
[20] N. K. Bridge and G. Porter, *Proc. Roy. Soc. (London) Ser. A* **244**, 259, 276 (1958).
[21] P.-S. Song and D. E. Metzler, *Photochem. Photobiol.* **6**, 691 (1967).
[22] K. Gollnick and G. O. Schenck, *in* "Organic Photochemistry," p. 507. Butterworth, London, 1965.
[23] G. R. Penzer, *Biochem. J.* **116**, 733 (1970).

to give both the $^1\Delta_g$ and $^1\Sigma_g^+$ oxygen species. The $^1\Delta_g$ oxygen reacts to produce a higher yield of the epoxy ketone.

Acknowledgments

Part of this work was supported by the Science and Medical Research Councils of the United Kingdom. One of us (M. B. T.) thanks the Science Research Council for a research studentship.

[154] Flavin-Sensitized Isomerizations

By ANN GORDON-WALKER and G. K. RADDA

Scope of the Reaction

Flavins, like many other substances,[1] can photosensitize the isomerization of olefinic compounds.[2,3] The inclusion of a description of this reaction in a methods volume is justified for a number of reasons. First, like photo-oxidation,[4] it provides a method for introducing energy of a given magnitude into a system using light. We feel that potentially the use of light as a "reagent" is the simplest way of initiating chemical reactions in a complex organized or compartmented system. Second, flavin-sensitized isomerizations may produce experimentally undesirable artifacts (e.g., in fluorescence studies where powerful light sources are used). Indeed, the discovery of these reactions stems from an investigation on the effect of the fungicide pimaricin on yeast cells.[2] Third, this reaction provides, as will be seen below, a method of characterizing the triplet levels of flavins. Finally, riboflavin is present in the eyes of many species,[5] sometimes even in a crystalline form,[6] and it can sensitize the isomerization of retinols and retinals.[3] Thus the process may be of importance in some photobiological systems.

[1] G. S. Hammond, J. Saltiel, A. A. Lamola, N. J. Turro, J. S. Bradshaw, D. O. Cowan, R. C. Counsell, V. Vogt, and C. Dalton, J. Am. Chem. Soc. **86**, 3197 (1964).

[2] J. Posthuma and W. Berends, Biochim. Biophys. Acta **112**, 422 (1966).

[3] A. Gordon-Walker and G. K. Radda, Nature **215**, 1483 (1967).

[4] M. B. Taylor and G. K. Radda, this volume [153].

[5] A. Pirie, in "Aspects of Comparative Ophthalmology. Proc. Symp. of the British Small Animal Veterinary Assoc." (O. Graham-Jones, ed.), p. 57. Macmillan (Pergamon), New York, 1965; J. H. Elliot and S. Futterman, Arch. Ophthalmol. **70**, 531 (1963).

[6] A. Pirie, Nature **183**, 985 (1959); H. J. A. Dartnall, G. B. Arden, H. Ikeda, C. P. Luck, M. E. Rosenberg, C. M. H. Pedler, and K. Tansley, Vision Res. **5**, 399 (1965).

Method

The illumination set-up is similar to that described elsewhere in this volume.[7] Generally, higher light intensities or longer illumination times are required for these sensitizations than for photoreductions.

This type of reaction occurs both in water and in organic solvents. The choice of the solvent depends on the solubility of the olefinic compound; e.g., the reaction of stilbene carboxylic acids has been carried out in water[2] whereas stilbene[8] and retinols[3] were dissolved in methanol. The presence of oxygen is undesirable because it inhibits isomerization and may also lead to photooxidation. Spectroscopic cuvettes are again suitable for deoxygenation[7] but should not be stoppered with rubber bungs when organic solvents are used. In addition, efficient deoxygenation of organic solvents generally requires longer periods of bubbling of the inert gas. Evaporation of solvent can be minimized by passing the inert gas through pure solvent before it enters the cuvette.

In general, the reactions can be followed spectrophotometrically. This method has been employed for observing the reaction of pimaricin, stilbene-4-carboxylic acid,[2] and retinols.[3]

The reaction of stilbenes provides a method of studying the triplet levels of sensitizers. This requires an accurate measure of the ratio of cis:trans stilbenes in the reaction mixture. This ratio can be most easily followed (at $10^{-3} M$ stilbene) by gas–liquid chromatography (GLC). In the sensitization experiments, the concentration of flavins ($\sim 10^{-4}$ to $5 \times 10^{-5} M$) are chosen to give optical densities between 0.5 and 1.0 in the reaction cuvette. The concentration of stilbenes should be between 10^{-2} and $10^{-3} M$, because at low stilbene concentration ($10^{-5} M$) appreciable photodestruction of the isomers occurs during long illuminations. Methanol or aqueous methanol are convenient solvents. One to two microliters of the reaction mixture are taken at intervals with a Hamilton microsyringe (Hamilton Co., Inc., Whittier, California), inserted through the deoxygenation needle left in the stopper of the cuvette, and injected directly into the chromatograph.

A Perkin-Elmer F11 gas chromatograph with a hydrogen flame ionization detector has been routinely used in the authors' laboratory. Complete resolution of isomers is achieved on a 2-meter column (2 mm i.d.) packed with Apiezon L/Chromosorb P (80–100 mesh) 15:85 w/w. The column is operated at 220° with nitrogen as the carrier gas. The retention times are in the range: *cis*-stilbene 15–20 minutes, *trans*-stilbene 40–50 minutes, depending on the carrier gas flow rate and column packing. The GLC

---

[7] G. R. Penzer and G. K. Radda, this volume [152].
[8] A. Gordon-Walker and G. K. Radda, submitted for publication.

should be standardized by making up known mixtures of pure stilbene isomers by weight and obtaining peak-area ratios. The method of triangulation is accurate to within $\pm 5\%$. In our system, the measured cis:trans peak-area ratios had to be multiplied by 0.82 to give correct weight ratios.

The reaction can also be followed spectrophotometrically. Because of the high stilbene concentrations necessary to avoid significant destruction, the course of the reaction can be followed only by a sampling technique involving dilution to give measurable optical densities. This also eliminates the interference from the flavin absorption. Cis:trans ratios are obtained by following optical densities at two wavelengths (295 and 321 mμ) and knowing the extinction coefficients of the pure stilbenes at these wavelengths.

The photostationary state is reached after ~ 6 hours of illumination, but it is best to follow the time course of the reaction until no more isomerization is observed. It is essential that the photostationary state should be approached from both the *cis*- and *trans*-isomer sides. Some of the flavins decompose significantly during the time required to reach the photostationary equilibrium. In these cases, an approximate value of the stationary state ratio is first obtained by illuminations of one stilbene isomer and flavin. Mixtures of the stilbene isomers are then made up around the equilibrium mixture, and by short illuminations the final equilibrium value can be found accurately.

The dependence of the photostationary state ratio on the triplet transition energies (E_t) for a number of flavins is shown in Fig. 1 (see footnote 9). More complicated correlations are obtained with different types of sensitizer.[1]

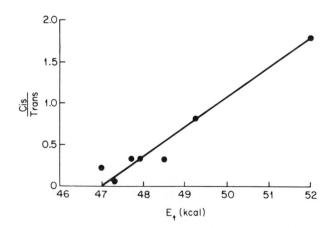

FIG. 1. Variation of *cis-trans* stilbene ratio with triplet energy of flavins.

Characteristics of the Reaction[3,9]

The quantum yields are generally low, e.g., for the reaction of retinol, 0.002 at concentration $10^{-5} M$; and for stilbenes, 0.05 at concentrations of $10^{-3} M$. The reactions are subject to quenching by O_2, KI, and paramagnetic metal ions. No ground-state complexes between flavins and the olefins are detectable by either fluorescence or absorption spectra. The reactions most probably involve triplet–triplet energy transfers by "nonvertical" transitions. (For a detailed theoretical description see footnote 1.)

Acknowledgments

One of us (A. G.-W.) thanks the Medical Research Council of Great Britain for a Scholarship. The work reported here was partly supported by the Science and Medical Research Councils of the United Kingdom.

[9] G. K. Radda, *Abstr. Vth Intern. Photobiol. Congr., Hanover* (1969).

[155] Photoreductants of FAD and of FAD-Dependent Oxidases

By C. Veeger, J. F. Koster, and D. B. McCormick

In general, for the photoreduction of FAD and flavoproteins one photoreductant was used, namely ethylenediaminetetraacetate (EDTA), and other possible photoreductants were not taken into account. This contribution is concerned with the rates and efficiencies of other reductants for the photoreduction of FAD and of FAD-dependent oxidases.[1,2]

Method

Reagents

Sodium pyrophosphate buffer, 0.1 M, pH 8.3 or pH 8.0
Sodium phosphate buffer, 0.1 M, pH 7.0
FAD, $10^{-3} M$
Photoreductant, 1 M
D-Amino acid oxidase (EC 1.4.3.3), prepared according to Massey, Palmer, and Bennett[3] and made benzoate-free with an excess of D-alanine[4]

[1] D. B. McCormick, J. F. Koster, and C. Veeger, *European J. Biochem.* **2**, 387 (1967).
[2] V. Massey and G. Palmer, *Biochemistry*, **5**, 3181 (1966).
[3] V. Massey, G. Palmer, and R. Bennett, *Biochim. Biophys. Acta* **48**, 1 (1961).
[4] K. Yagi and T. Ozawa, *Biochem. Z.* **338**, 330 (1963).

L-Amino acid oxidase (EC 1.4.3.2), prepared according to Wellner and Meister[5]

Glucose oxidase (EC 1.1.3.4), prepared according to Bentley[6]

Procedure. A Thunberg cuvette is filled with 0.1 ml of the compound to be reduced, which is either 4–5 mg of enzyme or 25 μM FAD, 0.1 ml of photoreductant, and either 0.1 M sodium pyrophosphate buffer, pH 8.3, or 0.1 M sodium phosphate, pH 7.0, to a total volume of 4 ml. The temperature of the solution is maintained at 5°. The cuvette is deoxygenated by repeated evacuation followed by readmission of oxygen-free nitrogen (obtained by passage through pyrogallate). The cuvette is placed in the thermostatted sample compartment of a monochromator (Zeiss M4QIII) equipped with a cuvette holder, magnetic stirrer, and 500-W xenon lamp unit (Zeiss LX501) with water bath, but without a blue filter. The wavelength is selected by the monochromator with a bandwidth of 25 mμ. The content of the cuvette is stirred by the magnetic stirrer to obtain a homogeneous reduction. The rate of the reduction is followed by registering the spectra with a spectrophotometer (Cary Model 14) equipped with a cell holder also controlled at the same temperature as the illumination. During the deoxygenation process the solution is slightly concentrated.

Comments

During the photoreduction of FAD-dependent oxidases like D- and L-amino acid oxidases, glucose oxidase, but less with free FAD, a lag period is observed quite often. This lag period is due to small amounts of oxygen still present in the cuvette. The reoxidation is much faster than the photoreduction; this means that the flavin will remain in the quinone form as long as oxygen is present. The lag period is determined by the rate of photoreduction.

EDTA, EDTP (ethylenediaminetetrapropionate), and L-methionine are good photoreductants for FAD (Table I). The efficiencies of dimethylaminopropanol and triethylamine depend on the pH. At higher pH the efficiency becomes larger. The rates of photoreduction of D-amino acid oxidase and glucose oxidase with several photoreductants are summarized in Table II. In all cases, the photoreduction of free FAD is much faster than that bound to the oxidases. Again EDTA, EDTP, L-methionine, and nicotine are good photoreductants for the enzymes, while dimethylaminopropanol and dimethylglycine ethyl ester are less so, and triethylamine is rather ineffective. All the rates of photoreduction for glucose oxidase with the different photoreductants are lower than for D-amino acid oxidase. The order of effectiveness is somewhat different.

[5] D. Wellner and A. Meister, *J. Biol. Chem.* **235**, 2013 (1960).
[6] R. Bentley, Vol. I, p. 340.

TABLE I
PHOTOREDUCTION OF FAD[a]

| Photoreductant | Time for half reduction (min) | |
	pH 7.0	pH 8.0
EDTA	1	1
EDTP	2	2
L-Methionine	2	3
Nicotine	4	4
Dimethylaminopropanol	5	2
Dimethylglycine ethyl ester	7	6
Triethylamine	60[b]	15

[a] Anaerobic solutions contained 25 μM FAD and 25 mM photoreductant in 0.1 M sodium phosphate (pH 7.0) or pyrophosphate (pH 8.0) buffer at 5°. Excitation wavelength, 450 mμ.

[b] Significant photodecomposition occurs during the prolonged times necessary for photoreduction.

The photoreduction of D- and L-amino acid oxidases with the photoreductant EDTA can be markedly enhanced by adding FAD or FMN (25 μM). The rate approaches that of free FAD or FMN. The rate is also enhanced in 6 M urea. Addition of free flavin or 6 M urea has little or no effect on glucose oxidase.

TABLE II
PHOTOREDUCTION OF D-AMINO ACID OXIDASE AND GLUCOSE OXIDASE[a]

| Photoreductant | Time for half reduction (min) | |
	D-Amino acid oxidase	Glucose oxidase
EDTA	9	30
EDTP	10	90
Nicotine	8	15
L-Methionine	8	60
Dimethylaminopropanol	25	90
Dimethylglycine ethyl ester	20	120
Triethylamine	30	180

[a] Anaerobic solutions contained 4–5 mg of enzyme and 25 mM photoreductant in 0.1 M sodium pyrophosphate buffer (pH 8.3) in 4 ml at 5°. Excitation wavelength for D-amino acid oxidase, 455 mμ; and for glucose oxidase, 450 mμ.

[156] Preparation of N^5-Acyl Leuco FAD Compounds
D- and L-Amino Acid Oxidases

By C. VEEGER and A. DE KOK

The method is based on a photochemical reaction between enzyme-bound FAD and an α-keto acid, the latter being a product of the enzymatic oxidation of the corresponding amino acid. The structure of the addition compounds has recently been established.[1] The reaction, based on the scheme of Hemmerich *et al.*[2] is tentatively given by:

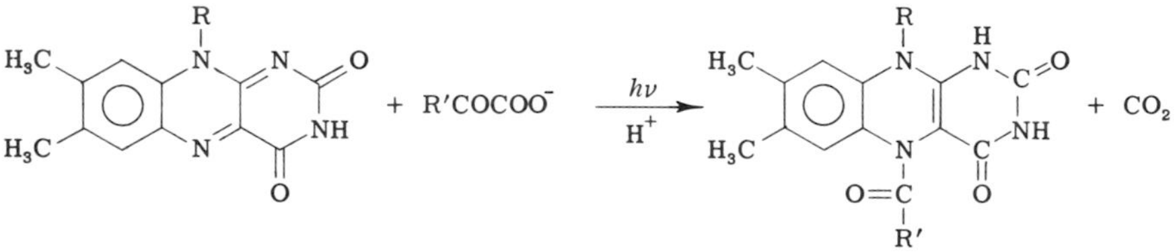

These structures represent active acyl compounds and hydrolyze easily to reform the original isoalloxazine compound. Ultraviolet light is also an active deacylating agent. It is advisable to work in dim light when preparing these compounds. The isolation procedure is described for N^5-acetyl-FADH. In this case D-amino acid oxidase (EC 1.4.3.3) has to be used rather than L-amino acid oxidase (EC 1.4.3.2) because of substrate specificity reasons.

Preparation

Equipment and Reagents

500-W xenon lamp with housing and heat filter (e.g., Zeiss No. 46 72 10)

Monochromator (e.g., Zeiss M4QIII or filter with zero transmission below 360 mμ)

Cuvette holder, thermostatable and equipped with magnetic stirring motor (e.g., Zeiss No. 50 74 88)

Anaerobic spectrophotometer cell, 1-cm pathlength

Freeze-drying apparatus and rotary evaporator

Equipment for thin-layer chromatography

Equipment for Sephadex chromatography

Sodium pyruvate, 0.3 M, in 0.5% ammonium bicarbonate

[1] A. de Kok, C. Veeger, M. Brüstlein and P. Hemmerich, to be published.
[2] P. Hemmerich, V. Massey, and G. Weber, *Nature* **213**, 728 (1967).

D-Amino acid oxidase, purified,[3] 10 mg/ml in 0.5% ammonium bicarbonate, made benzoate-free by the method of Yagi and Ozowa.[4] L-Amino acid oxidase is purified by the method of Wellner and Meister.[5]

Sephadex G-50

Dextran Blue

t-Butyl alcohol

Procedure. The cuvette is filled with 1.0 ml of the sodium pyruvate solution and 2.0 ml of the D-amino acid oxidase solution. A 5-mm stirring bar is added, and the cuvette is made anaerobic by flushing with pure nitrogen or argon. The cell is then placed in the cuvette holder, thermostatted at 0°–5°. A stream of dry air is passed along the cuvette to avoid condensation. The excitation wavelength is 390 mμ, with slit set on 2 mm, giving a bandwidth of 33 mμ. The course of the reaction can be followed by measuring the decline in absorbancy at 460 mμ. After about 80% bleaching, the illumination, which takes about 2 hours in the above-described apparatus, is stopped. The contents of the cell is dialyzed against three changes of 1 liter of 0.5% ammonium bicarbonate during 12 hours at 2°.

The solution is then deproteinized by heating at 100° in a waterbath. Care should be taken for effervescence. The heating time should be as short as possible. Small amounts of protein will still remain in solution, but are removed in the course of further purification on Sephadex. After heating, the mixture is rapidly cooled to 0° and centrifuged at 31,000 g for 10 minutes at 0°–5°. The supernatant solution is freeze dried. After drying, evacuation should be continued for 2 days at low temperature to remove the volatile buffer system.

A concentrated solution of the flavins in water is placed on top of a Sephadex G-50 column (1 × 30 cm) and eluted with water. The high molecular weight material, as indicated by Dextran Blue elution in a separate run, is discarded. The flavins are concentrated with a rotary evaporator at room temperature.

Preparative thin-layer chromatography is now carried out on cellulose. A neutral solvent system that will separate the flavins quite well is t-butyl alcohol–water, 60:40 (footnote 6). A dark band with an R_f value of about 0.4, which can be seen under 360 mμ light, moving ahead of FAD and a small band of FMN, is N⁵-acetyl-FADH. This latter compound can be identified further after most of the material has been scratched from the plate by exposure to HNO₂ fumes. This deacylates the leucoflavin very

[3] V. Massey, G. Palmer, and R. Bennett, *Biochim. Biophys. Acta* **48**, 1 (1961).

[4] K. Yagi and T. Ozawa, *Biochim. Biophys. Acta* **60**, 200 (1062).

[5] D. Wellner and A. Meister, *J. Biol. Chem.* **235**, 2013 (1960).

[6] G. L. Kilgour, S. P. Felton, and F. M. Huennekens, *J. Am. Chem. Soc.* **79**, 2254 (1957).

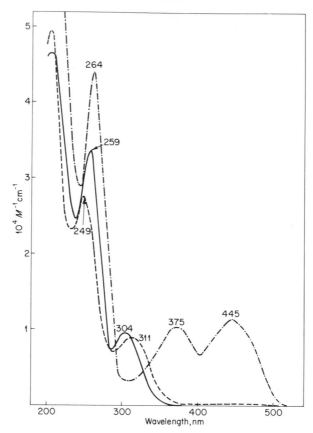

FIG. 1. Spectra of N^5-acetyl-FADH: (———) in 0.1 M phosphate buffer, pH 7.0, (- - - - -) in 0.1 M KCl–HCl buffer, pH 1.5, ($\cdot$ — $\cdot$ — $\cdot$ —) the addition of NaNO$_2$ and subsequent removal of HNO$_2$ by heating and evacuation.

rapidly to give back FAD. The material from the plate is eluted with water for 30 minutes.

After the final purification step, N^5-acetyl-FADH shows the spectrum given in Fig. 1.

Specificity

The reaction is only found with α-keto acids. The concentrations of the α-keto acid must be high, reflecting the high dissociation constants of the oxidase–α–keto acid complex (e.g., 38 mM for D-amino acid oxidase-pyruvate).[7] Pyruvate is apparently not bound to L-amino acid oxidase in the way longer-chain acids are bound, because upon 390-mμ irradiation,

[7] J. F. Koster and C. Veeger, *Biochem. Biophys. Acta* **151**, 11 (1968).

only semiquinone formation is observed. A reaction with free flavin coenzyme does not take place under the conditions mentioned above. At higher light intensities a slow reaction is found, but reversible photoreduction and degradation of FAD are the predominant pathways. An exception is benzoyl formate. It is assumed that this results from resonance stabilization of a reactive intermediate, as may be the case with the other keto acids when bound to the enzyme.

General Comments

The wavelength of irradiation is very critical. Above 420 mμ the reaction is very poor, whereas below 360 mμ reversible photoreduction is the predominant pathway.

When radioactive keto acids are used, radioautography and subsequent analysis of the spots can be performed. The method as described by Veen[8] is advised.

[8] H. Veen, *Acta Botan. Neerl.* **15,** 419 (1966).

[157] The Enzymatic Synthesis of Riboflavin[1]

By G. W. E. PLAUT and R. A. HARVEY

Intact cells of riboflavin-producing organisms have been shown to incorporate radioactivity from precursors of purines and from all atoms of purine molecules themselves, except carbon atom No. 8, into the corresponding positions of the pyrimidine portion of riboflavin.[1a-4] Studies with purineless mutants of *Escherichia coli* indicate that a purine derivative(s) is an obligatory intermediate in riboflavin synthesis.[5]

Two-carbon fragments are utilized for the synthesis of the *o*-xylene portion of riboflavin, and the location of radioactivity from ^{14}C-labeled glucose has been traced, especially in the ribityl group.[6]

Two substances structurally related to flavins, 6,7-dimethyl-8-ribityllumazine and 6-methyl-7-hydroxy-8-ribityllumazine, have been isolated from a number of riboflavin-producing organisms and prepared by chemical

[1] The experimental work from this laboratory presented here and the preparation of this article have been assisted by grants from the National Institute of Arthritis and Metabolic Diseases (AM 10501), National Institutes of Health, U.S. Public Health Service, and the National Science Foundation (GB 6915).

[1a] G. W. E. Plaut, *J. Biol. Chem.* **208,** 513 (1954).
[2] W. S. McNutt, *J. Biol. Chem.* **210,** 511 (1954).
[3] W. S. McNutt, *J. Biol. Chem.* **219,** 365 (1956).
[4] W. S. McNutt, *J. Am. Chem. Soc.* **83,** 2303 (1961).
[5] D. J. Howells and G. W. E. Plaut, *Biochem. J.* **94,** 755 (1965).
[6] G. W. E. Plaut, *Ann. Rev. Biochem.* **30,** 409 (1961).

synthesis. 6,7-Dimethyl-8-ribityllumazine was proved to be an intermediate in riboflavin synthesis by the demonstration that radioactivity
from earlier precursors can be recovered in the analogous position in the
lumazine derivative and flavin, and by the conversion of 6,7-dimethyl-8-
ribityllumazine to riboflavin by extracts and purified enzymes from microorganisms and plants (for review see footnotes 6, 7). Degradative methods
for the localization of isotopes from precursors in various portions of the
lumazine and flavin molecule have been described.[6]

Studies have been made with purified enzymes from *Ashbya gossypii*
and yeast (for review see footnotes 6, 7). The riboflavin synthetase reaction
(Fig. 1) involves removal of the 4-carbon portion containing carbons 6
and 7 and the adjacent methyl groups from one molecule of 6,7-dimethyl-8-
ribityllumazine and its transfer to a second molecule of the lumazine. Thus,
two molecules of 6,7-dimethyl-8-ribityllumazine are consumed for the
formation of one molecule of riboflavin and one of 4-ribitylamino-5-amino-
2,6-dihydroxypyrimidine. Radioactivity from 6,7-dimethyl-^{14}C-8-ribityl-
lumazine is recovered in equal amounts in the methyl groups and carbon
atoms 5 and 8 of the *o*-xylene ring of riboflavin, whereas label from position
2 or positions 4 + 8a of the lumazine is recovered in equimolecular amounts
in the pyrimidine portion of riboflavin and in 4-ribitylamino-5-amino-
2,6-dihydroxypyrimidine.

6,7-Dimethyl-8-ribityllumazine   Riboflavin   4-Ribitylamino-5-amino-
                                                            2,6-dihydroxypyrimidine

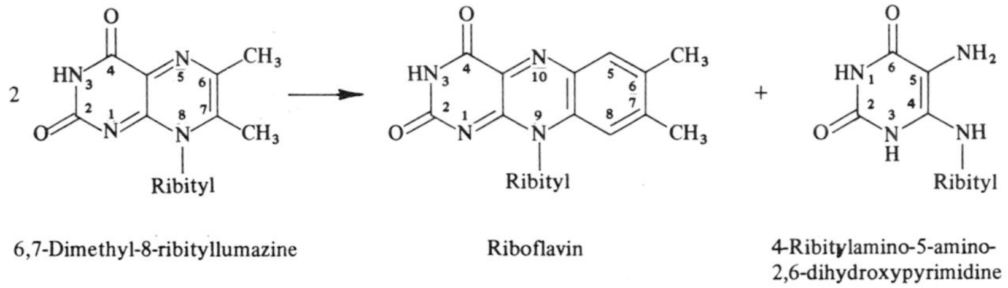

Fig. 1. The riboflavin synthetase reaction.

Chemical Syntheses

6,7-Dimethyl-8-(1'-D-ribityl)lumazine

Method of Synthesis

In the synthesis of intermediates described here, the batch sizes indicated have been found convenient in this laboratory.[8–11] In some cases
material is accumulated before proceeding with the next step. The method
of synthesis is outlined in Fig. 2.

---

[7] C. H. Winestock and G. W. E. Plaut, *in* "Plant Biochemistry" (J. Bonner and V. E.
Varner, eds.), p. 391. Academic Press, New York, 1965.

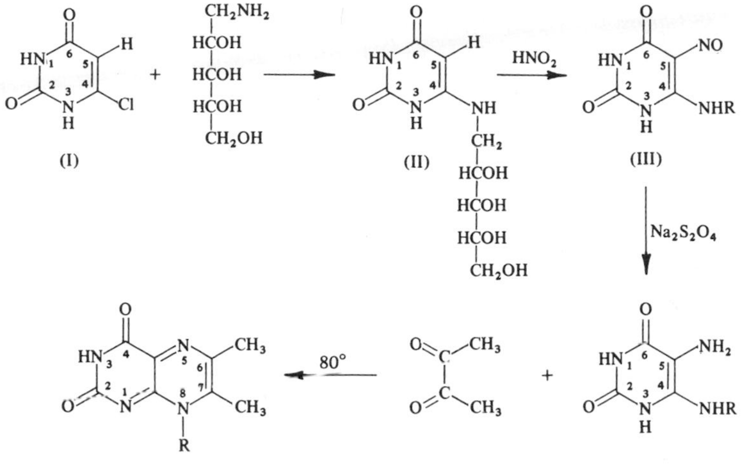

FIG. 2. Method of synthesis of 6,7-dimethyl-8-(1'-D-ribityl) lumazine.

4-Chloro-2,6-dihydroxypyrimidine. A suspension of 100 g of 4-chloro-2,6-dimethoxypyrimidine (Aldrich Chemical Co.) in a mixture of 300 ml of 48% HBr and 300 ml of water is heated at 80° for 3 hours under reflux; 700 ml of water is added, and crystallization occurs overnight at 5°. The solid is collected by filtration, dried at 60°, and recrystallized from 2 liters of water. The colorless crystals are dried in a vacuum at 60°; m.p. 301°–302° (decomposes); yield 60 g.

D-*Ribose Oxime.* Powdered hydroxylamine hydrochloride (59 g) is suspended in 300 ml of absolute ethanol containing 2 drops of 1% phenolphthalein in alcohol. A freshly prepared solution of 42.5 g of sodium methoxide in 400 ml of absolute ethanol is added dropwise to the vigorously stirred suspension of hydroxylamine hydrochloride until a pink color persists for about 1 minute. Excess sodium methoxide should be removed by the addition of a small amount of hydroxylamine hydrochloride. Precipitated sodium chloride is removed by filtration. The solution of hydroxylamine is heated to 70°, stirred, and 54.5 g of dry D-ribose is added in small portions. The reaction mixture is allowed to remain at room temperature overnight. The crystalline material formed is collected by filtration; additional ribose oxime can be recovered from the mother liquor. The solid is dried in a vacuum at 56°. Yield of D-ribose oxime from D-ribose is 80–90%; m.p. 138°–139°.

[8] G. F. Maley and G. W. E. Plaut, *J. Biol. Chem.* **234**, 641 (1959).

[9] G. W. E. Plaut, *J. Biol. Chem.* **238**, 2225 (1963).

[10] C. H. Winestock and G. W. E. Plaut, *J. Org. Chem.* **26**, 4456 (1961).

[11] C. H. Winestock, T. Aogaichi, and G. W. E. Plaut, *J. Biol. Chem.* **238**, 2866 (1963).

D-*Ribitylamine.* A suspension of 25 g (152 millimoles) of D-ribose oxime and 300 mg of platinum oxide in 250 ml of glacial acetic acid is treated with hydrogen (50 psi) and shaken until the white solid has dissolved (approximately 24 hours). The catalyst is removed by filtration, and the liquid is concentrated to a syrup in a vacuum. The residue is dissolved in 100 ml of water and passed through a 2 × 30 cm column of AG 50W-X8 hydrogen form (Bio-Rad Corporation; 200–400 mesh) with ice water circulated through the jacket of the column. The cation exchange resin is washed with 200 ml of water and then treated with 300 ml of 3 M NH$_4$OH. The pH of the effluent is monitored; the alkaline portion of the eluate is collected and concentrated to a syrup in a vacuum. Residual ammonia is removed by repeated dilution of the residue in water followed by concentration in a vacuum. The yield (65–75%) of D-ribitylamine is determined by titration with standard acid. The solution of D-ribitylamine is stored at −20°.

4-Ribitylamino-5-nitroso-2,6-dihydroxypyrimidine. A solution of 13.8 g (95 millimoles) of 4-chloro-2,6-dihydroxypyrimidine in 95 ml of 2 N D-ribitylamine (190 millimoles) is prepared in a round-bottom flask, and excess water is removed under reduced pressure in a flash evaporator. The material is heated at 95° for 1 hour, and then for 2 hours at 110° under high vacuum. The brown syrup is dissolved in 300 ml of water, and 19.6 g of NaNO$_2$ is added; the solution is adjusted to pH 4.6 with 3 N acetic acid, allowed to remain at room temperature for 2 hours, and evaporated to dryness in a vacuum.

The residue is dissolved in 300 ml of 0.15 M NH$_4$OH and placed on a 4.5 × 30 cm column of AG 1-X8 formate form (Bio-Rad Corporation; 200–400 mesh). The column is washed with 200 ml of water followed by 250 ml of 0.01 M formic acid. Formic acid, 0.1 M, is placed on the anion exchange resin, and the red percolate is collected and concentrated to dryness in a vacuum. The residue is crystallized from 100 ml of water. 4-(1'-D-Ribitylamino)-5-nitroso-2,6-dihydroxypyrimidine is obtained as a red solid in a yield of 50–70% from 4-chloro-2,6-dihydroxypyrimidine.

6,7-Dimethyl-8-ribityllumazine. A solution of 5 g of 4-ribitylamino-5-nitroso-2,6-dihydroxypyrimidine in 100 ml of water is heated to 90°, adjusted to pH 5.8 with 2 N KOH, and treated with 10.0 g of sodium hydrosulfite. The solution turns pale yellow within a few minutes, indicating reduction of the nitroso derivative to 4-ribitylamino-5-amino-2,6-dihydroxypyrimidine. All subsequent steps are done in the dark or in dim light. The solution is cooled, adjusted to pH 4.6 with 2 N HCl, treated with 10 ml of 2,3-butanedione and heated at 76° for 40 minutes. The reaction mixture is evaporated to dryness in a vacuum at a bath temperature of 40°.

The residue is dissolved in 80 ml of water, diluted with 320 ml of ethanol and placed on a 4.5 × 20 cm column of acid-washed alumina AG 4 (Bio-Rad Corporation) of 200–400 mesh (prepared from a suspension of 250 g of acid-washed alumina in absolute alcohol). The column is washed with 400 ml of 80% (v/v) ethanol, and the green fluorescent material is eluted with 50% (v/v) ethanol (approximately 2.5 liters of effluent). The solvent is removed under reduced pressure, and the residue is taken up in 125 ml of water and treated with twenty-five 25-ml portions of water-saturated benzyl alcohol. The benzyl alcohol layers are combined and filtered through Whatman No. 3 paper; an equal volume of ether is added. The compound is extracted from the solution with twenty-six 40-ml portions of water. Residual benzyl alcohol in the aqueous phase is removed by extraction into three 50-ml portions of ether. The water layer is evaporated to dryness in a vacuum. The residue is dissolved in 20 ml of water and crystallized by the addition of 30 ml of ethanol. Additional compound can be recovered from the mother liquor by concentration to smaller volumes. 6,7-Dimethyl-8-(1'-D-ribityl)lumazine is obtained as a yellow solid in a yield of 40–70% from 4-ribitylamino-5-nitroso-2,6-dihydroxypyrimidine; m.p. 270°–274° (decomposes). $[\alpha]_D^{22°}$, -180 ± 4 (c, 0.517 in water).

The compound can be recrystallized from 80% ethanol; analytical samples have been crystallized from water, but with substantial losses due to solubility.

Properties of 6,7-Dimethyl-8-(1-D-ribityl)lumazine

The light absorption[10] has been determined in 0.1 N H_2SO_4: maxima at 407 mμ (log ϵ = 4.01) and 256 mμ (log ϵ = 4.20), minima at 300 mμ (log ϵ = 3.22) and 220 mμ (log ϵ = 3.93); and in 0.1 N NaOH: maxima at 313 mμ (log ϵ = 4.00) and 279 mμ (log ϵ = 4.15), minima at 292 mμ (log ϵ = 3.92) and 257 mμ (log ϵ = 4.01).

The crystalline compound gives a single green fluorescent spot in several systems of paper chromatography.

6,7-Dimethyl-8-ribityllumazine and other substituted lumazines are light sensitive and unstable in solution. All steps in the preparation of these compounds should be carried out rapidly and in as little light as possible. Prolonged heating should be avoided in the recrystallization step, and the compounds should be dried in a vacuum at a temperature not exceeding 55°.

Syntheses of ¹⁴C-Labeled 6,7-Dimethyl-8-Ribityllumazines

The compound has been prepared with label in various positions by adaptations on a micro scale of the synthesis described above. The following compounds have been synthesized.[9]

6,7-Dimethyl-8-ribityllumazine-2-[14]C: Barbituric acid-2-^{14}C is converted to 2,4,6-trichloropyrimidine, which is hydrolyzed selectively[12] to 4-chloro-2,6-dihydroxypyrimidine-2-[14]C.

6,7-Dimethyl-8-ribityllumazine-4,8a-[14]C: Barbituric acid 4,6-^{14}C, prepared from diethylmalonate-1,3-^{14}C and urea, is converted to 4-chlorouracil by way of the trichloropyrimidine.

6,7-Dimethyl-8-ribityl-^{14}C-lumazine: D-Ribose-^{14}C is converted to ribitylamine by way of the oxime.

6,7-Dimethyl-^{14}C-8-ribityllumazine and 6,7-dimethyl-^{14}C-8-[1'-(5'-deoxy-D-ribityl)]lumazine[13]: Prepared by reduction of 4-ribitylamino-2,6-dihydroxy-5-nitrosopyrimidine and 4-(5'-deoxy-D-ribitylamino)-2,6- dihydroxy-5-nitrosopyrimidine,[11] respectively, followed by condensation with 2,3-butanedione-1,4-[14]C.

Discussion

In the synthesis described above the intermediates 4-ribitylamino-2,6-dihydroxypyrimidine and 4-ribitylamino-5-amino-2,6-dihydroxypyrimidine are not isolated. The characterization of 4-ribitylamino-2,6-dihydroxypyrimidine has been reported elsewhere[8]; the synthesis of 4-ribitylamino-5-amino-2,6-dihydroxypyrimidine is described below.

Various methods of chemical synthesis of 6,7-dimethyl-8-ribityllumazine have been described, varying in some detail from that reported here. Thus, ribitylamine and 4-chloro-5-nitro-2,6-dihydroxypyrimidine condense to form 4-ribitylamino-5-nitro-2,6-dihydroxypyrimidine[12]; the nitrogen function has been introduced at position 5 of 4-ribitylamino-2,6-dihydroxypyrimidine with a diazonium salt.[14] Either of these intermediates is then reduced to 4-ribitylamino-5-amino-2,6-dihydroxypyrimidine.

4-(1'-D-Ribitylamino)-5-amino-2,6-dihydroxypyrimidine[15]

4-(1'-D-Ribitylamino)-5-nitroso-2,6-dihydroxypyrimidine (112 mg) in 2.5 ml of hot water is reduced with 213 mg of sodium hydrosulfite. The solution is cooled in ice, 1 ml of concentrated HCl is added, and the precipitate is removed by filtration. The filtrate is treated with two 5-ml portions of carbon disulfide, the aqueous phase is stored in a refrigerator overnight, and the precipitate is filtered off. The solution is cooled in an ice bath, saturated with gaseous HCl, and the precipitate of sodium chloride is removed by filtration in the cold. The solution is warmed to room tem-

[12] J. Davoll and D. D. Evans, *J. Chem. Soc.* p. 5041 (1960).
[13] R. A. Harvey and G. W. E. Plaut, *J. Biol. Chem.* **241**, 2120 (1966).
[14] W. Pfleiderer and G. Nübel, *Ber. Deut. Chem. Ges.* **93**, 1406 (1960).
[15] H. Wacker, R. A. Harvey, C. H. Winestock, and G. W. E. Plaut, *J. Biol. Chem.* **239**, 3493 (1964).

perature, and 12 ml of ethanol is added dropwise with mixing. The suspension is allowed to stand in a refrigerator overnight and filtered; the white solid is washed with about 1 ml of ethanol, and dried in a vacuum at room temperature. 4-(1'-D-Ribitylamino)-5-amino-2,6-dihydroxypyrimidine·HCl is obtained in a yield of 50%; absorption spectrum in 0.1 N HCl, maximum at 268 mμ ($\epsilon = 24{,}500$ M^{-1} cm^{-1}). The compound is fairly stable in aqueous solutions of strong acids, but decomposes rapidly as the pH of solutions rises.

Synthesis of Other Substituted Lumazines

A large number of derivatives of lumazine have been prepared bearing different substituents at positions 2, 6, 7, or 8 of the pteridine molecule (for review see footnotes 10–12, 14, 16). The methods of synthesis are essentially those described above for 6,7-dimethyl-8-ribityllumazine. For example, for the preparation of 6,7-dimethyl-8-(1'-aldityl)lumazines, the intermediate 4-(1'-alditylamino)-2,6-dihydroxypyrimidines have been prepared by condensation of the appropriate glycamine with 4-chloro-2,6-dihydroxypyrimidine. The glycamines have been obtained by catalytic reduction of sugar oximes or N-benzylalditylamines.[17] Variations in substituents at positions 6 and 7 have been produced by condensation of 4-(1'-alditylamino)-5-amino-2,6-dihydroxypyrimidines with the α-diketones. For example, the naturally occurring substance, 6-methyl-7-hydroxy-8-ribityllumazine, has been prepared chemically by the condensation of 4-ribitylamino-5-amino-2,6-dihydroxypyrimidine with ethyl pyruvate.[18] The oxygen function in position 2 has been replaced with an amino group by using as the starting material the appropriate 2-amino pyrimidine derivative. For example, 2-amino-4-hydroxy-6,7-dimethyl-8-(1'-D-ribityl)-pteridine has been prepared by the interaction of 2-amino-4-chloro-6-hydroxy-5-nitropyrimidine with ribitylamine followed by reduction and condensation with 2,3-butanedione.[12]

Analyses

Assay of Riboflavin Synthetase

Direct Spectrophotometric Assay

Riboflavin synthetase activity is most conveniently measured by a spectrophotometric assay that makes use of the fact that the product of the

[16] W. Pfleiderer, Angew. Chem. **3**, 114 (1964).
[17] F. Kagan, M. A. Rebenstorf, and R. V. Heinzelman, J. Am. Chem. Soc. **79**, 3541 (1957).
[18] G. W. E. Plaut and G. F. Maley, Arch. Biochem. Biophys. **80**, 219 (1959).

reaction, riboflavin, retains appreciable light absorption at 470 mμ (ϵ = 9600) whereas the substrate, 6,7-dimethyl-8-ribityllumazine, exhibits only minor absorbance at this wavelength (ϵ = 66).[19] Both substances absorb at 405 mμ (riboflavin, ϵ = 6800; 6,7-dimethyl-8-ribityllumazine, ϵ = 10,300). Riboflavin can therefore be determined in the presence of 6,7-dimethyl-8-ribityllumazine by measuring the absorbance at 470 and 405 mμ.

The routine assay is performed in 3.0 ml of a medium containing 50 mM potassium phosphate at pH 7.0, 10 mM NaHSO$_3$, and 0.6 mM 6,7-dimethyl-8-ribityllumazine. The reaction is initiated by the addition of the enzyme, and the mixture is incubated for 60 minutes at 37°. At the start and end of the incubation period, 1-ml samples are withdrawn from the reaction mixture and added to 0.5 ml of 15% trichloroacetic acid in 12-ml centrifuge tubes. Coagulated protein is removed by centrifugation, and the color of the clear supernatant solution is measured at 470 and 405 mμ. The following equation is used to calculate riboflavin formation:

$$(104.6 \times OD_{470 \, m\mu}) - (0.6705 \times OD_{405 \, m\mu}) =$$
$$\text{millimicromoles riboflavin/ml deproteinized solution}$$

Multiplication of the equation by a factor of 1.5 gives the concentration of riboflavin in the original reaction medium. The enzyme in the assay mixture should be at a concentration that gives a formation of 10–50 millimicromoles of riboflavin per milliliter after 1 hour of incubation; enzyme concentrations that lead to riboflavin formation in excess of 50 μM result in nonlinearity in the assay due to product inhibition. The specific activity of the enzyme is expressed as millimicromoles of riboflavin formed per milligram of protein per hour at 37°.[20] Enzyme protein is determined by the method of Lowry et al.[22] The concentration of bovine serum albumin used as reference standard was calculated from its ultraviolet absorption at 280 mμ.[23]

Measurement of Initial Reaction Rates

The assay described above employs saturating concentrations of substrate and provides reaction rates that are essentially linear during the

---

[19] This molar extinction coefficient is revised from a value previously reported.[9]

[20] In the previous report on the properties of the enzyme from yeast, activity was expressed as formation of riboflavin per milliliter of assay mixture caused by the amount of enzyme introduced into 3 ml of reaction mixture.[13] The expression of enzyme units here is in accord with current usage.[21] Therefore, the enzyme units used in the earlier report should be multiplied by a factor of 3 to make them comparable with those used here.

[21] "Enzyme Nomenclature," International Union of Biochemistry, p. 6. Elsevier, New York, 1965.

[22] O. H. Lowry, N. J. Rosebrough, A. L. Farr, and R. J. Randall, *J. Biol. Chem.* **193**, 265 (1951).

[23] C. Tanford and G. L. Roberts, *J. Am. Chem. Soc.* **74**, 2509 (1952).

60-minute incubation period. This assay has been used routinely to follow the purification of the enzyme. However, detailed kinetic analyses of the riboflavin synthetase reaction require the measurement of initial reaction rates. Determination of initial velocities is performed at 25° with a Zeiss PMQII spectrophotometer coupled to a logarithmic recorder (Photovolt Corp.) or equivalent equipment. The incubation mixture (except enzyme) in 1.0-cm light path cells is equilibrated for 10 minutes in the thermostatted cell compartment of the spectrophotometer. The reaction is initiated by the addition of the enzyme, and the light absorption at 470 mμ due to the formation of riboflavin is recorded.[24] A molar extinction of 9468 rather than 9600 is used in calculating the formation of riboflavin to correct for the small decrease in absorbance at this wavelength due to substrate consumption.

Electrophoresis of Enzyme on Cellulose Acetate Strips

The method outlined below is useful for following the progress of the purification of riboflavin synthetase.

Electrophoresis

The enzyme solution is spotted on a cellulose acetate strip previously wetted with a buffer containing 0.10 M Tris·HCl at pH 8.0 and 0.01 M Na$_2$SO$_3$. The strip is placed in a Beckman Microzone apparatus and current (250 V, 3.5 mA) is applied for 30 minutes.

Detection of Riboflavin Synthetase

Detector Gel. To 0.43 g Bacto-Agar (Difco Laboratories) is added 20 ml of a solution containing 0.10 M Tris·HCl and 0.01 M Na$_2$SO$_3$ at pH 7.0. The mixture is stirred mechanically and boiled until the agar has dissolved. The solution is allowed to cool at room temperature for 1 minute, 0.66 ml of 6,7-dimethyl-8-ribityllumazine (3 mg/ml) is added, and the gel is poured immediately into four plastic boxes (approximately $1 \times 5 \times 8$ cm), covering the bottom of each container completely. Indicator agars not used immediately can be stored at 0° for up to 2 days.

Detection of Enzyme Activity and Protein. The detector gel is incubated in air at 37° for 15 minutes. The cellulose acetate strip is removed from the electrophoresis apparatus, trimmed to a length of 7 cm, and placed on the warmed detector agar with the side of the strip originally spotted with the enzyme in contact with the gel. The box is covered with an opaque lid to exclude light and incubated at 37° for 15 minutes or until a yellow fluorescent spot can just be detected on the strip under ultraviolet light.

---

[24] Because of the photosensitivity of the reaction components, care must be taken to reduce to a minimum the amount of light impinging on the cells during the spectrophotometric observations.

The strip is peeled away from the agar by carefully grasping one end with a pair of tweezers. The strip is exposed to vapors of ammonia to bleach the intense green fluorescence due to 6,7-dimethyl-8-ribityllumazine. Ultraviolet light reveals a yellow fluorescent spot on a colorless background, corresponding to the location of riboflavin synthetase activity. The yellow zone can be outlined for future references by a series of small pin holes. The strip can then be stained for protein in the usual manner.

The minimal amount of enzyme activity that can be detected (60 minutes of incubation at 37°) is 4.5 units, corresponding to about 0.3 μg of enzyme protein.

Methods of Separation of Reactants and Products

Column Chromatography

The methods of column chromatography described here have been used principally for the separation, identification, and quantitative determination of reactants and products of the riboflavin synthetase reaction.[9] Scaled-up versions of these columns have been used for preparative purposes.

Lloyd's Reagent

Cellulose Suspension. Two grams of cellulose powder (Whatman, No. CF11 standard grade ashless powder for chromatography) are washed by repeated suspension in 100-ml portions of water, allowed to settle, and decanted until essentially all fines have been removed. The suspension is stored in 500 ml of water. The residue is washed again immediately before use to obtain satisfactory flow rates.

Lloyd's Reagent Suspension. A suspension of 20 g of cellulose in 1 liter of distilled water is stirred mechanically, and 5 g of Lloyd's reagent (Hartman-Leddon Company, Philadelphia) in 100 ml of water is added. The suspension is permitted to settle for 5 minutes, and the fines are decanted. The washing is repeated just before use.

Preparation of Columns. A column (0.7 cm i.d.) is fitted with a glass wool plug, filled partially with water, and then with the cellulose suspension. The cellulose is packed to a height of 2 cm. Lloyd's reagent-cellulose mixture is suspended in 5% acetic acid and degassed in a vacuum for about 30 minutes. The suspension is poured into the column above the cellulose plug, packed to a height of 8 cm, and washed with 500 ml of 5% acetic acid.

Chromatography. Trichloroacetic acid filtrate (1.5 ml) from an incubation mixture (see Assay, p. 522) is mixed with 8.5 ml of 5% acetic acid and applied to the column. The column is then treated sequentially with 100-ml portions of aqueous solutions containing 5% acetic acid, 0.5% pyridine and 2% acetic acid, and 3% pyridine and 2% acetic acid to recover 6-methyl-7-hydroxy-8-ribityllumazine, 6,7-dimethyl-8-ribityllumazine, and riboflavin, respectively. A flow rate of 0.5 ml per minute is maintained by applying air pressure to the column. With the appropriate eluting solvents, a recovery of 95–100% of the individual components (up to 0.5 micromole of each compound) can be obtained.

Acid Alumina

Preparation of Column. A column (1 cm i.d.) is fitted with a porous glass plate and a clamp or stopcock at the outlet. A suspension of 10 g of acid alumina AG4 (Bio-Rad Laboratories; 200–400 mesh) is made in 50 ml of a solvent composed of equal volumes of *n*-butanol and absolute ethanol and poured into the column. A disk of filter paper is pressed on top of the adsorbent once it has settled. The level of solvent in the column is adjusted so that the top of the alumina is covered by 2–3 mm of liquid.

Chromatography. The sample is prepared in a solvent composed of 1 ml of water, 9 ml of ethanol, and 10 ml of *n*-butanol and applied to the column. The column is washed with 10 ml of the above solvent. Riboflavin is brought off the column with 50 ml of *n*-butanol–ethanol–H_2O (80:28:28); and 6,7-dimethyl-8-ribityllumazine subsequently with 50 ml of *n*-butanol–ethanol–H_2O (80:28:56). The column is then washed with 50 ml of 50% (v/v) ethanol and 6-methyl-7-hydroxy-8-ribityllumazine is eluted with 50 ml of 0.03 M NH_4OH.

Recoveries of 95 to 100% of the individual components can be obtained with the appropriate eluting solvents.

Other Adsorbents. Columns of Dowex AG 50 W-X4 in the hydrogen form have been used to separate the second product of the riboflavin synthetase reaction, 4-ribitylamino-5-amino-2,6-dihydroxypyrimidine, from riboflavin and 6,7-dimethyl-8-ribityllumazine[15] using dilute HCl (0.05–0.1 M) as the developing solvent.

Columns of Magnesol[9] and Florisil[8] have been used for the separation of riboflavin and the lumazines, but considerable amounts of inorganic solids are obtained in the effluents. Much less inorganic matter occurs in effluents with columns of acid alumina and Lloyd's reagent.

Paper Chromatography and Electrophoresis

Materials are chromatographed on Whatman No. 3 MM or Schleicher and Schuell No. 407 paper and developed by the ascending method usually for 16–18 hours at ambient temperatures. Compounds are detected by fluorescence or quenching of fluorescence under ultraviolet light (Mineralight, Model SL-2537).

TABLE I
SOLVENT SYSTEMS FOR PAPER CHROMATOGRAPHY AND ELECTROPHORESIS,
AND THIN-LAYER CHROMATOGRAPHY[a]

Compound	Solvent systems[b]						
	I	II	III	IV	V	VI	VII[c]
6,7-Dimethyl-8-ribityl-lumazine	0.21 (0.42)	0.48 (0.65)	0.51	0.22	0.15 (0.28)	0.78 (0.83)	−9
6,7-Dimethyl-8-(5′-deoxyribityl)lumazine	0.33	0.64	—	0.16	0.10	—	—
6-Methyl-7-hydroxy-8-ribityllumazine	0.20	0.39	0.38	0.23	0.15	0.69	+70
Riboflavin	0.37 (0.58)	0.61 (0.71)	0.41	0.33	0.28 (0.46)	0.41 (0.45)	−6
5′-Deoxyriboflavin	0.60	0.73	—	0.22	0.41	—	—

[a] The results quoted in the table represent the averages of values obtained in a number of experiments. The numbers in parentheses refer to values of R_f obtained in chromatography on thin layers of cellulose.

[b] The following solvent systems were used for paper chromatography: (I) 1-butanol–ethanol–H_2O (500:175:360); (II) isobutyric acid–1 N NH_4OH–0.1 M ethylenediaminetetraacetate (250:150:4); (III) 1-propanol–1 N NH_4OH–H_2O (6:3:1); (IV) isopropyl ether–88% to 90% formic acid (9:6); (V) 1-butanol–acetic acid–H_2O (200:30:75); (VI) 3% NH_4Cl. Chromatograms were developed by the ascending method for 15–18 hours excepting system VI, which required 6–8 hours.

[c] Electrophoresis on strips of Whatman No. 3 MM filter paper in a solvent containing 11.6 ml of pyridine and 9.8 ml of glacial acetic acid in 1 liter of aqueous solution at pH 4.85. Current was applied at 1000 V for 3 hours in a water-cooled (15°) electrophoresis apparatus. All results are expressed as millimeters of travel of the compound from the origin (center) either to the cathode (−) or to the anode (+) in 3 hours.

Values of migration on paper of a number of compounds are given in Table I.

Thin-Layer Chromatography

Cellulose on Glass. A suspension of 15 g of Machery Nagel Cellulose 300 G (Brinkmann Instruments, Inc.) in 70 ml of water is prepared by homogenization for a few minutes in a Waring blendor. The suspension is applied to a thickness of 250 μ on 5 × 20 cm or 20 × 20 cm glass plates. The plates are dried in air at room temperature and then at 110°.

The plates are developed with the appropriate solvent systems to a height of 10–12 cm within 1–2 hours. Values of R_f are shown in Table I.

Silicic Acid on Glass Cloth.[25] Satisfactory resolution of lumazines and

[25] T. Aogaichi and G. W. E. Plaut, unpublished observations, 1968.

riboflavin has been obtained in a development time of 10–90 minutes on sheets of Chrom AR 500 (Mallinckrodt Chemical Co.). The R_f values for riboflavin, 6,7-dimethyl-8-ribityllumazine, and 6-methyl-7-hydroxy-8-ribityllumazine with the solvent systems described in Table I are, respectively: (I) 0.86, 0.64, and 0.72; (II) 0.87, 0.81, and 0.84; (V) 0.75, 0.45, and 0.56; (VI) 0.60, 0.84, and 0.95.

Riboflavin Synthetase

Purification of Riboflavin Synthetase from Yeast[25a]

Riboflavin synthetase of high specific activity has been obtained from *Ashbya gossypii* and yeast[9,13,25a]; partially purified enzyme has been prepared from *E. coli*[9] and spinach.[26] The highest degree of purification, achieved with the enzyme from yeast, is described below.

General Procedures

The enzyme is inactivated by oxygen. Therefore, all buffers from steps 2 to 14 are purged with nitrogen gas to displace oxygen before use.

Dialysis tubing is washed by boiling in several batches of distilled water.

The operations between steps 1 and 4 are at room temperature except where specified. Subsequent steps are performed at lower temperatures as indicated in the text and in the dark or in dim light.

Ammonium sulfate saturation is calculated by the formulation of Noda and Kuby,[27] a saturated solution being a mixture of 707.4 g of ammonium sulfate and 1 liter of water.

Red Star active dry yeast can be purchased from Universal Foods Corporation, Milwaukee, Wisconsin.

Step 1. Extraction. Red Star active dry yeast (6 kg) is processed as follows: 2.5 liters of 0.24 M $NaHCO_3$, containing 0.01 M Na_2SO_3, and 25 ml of toluene are placed into the bowl of a 1-gallon Waring blendor (Model CB 5) with 375 g of yeast and mixed at low speed for 1 minute. A total of 16 such suspensions are combined: The resulting 40 liters are placed into a 60- to 70-liter capacity plastic vessel. The contents of the vat are stirred overnight with a Lightnin Model L mixer (or equivalent overhead stirrer) fitted with a stainless steel shaft and propeller.

The reaction of the mixture (initially about pH 7.0) is adjusted to pH 5.1 by the slow addition of 3 M acetic acid with mechanical stirring, requiring about 3 hours for the addition of around 4.2 liters of the acid. The suspension is allowed to stand for 1 hour and then is placed on 16 one-

[25a] G. W. E. Plaut, R. L. Beach, and T. Aogaichi, *Biochemistry* **9**, 771 (1970).
[26] H. Mitsuda, *Vitamins* **28**, 465 (1963).
[27] L. Noda and S. A. Kuby, *J. Biol. Chem.* **226**, 541 (1952).

gallon funnels fitted with 50-cm diameter fluted filter paper (E and D No. 512). The filtration requires 10–12 hours and is done most conveniently overnight.

Step 2. First Ammonium Sulfate Fractionation. To the acidified filtrate is added with stirring solid ammonium sulfate (Mallinckrodt AR) to a final concentration of 52% saturation. The resulting suspension is allowed to stand for 3 hours and then centrifuged at 20° in 1-liter plastic cups in the No. 276 rotor of the International PR-2 centrifuge at 2200 rpm for 15 minutes. The supernatant solution is discarded, and the residue is taken up in 2500 ml of a solution of 0.1 M ammonium acetate and 0.01 M Na_2SO_3. After standing overnight, the suspension is centrifuged in the GSA or GS-3 rotor of a Sorvall RC-2B centrifuge at 8000 rpm for 20 minutes at 20°. The supernatant fluid is retained, and the residue is suspended in 500 ml of the above buffer and centrifuged. The supernatant fractions are combined and stored at room temperature.

An additional 6 kg of yeast is processed through steps 1 and 2, and the solutions obtained from both batches at the end of step 2 are combined before proceeding with the next step.

Step 3. Second Ammonium Sulfate Fractionation. The concentration of ammonium sulfate in the solution from step 2 is determined by the Nessler method.[28,29] To the stirred solution is added solid ammonium sulfate to a final concentration of 50% saturation. The suspension is allowed to stand for 2 hours and then centrifuged at 20° in the stainless steel inserts of a No. 959 rotor of an International PR-2 centrifuge at 4500 rpm for 15 minutes. The supernatant fluid is discarded and the residue is taken up in 2.5 liters of a solution containing 0.1 M ammonium acetate, 0.01 M Na_2SO_3, and 0.0002 M riboflavin. The solution is allowed to stand overnight and clarified by centrifugation as above. The residue is discarded.

Step 4. Heat Step. The ammonium sulfate concentration of the solution from step 3 is determined by the Nessler method, and enough solid ammonium sulfate is added to bring the concentration to 21% saturation (0.8 M) to stabilize the enzyme for the subsequent treatment. One-liter quantities of the solution are placed in 2-liter suction flasks. The air above the solution is displaced with nitrogen, and closure is effected with a 1-hole silicone rubber stopper fitted with a thermometer whose bulb is immersed in the liquid. The flask is placed into a boiling water bath with occasional

---

[28] For the calculation of the concentration of ammonium sulfate, the value obtained by the Nessler method is corrected for ammonium ion from the 0.1 M ammonium acetate buffer.

[29] M. J. Johnson, *in* "Manometric Techniques and Related Methods for the Study of Tissue Metabolism," 2nd ed. (W. W. Umbreit, R. H. Burris, and J. F. Stauffer, eds.) p. 161. Burgess, Minneapolis, 1949.

gentle swirling until the solution reaches 55°. The container is then placed into ice water until its contents reach 25°. The treated preparations are combined and centrifuged at 20° in the No. 959 rotor of an International PR-2 centrifuge at 4500 rpm for 15 minutes.

Step 5. Dialysis. The supernatant fluid from step 4 is placed into 7 sections of 5-cm diameter dialysis tubing and dialyzed with stirring against 12 liters of a solution of 0.1 M ammonium acetate–0.01 M Na_2SO_3 for 2–3 hours. The dialysis fluid is replaced by an equal quantity of the same buffer, and the dialysis is continued overnight.

Step 6. Acetone Fractionation. Mercaptoethanol (Eastman Kodak, white label) is added to the solution from step 5 to a final concentration of 0.2 M (14 ml of mercaptoethanol per liter). The solution is adjusted to pH 8.3 with 8 M NH_4OH. About one-third of the preparation (1 liter) is placed into a 4-liter glass bottle fitted with a stirrer. The container is chilled in an ethylene glycol bath at $-10°$, and acetone is added slowly to a final concentration of 48% (v/v). The suspension is left at $-10°$ for 1 hour and centrifuged at $-10°$ in the No. 959 rotor of an International PR-2 centrifuge at 4500 rpm for 15 minutes. The residue usually contains little activity and is discarded. The supernatant solution is adjusted to a final acetone concentration of 70% by the procedure described above, allowed to stand for 1 hour at $-10°$ and centrifuged at this temperature in the No. 959 rotor of an International PR-2 centrifuge at 4500 rpm for 15 minutes. The supernatant fluid is discarded and the residue is taken up in 150 ml of 0.1 M ammonium acetate–0.2 M mercaptoethanol at 5°.

Acetone fractionation is repeated on the remaining portions from step 5 as described above. Suspensions of the 48–70% acetone fractions in buffer are combined, homogenized in a large plastic homogenizer (80-ml capacity), and centrifuged at 3° in the GSA rotor of the Sorvall RC-2B centrifuge at 8000 rpm for 15 minutes. The residue is discarded, and the supernatant solution is stored at 3°.

Step 7. Third Ammonium Sulfate Fractionation. A solution of 0.1 M ammonium acetate–0.2 M mercaptoethanol is added to the solution from step 6 to adjust the protein concentration to about 10 mg/ml. Saturated ammonium sulfate solution (Mann, enzyme grade) is added to a final concentration of 37% saturation. The preparation is allowed to stand for 20 minutes in an ice bath and centrifuged in the stainless steel inserts of a No. 959 rotor of an International PR-2 centrifuge at 3° and 4500 rpm for 10 minutes. The residue contains little activity and is discarded. The supernatant fluid is adjusted to 53% saturation by the addition of saturated ammonium sulfate solution with stirring. The suspension is centrifuged as above. The supernatant layer is discarded, and the residue is taken up in 30 ml of 0.1 M ammonium acetate–0.2 M mercaptoethanol and centrifuged in the SS 34 rotor of a Sorvall RC-2B centrifuge at 3° and 18,000 rpm for 10 minutes. The residue is discarded, and the supernatant fluid is stored at 5°.

Step 8. Acidification and Freezing-Thawing Step. To the solution from step 7 is added an equal volume of 0.1 M sodium acetate–acetic acid buffer at pH 5.2 containing 0.2 M mercaptoethanol. The resulting solution is at about pH 5.6 and is acidified further to pH 5.40 with 0.1 M sodium acetate–acetic acid buffer at pH 4.8 containing 0.2 M mercaptoethanol. Twenty-five-milliliter portions are placed into 30-ml capacity Corex tubes, which are then stored in a deep freezer at $-90°$ for 1.5 hours. The frozen solutions are removed from the freezer and allowed to stand in air at room temperature for 5–10 minutes and are then brought to room temperature in a water bath. A voluminous white precipitate is removed by centrifugation, and the freezing-thawing centrifugation cycle is repeated. The solution is placed at $-90°$ overnight and thawed and centrifuged in the morning before proceeding to the next step.

Step 9. Dialysis. Enough 0.1 M sodium acetate–acetic acid buffer at pH 5.2 containing 0.2 M mercaptoethanol is added to the yellow solution from step 8 to adjust the protein concentration to approximately 10 mg/ml. The diluted enzyme solution is dialyzed with stirring against 25 volumes of the above buffer contained in a 4-liter graduated cylinder at 5° for 2 hours. The dialysis fluid is replaced by an equal volume of fresh buffer, and the dialysis is continued for an additional 2 hours. The solution is clarified by centrifugation in an SS-34 rotor of the Sorvall RC-2B centrifuge at 5° and 15,000 rpm for 10 minutes.

Step 10. First Ethanol Fractionation. The dialyzed enzyme is brought to room temperature, and 95% ethanol is added with stirring to a final concentration of 7% (v/v) alcohol. The air in the flask containing the enzyme solution is displaced with nitrogen, the container is stoppered and incubated at 24° for 4 hours. A white precipitate is formed which is removed by centrifugation in the SS-34 rotor of the RC-2B centrifuge at 20° and 15,000 rpm for 10 minutes. The supernatant fluid is incubated at 24° overnight as above, and is then clarified by centrifugation (Step 10a).

The supernatant solution is chilled to $-3°$ and brought with stirring to 25% ethanol. After 2 hours at $-3°$, the suspension is centrifuged at $-10°$ in the SS-34 rotor of a Sorvall RC-2B centrifuge at 15,000 rpm for 10 minutes. The supernatant solution usually contains little activity and is discarded.[30] The residue is taken up in 30 ml of 0.1 M sodium acetate–

---

[30] In some cases considerable activity (and yellow color indicating the presence of the enzyme) has been found in the fraction above 25% ethanol. This activity can be recovered by raising the alcohol concentration to 30–40% and allowing the solution to stand overnight at $-3°$. The precipitate formed is collected by centrifugation and dissolved in 6 ml of the acetate–mercaptoethanol buffer.

acetic acid buffer at pH 5.2 containing 0.2 M mercaptoethanol and clarified by centrifugation (Step 10b).

Step 11. Second Freezing-Thawing Step. The protein concentration of the solution from Step 10b is adjusted to 10 mg/ml by the addition of 0.1 M sodium acetate–acetic acid buffer at pH 5.2 containing 0.2 M mercaptoethanol. The diluted enzyme solution is placed in 30-ml capacity Corex glass tubes and frozen in a deep freezer at $-90°$ for 1.5 hours. The white precipitate formed after thawing is removed by centrifugation, and the supernatant fluid is frozen overnight at $-90°$, thawed, and centrifuged. The supernatant fluid is retained.

Step 12. Second Ethanol Fractionation. The solution is brought to 7% alcohol by the dropwise addition of 95% ethanol at $-3°$. If a precipitate forms after 1 hour of standing at this temperature, it is removed by centrifugation. The solution is then brought to 20% ethanol, allowed to stand for 1 hour at $-3°$ and centrifuged at $-10°$ in the SS-34 rotor of the Sorvall RC-2B centrifuge at 16,000 rpm for 10 minutes. The supernatant fluid is discarded, and the residue is taken up in 10 ml of a solution of 0.02 M potassium phosphate–0.2 M mercaptoethanol–0.001 M dithioerythritol at pH 7.2. The enzyme solutions are stored in an ice bath.

Step 13. TEAE-Cellulose Chromatography. Ten 0.7 $\times$ 4 cm columns of TEAE-cellulose[31] are prepared. Each column is washed with 30 ml of 0.02 M potassium phosphate–0.2 M mercaptoethanol–0.001 M dithioerythritol at pH 7.2.

One-milliliter portions of the solution from step 12 are placed on each column, which is then washed with 30 ml of 0.04 M potassium phosphate–0.2 M mercaptoethanol–0.001 M dithioerythritol at pH 7.2 at a flow rate of 0.4–0.5 ml/min. The enzyme appears on the column at first as a sharp yellow band which becomes more diffuse as the column is treated with the above buffer. The effluent contains mainly inert protein. The enzyme activity and the yellow color is brought off the column with 0.1 M potassium phosphate–0.2 M mercaptoethanol–0.001 M dithioerythritol at pH 7.2 at a flow rate of 0.3 to 0.4 ml/minute. The effluents are collected in 2-ml portions.

Step 14. Concentration of TEAE-Cellulose Effluent. Fractions containing

[31] Forty grams of TEAE-cellulose (Bio-Rad) are suspended in 2 liters of 1 N NaOH. The large particles are allowed to settle by gravity, and the fines in the supernatant fluid are removed by decantation. The process is repeated, and the residue is washed with three 4-liter portions of water. After the last water wash, the cellulose is recovered by filtration, suspended in 2 liters of 2 N HCl, filtered within 5 minutes, and washed immediately with water to remove excess acid. The residue is suspended in 2 liters of 2 N NaOH with stirring for 1 hour and allowed to settle by gravity. Residual fines are removed by decantation. The cellulose is washed with water until excess alkali has been removed. The suspension of TEAE-cellulose in water is stored in an amber colored bottle in the refrigerator.

TABLE II
PURIFICATION OF RIBOFLAVIN SYNTHETASE FROM YEAST[a]

Step	Volume (ml)	Total protein (mg)	Activity (millimicromoles B₂/ml)	Total activity (millimicromoles B₂)	Activity yield (%)	Specific[20] activity (millimicromoles B₂/mg)
1. Extract	66,800	804,000	45.0	2,996,000	(100)	3.7
2. First ammonium sulfate	7180	182,000	480	3,441,000	115	18.9
3. Second ammonium sulfate	2975	103,000	978	2,910,000	97	28.3
4. Heat treatment	2800	69,000	948	2,656,000	89	38.5
5. Dialysis	3625	67,000	693	2,508,000	84	37.4
6. Acetone fractionation	460	9000	4140	1,907,000	64	212
7. Third ammonium sulfate	55	3680	31,800	1,756,000	59	477
8. Acidification and freezing-thawing	102	2450	15,630	1,595,000	53	651
9. Dialysis	222	2310	7920	1,755,000	59	760
10. First ethanol fractionation	40	655	26,730	1,074,000	36	1640
11. Second freezing-thawing	69	342	15,360	1,060,000	35	3100
12. Second ethanol fractionation	14	167	71,670	1,004,000	33	6010
13. TEAE-cellulose chromatography	92	—	8220	756,000	25	—
14. Concentrate	6.2	71	136,200	845,000	28	11,900

[a] Active dry yeast, 12 kg.

the activity from the TEAE-cellulose step are combined (40–70 ml) and concentrated under vacuum in an S. and S. collodion membrane filtration apparatus to a final volume of 5–6 ml. The collodion sac is bathed in a solution of 0.1 M potassium phosphate–0.2 M mercaptoethanol–0.001 M dithioerythritol at pH 7.2 during vacuum filtration. The final enzyme preparation is stored in liquid nitrogen.

A typical purification procedure is described in Table II.

Discussion

The procedure described above, an improvement of the method of Harvey and Plaut,[13] gives 20–30% recovery of activity from the extract in the purified enzyme. A purification from the extract of 3000- to 5000-fold has been achieved in a number of preparations. The final preparation contains firmly bound riboflavin.

Fractionation of the enzyme is highly dependent on the nature of batches of dried yeast used as the starting material. Special precautions at certain stages of the preparation help to reduce the effects of such batch variations. Thus, a protein concentration greater than 50 mg/ml at the heat step reduces the effectiveness of subsequent fractionation procedures; 20–25 mg/ml of protein has been used in recent preparations. The introduction of the freezing-thawing and alcohol denaturation treatments (steps 9, 10a, 11) appears to decrease the solubility of the enzyme in the alcohol at steps 10b and 12 (Table II), and greatly improves the effectiveness of the purification at these steps. Early stages of preparation are conducted at room temperature, since activity is lost more rapidly at 0° to 5°. The stability of the enzyme is improved by the inclusion of Na_2SO_3 into the buffers at steps 2–5.

With improved purity the enzyme becomes increasingly sensitive to atmospheric oxygen. It is stabilized by the exclusion of oxygen by nitrogen wherever possible, the addition of riboflavin (step 3), and relatively high concentrations of mercaptoethanol starting at step 6. Buffers containing both mercaptoethanol and dithioerythritol at and after the TEAE-cellulose chromatography step afford better protection of activity than either agent alone.

The final enzyme solution has been stored for several months in liquid nitrogen or at −90° without substantial loss in activity.

Properties of Riboflavin Synthetase from Yeast

Physicochemical Properties. Certain physicochemical studies of the purified enzyme have been made. Analytical ultracentrifugation[13] shows a main active component with a sedimentation constant of 4.3 S, suggesting a molecular weight of the protein of 70,000–80,000. More recent preparations with a specific activity of 9000 millimicromoles/mg revealed a single protein component in disk electrophoresis in the standard buffer and gel.[32] However, electrophoresis on acetylated cellulose strips indicated at least two protein components in about equal amounts, one of these containing the activity. Additional purification of this preparation by chromatography on TEAE-cellulose removed the inactive component, leaving an active enzyme with a specific activity of about 15,000. The preparation described in Table II had a final specific activity of 11,900 units/mg and still contained about 22% of the inactive protein component.

Substrate Specificity. Chemical changes catalyzed by riboflavin synthetase involve the heterocyclic ring system of 6,7-dimethyl-8-(1'-D-ribityl)lumazine. Only one other 6,7-dimethyllumazine substituted in position 8 (6,7-dimethyl-8-[1'(5'-deoxy-D-ribityl)]lumazine) of those examined is converted enzymatically to the corresponding flavin (5'-deoxyriboflavin).[11]

Compounds substituted at the D-ribityl group by simple substituents (hydrogen, methyl, β-hydroxyethyl), tetrahydroxypentyl groups (D- and L-arabityl, D- and L-lyxityl, D- and L-xylityl), and pentahydroxyhexyl groups (D-galactityl, D-glucityl, D-mannityl) have been studied; only 6,7-dimethyl-8-(1'-D-xylityl)lumazine combined with the purified enzyme. It is a competitive inhibitor (K_i, $9 \times 10^{-5}\ M$) of the conversion of the ribityl compound (K_m, $1 \times 10^{-5}\ M$) to riboflavin,[11] and direct evidence for binding to the purified yeast enzyme has been obtained.[13] It seems significant that of the epimeric forms of the pentyl derivatives tested, only those possessing hydroxyl groups in the D-configuration in positions 2' and 4' of the side chain (D-ribityl, 5'-deoxy-D-ribityl, and D-xylityl) are bound to the enzyme. The recent observation[33] that 6,7-dimethyl-8-[1'-(2'-deoxy-D-ribityl)]lumazine is neither substrate nor inhibitor for the enzyme from yeast is consistent with these requirements for the configuration of the side chain. Lengthening of the side chain to form the D-glucityl derivative leads to loss of activity, although this compound contains D-2' and D-4'-hydroxyl groups. Shortening the pentyl side chain by one carbon atom (D- and L-threityl and D- and L-erythrityl) produces compounds which are relatively ineffective competitive inhibitors (K_i, approx. 0.5 to $1 \times 10^{-3}\ M$), but not substrates, of the yeast enzyme.[33] The requirements of riboflavin synthetase for the configuration of the side chain of the substrate are more rigid than that of riboflavin kinase,[34,35] the next enzyme in the sequence of biosynthesis of flavin coenzymes.

[32] B. J. Davis, *Ann. N.Y. Acad. Sci.* **121,** 404 (1964).
[33] R. L. Beach and G. W. E. Plaut, unpublished observations, 1968.
[34] B. M. Chassy, C. Arsenis, and D. B. McCormick, *J. Biol. Chem.* **240,** 1338 (1965).
[35] D. B. McCormick and R. C. Butler, *Biochim. Biophys. Acta* **65,** 326 (1962).

Replacement of the methyl groups at positions 6 and 7 of 8-(1'-D-ribityl)lumazines[11] leads to substances that are either inactive (methyl, ethyl; methyl, n-propyl; methyl, n-butyl; diethyl; di-n-propyl; diphenyl; methyl, phenyl) or competitive inhibitors of the enzyme [methyl, n-pentyl $(K_i,\ 2 \times 10^{-4}\ M)$; 5,6,7,8-tetrahydro-9-(1'-D-ribityl)isoalloxazine $(K_i,\ 1.6 \times 10^{-4}\ M)$]. Among these, 6-methyl-7-hydroxy-8-ribityllumazine $(K_i,\ 2 \times 10^{-6}\ M)$ and 6,7-dihydroxy-8-ribityllumazine $(K_i,\ 9 \times 10^{-9}\ M)$ are particularly effective inhibitors. Inhibition by substances bearing substituents other than methyl groups at positions 6 and 7 depends on the presence of the proper group at position 8; replacement of the D-ribityl group of 6-methyl-7-hydroxy-8-ribityllumazine by methyl or β-hydroxyethyl substituents leads to inactive compounds.

Substitution of the oxy by an amino group at position 2 of 6,7-dimethyl-8-ribityllumazine produces a compound that is neither a substrate nor an inhibitor of riboflavin synthetase. Other pteridine derivatives with an amino group at position 2 are also ineffective,[11] except 2-amino-4,6-dihydroxy-8-D-ribityl-7(8H)-pteridinone; this is an inhibitor, though considerably less potent than the analogous 6,7-dihydroxy-8-ribityllumazine.

Riboflavin is a product inhibitor $(K_i,\ 5 \times 10^{-6}\ M)$ of the enzyme from yeast,[13] but less effective for that from *A. gossypii*. FMN or FAD are essentially inactive. Riboflavin-2-imine, a substrate for rat liver flavokinase,[35] shows doubtful inhibitory action with riboflavin synthetase.

Other Properties. Riboflavin synthetase from yeast exhibits maximal activity at pH 7.0 under the conditions of the standard assay. Enzyme activity decreases almost symmetrically as the pH is increased or decreased from the optimum; half-maximal activity is observed at pH 5.8 and pH 8.4. There is little loss in activity when the enzyme is stored for several hours in the presence of suitable reducing agents at 0° between pH 4 and pH 8.

The velocity of the reaction is increased 2.0-fold by raising the temperature from 25° to 37°.

The activity of the yeast enzyme is inhibited approximately 50% by $10^{-6}\ M$ p-chloromercuribenzenesulfonate. The inhibition can be reversed almost completely by subsequent addition of an excess of cysteine or mercaptoethanol.

Mechanism of Action. A general mechanism for the riboflavin synthetase reaction has been proposed which involves the addition of 6,7-dimethyl-8-ribityllumazine to two sites on the enzyme.[13] One site binds the substrate in such a way that it functions as a donor of the 4-carbon moiety; the other binds the lumazine which serves as acceptor of the 4-carbon fragment. Such a mechanism is supported both by the stoichiometry of the overall reaction and by the finding that 6,7-dimethyl-^{14}C-8-ribityllumazine is enzymatically converted to riboflavin labeled exclusively in carbon atoms 5 and 8 and the methyl groups.

Riboflavin synthetase catalyzes the conversion of 6,7-dimethyl-8-(5'-deoxyribityl)lumazine to 5'-deoxyriboflavin at a slow rate. However, rapid formation of riboflavin-^{14}C (but not 5'-deoxyriboflavin-^{14}C) occurs in the presence of a mixture of equimolecular amounts of 6,7-dimethyl-8-ribityllumazine and 6,7-dimethyl-^{14}C-8-(5'-deoxyribityl)lumazine. This indicates that 6,7-dimethyl-8-(5'-deoxyribityl)lumazine is an efficient donor of the 4-carbon fragment involved in the formation of the o-xylene portion of riboflavin, but functions poorly as an acceptor of the 4-carbon moiety. Thus, the donor and acceptor sites of riboflavin synthetase are not equivalent in that they show different substrate specificities with respect to flavin formation.

Enzyme–substrate complexes isolated by Sephadex chromatography at pH 4.5 appear to contain the lumazine derivatives at the site which leads to donation of the 4-carbon moiety. For example, reaction of equivalent amounts of enzyme-6,7-dimethyl-^{14}C-8-(5'-deoxyribityl)lumazine with free 6,7-dimethyl-8-ribityllumazine yields only riboflavin-^{14}C, but not 5'-deoxyriboflavin. Evidence for a second substrate binding site on the enzyme, presumably the acceptor site, has been obtained by measurement of changes of polarization of fluorescence caused by the addition of substrate to isolated enzyme–substrate complexes.

It has not been possible to demonstrate an intermediate in the riboflavin synthetase reaction; furthermore, no evidence has been obtained that the enzyme contains or requires an inorganic or an organic cofactor for activity.

The transfer of 4-carbon units from 1 molecule of 6,7-dimethyl-8-ribityllumazine to a second molecule of lumazine may be due to a spatial effect, wherein the formation of the C—C bonds of the product facilitates the rupture of the C—N bonds of the substrate. In this connection, riboflavin has been obtained in good yield when 6,7-dimethyl-8-ribityllumazine was refluxed anaerobically in phosphate buffer at pH 7 for a prolonged period of time.[36] Under the conditions of the chemical conversion, with an extremely high concentration of substrate, enough lumazine molecules in solution may be oriented in the configuration required for reaction to occur. If this mechanism is also applicable to riboflavin synthetase, the principal function of the enzyme may involve the proper alignment of the substrate molecules to permit more efficient formation of products.

Enzyme Complexes. Complexes of riboflavin synthetase with substrates or substrate analogs are prepared by displacement of enzyme-bound riboflavin by an excess concentration of the desired ligand. The new enzyme

---

[36] T. Rowan and H. C. S. Wood, *Proc. Chem. Soc.* p. 21 (1963).

complex thus formed is separated from the low molecular weight components by chromatography on Sephadex G-25.

In a typical experiment, 0.5 ml of a solution containing 40 millimicromoles of enzyme-bound riboflavin is mixed with 1–2 micromoles of 6-methyl-7-hydroxy-8-ribityllumazine and applied to a 1 × 20 cm column of Sephadex G-25 (fine grade, Pharmacia Co.) equilibrated with 0.10 M potassium phosphate–0.2 M mercaptoethanol–0.001 M dithioerythritol at pH 7.0. The column is eluted with the same buffer at a flow rate of 0.5 ml/min, and 1-ml fractions are collected. The enzyme complex containing 6-methyl-7-hydroxy-8-ribityllumazine emerges in the void volume of the column, while the low molecular weight components are retarded.

Complexes of the enzyme with substrates have been prepared similarly, but in buffers at pH 4.5 where the enzyme is inactive.[13] The activity can be restored by incubating such complexes at a pH where conversion to flavin occurs.

Complexes of the enzyme have been prepared containing 6-methyl-7-hydroxy-8-ribityllumazine, 6,7-dimethyl-8-ribityllumazine, 6,7-dimethyl-8-(5′-deoxyribityl)lumazine, 6,7-dimethyl-8-D-xylityllumazine, or 5′-deoxyriboflavin.[13] Only those substances are bound that have a kinetic effect on the enzyme. For example, 6,7-dimethyl-8-L-xylityllumazine does not inhibit activity and is not bound. The complexes contain riboflavin, substrate, or analogs of 6,7-dimethyl-8-ribityllumazine, respectively, in equimolecular amounts. The binding is not due to covalent bonding since riboflavin (or lumazines) can be removed from the protein with charcoal; the treated enzyme retains activity.

Characteristics of absorption and fluorescence spectra of riboflavin and various lumazine derivatives in the free and enzyme-bound form are given in Table III. The absorption spectrum of riboflavin bound to the enzyme is displaced toward longer wavelengths when compared to the free compound; it is thus similar to absorption spectra of flavoproteins containing bound phosphorylated forms of flavin. Binding of riboflavin to the synthetase does not result in displacement of the fluorescence emission maximum; however, it is accompanied by a 70–80% decrease in the intensity of fluorescence. The lumazine derivatives, also, exhibit a displacement of absorption spectra toward longer wavelengths on binding; the fluorescence emission maxima are positioned at slightly shorter wavelengths, but the intensity of fluorescence is the same in the free and bound forms.

Distribution. Riboflavin synthetase activity has been found in extracts of all riboflavin-producing organisms examined. Activity (2.4–4.8 millimicromoles riboflavin formed per milligram of protein per hour at 37°) has been found in extracts of *Aerobacter aerogenes, Pseudomonas, Bacillus*

TABLE III
ABSORPTION AND FLUORESCENCE EMISSION SPECTRA OF ENZYME-BOUND
RIBOFLAVIN AND LUMAZINES[a]

Compound	pH	Emission maximum[e]		Absorption maximum	
		Free	Enzyme bound	Free	Enzyme bound[b]
Riboflavin	7.0	522	522[c]	372	387
				445	467
6,7-Dimethyl-8-ribityllumazine	4.5	483	473[d]	407	412
6,7-Dimethyl-8-D-xylityllumazine	7.0	483	473[d]	407	412
6-Methyl-7-hydroxy-8-ribityllumazine	7.0	422	407[d]	340	345

[a] The preparation of the enzyme complexes (riboflavin synthetase, specific activity 7500) is described in the text. Potassium phosphate $(0.1 M)$–0.01 M NaHSO$_3$ was used as the buffer at pH 7.0 and 0.1 M sodium citrate–citric acid –0.1 M NaHSO$_3$, at pH 4.5.

[b] Molar absorbance at λ_{max} shows little or no change on binding.

[c] Fluorescence intensity at emission maximum decreased 70–80%.

[d] Fluorescence intensity at emission maximum shows no change on binding.

[e] Excitation at 366-mμ line of mercury lamp; temperature, 10°.

subtilis, Escherichia coli, Ashbya gossypii, Saccharomyces cerevisiae, Neurospora crassa, and the folate-requiring organism *Lactobacillus plantarum* ATCC 8014; whereas *L. casei* ATCC 7496, which requires riboflavin for growth, does not contain the enzyme.[37] Inability to form flavin does not necessarily indicate lack of the synthetase reaction, since extracts of the riboflavinless mutant *Neurospora crassa* FGSE No. 83[38] contain as much riboflavin synthetase activity as the wild strain.[25]

Addendum

A number of publications have appeared recently on the chemical[39,40,41] and enzymatic[25a,42] formation of flavin from precursor lumazines.

[37] G. W. E. Plaut, *in* "Metabolic Pathways" (D. M. Greenberg, ed.), Vol. II, p. 673. Academic Press, New York, 1961.

[38] L. Garnjobst and E. L. Tatum, *Am. J. Botany* **43,** 149 (1956).

[39] T. Paterson and H. C. S. Wood, *Chem. Commun.* 290 (1969).

[40] R. L. Beach and G. W. E. Plaut, *Tetrahedron Letters* **40,** 3489 (1969).

[41] R. L. Beach and G. W. E. Plaut, *Biochemistry* **9,** 760 (1970).

[42] R. L. Beach and G. W. E. Plaut, *J. Am. Chem. Soc.* **92,** 2913 (1970).

[158] Assay Methods, Isolation Procedures, and Catalytic Properties of Riboflavin Synthetase from Spinach

By HISATERU MITSUDA, FUMIO KAWAI, and YUZURU SUZUKI

2 (6,7-Dimethyl-8-ribityllumazine) →
   riboflavin + 4-ribitylamino-5-amino-2,6-dihydroxypyrimidine

In this conversion catalyzed by riboflavin synthetase, four carbon atoms from the methyl group substituents and carbons 6 and 7 of *one* molecule of 6,7-dimethyl-8-ribityllumazine are transferred to a *second* molecule of the lumazine to form the *o*-xylene ring of riboflavin. This mechanism is established both by the stoichiometry of the reaction and by the isotopic studies.[1-5] Riboflavin synthetase activity has been detected in extracts of a number of plants and flavinogenic microorganisms, and was found also in homogenates of some animal livers.[6] Partial purification of the enzyme has been performed with spinach,[7] *Escherichia coli*, and *Ashbya gossypii*.[8] Harvey and Plaut isolated it from bakers' yeast with 2000-fold purity.[9]

Assay Methods

Principle. 6,7-Dimethyl-8-ribityllumazine is mixed with the enzyme, and the riboflavin produced is measured spectrophotometrically by reading absorbancies at 470 and 405 nm, or fluorophotometrically after isolation by paper chromatography.

Procedure for Spinach Enzyme.[7] Assay medium is placed in a tube, which contains 0.2 micromole of 6,7-dimethyl-8-ribityllumazine, 20 micromoles of cysteine, 20 micromoles of ascorbate, 100 micromoles of phosphate buffer of pH 7.5, and enzyme in a total volume of 2.0 ml. The reaction is started by addition of enzyme and proceeds at 37° for 1–3 hours. Incubation is terminated by heating the medium at 100° for 2 minutes. The

[1] H. Mitsuda, F. Kawai, and S. Moritaka, *J. Vitaminol. (Kyoto)* **7**, 128 (1961).

[2] H. Mitsuda, Y. Suzuki, and F. Kawai, *J. Vitaminol. (Kyoto)* **9**, 121 (1963).

[3] G. W. E. Plaut, *J. Biol. Chem.* **235**, PC41 (1960).

[4] T. W. Goodwin and A. A. Horton, *Nature* **191**, 772 (1961).

[5] H. Wacker, R. A. Harvey, C. H. Winestock, and G. W. E. Plaut, *J. Biol. Chem.* **239**, 3493 (1964).

[6] G. W. E. Plaut, *in* "Metabolic Pathways" (D. M. Greenberg, ed.), Vol. II, p. 673. Academic Press, New York, 1961.

[7] H. Mitsuda, F. Kawai, S. Yoshimoto, and Y. Suzuki, *Annual Meeting of Japan Vitamin Society Yudanaka (Japan), April* (1963); *Vitamins (Kyoto)* **28**, 98, 465 (1963).

[8] G. W. E. Plaut, *J. Biol. Chem.* **238**, 2225 (1963).

[9] R. A. Harvey and G. W. E. Plaut, *J. Biol. Chem.* **241**, 2120 (1966).

coagulated protein is removed by centrifugation or filtration through gauze. A 0.1-ml aliquot of the supernatant solution is chromatographed (Toyo filter paper No. 51, 2 × 40 cm) using a solvent system of n-butanol–acetic acid–water (4:1:5 by volume, upper layer). The yellow band (riboflavin, R_f = 0.30) detectable under ultraviolet light on the chromatogram is cut out and eluted with 5 ml of distilled water at 80° for 15 minutes in the dark. The fluorescent intensity can be determined with a fluorophotometer. Controls lacking only enzyme or the substrate must be run parallel with all experiments. Initial concentration of the lumazine is determined by measuring absorbancy at 405 mμ, using a molar absorbancy index [a_M = absorbancy × molecular weight/grams per liter] of 10,300 in neutral solution. Protein is estimated by spectrophotometry.[10]

Units and Specific Activity. A unit of activity is defined as that amount of enzyme yielding 1 millimicromole of riboflavin per hour at 37°. Specific activity is expressed in units per milligram of protein.[11,12]

Purification Procedure for Spinach Enzyme[7]

Operations in all steps are carried out at 0–4°.

Step 1. Extraction. Fresh green leaves (5 kg) of spinach are washed with tap water and chopped fine with a knife. The broken materials (each 200 g) are ground with 50 ml of 0.1 M phosphate buffer at pH 7.0 with a cold mortar and pestle and homogenized with a blender. This is followed by pressing out the juice by hand through three layers of gauze. The juice (4.5 liters) obtained from 5 kg of the leaves is adjusted to pH 5.4 with 10% acetic acid and allowed to stand overnight. The resulting precipitate is removed by centrifuging at 3500 g for 20 minutes. The supernatant liquid is neutralized to pH 6.5 with 1 N NaOH.

Step 2. First Ammonium Sulfate Fractionation. Solid ammonium sulfate is slowly added with overhead stirring to the supernatant solution (3830 ml). The concentration of ammonium sulfate is increased to 38% saturation (100% saturation = 760 g per liter). The pH of the solution is maintained at pH 6.5 by addition of 1 N NaOH during this operation. The precipitate is collected by centrifugation and then suspended in a small volume of 0.02 M phosphate buffer of pH 7.0, followed by dialysis against the same buffer for 2 hours. At this stage, the enzyme is stable if kept frozen.

Step 3. Protamine Sulfate Treatment. To the dialyzate (330 ml), 26.4 ml of protamine sulfate solution (20 mg/ml) is added. After magnetic stirring

[10] H. M. Kalcker, *J. Biol. Chem.* **167,** 461 (1947).
[11] O. H. Lowry, N. J. Rosebrough, A. L. Farr, and R. J. Randall, *J. Biol. Chem.* **193,** 265 (1951).
[12] A. G. Gornal, C. J. Bardawill, and M. M. David, *J. Biol. Chem.* **177,** 751 (1949).

for 20 minutes, the precipitate is removed by centrifuging at 3500 g for 20 minutes.

Step 4. Second Ammonium Sulfate Fractionation. The supernatant solution (320 ml) is saturated to 23% by the addition of solid ammonium sulfate in the same manner as in step 2. The suspension is centrifuged. To the supernatant solution, an additional solid ammonium sulfate is added to 40% saturation, followed by centrifugation. The precipitate is dissolved in a small volume of 0.025 M phosphate buffer at pH 6.5, and dialyzed against the same buffer.

Step 5. CM-Cellulose Treatment. The dialyzate (75 ml) is divided into three parts of 25 ml each for convenience. Each 25-ml portion is put on a column (2.2 cm diameter) involving 8 g of CM-cellulose (Serva) previously washed with 0.025 M phosphate buffer of pH 6.5. The column is developed with 150 ml of the same buffer. A saturated solution of ammonium sulfate is added to the column effluents to give a 55% saturated solution. The precipitate is recovered by centrifuging at 9700 g for 20 minutes, and suspended in 0.025 M phosphate buffer at pH 6.5. The suspension is dialyzed against the same buffer for 5 hours.

Step 6. First DEAE-Cellulose Column Chromatography. A 12.5-ml aliquot of the dialyzate (75 ml) is applied to a column (2.1 cm diameter) prepared from 5.0 g of DEAE-cellulose (Serva) previously washed and equilibrated with 0.01 M Tris–malate NaOH buffer of pH 6.2. The column is first washed with 60 ml of the Tris buffer, followed by 80 ml of the buffer containing 0.2 M NaCl. Most of the enzyme can be eluted by increasing the concentration of NaCl to 0.5 M in the same buffer. The eluate is brought to 60% saturation with a saturated solution of ammonium sulfate. This step is repeated five more times. The precipitate is collected by centrifugation, and then suspended in 0.01 M Tris–malate–NaOH buffer (pH 6.2), followed by dialysis against the same buffer for 5 hours.

Step 7. Second DEAE-Cellulose Column Chromatography. The dialyzate is immediately put on a column (2.1 cm diameter) containing 4.2 g of DEAE-cellulose which is previously equilibrated with 0.01 M Tris–malate–NaOH buffer at pH 6.0. The chromatogram is developed by stepwise elution with NaCl solution buffered by 0.01 M Tris–malate–NaOH at pH 6.0 (60 ml of the Tris buffer containing no NaCl, 120 ml with 0.2 M NaCl, 80 ml with 0.3 M NaCl, 90 ml with 0.4 M NaCl, and 120 ml with 0.7 M NaCl). Fractions of 10 ml each are collected. The major activity of the enzyme can be eluted from the column by 0.3 M salt concentration. These fractions are combined. In steps 6 and 7, one equivalent weight of cellulose to ten equivalent weights of protein present is usually used.

An example of purification is summarized in the table.

PURIFICATION OF SPINACH RIBOFLAVIN SYNTHETASE[a]

Step	Total protein (mg)	Total activity (units)	Specific activity (units per mg protein)	Recovery (%)	Purification degree
1	432,000	9520	0.022	100	1
2	18,100	7060	0.39	74.4	17.7
3	10,500	5830	0.55	61.3	25.0
4	7,180	5350	0.75	56.3	34.1
5	5,300	4140	0.78	43.5	35.5
6	522	2640	5.05	27.7	230
7	117	1760	15.1	18.5	648

[a] Fresh leaves of spinach; 5 kg is used.

Properties

Stability.[7,9] Purified enzyme of spinach retains initial activity at 0°–4° for 24 hours and is stable at least for 1 week at 0°–4° in the presence of saturated ammonium sulfate. Freezing leads to complete loss of the activity. Spinach and yeast enzymes are more labile at a pH higher than 8.0. Yeast enzyme shows appreciable lability to cold (0°–2°), and its stability is enhanced at lower pH (<7.5). Heating at 80° for only 1 minute completely destroys the yeast enzyme. Both enzymes are stabilized by reducing agents such as cysteine, ascorbate, and Na_2SO_3. Higher ionic strength protects the yeast enzyme. Ten percent saturated ammonium sulfate at pH 6.0 containing 0.01 M Na_2SO_3 is particularly effective for storage. Oxidizing substances readily and irreversibly inactivate the yeast enzyme, especially dissolved O_2, requiring the use of degassed buffer in the latter steps of purification.

pH Optimum.[7–9] The enzymes of *A. gossypii*, yeast, and spinach exhibit optimal activity at pH 6.9, 7.0, and 7.5, respectively, with rather broad pH activity curves.

Effect of Temperature.[7,8] The apparent activation energies of spinach (25°–45°) and yeast (12°–37°) enzymes are approximately 15,000 and 10,000 calories per mole, respectively.

Stoichiometry.[1–3,5,7,9] Crude or purified enzymes of spinach, *E. coli*, *A. gossypii* and yeast catalyze the conversion of 2 molecules of 6,7-dimethyl-8-ribityllumazine to yield 1 molecule of riboflavin. The second product, 4-ribitylamino-5-amino-2,6-dihydroxypyrimidine, has been proven in the case of the yeast enzyme to be equivalent in amount to the riboflavin formed.

Apparent K_m and V_{max}.[7–9] The reaction catalyzed by yeast and spinach enzymes proceed under zero to first-order kinetics over 2×10^{-4} to $6 \times$

$10^{-7} M$ of the substrate, giving only one K_m of $1.0 \times 10^{-5} M$ (pH 7.0) and $4.5 \times 10^{-5} M$ (pH 7.5), respectively. K_m of $A.$ *gossypii* enzyme is $2.9 \times 10^{-5} M$ (pH 6.9). K_m and V_{max} values of yeast enzyme are influenced by change of pH: K_m, $3.8 \times 10^{-5} M$ at pH 5.8, $4.4 \times 10^{-5} M$ at pH 6.4, $1.0 \times 10^{-5} M$ at pH 7.0, $1.1 \times 10^{-5} M$ at pH 7.6, $4.1 \times 10^{-5} M$ at pH 8.2, and $47.0 \times 10^{-5} M$ at pH 8.9; V_{max}, 41% at pH 5.8, 62% at pH 6.4, 100% at pH 7.0, 61% at pH 7.6, 32% at pH 8.2, and 6% at pH 8.9.

Inhibitors (SH-Reagents and Avidin).[1,7-9] p-Chloromercuribenzoate, $HgCl_2$, and $CuSO_4$ (each $10^{-3} M$) inhibit completely the spinach enzyme. Although p-hydroxymercuribenzoate ($10^{-4} M$) does not inhibit $E.$ *coli* and $A.$ *gossypii* enzymes, p-chloromercuribenzoate and p-chloromercuribenzene sulfonate ($>10^{-4} M$) block completely yeast enzyme activity. This inhibition is completely reversed by subsequent addition of excess cysteine or thioethanol. At $3 \times 10^{-5} M$ of the p-mercuribenzene sulfonate (50% inhibition), the substrate can afford some degree of protection if added before the mercurial. A similar case is shown with 6,7-dimethyl-8-(1'-D-xylityl)lumazine, but not with the L-xylityl derivative. Avidin (1.7–8.4 units/ml) is moderately inhibitory to the yeast enzyme. This inhibition cannot be prevented by biotin.

Side Reaction (Formation of 6-Methyl-7-hydroxy-8-ribityllumazine). 6-Methyl-7-hydroxy-8-ribityllumazine existing in a number of riboflavin-producing organisms[6] is proved to be directly formed from 6,7-dimethyl-8-ribityllumazine by extracts of spinach and $A.$ *gossypii*.[13,14] This reaction is characterized as dehydrogenic demethylation of the substrate by quinones, which can be regenerated by coupling with the polyphenol oxidase system in plants.[15] This conversion is inhibited by reducing agents such as cysteine, ascorbate, mercaptoethanol, Na_2SO_3, and dithiothreitol, and completely blocked under anaerobic conditions.[2,16] Assay of riboflavin synthetase must be carried out in the reduced condition for this reason and because of lability of the enzyme in the presence of O_2, particularly in the case of plants and crude samples. The possible role of this side reaction in the control of riboflavin synthesis has been presented.[17,18]

[13] H. Mitsuda, F. Kawai, and Y. Suzuki, *J. Vitaminol. (Kyoto)* **7**, 243 (1961).
[14] G. W. E. Plaut, *Federation Proc.* **19**, 312 (1960).
[15] H. Mitsuda, Y. Suzuki, and F. Kawai, *J. Vitaminol. (Kyoto)* **9**, 125 (1963); *Vitamins (Kyoto)* **27**, 234 (1963).
[16] H. Mitsuda, F. Kawai, and Y. Suzuki, *J. Vitaminol. (Kyoto)* **7**, 247 (1961).
[17] C. H. Winestock, T. Aogaichi, and G. W. E. Plaut, *J. Biol. Chem.* **238**, 2866 (1963).
[18] H. Mitsuda, *Proc. Japan Acad.* **42**, 940 (1966).

[159] Flavokinase (ATP:Riboflavin 5'-Phosphotransferase, EC 2.7.1.26) from Rat Liver

By Donald B. McCormick

Flavin + ATP → flavin monophosphate + ADP

Assay Method

Principle. The flavin phosphate formed, usually FMN, is measured spectrophotometrically at 450 nm in a protein-free filtrate from which most nonphosphorylated flavin has been removed by extraction into benzyl alcohol according to the method of Burch *et al.*[1] as modified by Kearney and Englard.[2]

Reagents

> Flavin, $2 \times 10^{-4} M$. Fresh solutions are made by solubilization in a minimum volume of 0.1 N NaOH followed dy dilution.
> ATP, 0.01 M, pH 8
> ZnSO$_4$, 0.001 M
> Potassium phosphate buffer, 0.75 M, pH 8

Procedure. Mixtures contain 2.5 ml of riboflavin or analog, 0.5 ml each of ATP, ZnSO$_4$, and phosphate buffer, and enzyme plus water to a total volume of 5 ml. Final concentrations are 0.1 mM flavin, 1 mM ATP, 0.1 mM Zn^{2+}, 75 mM phosphate, and sufficient enzyme, usually 0.5–2 mg of protein, to ensure measurable activity. Incubation is carried out in the dark for 1 hour at 37°. The reaction is terminated by the addition of 2 ml of 17.5% trichloroacetic acid. A control is run by addition of this acid before the enzyme. Analysis is done according to Kearney (see Vol. II [108]) by measuring the absorbance at 450 nm of total flavin (A) in a neutralized filtrate and the flavin phosphate (B) in an aqueous phase after extractions with benzyl alcohol and chloroform. From measurements of the distribution coefficients of free and phosphorylated flavins between benzyl alcohol and aqueous phases, the fraction of free (x) and of phosphorylated (y) flavin remaining in the final aqueous phase is known. The quantity of flavin phosphate produced is then calculated as $(B - xA)/(y - x)$.

Definition of Unit and Specific Activity. One unit is that amount of enzyme which catalyzes the synthesis of 1 millimicromole of flavin mono-

[1] H. B. Burch, O. A. Bessey, and O. H. Lowry, *J. Biol. Chem.* **175**, 457 (1948).
[2] E. B. Kearney and S. Englard, *J. Biol. Chem.* **193**, 821 (1951).

phosphate in 1 hour at 37° in the above assay. Specific activity is defined as units per milligram of protein determined by the method of Lowry et al.[3]

Comments. The assay is generally applicable to extracts where only moderate contamination by phosphatases occurs. However, the levels of acid phosphatases are so high in several tissues from the rat[4] and other animals[5] that much of the flavokinase is masked in crude preparations. The present use of Zn^{2+} and phosphate seems to obviate some of the difficulty, since the conventional use of Mg^{2+} and Tris, even with inclusion of fluoride, often leads to lower values for kinase activity. A test for the level of interfering phosphatase can be made by performing the above incubation with FMN instead of riboflavin as substrate.

Purification Procedure

Flavokinase from liver cells is located in the true supernatant solution and is enriched through classic fractionation.[6]

Step 1. Homogenization of Tissue. Rats are killed by rapid decapitation and exsanguination. The livers are extirpated, rinsed, blotted, and homogenized in 4 volumes of cold, 0.05 M potassium phosphate buffer, pH 7, with the TenBroeck apparatus. All subsequent operations are performed at 0°–4°.

Step 2. Preparation of Supernatant Solution. A supernatant solution is prepared from the 20% homogenate by centrifuging at 18,500 g for 30 minutes and discarding the debris.

Step 3. Precipitation with Ammonium Sulfate. The supernatant solution is brought to 55% saturation with ammonium sulfate by addition of a 100% saturated solution of the salt over a 1-hour period. The mixture is centrifuged at 12,500 g for 15 minutes, and the precipitate is discarded. The supernatant solution is then brought to 75% saturation with ammonium sulfate over a similar period of time, and the precipitate is collected by centrifugation.

Step 4. Dialysis. The precipitate is dissolved in a small volume of 0.02 M glycine, pH 6.8, and dialyzed overnight against a large volume of this solution; the dialyzate is clarified by centrifugation at 12,500 g for 15 minutes. This solution is fairly stable when frozen and may be used as a supply of the crude enzyme.

Step 5. Chromatography on DEAE-Cellulose. The clear supernatant

[3] O. H. Lowry, N. J. Rosebrough, A. L. Farr, and R. J. Randall, J. Biol. Chem. **193**, 265 (1951).
[4] D. B. McCormick, Proc. Soc. Exptl. Biol. Med. **107**, 784 (1961).
[5] D. B. McCormick and M. Russell, Comp. Biochem. Physiol. **5**, 113 (1962).
[6] D. B. McCormick, J. Biol. Chem. **237**, 939 (1961).

solution is poured over a column of DEAE-cellulose (1 g per 100 mg of protein) which has been previously equilibrated with 0.02 M glycine, pH 6.8. Protein is eluted fractionally in a linear gradient established between 0.02 M glycine, pH 6.8, and a pH 6.8 buffer which is 0.05 M in potassium phosphate and 0.02 M in glycine. Fractions that contain most flavokinase are combined. The enzyme is precipitated by 75% saturation with ammonium sulfate, collected by centrifugation at 12,500 g for 15 minutes, and dissolved in 0.05 M potassium phosphate buffer, pH 7.

A typical partial purification of flavokinase from liver is summarized by the data in Table I.

TABLE I
PURIFICATION OF FLAVOKINASE FROM RAT LIVER

Fraction	Total protein[a] (mg)	Total activity (units)	Specific activity (units/mg)	Yield (%)
1. Homogenate	1590	1590	1	100
2. Supernatant solution	750	4500	6	280[b]
3. Ammonium sulfate precipitate	150	2100	14	130
4. Dialyzate	125	2500	20	160
5. DEAE-cellulose eluate	10	800	80	50

[a] From 7 g (wet weight) of liver.
[b] High apparent yields are primarily due to removal of considerable particulate phosphatase.

Comments. These procedures readily effect an apparent purification of approximately 80-fold over the activity shown by crude homogenates, and 15-fold or greater over the protein of the first supernatant solutions. More extensive purification can be accomplished by specific absorption on flavin-cellulose compounds which utilize complexing propensities of the kinase for substrate or competitive inhibitor.[7] An example of this technique with 7-celluloseacetamido-6,9-dimethylisoalloxazine is given in another section (see chapter by Arsenis, this volume [160]).

Properties

Optima for Temperature and pH. The optimal temperature for phosphorylation of riboflavin by the liver kinase is near 50° for a 30-minute incubation, and activity decreases markedly at higher temperatures or longer times.[6] Optima for pH are approximately 7.5 and 8.0 with $10^{-4} M$ Mg^{2+} and Zn^{2+}, respectively.[6]

Requirements for Metal Ions and Nucleoside Triphosphate. Zn^{2+} is more effective than Mg^{2+} over a considerable range of added cation with kinase

[7] C. Arsenis and D. B. McCormick, *J. Biol. Chem.* **239**, 3093 (1964).

which still contains some phosphatase after the DEAE-cellulose step.[6] However, Mg^{2+} becomes as effective an activator upon extensive purification over flavin-cellulose compounds.[7] ATP $(K_m = 2 \times 10^{-4} M)$ is the most effective phosphorylating agent, but some activity with GTP and, to a lesser extent, CTP is found with partially purified preparations of the kinase.[6]

Specificity for Flavins. The substrate requirements of the liver flavokinase appear similar for those analogs of riboflavin which were tested with the yeast kinase.[8] Much more information exists for the liver kinase,

TABLE II
RELATIVE SUBSTRATE EFFICIENCIES OF FLAVINS
WITH FLAVOKINASE FROM RAT LIVER

Substituent changes in D-riboflavin[a]	Relative efficiency for phosphorylation
None	100
9-Substituent	
D-Erythrityl	33
D-Allityl	26
D-Arabityl	25
2′-Deoxy-D-ribityl	18
6,7-Substituents	
Dichloro	94
Dibromo	51
Diiodo	47
Diethyl	40
6-Methyl-7-fluoro	38
Dimethoxy	2
7-Substituent	
Amino	31
6-Substituent	
Methyl	24
6-Methyl-8(N)-pyrido	22
Chloro	20
Ethyl	16
Ethoxy	13
Methoxy	11
Amino	11
2-Substituent	
Thio	62[b]
Imino	31
Benzylazino	31
Methylmercapto	9
Deoxy	4

[a] 6,7-Dimethyl-9-(1′-D-ribityl)isoalloxazine.
[b] Partially decomposes to riboflavin during incubation.

[8] E. B. Kearney, *J. Biol. Chem.* **194**, 747 (1952).

the activity of which has been examined for a large number of analogs with substituent changes in various parts of the flavin structure.[9-12] The substrate efficiencies of numerous flavins compared to riboflavin ($K_m = 1.2 \times 10^{-5}\ M$) is summarized by the data in Table II.

Flavins found active as competitive inhibitors include lumiflavin (6,7,9-trimethylisoalloxazine) and its 7-amino analog, the 4-imino derivative of riboflavin, L-fucoflavin [6,7-dimethyl-9-(1'-L-fucityl)isoalloxazine], and 6,7-dimethyl-9-(ω-hydroxyalkyl)isoalloxazines where the alkyl side chain varies from 2 to 6 carbons in length.

Inactive flavins result from substitution of D- and L-lyxityl, D-xylityl, L-arabityl, L-rhamnityl, D-sorbityl, D-ducityl, and D-mannityl in the side chain; 6,7-demethyl and iso(5,6-dimethyl) in the benzenoid portion; 3-methyl, 2-β-hydroxyethylamino, 2-phenylamino, and 2-morpholino in the pyrimidinoid portion.

[9] D. B. McCormick and R. C. Butler, *Biochim. Biophys. Acta* **65**, 326 (1962).
[10] D. B. McCormick, C. Arsenis, and P. Hemmerich, *J. Biol. Chem.* **238**, 3095 (1963).
[11] C. S. Yang, C. Arsenis, and D. B. McCormick, *J. Nutr.* **84**, 167 (1964).
[12] B. M. Chassy, C. Arsenis, and D. B. McCormick, *J. Biol. Chem.* **240**, 1338 (1965).

[160] Preparation and Use of Flavin-Cellulose Compounds

By CHARALAMPOS ARSENIS

Since enzymes enter into temporary combination with their substrates, purification of enzymes by column chromatography can be achieved by (a) coupling the substrate, inhibitor, or coenzyme to the surface of an insoluble matrix, to which the enzyme can be temporarily adsorbed and later be released by a change in the medium to conditions unfavorable for combination, or (b) adsorbing the enzyme in an unspecific manner onto the insoluble matrix and specifically eluting the enzyme by a solution of its substrate.[1,2] The former technique was used for the purification of rat liver flavokinase[3] and spinach glycolate apooxidase.[4]

Preparation of Cellulose Derivatives

Syntheses of Flavin-Cellulose Compounds. Dry carboxymethyl cellulose is suspended in 9 volumes per gram of anhydrous pyridine. Twice the

[1] B. M. Pogell, Vol. IX, p. 9.
[2] C. Arsenis and O. Touster, *J. Biol. Chem.* **242**, 3400 (1967).
[3] C. Arsenis and D. B. McCormick, *J. Biol. Chem.* **239**, 3093 (1964).
[4] C. Arsenis and D. B. McCormick, *J. Biol. Chem.* **241**, 330 (1966).

equivalent amount of thionyl chloride, based on 0.7 meq of carboxymethyl groups per gram of cellulose, is stirred in. The chlorocarbonylmethyl cellulose is filtered off immediately, washed with another 9 volumes of pyridine, and dried over P_2O_5 in a vacuum. Chlorocarbonylmethyl cellulose is suspended in 9 volumes of anhydrous pyridine which contains an amount of flavin equivalent to the initial carboxymethyl groups of the cellulose. The mixture is kept in the dark at 45°–50° for 24 hours with occasional stirring, and the flavin-cellulose derivative is filtered off. The product is thoroughly washed with pyridine, water, 0.1 N HCl, water again, ethyl alcohol, and diethyl ether. When wet, the flavin-cellulose compounds have a faint reddish-brown color. They are dried at 60° in a vacuum oven.

Syntheses of FMN-Cellulose Compounds. The FMN derivative of cellulose phosphate is prepared as follows: To 10 g of dry riboflavin in 50 ml of anhydrous pyridine, 40 ml of phosphorus oxychloride is added dropwise. The mixture, containing riboflavin dichlorophosphate and excess phosphorus oxychloride, is kept at 50° for 3 hours and then added to a suspension of 80 g of dry cellulose in 600 ml of anhydrous pyridine. The suspension is kept at 50° for 24 hours with occasional stirring, and the FMN-cellulose phosphate is filtered off. This material is thoroughly washed with cold pyridine, dilute HCl, aqueous 5% NaCl, water, ethyl alcohol, and diethyl ether until no flavin can be detected in the washings. The FMN derivative of DEAE-cellulose is prepared in the same manner, but more extensive washing is necessary to remove the last traces of noncovalently-bound flavin.

The FMN derivative of cellulose is prepared by adding 10 g of FMN and 25 g of dicyclohexylcarbodiimide to 60 g of dry cellulose in 400 ml of anhydrous pyridine. The suspension is kept at 30° for 5 days with shaking, and the FMN-cellulose is filtered off and washed as before.

Yields. Yields of 19–36% for the coupling of flavins to chlorocarbonyl methyl cellulose are calculated on the basis of analyses for Cl and N. These represent overall yields of 3–7% linkage of flavin to initial carboxymethyl substituents. When 7-amino-6,9-dimethylisoalloxazine was coupled to carboxymethyl cellulose to form 7-celluloseacetamido-6,9-dimethylisoalloxazine, there was 0.32% N found, which corresponds to 1.17% flavin and, therefore, a 1:135 mole ratio of flavin to glucose units. On the other hand, when 6-amino-9-(1'-D-ribityl)isoalloxazine is coupled to carboxymethyl cellulose to give 6-celluloseacetamido-9-(1-D-ribityl)isoalloxazine, 0.12% N is found, corresponding to 0.66% flavin, and 1:355 mole ratio of flavin to glucose units.

Yields for coupling of FMN are based on the amount of FMN, measured as absorbance at 450 nm, liberated after acid hydrolysis. Approximately 20–100 mg of FMN per 100 g of cellulose derivative are bound.

Assays

Flavokinase or FMN-Phosphatase Activity

This activity was followed by measuring FMN formation or disappearance spectrophotometrically.[4a]

Glycolate Apooxidase

Glycolate oxidation is followed by measuring the rate of oxygen uptake in Warburg vessels as described by Clapett et al.[5]

Reagents

Potassium glycolate, 2.4×10^{-4} M
Potassium phosphate buffer, 0.1 M, pH 8.0
Flavin mononucleotide, 6×10^{-4} M
Enzyme

Procedure. Mixtures for assaying glycolate apooxidase are placed in Warburg manometric flasks that contain 100 micromoles of potassium phosphate buffer (pH 8), 60 millimicromoles of FMN, and apooxidase in an incubation volume of 2.75 ml. The contents are equilibrated at 37° for 15 minutes. The reactions are initiated by tipping in 60 millimicromoles of potassium glycolate in 0.25 ml. Uptake of O_2 is followed for 10 minutes.

Definition of Unit and Specific Activity. For the flavokinase or FMN phosphatase, 1 unit is that amount of enzyme which catalyzes the synthesis or hydrolysis of 1 millimicromole of flavin monophosphate in 1 hour at 37°, respectively. For the glycolate apooxidase one unit is that amount of enzyme which takes up 1 microliter of O_2 in 10 minutes at 37°. Specific activity is defined as units per milligram of protein determined by the method of Lowry et al.[6] or by measurement of absorbance at 280 nm.

Column Chromatography

Flavokinase Chromatography on Flavin Cellulose (and CM-Cellulose for Comparison). The partially purified preparations of flavokinase, obtained after dialysis,[6a] are poured over columns (1.4 × 65 cm) containing approximately 25 g of flavin- (or CM-)cellulose derivative equilibrated with 0.005 M potassium phosphate buffer. Approximately 80% of the protein is eluted with a linear gradient established between 125 ml of 0.005 M potassium phosphate buffer, pH 7.0, and 125 ml of 0.1 M potassium phos-

---

[4a] H. B. Burch, O. A. Bessey, and O. H. Lowry, *J. Biol. Chem.* **175,** 457 (1948). This volume [159].

[5] C. O. Clapett, N. E. Tolbert, and R. H. Burris, *J. Biol. Chem.* **178,** 977 (1949).

[6] O. H. Lowry, N. J. Rosebrough, A. L. Farr, and R. J. Randall, *J. Biol. Chem.* **193,** 265 (1951).

[6a] D. B. McCormick, *J. Biol. Chem.* **237,** 959 (1962). This volume [159].

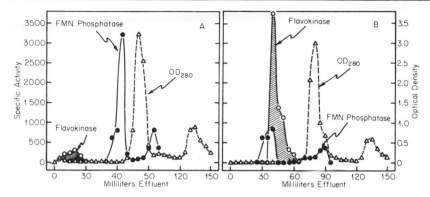

FIG. 1. Chromatography of liver flavokinase on CM-cellulose (A) and on 7-cellulose-acetamido-6,9-dimethylisoalloxazine (B).

phate, pH 7.0. A typical chromatographic elution pattern of flavokinase on CM-cellulose or 7-celluloseacetamido-6,9-dimethylisoalloxazine is presented in Fig. 1.

Glycolate Apooxidase Chromatography on FMN-Cellulose (and Cellulose for Comparison). Glycolate apooxidase from spinach leaves is obtained as a supernatant solution following a second acid precipitation.[7] Chromatography of the apooxidase on cellulose or FMN-cellulose is accomplished by pouring the partially purified preparations of the enzyme over

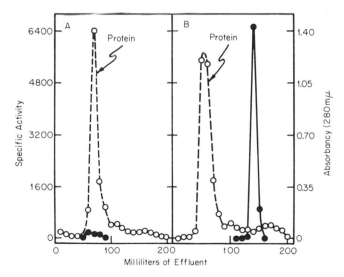

FIG. 2. Chromatography of glycolate apooxidase on cellulose (A) and FMN-cellulose (B).

[7] I. Zelitch and S. Ochoa, *J. Biol. Chem.* **210,** 707 (1953).

columns (1.5 × 30 cm) and eluting with a linear gradient established between 100 ml of 0.005 M and 100 ml of 0.05 M potassium phosphate buffer, pH 8.0.

The elution pattern from chromatography of crude glycolate apooxidase on a control column of cellulose, A, and on an FMN-cellulose, B, is illustrated in Fig. 2.

The recoveries of flavokinase and glycolate apooxidase from chromatography on flavin-cellulose compounds is shown in the table. It is evident that extensive purification can be accomplished by specific adsorption of the enzyme on the flavin-cellulose derivatives.

RECOVERIES OF ENZYMES FROM COLUMN CHROMATOGRAPHY
ON DERIVATIVES OF CELLULOSE

Enzyme	Protein (mg)	Total activity (units)	Specific activity (units/mg)
Flavokinase			
Before chromatography	72	1152	16
After CM-cellulose chromatography	0.3	60	200
After flavin-cellulose chromatography	0.3	348	1160
Glycolate apooxidase			
Before chromatography	36	470	13
After cellulose chromatography	34	475	14
After FMN-cellulose chromatography	0.2	720	3600

The use of biochemically specific adsorbents for selective purification of enzymes that are not readily amenable to other fractionation procedures is beginning to be of considerable importance. The general principle of attachment of specific reagents or natural substances to cellulose or cellulose derivatives has already been applied with success for selective removal of antibodies, nucleosides, and ribonucleotides. The relative ease of synthesis of many such specific adsorbents, the small number of reactive sites generally needed, and the simplicity of the overall method are advantages that should lead to further development of the general technique.

[161] GTP-Dependent Flavin Phosphate Synthetase

By Sei Tachibana

Assay Method

Principle. GTP-dependent riboflavin phosphate synthetase[1] can be obtained from *Rhizopus javanicus*. The enzyme requires GTP, ITP, UTP, or some unknown cofactor; ATP and CTP are inactive. *Rhizopus oryzae*, *R. hangchao*, *R. nigricans*, *R. javanicus* var. *kawasakiensis*, *R. delemer*, *R. formosensis*, *R. batatus*, and *R. tritici* are also found to be available for the enzyme preparation. Other species of *Rhizopus* (except at least two: *R. chinensis* and *R. oligosporus*) may be available for the enzyme preparation.

Preparation of Boiled Cell-Free Extract. One part of the dried mycelial powder of *R. javanicus* is extracted with a 10-fold volume of 0.05 M phosphate buffer (pH 5.2), followed by heating for 10 minutes in boiling water. The resulting supernatant solution obtained by centrifugation at 8000 g for 20 minutes is concentrated to one-tenth volume under reduced pressure at 50°, and the solution is kept in a deep freeze to use as the unknown cofactor.

Enzyme Reaction. A reaction mixture of total volume 1.5 ml containing 1.06 × 10⁻⁴ M riboflavin, 2 × 10⁻³ M GTP, or 0.03 ml of boiled cell-free extract, 2 × 10⁻³ M MgSO₄, 2 × 10⁻² M phosphate buffer (pH 5.5), and 0.3 ml of enzyme solution is incubated for 2 hours at 37°.

Assay of Reaction Product. Paper partition chromatography using the upper layer of *n*-butanol–acetic acid–water (4:1:5), water, saturated with isoamyl alcohol, 5% Na₂HPO₄ solution, *n*-butanol–pyridine–water (6:4:3), or the lower layer of phenol *n*-butanol–water (16:3:10) is used. When the enzyme preparation is contaminated with riboflavin kinase (ATP:riboflavin 5′-phosphotransferase, EC 2.7.1.26), the last solvent system is preferable, because in the cases of the three former solvent systems, the R_f values of the product FMN′ (riboflavin cyclic-2′,5′-mononucleotide) are almost the same as those of FMN.

In the case of paper electrophoresis[2,3] carried out at pH 5.9 in 0.05 M phosphate buffer (KH₂PO₄ and Na₂HPO₄) at a constant voltage of 200 V (5–14 mA) for 4–5 hours, the reaction product (FMN′) spotted on the center of Toyo filter paper No. 50 migrates 40–45 mm toward the anode,

[1] S. Tachibana, *Vitamins (Kyoto)* **33**, 533 (1966); *J. Vitaminol.* **13**, 89 (1967).

[2] S. Tachibana, *Vitamins (Kyoto)* **22**, 291 (1961); *J. Vitaminol.* **7**, 294 (1961).

[3] S. Tachibana, J. Siode, and S. Matsuno, *Vitamins (Kyoto)* **27**, 210 (1963); *J. Vitaminol.* **9**, 197 (1963).

being separated from FMN, which migrates 28–30 mm toward the anode. The amount of flavin is determined fluorometrically on a paper strip or after elution by water at pH 3.0.

Definition of Unit and Specific Activity. One unit of the enzyme is defined as that amount which transforms 1 millimicromole of substrate per hour. The specific activity is presented as units per milligram of protein.

Growth and Harvest of R. javanicus. Subculture of *R. javanicus* is kept on agar slants containing 3% glucose, 0.2% NH_4NO_3, 0.1% KH_2PO_4, 0.05% $MgSO_4 \cdot 7H_2O$, 0.05% KCl, 0.2% peptone, 1.2% agar, and tap water. A liquid medium, containing 3% glucose, 0.1% $NaNO_3$, 0.08% NH_4NO_3, 0.1% KH_2PO_4, 0.05% $MgSO_4 \cdot 7$ H_2O, and 0.05% KCl, per 200 ml of tap water in a 500-ml Fernbach flask, is sterilized; previously sterilized 1% $CaCO_3$ is added to the medium just before inoculation. After inoculation by transferring three loops of spores from the agar slant subculture, the flask is incubated for 6 days at 28° without agitation. Growth conditions may be varied over a wide range of temperature, media, and population density without affecting the efficiency of the method described. At the end of the culture period, the mycelia are harvested, washed, pressed in filter paper layers, and dried with a fan at room temperature. The drying is further carried out *in vacuo* in a $CaCl_2$-desiccator for at least 2 weeks in a cold room at 5°. The dried mycelia may be stored for years without significant loss of activity of the enzyme.

Preparation of Enzyme Solution

The dried mycelia are powdered in a homogenizer. Twenty grams of the mycelial powder is extracted with some glass powder in 200 ml of 0.05 M phosphate buffer (pH 5.2) in a chilled mortar which is kept in ice throughout the grinding procedure. The homogenate is centrifuged at 8000 g for 20 minutes, and the resulting residue is discarded. To the supernatant solution, crystalline $(NH_4)_2SO_4$ is added up to 0.7 saturation, followed by centrifugation at 8000 g for 30 minutes. To the resulting supernatant, $(NH_4)_2SO_4$ is added up to 0.8 saturation, and the mixture is kept for 14 hours at 5°. The precipitate is collected by centrifugation at 8000 g for 30 minutes and dissolved in a small amount of 0.005 M phosphate buffer (pH 5.5). The solution is dialyzed in a Cello-Tube (cellophane bag) against 3 liters of pure water for 16 hours at 5°. The resulting dialyzate is centrifuged at 15,000 g for 30 minutes to remove the precipitate and then kept in a deep freeze for use. The protein concentration is determined to be about 0.1–0.2% by the spectrophotometric method of Warburg and Christian.[4] The specific activity of the enzyme solution is about 200, which is 70- to 90-fold that of the starting cell-free extract.

[4] O. Warburg and W. Christian, *Biochem. Z.* **310**, 384 (1941).

This enzyme preparation contains neither riboflavin kinase nor phosphotransferase for which dinitrophenylphosphate, glycerophosphate, glucose monophosphate, AMP, CMP, GMP, IMP, or UMP is active.

At 37°, the optimal pH range is around 5.5. At pH 5.5, the apparent temperature optimum is 37°. If the enzyme solution is heated for 5 minutes at 80°, the activity of enzyme is almost lost. The K_m for riboflavin as determined from the Lineweaver-Burk type of plot is $1.5 \times 10^{-5}\ M$. The K_m for Mg^{2+} is $2.5 \times 10^{-5}\ M$. Mn^{2+} is as effective as Mg^{2+}. The K_m for GTP is $1 \times 10^{-4}\ M$. UTP and ITP are active with higher K_m values than that of GTP. ATP and CTP are inactive.

[162] Specificity of FAD Pyrophosphorylase (ATP:FMN Adenyltransferase, EC 2.7.7.2) from Rat Liver

By DONALD B. McCORMICK

$$\text{Flavin monophosphate} + \text{ATP} \xrightarrow{Mg^{2+}} \text{flavin adenine dinucleotide} + \text{PP}$$

Assay Method

Principle. The particular flavin adenine dinucleotide formed is measured fluorometrically according to the method of Yagi[1] after elution from paper chromatograms following phenol extraction of flavin material from boiled filtrates saturated with ammonium sulfate.[2]

Reagents

Flavin monophosphate, 0.001 M, neutralized
ATP and other nucleotides, 0.01 M, pH 7.5
$MgCl_2$, 0.01 M
Potassium phosphate buffer, 0.25 M, pH 7.0 and 7.5
Ammonium sulfate
Liquid phenol. Solution is effected with a minimum of water.

Enzyme. FAD pyrophosphorylase from rat liver is partially purified as described by DeLuca and Kaplan.[3] The fraction obtained after precipitation by 40% saturation of ammonium sulfate is sufficiently active for specificity studies, although some nonspecific FAD destroying activity is still present (see Vol. VI [43]).

[1] K. Yagi, *J. Biochem.* (*Japan*) **38**, 161 (1951).
[2] D. B. McCormick, *Biochem. Biophys. Res. Commun.* **14**, 493 (1964).
[3] C. DeLuca and N. O. Kaplan, *Biochim. Biophys. Acta* **30**, 6 (1958).

Procedure. Mixtures for assessing substrate reactivity contain 1.5 ml of FMN or other flavin phosphate, 0.3 ml each of ATP or other nucleotide, $MgCl_2$, and phosphate buffer (pH 7.5), and enzyme plus water to a total volume of 3 ml. Final concentrations are 0.5 mM flavin phosphate, 1 mM each of ATP and Mg^{2+}, 25 mM phosphate, and sufficient enzyme, usually 1–2 mg of protein, to ensure measurable activity. Incubation is carried out in the dark for 1 hour at 37°. The reaction is terminated by heating the mixture in a boiling water bath for 3 minutes. The solutions are saturated with solid ammonium sulfate, the mixtures are filtered, and the flavin compounds in the filtrates are extracted by shaking with 0.25 ml of liquid phenol. Aliquots (50 μl) of the top phenol layer are chromatographed on Whatman No. 1 paper by the method of Giri and Krishnaswamy,[4] with *n*-butyl alcohol–acetic acid–water (4:1:5, top phase) as solvent.[5] The chromatograms are examined under an ultraviolet light to locate fluorescent flavin bands, and the paper strips containing the slower moving FAD compounds are cut out and eluted with 5 mM phosphate buffer, pH 7. FAD compounds in the eluates are determined fluorometrically by the method of Yagi.[1]

Specificity

For Nucleotides. ATP is the only effective nucleotide substrate for the liver pyrophosphorylase. Ribo- and deoxyribofuranoside 5′-triphosphates with inosine, guanine, cytosine, uracil, and thymine as the base are all inactive.[2] Some activity is found with ADP and the liver enzyme, probably due to contamination with adenylate kinase[3]; this is not the case with the pyrophosphorylase from yeast (see Vol. II [117]).

SUBSTRATE BEHAVIORS OF FLAVIN PHOSPHATES WITH
FAD PYROPHOSPHORYLASE FROM RAT LIVER

Flavin monophosphate added (500 micromoles)	Flavin adenine dinucleotide formed (micromoles)
Riboflavin 5′-phosphate	10.3
Isoriboflavin 5′-phosphate	28.8
6-Methylriboflavin 5′-phosphate	7.5
6,7-Dibromoriboflavin 5′-phosphate	3.2
2′, 3′, 4′-Trideoxyriboflavin 5′-phosphate	2.2
2′-Deoxyriboflavin 5′-phosphate	1.0
D-Araboflavin 5′-phosphate	1.2
D-Erythroflavin 4′-phosphate	1.2

[4] K. V. Giri and P. R. Krishnaswamy, *J. Indian Inst. Sci.* **38**, 232 (1956).
[5] G. A. Kimmich and D. B. McCormick, *J. Chromatog.* **12**, 394 (1963).

For Flavin Phosphates. Several flavin monophosphates have been observed to serve as substrates for the liver pyrophosphorylase.[2] Alterations in the benzenoid portion of riboflavin [6,7-dimethyl-9-(1'-D-ribityl)isoalloxazine] are tolerated to varying degrees, as 5'-phosphates of iso(5,6-dimethyl), 6-methyl, and 6,7-dibromo analogs are active. Less reactivity is observed with the phosphate esters of side-chain altered compounds such as 2',3',4'-trideoxyribo, 2'-deoxyribo, D-arabo, and D-erythro analogs. The substrate behavior of flavin phosphates is summarized by the data in the table.

[163] Riboflavin Degradation

By L. Tsai and E. R. Stadtman

General

Detailed information on the mechanism of riboflavin degradation is still lacking; however, general pathways of metabolism have been established in several organisms. The information available was derived mostly from studies with microorganisms that were isolated from soil enrichment cultures in which riboflavin was supplied as the major source of carbon and energy. These studies show that the pathway of riboflavin degradation is not the same for all organisms. Figure 1 summarizes the results from different laboratories illustrating three separate pathways of riboflavin degradation.

In one pathway riboflavin is converted to lumichrome. This was first demonstrated by Foster[1] for an organism *Pseudomonas riboflavina* which he isolated from riboflavin-supplemented enrichment cultures. Later, Yanagita and Foster[2] showed that lumichrome is formed by a hydrolytic cleavage of the riboflavin side chain to produce ribitol as the other product. The ribitol is subsequently oxidized to CO_2 to provide energy and carbon for growth, but lumichrome accumulates as a fairly stable, relatively insoluble end product of metabolism, which crystallizes out of the culture medium. Lumichrome has also been identified as the major end product of riboflavin by another organism, tentatively identified as a *Nocardia* sp.[3] Although lumichrome is resistant to further degradation by *Pseudomonas riboflavina*, it is metabolized to unidentified products by an aerobic bac-

[1] J. W. Foster, *J. Bacteriol.* **47**, 27 (1944); *ibid.*, **48**, 98 (1944).

[2] T. Yanagita and J. W. Foster, *J. Biol. Chem.* **21**, 593 (1956).

[3] E. R. Stadtman, *Proc. Intern. Congr. Biochem. 4th Vienna*, 1958, Symp. XI, Vitamin Metabolism, Reprint No. 15.

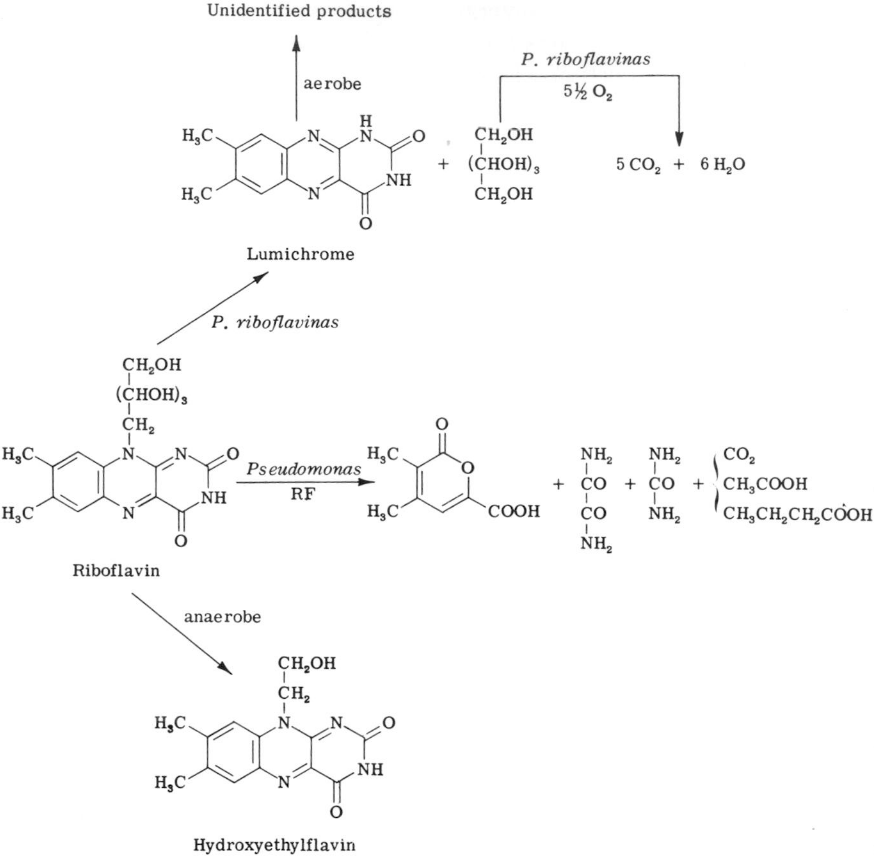

FIG. 1. Three pathways of riboflavin degradation based on results from different laboratories.

terium that was isolated from soil enrichment cultures in which lumichrome degradation by this organism has not been investigated.

A second pathway of riboflavin degradation has been demonstrated to occur in anaerobic enrichment cultures. This involves cleavage of the ribityl side chain between carbons 2' and 3' to produce 6,7-dimethyl-9-(2'-hydroxyethyl)isoalloxazine (hydroxyethylflavin). This accumulates as a crystalline, green, quinhydrone-like complex consisting of 1 mole of half-reduced and 1 mole of oxidized flavin.[4] The specific organism responsible for this conversion has not been isolated in pure culture, and further degradation of the hydroxyethylflavin produced in the crude enrichment

[4] H. T. Miles and E. R. Stadtman, *J. Am. Chem. Soc.* **77**, 5746 (1955).

cultures has not been investigated. The hydroxyethylflavin has also been isolated from urine of goats that had eaten riboflavin.[5]

Finally, a third pathway of riboflavin metabolism, involving complete degradation of the isoalloxazine ring has been demonstrated in another aerobic organism, *Pseudomonas* RF.[6-12] Acetate, propionate, butyrate, CO_2, oxamide, urea, and 3,4-dimethyl-6-carboxy-α-pyrone have been established as the ultimate end products of riboflavin degradation by this organism.[9,11,12] The pathway of this degradation has been extensively studied and will be the subject of this report.

Riboflavin Degradation by *Pseudomonas* RF

Present knowledge concerning the pathway of riboflavin degradation by *Pseudomonas* RF is summarized in Fig. 2. The first intermediate detected is 1-ribityl-2,3-diketo-1,2,3,4-tetrahydro-6,7-dimethylquinoxaline (compound I).[12a] Studies with 2-^{14}C-riboflavin[7] and with soluble enzyme preparations[10] have shown that this formation of compound I involves an oxygen-dependent mechanism in which ring C of the isoalloxazine nucleus is cleaved to yield urea, CO_2, and compound I. The next step involves oxidative cleavage of compound I to 6,7-dimethylquinoxaline-2,3-diol (compound II) and ribose.[8] In a third oxygen-dependent step, compound II is converted to oxamide plus either 3,4-dimethyl-6-carboxy-α-pyrone (compound III)[5,9,11] or a mixture of small metabolites, including butyrate, propionate, acetate, and CO_2.[11]

Unfortunately efforts to establish the detailed mechanism of the various reactions involved have been hampered by the inability to demonstrate the reactions in cell-free extracts. Only the first step shown in Fig. 2 (i.e., the conversion of riboflavin to compound I, urea, and CO_2) has been demonstrated in cell-free extracts.[10] Knowledge concerning the remaining steps was obtained from studies with suspensions of dried cells which were facilitated by the unique spectral characteristics and chromatographic behavior of the various intermediates and by the use of metabolic inhibitors

[5] E. C. Owen, *Biochem. J.* **84**, 96P (1962).

[6] P. Z. Smyrniotis, H. T. Miles, and E. R. Stadtman, *J. Am. Chem. Soc.* **80**, 2541 (1958).

[7] H. T. Miles, P. Z. Smyrniotis, and E. R. Stadtman, *J. Am. Chem. Soc.* **81**, 1946 (1959).

[8] L. Tsai, P. Z. Smyrniotis, D. Harkness, and E. R. Stadtman, *Biochem. Z.* **338**, 561 (1963).

[9] D. R. Harkness, L. Tsai, and E. R. Stadtman, *Arch. Biochem. Biophys.* **108**, 323 (1964).

[10] D. R. Harkness and E. R. Stadtman, *J. Biol. Chem.* **240**, 4089 (1965).

[11] W. Barz, L. Tsai, and E. R. Stadtman, unpublished results.

[12] D. R. Harkness, L. Tsai, and E. R. Stadtman, unpublished results.

[12a] K. Hotta and O. Ando, *J. Vitaminol.* (*Kyoto*) **7**, 196 (1961), reported the isolation of this compound as a degradation product of riboflavin in plant tissues.

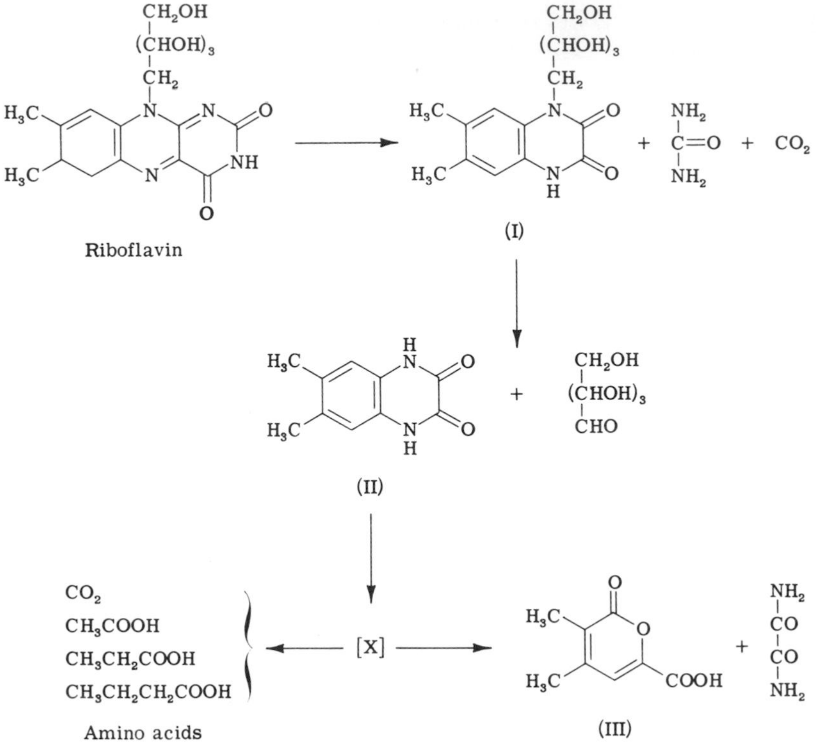

Fig. 2. Pathway of riboflavin degradation by *Pseudomonas* RF.

and isotopically labeled substrates. Figure 3 summarizes properties of the various intermediates that have been useful in establishing the individual steps involved. The decomposition of riboflavin can be readily followed by measuring the absorbancy at 445 nm since none of the ultimate end products or the intermediates absorb at this wavelength. In the presence of 5 mM arsenite, which inhibits the conversion of compounds I and II to compound III, the products of riboflavin decomposition are urea and a mixture of compounds I and II.[7] Although compounds I and II have identical spectra at pH 6.0–7.0,[7] they can be readily distinguished spectroscopically by virtue of the fact that the spectrum of compound II is shifted by complexing with mercuric ions, whereas no such shift in spectrum is observed with compound I.[8] The decrease in absorbancy at 323 nm or the increase in absorbancy at 380 nm resulting from the addition of mercuric chloride is therefore a specific measure of the concentration of compound II. The conversion of riboflavin or of compounds I and II to compound III occurs readily in the absence of arsenite and may be followed by the decrease

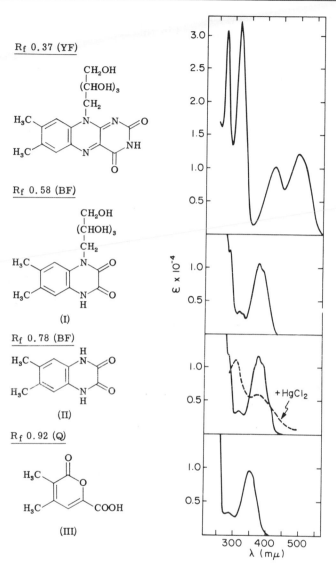

Fig. 3. Properties of riboflavin and compounds I–III. YF = yellow fluorescent spot; BF = bright blue fluorescence; Q = dark purple quenching spot.

in absorbancy at 450 nm (for riboflavin) and 350 nm (for compounds I and II) and by the increase in absorbancy at 300 nm (for compound III).[7] The various transformations are further established by paper chromatographic analysis of the reaction mixtures in a solvent system composed of n-butanol–acetic acid–water, 160:40:75. Riboflavin and compounds I, II, and III are readily separated in this chromatographic system (R_f = 0.37, 0.58, 0.78, and 0.92, respectively) and are easily located on paper chromatograms by their appearance under ultraviolet light.[7] Riboflavin appears as a bright yellow fluorescent spot (YF), compounds I and II exhibit a bright blue fluorescence (BF), and compound III appears as a dark purple quenching spot (Q). With some cell preparations, compound III is the major end product of riboflavin degradation, but with other preparations it may account for only 20–50% of the total substrate decomposed. Once formed compound III is fairly resistant to further degradation; however, it is slowly metabolized by suspensions of cells that have become adapted to compound III.[9] Studies with methyl-^{14}C-labeled compound II indicates that its further degradation to oxamide, [14]CO$_2$, and ^{14}C-labeled short-chain fatty acid does not necessarily involve compound III as an intermediate.[9,11,12] On the other hand, compound III degradation by adapted cells also yields CO$_2$ and lipids.[9] At the present state of knowledge it seems likely that compound II is converted to oxamide plus a derivative that may be converted either to compound III by a relatively irreversible step or may be further degraded to CO$_2$ and fatty acids. No information is presently available on the detailed mechanism of these latter steps in riboflavin metabolism.

Riboflavin Mixed Function Oxygenase: Conversion of Riboflavin to Urea, CO$_2$, and 1-Ribityl-2,3-diketo-1,2,3,4-tetrahydro-6,7-dimethylquinoxaline (Compound I)

Riboflavin + O$_2$ + DPNH + H$^+$ + H$_2$O → compound I + urea + CO$_2$ + DPN$^+$

Assay Method

Principle. The method is based on the fact that riboflavin has a strong absorption band at 445 nm whereas compound I does not.[7] Therefore, under the specified conditions the decrease in absorbancy at 445 nm is a measure of the enzyme activity. DPNH is supplied by a generating system consisting of ethanol, DPN$^+$, and alcohol dehydrogenase, and the mixture is strongly aerated to prevent accumulation of reduced riboflavin which is not a substrate for this enzyme. Catalase is added to decompose hydrogen peroxide that might be produced as a result of autoxidation of any reduced riboflavin that may be generated by means of dehydrogenases that are present in the cell-free extract.

Reagents

Potassium pyrophosphate buffer, 1.0 M, pH 8.0
Riboflavin (freshly prepared), 0.005 M

2-Mercaptoethanol, 0.1 M
Horse liver catalase, 10 mg/ml
DPN, 0.1 M, pH 6.0
Ethanol, 0.05 M
Yeast alcohol dehydrogenase, 10 mg/mole
Cell-free extracts of *Pseudomonas* RF containing 30 mg of protein per
　　mole

Procedure. The procedure was previously described by Harkness and
Stadtman.[10] Eight-tenths milliliter of water, 0.2 ml of potassium pyrophosphate buffer, 0.4 ml of cell-free extract of *Pseudomonas* RF, and 0.1-ml
aliquots of each reagent, ethanol, 2-mercaptoethanol, catalase, alcohol
dehydrogenase, riboflavin, and DPN, are placed in a 10-ml Erlenmeyer
flask. This is shaken on a rotary shaker (New Brunswick Scientific Inc.,
Model 53, 240 cycles/min, excursion 2 cm) in air at 26° for 15 minutes. A
sample without added cell-free extract serves as a control. After incubation,
the samples are diluted 1:5 in the potassium phosphate buffer, and the
absorbancy is measured at 445 nm. The difference in absorbancy between
a sample containing no cell-free extract and that containing cell-free extract
is a measure of riboflavin disappearance catalyzed by the mixed-function
oxygenase. Crude cell-free extracts do not catalyze the further conversion
of compound I to compound II; therefore, compound I accumulates in an
amount stoichiometric with respect to the amount of riboflavin that disappears. Its presence can be confirmed by paper chromatography of the
reaction products in a solvent system composed of n-butanol–acetic acid–
H_2O, 160:40:75 ($R_f = 0.53$). If the reaction is allowed to proceed until
all the riboflavin is decomposed, the stoichiometric accumulation of compound I can be confirmed directly by diluting the reaction mixtures 1:5
with 0.1 N NaOH and measuring the spectrum of the resultant solution.[10]
Stoichiometric production of urea and CO_2 as the other products of riboflavin degradation by cell-free extracts has been established by direct
analysis.[10]

Preparation of Cell-Free Extracts

The isolation of *Pseudomonas* RF from soil enrichment cultures and the
conditions for its growth in large culture have been described previously.[7]
The bacteria were harvested in a Sharples centrifuge as close to the time of
riboflavin exhaustion as possible; this was usually 16–24 hours after inoculation. The cell paste was frozen in liquid nitrogen and stored at −80°. A
suspension of 25 g of cell paste in 25 ml of 0.05 M potassium phosphate
buffer, pH 7, containing 5 mM 2-mercaptoethanol were ruptured in a
French pressure cell at 8000 psi. The viscous extract was treated with
DNase (5 μg/10 ml) under H_2 for 10 minutes at 26° and then centrifuged

for 30 minutes at 26,000 g. Aliquots of the supernatant fluid were frozen in an alcohol–dry ice bath under hydrogen and were stored at $-80°$. Such extracts retained most of their activity for 2–3 weeks. Extracts thawed and refrozen in the same manner retained full activity, but once thawed the activity declines rapidly during storage at $0°–3°$; no activity remains after 12 hours at pH 8.0, or after 1 hour at pH 5.0.[10]

Attempts to purify the enzyme have been unsuccessful. It precipitates between 40 and 50% ammonium sulfate saturation with no increase in specific activity and with only 50% recovery of total activity, which could not be restored by combinations of fractions or by addition of boiled extracts. Other procedures, including solvent fraction at $0°$ to $-10°$, and chromatography on DEAE-cellulose resulted in almost complete loss of catalytic activity.[10]

Properties of the Catalytic System

Specificity. Of a variety of riboflavin analogs tested, lyxoflavin is the only compound oxidized as rapidly as riboflavin. Lumiflavin and FMN are degraded at moderate rates, whereas galactoflavin and hydroxyethylflavin are oxidized very slowly, and lumichrome not at all. All compounds attacked are oxidized to the respective 1-substituted 2,3-diketo-1,2,3,4-tetrahydro-6,7-dimethylquinoxalines.[10]

TPNH is only one-fourth as active as DPNH as a cosubstrate; no degradation of substrate occurs in the absence of either oxygen or DPNH or TPNH.[10]

Activators and Inhibitors. The oxygenase activity is stimulated by catalase, by various monomercaptans (but not by dihydrolipoic acid or by 2,3-dimercaptopropanol), and by other reducing compounds including stannous chloride, sodium dithionite, and sodium bisulfite.[10] The enzyme system is inhibited by Hg^{2+}, Cu^{2+}, α,α-dipyridyl, and o-phenanthroline, but not by EDTA. The activity is inhibited by high concentrations of substrates (0.1–0.8 mM).[10]

Effect of pH. The pH optimum is about 7.8–8.0. At pH 8.0, equal activities are obtained with 0.1 M potassium phosphate or potassium pyrophosphate buffers. One-fourth as much activity is obtained with 2-aminopropanol buffer.[10]

Conversion of 1-Ribityl-2,3-diketo-1,2,3,4-tetrahydro-6,7-dimethylquinoxaline to 6,7-Dimethylquinoxaline-2,3-diol and Ribose (Conversion of Compound I to Compound II)

$$\text{Compound I} + O_2 + [2H] \rightarrow \text{compound II} + \text{ribose} + H_2O$$

Principle. In the presence of 0.01 M arsenite, cell suspensions of *Pseudomonas* RF catalyze the stoichiometric oxidation of compound I to compound II and ribose.[8] (Arsenite inhibits the further conversion of compound II to other products.) Although the ultraviolet absorption spectra of compounds I and II are identical, the spectrum of compound II is markedly changed by the addition of mercuric ions, whereas the spectrum of compound I is unchanged.[8] The decrease in absorbancy at 323 nm ($\Delta\epsilon = -5750\ M^{-1}\ cm^{-1}$) or the increase in absorbancy at 380 nm ($\Delta\epsilon = +2760\ M^{-1}\ cm^{-1}$) caused by the addition of 1.2 mM $HgCl_2$ is therefore a specific measure of compound II.[8]

Reagents

Potassium phosphate buffer, 0.1 M, pH 6.2
Potassium arsenite, 0.1 M
Compound I, 10 mg/ml, brought into solution by the addition of small amounts of 0.1 M potassium hydroxide
Hydrogen peroxide, 10%
Potassium phosphate buffer, 1.0 M, pH 6.85
$HgCl_2$, 0.2 M
Suspension of dried cells of *Pseudomonas* RF, 25 mg/ml

Procedure. The procedure was previously described.[8] Eight-tenths ml water, 0.035 ml potassium phosphate buffer (pH 6.2), 0.1 ml potassium arsenite, and 0.1 ml of cell suspension are placed in a 10-ml Erlenmeyer flask. This is shaken in air at 25° for 4 hours. After incubation, the samples are transferred to centrifuge tubes and centrifuged at 12,000 g to remove the bacterial cells. Two-tenths milliliter of the supernatant fluid is diluted with 0.88 ml of water containing 0.01 ml of H_2O_2 (to decompose the arsenite). After 4 minutes (pII 5.0–5.5), 0.03 ml of potassium phosphate buffer (pH 6.85) is added and the spectrum is determined; then 0.03 ml of $HgCl_2$ is added, and the spectrum is determined again. The concentration of compound II is calculated from the increase in absorbancy at 380 nm (+0.276 per 0.1 micromole) or by the decrease in absorbancy at 323 nm (−0.575 per 0.1 micromole) caused by the addition of $HgCl_2$.

Properties of the Catalytic System

Specificity. The enzyme responsible for the conversion of compound I to compound II is not specific. A number of 1-alkyl-2,3-diketo-1,2,3,4-tetrahydroquinoxalines are oxidized. Thus analogs of compound I in which the ribityl side chain is replaced by either a hydroxyethyl group, a formylmethyl group, or a methyl group are almost as active substrates as is compound I.[8]

The reaction of compound II with mercuric ions is highly specific. Other divalent ions including: Pb^{2+}, Cd^{2+}, Zn^{2+}, Co^{2+}, Mg^{2+}, Ba^{2+}, Ca^{2+}, and Fe^{2+}

in concentrations of $10^{-3} M$ do not cause a detectable change in the absorption spectrum of compound II; neither do these ions inhibit the reaction of Hg^{2+} with compound II.[8] With stoichiometric amounts of mercuric acetate, no complex occurs at pH 4.3 or below; at pH 4.8 complex formation is 67% of maximal; whereas over the pH range of 5.0 to 11.0 complex formation is essentially complete.[8]

Oxygen Requirement. No decomposition of compound I occurs in the absence of molecular oxygen. Studies with $1'$-^{14}C-compound I have shown that the oxidation leads to the formation of 1-^{14}C-ribose.[8] The inability to obtain the oxidase in cell-free extracts has precluded efforts to demonstrate requirements for DPNH or TPNH. However, it seems most likely that the conversion of compound II is catalyzed by a mixed function oxygenase.

Metabolism of 6,7-Dimethylquinoxaline-2,3-diol (Compound II)

In studies with methyl-^{14}C-compound II, it has been established that the oxidation of compound II leads to the formation of an equivalent amount of oxamide[9] and a mixture of ^{14}C-labeled products including 3,4-dimethyl-6-carboxy-α-pyrone,[9,11] CO_2, and organic acids.[11] A significant amount of the labeled carbon is also assimilated by the cells and is found largely in the alanine, phenylalanine, methionine, and glutamic acid residues of the cellular protein.[9,11,12] Acetate, propionate, and butyrate have been identified as the major organic acid derivatives, and it has been established that the ^{14}C is located almost exclusively in the methyl carbon atoms of these fatty acids.[11] Alanine is the major amino acid formed, and the ^{14}C in this compound is found almost entirely in the methyl carbon atom.[11]

Although oxamide always accumulates as a stoichiometric end product of compound II oxidation, the distribution of other products produced by different batches of cells is highly variable and is a function of the conditions used in growth of the bacteria. Suspensions of cells derived from highly aerobic cultures produce relatively small amounts of the α-pyrone, whereas cells derived from poorly aerated cultures produce α-pyrone as the major product other than oxamide.[11] It is evident that the degradation of compound II can occur by two different pathways. One pathway leads to the formation of equivalent amounts of compound III and oxamide, whereas the other leads to production of oxamide and a mixture of CO_2, fatty acids, and assimilated products.[9,11] Since compound III is a stable end product of compound II (it is not metabolized by cell suspensions),[9,11] it is evident that compound III is not an intermediate in the conversion of compound II to other products; however, the possibility remains that it is derived from compound II via an intermediate that is common to both pathways of degradation (see Fig. 2).

Oxidation of Methyl-^{14}C Compound II

Reagents

Potassium phosphate buffer, 0.05 M, pH 7.5
Ethanol solution, 1.0%
Lyophilized cells of *Pseudomonas* RF
Methyl-^{14}C-compound II, 0.05 M, dissolved in 0.05 M NaOH

Procedure.[11] Five milliliters of phosphate buffer, 0.2 ml of ethanol solution, 20 mg of lyophilized cells, and 0.02 ml of methyl-^{14}C-compound II are mixed in a 25-ml Erlenmeyer flask and shaken in air at 30°. At 1-hour intervals, 1-ml aliquots are pipetted into 1.0 ml of 0.05 M NaOH; the mixture is centrifuged for 10 minutes at 25,000 g, to remove the cells, and the absorbancy of the supernatant fluid is measured at 333 nm. If the absorbancy decreases to less than 0.05 absorbancy unit, the yield of compound III is determined by the absorbancy at 300 nm ($\epsilon = 9850\ M^{-1}$ cm^{-1}). The amount of ^{14}C assimilated can be determined by measuring the radioactivity of the washed cells. The concentrations of ^{14}C-labeled fatty acids and ^{14}CO$_2$ produced can be determined by standard procedures.

Inhibitors and Activators. In the presence of $5 \times 10^{-4}\ M$ KCN, assimilation of compound II and its degradation to organic acids is almost completely inhibited, but the conversion to compound III is not. Therefore, in the presence of KCN, compound II is converted to stoichiometric amounts of oxamide and compound III.[11] In the presence of $5 \times 10^{-3}\ M$ NaF, the synthesis of compound III is slightly inhibited and the synthesis of ^{14}CO$_2$ is slightly stimulated. Iodoacetamide and iodoacetate at concentrations of $5 \times 10^{-4}\ M$ cause substantial inhibition (30–90%) of both degradation mechanisms.[11]

No decomposition of compound II occurs in the absence of molecular oxygen.

Synthesis of 1-Ribityl-2,3-diketo-1,2,3,4-tetrahydro-6,7-dimethylquinoxaline

Principle. The synthesis of quinoxaline-2,3-diols involves the condensation of the appropriately substituted *o*-phenylenediamine with oxalic acid or its derivative. Thus, the preparation of 1-ribityl-2,3-diketo-1,2,3,4-tetrahydro-6,7-dimethylquinoxaline (I) can readily be achieved by the reaction of diethyl oxalate with *N*-ribityl-2-amino-4,5-dimethylaniline, which is the key intermediate in several processes for the production of riboflavin and related compounds. The synthesis of (I), outlined in Fig. 4, is a modification of a method for synthesis of lyxoflavin by Heyl *et al.*[13]

---

[13] D. Heyl, E. C. Chase, F. Koniuszy, and K. Folkers, *J. Am. Chem. Soc.* **73**, 3826 (1951).

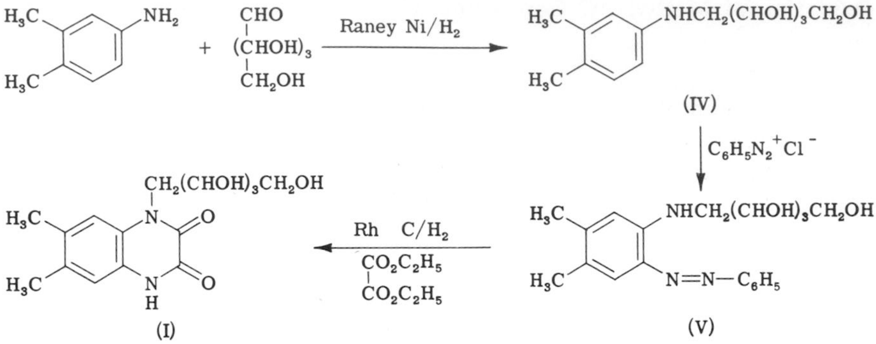

$$H_3C-C_6H_2(CH_3)-NH_2 + \begin{array}{c} CHO \\ | \\ (CHOH)_3 \\ | \\ CH_2OH \end{array} \xrightarrow{\text{Raney Ni}/H_2} H_3C-C_6H_2(CH_3)-NHCH_2(CHOH)_3CH_2OH$$

(IV)

$$\downarrow C_6H_5N_2{}^+Cl^-$$

$$\begin{array}{c} CH_2(CHOH)_3CH_2OH \\ | \\ N \\ \diagdown \diagup O \\ \diagup \diagdown O \\ N \\ | \\ H \end{array} \xleftarrow[\substack{CO_2C_2H_5 \\ | \\ CO_2C_2H_5}]{Rh \quad C/H_2} \begin{array}{c} NHCH_2(CHOH)_3CH_2OH \\ \\ N{=}N{-}C_6H_5 \end{array}$$

(I)   (V)

FIG. 4. Synthesis of 1-ribityl-2,3-diketo-1,2,3,4-tetrahydro-6,7-dimethylquinoxaline (compound I).

This method is adaptable for the preparation of radioactively labeled compounds.[8]

Procedures. (a) *N-Ribityl-3,4-dimethylaniline (IV).* To a reaction bottle of the Parr hydrogenation apparatus, the following reagents are introduced: 1.25 g of D-ribose in 5 ml of methanol; 1.0 g of xylidine in 2.5 ml of methanol; and approximately 0.5 ml of a suspension of Raney nickel. This mixture is shaken under H_2 of 36 psi for 9 hours at room temperature; the reaction is stopped, and 40 ml of methanol is added. The mixture is warmed slightly, and the catalyst is filtered off. The filtrate is concentrated to about half its volume and allowed to crystallize. Yield: 1.58 g of *N*-ribityl-3,4-dimethylaniline; m.p. 139°–141°.

(b) *N-Ribityl-2-phenylazo-4,5-dimethylaniline (V).*[14] Benzene diazonium salt solution: To a solution of 800 mg of aniline in 7.85 ml of 3.5 N HCl at 0°, 600 mg of sodium nitrite is added portionwise at such a rate that the temperature of the mixture does not rise above 5°.

A mixture of 177 mg of *N*-ribityl-3,4-dimethylaniline in 1.4 ml of H_2O, 0.23 ml of conc. HCl, and 228 mg of sodium acetate is cooled in an ice–salt bath to −5°. To this mixture, 0.9 ml of the benzene diazonium salt solution is added. The entire reaction mixture in the ice–salt bath is then put in a cold room at 5° for 3 hours. Upon warming to room temperature, a solution of 215 mg of sodium acetate in 1.75 ml of water is added. The mixture turns brown and precipitates begin to form. It is allowed to stand at room temperature for 14 hours. The crude azo compound (V) is then collected on a filter, washed thoroughly with water, and dried in a vacuum desiccator overnight. Yield: 232 mg; m.p. 155°–159°. This crude material is suitable for use in the subsequent step. Purification of the azo compound can be

[14] This procedure is essentially a scaled-down version of that of Heyl *et al.*[13] It is given here for the sake of continuity.

achieved by recrystallizations from methanol; pure samples melt at 176°–177°.

(c) *1-Ribityl-2,3-diketo-1,2,3,4-tetrahydro-6,7-dimethylquinoxaline* (*I*). To a solution of 90 mg of the azo compound (V) in 13 ml of ethanol is added 2 ml of diethyl oxalate and 40 mg of 5% Rh-C catalyst. This mixture is shaken under an atmosphere of hydrogen at 21°. Hydrogen uptake is followed from the reading of a gas-burette. The reaction is stopped after the theoretical amount of hydrogen uptake is observed. The reaction mixture is heated at 110° under a stream of He (or N_2) for 2 hours during which time the solvent is entirely evaporated. The residue is taken up in 2 ml of ethanol and filtered to move the excess diethyl oxalate. The mixture is then extracted with 2 ml of warm 0.1 N NaOH and the catalyst filtered off and washed three times with 3 ml of 0.1 NaOH. The combined filtrate and washings are concentrated at room temperature under a stream of N_2. Upon exposure of the concentrated solution to a carbon dioxide atmosphere, 49 mg of the quinoxaline compound (I), m.p. 250°–260°, is precipitated. The precipitate is collected on a filter and dried in a vacuum desiccator at 100° for 6 hours.

Synthesis of 6,7-Dimethylquinoxaline-2,3-diol-methyl-^{14}C

Principle. In order to introduce ^{14}C-labeling into the benzenoid moiety of a quinoxaline molecule, an appropriately labeled benzene derivative is required as starting material. *o*-Xylene is eminently suitable for this purpose, since reaction between *o*-xylene and oxalylchloride yields 3,4-dimethylbenzoic acid (VI),[15] the carboxyl group of which serves as a starting point for the construction of the pyrazine ring. The synthesis of 6,7-dimethylquinoxaline-2,3-diol-methyl-^{14}C (II) from *o*-xylene-methyl-^{14}C is outlined in Fig. 5.

Procedures.[16] (a) *3,4-Dimethylbenzoic acid-methyl-^{14}C* (*VI*).[17] A solution of 0.9 g of *o*-xylene (containing *o*-xylene-methyl-^{14}C, Tracerlab) and 3 ml of oxalyl chloride in 26 ml of carbon disulfide is added dropwise to a suspension of 2.6 g of aluminum chloride in 10 ml of carbon disulfide. The addition requires 7 hours; during this time, the temperature of the mixture is kept at 2° by cooling in an ice bath. It is then stirred for 5 hours at room temperature. At the end of the reaction, 25 ml of ice-water is added and the mixture is warmed over a steam bath until all carbon disulfide is evaporated. Upon cooling the brown solid is collected on a filter, then suspended in 10% Na_2CO_3 solution and heated to boiling for a few minutes. The mixture is filtered while hot; the alkaline filtrate is treated with charcoal,

[15] S. Coffey, *Rec. Trav. Chim.* **42**, 426 (1923).
[16] L. Tsai, unpublished results.
[17] This is a modification of a procedure reported by Coffey.[15]

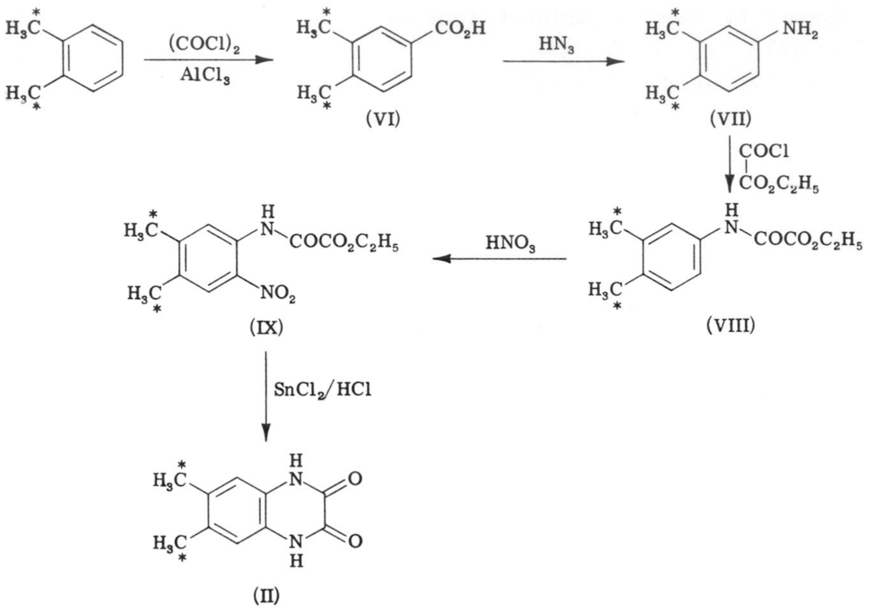

FIG. 5. Synthesis of 6,7-dimethylquinoxaline-2,3-diol-methyl-^{14}C (compound II) from o-xylene-methyl-^{14}C.

filtered, and acidified with 6 N HCl. 3,4-Dimethylbenzoic acid-methyl-^{14}C (VI) is precipitated. This is filtered, washed with water, and dried in a vacuum desiccator overnight; m.p. 155°–158°, 0.58 g.

(b) *3,4-Dimethylaniline-methyl-^{14}C* (*VII*). To a mixture of 450 mg of 3,4-dimethylbenzoic acid-methyl-^{14}C (VI), 4 ml of conc. H$_2$SO$_4$, 0.1 ml of fuming H$_2$SO$_4$, and 8 ml of chloroform, 210 mg of solid sodium azide is added portionwise[18] under vigorous stirring over a period of about 30 minutes while the temperature is kept at 50°. The mixture is stirred for another 30 minutes at 50° when gas evolution has practically ceased. It is cooled in an ice-bath and about 20 ml of ice-water is added, extracted with two 10-ml portions of chloroform. This strongly acidic solution is made basic by careful addition of 20% KOH, then extracted three times with 10 ml of chloroform each. The combined chloroform extract is dried over anhydrous MgSO$_4$, filtered, and evaporated under reduced pressure to yield colorless crystals of 3,4-dimethylaniline-methyl-^{14}C (VII); m.p. 45°–47°, 215 mg.

(c) *Ethyl N-3,4-Dimethylphenyloxamate-Methyl-^{14}C* (*VIII*). An ice-cooled mixture of 205 mg of 3,4-dimethylaniline-methyl-^{14}C (VII), 250 mg

---

[18] The addition of sodium azide should be at such a rate as to maintain a steady rate of gas evolution; too rapid addition may cause a violent reaction.

of anhydrous Na_2CO_3, and 3.5 ml of ethyl ether is stirred vigorously while 0.3 ml of ethyl oxalyl chloride is added slowly. Stirring is continued for 20 minutes under ice-cooling, then for 5 hours at room temperature. Ice-water, 20 ml, is added, and ethyl ether is evaporated under a stream of nitrogen. The solid material is filtered and washed thoroughly with a small volume of dilute HCl and then with water. Recrystallization from dilute ethanol gives colorless crystals of the oxamate (VIII); m.p. 68°–69°, 373 mg.

(d) *Ethyl N-3,4-Dimethyl-6-nitrophenyloxamate-*^{14}*C-methyl* (*IX*). To 15 ml of conc. HNO_3, cooled to 5° in an ice bath, is added portionwise over a period of 25 minutes 330 mg of ethyl N-3,4-dimethylphenyloxamate-methyl-^{14}C (VIII). Temperature of the mixture rises to 25°. After completion of addition the mixture is warmed to 35° and stirred for 40 minutes. It is then poured over ice, upon which pale yellow precipitates are formed. Recrystallization from ethanol furnishes the crystalline nitro compound (IX); m.p. 108°–110°, 270 mg.

(e) *6,7-Dimethylquinoxaline-2,3-diol-Methyl-*^{14}*C* (*II*). A mixture of 245 mg of the nitro compound (IX), 750 mg of $SnCl_2 \cdot 2H_2O$, 2.5 ml of glacial acetic acid and 2.5 ml of conc. HCl is stirred at 40° for 4 hours. At the end of the reaction, 25 ml of ice-water is added and the white precipitates of (II) are collected on a filter, washed with small volumes of water and ethanol and finally dried in a vacuum desiccator to yield 155 mg of product. Purification is effected by recrystallization either from glacial acetic acid or from gradual acidification of an alkaline solution of the compound.

[164] Riboflavin Hydrolase (EC 3.5.99.1) from *Pseudomonas riboflavina*

By C. S. YANG and DONALD B. McCORMICK

Assay Method

$$\text{Riboflavin} + H_2O \rightarrow \text{lumichrome} + \text{ribitol}$$

Principle. The disappearance of flavin, usually riboflavin, during the reaction is assayed fluorometrically. By use of riboflavin-2-^{14}C, both the disappearance of riboflavin and appearance of lumichrome are also determined by measuring the radioactivities after their separation by paper chromatography.[1]

---

[1] C. S. Yang and D. B. McCormick, *Biochim. Biophys. Acta* **132**, 511 (1967).

Reagents

Riboflavin or other flavins, $1 \times 10^{-5} M$ in water
Potassium phosphate buffers, 0.1 M and 0.04 M, pH 7.0
Acetic acid, 5%, in ethanol
n-Butanol–acetic acid–water (4:1:5, v/v/v, upper phase)

Enzyme. The enzyme is prepared from *P. riboflavina* (ATCC 9526) which has been cultured aerobically for 24 hours at 30° on a riboflavin–yeast–salts medium.[2] Cells are suspended in 20 volumes of 0.1 M phosphate buffer (pH 7.0) and ruptured by sonic oscillation for 5 minutes in a Branson sonifier. The mixture is then centrifuged at 900 g for 30 minutes to remove the unbroken cells and large fragments. The turbid supernatant solution is centrifuged again at 10,000 g for 30 minutes to obtain most of the enzyme in the sedimented particles. This particulate fraction is washed three times and resuspended in the phosphate buffer for use as the enzyme preparation. Aliquots are diluted as required for assay.

Procedure. To 2.3 ml of 0.04 M phosphate buffer in a 25-ml Erlenmeyer flask, 0.1 ml of $1 \times 10^{-5} M$ flavin is added. After equilibration at 30°, 0.1 ml of enzyme is added and the mixture is incubated at 30° with rotary shaking for 30 minutes. The reaction is stopped by adding 2.5 ml of ethanol–acetic acid solution, and the mixture is centrifuged to remove the particulate matter. A control is run in the same manner except that the ethanol–acetic acid solution is added before the enzyme preparation. The amount of flavin remaining is measured with a suitable spectrophotofluorometer, in which the activating wavelength is set at 450 nm and the fluorescence read at 520 nm. Under these assay conditions the fluorescence is directly proportional to the amount of flavin. The amount of flavin hydrolyzed is used as a measure of enzyme activity. The amount of riboflavin-^{14}C remaining and the formation of lumichrome-^{14}C can also be determined by applying aliquots of the ethanol–acetic acid extracts to Whatman No. 1 paper, developing the chromatograms in n-butanol–acetic acid–water, and quantitating the radioactivity with a radiochromatogram scanner.

The concentrations of substrate and enzyme, as well as the incubation time, are subject to modification as needed. No significant alteration of the result has been found when the steps of adding ethanol–acetic acid solution and centrifugation are omitted.

Properties

The enzyme preparation is relatively unstable and attempts to purify the enzyme further have not been successful.[3] The enzyme does not func-

[2] T. Yanagita and J. W. Foster, *J. Biol. Chem.* **221,** 593 (1956).
[3] C. S. Yang, Ph.D. thesis, Cornell University, Ithaca, New York, 1967.

TABLE I
RELATIVE SUBSTRATE ACTIVITIES OF FLAVINS

Flavin	Relative activity	Flavin	Relative activity
D-Riboflavin	100	D-Alloflavin	5
Isoriboflavin	31	D-Erythroflavin	6
Dichlororiboflavin	16	DL-Glyceroflavin	5
3-Methylriboflavin	2	6'-Hydroxyhexylflavin	9
L-Lyxoflavin	26	5'-Hydroxypentylflavin	14
L-Deoxylyxoflavin	8	4'-Hydroxybutylflavin	17
D-Lyxoflavin	3	3'-Hydroxypropylflavin	7
D-Araboflavin	13	2'-Hydroxyethylflavin	6
D-Galactoflavin	4	Formylmethylflavin	5
D-Sorboflavin	1	Lumiflavin	7
D-Dichlorosorboflavin	3		

tion under anaerobic conditions and is inhibited by higher levels of substrate. Inorganic phosphate or ATP is not involved in the reaction.[2] The substrate specificity of this bacterial enzyme has been extensively studied,[1] and some of the data are listed in Tables I and II. The significant but

TABLE II
SUBSTRATE PROPERTIES OF REPRESENTATIVE FLAVINS

Flavin	V_{max} (millimicromoles decomposed per 30 min)	K_m ($M \times 10^7$)
D-Riboflavin	8.4	16.0
Isoriboflavin	1.0	1.3
L-Lyxoflavin	0.9	0.6
5'-Hydroxypentylflavin	0.4	0.7
D-Erythroflavin	0.2	0.8

decreased activities of iso- and dichlororiboflavin point up the preference for the 6,7-dimethylisoalloxazine structure of the natural vitamin. Also, the system is more reactive toward a 9-D-ribityl side chain as shown by the lower substrate reactivities of the flavins with substituents on the 9-position of the isoalloxazine ring system. The mechanism of the reaction is not known. A similar hydrolase has also been isolated from plant tissues and shown to require reduced glutathione and Mg^{2+} for maximal activity.[4]

[4] S. A. Kumar and C. S. Vaidyanathan, *Biochim. Biophys. Acta* **89**, 127 (1964).

[165] Isolation and Identification of 7,8-Dimethyl-10-(2'-hydroxyethyl)isoalloxazine from Natural Sources

By E. C. OWEN and D. W. WEST

A ruminant, such as a cow or goat, produces minute amounts of 7,8-dimethyl-10-(2-hydroxyethyl)isoalloxazine in its urine under normal feeding regimes. However, when large doses (10–20 mg/kg body weight) of riboflavin are fed to these animals, amounts adequate for isolation of this metabolite appear in the milk and urine within 24 hours. The production of this metabolite is due to the degradative action of the intestinal bacteria, and there are thus three ways in which the metabolite may be obtained: (a) by extraction from media in which rumen bacteria have been incubated; (b) by extraction from milk; (c) by extraction from urine (Fig. 1).

Extraction from Culture Medium

Step 1. Rumen contents (100 ml) obtained by means of a fistula are centrifuged at 800 *g* for 30 minutes to remove food particles and protozoa. The residue is discarded, and the supernatant is centrifuged further at 27,500 *g* for 10 minutes. The bacterial residue from this high-speed centrifugation is then inoculated into a culture medium containing 15 g of proteose peptone, 5 g of yeast extract, 2.5 g of liver digest (all Oxoid materials), 5 g of NaCl, and 0.5 g of riboflavin per liter of tap water. After 3–5 days of incubation at 37° in the dark, most of the riboflavin has been converted to the required metabolite. The culture is then acidified to pH 4.0 with concentrated sulfuric acid and extracted with chloroform for 24 hours in a continuous extractor.

Step 2. This chloroform extract is concentrated to about 20 ml under vacuum in a rotatory evaporator after filtration through phase-separating filter paper (Whatman) to remove any aqueous phase carried over in the extraction process. A slurry of 75 g of silica gel (Camag. D5) with 170 ml of water is spread on glass plates (20 × 20 cm), dried overnight at room temperature, and then further dried at 90° for 1 hour. This gives a 0.75-mm layer of dried silica gel. The concentrated chloroform extract (ca. 1 ml of extract per plate) is applied to these plates as a band at the origin, and the plates are developed twice in a solution of chloroform and methanol (9:1, v/v).

Step 3. The band of silica gel containing the 7,8-dimethyl-10-(2'-hydroxyethyl)isoalloxazine, located by its yellow fluorescence under long wavelength ultraviolet light, is scraped from the plate, and the metabolite is eluted from the silica by methanol. Elution is most effectively carried out by the double centrifuge tube method of Owen.[1a] The apparatus con-

[1a] E. C. Owen, *Lab. Pract.* **17**, 1137 (1968).

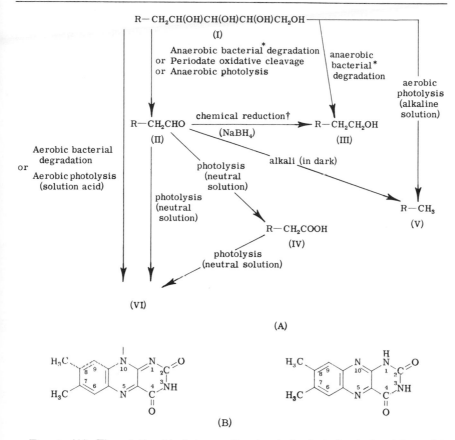

FIG. 1. (A). The relationship between the chemical, photochemical and bacterial degradation products of the ribityl side-chain of riboflavin. I is riboflavin; II is 7,8-dimethyl-10-formylmethyl-isoalloxazine; III is 7,8-dimethyl-10(2′-hydroxyethyl)-isoalloxazine; IV is 7,8-dimethyl-isoalloxazine-10-acetic acid[1b]; V is 7,8,10-trimethyl-isoalloxazine (lumiflavin); VI is 7,8-dimethyl-alloxazine (lumichrome). (* Of bacteria, isolated from rumen contents by the authors,[1c] some produced II and others III from I. † Since their submission of the original typescript the authors have found that this reduction occurs in the intact animal *in vivo* and that ruminant liver or kidney slices can cause this reduction *in vitro*.)[1c]

FIG. 1. (B). Structural formula of the 7,8-dimethyl-isoalloxazine nucleus (R of diagram A above) and on its right the structural formula of lumichrome.

sists of a conical centrifuge tube with a small hole in the bottom supported concentrically inside a larger round-bottom centrifuge tube by means of a flanged Perspex washer. The washer is of sufficient thickness to ensure that the bottom of the inner tube is 2–3 cm from the bottom of the outer tube. A layer of glass wool covered with Celite (Johns Manville, London) supports the silica gel in the inner centrifuge tube. The silica gel is "wetted"

[1b] E. C. Owen and D. W. West, *Chem. & Industr.* **881** (1968).
[1c] E. C. Owen and D. W. West, *Brit. J. Nutr.* **24, 45** (1970).

with the required solvent, and the whole apparatus is centrifuged to collect the eluate in the outer tube. This method is rapid and requires a minimal amount of eluant.

Step 4. The methanol solution thus obtained, concentrated almost to dryness on a rotatory evaporator, is redissolved in chloroform (20 ml), and the metabolite is purified by column chromatography on neutral alumina (Woelm). A 2-ml portion of the chloroform solution is applied to the top of a column (2 × 50 cm) containing 100 g of neutral alumina in chloroform. The column is washed through with 50 ml of chloroform, and then elution is continued with chloroform containing 2% acetone (v/v) until no more fluorescent material is eluted. Elution with chloroform containing 2% acetic acid (v/v) then removes several more fluorescent impurities before the 7,8-dimethyl-10-(2'-hydroxyethyl)isoalloxazine. The metabolite slowly crystallizes from this solvent mixture as yellow needles (m.p. 300°–301°).

Alternative Procedure to Steps 3 and 4. It is not always easy to obtain a pure crystalline sample of the metabolite by this procedure, particularly when it is isolated from urine samples; for identification purposes it is easier to prepare and purify the acetyl derivative which crystallizes much more readily. Thus the band of silica containing the 7,8-dimethyl-10-(2'-hydroxyethyl)isoalloxazine from the preparative layer chromatoplates is extracted with pyridine instead of methanol, and the resulting pyridine solution is treated with 2 or 3 ml of freshly distilled acetic anhydride and heated on a water bath for 30 minutes. Methanol is then added to inactivate excess acetic anhydride, and the solution is concentrated to small bulk on a rotatory evaporator. This concentrate is repurified by preparative layer chromatography using the same solvent system as before and the 7,8-dimethyl-10-(2'-acetoxyethyl)isoalloxazine is eluted in methanol. The methanol solution is evaporated to dryness on a rotatory evaporator, and the residue is redissolved in chloroform. This chloroform solution is purified by column chromatography on neutral alumina using chloroform containing 2% ethanol (v/v) as the eluting solvent. The acetoxy compound crystallizes from this mixture of solvents and can be recrystallized from a mixture of ethanol and light petroleum (b.p., 80°–100°; m.p., 230°–232°, decomp.). If necessary, the acetoxy compound is easily hydrolyzed to the parent alcohol by treatment with concentrated hydrochloric acid at room temperature.

Extraction from Milk

Milk (1 liter), obtained from a goat or cow to which riboflavin has been fed 24 hours previously, is treated with 250 ml of a 50% (w/v) aqueous solution of trichloroacetic acid, and this mixture is allowed to stand in the dark for 15 minutes. The solution is then centrifuged at 800 g for 15 minutes, and the supernatant fluid is decanted off and retained. The residue is washed once with 200 ml of a 20% (w/v) aqueous solution of trichloroacetic acid and then several times with equivalent amounts of water, the solution being centrifuged at 800 g; the supernatant portion is retained after each wash. The residue is washed with water until no further fluorescence is observable in the precipitate. The original extract and the washings are combined and neutralized by the addition of 4 M K_2HPO_4. After the extract has stood in the cold for 1 hour, the flocculent precipitate that has settled out is filtered off. This solution is extracted three times with molten phenol (25 ml for every 100 ml of milk extract), and the combined phenol extracts are treated with 3–4 times their volume of diethyl ether. For quantitative estimations the ethereal solution is washed repeatedly with small amounts of water until all the fluorescence has been recovered in the aqueous phase. Three or four extractions suffice for qualitative work. The aqueous phase varies in volume from 100 to 400 ml, the more fluorescent milk samples producing greater volumes of extract. This concentrate is acidified to approximately pH 4.0 with concentrated sulfuric acid and extracted with chloroform in a continuous extractor. This chloroform extract is then treated in exactly the same way as the chloroform extract obtained from the bacterial incubation (step 2).

Extraction from Urine

Urine, obtained from a goat or cow to which riboflavin has been fed 24 hours previously, is acidified to pH 4.0 with concentrated sulfuric acid and extracted with chloroform in a continuous extractor. The chloroform extract is then treated in either of the two alternative ways already described for chloroform extracts from rumen bacterial incubations (see step 2 above).

Identification

7,8-Dimethyl-10-(2'-hydroxyethyl)isoalloxazine has the same fluorescence characteristics and ultraviolet absorption spectrum as riboflavin, and these physical properties are thus of limited usefulness in distinguishing this material from other isoalloxazines. However, because the difference between these isoalloxazines is in the length and substitution of the side chain, nuclear magnetic resonance (NMR) spectroscopy is applicable. The sparing solubility of the isoalloxazines in most solvents presents a difficulty that can be avoided by the use of deuteriodimethyl sulfoxide as solvent or by use of the computer-average-transient method for very dilute solutions. The NMR characteristics of 7,8-dimethyl-10-(2'-acetoxyethyl)isoalloxazine in deuteriochloroform are given in Table I.

The solubility of 7,8-dimethyl-10-(2'-hydroxyethyl)isoalloxazine in

TABLE I

PROTON MAGNETIC RESONANCE DATA FOR 7,8-DIMETHYL-10-
(2'-ACETOXYETHYL)ISOALLOXAZINE IN DEUTERIOCHLOROFORM
WITH TETRAMETHYLSILANE AS INTERNAL STANDARD[a]

τ Values	Intensity	Splitting (J, cps)	Assignment
(7.34, 7.45)	3, 3	s, s	Two aromatic methyls
(1.83, 2.28)	1, 1	s, s	Two isolated aromatic protons
4.94	2	t, d; 8, 1	CH$_2$—N<
5.27	2	d, 5	CH$_2$
			O
			C = O
7.9	3	s	CH$_3$ Methyl of acetate group

[a] Obtained by use of the computer-average-transient method, 225 scans. The authors wish to thank Dr. Scheidegger of Varian Associates Zurich, Switzerland, for this spectrum of the material, which was isolated at the Hannah Institute.

many organic solvents is greater than that of riboflavin, so that chromatography on paper or thin layers of silica gel in several solvent systems is the quickest routine method of identification (Table II). In addition, valuable characteristics are that the material is easily acetylated and that the acetyl derivative has a completely different R_f. If an unknown compound has the same R_f as 7,8-dimethyl-10-(2'-hydroxyethyl)isoalloxazine in several solvent systems and its acetyl derivative has the same R_f as

TABLE II

R_f VALUES OF ISOALLOXAZINE DERIVATIVES ON SILICA GEL THIN-
LAYER CHROMATOGRAPHY IN VARIOUS SOLVENT SYSTEMS

Flavin	R_f values in solvent system[a]				
	(a)	(b)	(c)	(d)	(e)
Riboflavin	0.10	0.50	0.50	0.64	0.02
Tetraacetoxyriboflavin	0.85	—	—	—	0.80
7,8-Dimethyl-10-formylmethylisoalloxazine	0.36	0.78	0.47	0.72	—
Lumiflavin	0.45	0.51	0.32	0.54	—
7,8-Dimethyl-10-(2'-hydroxyethyl)isoalloxazine	0.31	0.54	0.44	0.61	0.25
7,8-Dimethyl-10-(2'-acetoxyethyl)isoalloxazine	0.51	0.51	0.36	—	0.50

[a] Solvent systems: (a) chloroform–methanol (9:1); (b) butanol–ethanol–water (7:2:2); (c) water saturated with isoamyl alcohol; (d) butanol–pyridine–water (6:4:4); (e) benzene–pyridine–acetic acid (4:1:1).

7,8-dimethyl-10-(2′-acetoxyethyl)isoalloxazine, it is a safe assumption that the unknown is in fact 7,8-dimethyl-10-(2′-hydroxyethyl)isoalloxazine.

Urine or bacterial culture samples can be applied directly onto thin-layer chromatographic plates, but in this case it is advisable to develop the plates with both internal and external marker spots of authentic materials, since the presence of impurities can alter the R_f of the metabolite by acting as additions to the developing solvent system. The periodate oxidative procedure of Fall and Petering[2] provides a convenient source of authentic material for comparison purposes.

[2] H. H. Fall and H. G. Petering, *J. Am. Chem. Soc.* **78**, 377 (1956).

[166] Isolation and Identification of 7,8-Dimethyl-10-formylmethylisoalloxazine as a Product of the Bacterial Degradation of Riboflavin

By E. C. OWEN and D. W. WEST

Isolation

Step 1. Rumen contents (500 ml) obtained from a sheep, goat, or cow by means of a fistula or of a stomach pump are centrifuged at 800 g for 15 minutes to remove food particles and protozoa. The residue is discarded, and the supernatant is further centrifuged at 27,500 g for 10 minutes. The supernatant liquid from this second centrifugation (ca. 400 ml) is poured into an amber screw-top bottle containing riboflavin (1 g) and then inoculated with 300–400 mg of the bacterial residue from the high-speed centrifugation. The contents of the bottle are incubated at 37° for 4–7 days during which time some of the undissolved riboflavin forms a dark green complex with the required product, which collects at the bottom of the bottle. After this period the culture is acidified to approximately pH 4.0 with concentrated sulfuric acid and extracted with chloroform (500 ml) in a continuous extractor for 24 hours. This chloroform extract is filtered through phase-separating filter paper (Whatman) to remove any aqueous phase present and is then concentrated to ca. 20 ml under vacuum on a rotatory evaporator.

Step 2. Chloroform (75 ml) is added to a chromatography column (3 × 50 cm) fitted with a Teflon tap and a sintered-glass disk as a support for the adsorbent. A disk of glass fiber filter paper is placed on the top of the sintered-glass disk to prevent the pores of the glass becoming clogged. Silica gel (Mallinckrodt Chemical Co., Grade C.C.7, 200–325 mesh) (130 g) is made into a slurry with chloroform (200 ml), and the slurry is added to the column via an extension tube fitted to the top. The silica gel is allowed to settle for approximately 5 minutes before the tap is opened to permit the solvent to flow slowly until the silica has completely settled. The column is washed with chloroform (100 ml) while a slight pressure of nitrogen is applied and 2 ml of the concentrated chloroform extract is then absorbed onto the top of the silica. After the material has been washed into the column with chloroform, elution is started by applying a 10% (v/v) solution of diethyl ether in chloroform again under slight pressure of nitrogen. After approximately two column volumes of this eluent has been used, the diethyl ether concentration is increased to 50% (v/v) and elution is continued. After a further two-column volumes of this mixture have passed through, the washing is continued with pure diethyl ether until no further yellow- or blue-fluorescent materials are eluted. The 7,8-dimethyl-10-formylmethylisoalloxazine is then obtained by elution with diethyl ether containing 10% absolute ethanol (v/v). The product crystallizes from a mixture of ethanol and light petroleum (b.p. 80°–100°) as the hemiethanolate (m.p. 296°–297°, decomp.).

Identification

This metabolite is distinguishable from most other isoalloxazines by the manner in which its ultraviolet absorption spectrum varies with the pH of the solution. Most isoalloxazines, such as riboflavin, show a four-banded spectrum (Table I) from pH 1 to 13. Below pH 1 the 350 nm and

TABLE I

ULTRAVIOLET SPECTRAL CHARACTERISTICS OF ISOALLOXAZINE DERIVATIVES

Flavin	λ_{max}(nm) and ϵ (in parentheses) in H_2O				pH
Riboflavin	223 (35,500)	267 (35,500)	376 (10,700)	445 (22,500)	1
	222 (27,700)	270 (31,700)	356 (10,600)	447 (10,600)	13
7,8-Dimethyl-10-formylmethyl-isoalloxazine	216 (17,200)	263 (21,300)	387 (12,000)	—	1
	220 (23,100)	266 (23,500)	350 (6560)	445 (6560)	13[a]

[a] ϵ values low due to decomposition at high pH.

455 nm peaks coalesce to form a single peak with maximum around 387 nm. With 7,8-dimethyl-10-formylmethylisoalloxazine this change occurs around pH 3.5, so that in normal acid solution it shows only the three-banded spectrum typical of the alloxazines (Table I).

The product obtained by the purification procedure outlined is in fact the hemiethanolate of 7,8-dimethyl-10-formylmethylisoalloxazine,

TABLE II

PROTON MAGNETIC RESONANCE DATA[a] FOR THE HEMIETHANOLATE OF
7,8-DIMETHYL-10-FORMYLMETHYLISOALLOXAZINE IN DEUTERIODIMETHYL
SULFOXIDE WITH TETRAMETHYLSILANE AS INTERNAL STANDARD

τ Values	Intensity	Splitting (J, cps)	Assignment
5.21	2	d, 5	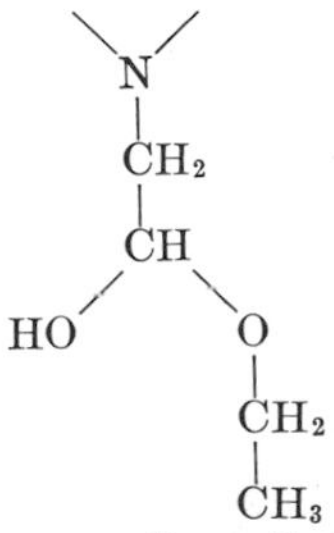
4.72	1	q, d; 6, 1	
3.28	1	d, 7	
ca. 6.3	2	m	
8.82	3	t	
7.34, 7.45	3, 3	s, s	Two aromatic methyls
1.96	2	d, 1	Two p-aromatic protons
−1.56	1	s, b	NH

[a] The authors wish to thank Dr. Karl Overton, of the Chemistry Department of Glasgow University, for the NMR data of this compound.

and Table II shows the nuclear magnetic resonance data obtained for this compound dissolved in deuteriodimethyl sulfoxide.

However the quickest way of preliminary identification is undoubtably by thin-layer chromatography. The appropriate R_f values are shown in Table III, together with those of a number of other isoalloxazines for comparison.

TABLE III

R_f VALUES OF ISOALLOXAZINE DERIVATIVES ON SILICA GEL THIN-LAYER
CHROMATOGRAPHY IN VARIOUS SOLVENT SYSTEMS

Flavin	R_f in solvent system[a]			
	(a)	(b)	(c)	(d)
Riboflavin	0.10	0.50	0.50	0.64
7,8-Dimethyl-10-formylmethylisoalloxazine	0.36	0.78	0.47	0.72
Lumiflavin	0.45	0.51	0.32	0.54
7,8-Dimethyl-10-(2′-hydroxyethyl)isoalloxazine	0.31	0.54	0.44	0.61

[a] Solvent systems: (a) chloroform–methanol (9:1); (b) butanol–ethanol–water (7:2:2); (c) water saturated with isoamyl alcohol; (d) butanol–pyridine–water (6:4:4).

[167] Preparation, Properties, and Conditions for Assay of Lipoamide Dehydrogenase Apoenzyme (NADH:Lipoamide Oxidoreductase, EC 1.6.4.3) from Pig Heart

By C. Veeger and J. Visser

The purification method used for lipoamide dehydrogenase from pig heart is only slightly different from that which has been described elsewhere.[1,2] This procedure is rather a general one and has been basically used for lipoamide dehydrogenases from various sources.[3-8] The apoenzyme, which is obtained by removal of the FAD, has been found, in addition to the D-amino acid oxidase apoenzyme (EC 1.4.3.3), useful in testing flavin analogs for their coenzymatic properties and their affinities for the protein.

Assay

Principle. The enzyme activity of the overall reaction:

$$\text{NADH} + \text{H}^+ + \text{oxidized lipoate (lipS}_2) \rightleftharpoons \text{NAD}^+ + \text{reduced lipoate [lip(SH)}_2] \quad (1)$$

is assayed spectrophotometrically by recording the decrease in absorbancy at 340 nm. The diaphorase activity is determined spectrophotometrically recording the reduction of 2,6-dichlorophenolindophenol (DCIP) at 600 nm in the reaction

$$\text{NADH} + \text{H}^+ + \text{DCIP} \rightarrow \text{NAD}^+ + \text{DCIPH}_2 \quad (2)$$

The specific activity with lipoate expressed as change in absorbancy units per milligram of protein per minute can be calculated from

$$\frac{\Delta A}{\text{addition}} \times \frac{\text{dilution factor}}{\text{protein conc.}}$$

The activity with DCIP is obtained the same way but multiplied by a factor of 100 (see footnote 9).

[1] V. Massey, *Biochim. Biophys. Acta* **37**, 310 (1960).
[2] V. Massey, Q. H. Gibson, and C. Veeger, *Biochem. J.* **77**, 341 (1960).
[3] E. P. Channing, A. Eberhard, A. H. Guindon, C. Kepler, V. Massey, and C. Veeger, *Biol. Bull.* **123**, 480 (1962).
[4] M. Koike, L. J. Reed, and W. R. Carroll, *J. Biol. Chem.* **238**, 30 (1963).
[5] A. Wren and V. Massey, *Biochim. Biophys. Acta* **110**, 329 (1965).
[6] E. Misaka and R. Nakanishi, *J. Biochem.* **53**, 465 (1963).
[7] D. R. Basu and D. P. Burma, *J. Biol. Chem.* **235**, 509 (1960).
[8] J. Matthews and L. J. Reed, *J. Biol. Chem.* **238**, 1869 (1963).
[9] N. Savage, *Biochem. J.* **67**, 146 (1957).

Reagents

Citrate buffer, 1 M, pH 5.65

Sodium phosphate buffer, 0.3 M, pH 7.2

2,6-Dichlorophenolindophenol (DCIP), 0.001 M in water

Bovine serum albumin, 0.03 M, 2% (w/v) in EDTA

DL-Dithio-n-octanoic acid (lipoic acid), 0.02 M, dissolved in methanol (2% final concentration) and further diluted in water containing an equivalent amount of NaOH

NAD⁺, 0.01 M in water

NADH, 0.01 M in water, freshly prepared every day and kept on ice

Procedure. (a) *Lipoate Activity.* To a spectrophotometer cuvette thermostated at 25° is added: water to a final volume of 3 ml, 2.5 ml of citrate buffer, 0.1 ml of bovine serum albumin, 0.1 ml of lipoate, 0.03 ml of NAD⁺, and 0.03 ml of NADH. The reaction is started by adding enzyme in an appropriate dilution giving an initial change in absorbancy at 340 nm, not exceeding 0.1 per minute.

(b) *Diaphorase Activity.* To a spectrophotometer cuvette thermostated at 25° is added: water to a final volume of 3.0 ml, 0.5 ml of sodium phosphate buffer, 0.1 ml of bovine serum albumin, 0.12 ml of DCIP, and 0.03 ml of NADH. The reaction is started by adding that amount of enzyme which gives an initial change in absorbancy at 600 nm not exceeding 0.1 per minute.

All activities are based upon the initial rate.

Preparation of the Enzyme

Reagents

Sodium phosphate buffer, 0.02 M, pH 7.4

Acetic acid, 1 N

Ammonium sulfate, 1% (w/v) in twice-distilled water

Disodium hydrogen phosphate, 0.3 M, pH 7.6

Sodium phosphate buffer, 0.3 M and 0.03 M, pH 7.6, containing 0.3 mM EDTA

Cellulose powder, Whatman, standard grade suspension, 10% (w/v) in twice-distilled water

Calcium phosphate gel, prepared according to Swingle and Tiselius[10] (70 mg/ml dry weight)

Calcium phosphate gel column (4.5 cm diameter, 15 cm height) packed with a flow rate of 50–80 ml per hour after pouring in a

[10] S. M. Swingle and A. Tiselius, *Biochem. J.* **48**, 191 (1951).

slurry of 400 ml of cellulose suspension and 45 ml of calcium phosphate gel. After packing, the column is washed with 1 liter of water followed by 1 liter of 0.03 M phosphate buffer, pH 7.6, containing 0.3 mM EDTA.

Dialysis bags are soaked in 1 mM EDTA for 20 minutes, then boiled for 5 minutes. After cooling, they are rinsed with twice-distilled water.

Procedure

Step 1. Preparation of Particles According to Keilin and Hartree.[11] Ten kilograms of pig hearts, cleaned of fat and connective tissue, are minced, and the mince is washed with tap water. The washings are repeated until they are almost colorless. The mince is squeezed through cheese cloth. Aliquots (800 g) are homogenized in a mortar with 600 g of sand and 400 ml of cold sodium phosphate buffer, 0.02 M, pH 7.4, for 40 minutes. An additional 600 ml of buffer is added. The homogenate is centrifuged for 30 minutes at 1350 g. The supernatant solution is decanted, cooled to 0°, and adjusted to pH 5.4 by the addition of 1 N acetic acid; it is then centrifuged for 20 minutes at 2000 g. The precipitate is suspended and homogenized with a Waring blendor in 1 liter of a cold 1% ammonium sulfate solution to which 5 ml of EDTA, 0.1 M, is added. The homogenate is stored in the cold until used.

Step 2. Two portions of pig heart particles are stirred for 12 hours at 0° before an ethanolic heat treatment is applied.[2] Commercial ethanol (96%) is added carefully to 4.5% (v/v) under constant stirring at 5°. After 15 minutes the suspension is heated to 42°–44° and extracted for another 15 minutes. After cooling to 5° the solution is centrifuged for 20 minutes at 10,000 g, and the precipitate is discarded. Then 0.3 M Na$_2$HPO$_4$ (1:30, v/v) is added, after which the solution is brought to 0.40 saturation by addition of (NH$_4$)$_2$SO$_4$ (240 g per liter) followed by centrifugation for 15 minutes at 32,000 g. The supernatant solution is brought to 0.80 saturation by the addition of (NH$_4$)$_2$SO$_4$ (280 g per liter); the precipitate obtained upon centrifugation at 30,000 g for 20 minutes is dissolved in 0.03 M sodium phosphate buffer, pH 7.6, containing 0.3 mM EDTA and dialyzed against twice-distilled water.

Step 3. After the removal by centrifugation of denatured protein, obtained on dialysis, 200 ml of enzyme solution (20 mg/ml) is placed on a cellulose-calcium phosphate gel column, then washed with 500 ml 0.03 M sodium phosphate and 0.1 M sodium phosphate pH 7.6 until the absorbancy of the eluate is less than 0.05. The yellow, fluorescent band of lipoamide

[11] D. Keilin and E. F. Hartree, *Biochem. J.* **41,** 500 (1947).

dehydrogenase is eluted with 0.1 M sodium phosphate, pH 7.6, containing 5% ammonium sulfate (w/v).

Step 4. The eluate is fractionated by $(NH_4)_2SO_4$ precipitation, and the precipitate from the 0.45 to 0.8 saturation fraction (between 280 and 560 g per liter) dissolved in 0.03 M sodium phosphate buffer, pH 7.2, containing 0.3 mM EDTA. The solution is stored at $-15°$ at a protein concentration of 20–30 mg/ml.

Step 5. Ten milliliters of enzyme solution is thawed and mixed with 650 mg of $(NH_4)_2SO_4$, 10 ml of 0.3 M phosphate buffer, pH 7.6, and 25 ml of twice-distilled water. The solution is heated to 70° and maintained at that temperature for 5 minutes. After cooling to 5°, the denatured, coagulated protein is removed by centrifugation at 25,000 g for 10 minutes. The supernatant solution is fractionated by ammonium sulfate precipitation, and the precipitate from 0.45 to 0.80 saturation is dissolved in 0.03 M sodium phosphate buffer pH 7.2 with 0.3 mM EDTA. The enzyme solution is extensively dialyzed overnight against 4 changes of 1 liter of 0.03 M phosphate buffer containing 0.3 mM EDTA.

Step 6. The calcium phosphate gel column chromatography is repeated exactly as described under step 3.

Step 7. The eluate is fractionated again with ammonium sulfate. Up to 0.50 saturation (312 g per liter), there is hardly any precipitate; enzyme fractions are collected between 0.50 and 0.62 saturation (80 g per liter) and between 0.62 and 0.80 saturation (90 g per liter). The fraction collected between 0.55 and 0.62 saturation has, in general, an absorbancy ratio 280 nm:455 nm of 5.0–5.3 and specific activities with lipoate and DCIP of 38–44 and 150–250, respectively. For the enzyme obtained between 0.62 and 0.80 saturation, the absorbancy ratio is 5.2–5.5, while the activities are 36–40 and 150–250, respectively. The enzyme preparations are dissolved in 0.03 M phosphate buffer containing 0.3 mM EDTA (pH 7.2), without removal of $(NH_4)_2SO_4$ in a concentration of 20–30 mg/ml and stored at $-10°$.

Preparation of the Apoenzyme

Apoenzyme preparations of bacterial lipoamide dehydrogenases have been reported before.[12,13] The removal of the flavin without denaturation of the apoenzyme is more difficult in the case of the pig heart enzyme.[14,15]

[12] M. Koike, L. J. Reed, and W. R. Carroll, *Biochem. Biophys. Res. Commun.* **7,** 16 (1962).
[13] C. H. Williams, *J. Biol. Chem.* **240,** 4793 (1965).
[14] C. Veeger, D. V. DerVartanian, J. F. Kalse, A. de Kok, and J. F. Koster, *in* "Flavins and Flavoproteins" (E. C. Slater, ed.), Vol. 8, p. 242. Biochim. Biophys. Acta Library, Elsevier, Amsterdam 1966.
[15] J. F. Kalse and C. Veeger, *Biochim. Biophys. Acta* **159,** 244 (1968).

The procedure is basically the method used by Strittmatter[16] for DPNH cytochrome b_5 reductase, i.e., an acid ammonium sulfate treatment.

Reagents

Tris-acetate, 1 M, pH 8.1, containing 0.1 M EDTA

KBr, 3 M in water

Ammonium sulfate, solution saturated at 20° and acidified with concentrated H_2SO_4, to pH 1.9 (1:10 dilution in H_2O; pH meter reading of concentrated solution pH 1.4–1.5), prepared freshly every week

Sodium phosphate buffer, 0.3 M, pH 7.6

Sodium phosphate buffer, 0.03 M, pH 7.2 containing 3 mM EDTA

EDTA, 0.1 M

Procedure. Enzyme, 2–3 mg, from a 2% stock solution prepared in 0.03 M sodium phosphate, pH 7.2, with 3 mM EDTA, is diluted with cold Tris buffer to a final volume of 1 ml. One milliliter of KBr solution is added, and the mixture is kept on ice in a 25-ml beaker. The pH of the saturated ammonium sulfate solution is checked before use. The most critical point is the pH of this solution[15]; if it is too low the protein easily denatures, if it is too high there is no efficient removal of the flavin. One milliliter of this saturated ammonium sulfate is added dropwise within 20 seconds, swirling slightly. This addition results in a slight turbidity. After a 40-second interval, an additional 4 ml of ammonium sulfate is added which precipitates the apoenzyme. Immediately the solution is centrifuged for 6 minutes at 18,000 g (temperature 5°).

The yellow supernatant solution is decanted, after which the centrifuge tubes are wiped out carefully with absorbent paper. One or two drops EDTA are added to the precipitate, which is dissolved in 0.6 ml of 0.3 M sodium phosphate buffer, pH 7.6, kept between 20° and 25°. When the apoenzyme has dissolved, the tube is immediately placed on ice and the contents diluted with 0.4 ml of a cold 0.03 M sodium phosphate buffer, pH 7.2, containing 3 mM EDTA to prevent buffer crystallization. The apoenzyme is stored on ice. It has been pointed out previously[15] that it is critical to use a buffer of a high ionic strength to dissolve the apoenzyme. The temperature of this buffer is important as well; no differences, however, were found in varying the pH from 7.2 to 7.6.

If the precipitate is still yellow, it is dissolved in half its original volume of Tris buffer and KBr, after which a second $(NH_4)_2SO_4$ treatment is carried out.

[16] P. Strittmatter, *J. Biol. Chem.* **236**, 2329 (1961).

Properties

The individual preparations obtained may vary considerably. The protein content of the apoenzyme is calculated according to $A^{1\,cm}_{1\,mg/ml} = 0.8$ at 280 mμ and varies between 1.2 and 2 mg/ml.[14] The FAD content is approximately 5% or less; part of it is not enzyme bound. The remaining activity with lipS$_2$ is 0.3–3%, and the DCIP-activity varies between 80 and 300% of the original activity of the holoenzyme.

Stability. The apoenzyme is labile at room temperature. Kept on ice for 24 hours, 70% of the apoenzyme is still able to recombine as judged by the increase of the DCIP activity upon the addition of FAD at 0° (footnote 17). If prepared from a slightly impure enzyme, which has a ratio $A\ 280 : A\ 455 = 6.0$–6.2 (for pure enzyme this ratio is 5.0–5.2), the apoenzyme obtained has a somewhat greater stability. The stability is also enhanced by small amounts of free FAD; the flavin probably maintains the right conformation of the apoenzyme by a reversible association-dissociation.[15]

Physical Properties. Apoenzyme (2.3 mg/ml) and reconstituted enzyme (at 0°, 2.6 mg/ml) have $s_{20,w}$ values of 4.2 S and 3.8 S, respectively; $D_{20,w}$ values are 6.8–6.9 $\times$ 10^7 cm^2 sec^{-1} and 6.4 $\times$ 10^7 cm^2 sec^{-1}. The molecular weight is half that of the holoenzyme, i.e., 52,000 (see footnote 18).

Recombination

The binding of FAD occurs to the monomer, which then dimerizes in a series of temperature-dependent conformational changes induced by the bound flavin molecule. The FAD binding process to the monomer is exothermic ($\Delta H = -8300$ cal mole^{-1}, $\Delta S = -4$ esu); detectable dimerization occurs at temperatures higher than 5°; the process has an activation energy of 21,000 cal mole^{-1}. The DCIP activity of the apoenzyme–FAD complexes is at the maximum strongly enhanced with respect to the activity of the holoenzyme (maximum values are 4400–5000 and 180–250 units, respectively), but is dependent on the time of incubation (Fig. 1). The lipoate activity is connected with the dimer structure and returns in a second-order rate process. K_{ass} values for FAD binding have been obtained using the DCIP activities obtained with flavin-saturation curves and with fluorescence polarization techniques.[15,19] K_{ass} values for FAD vary between 5.3 and 1.5 $\times$ 10^5 l-mole^{-1} depending on the temperature used. The FAD binding is specific and amounts to 1 mole of FAD per 52,000 g of protein.

---

[17] J. Visser and C. Veeger, unpublished results.
[18] J. Visser and C. Veeger, *Biochim. Biophys. Acta* **159,** 265 (1968).
[19] J. Visser, D. B. McCormick, and C. Veeger, *Biochim. Biophys. Acta* **159,** 257 (1968).

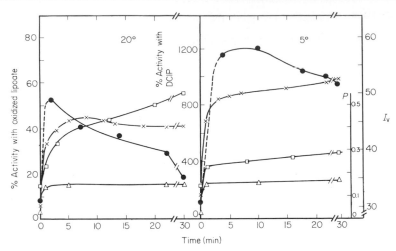

Fig. 1. The effect of temperature and time on the development of the activity with oxidized lipoate upon the addition of FAD to the apoenzyme of lipoamide dehydrogenase. The cuvettes contained in a volume of 2.5 ml: 0.05 M phosphate buffer *plus* 1 mM EDTA (pH 7.6), 8 μM apoprotein and 16 μM FAD. Samples were withdrawn from the cuvettes at the times indicated and the activities of the enzyme determined. ● ● ●, DCIP-activity; □ □ □, activity with oxidized lipoate; △ △ △, fluorescence polarization (P); $\times \times \times$, vertical component (I_v) of polarized light in arbitrary values. The fluorescence and fluorescence polarization values at zero time are those of the added free FAD. From *Biochim. Biophys. Acta* **159**, 244 (1968).

The flavin spectrum shifts to a longer wavelength during the dimerization; this indicates creation of a more apolar environment.

Conditions. Binding studies have been performed in 0.04 M to 0.08 M sodium phosphate buffers with 1 mM EDTA while the pH values varied from 7.2 to 7.6. Protein concentration was in the range of 2–8 μM based upon a molecular weight of 50,000. In Fig. 1 a typical binding experiment is shown. The activities with both DCIP and lipS$_2$ were determined by withdrawing samples from the fluorescence cuvette; the activities were determined at 25°. The binding of FAD is a fast process, as can be concluded from the rapid increase of the fluorescence polarization. Especially at lower temperatures, the fluorescence intensity keeps increasing, even if the polarization is constant. The DCIP activity passes a maximum which appears to be time and temperature dependent; it maintains a high level at low temperatures. At 0°–5° the lipoate activity returns very slowly in contrast to the fast return at higher temperatures (20°–25°).

Flavin saturation curves have been made by incubating apoenzyme (2–6 μM) with varying concentrations of FAD on ice during 20 minutes, after which the DCIP activities are determined. K_m values of Lineweaver-Burk plots of the FAD-dependent return of the activity correspond well with the value of the dissociation constant of the flavin.

Flavin Analogs. The following FAD analogs are bound to the apoenzyme: 3-N methyl FAD (K_{ass} 1.1 $\times$ 10^5 l-mole^{-1} at 10°), 3-N carboxymethyl FAD (K_{ass} 3 $\times$ 10^4 l-mole at 10°) and flavin 8-bromoadenine dinucleotide F 8-Br AD (K_{ass} 1.5 $\times$ 10^4 l-mole^{-1} at 0°) (footnotes 19 and 20) restore the DCIP activity to some extent. 3-N methyl FAD restores 17% of the original activity with lipoate. The 2-morpholino-FAD does not restore any activity. Although FMN is not able to restore the DCIP

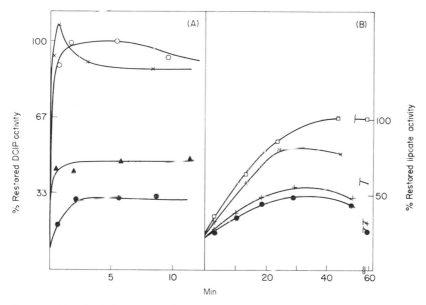

Fig. 2. (A) The influence of FMN preincubation with lipoamide dehydrogenase apoenzyme on the restoration of activity with DCIP by FAD. Apoenzyme (12 μM) was incubated on ice with FMN (200 μM) in 0.03 M sodium phosphate buffer (pH 7.2) with 0.3 mM EDTA. Samples were withdrawn at 5 (O—O), 30 (X—X), 90 (△—△) and 180 (O—O) min, incubated with 100 μM FAD on ice and the DCIP activity determined with time. (B) Influence of FMN on the restoration of the activity with lipoate. Recombination of 9 μM apoenzyme of lipoamide dehydrogenase at 20° with 20 μM FAD in 0.03 M sodium phosphate buffer (pH 7.2) with 0.03 mM EDTA. □—□, apoenzyme not preincubated with FMN; X—X, apoenzyme preincubated for 30 min with 10 μM FMN; +—+, with 50 μM FMN; ●—●, with 90 μM FMN. From *Biochim. Biophys. Acta* **206**, 224 (1970).

[20] J. Visser, J. F. Koster, D. B. McCormick, and C. Veeger, unpublished results.

activity, it is bound to the apoenzyme. This binding is observed by means of protein-fluorescence quenching and equilibrium dialysis; preincubation of the apoenzyme with this aspecific flavin leads to competition with FAD-binding (Fig. 2) (see footnote 21). K_{ass} values increase with incubation time from 0.1–0.3×10^5 l-mole^{-1} after 30 minutes to 2×10^5 l-mole^{-1} after several hours of preincubation on ice. FMN analogs are also bound without restoration of activity, e.g., iso-FMN, 2-thio-FMN, 2-N CH$_2$COOH FMN, 3-N CH$_2$COOH FMN, and 3-N methyl FMN (footnote 21).

Acknowledgments

We thank Drs. D. B. McCormick and P. Hemmerich for the supply of FAD- and FMN analogs.

[21] J. Visser and C. Veeger, *Biochim. Biophys. Acta* **206**, 224 (1970).

[168] Determination of the Xanthine Oxidase

By E. C. OWEN

Xanthine:oxygen oxidoreductase (EC 1.2.3.2) is an enzyme that contains riboflavin in its molecule in the form of FAD. Xanthine oxidase is determined aerobically[1] by Warburg manometry and anaerobically by using triphenyltetrazolium chloride (TTC). The TTC method, being colorimetric, is more readily adapted for measuring very low titers of the enzyme and can be made even more sensitive merely by using a longer time of incubation. Titers of xanthine oxidase in human milk or in the milk of animals which have been given tungstate are very low but can be measured by the TTC method.

Anaerobic Procedure for Estimating Xanthine Oxidase in Milk Using Triphenyltetrazolium Chloride (TTC) (Owen and Proudfoot, 1968)[2]

The conditions of Zittle *et al.*[3] are used with Thunberg tubes replacing open ones. Xanthine is omitted to obtain blanks. These blanks are easily measureable in cows' milk but are often negligible in goats' milk. Blanks on boiled cows' milk show color because the enzyme is not completely destroyed by boiling. Large blanks may be encountered in internal organs,

[1] L. I. Hart, E. C. Owen, and R. Proudfoot, *Brit. J. Nutr.* **21**, 617 (1967).

[2] E. C. Owen and R. Proudfoot, *Brit. J. Nutr.* **22**, 331 (1968).

[3] C. A. Zittle, E. S. Dellamonica, J. H. Custer, and R. K. Rudd, *J. Dairy Sci.* **39**, 552 (1956).

such as liver, because of the well-known nonspecific reduction due to endogenous respiration. The enzyme, which in this type of assay is usually referred to as xanthine dehydrogenase, is measured by the amount of formazan it produces in a given time.

Reagents

Xanthine (Sigma Chemical Company, St. Louis, Missouri): 0.380 g of xanthine is dissolved in 50 ml of 0.05 N KOH. This solution is diluted daily to 0.005 N.

Triphenyltetrazolium chloride (TTC) (Sigma): 334.5 mg dissolved in 20 ml of 0.1 M phosphate buffer

Phosphate buffer, 0.5 M: 41.2 g of disodium hydrogen orthophosphate, $Na_2HPO_4 \cdot 12H_2O$ and 2.96 g of sodium dihydrogen orthophosphate, $NaH_2PO_4 \ 2H_2O$, dissolved in 250 ml of distilled water

Procedure. Into the tube of the Thunberg apparatus are placed 1 ml of 0.5 M buffer solution, 1 ml of distilled water, 0.2 ml of TTC solution, and 1 ml of sample. The sample is mixed by means of a piston homogenizer of household size. The homogenate is diluted 1:5 with distilled water; 1 ml of this diluted sample is added, and the contents are mixed. Into the lids are placed 0.2 ml of the dilute xanthine solution; or for the blanks, 0.2 ml of distilled water.

The tube is stoppered and evacuated by means of a water pump. It is then filled with nitrogen from a cylinder and again evacuated. Nitrogen is readmitted, and the contents of the tube and lid are mixed by shaking in the hand. The tube is then placed with others in a rack in a shaking incubator for half an hour at 37°. Five milliliters of acetic acid is added to each tube, followed by 5 ml of toluene. The tubes are then shaken vigorously to dissolve the red formazan in the toluene. The tubes are allowed to stand until the toluene layer separates. The red color is read in a spectrophotometer at 495 nm. The amount of enzyme present is proportional to the rate of production of the formazan.

By dissolving weighed amounts of triphenylformazan, prepared from TTC, in toluene, the $E^{1\,\%}_{1\,cm}$ at 495 mμ was found to be 511 (510–512); this is comparable with the E value calculable from the results of Zittle et al.[3]

Estimation of the chloride in the TTC by titration with silver nitrate indicated a purity of 99.7 (98.9–100.2). Experiments showed that on the same milk samples the conditions of Zittle *et al.* gave the same result when Thunberg tubes replaced the open tubes used by them.

Using phosphate buffer we found that the optimal rate of production of the formazan is at pH 7.4. Aerobically the optimal pH of milk xanthine oxidase is 9.0.[1] This change of pH optimum with change from aerobic to anaerobic conditions has also been noted by Muraoka *et al.*[4]

The milk samples were stored in a refrigerator for 1–2 days before analysis; they were then thawed and passed through a hand-operated homogenizer to improve the sampling of duplicates before being diluted 1:5 for testing.

[4] S. Muraoka, H. Enomoto, M. Sugiyama, and H. Yamasaki, *Biochim. Biophys. Acta* **143**, 408 (1967).

[169] Flavodoxin of *Clostridium pasteurianum*

By E. KNIGHT, JR. and R. W. F. HARDY

Flavodoxin (Fld) is an electron-transferring flavoprotein isolated from extracts of *Clostridium pasteurianum.*[1-3] It has been suggested that it replaces the electron-transferring protein, bacterial ferredoxin (Fd), in the iron-deficient cell.[2] Flavodoxin has also been isolated from extracts of *Peptostreptococcus elsdenii*[4] and *Desulfovibrio gigas.*[5,6] A flavoprotein with similar properties to flavodoxin and named phytoflavin has been obtained from *Anacystis nidulans.*[7-9]

Assay Methods

A phosphoroclastic[2] and a hydrogenase[1] assay for flavodoxin will be described. The method of choice depends primarily on the available equipment.

Phosphoroclastic Assay

Flavodoxin catalyzes the oxidation of pyruvate by the clastic system of *C. pasteurianum* with the subsequent formation of acetyl phosphate.

[1] E. Knight, Jr., A. J. D'Eustachio, and R. W. F. Hardy, *Biochim. Biophys. Acta* **113**, 626 (1966).
[2] E. Knight, Jr. and R. W. F. Hardy, *J. Biol. Chem.* **241**, 2752 (1966).
[3] E. Knight, Jr. and R. W. F. Hardy, *J. Biol. Chem.* **242**, 1370 (1967).
[4] S. G. Mayhew, *Federation Proc.* Abstr. No. 3228 (1968).
[5] J. LeGall and E. C. Hatchikian, *Compt. Rend. Acad. Sci. Paris* **264**, 2580 (1967).
[6] L. J. Guaraia, E. J. Laishley, N. Forget, and H. D. Peck, Jr., *Microbiol. Proc.*, Abstr. No. P129 (1968).
[7] R. M. Smillie, *Biochem. Biophys. Res. Commun.* **20**, 621 (1965).
[8] A. Trebst and H. Bothe, *Ber. Deut. Botan. Ges.* **79**, 44 (1966).
[9] Herman Bothe, Ph.D. dissertation, "Ferredoxin und Phytoflavin in Photosynthetischen einer Preparation aus der Blaualge *Anacystis nidulans*," Universität Göttingen, Germany, 1968.

Flavodoxin activity is related to the generation of acetyl phosphate as measured by the method of Lipmann and Tuttle.[10] A ferredoxin-free extract of *C. pasteurianum* is used as the source of clastic enzymes, phosphotransacetylase, and hydrogenase.[11]

$$CH_3\text{—}\overset{O}{\overset{\|}{C}}\text{—}\overset{O}{\overset{\|}{C}}\text{—OH} + CoASH \xrightarrow[\text{clastic enzymes} \atop \text{H}_2\text{ase}]{\text{Fld}} CH_3\text{—}\overset{O}{\overset{\|}{C}}\text{—SCoA} + H_2 + CO_2$$

$$CH_3\text{—}\overset{O}{\overset{\|}{C}}\text{—SCoA} + P_i \xrightarrow{\text{phosphotransacetylase}} CH_3\text{—}\overset{O}{\overset{\|}{C}}\text{—O—P} + CoASH$$

Reagents

Sodium phosphate buffer, 0.5 M, pH 6.5

Sodium pyruvate, 1 M

Coenzyme A, 0.0004 M

Hydroxylamine, 4 M. The hydroxylamine is neutralized with an equal volume of 4 M KOH immediately before use.

Ferredoxin-free extract of *C. pasteurianum*, 20 mg/ml. A cell-free extract of *C. pasteurianum*, grown in an iron-sufficient medium, is treated with DEAE-cellulose (1 g of damp DEAE/0.25 g of protein) in a batch process for 15 minutes at 5°, pH 7. The DEAE-cellulose is removed by filtration, and the filtrate is the ferredoxin-free extract.

Flavodoxin. Dilute to less than 40 units/ml with water.

FeCl₃ reagent. Mix equal volumes of 5% FeCl₃, 3.0 N HCl, and 12% TCA.

Procedure. To each reaction tube add 0.05 ml of phosphate buffer, 0.05 ml of pyruvate, 0.1 ml of coenzyme A, 1–10 units of flavodoxin, and 0.7 ml of the ferredoxin-free extract; total volume is 1.1 ml. Incubate 15 minutes at 30°. Add 1.0 ml of neutralized hydroxylamine and incubate 10 minutes more at 30°. The reaction is terminated by adding 3.0 ml of FeCl₃ reagent. Precipitate is removed by centrifugation, and the color of the supernatant is read at 540 nm in a Klett or Lumetron colorimeter. A standard curve may be obtained with synthetic acetyl phosphate. With the assay conditions described, 1 micromole of acetyl phosphate gives an optical density of 0.06 at 540 nm in a Lumetron colorimeter.

Definition of Unit and Specific Activity. One unit of flavodoxin is that amount which catalyzes the formation of 1 micromole of acetyl hydroxamate under the conditions of the assay. The assay is linear below 10 units per reaction tube. Specific activity is defined as units per milligram of protein.

[10] F. Lipmann and L. C. Tuttle, *J. Biol. Chem.* **159**, 21 (1945).

[11] L. E. Mortenson, R. C. Valentine, and J. E. Carnahan, *J. Biol. Chem.* **238**, 794 (1963).

Hydrogenase Assay

This assay measures the capacity of flavodoxin to catalyze the evolution of hydrogen gas from dithionite solutions in the presence of the enzyme hydrogenase. A ferredoxin-free extract of *C. pasteurianum* is used as the source of hydrogenase.

$$2H^+ + 2e \xrightarrow[H_2ase]{Fld} H_2$$

Reagents

Sodium phosphate buffer, 0.5 M, pH 6.5
Sodium dithionite, dry powder
Potassium hydroxide, 20%
Ferredoxin-free extract of *C. pasteurianum*, 20 mg/ml
Flavodoxin. Dilute with water to give a concentration that catalyzes the evolution of less than 15 microliters of H_2 per minute.

Procedure. Each Warburg flask contains in the center compartment 0.2 ml sodium phosphate buffer, 0.1 ml of ferredoxin-free extract, flavodoxin, and water to 3 ml. The side arm contains 7 mg of solid sodium dithionite, and the center well contains 0.2 ml of 20% KOH. The flasks are flushed for 15 minutes with argon, then equilibrated for 10 minutes at 30° with argon as the gaseous phase. The reaction is initiated by tipping in the dithionite. Manometer readings are made every 30 seconds and continued for 6–10 minutes. Activity determinations are usually made in the 3- to-5-minute region.

Definition of Unit and Specific Activity. One unit is defined as that amount of flavodoxin which catalyzes the evolution of 1 microliter of hydrogen gas per minute under the conditions of the assay. Specific activity is defined as units per milligram of protein. One unit of activity in the hydrogenase assay is equal to 1.5 units in the phosphoroclastic assay.

Growth of Cells

C. pasteurianum is grown on nitrogen gas according to the procedure of Carnahan *et al.*,[12] with one modification: The growth medium contains 0.44 μg of iron per milliliter (low iron medium) instead of 8.8 μg of iron per milliliter (normal iron medium). The final yield of cells grown in the low iron medium is 75% of the yield of cells grown in the normal iron medium. The "low iron" cells are much lighter in color than the "normal iron" cells. These cells contain flavodoxin and very little ferredoxin. The "low iron" cells may be easily converted to "normal iron" cells containing

[12] J. E. Carnahan, L. E. Mortenson, H. F. Mower, and J. E. Castle, *Biochim. Biophys. Acta* **44**, 520 (1960).

ferredoxin by addition to the medium of iron to 8.8 μg/ml. Cells are harvested in a Sharples centrifuge, and the wet cell paste is dried under vacuum at ca. 30° in a Rinco rotary evaporator. The dried cells may be stored at -20° for at least one year without loss in flavodoxin activity.

Purification Procedure

Autolysis of Cells. The dried *C. pasteurianum* cells are suspended in distilled water (50 g of cells in 500 ml of water) and stirred for 1 hour at room temperature. All operations hereafter are performed at 4°. The autolyzate is centrifuged for 30 minutes at 34,000 g and the precipitate is discarded. The light yellow supernatant constitutes the crude extract. Flavodoxin activity is determined using the phosphoroclastic assay.

Batch DEAE-Cellulose Chromatography. To the crude extract is added 17 g of DEAE-cellulose (damp) and the resulting mixture adjusted to pH 7 and stirred for 15 minutes. The DEAE-cellulose is removed by filtration, and the filtrate is discarded. The DEAE-cellulose is washed with 150 ml of 0.05 M Tris-HCl, pH 7.3, and the filtrate is discarded. Flavodoxin is eluted by stirring the DEAE-cellulose with 140 ml of 0.8 M NaCl in 0.05 M Tris-HCl for 15 minutes. The eluate is dialyzed for 16 hours against 6 liters of 0.02 M Tris-HCl, pH 7.3.

Column DEAE-Cellulose Chromatography. The batch DEAE-cellulose eluate (125 ml) is applied to a DEAE-cellulose column (2.5 × 24 cm) that has been previously equilibrated with 0.05 M Tris-HCl, pH 7.3. The column is washed with 200 ml of the same buffer, then eluted stepwise with 200 ml of 0.2 M NaCl, 0.3 M NaCl, and 0.4 M NaCl (all in 0.05 M Tris-HCl, pH 7.3), respectively. Flavodoxin is eluted in the 0.4 M NaCl fraction. The fraction is dialyzed against 6 liters of 0.02 M Tris-HCl, pH 7.3, for 16 hours, then concentrated by vacuum ultrafiltration to 10 ml.

As an alternative to the DEAE-cellulose column chromatography step, the batch eluate may be further purified by electrophoresis in an Elphor electrophoresis apparatus with a buffer of 0.033 M Tris, 0.008 M EDTA, 0.04 M boric acid, pH 8.6. After electrophoresis, the fractions containing flavodoxin are concentrated as above.

Bio-Gel P-60 Chromatography. The eluate from the DEAE-cellulose column, or the fraction containing the flavodoxin after electrophoresis, is passed through a Bio-Gel P-60 column (2.8 × 44 cm) with 0.02 M Tris-HCl, pH 7.3, as the buffer. Flavodoxin is eluted after about 210 ml of buffer has passed through the column. The most active fractions are pooled and dialyzed against 2 liters of 0.02 M Tris-HCl, pH 7.3, for 16 hours.

First Crystallization. The Bio-Gel column eluate is concentrated by ultrafiltration to 3–5 mg of protein per milliliter. The yellow flavodoxin solution is brought to pH 7 with 0.1 N HCl and solid ammonium sulfate

TABLE I
PURIFICATION OF FLAVODOXIN FROM *C. pasteurianum*

Fractionation step	Volume	Total protein	Total units[a]	Specific activity	Yield
Crude extract	282	5499	550	0.10	100
DEAE-cellulose batch, 0.8 NaCl eluate	125	275	440	1.6	80
DEAE-cellulose column, 0.4 M NaCl eluate	50	70	315	4.5	57
Bio-Gel P-60 column	30	6	102	17.0	20
First crystallization	1	3	66	22.0	12
Second crystallization	1	2.5	55	22.0	10

[a] Phosphoroclastic assay.

is slowly added to 66% saturation. At this point a white precipitate forms and is collected by centrifugation and discarded. Ammonium sulfate is added to the supernatant solution to 76% saturation, causing a yellow turbidity. The precipitate is removed by centrifugation and allowed to stand at 4° for 16–24 hours. After this period, yellow crystals of flavodoxin have formed. The crystals are collected by centrifugation, washed with a cold solution of saturated ammonium sulfate, and then dissolved in 1 ml of 0.02 M Tris-HCl, pH 7.

Second Crystallization. To the solution containing the dissolved crystals (first crystallization) solid ammonium sulfate is added to 76% saturation. The crystals are allowed to form over a period of 24–48 hours. The crystals are stored in 76% ammonium sulfate and retain full activity for at least 2 weeks.

A summary of the purification data for flavodoxin is presented in Table I.

The crystalline flavodoxin migrates as a single band on starch-gel electrophoresis (pH 8.6) and sediments as a single symmetrical component on ultracentrifugation in 0.05 M Tris-HCl, pH 7.3, containing 0.15 M NaCl.

Properties

Physical. Oxidized flavodoxin has a typical flavoprotein absorption spectrum with maxima at 272, 372, and 443 nm with a shoulder at 472 nm (Fig. 1). Formamidine sulfinic acid, sodium dithionite, and potassium borohydride reduce flavodoxin. On slow reduction of flavodoxin with formamidine sulfinic acid, a grayish blue semiquinone form may be observed. The nuclear magnetic resonance spectrum of oxidized flavodoxin has been obtained.[13] The spectrum of the oxidized form is not substantially

[13] C. C. McDonald and W. D. Phillips, unpublished results.

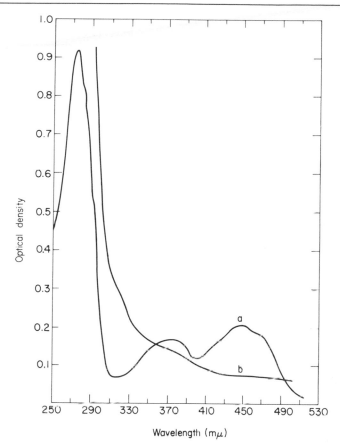

FIG. 1. Spectra of (a) oxidized flavodoxin and (b) flavodoxin reduced with form-amidine sulfinic acid (0.0006 M). These spectra were obtained in 0.02 M Tris-HCl, pH 7.3, at 0.25 mg of flavodoxin per milliliter.

changed on reduction of flavodoxin. Flavodoxin has a molecular weight of 14,600.

Chemical. The prosthetic group of flavodoxin is flavin mononucleotide (FMN). The FMN may be dissociated from the apoprotein by allowing flavodoxin to stand in 6 M guanidine-HCl, pH 7, for 24 hours. The FMN may also be dissociated by reacting flavodoxin with sodium mersalyl in a 1:80 molar ratio. Flavodoxin contains no iron, cobalt, or molybdenum. An E_0' of -0.28 V has been calculated for flavodoxin. The amino acid content of flavodoxin is presented in Table II.

Biological. Flavodoxin will replace ferredoxin in numerous biological reactions. It can replace ferredoxin in the following reactions: (1) the oxidation of pyruvate and the subsequent evolution of hydrogen gas by

TABLE II
AMINO ACID CONTENT OF FLAVODOXIN

Amino acid	Amount (g/100 g flavodoxin)	Residues (No./molecule flavodoxin)
Aspartic acid	15.9	18
Glutamic acid	15.7	19
Threonine[a]	2.5	4
Serine[a]	8.3	14
Proline[a]	2.5	4
Glycine	6.7	17
Alanine	6.8	14
Half-cystine[b]	1.0	1
Valine	10.3	15
Methionine	3.4	4
Isoleucine	3.7	5
Leucine	9.2	12
Histidine	ND[c]	—
Tryptophan[d]	4.6	4
Tyrosine	2.6	2
Phenylalanine	3.4	3
Lysine	8.5	10
Arginine	2.0	2

[a] Corrected for destruction during hydrolysis.
[b] Determined as cysteic acid after performic acid oxidation and hydrolysis.
[c] Not detected.
[d] Determined colorimetrically.

extracts of *C. pasteurianum;* (2) the fixation of nitrogen with pyruvate as the electron donor by nitrogen-fixing extracts of *C. pasteurianum;* (3) the reduction of TPN by extracts of *C. pasteurianum* using hydrogen gas as the electron donor; and (4) the reduction of TPN by illuminated chloroplasts.

The flavodoxin of *P. elsdenii* will replace ferredoxin in the phosphoro-clastic and hydrogenase reactions.[4] The flavodoxin of *D. gigas* functions as an electron-transferring protein in the reduction of sulfite,[5,6] and has been implicated in the reduction of thiosulfate,[6] and in the ATP-dependent reduction of sulfate.[6]

Section IX

Pteridines, Analogs, and Pterin Coenzymes

[170] Pteridine Content of Bacteria Including Photosynthetic Bacteria

By H. S. FORREST

Principle

The 2-amino-4-hydroxypteridinyl radical is resistant to oxidation by alkaline permanganate. Most naturally occurring pteridines contain this radical in more-or-less modified form. Compounds with substituents at the 6-position (e.g., biopterin, neopterin) yield the 6-carboxylic acid on oxidation; compounds in which the pyrazine ring of the pteridine moiety is reduced, are oxidized to the stable aromatic form; compounds with substituents on the pyrazine nitrogens probably lose these by oxidation and decarboxylation. Thus the procedure converts all pteridines in the cell or culture fluid into a few derivatives that are easily separable from each other and other fluorescent permanganate oxidation products; since most naturally occurring pteridines have carbon substituents at the 6-position, 2-amino-4-hydroxy-6-carboxylic acid is the major oxidation product. Synthetic material is readily available for comparison, quantitatively and qualitatively.

Method

Cultures (using 1 liter of culture fluid) at appropriate stages are harvested by centrifugation, and the cells are washed three times with distilled water. The combined supernatant solution, and/or the cells are again suspended in distilled water, made alkaline with KOH, and heated on a steam bath; a saturated solution of $KMnO_4$ is added, about 10 ml at a time. Oxidation is complete when the solution remains red for about 3 hours. The volume of $KMnO_4$ solution needed for complete oxidation may vary from 10 to 150 ml, and the time required for oxidation, from 2 to 20 hours. Excess $KMnO_4$ is destroyed by adding ethanol or methanol, and MnO_2 is removed by filtration and washed with ethanol–concentrated ammonia (2:1). The washings are evaporated to remove ethanol and ammonia, and the concentrated solution is combined with the original filtrate. The whole is brought to pH 3 with acetic acid, and the solution is treated with sufficient charcoal (Darco G60) to absorb all fluorescent materials. The charcoal is collected by centrifugation (at which time it is convenient to check the supernatant solution for fluorescence using a long wavelength UV light) and washed three times with 0.1 N acetic acid. If the amount of charcoal is small, it may be suspended in water and applied directly as a streak to Whatman

No. 3 MM filter paper. With large amounts, it is more convenient to elute the charcoal with ethanol–concentrated ammonia (2:1), concentrate the eluate, and apply it as a streak to the filter paper. The chromatogram is developed with n-propanol–1% ammonia (2:1); the resultant blue-fluorescent bands are cut out from the dried chromatogram, the fluorescent materials are eluted from them with 1% ammonia, and small samples of the eluates are rechromatogramed against authentic standards (2-amino-4-hydroxypteridine and 2-amino-4-hydroxy-6-carboxypteridine) in three solvent systems: n-propanol–1% ammonia (2:1); n-butanol–acetic acid–water (4:1:5), and 4% aqueous ammonia. The amounts of pteridines obtained are then estimated on the original eluate by fluorescence in a standard fluorometer [appropriate filters, for example, are (Corning numbers): primary, 5860; secondary, 3389].

In most cases, the amounts obtained are not sufficient to allow estimation by ultraviolet absorption (using the characteristic 360-nm peak of such compounds in 0.1 N sodium hydroxide); but when this is possible, the results are in good agreement with those obtained by fluorescence.

[171] Radioisotope Assay for Reduced Pteridines: the Phenylalanine Hydroxylase Cofactor

By GORDON GUROFF

Reduced pteridines can serve as cofactors for a number of oxygenase reactions.[1] In two cases it has been shown that the natural cofactor is indeed a reduced pteridine. In rat liver the cofactor for phenylalanine hydroxylation is reduced 2-amino-4-hydroxy-6-(L-erythro-1,2-dihydroxypropyl)pteridine (L-erythrobiopterin).[2] In *Pseudomonas* the same reaction is catalyzed by reduced 2-amino-4-hydroxy-6-(L-threotrihydroxypropyl)-pteridine (L-threobiopterin).[3] Because of the wide distribution of pteridines in nature, it is reasonable to assume that most of the reactions for which pteridines can serve as cofactors will be found to be pteridine-dependent in their native state.

In view of the increasing awareness of and interest in pteridine-requiring reactions, methods for the determination of pteridine levels in biological materials have become valuable. The following method takes advantage of the ability of reduced pteridines to serve as cofactors for the phenyl-

[1] S. Kaufman, *Ann. Rev. Biochem.* **36,** 171 (1967).
[2] S. Kaufman, *Proc. Natl. Acad. Sci. U.S.* **50,** 1085 (1963).
[3] G. Guroff and C. A. Rhoads, *J. Biol. Chem.* **244,** 142 (1969).

alanine hydroxylase enzyme of *Pseudomonas* sp. (11299a).[4-6] This method, previously published in some detail,[7] is based on an earlier method for phenylalanine hydroxylase itself[8] and utilizes the *Pseudomonas* enzyme as a reagent.

Assay Method

Basis of the Method. Phenylalanine hydroxylase [L-phenylalanine, tetrahydropteridine:oxygen oxidoreductase(hydroxylating), EC 1.14.3.1] catalyzes the hydroxylation of *p*-tritio-L-phenylalanine to *m*-tritio-L-tyrosine[9] as shown in the following diagram:

$$H_2C-\underset{\underset{H}{|}}{\overset{\overset{NH_2}{|}}{C}}-COOH \quad (p\text{-T-phenyl ring}) \longrightarrow H_2C-\underset{\underset{H}{|}}{\overset{\overset{NH_2}{|}}{C}}-COOH \quad (m\text{-T, OH phenyl ring})$$

The *m*-tritiotyrosine produced can be iodinated with concomitant release of tritium. The combination of hydroxylation and iodination yields tritium in the form of tritiated water in amounts stoichiometric with the hydroxylation.

Since the phenylalanine hydroxylase can be obtained as a stable preparation in substantial amounts, it can be used as a reagent. Then, by appropriate modification, the method can be used to measure small amounts of reduced pteridine, upon which the hydroxylation is absolutely dependent. The amount of tritium released can be related to the amount of pteridine present, since the enzyme and all the other components are in excess.

Preparation of the Reagents. The purification of phenylalanine hydroxylase from *Pseudomonas* has been presented in an earlier volume in this series.[10] The properties of this enzyme and the details of measuring its activity are also included in that discussion as well as in recent publications.[4,5,10] In addition to the hydroxylase, glucose dehydrogenase and dihydropteridine reductase are added. Glucose dehydrogenase from beef liver is purified through the second ammonium sulfate step[11] and added

[4] G. Guroff and T. Ito, *J. Biol. Chem.* **240**, 1175 (1965).

[5] G. Guroff and C. A. Rhoads, *J. Biol. Chem.* **242**, 3641 (1967).

[6] G. Guroff and C. A. Strenkoski, *J. Biol. Chem.* **241**, 2220 (1966).

[7] G. Guroff, C. A. Rhoads, and A. Abramowitz, *Anal. Biochem.* **21**, 273 (1967).

[8] G. Guroff and A. Abramowitz, *Anal. Biochem.* **19**, 548 (1967).

[9] G. Guroff, J. Daly, D. Jerina, J. Renson, B. Witkop, and S. Udenfriend, *Science* **157**, 1524 (1967).

[10] G. Guroff, see Vol. XVII [74].

[11] H. J. Strecker, see Vol. I [44].

to the incubation to regenerate NADPH. Dihydropteridine reductase is prepared from sheep liver and purified through the calcium phosphate gel procedure.[12] This enzyme is added to recycle the reduced pteridine as it is oxidized by the hydroxylase and effectively magnifies the response to small amounts of pteridine.

In addition to these enzymes, the assay also requires p-tritio-L-phenyl-alanine. Commercial material is diluted with carrier L-phenylalanine to a specific activity of 1 micromole per millicurie and then chromatographed on Whatman No. 3 MM paper with 2-propanol–NH$_3$–H$_2$O (80:10:10) as solvent. The phenylalanine is located as an ultraviolet absorbing spot, eluted from the paper, acidified, and placed on a 0.5-ml column of Dowex 50, H$^+$. The column is washed with 10 ml of water, and then the amino acid is eluted with 20 ml of 1 N HCl. The eluate is evaporated to dryness on a rotary evaporator. The residue is taken up in 1 ml of water. Finally 9 ml of alcohol is added, and the isotope is stored at $-20°$.

2-Amino-4-hydroxy-6,7-dimethyltetrahydropteridine from Aldrich Chemical Corporation is used as a standard. Biopterin is reduced with hydrogen gas in the presence of PtO$_2$ and 0.1 N HCl.[7] (The biopterin used in our studies was a gift from Drs. Joseph Weinstock and Alfred Maas of Smith, Kline, and French Laboratories.) N-Iodosuccinimide is recrystallized from hot dioxane by dropwise addition of chloroform.

Assay Procedure. Samples to be assayed are homogenized in 2 volumes of cold 0.01 M mercaptoethanol and centrifuged at 15,000 g for 10 minutes in a refrigerated centrifuge. Three-milliliter portions of the supernatant fractions are placed in chilled glass centrifuge tubes, and nitrogen is bubbled through the samples for 2 minutes. During the nitrogen treatment, the samples are held at 0°. Then they are stoppered, heated at 60° for 3 minutes, and centrifuged at 15,000 g for 20 minutes at 0°. The supernatant portions are used for assay.

For each assay, 20 μl of purified *Pseudomonas* phenylalanine hydroxylase (approximately 250 μg of protein) is incubated with 20 μl of 12.5 mM ferrous ammonium sulfate for 10 minutes at 30°. The tubes containing this activated enzyme are cooled, and Tris buffer, pH 7.3, 30 micromoles (1.5 M, 20 μl); L-phenylalanine, 1 micromole (0.1 M, 10 μl); p-tritio-L-phenyl-alanine, 200,000 cpm (10 μl); NADPH, 1 micromole (0.1 M, 10 μl); glucose, 1 micromole (0.1 M, 10 μl); glucose dehydrogenase, 1.2 mg of protein; dihydropteridine reductase, 35 μg of protein; and sample and water are added in a final volume of 0.25 ml. When the sample is omitted, an equivalent amount of 0.01 M mercaptoethanol is added. When standard 2-amino-4-hydroxy-6,7-dimethyltetrahydropteridine is added in place of the sample, it is made up in 0.1 M mercaptoethanol at a concentration of 1 mg/ml.

[12] S. Kaufman, see Vol. V [109].

The tubes are incubated at 30° for 60 minutes in air and then heated at 100° for 1 minute. After the heat step, the tubes are cooled on ice. Acetate buffer, 0.5 ml, pH 5.5, 0.2 M, and N-iodosuccinimide, 0.2 ml of a freshly prepared 1% solution, are added, and the samples are kept at 0° for 5 minutes. Trichloroacetic acid, 0.05 ml of a 30% solution, is added, and the tube contents are applied to small columns containing 1 ml of Dowex 50, H$^+$, with 10 mg of charcoal layered on top. These columns are contained in cotton-plugged Pasteur pipettes. After the sample has run through, the columns are washed with 1-ml portions of water. The eluate and wash are collected in counting vials, 10 ml of Bray's solution[13] is added, and the samples are counted.

Sample Data. Data on the dependence of the method on the several components have been presented previously, as have experiments relating to the recovery of various reduced pteridines when carried through the homogenization and heating.[7] The recovery experiments showed that both the model cofactor, 2-amino-4-hydroxy-6,7-dimethyltetrahydropteridine, and the natural cofactor from rat liver,[2] tetrahydrobiopterin, would be completely recovered by this procedure.

Studies on the amounts of cofactor in various tissues are presented in the table along with values for the model pteridine, 2-amino-4-hydroxy-6,7-dimethyltetrahydropteridine, and for tetrahydrobiopterin, the natural cofactor from rat liver. Based on the response to standard amounts of tetrahydrobiopterin, values for cofactor levels in various tissues can be calculated.[7] Those values range from 13 μg per gram for rat liver to 0.6 μg per gram for guinea pig brain. By way of comparison, a value of 8.1 μg per gram of rat liver was found using an isotope dilution method.[14] For routine work 2-amino-4-hydroxy-6,7-dimethyltetrahydropteridine seems to be the most convenient standard because of its greater stability and its availability from commercial sources.

Comparison with Other Methods

Tetrahydrobiopterin has been measured by isotope dilution.[14] This method is, of course, quite specific but will not detect other pteridine cofactors if they are present. Also, the isotope dilution method is considerably more time consuming than the present method.

Cofactor assays have been done using rat liver phenylalanine hydroxylase and a colorimetric method for tyrosine determination.[15] The liver enzyme has some advantage over the bacterial enzyme because it is readily available in most laboratories. On the other hand, the bacterial enzyme is somewhat more stable than the rat liver enzyme. Also, the

[13] G. A. Bray, *Anal. Biochem.* **1**, 279 (1960).
[14] H. Rembold and H. Metzger, *Z. Naturforsch.* **22b**, 827 (1967).
[15] S. Kaufman, *J. Biol. Chem.* **230**, 931 (1958).

COFACTOR LEVELS IN SEVERAL ANIMAL TISSUES[a]

Source of cofactor	Amount	Counts per minute in expt. no.			
		1	2	3	4
None	—	170	201	189	157
Rat liver	0.02 ml	1214	2163	2373	2492
	0.04	1842	3449	4184	3779
Rat brain	0.04 ml	310	367	447	371
	0.08	473	486	683	553
Rat kidney	0.04 ml	472	778	575	587
	0.08	656	1171	930	995
Guinea pig liver	0.04 ml	1560	2260	2057	2666
	0.08	3016	3923	3950	5039
Guinea pig brain	0.04 ml	277	324	333	339
	0.08	380	432	466	429
Guinea pig kidney	0.04 ml	389	628	609	493
	0.08	722	991	1000	720
Tetrahydrobiopterin	0.1 μg	2247	2125	2642	2669
	0.2	3830	4667	4914	4413
2-Amino-4-hydroxy-6,7-dimethyltetra-hydropteridine	0.2 μg	3596	—	—	—

[a] Tissue extracts were prepared by homogenizing the tissue in two volumes of 0.01 M mercaptoethanol and then proceeding as described in the text. Each experiment represents assays done on different days with tissues from different animals.

bacterial enzyme is much easier to obtain in a cofactor-free form. Both enzymes are specific for reduced pteridines but will measure cofactor only, and the assay tells nothing about the overall level of total pteridines in the sample. None of the other pteridine-requiring enzymes in the literature are currently available at an appropriate stage for such an assay.

Use of the isotope method presents some distinct advantages over the colorimetric or fluorimetric measurement of tyrosine formation. First, endogenous tyrosine does not interfere. Second, tyrosine produced during the incubation by other routes such as by proteolysis will not be measured. Third, further metabolism of tyrosine will not affect the results. These advantages make the isotope method eminently applicable to crude tissue extracts.

Conclusion

Appropriate modifications of the isotope assay for phenylalanine hydroxylase lead to a method for the measurement of phenylalanine hydroxylase cofactors in tissue extracts. *Pseudomonas* phenylalanine hydroxylase can be used conveniently as a reagent in this method. Recovery studies indicate that the method will quantitatively measure tetrahydrobiopterin, the hydroxylase cofactor of rat liver. The use of the isotope measurement has clear advantages over other methods of assay, especially for the estimation of cofactor levels in crude tissues.

[172] Enzymatic Determination of Folate Compounds

By L. JAENICKE

Tetrahydrofolic acid and its one-carbon derivatives are substrates or products of numerous enzymatic reactions. Therefore, exact, quick, and not too cumbersome methods for their determination are desired. For this purpose enzymatic procedures are useful, if the preparation and the properties of the enzymes required allow their use.

In the following are compiled some procedures for the enzymatic analysis of folate compounds which have proved useful. They are more specific than the chemical and spectrophotometric methods, and quicker than the microbiological assays. However, none of them attains the sensitivity of the latter, which, therefore, seem to be the only useful methods so far for determining folate derivatives in biological material.

Determination of Tetrahydrofolic Acid and 5,10-Methylenetetrahydrofolic Acid by Methylenetetrahydrofolate Dehydrogenase (5,10-Methylenetetrahydrofolate:NADP Oxidoreductase, EC 1.5.1.5)

The reaction catalyzed by methylenetetrahydrofolate dehydrogenase follows Eq. (1)

$$5,10\text{-Methylenetetrahydrofolate} + TPN^+ \rightleftharpoons$$
$$5,10\text{-methenyltetrahydrofolate} + TPNH \quad (1)$$

However, at the pH of the incubations and by action of the enzyme cyclohydrolase, which is contained in the impure dehydrogenase preparations used, the end product is 10-formyltetrahydrofolate instead of the anhydroformyl compound. Thus, methylenetetrahydrofolate dehydrogenase may be used for the determination of tetrahydrofolic acid and of 5,10-methylenetetrahydrofolic acid formed by condensation of the former with formaldehyde (enzymatically from serine). The reaction can be followed photometrically either directly by measuring TPNH formation or indirectly, after acidification, by measuring 5,10-methenyltetrahydrofolic acid.

Direct Assay

Solutions

Potassium phosphate buffer, 0.05 M, containing 0.001 M mercapto-
ethanol, pH 7.4
TPN⁺, 0.1 M, (Boehringer-Mannheim) in water
Formaldehyde, 0.01 M
Perchloric acid, 0.5 M (ca. 5% v/v)

Preparation of Methylenetetrahydrofolate Dehydrogenase. (a) FROM
YEAST.[1] Bakers' yeast is lysed by twice freezing in liquid air and thawing
overnight. The paste is mixed with an equal volume of 0.01 M potassium
phosphate buffer pH 7.3, containing 0.001 M ethylenediaminetetraacetate
and centrifuged for 60 minutes at 24,000 g. The supernatant is cooled to
0° in a dry-ice methyl cellosolve bath, stirred, and mixed with precooled
acetone to 30% v/v, the temperature being simultaneously lowered to
−5°. The precipitate is centrifuged off at −10°, and the supernatant
solution is treated at −5° with a further 10% acetone. The protein is
collected by centrifugation at −10° and redissolved in a small volume of
the above buffer. To this solution is added saturated ammonium sulfate
solution (pH 7.2) to 0.45 saturation. After centrifugation, the supernatant
is fractionated in the same way to 0.60 saturation and the protein is separated and redissolved in a small amount of the phosphate buffer. A TEAE-
cellulose column is prepared from a weight of the exchanger approximately
10 times that of the protein. The extract is poured onto the top of the
column, which is then washed with 0.01 M potassium phosphate buffer
containing 0.001 M ethylenediaminetetraacetate, pH 7.3, until the optical
density falls below 0.025. After washing with a further 100 ml of this buffer,
the enzyme is eluted with 0.05 M potassium phosphate containing 0.001 M
ethylenediaminetetraacetate, pH 7.45. The enzyme, which forms a yellow
band, is purified about 100- to 120-fold. Enzyme preparations from which
the TEAE-cellulose chromatography step has been omitted may also be
used for the determinations.

(b) FROM *Clostridium cylindrosporum.*[2] A suspension of 1 g of lyophilized
cells in 20 ml of 0.05 M potassium maleate containing 0.02 M mercaptoethanol, pH 7.0, is stirred for 30 minutes at 37° and then centrifuged for
15 minutes at 144,000 g. Then 2.5 ml of 2% protamine sulfate is added to
the supernatant, and the mixture is stirred for 5 minutes at 25° and centrifuged for 10 minutes at 30,000 g. The supernatant solution is brought to

[1] B. V. Ramasastri and R. L. Blakley, *J. Biol. Chem.* **239**, 106 (1964).
[2] K. Uyeda and J. C. Rabinowitz, *J. Biol. Chem.* **242**, 4378 (1967).

pH 6.5 and then to 0.45 saturation with respect to solid ammonium sulfate. After 5 minutes of centrifugation at 30,000 g, the supernatant solution is adjusted to pH 7 and heated with stirring for 15 minutes at 55°. Denatured protein is centrifuged off, and the supernatant is mixed with 0.02 ml of 1 M MgCl$_2$ and then brought to 0.60 saturation with ammonium sulfate. After standing for 75 minutes at 2°, the precipitated protein, containing the *tetrahydrofolate formylase* (see below) is centrifuged off. To the supernatant is added ammonium sulfate to 0.68 saturation at pH 7.0. After 1 hour of stirring at 20°, the protein is centrifuged off, redissolved in 1.5 ml of potassium maleate–mercaptoethanol buffer, pH 7.0, and dialyzed against 125 ml of potassium maleate buffer, pH 7.0. One enzyme unit forms 4.23 millimicromoles of methenyltetrahydrofolate per minute.

Procedure. For the determination of tetrahydrofolic acid, half-micro quartz cuvettes of 1-cm light path are used. The assay mixture is composed of: 0.40 ml of potassium phosphate–mercaptoethanol buffer, pH 7.4 (20 micromoles); 0.05 ml of TPN$^+$ (0.5 micromoles); 0.05 ml of formaldehyde (0.5 micromoles), and up to 0.50 ml of sample (containing 0.05–0.3 micromoles of tetrahydrofolate in clear solution; extraction of tissue, see below). The reaction is started with 0.01 ml of enzyme solution (2 units). A blank determination without TPN$^+$ is made concurrently.

The increase in optical density at 340 nm is measured spectrophotometrically at 25° (thermostated cuvette holder). $\epsilon_{340 \text{ nm}} = 7100$.

Calculation

Micromoles (methylene) tetrahydrofolate/ml =
$$\Delta E/7.1 \times \text{volume of sample}$$

In the same way, methylene tetrahydrofolate is determined, replacing formaldehyde solution by water.

Indirect Assay

With turbid solutions or solutions containing TPN$^+$- or TPNH-consuming reactions, the direct optical test is unsuitable. Instead, the more sensitive (0.005 to 0.2 micromole of tetrahydrofolate per sample) indirect assay must be used; in this formyltetrahydrofolic acid is converted to its anhydroformyl derivative. After being incubated in test tubes for 30 minutes at 37°, the mixtures are acidified with an equal volume (1.0 ml) of 5% perchloric acid, held 20 minutes at room temperature, and then centrifuged. The supernatant solution is measured in 1-cm glass cells at 350 nm in a spectrophotometer or at 366 nm in a filter photometer: $\epsilon_{350 \text{ nm}} = 25,100$; $\epsilon_{366 \text{ nm}} = 22,600$.

Calculation

Micromoles (methylene) tetrahydrofolate/ml =

$$\Delta E_{350}/12.5 \times \text{volume of sample}$$
$$\text{resp.} = \Delta E_{366}/11.3 \times$$
$$\text{volume of sample}$$

The sensitivity of this assay is 0.005 micromoles folate compound per milliliter.

Samples from Biological Material

Biological samples are prepared as follows: The tissue is homogenized in 0.05 M potassium phosphate buffer, pH 7.3, containing 0.05 M ascorbate, heated for 5 minutes in a boiling water bath, and centrifuged. Although as little as 0.25 micromole of folate per gram of tissue (ca. 100 $\mu g/g$) may be assayed by this method, this is still far in excess of amounts usually found in biological materials.

Tetrahydrofolate Determination with Tetrahydrofolate Formylase

Tetrahydrofolate formylase formylates tetrahydrofolate in an ATP-requiring reaction to 10-formyltetrahydrofolic acid according to Eq. (2)

$$\text{Tetrahydrofolate} + \text{HCOO}^- + \text{ATP} \overset{\text{Mg}^{2+}}{\rightleftharpoons} \text{10-formyltetrahydrofolate} + \text{ADP} + \text{P}_i \quad (2)$$

The product, 10-formyltetrahydrofolic acid is cyclized to 5,10-methenyl-tetrahydrofolic acid by addition of acid. Its characteristic absorption at 350 nm (resp. 366 nm) serves for the spectrophotometric analysis. The enzyme is widespread although usually found only in small quantities. An excellent source was found by J. C. Rabinowitz[3] in *Clostridium cylindrosporum*, where it constitutes about 10% of the total extractable protein. Enzyme preparations sufficiently pure for the assay are obtained during the above-mentioned purification of methylenetetrahydrofolate dehydrogenase (see pp. 606, 607). Like all the enzymatic assays, the method here described is useful for the determination of folate concentrations > $5 \times 10^6\ M$; these, however, are rarely reached in biological materials.

Reagents

Sodium formate, 0.1 M
ATP-Na$_4$, 0.05 M (Boehringer-Mannheim), pH 7.0 (stable at $-20°$)
MgCl$_2$, 0.1 M, containing 0.1 M KCl
Triethanolamine, 1 M, containing 0.05 M mercaptoethanol buffer, pH 8.0

---

[3] J .C. Rabinowitz and W. E. Pricer, Jr., *J. Biol. Chem.* **237**, 2898 (1962).

Perchloric acid, 2%

Tetrahydrofolate formylase from *C. cylindrosporum* (see pp. 606, 607), 10–15 mg protein/ml (stable at 0° for 4–6 weeks)

For quicker pipetting, equal volumes of the first four solutions may be premixed (FAST solution).

Procedure. Pipette into small test tubes (10 × 100 mm): 0.2 ml FAST solution and a 0.25-ml sample (containing 0.005–0.2 micromole of tetrahydrofolate; preparation from tissue see pp. 606, 607). The reaction is started with 0.05 ml of enzyme. The incubation mixtures may be layered with petroleum ether (b.p. 40°–60°). After 30 minutes incubation at 37°, 0.5 ml of perchloric acid is added. The mixture is kept for 15 minutes at room temperature, then centrifuged. The clear supernatant solution is measured photometrically in 1-cm half-microglass cuvettes at 366 nm.

Calculation

$$\text{Micromoles of tetrahydrofolate/ml} = \Delta E_{366}/22 \times \text{volume of sample}$$
$$\text{resp.} = \Delta E_{350}/25 \times \text{volume of sample}$$

Determination of Dihydrofolic Acid

Dihydrofolic acid is reduced to tetrahydrofolic acid by the enzyme dihydrofolate reductase, which requires as reducing cofactor either DPNH or TPNH, according to its specificity. The reaction is represented by Eq. (3):

$$\text{Dihydrofolate} + \text{TPNH (DPNH)} + \text{H}^+ \rightleftharpoons \text{tetrahydrofolate} + \text{TPN}^+ (\text{DPN}^+) \quad (3)$$

The amount of tetrahydrofolic acid formed can be measured directly by the change in extinction at 280 nm and 300 nm, resp., or indirectly by the decrease in extinction at 340 (366) nm. The latter method cannot be used with impure enzymes containing other TPNH-oxidases. The most favorable wavelength for the direct optical test is 300 nm, the isosbestic point of TPN$^+$ and TPNH, so that the extinction difference in the reaction mixture is due merely to the transformation of dihydrofolic acid to tetrahydrofolic acid: $\Delta \epsilon_{300 \text{ nm}} = -5000$.

Reagents

Potassium phosphate, 0.1 M, containing 0.003 M mercaptoethanol buffer, pH 7.5

TPNH, 0.02 M (Boehringer-Mannheim) in 1% KHCO$_3$ (stable at $-20°$)

Preparation of Dihydrofolate Reductase. (a) PROCEDURE OF SCRIMGEOUR

AND HUENNEKENS.[4] Chicken liver acetone powder (20 g) is stirred mechanically with 120 ml of 0.05 M potassium phosphate buffer, pH 7.5, for 60 minutes in the cold. The suspension is centrifuged, and the extract is mixed with 0.3 volume of 2% protamine sulfate. After 5 minutes of stirring, the turbid fluid is centrifuged. The supernatant solution is readjusted to pH 7.5. The protein content of this extract is adjusted to 20 mg/ml, and the solution is cooled down close to its freezing point. Per each 10 ml of extract 3.2 ml of ethanol, precooled to $-20°$, is added, followed by 1.3 ml of chloroform. The mixture is shaken vigorously for 1 minute, then centrifuged. The aqueous layer, separated from the sediment, is dialyzed for 4 hours against distilled water and fractionated with solid ammonium sulfate between 0.60 and 0.80 saturation. The protein is dissolved in 20 ml of potassium phosphate buffer, pH 6.5, dialyzed 3–4 hours against the same buffer and adsorbed on a hydroxylapatite column (3 × 16 cm), equilibrated with 0.05 M potassium phosphate buffer, pH 6.5. The protein is eluted with a phosphate gradient (mixing chamber: 500 ml of 0.05 M potassium phosphate buffer, pH 6.5; reservoir: 500 ml of 0.3 M potassium phosphate buffer, pH 6.5, under slight nitrogen pressure. The active fractions are pooled. The purification is about 500-fold, the yield approximately 40%. The enzyme may be kept at $-20°$.

(b) PROCEDURE WITH BAKERS' YEAST. A useful enzyme preparation can also be obtained from bakers' yeast. The yeast is suspended in 0.5 volume of 0.1 M phosphate buffer, pH 7.5, and shaken with glass beads (Ballottini No. 8½) for 10 minutes in a Vibrogen cell mill (Bühler & Co., Tübingen). The homogenate is stirred for 30 minutes at 2°, filtered to remove the glass beads, and centrifuged for 1 hour at 40,000 g. The supernatant solution is brought, with stirring, to pH 4.5 with 2 M acetate buffer, pH 4.0, and centrifuged for 10 minutes at 27,000 g; the supernatant solution is immediately readjusted to pH 7.5. The protein which is precipitated between 0.4 and 0.9 ammonium sulfate saturation is centrifuged and redissolved in 0.1 M phosphate buffer, pH 7.5, to a concentration of ca. 15 mg/ml. After dialysis overnight against 0.05 M phosphate buffer, pH 7.5, containing 0.001 M mercaptoethanol and 0.001 M ethylendiaminetetraacetate, the extract is applied to a hydroxylapatite column (1 ml/8 mg protein), which is then washed with 0.001 M sodium phosphate buffer, pH 7.0. The active protein is eluted by a phosphate gradient at pH 7.0.

Procedure. Pipette into 1-cm half-micro quartz cuvettes 0.2 ml of phosphate buffer (20 micromoles), 0.5-ml sample (containing > 0.05 micromole of dihydrofolic acid; extraction from tissue see p. 608), 0.005 ml TPNH

[4] K. G. Scrimgeour and F. M. Huennekens, *in* Hoppe-Seyler-Thierfelder, "Handbuch der physiologisch und pathologisch-chemischen Analyse," 10th ed., Vol. VIB, p. 198. Springer, Berlin, 1966.

(0.1 micromole); place in the spectrophotometer, which is set at 300 nm, and adjust extinction to 0.500. Start reaction with 0.10 ml of enzyme. Let the reaction go to completion, and determine the extinction difference. Subtract the blank value obtained by omitting the sample.

Calculation

Micromoles of dihydrofolic acid/ml $= \Delta E / -6200 \times$ volume of sample

Enzymatic Determination of 5-Methyltetrahydrofolic Acid

5-Methyltetrahydrofolic acid is the folate-activated one-carbon intermediate in methionine biosynthesis according to Eq. (4).

$$\text{5-Methyltetrahydrofolate} + \text{homocysteine} \xrightarrow[\text{system}]{\text{reducing}} \text{methionine} + \text{tetrahydrofolate} \quad (4)$$

5-Methyltetrahydrofolate is formed in enzymatic reactions from the formylated tetrahydrofolates or from 5,10-methylenetetrahydrofolic acid, or these compounds may be reduced chemically by boranates to 5-methyltetrahydrofolic acid. It is isolated and purified by TEAE-cellulose chromatography. The enzymatic reactions yield directly the biologically active diastereomers, whereas the chemical condensation of formaldehyde with *dl*-tetrahydrofolic acid gives the racemic mixture of *d*- and *l*-5,10-methylenetetrahydrofolic acid which may be separated into the two diastereomers by TEAE-cellulose exchange chromatography. The more acidic form is the (+)-diastereomer, which subsequently, is reduced to (−)-5-methyltetrahydrofolic acid, the active intermediate in methionine formation. For the enzymatic assay of the biologically active (−)-stereomer of 5-methyltetrahydrofolic acid, the methionine synthetase from *Escherichia coli* B is suitable. This reaction is practically irreversible ($K_{eq} = 1.4 \times 10^7$, see footnote 5), and tetrahydrofolic acid is liberated from the 5-methyl derivative in the complete reaction according to Eq. (4). Tetrahydrofolic acid is determined in the range of 0.1–2 micromoles/ml by chemical formylation to 10-formyltetrahydrofolic acid,[6] which cyclizes spontaneously in the acidic milieu of the assay mixture to anhydroformyltetrahydrofolic acid (5,10-methenyltetrahydrofolic acid). The latter can be measured photometrically by its characteristic and high absorption at 350 nm or 366 nm, respectively. The molar extinction coefficient is 25,000 at 350 nm and 22,600 at 366 nm. In spite of the somewhat lower sensitivity, measuring in simple filter-photometers at 366 nm is advantageous. An enzyme preparation sufficient for the enzymatic assay is obtained by initial gross fractionation of *E. coli* extracts.

[5] H. Rüdiger and L. Jaenicke, *FEBS Letters* **4**, 316 (1969).
[6] S. Rosenthal, L. C. Smith, and J. M. Buchanan, *J. Biol. Chem.* **240**, 836 (1965).

Reagents

    dl-Homocysteine (Fluka AG, Buchs SG), 0.05 M, in 0.25 M phosphate buffer, pH 7.2, freshly prepared

    S-Adenosylmethionine-Cl (BDH, Dorset), 0.01 M, in water; stable if kept frozen

    FMN (E. Merck AG, Darmstadt), resp. FAD (Sigma Chemicals, St. Louis), 0.001 M, in water; stable if kept frozen

    NADH, (Boehringer-Mannheim), 0.028 M, in 0.5 M phosphate buffer pH 7.2, freshly prepared daily, resp.: Aquocobalamin (E. Merck AG, Darmstadt), 0.1 mM in water; stable if kept frozen

    Dithiothreitol (Calbiochem, Lucerne), 0.05 M in water; stable if kept frozen

Methionine Synthetase.[7] *E. coli* B (phage resistant, grown in 10^{-8} g/ml vitamin B_{12}-containing medium) is disintegrated with twice its weight of alumina (Alcoa 303 or FPTO, Martinswerk, Bergheim/Erft), taken up in 10 volumes 0.1 M Tris-HCl buffer, pH 8.0, containing 0.01 M mercaptoethanol, and the suspension is stirred for 30 minutes. After centrifugation, 0.05 part of 1.0 M MnCl$_2$ is added slowly with stirring to the crude extract, and the mixture is added with stirring to 0.1 part 4 M K$_2$HPO$_4$. The suspension is centrifuged and the supernatant solution is fractionated at 2° with finely powdered ammonium sulfate between 0.3 and 0.6 saturation. The protein is separated by centrifugation, redissolved to a concentration of about 30 mg/ml in 0.1 M Tris-mercaptoethanol buffer, pH 8.0, and dialyzed against the same buffer. This enzyme preparation contains sufficient amounts of NADH/FAD oxidoreductase, required for the activation of the synthetase. It may be replaced by platinum/hydrogen, by FADH$_2$ or FMNH$_2$, or, most conveniently, by a reduced flavins generating system, e.g., FMN$^+$ aquocobalamin/dithiothreitol. Particularly, in the former cases, strictly anaerobic conditions are necessary.

    An enzyme preparation that is greatly independent upon S-adenosylmethionine may be prepared as follows[8]: 500 g frozen *E. coli* B cells are broken into small lumps, thawed in 4 liters Tris-buffer (0.05 M Tris-HCl, pH 8.0, containing 0.01 M mercaptoethanol and 0.02% Na-azide), and immediately passed through a Manton-Gaulin Laboratory Homogenizer, model 15M-8 TA (Manton-Gaulin Manufacturing Co., Inc., Everett, Mass.) at 500 atm. The temperature is kept below 15°. Into the suspension of disintegrated cells is stirred 250 g DEAE cellulose (Serva Entwicklungslabor, Heidelberg), preequilibrated with the same buffer. The mixture is made 0.1 M in NaCl and stirred for one hour. Then, the slurry is filtered

[7] J. Stavrianopoulos and L. Jaenicke, *European J. Biochemistry* **3**, 95 (1967).
[8] H. Rüdiger and L. Jaenicke, *European J. Biochemistry* **10**, 557 (1969).

by suction through a 24-cm Büchner funnel and the filter cake washed twice with 0.05 M Tris-buffer, pH 8.0/0.1 M NaCl. The ion exchanger is resuspended in the same buffer, poured into a column, 10 cm wide, and allowed to settle. The height of the column is about 14 cm. The column is developed by a NaCl-gradient in 0.05 M Tris-buffer, pH 8.0. Methionine synthetase is eluted between 0.15 and 0.19 M NaCl, estimated conductometrically. This procedure removes practically all nucleic acids, enriches the enzyme 8- to 18-fold, and concentrates it in one fifth of the original volume. The enzyme may be fractionated further by repeated DEAE, Sephadex, and hydroxyl-apatite chromatographies, according to Rüdiger and Jaenicke,[8] but is essentially sufficient for the given purpose at this stage of purification.

Procedure.[9] Calibrated test tubes (10 × 100 mm) are used which bear a mark so that for each measurement they can be aligned in the same way to the light path of the photometer. The photometer is fitted with a holder for the test tubes and a hole, bored through the lid of the instrument, for easy changing of the test tubes.

The incubation mixture (total volume 0.25 ml) consists of: 0.025 ml of sample (containing 0.005–0.07 micromole of 5-methyl tetrahydrofolate), 0.050 ml of *dl*-homocysteine (2.5 micromoles in sodium phosphate buffer, pH 7.2; 12.5 micromoles), 0.025 ml of S-adenosylmethionine (0.25 micromole), 0.025 ml of FAD resp. FMN (0.025 micromole), 0.025 ml of NADH (0.7 micromole) resp. 0.025 ml of aquocobalamin (0.25 millimicromole), and 0.050 ml dithiothreitol (2.5 micromoles), and 3 mg of methionine synthetase in 0.100 ml. The solution is incubated at 30° under nitrogen in the dark. After 2 hours of incubation 1.0 ml 98% formic acid containing 1 M mercaptoethanol is added, and the mixture is heated for 5 minutes in a boiling water bath. After cooling, the 5,10-methenyltetrahydrofolic acid formed is measured at 366 nm against a blank that contains no homocysteine or that has been kept at 0°.

Calculation

Micromoles 5-methyltetrahydrofolic acid/ml =
$$\Delta E_{366} \times 1.25/22.6 \times \text{volume of sample} \times \text{light path (cm)}$$

This procedure is less sensitive than the microbiological determination with *Leuconostoc mesenteroides* for the methionine formed (see below). However, the values are in stoichiometric agreement, and the method is much quicker. The same method, incidentally, may be used for the determination of free tetrahydrofolic acid or of that liberated in other tetrahydrofolate-dependent transfer reactions.

Microbiological Procedure. For the assay of smaller concentrations

[9] H. Rüdiger and L. Jaenicke, *European J. Biochemistry* **10**, 557 (1969).

(ca. 10^{-2} micromole/ml) of 5-methyl tetrahydrofolate, the microbiological procedure has to be employed. One-tenth milliliter of the sample solution is mixed with 1.0 ml of Difco methionine assay medium in small test tubes (10 × 100 mm), which are plugged, steam sterilized for 45 minutes, and, after being cooled, inoculated with 0.025 ml of a fresh 12-hour culture of *Leuconostoc mesenteroides* P-60 (ATCC 8042). After 16 hours of incubation at 37°, the turbidity is measured nephelometrically, using a 546 nm filter. The values are read from a standard curve with 0 to 0.10 micromole/ml L-methionine.

Discussion and Conclusions

The procedures described here allow the analysis of some folate compounds in concentrations down to ca. 10^{-7} M, as may occur or be produced in biochemical reactions. Other procedures for the determination of formylated tetrahydrofolates may be developed, based on known folate-dependent enzymatic reactions. However, they will have little practical value, since these compounds are much more easily measured by their pH-dependent cyclization or decyclization reactions, provided the samples do not contain interfering substances. This, however, is also true, at least to some extent, for the enzymatic methods.

For the determination of 10-formyltetrahydrofolic acid, it would be possible to utilize the TPN-dependent 10-formyltetrahydrofolate dehydrogenase reaction, recently found by Kutzbach.[10] However, this method, as well as coupled reactions and regenerating or relay reactions in which folate may be used catalytically, require highly purified enzymes in relatively great quantities. Thus, they would be only of academic interest. Obviously the highest amplifying effects in this sense are obtained with the known microbiological assay procedures, which reach down into the range of 10^{-12} M. In spite of their known weaknesses, they operate to a certain extent selectively, since microorganisms having different growth requirements for folate compounds are at hand.

[10] C. Kutzbach and E. L. R. Stokstad, this volume [199].

[173] Enzymatic Assay for 7,8-Dihydrobiopterin

By SEYMOUR KAUFMAN

Principle. 7,8-Dihydrobiopterin (7,8-dihydro-2-amino-4-hydroxy-6-[1,2-dihydroxypropyl-(L-*erythro*)]-pteridine) is the form of the naturally occurring hydroxylation cofactor that has been isolated from rat liver extracts.[1] It functions in the phenylalanine hydroxylating system as shown in Eqs. (1–3).[2,3]

$$\text{7,8-Dihydrobiopterin} + \text{TPNH} + \text{H}^+ \rightarrow \text{tetrahydrobiopterin} + \text{TPN}^+ \quad (1)$$

$$\text{Tetrahydrobiopterin} + \text{O}_2 + \text{phenylalanine} \rightarrow$$
$$\text{tyrosine} + \text{H}_2\text{O} + \text{quinonoid dihydrobiopterin} \quad (2)$$

$$\text{Quinonoid dihydrobiopterin} + \text{TPNH} + \text{H}^+ \rightarrow \text{tetrahydrobiopterin} + \text{TPN}^+ \quad (3)$$

It should be noted that reaction (1), catalyzed by dihydrofolate reductase (5,6,7,8-tetrahydrofolate:NADP oxidoreductase, EC 1.5.1.3), serves only to convert the 7,8-dihydrobiopterin to the tetrahydro form; this reaction plays no further role in the catalytic functioning of the pteridine in the hydroxylation reaction.[4] Reaction (2) is catalyzed by phenylalanine hydroxylase, [L-phenylalanine, tetrahydropteridine:oxygen oxidoreductase (4-hydroxylating), EC 1.14.3.1], and reaction (3) by dihydropteridine reductase.

Dihydrobiopterin can be assayed either by its ability to stimulate the rate of conversion of phenylalanine to tyrosine in the presence of the three enzymes shown above (catalytic assay),[1,5] or by following its conversion to tetrahydrobiopterin in the presence of TPNH and dihydrofolate reductase (noncatalytic assay).[4] In the latter assay, the decrease in absorbance at 340 nm is a measure of the amount of dihydrobiopterin present.

The catalytic assay is useful in the range of 0.1–3.0 millimicromoles of dihydrobiopterin; the noncatalytic assay is useful in the range of 1.0–10.0 millimicromoles of dihydrobiopterin. 7,8-Dihydrobiopterin can be prepared from biopterin by reduction with Zn and alkali[4] or with sodium hydrosulfite.[6]

Catalytic Assay

Reagents

L-Phenylalanine, 0.02 M
Potassium phosphate, 1.0 M, pH 6.8
TPN or TPNH, 0.0025 M
Glucose, 2.5 M
Phenylalanine hydroxylase, sufficient to form 0.05–0.1 micromole of tyrosine in 30 minutes
Glucose dehydrogenase in excess

[1] S. Kaufman, *Proc. Natl. Acad. Sci. U.S.* **50**, 1085 (1963).
[2] S. Kaufman, *J. Biol. Chem.* **234**, 2677 (1959).
[3] S. Kaufman, *J. Biol. Chem.* **239**, 332 (1964).
[4] S. Kaufman, *J. Biol. Chem.* **242**, 3934 (1967).
[5] S. Kaufman, *J. Biol. Chem.* **230**, 931 (1958).
[6] M. Nagai (Matsubara), *Arch. Biochem.* **126**, 426 (1968).

Dihydropteridine reductase in excess
Dihydrofolate reductase in excess[7]
Dihydrobiopterin solution to be assayed

Procedure. The reaction mixtures are prepared by the addition of 0.1 ml of each solution to test tubes cooled in ice. Water is added so that the final volume will be 1.0 ml. The last two additions are the hydroxylase and the pteridine, respectively. A reaction mixture without any added pteridine serves as a control. In addition, a tube containing 0.1 ml of a freshly prepared solution of 0.001 M 2-amino-4-hydroxy-6,7-dimethyl-tetrahydropteridine (dissolved in 0.005 M HCl) in place of the dihydrobiopterin, is included as a standard. The mixtures are incubated at 25° for 30 minutes with shaking. The reaction is stopped by the addition of 2.0 ml of 12% trichloroacetic acid. The precipitated protein is removed by centrifugation, and tyrosine is determined[8] on a 2.0-ml aliquot of the supernatant fluid. A tyrosine standard containing the same amount of TCA as the experimental tubes is carried through each assay.

The amount of dihydrobiopterin in the unknown is calculated from the amount of tyrosine formed in the presence of the 6,7-dimethylpteridine and the amount formed in the presence of dihydrobiopterin. In the range of 0.1–3.0 millimicromoles of dihydrobiopterin per milliliter of reaction mixture, dihydrobiopterin is about 15 times more active on a molar basis than 100 millimicromoles of the dimethyl compound in supporting tyrosine formation.[4] Therefore, if x is the amount of tyrosine formed in the presence of 100 millimicromoles of the dimethylpteridine, and y is the amount of tyrosine formed in the presence of z millimicromoles of dihydrobiopterin, $z = 100 \ x/15 \ y$. Several concentrations of dihydrobiopterin must be included in the assay to make certain that the concentration of the pteridine is in the range where tyrosine formation is proportional to dihydrobiopterin concentration.

An alternative method of following the hydroxylation reaction is to measure the phenylalanine-dependent oxidation of TPNH.[5] In this case, TPNH is used in place of the TPN, and the TPNH-regenerating system (glucose and glucose dehydrogenase) is omitted.

Specificity. As far as naturally occurring compounds are concerned, the only ones that show cofactor activity in the above assays are tetrahydro- and dihydrobiopterin, tetrahydro- and dihydroneopterin[4] (2-amino-4-hydroxy-6-[1,2,3-trihydroxypropyl-(L-*erythro*)]-pteridine),   and   tetrahydro-

---

[7] Partially purified preparations of dihydropteridine reductase from sheep liver contain adequate amounts of dihydrofolate reductase.

[8] S. Udenfriend and J. R. Cooper, *J. Biol. Chem.* **196,** 227 (1952).

folate.[9] Tetrahydropterins can be distinguished from 7,8-dihydropterins by the fact that dihydrofolate reductase is not an essential component of the hydroxylase system with the tetrahydropterins, whereas it is essential with the 7,8-dihydro compounds. The cofactor activity of tetrahydrofolate can be readily distinguished from that of the biopterin derivatives by a study of tyrosine formation as a function of time of incubation. Because of the instability of tetrahydrofolate under the conditions of the assay, the rate of tyrosine formation in the presence of this pteridine falls off rapidly with time.[9] By contrast, the rate of tyrosine formation is essentially constant with either dihydro- or tetrahydrobiopterin. Sepiapterin, another naturally occurring pteridine with potential cofactor activity, must first be converted to dihydrobiopterin before it is active.[10] This conversion requires a separate enzyme, sepiapterin reductase.[10]

Noncatalytic Assay

Reagents

> Potassium phosphate, 1.0 M, pH 6.8
> TPNH, 0.4 mM
> Glucose, 1.25 M
> Dihydrofolate reductase
> Glucose dehydrogenase, in excess
> Dihydrobiopterin, to be assayed

Procedure. The reaction mixture is prepared by the addition of 0.1 ml of each solution to a cuvette. Water is added to a final volume of 1.0 ml. A mixture in which the dihydrobiopterin has been omitted serves as a control. The reaction is started by the addition of dihydrofolate reductase, and the decrease in absorbance at 340 nm is measured. The amount of dihydrobiopterin present can be calculated from the fact that the change in extinction coefficient for the reaction, 7,8-dihydrobiopterin $\rightarrow$ tetrahydrobiopterin, is 6120 $\pm$ 170 M^{-1} cm^{-1}.[4]

The assay can be made twice as sensitive by omission of the TPNH-regenerating system (glucose, glucose dehydrogenase). Under these conditions the observed decrease in absorbance at 340 nm is due to the oxidation of TPNH to TPN as well as to the reduction of the pteridine [see Eq. (1)], and the change in extinction coefficient for the reaction is equal to the sum of 6.22×10^3 (extinction coefficient for TPNH at 340 mμ) and 6.12×10^3

[9] S. Kaufman, *Biochim. Biophys. Acta* **27**, 428 (1958).
[10] M. Matsubara, S. Katoh, M. Akino, and S. Kaufman, *Biochim. Biophys. Acta* **122**, 202 (1966).

(extinction coefficient for dihydrobiopterin reduction) or $12.3 \times 10^3 \, M^{-1}$ cm^{-1}.[4]

Specificity. The tetrahydropteridines will not be determined in this assay if the time of incubation is short. During long incubations, some oxidation of the tetrahydro- to the dihydropteridines will occur. The oxidation can be eliminated easily by carrying out the reaction anaerobically. The assay will not distinguish between the L-*erythro* isomers of dihydrobiopterin and dihydroneopterin.[4]

[174] Assay of Unconjugated Pteridines

By VIRGINIA C. DEWEY and G. W. KIDDER[1]

The only organisms at present known to have a dietary requirement for an unconjugated pteridine are members of the flagellate order Kinetoplastida (trypanosomids). This requirement was discovered by Cowperthwaite *et al.*[1a] and shown to be an unconjugated pteridine by Patterson *et al.*[2] for *Crithidia fasciculata.* Growth of the organism in the media used in that work is, however, rather slow (7–15 days) and scanty. The medium of Kidder and Dutta,[3] which gives heavier growth in a shorter time, is the basis for the media used in this laboratory. A discussion of the use of crithidias for assays of unconjugated pteridines has been given by Guttman (Nathan) *et al.*[4]

Media

Composition of Media

Stock cultures are maintained by weekly transfer in a peptone medium (Table I).

The medium used for the depletion of the organisms of folate is given in Table II (Medium D). Growth in this medium reaches only about half that

[1] Supported by research grants AM01005 and CA02924 from the National Institutes of Health, U.S. Public Health Service.
[1a] J. Cowperthwaite, M. M. Weber, L. Packer, and S. H. Hutner, *Ann. N.Y. Acad Sci.* **56,** 972 (1953); H. A. Nathan and J. Cowperthwaite, *J. Protozool.* **2,** 37 (1955).
[2] E. L. Patterson, II. P. Broquist, A. M. Albrecht, M. H. v. Saltza, and E. L. R. Stokstad, *J. Am. Chem. Soc.* **77,** 3167 (1955).
[3] G. W. Kidder and B. N. Dutta, *J. Gen. Microbiol.* **18,** 622 (1958).
[4] H. A. Nathan and J. Cowperthwaite, *Proc. Soc. Exptl. Biol. Med.* **85,** 117 (1954); H. A. Nathan, S. H. Hutner, and H. L. Levin, *J. Protozool.* **5,** 134 (1958); H. N. Guttman, *in* "Pteridine Chemistry" (W. Pfleiderer and E. C. Taylor, eds.), Pergamon Press, Oxford, 1964.

TABLE I
STOCK MEDIUM

Proteose peptone	20.0 g
Liver fraction L	1.0 g
Glucose	10.0 g
Folic acid	1.0 mg
Hemin[a]	25.0 mg
Water	1.0 liter
pH 8.0	

[a] Eastman Organic Chemicals Co.

obtained in a similar medium containing folate, but it can be transplanted indefinitely.

For depletion of the organisms of all pteridines (Medium DD), biopterin is omitted from the medium given in Table II. Growth is scanty and not transplantable.

The assay medium is the same as Medium DD, with the addition of folate at 0.001 μg/ml.

Preparation of Media

It has been found convenient to make up the ingredients of the experimental media (Table II) in the form of several solutions which are mixed in the appropriate proportions to give the final medium.

Solution A: amino acids at 10-fold concentration
Solution B: tyrosine ethyl ester at 100-fold concentration
Solution C: vitamins at 200-fold concentration
Solution D: sodium chloride and the phosphates at 100-fold concentration (keep at room temperature)
Solution E: tetrasodium ethylenediaminetetraacetate plus $MgSO_4$, $FeSO_4$, $CaCl_2$, and $NaMoO_4$ at 100-fold concentration
Solution F: trace metals at 200-fold concentration
Solution G: thymine at 100-fold concentration
Solution H: glucose, 25% (w/v), 25-fold concentration
Solution I: hemin dissolved in 50% triethanolamine at a concentration of 5 mg/ml. This must be diluted with water to a concentration of 1 mg/ml, before sterilizing by filtration through a Millipore filter, in order to reduce the viscosity of the solution. Make fresh at least once a month.

All solutions except D are stored in the refrigerator under toluene. Because of its relative insolubility, a solution of adenine is not prepared as a stock. Instead adenine is weighed out for each batch of medium.

TABLE II
MEDIUM D[a]

Component	Amount (μg/ml)	Component	Amount (μg/ml)
L-Arginine HCl	430	Adenine[c]	50
L-Histidine HCl·H$_2$O	210	Hemin[d]	5
DL-Isoleucine	630	Triethanolamine[d]	10
L-Leucine	970	Glucose	10,000
L-Lysine HCl	760	Ethylenediaminetetraacetic acid	700
DL-Methionine	340	tetrasodium salt	
L-Phenylalanine	500	Thymine	40
DL-Threonine	440	NaCl	4000
L-Tryptophan	120	Na$_2$HPO$_4$	1250
DL-Valine	660	KH$_2$PO$_4$	500
L-Tyrosine ethyl ester·HCl	200	MgSO$_4$·7 H$_2$O	400
Biotin	0.02	FeSO$_4$·7 H$_2$O	4.9
Nicotinamide	5	CaCl$_2$·2 H$_2$O	7.3
Pyridoxamine·2 HCl	2	Na$_2$MoO$_4$·2 H$_2$O	10.1
Ca pantothenate	8	ZnSO$_4$	68
Riboflavin	2	MnSO$_4$·H$_2$O	77
Thiamine	2	CoSO$_4$·7 H$_2$O	2.4
Biopterin[b]	0.001	CuSO$_4$·5 H$_2$O	1
		H$_3$BO$_3$	0.07
		KI	0.03
		Tween 80[e]	5000
		Adjust to pH 8.0	

[a] The amino acids omitted (serine, glycine, alanine, proline, aspartate, and glutamate) are not required by the organism, and their presence in the medium makes no significant improvement in growth. Although it has been shown that methionine can be dispensed with,[4] this is true only when the medium contains large amounts of folate as well as cysteine or homocysteine. Such a medium cannot be used for pteridine assay. It has been stated[5] that there is no threonine requirement, but this again is true only when the folate concentration is relatively high.[3] Various commercial preparations of acid-hydrolyzed casein, supplemented with tryptophan, may be substituted for the amino acid mixture at a final concentration of 10 mg/ml if tested and found to be free of pteridine. This would permit a saving of time.

[b] Folate is omitted from the medium of Kidder and Dutta and added in the indicated quantities in other media. For the assay medium, biopterin is also omitted and folate added at 0.001/μg/ml. Biopterin is 2-amino-4-hydroxy-6-L-*erythro*-1',2'-dihydroxypropylpteridine.

[c] Adenosine or adenylic acid in equimolar amounts may be substituted for adenine, but only if they are found to be free of pteridines.

[d] To be added after sterilization by filtration. Reduced from the 25 μg/ml recommended by Kidder and Dutta.

[e] Although Tween is not an absolute requirement, its inclusion results in about 50% better growth. It can be replaced by serum albumin, but this could be a potential source of pteridines. It can also be replaced by Triton WR 1339 at a concentration of 0.5 mg/ml. This is particularly desirable if the cells are to be harvested for any purpose. The effect of the detergents is particularly marked at the low concentration of pteridines with which one works in performing assays.

Adenosine or adenylic acid in equimolar amounts may be substituted for adenine, but only if they are free of pteridine (nonfluorescent). Most commercial preparations contain pteridine.

Tubes for assay are 15 × 125 mm Pyrex. The total volume per tube is 4 ml. The standard and the material to be assayed are dispensed in volumes of 0.2–2.0 ml per tube to cover the desired concentration range. Duplicate or triplicate sets should be prepared. The volume in all tubes is made up to 2.0 ml with water, (see below) including a blank containing only water. In preparing the medium, sufficient amounts of solutions A through G to make the final amount of medium desired are mixed and brought to one-fourth the final volume and added at 1 ml per tube. Loosely fitting aluminum caps are placed on the tubes and they are autoclaved for 6 minutes at 15 psi. Cotton plugs are undesirable, since cotton contains pteridines.

In a separate container are placed the desired amounts of Tween (or Triton) and glucose with a small amount of water. After heating to bring the detergent into solution and cooling with agitation, the volume is brought to one-fourth the final volume. The solution is then autoclaved and again cooled immediately under running water with swirling; the appropriate amount of filtered hemin is added aseptically.

Inoculation

To the solution of glucose, detergent, and hemin, the inoculum is added. This is usually 1/200th of the final volume of medium. Increasing the size of the inoculum to 1/50th permits reading of the cultures at 3 days instead of the usual 4, but may lead to high blank readings if the inoculum is not thoroughly depleted. One milliliter of the inoculum solution is then added aseptically to each tube, most conveniently with an automatic pipetting apparatus.

The tubes are then incubated at 25° for 4 (or 3) days in such a way as to provide sufficient aeration. This is done in this laboratory by placing the tubes on a slant just great enough to prevent medium from spilling out. An alternative to this procedure is to use small (25–50 ml) Erlenmeyer flasks. These are awkward to handle in large numbers and are more expensive in terms of space as well as requiring the cultures to be poured into tubes before reading.

At the completion of incubation the absorbance of the cultures is read. In this laboratory a Lumetron colorimeter with a 650 filter has been found

---

[5] S. H. Hutner and L. Provasoli, *in* "Biochemistry and Physiology of Protozoa" (S. H. Hutner and A. Lwoff, eds.), Academic Press, New York, 1955.

most convenient and rapid. The sample carrier is masked to accommodate the 4-ml volume. A top plate to fit the 15-mm tubes must also be used.

The dose-response curve to biopterin is a straight line when plotted logarithmically.

Since crithidias cling to the walls of the tubes, cultures must be shaken to disperse the flagellates evenly through the medium. This is done using a Super-Mixer with a touch plate switch after a few preliminary shakes by hand to wet down the areas covered with cells. The speed of the mixer should be adjusted and the tube positioned so as to avoid excessive frothing. After wiping with a soft cloth, the tube is placed in the light path of the colorimeter.

Special Precautions

All work with pteridines must be done in red or very dim light to avoid photodecomposition. The cultures must be incubated in the dark.

Glassware must be carefully cleaned because of the high degree of activity of the pteridines. We have found that sulfuric acid–dichromate soaking followed by rinsing of tubes in an automatic test tube washer[6] and of pipettes in an automatic pipette washer for 1.5 hours is satisfactory.

Preparation of Inoculum

The organism used is *Crithidia fasciculata*. By the use of Medium D (Table II), the crithidia are depleted of folate. If, however, it is desired to use crithidia for folate assay, they must be grown for one transplant in Medium D from which thymine has been omitted. Cultures can be maintained indefinitely in Medium D by loop transplant.

Medium DD is the same as Medium D except that the biopterin is omitted. This medium is used to deplete the cells of all pteridines. Since growth in this medium occurs in the first transplant only, cultures are not maintained in it but are grown for one transplant starting with cells from Medium D, as diagrammed below. The cultures in Medium DD are used to inoculate the assay media.

$$D \rightarrow DD$$
$$\downarrow$$
$$D \rightarrow DD$$
$$\downarrow$$
$$D \rightarrow DD$$

Standards

With each assay a series of dilutions of a standard pteridine must be carried. Although the amount of 6-biopterin required for half-maximal

[6] R. E. Parks, V. C. Dewey, and G. W. Kidder, *Anal. Chem.* **23,** 1193 (1951).

growth averages 0.018 ± 0.008 mμg/ml, in good agreement with the figure given by Rembold,[7] the spread is fairly large (0.014–0.041 mμg/ml).

For many years we used L-neopterin (2-amino-4-hydroxy-6-L-*erythro*-1′,2′,3′-trihydroxypropylpteridine) as a standard. This was obtained from the Lederle Division of American Cyanamid and has about the same biological activity as does 6-biopterin.[7] It is, however, no longer available from this source but may be synthesized according to the method of Patterson *et al.*,[8] Rembold and Metzger[9] or of Petering and Schmitt.[10] Although the synthesis is relatively simple the isolation of the 6-isomer

TABLE III

RESPONSE OF *Crithidia fasciculata* TO PTERIDINES:
AMOUNTS REQUIRED FOR HALF MAXIMAL GROWTH

Component	Amount (mμg/ml)	Component	Amount (mμg/ml)
Biopterin (R_1)[a]	0.018 ± 0.008[g]	Pteroyl glutamate[b]	70
L-Neopterin (R_2)[b]	0.126 ± 0.024	Pteroyl D-glutamate[b]	75
6-Hydroxymethylpterin (R_3)[c]	0.55 ± 0.17	Pteroyl asparate[b]	150
Sepiapterin (R_1)[d]	0.080	Pteroylglycine[b]	200
Pterin (R_4)[e]	4000	Pteroylalanine[b]	200
Tetrahydroxybutylpterins (R_5)[e]		N^5-Formylpteroylglutamate	70
D-*lyxo*	35	Aminopterin[b] (free of pteroyl	375
L-*lyxo*	0.35	glutamate)	
D-*arabino*	45		
L-*arabino*	2.0		
Trihydroxybutylpterin (R_6)[e]		Inactive:	
L-*arabino*	0.40	Pterin-6-carboxylate[e]	—
		Ichthyopterin[f]	—
2,4,5-Triamino-6-hydroxy-pyrimidine	240 ± 17	For data on other compounds see Rembold[7]	

[a] Smith, Kline, and French.
[b] Lederle.
[c] Dr. G. M. Brown.
[d] Isolated from a *Drosophila melanogaster* mutant in this laboratory.
[e] Prepared in this laboratory.
[f] Dr. M. Viscontini.
[g] Standard error.

[7] H. Rembold, *Vitamins Hormones* **23**, 359 (1965).
[8] E. L. Patterson, R. Milstrey, and E. L. R. Stokstad, *J. Am. Chem. Soc.* **78**, 5868 (1956).
[9] H. Rembold and H. Metzger, *Ber.* **96**, 1395 (1963).
[10] H. G. Petering and J. A. Schmitt, *J. Am. Chem. Soc.* **71**, 3977 (1949).

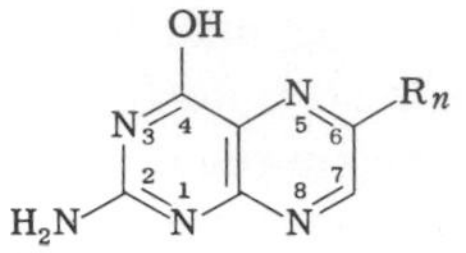

R_1 = —CHOHCHOHCH$_3$
R_2 = —CHOHCHOHCH$_2$OH
R_3 = —CH$_2$OH
R_4 = H
R_5 = —CHOHCHOHCHOHCH$_2$OH
R_6 = —CHOHCHOHCHOHCH$_3$

from the large quantity of the 7-isomer and other products by repeated rechromatography is somewhat laborious.[9]

More recently we have been using as a standard 6-biopterin (2-amino-4-hydroy-6-L-*erythro*-1',2',-dihydroxypropylpteridine) synthesized by Dr. J. Weinstock and obtained through the kindness of Dr. A. Maass of the Smith, Kline, and French Co. After chromatography on phosphocellulose[9] to remove 7-biopterin (30% present), this material gave results comparable to those of Rembold.[7] Since the synthesis of biopterin is somewhat more complicated than that of neopterin and the synthesis of either gives low yields of the 6-isomer, it might be better to choose some other standard material. Commercial biopterin (Sigma, K & K) is very impure but can presumably be purified on phosphocellulose. If such material is to be used as a standard, its activity relative to pure biopterin should be determined.

A more simply prepared pteridine of high activity is 2-amino-4-hydroxy-6-D-*lyxo*-1',2',3',4'-tetrahydroxybutylpteridine. Although this compound has only about 1/20 of the activity of biopterin[11] (Table III), it has the advantage that its synthesis by condensation of 2,4,5-triamino-6-hydroxypyridimidine and D-galactose leads to a large proportion of the 6-isomer as compared to the 7-isomer. The isomers can be separated on phosphocellulose.[7]

[11] V. C. Dewey, G. W. Kidder, and F. P. Butler, *Biochem. Biophys. Res. Commun.* **1**, 25 (1959).

[175] Simplified *Lactobacillus casei* Assay for Folates

By HERMAN BAKER, OSCAR FRANK, and S. H. HUTNER

Lactobacillus casei as an assay organism has the advantages, unlike *Streptococcus faecalis*, of responding to (a) those folic polyglutamates including, and below, the triglutamyl derivatives occurring in foodstuffs, and (b) the N^5-methyltetrahydrofolic acid, which may be the folate predominating in blood and tissues. A disadvantage shared with *S. faecalis* is that the folic acid requirement may be bypassed by thymine plus purines. The assay was designed primarily for assessing clinical folate status by measuring folate activity in serum. The poorer results with red cells or whole blood may reside in red cells being high in folate-bypassing compounds; this is not definitely established. Folate assays are covered in recent reviews[1,2] and in a brief commentary on its rationale.[3]

Conservation and Basal Media

These are given in Tables I and II.

TABLE I
CONSERVATION AGAR

Constituent	Amount (mg)
Proteose peptone (Difco)	750
Yeast hydrolyzate[a]	750
Glucose	1000
KH$_2$PO$_4$	200
Tomato juice filtrate[b]	10 ml
Tween 80[c]	10
L-Cysteine HCl	100
Agar	1500
Distilled water	To 100 ml

[a] "Yeast hydrolyzate" (N.B.C.) is as satisfactory as the yeast extract generally specified.

[b] Adjust filtrate to pH 7.0 with KOH before addition to the medium.

[c] Polyoxyethylene sorbitan monooleate (Atlas Powder Co.). Obtainable from Sigma Chemical Co. and other supply houses. Dispense from a 10% solution in 50% ethanol kept frozen and tightly stoppered to minimize autoxidation. Adjust pH to 6.6–6.9. Incubate stabs for 24–48 hours at 37°, store at 0°–6°, transfer monthly.

Preparation of Samples

Serum or plasma (unhemolyzed) may be stored frozen. Put 1-ml samples into 35-ml centrifuge tubes and dilute with 9 ml of phosphate–ascorbate buffer. Dissolve 27.8 g of NaH$_2$PO$_4$·H$_2$O in 1000 ml of water (solution A), and 71.7 g of Na$_2$HPO$_4$·12 H$_2$O in 1000 ml (solution B). Dilute 212.5 ml

[1] V. Herbert and J. R. Bertino, *in* "The Vitamins" (P. György and W. N. Pearson, eds.), 2nd ed., Vol. VII, p. 243. Academic Press, New York, 1968.

[2] H. Baker and O. Frank, "Clinical Vitaminology: Methods and Interpretation." Wiley (Interscience), New York (1968).

[3] O. Frank, H. Baker, and S. H. Hutner, *Am. J. Clin. Nutr.* **21**, 327 (1968).

TABLE II
BASAL MEDIUM FOR FOLATE ASSAY (FOR 500 ML OF 2× MEDIUM;
= 1 LITER OF FINAL MEDIUM)

Constituents	Amount (mg)
Hycase SF[a]	5000
L-Tryptophan[b]	100
L-Asparagine·H$_2$O	300
L-Cysteine HCl	250
Adenine[c,d]	5
Guanine[c,d]	5
Uracil[c,d]	5
Xanthine[c,d]	10
Riboflavin[d]	0.5
p-Aminobenzoic acid[d]	1.0
Pyridoxine HCl[d]	2.0
Thiamine HCl[d]	0.2
Ca pantothenate[d]	0.4
Nicotinic acid[d]	0.4
Biotin[c]	0.01
Glucose	20,000
K$_2$HPO$_4$	500
Glutathione (reduced)	2.5
Salt mix[e]	5 ml
Sodium acetate (anhydrous)	20,000
KH$_2$PO$_4$	500
MnSO$_4$·H$_2$O	100
Distilled water[f]	To 500 ml

[a] A low-salt acid hydrolyzate of "vitamin-free" casein (Sheffield Chemical Division, National Dairy Products Corp., 2400 Morris Ave., Union, New Jersey 07083).

[b] Tryptophan is largely destroyed upon acid hydrolysis of casein.

[c] Dissolve by boiling in 200–250 ml of dilute KOH before adding to basal medium.

[d] Added from a frozen-stored stock solution. Recent experiments indicate it may be preferable to use the far more soluble Na riboflavin PO$_4$·2 H$_2$O, which has equivalent activity on a molar basis. For ease in compounding ingredients as dry mixes (stored frozen), the vitamins, purines, and pyrimidines may be dispensed as triturates in mannitol or the metabolically inert pentaerythritol, thus avoiding the need for a microbalance. For example, biotin may be stored at room temperature and dispensed as 1:000 and 1:10,000 triturates.

[e] For 100 ml: MgSO$_4$·7 H$_2$O, 4 g; NaCl, 0.2 g; FeSO$_4$·7 H$_2$O, 0.2 g; MnSO$_4$·4 H$_2$O, 0.2 g; conc. HCl, 0.4 ml. For compounding as a dry mix, the following should be used: MgSO$_4$ (anhyd.), Fe(NH$_4$)$_2$(SO$_4$)$_2$·6 H$_2$O, and MnSO$_4$·H$_2$O, to minimize loss of water by efflorescence on frozen storage.

[f] Boil to dissolve the constituents and drive off CO$_2$. Adjust pH to 6.6–6.8 with KOH and bring to a 500-ml volume.

of solution A and 37.5 ml of solution B to 1000 ml with distilled water; the pH should be 6.1. Before use add 2.0 mg/ml ascorbic acid to help precipitate the proteins and protect serum folates from destruction by heat and oxidation. Cover the centrifuge tubes with polypropylene caps. Autoclave 10 minutes.

This procedure and the final autoclaving destroys thermolabile antibiotics in serum. For a reliable determination, 4–5 days should elapse after discontinuation of antifolates. Centrifuge and add 0.5, 1.0, and 1.5 ml of supernatant solution to separate flasks according to the following protocols (Tables III and IV). Double-strength basal medium is distributed as 5-ml portions in micro-Fernbach flasks (10 ml nominal volume; Bellco or Kimble) deployed in Pyrex utility trays. The inoculum is a suspension of an overnight or 6–8-hour culture (the stock culture medium minus agar) diluted aseptically 1:10 with distilled water. One drop is added per flask.

When the additions to individual flasks are completed, the flasks are covered with No. 6 polypropylene caps (sold as "stoppers") and autoclaved 30 minutes. Longer autoclaving or autoclaving above 121° may badly caramelize the glucose in the basal medium; hence, standardization of pH, time, and temperature of autoclaving is critical. Allow pressure in the autoclave to fall gradually to prevent boiling over. Upon removal from the autoclave, the trays with sterile flasks are covered with inverted trays. Tray covers are

TABLE III
STANDARD CURVE

Flask no.	Concentration (ng/ml)	Folic acid (per flask)	Distilled water (ml)
1	—	—	5.0
2	0.01	1 ml of 0.1 ng/ml	4.0
3	0.03	3 ml of 0.1 ng/ml	2.0
4	0.1	1 ml of 1.0 ng/ml	4.0
5	0.3	3 ml of 1.0 ng/ml	2.0
6	1.0	1 ml of 10.0 ng/ml	4.0
7	3.0	3 ml of 10.0 ng/ml	2.0

TABLE IV
SAMPLE ADDITION

Flask no.	Sample addition (ml)	Distilled water (ml)
8	0.5	4.5
9	1.0	4.0
10	1.5	3.5

removed for inoculation. After inoculation the covers are replaced, the joint between trays is sealed with freezer tape (Sears, Roebuck), and the flasks are incubated 14–18 hours at 37°.

Calculation of Results

Plot concentration of folic acid in the standard curve as ng/ml on the log scale and absorbance (formerly "optical density") on the linear axis. Absorbances of the individual assay flasks are plotted against concentration of folic acid. Folate activity of the sample is calculated by the formula:

$$\frac{\text{Folic acid conc. (ng/ml) derived from sample absorbance}}{\dfrac{\text{ml sample used}}{\text{ml final vol. of sample}} \times \dfrac{\text{ml supernatant added}}{\text{10 (vol. in assay flask)}}} = \text{folic acid activity (in ng/ml)}$$

Comment

Modifications of the *L. casei* method for tissues have been given along with more detail.[4] Juxtaposition in the basal medium of high concentrations of glucose, phosphate, and amino acids at neutral or near-neutral pH spells, as noted, hazard from overheating, especially at pH values above neutrality: This favors aldolization and Maillard reactions. An overhaul of the medium is badly needed to eliminate these hazards. The method yields consistently good results once these precautions become routine. The dehydrated basal medium is available commercially, but most workers prepare their own, as some batches have been poor. If the object is to detect folic acid or, more properly, its reduced forms as the prosthetic group of an enzyme, it might be preferable either to use the proteolytic ciliate *Tetrahymena*[5] (the protozoological literature should be consulted for improvements in basal media[6]) or to apply chemical hydrolysis followed by microbiological assay of the liberated *p*-aminobenzoic acid.

The commercial availability of biopterin (∼6% pure but nevertheless satisfactory) may make practical an exceptionally sensitive assay for folic acid by means of the flagellate *Crithidia fasciculata*.[7] Media and methods for handling *C. fasciculata* are under scrutiny.[8] As far as known, folic acid requirements in protozoa can be spared but not bypassed. Biopterin withstands conditions of acid hydrolysis that destroy folic acid, enabling a differential assay for biopterin and other microbiologically

[4] H. Baker and O. Frank, *in* "The Vitamins" (P. György and W. H. Pearson, eds.), 2nd ed., Vol. VII, p. 269. Academic Press, New York, 1968.

[5] T. H. Jukes, *Methods Biochem. Anal.* **2**, 121 (1955).

[6] D. Cox, O. Frank, S. H. Hutner, and H. Baker, *J. Protozool.* **15**, 713 (1968). [The amino acids in this medium (except for tryptophan) can probably be replaced by Hycase SF.]

[7] H. A. Nathan, S. H. Hutner, and H. L. Levin, *J. Protozool.* **5**, 134 (1958).

[8] K. M. O'Connell, S. H. Hutner, H. Fromentin, O. Frank, and H. Baker, *J. Protozool.* **15**, 719 (1968).

active unconjugated pteridines. Responses of *C. fasciculata* to polyglutamyl folates are unexplored; so far *Tetrahymena* behaves like higher animals.[5] Charting the applicability of the *L. casei* assay to enzymological problems must come from parallel assays with the aforementioned organisms. For further discussion of protozoological aspects of this topic, the reader is referred to pertinent articles in a recent work.[9]

[9] G. M. Kidder (ed.), "Chemical Zoology," Vol. I: Protozoa. Academic Press, New York, 1968.

[176] Microbiological Assay of Folic Acid Activity in Serum and Whole Blood[1]

By Jack M. Cooperman

The microbiological assay for serum and whole blood folic acid activity is a good example of how development in the microbiological assay of a vitamin led to advancement in the knowledge of the coenzyme forms of the vitamin.

When initial attempts were made to determine the folic acid activity of serum and whole blood, little or no activity could be detected using the microbiological assays available.[1a-4] In these techniques, the samples to be assayed were diluted with water and then heated to coagulate proteins prior to assay. It was subsequently observed that if whole blood is dialyzed and the dialyzate assayed without further heat treatment, considerably more folic acid activity is obtained than when the blood is diluted with water and then heated.[5] It was thus apparent that the blood folic acid activity is heat labile. Further investigations revealed that if blood is diluted with a phosphate buffer containing 0.05% ascorbic acid and then heated to obtain a protein-free filtrate, no loss in folic acid activity occurs.[6] These investigators then demonstrated that whole blood contains various reduced folates that are heat labile and that ascorbic acid is effective in preventing the heat destruction of these folates.[7]

[1] Studies reported herein were supported in part by National Institutes of Health Grants AM-01778 and HD-01162.
[1a] B. S. Schweigert and P. B. Pearson, *Am. J. Physiol.* **148,** 319 (1947).
[2] B. S. Schweigert, *J. Lab. Clin. Med.* **33,** 1271 (1948).
[3] G. Toennies and D. L. Gallant, *J. Lab. Clin. Med.* **34,** 501 (1949).
[4] I. Chanarin, B. B. Anderson, and D. L. Mollin, *Brit. J. Haematol.* **4,** 156 (1958).
[5] G. Toennies, H. G. Frank, and D. L. Gallant, *J. Biol. Chem.* **200,** 23 (1953).
[6] G. Toennies, E. Usdin, and P. M. Phillips, *J. Biol. Chem.* **221,** 855 (1956).
[7] E. Usdin, P. M. Phillips, and G. Toennies, *J. Biol. Chem.* **221,** 865 (1956).

These findings led to the development of several microbiological assays for serum and whole blood folic acid activities, and these were previously reviewed and critically discussed.[8]

In this paper, the various phases of the microbiological assay are presented and recommended procedures given in detail.

Response of Assay Microorganisms to Different Folates

Three microorganisms are commonly used to determine the folic acid activity of serum and whole blood. These are *Lactobacillus casei* [American Type Culture Collection (ATCC) 7469], *Streptococcus faecalis* (ATCC 8043), and *Pediococcus cerevisiae* (ATCC 8081). These differ in their pattern of response to the naturally occurring folates. The growth-promoting ability of different folates for these microorganisms is shown in Table I. In

TABLE I

The Microbiological Activity of Folates Compared on a Mole for Mole Basis with That of Pteroylglutamic Acid[a]

Folate	*Lactobacillus casei*	*Streptococcus faecalis*	*Pediococcus cerevisiae*
PteGlu[b]	100	100	<1
PteGlu$_2$	100	100	<1
PteGlu$_3$	100	<1	<1
PteGlu$_7$	1	<1	<1
H$_4$PteGlu[c]	100	100	100
5-CHO—H$_2$PteGlu	100	100	<1
5-CHO—H$_4$PteGlu	100	100	100
10-CHO—H$_4$PteGlu	100	100	100
5,10-CH=H$_4$PteGlu	100	100	100
5-CH$_3$—H$_4$PteGlu	60–80	<1	<1

[a] For *P. cerevisiae*, standard (100%) is 5-CHO—H$_4$PteGlu.
[b] Pteroylglutamic acid = 100%.
[c] Tetrahydropteroylglutamic acid.

general, *L. casei* can utilize for growth the widest spectrum of folates, and these include the mono-, di-, and triglutamates of pteroylglutamic acid and reduced derivatives. *Pediococcus cerevisiae*, on the other hand, is most specialized in its requirements, being able to utilize only the monoglutamate of tetrahydropteroylglutamic acid and its derivatives with the exception of 5-methyltetrahydropteroylglutamic acid.

Choice of Assay Microorganism

Assays of serum and whole blood with *L. casei* usually result in higher folic acid activity levels than assays with the two other microorganisms

[8] J. M. Cooperman, *Am. J. Clin. Nutr.* **20**, 1015 (1967).

mentioned above. For this reason, the *L. casei* assay has been most widely used. However, assay with all three microorganisms is useful to differentiate the folates occurring in blood.[7,9] In addition, the *S. faecalis* assay of serum is used to determine the gastrointestinal absorption of orally administered pteroylglutamic acid[10] and food folates[11] and to determine the plasma clearance of intravenously administered pteroylglutamic acid.[12]

Maintenance of Stock Culture

It has been found that when the assay microorganisms are maintained on an enriched agar with minimum transfers, a stable population of each culture results.[8] Under these conditions, the growth response patterns of these assay bacteria to folates will remain relatively constant for long periods of time.

On the other hand, when cultures are transferred daily through liquid broth,[13,14] the faster growing strains will soon predominate, resulting in cultures unsuitable for assay purposes.

A liver-tryptone agar has been devised and used successfully to maintain all three assay microorganisms mentioned above.[8] The composition and means of preparing the agar medium is given in Table II.

The bacteria should be transferred according to the scheme shown in Fig. 1. Initially, stabs of each culture are made into 3 tubes of enriched agar. These are allowed to incubate for 24 hours at 37° and 2 are then refrigerated. These are designated as the monthly stabs. From the third monthly stab, 4 new subcultures are made, and similarly incubated and refrigerated. These are the weekly stabs. Each week, 3–5 daily stabs are prepared from a weekly stab. On the day before assay, a transfer is made from a daily stab into liquid inoculum broth which is then allowed to incubate at 37° for 16–24 hours, and then treated further as discussed below before being used to inoculate the assay.

At the end of each month, one of the two refrigerated monthly stab cultures is used to prepare 3 new monthly cultures and the scheme is repeated. The third monthly stab is used as a spare in the event that a transfer does not take.

[9] N. Grossowicz, J. Aronovitch, M. Rachmilewitz, G. Izak, A. Sadovsky, and B. Bercovici, *Brit. J. Haematol.* **6**, 296 (1960).
[10] I. Chanarin, B. M. MacGibbon, W. J. O'Sullivan, and D. L. Mollin, *Lancet* **2**, 634 (1959).
[11] J. M. Cooperman and A. L. Luhby, *Israel J. Med. Sci.* **1**, 704 (1965).
[12] I. Chanarin, D. L. Mollin, and B. B. Anderson, *Brit. J. Haematol.* **4**, 435 (1958).
[13] H. Baker, V. Herbert, O. Frank, I. Pasher, S. H. Hutner, L. R. Wasserman, and H. Sobotka, *Clin. Chem.* **5**, 275 (1959).
[14] V. Herbert, R. Fisher, and B. J. Koontz, *J. Clin. Invest.* **40**, 81 (1961).

TABLE II

Composition and Means of Preparing Agar Medium in Which
the Three Assay Microorganisms Are Maintained

Ingredient[a]	Amount per liter
Bacto-Tryptone (Difco B123)	10 g
Glucose	10 g
K_2HPO_4	2 g
$CaCO_3$	3 g
Liver powder[b]	10 g
Salts A[c]	5 ml
Salts C[d]	5 ml
Agar	15 g

[a] The ingredients are added to distilled water which is then brought to a boil and stirred until the agar dissolves. The stirring is continued while 10-ml portions are doled out into 16 × 150 mm Pyrex lipless tubes with an agar pipette. The tubes are capped with stainless steel test tube closures, autoclaved at 121° for 15 minutes, cooled, and then stored in the refrigerator.

[b] Liver concentrate NF, Pharmaceutical Division, Wilson and Co., Chicago, Illinois.

[c] Salts A, 50 g K_2HPO_4 and 50 g KH_2PO_4 in 500 ml of distilled water.

[d] Salts C, 10 g $MgSO_4 \cdot 7 H_2O$, 0.5 g NaCl, 0.5 g $FeSO_4 \cdot 7 H_2O$, 1.5 g $MnSO_4 \cdot H_2O$, and 1 ml of concentrated HCl in 250 ml of distilled water.

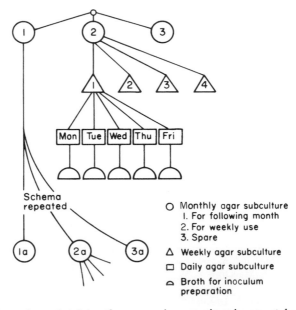

Fig. 1. Scheme for maintaining the assay microorganisms in agar stabs. This method using minimal transfers ensures stability of the cultures. From J. M. Cooperman, A. L. Luhby, and C. M. Avery, *Proc. Soc. Exptl. Biol. Med.* **104,** 536 (1960).

Preparation of Inoculum Broth

A transfer is made from a daily agar culture into liquid inoculum broth on the day before an assay is set up, and the inoculum is allowed to incubate at 37° for 20–24 hours. From the latter, the bacterial suspension used to inoculate the assay is prepared. The liquid inoculum broth used is an enriched medium. This consists of rehydrated riboflavin assay medium (Difco 0325) to which is added the equivalent of 1% skim milk powder.[15] To about 50 ml of distilled water is added 2.4 g of riboflavin assay medium, and the mixture is stirred until dissolved.

The clarified skim milk is prepared by dissolving skim milk powder in 10 volumes of distilled water, adjusting the pH to 4.2 with 6 N HCl, and centrifuging. The clear supernatant solution is collected and adjusted to pH 7.0 with dilute NaOH and recentrifuged. A volume of supernatant solution containing the equivalent of 1 g of original skim milk is added to the dissolved riboflavin medium, and distilled water is added to make 100 ml of final solution. Ten milliliters of this solution is added to 16 × 125 mm lipless test tubes. The tubes are capped with stainless steel test tube enclosures and autoclaved at 121° for 15 minutes. These tubes may be stored in the refrigerator for 3–6 months and still be suitable for assay.

Preparation of the Bacterial Suspension Used to Inoculate the Assay

Streptococcus faecalis and Pediococcus cerevisiae. The liquid broth cultures prepared as described above require further treatment before they can be used to inoculate the assay. For *S. faecalis* and *P. cerevisiae* assays, the thick bacterial suspension is centrifuged at 3000 rpm in an International clinical centrifuge for 5 minutes, after which the supernatant solution is discarded, and the bacterial pellet is resuspended in about 10 ml of sterile 0.9% saline solution. The suspension is recentrifuged and resuspended in sterile saline twice more. This is necessary to prevent carry-over of folic acid activity into the tubes to be inoculated.

After the third resuspension, an aliquant of the latter is diluted with 9 volumes of sterile saline. The amount prepared will, of course, depend upon the number of tubes to be inoculated. One drop of the latter, dispensed from a sterile 5-ml pipette plugged with cotton, is used to inoculate each assay tube.

Lactobacillus casei. For the *L. casei* assay, the inoculum suspension is prepared differently. Toennies *et al.*[5] showed that when a light suspension of *L. casei* in the logarithmic phase of growth is used to inoculate the assay, the incubation period of the assay can be reduced from 48 to 16 hours. This method is now widely used in *L. casei* assays for folic acid activity.

[15] J. M. Cooperman, R. Drucker, and B. Tabenkin, *J. Biol. Chem.* **191**, 135 (1951).

TABLE III

COMPOSITION AND MEANS OF PREPARING THE BASAL MEDIUM USED FOR
MICROBIOLOGICAL ASSAY WITH EACH OF THE THREE ASSAY MICROORGANISMS

Ingredient	Amount per liter
Acid-hydrolyzed casein[a]	12.7 g
Anhydrous sodium acetate	41.0 g
Glucose	40.0 g
K_2HPO_4	6.96 g
KH_2PO_4	6.80 g
Ascorbic acid	500 mg
DL-Alanine	400 mg
DL-Tryptophan	400 mg
L-Cystine[b]	400 mg
L-Asparagine	200 mg
Adenine[c]	20 mg
Guanine[c]	20 mg
Uracil[c]	20 mg
Xanthine[d]	20 mg
Salts C[e]	10 ml
Vitamin mixture[f]	5 ml
Pyridoxamine[g]	0.4 mg
Adjust pH to 6.4–6.5[h]	

[a] Hycase amino, Sheffield Co., Norwich, New York.

[b] Keep as a solution containing 25 mg/ml. Add weighed amount to distilled water, add concentrated HCl with stirring until dissolved, and make up to volume.

[c] Keep as a solution containing 1 mg/ml of each. To prepare, add 250 mg each of adenine, guanine, and uracil to about 100 ml of distilled water. Add concentrated HCl with stirring until dissolved and make up to 250 ml with distilled water. Add a layer of toluene and store in the refrigerator.

[d] Keep as a solution containing 1 mg/ml. To prepare, add 250 mg of xanthine to about 100 ml of distilled water, add sufficient concentrated NH_4OH to dissolve with stirring, and make up to 250 ml with distilled water. Add a layer of toluene and store in refrigerator.

[e] For composition of salts C see Table II.

[f] Vitamin mixture, add 50 mg of niacin, 50 mg of riboflavin, 50 mg of thiamine, 50 mg of calcium pantothenate, 10 mg of p-aminobenzoic acid, and 0.5 mg of biotin to 200 ml of distilled water. Add a layer of toluene, and store in refrigerator. Shake well before using. Prepare fresh every 2 months.

[g] Pyridoxamine is prepared as a solution containing 0.1 mg/ml, layered with toluene, and stored in the refrigerator.

[h] To prepare medium, dissolve all constituents in distilled water except vitamins, adjust pH to 6.4–6.5, and make up to volume. Add 2 drops of Tween 80 to prevent clumping of bacteria. Approximately 200-ml portions are placed in Erlenmeyer flasks and frozen at $-20°$. These are thawed on the day of assay, and the required amount of vitamin mixture is added.

A transfer is made from a daily agar stab into the skim milk–riboflavin assay medium broth the day before an assay is run and allowed to incubate 20–24 hours. During the next morning, the heavy bacterial growth is thoroughly suspended, and 0.1–0.2 ml of the suspension is transferred aseptically to a tube containing 6 ml of fluid consisting of 3 ml of basal medium (see below) and 3 ml of distilled water containing 200 $\mu\mu$g (pg) of pteroylglutamic acid. These tubes are prepared in advance and, after sterilization, stored in a refrigerator.

The transfer is allowed to incubate for 6 hours. A light growth occurs during this period. The tube is centrifuged, the supernatant solution is discarded, and the small pellet is resuspended in 6 ml of sterile 0.9% saline solution. This washing process is repeated twice, and then an aliquant of the suspension is diluted 10-fold with sterile 0.9% saline solution. One drop, from a sterile 5-ml pipette plugged with cotton, is used to inoculate the assay tubes.

Composition and Preparation of Basal Medium

A basal medium based on that of Toennies *et al.*,[6] has been found most useful for assays for folic acid activity with either *L. casei*, *S. faecalis*, or *P. cerevisiae*. Each of these assay bacterium grows in a manner suitable for assay purposes when the medium is supplemented with a utilizable folate. The composition and means of preparation of the basal medium is given in Table III.

In addition to the constituents usually found in media for microbiological assay, this medium contains ascorbic acid and high levels of phosphate. The ascorbic acid protects labile folates during the heat sterilization and incubation period. Other reducing agents, such as glutathione and cysteine, are less effective in this respect.[6] It has also been shown that, under certain conditions such as exist in the present assay, *L. casei* requires ascorbic acid for growth.[16] If ascorbic acid were omitted from the medium, it would be difficult to differentiate the folic acid activity of serum and whole blood, both of which are diluted with an ascorbic acid solution before assay.

The higher phosphate content results in superior growth for all three assay bacteria, since it has a better buffering capacity at near neutral pH than sodium acetate, which is usually the sole buffer in some of these assays.

The Reference Standard

A standard curve relating growth of bacteria to concentration of a reference standard is plotted for each assay. From this curve, calculations of the folic acid activity of the sample are made.

[16] H. Kihara and E. E. Snell, *J. Biol. Chem.* **235**, 1409 (1960).

For the *L. casei* and *S. faecalis* assays, synthetic pteroylglutamic acid is used as the reference standard. It is readily available and relatively stable in solution.

Pteroylglutamic acid cannot be used in the *P. cerevisiae* assay since this organism can utilize only reduced folates for growth. The calcium salt of leucovorin, 5-formyltetrahydropteroylglutamic acid, is used as the reference standard for assay with *P. cerevisiae*. It is also readily available and is the most stable of the reduced folates.

Stock solutions of the above two compounds are prepared to contain 1 $\mu g/ml$ each.

One hundred milligrams of synthetic pteroylglutamic acid are carefully weighed out on an analytical balance and suspended in about 100 ml of glass-distilled water. The suspension is dissolved by adding 1 N NaOH dropwise with stirring until a clear yellow solution results. The pH is then adjusted to 7.0–7.4 and diluted in a volumetric flask to 1 liter with distilled water. This solution contains 100 $\mu g/ml$. One milliliter of the latter is diluted in a volumetric flask to 100 ml with distilled water. This solution, containing 1 $\mu g/ml$, is stored in an opaque plastic bottle in the freezer.

Preparation of the standard solution containing 1 μg per ml of L-5-formyltetrahydropteroylglutamic acid presents no difficulty since the calcium salt is readily soluble. The synthetic salt is the DL-isomer, and this should be taken into consideration when preparing the stock solution.

When an assay is run, the appropriate stock solution is thawed and diluted further as necessary. A solution containing 500 pg per milliliter of pteroylglutamic acid is prepared for the *L. casei* assay; for the *S. faecalis* assay, a solution containing 1000 pg/ml. A solution containing 500 pg of L-5-formyltetrahydropteroylglutamic acid is prepared for the *P. cerevisiae* assay.

Further dilutions of the stock solutions containing 1 $\mu g/ml$ are made with 0.05 M phosphate buffer, pH 6.1. The latter is prepared by adding 5.85 g of KH_2PO_4 and 1.22 g K_2HPO_4 to 1 liter of water. This is done to equalize the phosphate content of the tubes containing the reference standard with those containing serum or whole blood samples, both of which are diluted with 0.05 M phosphate buffer.

Preparation of Samples

Serum. Venous blood (5–10 ml) is collected from patients who have fasted overnight. It is necessary that fasted blood be collected, since serum folates in humans will be elevated from 4 to 6 hours after a meal.[17]

[17] A. L. Luhby and J. M. Cooperman, *Advan. Metab. Disorders* **1**, 263 (1964).

The blood is allowed to clot for about an hour at 37°, and the tube containing the clotted blood is then centrifuged for 15 minutes at 3000 rpm in a desk-top centrifuge. The clear serum is collected and used for assay. The sample may be stored for about a month at −20° without loss of activity.

It is necessary to prepare a protein-free filtrate of serum for assay, since otherwise the lactic acid produced by the bacteria during the incubation period will precipitate these proteins. The latter will interfere with the turbidimetric reading of the bacterial growth, which is the end point for this assay. The method of Toennies et al.[6] in which serum is heated in an ascorbic acid–phosphate solution, provides a protein-free extract of serum which retains all the folic acid activity.

The extracting solution is prepared on the day sera are to be treated by adding 50 mg of ascorbic acid powder to each 100 ml of 0.05 M phosphate, pH 6.1 (see above for preparation of phosphate buffer). The pH of the solution is not readjusted after the addition of the ascorbic acid.

In a screw-top test tube, 0.5 ml of clear serum and 9.5 ml of the ascorbic acid–phosphate solution are added. The stopper is affixed to the tube, the contents are mixed, and the tube is then autoclaved for 10 minutes at 121°. The tube is then centrifuged, and the clear supernatant solution, which represents a 1:20 dilution of original serum, is used for assay. This dilution is suitable for sera containing between 3 and 15 ng/ml. Sera containing less than 3 ng/ml should be diluted 1:10 with the ascorbic acid–phosphate solution, and those above 15 ng/ml should be diluted to contain between 200 and 500 pg per milliliter of final solution.

Whole Blood. Toennies et al.[6] demonstrated that the folates in red blood cells are in a bound form that is not utilizable for growth by the common assay microorganisms. However, when the red blood cells are incubated in the presence of plasma, microbiologically available folic acid activity is released. They postulated that blood plasma contains an enzyme that hydrolyzes the red blood cell folates into forms that can be measured by the microbiological assay.

In their method, Toennies et al.[6] prepared whole blood for assay by diluting it with an ascorbic acid–phosphate solution, incubating it for 90 minutes so that the plasma enzyme could release folate activity, and then heating it to precipitate proteins. When this procedure is used, complete lysis of red blood cells does not occur during the initial dilution with phosphate buffer. Some investigators[18,19] subsequently altered this method by diluting whole blood with a hypotonic solution so that complete hemol-

---

[18] N. Grossowicz, F. Mandelbaum-Shavitt, R. Davidoff, and J. Aronovitch, *Blood* **20**, 609 (1962).

[19] A. V. Hoffbrand, B. F. A. Newcombe, and D. L. Mollin, *J. Clin. Pathol.* **19**, 17 (1966).

ysis was attained. Bird and McGlohon[20] compared these methods for preparing whole blood for assay. They found that if blood is completely hemolyzed initially, maximum microbiological folic acid activity is not released if the solution is subsequently incubated with the endogenous enzyme in plasma. Under these conditions, it is necessary to add an exogenous source of enzyme, namely, hog kidney conjugase preparation, in order to achieve maximum *L. casei* folic acid activity.

The method that Bird and McGlohon[20] recommended for whole blood treatment, consists of diluting heparinized blood 10-fold in 1.1% ascorbate containing 0.5% NaCl, heating 10 minutes in a 95° water bath, cooling, clarifying, and treating the resulting extract with hog kidney conjugase.

However, these authors pointed out that another method[8] in which only partial hemolysis occurred during the initial dilution of whole blood also releases maximal *L. casei* folic acid activity and is simpler to perform. The latter procedure is described below.

Heparinized blood is obtained from an intravenous puncture, and 0.1 ml is pipetted into a tube containing 7.4 ml of the ascorbic acid–phosphate buffer described above in the section on serum. The contents are mixed and allowed to incubate at 37° for 90 minutes; the test tube is then autoclaved for 10 minutes at 121°. After cooling, the tubes are centrifuged for 15 minutes at 3000 rpm in a desk-top centrifuge, and the clear supernatant, a 1:75 dilution of original whole blood, is collected. For whole bloods containing between 50 and 125 ng/ml, an aliquant of the supernatant solution is diluted with 2 volumes of 0.05 M phosphate buffer, pH 6.1. This represents a dilution of 1:225 of original whole blood. For whole bloods containing levels of folic acid activity above or below this range, the 1:75 dilution of whole blood should be diluted with sufficient 0.05 M phosphate buffer, pH 6.1, to provide a solution containing per milliliter between 200 and 500 pg of folic acid activity.

Mechanics for Setting Up the Assay

All three of the assay microorganisms described above are microaerophilic anaerobes and have similar environmental requirements. Thus assays with all three microorganisms are set up in a similar manner.

The test tubes used for these assays should be calibrated for equal light transmission so that turbidimeter readings of the contents can be made directly in the test tubes without transferring to a cuvette. The test tubes should be of good quality glass and 13 × 100 mm in size. The assay should not be carried out in Erlenmeyer or micro-Fernbach flasks, as is done in

[20] O. D. Bird and V. M. McGlohon, *in* "Analytical Microbiology" (F. Kavanagh, ed.), Vol. II. Academic Press, New York, in press.

some *L. casei* folic acid activity assays.[13,14] The latter containers are more suitable for aerobic organisms.

The final volume for these assays is 2 ml, consisting of 1 ml of basal medium and 1 ml of sample in phosphate buffer.

For the tubes containing the reference standard, 5 concentrations of appropriate reference material (see section on reference standard) are used in duplicate. The volume is brought to 1 ml where necessary with 0.05 *M* phosphate, pH 6.1. Two sets of tubes are included which contain 1 ml of 0.05 *M* phosphate and no standard solution. These are control tubes, and their use is explained below.

Five concentrations of each sample are also run in duplicate, and the volume is adjusted where necessary to a volume of 1 ml with 0.05 *M* phosphate, pH 6.1. Under these conditions, the concentration of phosphate is the same in all the standard and sample tubes.

The method of setting up the assay is illustrated in Table IV.

TABLE IV

METHOD OF SETTING UP THE ASSAY TUBES FOR THE THREE MICROBIOLOGICAL ASSAYS FOR FOLIC ACID ACTIVITY

Constituent	Tube number						
	0	1	2	3	4	5	6
Reference Standard Solution[a] (ml)	0	0	0.1	0.2	0.4	0.8	1.0
0.05 *M* Phosphate (ml)	1.0	1.0	0.9	0.8	0.6	0.2	0
Prepared sample (ml)	—	0.2	0.4	0.6	0.8	1.0	—
0.05 *M* Phosphate (ml)	—	0.8	0.6	0.4	0.2	0	—

[a] For *Lactobacillus casei* assay, Reference Standard Solution = pteroylglutamic acid, 500 pg/ml. For *Streptococcus faecalis* assay, Reference Standard Solution = pteroylglutamic acid, 1000 pg/ml. For *Pediococcus cerevisiae* assay, Reference Standard Solution = L-5-formyltetrahydropteroylglutamic acid, 500 pg/ml.

One milliliter of basal medium is then added to each tube which is then covered with an aluminum cap. Cotton may be used, but loose-fitting aluminum caps are easier to use and work just as well. The racks containing the tubes are shaken several times to mix the contents. The tubes are then autoclaved for 2 minutes at 121°. The tubes are then placed in a cold water bath for several minutes until the contents reach room temperature.

Each tube except one series containing 1 ml of 0.05 *M* phosphate buffer and 1 ml of medium, are then inoculated with one drop of inoculum prepared as described above. The tubes are shaken several times and then placed in a 37° incubator for 16–20 hours.

Reading the Assay Results

The end point for these assays is a turbidimetric reading in a colorimeter of the suspended bacteria in each assay tube. The colorimeter is adjusted to 660 nm, which tends to minimize the effect of the yellow color of the basal medium. The instrument is set to 100% transmission with the uninoculated color blank (test tube containing 1 ml of basal medium and 1 ml of 0.05 M phosphate buffer, pH 6.1). The inoculated blank serves to check that the reagents used in the assay contain no folic acid and that the proper bacterial inoculum is used. It should have a transmittance of 95–100%. The assay is ready to be read if the tube containing the highest concentration of standard has a transmittance of 50% or less.

The bacteria in each tube are thoroughly suspended using a Vortex Jr. mixer. The transmittance of each tube is then read and recorded.

Calculating the Assay Results

A standard curve, relating concentration of reference standard to transmittance of light through the bacterial suspension is plotted. For ease and accuracy of calculation, this plot should be a straight line for its useful portion.[21] For the *Lactobacillus* assay, this is achieved through the use of semilogarithmic graph paper. Percent transmittance is plotted on the linear ordinate, and concentration of pteroylglutamic acid on the semilogarithmic abscissa. The linear portion of this curve is shown in Fig. 2.

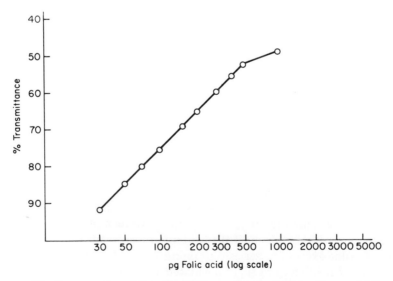

FIG. 2. The linear portion of the standard curve for the *Lactobacillus casei* folic acid activity assay.

[21] E. C. Wood and D. J. Finney, *Quart. J. Pharm. Pharmacol.* **19,** 112 (1946).

The useful linear portion of the standard curve in the *L. casei* assay ranges from a concentration of 25 to 500 pg of pteroylglutamic acid per tube; for the *S. faecalis* assay, from a concentration of 100 to 1000 pg of pteroylglutamic acid per tube; and for the *P. cerevisiae* assay, from 30 to 500 pg of L-5-formyltetrahydropteroylglutamic acid per tube.

For the *P. cerevisiae* and *S. faecalis* assays, a straight line is usually achieved for the reference standard with linear paper. If this does not occur, semilogarithmic graph paper should be employed.

The potency of each dilution of sample is calculated from the standard curve by determining the folic acid concentration for each transmittance value. The value for the sample calculated from each dilution of sample should not vary by more than 10%. If at least eight of the ten values from the different dilution of each sample do not agree within 10%, the sample is rerun.

If the results are to be statistically significant, it is essential that as many dilutions of either serum of whole blood be run as of reference material.[21-25] If this procedure is followed, the coefficient of variation, (1 standard deviation/mean), of the average serum or whole blood will be in the order of 4%. Using many dilutions of standard, but only one or two dilutions of unknown serum or whole blood, will decrease the accuracy and precision of the assay.

Choosing an Incubator

The maintenance of an exact temperature by the incubator to be used is not as crucial as the maintenance of a constant temperature throughout the box.[26] This cannot be attained in an incubator that heats by convection. In the latter type, variations of several degrees may exist between racks of tubes at different levels of the incubator, and this affects the assays adversely, especially those with short (16–24 hour) incubation periods.

It is essential that a forced draft incubator be used which contains a blower device to circulate the air. A vibrationless stirred constant-temperature water bath may also be used.

Cleaning Glassware

Chromic acid mixtures should be avoided if possible, since chromate ions often adhere to glassware and are later eluted into the assay medium causing inhibition of the growth of the assay microorganisms.

[22] E. C. Wood, *Nature* **155**, 632 (1945).
[23] D. J. Finney, *Quart. J. Pharm. Pharmacol.* **18**, 77 (1945).
[24] E. C. Wood, *Analyst* **72**, 84 (1947).
[25] E. E. Snell, *Physiol. Rev.* **28**, 255 (1948).
[26] M. H. Thomas, *in* "Methods of Vitamin Assay," 3rd ed., p. 43. Wiley (Interscience), New York, 1966.

A good laboratory detergent is preferred, and a solution is prepared according to the direction of the manufacturer. Tubes and beakers should be soaked for several hours in the detergent solution. They should then be rinsed ten times with hot water and three times with distilled water. No trace of detergent should remain, since detergents often affect the growth of the assay bacteria. To see whether or not the glassware is clean, add distilled water and then drain the water out. A smooth film of water should form. If droplets appear on the surface or the film is not smooth, rewash the glassware.

Pipettes are soaked for several hours in a tall cylinder containing detergent solution. They are rinsed for 4–5 hours in an automatic pipette washer. They are then drained and placed in a tall cylinder filled with distilled water and allowed to remain overnight. After draining, the pipettes are placed in a drying oven until dry.

Conclusions

The microbiological assays for serum and whole blood folic acid activity are superbly sensitive analytical techniques which are capable of great precision when properly set up. Three microorganisms are used for these assays, *Lactobacillus casei*, *Streptococcus faecalis*, and *Pediococcus cerevisiae*. Although the *L. casei* assay is employed most frequently by investigators, the other assay microorganisms are useful to differentiate serum and whole blood folates.

[177] Microbiological Assay of Folate by Thin-Layer Cup-Plate Method with *Streptococcus faecalis*

By Tohru Tsukahara and Masako Yamada

Nowadays, agar plate diffusion methods are not generally employed for the assay of vitamins by reason of their relative insensitivity to vitamins. The primary application of the assay methods has been only for the control of pharmaceutical preparations. However, the methods have such main advantages as their simplicity and high precision. So far as we know, the use of agar plate methods for folic acid assay has not yet been reported. In view of these advantages, we have developed the application of a cup-plate method with *Streptococcus faecalis* as test organism to the estimation of folic acid. In order to enhance the sensitivity of the method, a thin-layer agar plate technique was successfully adopted. Our experience has shown that a small amount of folacin contained in an agar added to assay medium causes the appearance of clear, sharp, and reproducible zones on the assay plates. Thus, in this procedure it is important to use an agar from which folic acid has been extracted as completely as possible. Further, we attempted to apply the cup-plate method using a thin-layer plate to several samples from animal and vegetable sources, and parallel assay was made with the turbidimetric method. In the assay of most samples used, the method gave a satisfactory result and was in accord with that given by the tube method. In some samples, however, the results by the plate and tube methods differed widely, and the recovery rates in the plate technique were unsatisfactory. For general acceptance of the method, it will be necessary to apply it to a wide variety of sample types and more comparison studies must be made with other techniques.

Test Organism

The test organism is *Streptococcus faecalis*, strain ATCC 8043. Stock cultures of this organism are maintained by monthly transfers to a nutrient agar cylinder medium of the following composition: yeast extract, 5 g; peptone, 12.5 g; glucose, 10 g; KH_2PO_4, 0.25 g; K_2HPO_4, 0.25 g; $CH_3COONa \cdot 3 H_2O$, 10 g; $MgSO_4 \cdot 7 H_2O$, 0.1 g; $MnSO_4 \cdot H_2O$, 5 mg; $FeSO_4 \cdot 7 H_2O$, 5 mg; agar, 20 g; and distilled water to make 1000 ml. The fresh cylinders inoculated with the organism are incubated at 30° for 18–24 hours and are stored at 4°.

Standard Solution

Ten milligrams of crystalline folic acid, which has been dried to constant weight and stored over $CaCl_2$ in a desiccator, is accurately weighed and dissolved in approximately 50 ml of 0.01 N NaOH containing 10% ethanol. The solution is adjusted to pH 7–8 with 0.1 N HCl solution and diluted with additional 10% ethanol solution to make the folic acid concentration exactly 100 μg/ml. This stock solution is stored in the dark at 5°.

A standard solution of folic acid may be obtained by an appropriate dilution of stock solution with distilled water in a wide range (1.25–160 mμg/ml). Fresh standard solution should be prepared from stock solution for each assay.

Preparation of Samples for Assay

Free Folate Extract

A weighed sample of natural products is chopped in small pieces and placed in a flask. A volume of 0.1 M phosphate buffer solution at pH 7 equal in milliliters to the wet weight of the sample in grams is added, and the mixture is boiled in a water bath for 5 minutes. Homogenize the mixture in a Potter-Elvehjem homogenizer for about 5 minutes; add a volume of 0.1 M phosphate buffer solution, pH 7.5, equal to 4 times the sample weight so that free folic acid may be extracted. Centrifuge the mixture, and filter the supernatant solution. Take an aliquot of the clear solution and dilute with phosphate buffer solution at pH 7–7.5 to a measured volume containing 1.25–160 mμg of folic acid per milliliter. This is the assay solution for free folic acid content.

Free and Bound Folate Extract

For the determination of the total folic acid content in biological materials, the combined types of folic acid (called conjugates) that cannot be utilized by the test organism must be released. To liberate free folic acid from these conjugates, enzymatic treatment of samples of animal and vegetable sources is frequently performed using a conjugase preparation from chick pancreas and hog kidney. Two types of conjugases have been described. One of them, a carboxypeptidase, occurs widely in nature.[1] The other one, a γ-glutamic acid carboxypeptidase, was isolated from chick pancreas.[2] The hog kidney and chick pancreas conjugases may be prepared as an aqueous suspension in the following manner.

Homogenize fresh defatted hog kidney with 3 parts of distilled water, centrifuge the mixture at 670 g for 20 minutes, and filter the supernatant solution through Hyflo Super-Cel.[3] The clarified filtrate can be stored in the frozen state.

The chick pancreas conjugase is prepared according to the method of Laskowski *et al.*[4,5] Homogenize fresh chicken pancreas in a Potter-Elvehjem homogenizer with 2 volumes of 0.1 M, pH 7, phosphate buffer. After autolysis for 24 hours at 37° under toluene, the autolyzate is centrifuged at 10,700 g for 30 minutes, and the middle brownish layer is taken out by capillary pipette from the upper fat layer and centrifuged residue. Add to this brownish liquid an equal volume of 0.1 M tricalcium phosphate solution and stir thoroughly. After centrifugation of the mixture for 20 minutes at 670 g, the supernatant is chilled to 5°, mixed with an equal amount of ice-cold absolute ethanol added drop by drop, and maintained at 5° for about an hour. The mixture is centrifuged, and the supernatant solution is discarded. Suspend the gray-white precipitate in 0.1 M, pH 7,

[1] O. D. Bird, S. B. Binkley, E. S. Bloom, A. D. Emmett, and J. J. Pfiffner, *J. Biol. Chem.* **157**, 413 (1945).

[2] A. Kazenko and M. Laskowski, *J. Biol. Chem.* **173**, 217 (1948).

[3] O. D. Bird, B. Bressler, R. A. Brown, C. J. Campbell, and A. D. Emmett, *J. Biol. Chem.* **158**, 631 (1945).

[4] M. Laskowski, V. Mims, and P. L. Day, *J. Biol. Chem.* **157**, 731 (1945).

[5] V. Mims and M. Laskowski, *J. Biol. Chem.* **160**, 493 (1945).

phosphate buffer and stir for 1 hour in the cold. The almost clear super-natant solution obtained by the final centrifugation contains the conjugase, which can be stored in the frozen state.

The assay solution for total folic acid can be obtained by the following procedure. A measured quantity of the assay solution for free folic acid prepared as described above is placed in a flask, mixed with 1 ml each of conjugase from chicken pancreas and hog kidney, and further mixed with 0.7 mg of cysteine hydrochloride per milliliter of the mixture. Incubate the mixture overnight at 37° under a small amount of toluene. Remove the toluene layer by capillary pipette, and dilute the mixture to a convenient volume. The assay solution will be obtained by filtration of the mixture.

Purification of Agar

A small amount of free folic acid contained in an agar added to the assay medium must be eliminated because of its effect on the sharpness of the zones of exhibition. Otherwise, a sharply defined growth zone can never be obtained owing to the clouding of the whole surface of the assay plate by a propagation of the test organism. Therefore, for the determination of folic acid by the plate technique, a vitamin-free agar must always be used. The agar, from which folic acid has been extracted as completely as possible, may be prepared in the following manner.

Gelatinized agar chopped in small pieces is suspended in a 0.1 M phosphate buffer solution, pII 7.5. After stirring by magnetic mixer for about 2 hours at 40°, incubate the mixture overnight at room temperature and exchange the buffer for a fresh solution. Repeat this stirring procedure several times, and collect the chopped agar by light centrifugation. Finally wash the agar repeatedly in running water for 1 hour to obtain an agar relatively free from folic acid activity.

Mechanics of the Assay

Design

In our laboratory the petri dish method is principally used, because of its simplicity, in the assay design. The petri dishes must be flat-bottomed so as to obtain an agar plate with uniform thickness; diameters of 90–100 mm are of convenient size.

Four dose levels of both standard and sample are used. Standard solutions of folic acid and test solutions from samples must be diluted to fall within the range 1.25–160 mμg of folacin per milliliter, over which the relationship between log dose and zone diameter has been demonstrated to be linear. Two sets of plates, each one-quarter full, are prepared; the sets are tied together to obtain the dose-response line for standard and the line for sample. The sample plates are replicated 5 times, each plate accommodating 4 stainless steel cylinders having an outside diameter of 8 mm (±0.1 mm), an inside diameter of 6 mm (±0.1 mm), and a length of 10 mm (±0.1 mm). Two alternate cylinders on each sample plate are used for the application of 2 dose levels in the ratio of 1:4 of the unknown solution, and the remaining 2 cylinders are used for the application of 2 dose levels of the already known concentrations of standard. The arithmetic mean response for each dose level is used to construct the dose-response line.

Assay Media

For the assay medium, one can employ the basal medium which has been recommended by the Association of Official Agricultural Chemists (A.O.A.C.).[6] The composition of the assay medium is given in the table. In practice it is convenient to prepare the stock solution of the various salts and vitamins and to store them under refrigeration until needed.

ASSAY MEDIUM

Acid-hydrolyzed casein	5 g
L-Cysteine	250 mg
DL-Tryptophan	200 mg
L-Asparagine	100 mg
Adenine sulfate	5 mg
Guanine hydrochloride	5 mg
Uracil	5 mg
Xanthine	10 mg
Glutathione	2.5 mg
Thiamine hydrochloride	200 μg
Riboflavin	500 μg
Pyridoxine hydrochloride	2 mg
Niacin	400 μg
Calcium pantothenate	400 μg
p-Aminobenzoic acid	500 μg
Biotin	10 μg
Potassium phosphate, monobasic	3.1 g
Magnesium sulfate, heptahydrate	200 mg
Ferric sulfate, heptahydrate	10 mg
Manganese sulfate, monohydrate	110 mg
Sodium citrate, dihydrate	25 g
Glucose, anhydrous	20 g
Polysorbate 80	50 mg
Agar (purified)	10 g
Distilled water to make	1000 ml
Final pH	7 ± 0.2

[6] "Official Methods of Analysis of the Association of Official Agricultural Chemists" (W. Horwitz, ed.), 9th ed. A.O.A.C., Washington, D. C., 1960.

Dispense this assay medium into a glass tube (27 ml in a 100-ml tube), plug with gauze-covered nonabsorbent cotton, and sterilize by steaming at 120° for 5 minutes. Five assay plates can be prepared from one glass tube. The medium is stored under constant refrigeration. Our experience has shown that the medium is satisfactory if not more than 1 week old. Some commercial sources of the basal media, e.g., Difco-Bacto-folic acid A.O.A.C. medium and Nissan-folic acid A.O.A.C. medium made in Japan, are available for a simple procedure.

Inoculum

Make a transfer from the stock culture of *Streptococcus faecalis* to a tube containing 10 ml of sterile inoculum medium of the following composition: glucose, 11 g; yeast extract, 5.5 g; peptone, 12.5 g; KH_2PO_4, 0.25 g; K_2HPO_4, 0.25 g; $CH_3COONa\cdot3\ H_2O$, 10 g; $MgSO_4\cdot7\ H_2O$, 0.1 g; $MnSO_4\cdot H_2O$, 5 mg; $FeSO_4\cdot7\ H_2O$, 5 mg; and distilled water to make 1000 ml. The culture is incubated at 30° for 16–18 hours. Centrifuge the culture of the third transfer, decant the supernatant liquid, and wash the cells 3 times with sterile saline. Resuspend the washed cells in sterile 0.9% NaCl solution. Adjust the washed cell suspension to an optical density of 0.25 with 0.9% NaCl solution using a photometer set at a specific wavelength between 540 and 660 nm. This suspension is used for the seeding of the assay plates.

Standard

Dilute aliquots of the stock solution of folic acid standard with phosphate buffer solution, pH 7–7.5, to a range of 1.25–160 mμg/ml. In this wide range, the folic acid assay gives a linear semilog response. Fresh standard solution should be prepared prior to use on the day of the assay. Four doses that are evenly spaced on the logarithmic scale of the standard response line must be prudently selected by the assayer in the light of his knowledge of the natural products which he is assaying.

Samples

Dilute all sample solutions with phosphate buffer solution to the assay range of 1.25–160 mμg of folacin per milliliter. The pH of the test solutions used has been found to affect the growth zone sizes to some extent. For this reason the pH of all final sample diluents must be the same as the pH of the standard solutions.

Preparation of Assay Plates

The assay medium of 27 ml in a 100-ml tube is melted by immersing the tubes in a boiling water bath. The tubes are cooled to approximately 48° and maintained at that temperature, and 3 ml of the inoculum is added to each tube. The contents of each tube are thoroughly mixed by rotation and stirred with a pipette to distribute the bacterial suspension evenly throughout the medium. Five milliliters of the contents of the tube is applied to each sterile petri dish, which is kept level before plating, and the contents are allowed to harden. It is necessary to use perfectly flat petri dishes in order to obtain a thin layer of agar of uniform thickness.

Four cups are placed on the seeded agar surface so that they are at approximately 60° intervals on a 2.8-cm radius. The plates are allowed to stand half an hour at room temperature to ensure proper seating of the cups on the agar surface. Cups are then filled with the appropriate concentration of standard and samples, respectively.

Assay Conditions

It has been indicated by Lees and Tootill[7] that various physicochemical and biological factors affect the size, density, and definition of the zones of exhibition on petri dishes. Therefore, for obtaining a clear and well-defined growth zone of desirable size, the assay procedure must be conducted under defined conditions. In folic acid assay, the following conditions are preferred.

pH of Assay Medium. The pH of the assay medium must be decided from the standpoint of both pH optimum for the growth of test organism and pH effective for the stability of folic acid. Our experiment has shown that in the pH range 6.8–7.5 the assay medium gives best results.

Agar Content of Assay Medium. The agar content of the assay medium must be considered in the light of its influence on the diffusion rate of folic acid under assay and the solidity of the assay plate which does not interfere with the assay procedures. Our experiment has also shown that 1% agar concentration is satisfactory for the purposes.

Depth of Seeded Agar in Assay Plate

The size of the zones of exhibition in a vitamin assay by plate technique varies inversely as the depth of seed layer in the assay plate. A thin-layer plate has the effect of inflating the zone size, and the layers must be uniform in thickness.

pH of Test Solution

The size of the growth zones is to some small extent related to the pH of the test solution applied to the plates. The pH of all final sample diluents should be adjusted with phosphate buffer solution to the same pH as the standard solutions.

[7] K. A. Lees and J. P. R. Tootill, *Analyst* **80,** 95 (1955).

Density of Seeding of Test Organism

The density of seeding of the test organism exerts a great influence on the size and sharpness of the zones of exhibition in vitamin assays. The final density of inoculum must be chosen to give convenient zone sizes while retaining satisfactory clarity and sharpness of the zones. This final density of inoculum in folic acid assay by the method is 0.025 OD (optical density) determined by a photometer set at 600 nm specific wavelength.

Relative Times for Diffusion of Folic Acid before Test Organism Begins to Grow

If the assay plates are allowed to stand at low temperature for any length of time after application of the test solutions, diffusion of the test solution proceeds but the test organism does not begin to grow, and consequently the growth zone sizes may be inflated. A prolonged time for diffusion of the test solution, however, will yield an ill-defined zone of exhibition. A diffusion time of 2–3 hours for the test solution has been found to be satisfactory for adequate development of the assay zones in the thin-layer technique.

Temperature and Length of Time of Incubation

Incubate the assay plates at 30° for 16–18 hours. This incubation time is sufficient to attain exhibition zones of a desirable size and definition. Further incubation does not alter the size of the zones.

The zones of exhibition that develop under the assay condition described above appear as an opaque and double circular area surrounding the cylinder containing the vitamin; the internal one is thick in bacterial growth; the external one is thinner.

Measuring the Response

Remove the assay plates from the incubator and take away the cups from their surface. The diameters of the zones of growth may be measured with calipers or a transparent celluloid ruler. The use of an illuminated reading box will be of great help for measuring the diameters. We prefer to read the zone diameters to the nearest 0.1 mm using a needle-point calipers. This method of reading is useful for avoiding the errors due to parallax.

Calculation of Results

Dose-Response Line and Assay Scale

The standard dose-response line is constructed by plotting the average response value (diameter of zones of exhibition in millimeters) for each of the standard levels against the logarithms of the concentrations (mµg/ml)

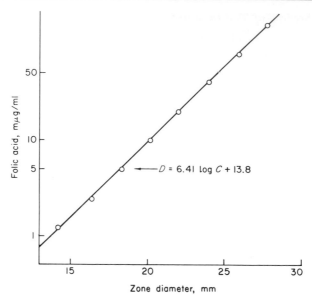

FIG. 1. Response of *Streptococcus faecalis* to folic acid as measured by the thin-layer cup-plate method and the "line of best fit" for the plotted points.

of the standard folic acid and by joining these plotted points. As indicated in Fig. 1, the plotted point that is the average from 5 cups is proportional to the logarithms of the amounts of folic acid placed in the cups. Under our conditions, this relationship holds with concentrations of folic acid over a wide range of 1.25–160 mμg/ml. For this technique, it is essential to ascertain the situation of these observed points on a straight line of the dose-response. Figure 1 also shows the "line of best fit" for the observed data drawn by the method of least squares. The observed points agree well with the computed ones. This fact may be expressed by the following equation: $D = \alpha \log C + \beta$, where α and β are constants, D is the zone diameter in millimeters, and C the folic acid concentration in millimicrograms per milliliter.

The application of both a thin-layer plate technique and a time of standing before incubation to the cup-plate method for the assay of folacin is fairly effective for extending the lower limit of sensitivity of the assay. In the thin-layer method the lower limit of sensitivity is 1.25 mμg/ml concentration, whereas it is 10 mμg/ml in the conventional plate method without any time for diffusion.

Calculation of Sample Potency

For this purpose, 4 dose levels of both standard and sample are used. Ten plates are required for each sample and they are separated into 2 sets. Two alternate cups on each plate are filled with the sample dilutions of 2 doses in the ratio of 1:4, and the remaining 2 cups with 2 standard dilutions in the same ratio. The sample plates are replicated 5 times. After incubation of the plates for 18 hours at 30°, the zones of exhibition are measured. The zone diameters for 4 dilutions of both sample and standard from 5 cups are each averaged. Then the theoretical values for the observed data of the standard are computed by the method of least squares, and the "line of best fit" is plotted on a semilogarithmic graph paper. The observed values for the sample are read from the dose-response line to obtain the corresponding folic acid concentration ($m\mu g/ml$) in each dilution of the sample. This concentration is multiplied by the appropriate dilution factor to obtain the potency of the original sample. In this way, for each level of assay solutions used, calculate the vitamin content of the sample. The potency of the sample is determined from the average of the values obtained from dose levels that do not vary by more than $\pm 10\%$ from this average. If the number of acceptable values remaining is fewer than 3, sufficient data are not available for calculating the potency of the sample.

Validity of Assay

The possibility may exist that the test samples, after treatment, contain some factors affecting the valid assay of folic acid by the plate method. To prove the validity of the method, the ability of the assay technique to give theoretical results for folic acid added to each sample should be examined. This is tested by adding different amounts of folacin to a sample solution and assaying as previously described. As a general rule, the recovery rate of the added folacin must fall within $100 \pm 5\%$.

The validity of the method is also confirmed by observing the parallelism between the standard and sample response lines. A remarkable drift of this line for sample from that for standard may indicate an interference with the reliable assay of folacin by the method.

The precision of the thin-layer plate method is further demonstrated by the applications of statistical method. Our statistical analysis of the assay results has shown that when using 5 plates, the probable errors mount to 20% at highest. These errors in the conventional plate method, however, are only 10%. The application of a thin-layer plate technique to the plate method for the assay of folate results in a marked increase in the sensitivity, but incurs a relative lowering of the reproducibility, that is the precision of the method.

Limitations of the Thin-Layer Plate Method

The method has limitations owing to mechanical machinery; particularly it is relatively low in sensitivity in relation to the turbidimetric assay method. This makes it impossible to assay the folic acid content of a natural product containing less than 50 mμg of the vitamin per gram of the original sample.

According to our investigations, the thin-layer plate method is effective for improvement of the recovery rates of the added folacin to the samples used, and it gives results in good accord with those given by the tube method. Some of the samples, however, are unsatisfactory in the recoveries, and the dose-response lines for the samples are not parallel with that for standard. In order to eliminate such disadvantages encountered in the plate method, test solutions from biological materials must be free from the factors affecting the assay results.

[178] Isolation of Pteridines by Ion-Exchange Chromatography

By H. REMBOLD

Principle

Unconjugated pterins and lumazines are heterocycles with considerable differences in base strength, depending upon the polarity of their ring substituents. For this reason chromatography by ion exchange is especially useful for the separation of pteridine mixtures and for their isolation from biological extracts.[1] By a standardized chromatography on the strong anion exchanger Dowex 1-X8 formate, they can first be divided into three main groups. The first one is eluted by water, the second by weak formic acid, and the third by formic acid–formate buffers. Owing to the good reproducibility of this method, the specific retention volume of every pteridine can be determined unequivocally. For this reason an expected compound can be reproducibly localized in chromatographies of biological extracts even when it is present in a very low amount. As an example, the isolation of 0.03 mg of neopterin contained in a TCA-extract from 300 g of bee pupae is described.[2]

The preliminary separation of a pteridine mixture or of a biological extract on Dowex 1-X8 is followed by a purification step on a cellulose ion-exchanger. Pteridines that are eluted from Dowex with water or with formic acid concentrations lower than 0.5 N can be separated with good success on phosphocellulose, owing to their amphoteric properties. Acidic pteridines are fractionated by anion exchange on Ecteola-cellulose. By

---

[1] H. Rembold and L. Buschmann, *Z. Physiol. Chem.* **330**, 132 (1962).
[2] H. Rembold and L. Buschmann, *Ann.* **662**, 72 (1963).

this means concomitant substances can be removed in most cases. Pteridines exhibit a typical fluorescence when illuminated with ultraviolet light (260 and 350 nm, respectively) in solution or after separation by paper or thin-layer chromatography. Quenching of fluorescence by acids or by strong UV absorption must be borne in mind.

The separation of 6- and 7-polyhydroxyalkylpterins[3] and of radioactive pteridine metabolites[4] by ion-exchange chromatography is described in this volume.

Preparation of the Ion-Exchange Columns

Dowex 1 Formate. Dowex 1-X8 chloride, 200–400 mesh, is washed with water by decantation until the fines are removed. The resin is placed in a glass tube (2 cm i.d., 60 cm length), equipped with a sintered-glass disk, a stopcock, and a ground joint, and washed with a solution of 3 *M* sodium formate until the resin is free from chloride ions (about 0.7 liter). The resin is then suspended in 500 ml of 7 *N* formic acid/1 *N* sodium formate, degassed, poured as a suspension into the column, and allowed to settle by gravity. The resin is washed with 200 ml of 88% formic acid, then with degassed water until the eluent is about pH 5. The column is finally adjusted to a length of exactly 45 cm. For regeneration, the resin is taken out of the glass tube and washed in the same way as above with 500 ml of 7 *N* formic acid/1 *N* sodium formate, 200 ml of 88% formic acid, and water.

Phosphocellulose. Phosphocellulose must have a capacity of 0.8–1.0 meq/g and must settle from a degassed water suspension almost completely within 30 minutes. The resin is suspended in 20 volumes of water and is decanted after 20–30 minutes. Sedimentation is repeated as often as fines remain in the supernatant solution after 20 minutes of standing (4 or 5 times). A glass tube (i.d. 2–5 cm, 15–30 cm in length) is filled one-third with water and is provided, instead of the sintered-glass disk, with a pad of cotton wool free from air bubbles. Then the resin is poured in as a thin slurry to obtain a bed 2–3 cm high. When the water has just disappeared from the top of the resin the material is packed by slight pressure applied by glass rod with a flattened end. Then water is filled in carefully, and another layer is piled up. This stepwise packing procedure is a prerequisite for preparative-scale separations with columns 2–5 cm i.d. and 15–30 cm high. To obtain homogeneously packed beds it is of equal importance to regenerate the resin in the column. It is first washed with a 10-fold bed volume of 5% formic acid/5% sodium formate (v/w), then with the same

[3] See chapter by Rembold and Eder, this volume [181].
[4] See chapter by Rembold and Gutensohn, this volume [200].

volume of 5% formic acid, and finally with water until neutrality. For analytical separations a column with 1.5 cm i.d. and 20 cm length is sufficient.

Ecteola-Cellulose. As described for phosphocellulose, the resin (capacity about 0.4 meq/g) is suspended in water several times and freed from fines. The column (i.d. 1.5 cm, length 20 cm) is successively washed with 500 ml of 1% sodium hydroxide, 500 ml of 1% formic acid, and water almost to neutrality. By the same procedure the column is regenerated without taking out the resin from the tube.

Chromatographic Separations

Standard Chromatography on Dowex 1-X8. The separations described below have to be carried out in an orange or, better, dark red light because of the light sensitivity of many pteridines. The extract (pH 6, 50–80 ml) is layered on the column, and the flow rate is adjusted to about 60 ml/hour. After a 3-fold wash with each 10 ml of water, a 5-cm layer of water is put on the column, which is subsequently connected to a reservoir containing degassed water. Elution is started with about 1.5 liters of water, and 10 ml fractions are collected. For the subsequent gradient elution a mixing chamber (1-liter round-bottom flask with magnetic stirrer) is put between the column and the reservoir. The mixing flask is filled with 500 ml of water. The reservoir is then successively filled with 500 ml each of 0.4 N, 1 N, and 2 N formic acid. These solutions are followed by 250 ml of 4 N formic acid and then 500 ml each of the following: 4 N formic acid/0.2 M sodium formate; 4 N formic acid/0.8 M sodium formate; and 8 N formic acid/3 M sodium formate. The intensity of fluorescence and the optical density at 260 nm are measured for every fraction.

Neutral solutions are concentrated by means of a rotary evaporator. Formic acid is removed from acidic fractions by continuous extraction with ether for 1–2 hours. To remove sodium formate, the substances are adsorbed on charcoal, which has been partially deactivated with 4% stearic acid according to the procedure of Asatoor and Dalgliesh.[5] To the solution freed from formic acid by ether extraction (pH 4), some charcoal is added. After standing for 30 minutes with occasional shaking, the solution is filtered and a sample of the clear fluid is tested for fluorescence. If necessary, the adsorption is repeated with another portion of charcoal. After a thorough wash with about 100 ml of water, the combined charcoals are treated with 100 ml of a mixture of water saturated with phenol and of concentrated ammonia (5:1). The solution is evaporated to dryness *in vacuo*, 50 ml of water is added, and the evaporation is repeated in order to

---

[5] A. Asatoor and C. E. Dalgliesh, *J. Chem. Soc.* p. 2291 (1956).

remove traces of phenol. The residue is dissolved in water and chromatographed on Ecteola-cellulose.

Rechromatography on Cellulose Exchangers

This procedure is carried out with water and then with formic acid solutions of rising concentration (1–5%) without use of a gradient. Fractions are collected in 5-ml portions.

Permanganate Oxidation for the Detection of 6- or 7-Substituted Pteridines

The sample to be oxidized is concentrated to about 0.5 ml and is filled into a test tube together with 0.1 ml of 1% sodium hydroxide. Without heating, a concentrated $KMnO_4$ solution is added dropwise until the violet color persists for some minutes. The tube is then placed in a boiling water bath, and additional permanganate is added if necessary. After about 5–10 minutes the excess oxidant is destroyed with some drops of

TABLE I

R_f Values of Some Pteridines Using Ascending Paper Chromatography[a]

| Pteridine | Solvent system[b] | | | |
	a	b	c	d
Biopterin	0.28	0.32	0.59	0.62
7 Hydroxybiopterin	0.15	0.24	0.44	0 52
Isoxanthopterin	0.14	0.18	0.27	0.27
Isoxanthopterin carboxylic acid	0.04	0.02	0.57	0.38
6-Methylisoxanthopterin	0.18	0.18	0.23	0.24
Lumazine	0.28	0.31	0.53	0.62
6-Hydroxylumazine	0.28	0.15	0.46	0.50
7-Hydroxylumazine	0.14	0.26	0.30	0.38
6,7-Dihydroxylumazine	0.04	0.07	0.29	0.33
Lumazine-6-carboxylic acid	0.09	0.13	0.48	0.62
Neopterin	0.13	0.21	0.55	0.60
Pterin	0.28	0.29	0.44	0.46
6-Hydroxymethylpterin	0.21	0.25	0.44	0.48
Pterin-6-carboxylic acid	0.09	0.09	0.40	0.45
Tetrahydroxybutylpterin	0.05	0.16	0.59	0.60
Xanthopterin	0.21	0.15	0.35	0.33

[a] Paper: Schleicher and Schüll 2043b.

[b] Solvent systems: a = *n*-butanol–glacial acetic acid–water (20:3:7); b = propanol-1–1% aqueous ammonia (2:1); c = 4% Na-citrate solution; d = 3% NH_4Cl solution.

methanol, the solution is neutralized with diluted formic acid, and filtered by the use of a filter aid (e.g., Celite). When fluorescence can be observed, the clear solution is diluted to about 75 ml and filtered through a small Ecteola-column (7 × 80 mm), followed by a water wash to remove salts. The pterin carboxylic acids are eluted by 1% formic acid with a retention volume of about 35 ml. The fractions are brought to dryness in a rotary evaporator, and the residue is tested by paper chromatography.

Paper Chromatography

A list of R_f values for different pteridines is given in Table I. The spots are made visible by their fluorescence in UV light (260 and 350 nm).

Behavior of Pteridines on the Anion Exchanger Dowex 1-X8

The pterins can be divided into three groups depending upon their acidity and their behavior on the ion-exchange column (Table II):

Group 1: Pterins with basic, neutral, or extremely weakly acidic prop-

TABLE II

BEHAVIOR OF PTERINS ON DOWEX 1-X8[a,b]

Pterin	Amount (mg)
Eluted by water:	
Drosopterin (1)	Trace
6-Tetrahydroxybutylpterin (2)	1
6-Methylisoxanthopterin-O-glucoside (3)	0.5
Biopterin (4)	1
6-Hydroxymethylpterin (5)	1
Pterin (6)	1
Eluted by diluted formic acid:	
Lumazine (7)	1
7-Hydroxybiopterin (8)	1
Xanthopterin (9)	4
6-Methylisoxanthopterin (10)	1
Isoxanthopterin (11)	1
Ekapterin (12)	Trace
The same after addition of HCO_2NH_4:	
7-Hydroxylumazine (13)	1
Pterin-6-carboxylic acid (14)	2
Pterin-7-carboxylic acid (15)	0.5
Isoxanthopterin carboxylic acid (16)	1
Leucopterin (17)	5
Xanthopterincarboxylic acid (18)	1
7-Hydroxylumazinecarboxylic acid (19)	1
Erythropterin (20)	1

[a] The quantities used in the chromatography are given. For the numbers in parentheses compare Fig. 1.

[b] Reproduced from Reference 1.

erties. They are eluted from the column with water (e.g., 6-polyhydroxy-alkylpterins and pterin).

Group 2: Weakly acidic pterins are eluted with diluted formic acid (e.g., xanthopterin, isoxanthopterin, and their derivatives).

Group 3: Pterins with strongly acidic functions are eluted by formic acid only after addition of formate. In this group are located pterins with carboxyl groups and leucopterin.

The behavior of a series of pterins is shown in Fig. 1. The curves demonstrate that the substances are separated very accurately by the gradient elution. By use of the flat formic acid gradient described above, the peaks are equally scattered throughout the chromatogram. The separation effect for the pteridines contained in group 1 is dependent upon the amount of concomitant substances. Biopterin (4), for example, exhibits an appreciable tailing effect. This may be prevented by guanine (17), which migrates

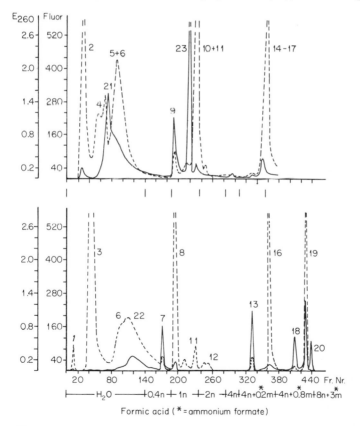

Fig. 1. Chromatographic behavior of pterins on Dowex 1-X8 formate. (————) OD at 260 nm; (- - - - - -) fluorescence. Substances 1–16 are compounds listed in Table II. Guanine (17), 10 mg; riboflavin (18), 1 mg; uric acid (19), 10 mg. Reproduced from Reference 1.

behind this pterin, pushing it out as a double peak. This effect is not found in the area of gradient elution. Here uric acid (19) is eluted as a sharp peak without interference with the other pterins. Decomposition of compounds tested have not been observed with the procedure described.

Fine Separation of Pterin Fractions on Cellulose Exchangers

After the group separation on Dowex 1-X8 several ways are possible for a further fractionation. Pterins which are eluted from Dowex 1-X8 with water already or with formic acid concentrations lower than 0.5 N should be rechromatographed on Dowex 1-X8 formate, eventually by use of another type of gradient. For further fractionation phosphocellulose is used. Table III gives the retention volumes of a series of pterins which are

TABLE III
RETENTION VOLUMES OF PTERINS ON ION EXCHANGERS (BIOPTERIN = 100)

Pterin	Phosphocellulose, water-elution	Dowex 1-X8, standard chromatogram
Biopterin	100	100
6-Tetrahydroxybutylpterin	81	42
6-Hydroxymethylpterin	106	140
Pterin	110	160
Xanthopterin	54	322
6-Methylisoxanthopterin	14	382
Isoxanthopterin	14	387
7-Hydroxybiopterin	10	327
Pterin-6-carboxylic acid	10	600
Pterin-7-carboxylic acid	10	585
Isoxanthopterin carboxylic acid	9	605

eluted from phosphocellulose with water compared with their behavior on Dowex 1. Acidic pterins are separated by the use of Ecteola-cellulose. A combination of Dowex 1 and a cellulose exchanger will allow the separation of a specific pteridine in every case.

Isolation of Pteridines from Biological Sources

The advantage of a fine separation by use of ion-exchange celluloses becomes evident when byproducts have to be removed from biological extracts. As an example, the fractionation of a TCA-extract from 2000 bee pupae (300 g fresh weight) is shown in Fig. 2. A pattern of fluorescing and of UV-absorbing zones is obtained. For the subsequent fractionation, the chromatogram is divided completely into distinct zones as follows:

1. Up to fraction 65 the substances are rechromatographed on Dowex 1-X8 and then on phosphocellulose.

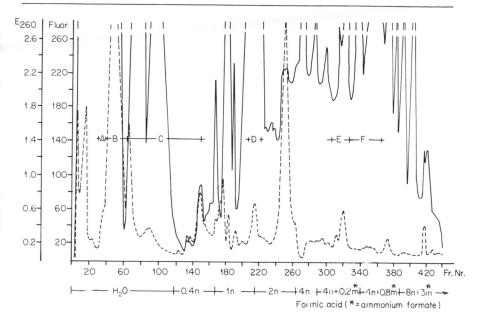

Fig. 2. Chromatography of a TCA extract on Dowex 1-X8 formate from 2000 bee pupae. Ten-milliliter fractions are collected. (——) OD at 260 nm; (- - - - -) fluorescence. A–F: Fractions from which a pteridine has been isolated (see also Table IV). Reproduced from Reference 2.

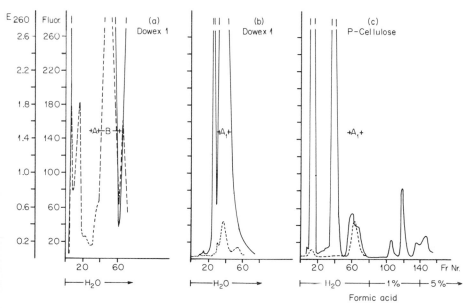

Fig. 3. Chromatographic enrichment of neopterin. (——) OD at 260 nm; (- - - - -) fluorescence. Reproduced from Reference 2.

2. Fractions 65–186 are further separated on phosphocellulose. If necessary, formic acid is first extracted with ether.

3. The zones following fraction 186 are freed from formic acid and from formate and are further fractionated on Ecteola-cellulose.

In control experiments, the behavior of such pteridines, which are on hand, can be determined on Dowex 1 as well as on cellulose exchangers. In such a case even traces of pteridines can be detected or be excluded by the use of paper chromatography.

Figure 3 presents the enrichment of 30 μg of neopterin as an example of the fractionation procedure described above. The compound is contained in zone A of the standard chromatogram on Dowex 1 (Figs. 2 and 3a). This fraction is rechromatographed on Dowex 1 in order to remove traces of concomitant biopterin, which is mainly located in zone B (Fig. 3b). UV-absorbing byproducts are finally separated by chromatography on phosphocellulose (Fig. 3c). By this procedure even traces of a certain pteridine may be highly enriched by use of only three chromatographic steps. The efficiency of this separation procedure can be seen from Table IV. All the pteridines isolated from this biological extract could be identified and quantitatively recovered in a quantity of even a few micrograms.

TABLE IV

PTERIDINES ISOLATED FROM A TCA EXTRACT OBTAINED FROM 2000 HONEY BEES

Pteridine[a]	Bee pupae (mg)	Flight bees (mg)
Biopterin (B)	0.4	0.4
Neopterin (A)	0.03	0.01
7-Hydroxylumazine (E)	0.5	0.8
Isoxanthopterin (D)	0.018	0.45
Pterin (C)	0.04	0
Pterin-6-carboxylic acid (F)	0.025	0.020
Isoxanthopterin carboxylic acid	0	0.007

[a] A–F, see Fig. 2.

[179] Separation and Identification of Folate Coenzymes on DEAE-Sephadex[1]

By P. F. Nixon[1a] and J. R. Bertino[1b]

The separation of naturally occurring folates has been achieved by column chromatography on DEAE-cellulose[1c,2] and on TEAE-cellulose,[3] and the separated fractions have been quantitated and identified by microbiological assay. However, the resolution of such columns is insufficient for adequate separation of several of the folate coenzymes. For the purpose of following the interconversion of a number of folate coenzymes by an alternative means, such as by radiolabel techniques, a chromatographic system is required by which the identity of compounds can be unambiguously assigned from their elution position relative to known markers.

A general method is now described by which folate coenzymes can be more satisfactorily resolved by column chromatography, and by which their elution can be more simply monitored, despite the presence of reducing agents which interfere with the ultraviolet absorbance spectra of folate coenzymes but which may be necessary to prevent their oxidation during chromatography. This general method does not replace those special conditions of elution from DEAE-cellulose which are designed to separate specific pairs of folate coenzymes (for example THF[4] and dihydrofolate[5]) or the diastereoisomers of 5,10-methylene-THF.[6]

Procedure

DEAE-Sephadex A-25, 40–120 μ bead size, is swelled in water for 24 hours, during which time any fines are removed by repeated decantation. The material is then stirred in 0.5 N KOH for 2 hours, washed with three changes of distilled water, washed with 0.5 M potassium dihydrogen phosphate, washed twice with distilled water, then washed and suspended in pH 6.0 potassium phosphate buffer, 0.1 M with respect to phosphate. All washes are removed by decantation. The DEAE-Sephadex is poured

[1] Supported by Grants CA08010 and CA08341 from the U.S. Public Health Service.
[1a] Merck Sharp and Dohme International Fellow in Clinical Pharmacology.
[1b] Career Development Awardee of the National Cancer Institute.
[1c] M. Silverman, L. W. Law, and B. T. Kaufman, *J. Biol. Chem.* **236**, 2530 (1961).
[2] O. D. Bird, V. M. McGlohon, and J. W. Vaitkus, *Anal. Biochem.* **12**, 18 (1965).
[3] E. Usdin, *J. Biol. Chem.* **234**, 2373 (1959).
[4] The abbreviation THF is used for 5,6,7,8-tetrahydrofolate.
[5] C. K. Mathews and F. M. Huennekens, *J. Biol. Chem.* **235**, 3304 (1960).
[6] B. T. Kaufman, K. O. Donaldson, and J. C. Keresztesy, *J. Biol. Chem.* **238**, 1498 (1963).

according to the manufacturer's recommendations[7] to form a column 0.9 ×
27 cm and is washed by 100 ml of "starting" buffer (0.1 M potassium
phosphate, pH 6.0, containing 0.2 M 2-mercaptoethanol) before use. The
column is run at 4° in the dark.

The extract containing folates is prepared in 0.2 M 2-mercaptoethanol,
so as to prevent the oxidation of reduced folates. Before application to the
column, phosphate buffer (pH 6) is added to the sample to achieve a concentration of approximately 0.1 M. Together with suitable markers, the
sample is applied to the column, then elution is commenced by "starting"
buffer. Fractions of 3–4 ml are collected at a flow rate of 40 ml per hour.
After collection of the first 10 fractions, the remaining fractions are eluted
by a linear gradient accomplished by use of 250 ml of starting buffer in
the mixing chamber, and, in the reservoir, 250 ml of 0.8 M potassium
phosphate buffer, pH 6.0, containing 0.2 M 2-mercaptoethanol.

Identification of Folate Coenzymes

For the identification of those folate coenzymes which may be radiolabeled during studies of the interconversion of the reduced folates, satisfactory markers have been found to be 0.5 micromole of p-aminobenzoyl-
glutamate and 2 micromoles each of THF and 5-formyl-THF. The elution
position of these markers, in the presence of 2-mercaptoethanol, may be
monitored by their fluorescence[8] emission at 365 nm on excitation at 305
nm. After measurement of the fluorescence of a fraction at pH 6, a 1-ml
portion of the fraction is adjusted to approximately pH 3 by the addition
of 0.2 ml of glacial acetic acid, and the fluorescence is again measured. The
above three markers are identified by the relative intensities of their fluorescence at these two pH values. The ratio of the fluorescence at pH 3 to the
fluorescence at pH 6 is approximately 0.8 for p-aminobenzoylglutamate, 1.1
for 5-formyl-THF, and 15 for THF.

The relative elution positions of the folate coenzymes are shown idealized in Fig. 1. The assigned position of 10-formyl-THF has been verified
by the absorbance spectral peak of 5,10-methenyl-THF at about 350 nm,
which can be generated by adjustment of those fractions which contain
10-formyl-THF to pH 1. The position of 10-formyl-THF has also been
verified by the use of radiolabeled material. It does not show appreciable
fluorescence under the conditions used above. The position of 5-methyl-
THF has been verified by radiolabeled material. Under the above conditions, its fluorescence is similar to that of THF. The identity of each eluted
material has also been confirmed by ultraviolet absorbance spectra, which

[7] Sephadex-gel filtration in theory and practice. Pharmacia Co.
[8] K. Uyeda and J. C. Rabinowitz, *Anal. Biochem.* **6**, 100 (1963).

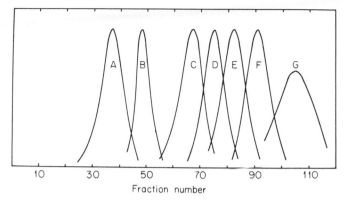

Fraction number

FIG. 1. Idealized diagram of the elution of folate coenzymes from a column of DEAE-Sephadex under the conditions described in the text. Fractions of 3.8 ml were collected, and the elution positions were as follows: A (peak fraction 37), 10-formyl-THF; B (48), p-aminobenzoylglutamate; C (67), 5-formyl-THF; D (75), 5-methyl-THF; E (82), THF; F (91), dihydrofolate; G (105), folate. The elution of (+),L,5,10-methylene-THF lies between D and E.

may be obtained by dilution of the peak fraction sufficient to reduce the concentration of 2-mercaptoethanol to 10 mM. The positions of (+),L-5,10-methylene THF, of DHF, and of folate have been assigned by use of radiolabeled compounds, and by absorbance spectra of the peak fractions.[9]

For specific purposes, other markers and other conditions of monitoring may be used. In our laboratory, the above conditions have consistently given the separations shown and have made possible determinations of the interconversion of radiolabeled folate coenzymes *in vivo*.

[9] The stability of 5,10-methylene-THF can be increased by the inclusion of 25 mM formaldehyde in the eluting buffers. It should be noted that this will result in the formation of 5,10-methylene-THF from THF.

[180] Synthesis and Assay of Tetrahydrohomofolic Acid and Tetrahydrohomopteroic Acid[1]

By R. L. KISLIUK

Tetrahydrohomofolic acid differs structurally from tetrahydrofolic acid only in having an additional methylene group between the pteridine and p-aminobenzoic acid portions of the molecule. It acts as a folic acid an-

[1] Supported by NIH Grant GM-1187102 and NSF Grant GB-1890.

tagonist in preventing the growth of *Streptococcus faecalis* (ATCC 8043)[2] and in prolonging the survival of mice inoculated with an amethopterin-resistant strain of L1210 leukemic cells.[3] Tetrahydrohomopteroic acid showed antimalarial activity against a strain of *Plasmodium cynomolgi* resistant to the antifolate pyrimethamine.[4]

Paper Chromatographic Analysis of Homofolic Acid and Homopteroic Acid[5]

Homofolic acid has been synthesized by three different routes.[6–8] Homopteroic acid is an intermediate in these syntheses and occurred as a contaminant in certain commercial preparations of homofolic acid. The two compounds can be separated by ascending chromatography on Whatman No. 3 filter paper.[9] Three μl of a 1% solution (pH brought to 11 with 1 N NaOH) were spotted on the paper, and 0.1 M ammonium bicarbonate was the solvent. The compounds were visualized as quenching spots under ultraviolet light. The R_f values for homofolic acid and homopteroic acid are 0.84 and 0.47, respectively.

One can estimate the amount of homopteroic acid in homofolic acid preparations by determining the ratio of absorption at 300 nm in 0.1 N KOH solutions. The value of this ratio for homopteroic acid is 0.60; for homofolic acid it is 0.88.

Folic acid can be separated from homofolic acid by ascending chromatography on Whatman No. 1 filter paper (1.5 $\times$ 57 cm strips) using 4% sodium citrate as the solvent.[9] One milligram per milliliter solutions of the compounds are prepared in the solvent, and 10 μl is spotted. The R_f values are folic acid, 0.28; homofolic acid, 0.53; pteroic acid, 0.11; and homopteroic acid, 0.37.

No folic acid (less than 0.01%) was observed in preparations of homofolic acid when 2-cm strips of the chromatograms were assayed micro-

[2] L. Goodman, J. DeGraw, R. L. Kisliuk, M. Friedkin, E. J. Pastore, E. J. Crawford, L. Plante, A. Nahas, J. F. Morningstar, G. Kwok, L. Wilson, E. Donovan, and J. Ratzan. *J. Am. Chem. Soc.* **86**, 308 (1964).

[3] J. A. R. Mead, A. Goldin, R. L. Kisliuk, M. Friedkin, L. Plante, E. J. Crawford, and G. Kwok, *Cancer Res.* **26**, 2374 (1966).

[4] R. L. Kisliuk, M. Friedkin, L. H. Schmidt, and R. Rossan, *Science* **156**, 1616 (1967).

[5] Homofolic acid and homopteroic acid were obtained through the courtesy of Drs. Harry Wood and Robert Engel of the Cancer Chemotherapy National Service Center, National Cancer Institute, Bethesda, Maryland.

[6] J. I. DeGraw, J. P. Marsh, Jr., E. M. Acton, O. P. Crews, C. M. Mosher, A. N. Fujiwara, and L. Goodman, *J. Org. Chem.* **30**, 3404 (1965).

[7] C. W. Mosher, E. M. Acton, O. P. Crews, and L. Goodman, *J. Org. Chem.* **32**, 1452 (1967).

[8] Y. Kim, V. Grubliauskas, and O. M. Friedman, *Abstr. Am. Chem. Soc.* **154**, 41 (1967).

[9] L. T. Plante, E. J. Crawford, and M. Friedkin, *J. Biol. Chem.* **242**, 1466 (1967).

biologically with *S. faecalis* and *Lactobacillus casei*. Homofolic acid will support the growth of *S. faecalis*, but not that of *L. casei*. Half-maximal growth occurs at $8 \times 10^{-9} M$ homofolic acid. The corresponding value for folic acid is $0.6 \times 10^{-9} M$.

Pteroic acid was not observed on paper chromatograms of homopteroic acid. An amethopterin-resistant variant of *S. faecalis* obtained in this laboratory by the procedure of Hutchison[10] was used to assay these chromatograms. Homopteroic acid supports half-maximal growth of this organism at $0.3 \times 10^{-6} M$, but does not support the growth of *S. faecalis* ATCC 8043. Pteroic acid supports growth of both strains at levels comparable to folic acid.

Preparation of Tetrahydrohomofolic Acid and Tetrahydrohomopteroic Acid

Eight hundred milligrams of PtO_2 catalyst (J. Bishop Co. Malvern, Pennsylvania) is suspended in 10 ml of H_2O, and allowed to settle. The fine material that remains suspended in the supernatant solution is decanted. This process is repeated twice, and the catalyst is suspended in 25 ml of H_2O and transferred to a 500-ml 3-neck, round-bottom flask containing a ¾-inch magnetic stirring bar. One side neck is stoppered with a ground-glass stopper (℥ 24/40). The central neck is stoppered with a ground-glass tube (℥ 34/45) which allows the contents of the flask to flow out when inverted. The other side neck is closed with a rubber serum cap. The flask is gassed 5 times with H_2 through the central neck, which is connected with rubber pressure tubing to the H_2 gas container, water aspirator, and manometer. The catalyst is then hydrogenated to completion (until it takes up 110 ml H_2) by stirring for 30 minutes.

Two grams of homofolic acid are suspended in 25 ml of H_2O and dissolved by adding 1 N NaOH dropwise to obtain a pH of 7.0 (7 ml required). This solution is taken up in a 50-ml syringe and injected into the hydrogenation flask through the rubber serum cap. This cap and the ground-glass connections are secured with adhesive tape. After 2–3 hours of stirring, 2 moles of H_2 are taken up per mole of homofolate, and hydrogen uptake ceases. The reaction flask is clamped off and inverted over a sintered-glass funnel (150 ml, medium porosity), the upper opening of which is fitted with a rubber stopper (size 13½) penetrated by a glass tube. The bottom of the funnel is connected through a second rubber stopper to a 2-liter lyophilizing bulb. The apparatus is flushed with hydrogen entering the rubber tubing above the funnel through a syringe needle and evacuated through a second opening in the rubber stopper in the lyophilizing flask. The con-

[10] D. J. Hutchison, *Cancer Res.* **18**, 214 (1958).

tents of the hydrogenation flask are then filtered directly into the lyophiliz-ing bulb *in vacuo*. The vacuum is released with H_2, and the contents of the bulb are lyophilized immediately. The sodium tetrahydrohomofolate (95% yield, 1.7 moles of Na per mole of tetrahydrohomofolate) is stored *in vacuo*.

A similar procedure is used to prepare sodium tetrahydrohomopteroate. Homopteroic acid is less soluble than homofolic acid; 6.2 ml of 1 *N* NaOH are added dropwise to 2 g of homopteroic acid in 25 ml H_2O. After 3 hours of stirring, the resulting slurry (pH 8.8) is injected into the hydrogenation vessel. As the hydrogenation proceeds, complete solution is attained.

Preparation of Dihydrohomofolic Acid and Dihydrohomopteroic Acid

Dihydrohomofolic acid is prepared by three methods.

Dithionite Reduction.[11] Thirty milligrams of homofolic acid is dissolved in 2.5 ml of 1 *M* 2-mercaptoethanol by adjusting the pH to 7.5 with KOH; 200 mg of sodium dithionite is added. After 30 minutes, the dihydro com-pound is precipitated by adjusting the pH of the mixture to 3.0 with HCl. The precipitate is washed 2 times with 2.5 ml of 1 *M* 2-mercaptoethanol (adjusted to pH 3.0 with HCl), lyophilized, and stored *in vacuo*. The yield is 70%.

Catalytic Reduction in Alkali.[12] Thirty milligrams of PtO_2 catalyst is hydrogenated in 1 ml of 0.1 *N* NaOH in a 20-ml Erlenmeyer flask capped with a rubber serum cap. Hydrogen is passed through the vessel via syringe needles. The mixture is stirred for 1 hour with a magnetic stirrer. One hundred milligrams of homofolic acid dissolved in 4 ml of 1 *N* NaOH is in-jected into the flask, and hydrogenation is continued for 2 hours. Then, 0.35 ml of 100% 2-mercaptoethanol (14.7 *M*) is injected into the flask to make a final concentration of 1 *M*. The preparation is stirred for 10 minutes, then the catalyst is removed by centrifugation. The supernatant solution is brought to pH 7.1 by adding 2 g of Amberlite CG50 (Mallinckrodt, 100–200 mesh; 50 g is washed with 2 liters of 1 *N* acetic acid and 5 liters of water, then dried *in vacuo*). After filtration on a Büchner funnel, the resin was washed with 20 ml of 1 *M* 2-mercaptoethanol. The combined filtrate and washings are lyophilized and stored *in vacuo*. The yield, determined by absorption at 282 nm, is 80%. Forty percent of the weight of the final product is due to material derived from the Amberlite.

Column chromatography shows a UV-absorbing impurity that amounts to 8% of the total absorption at 282 nm emerging from the column in

[11] M. Friedkin, E. J. Crawford, and D. Misra, *Federation Proc.* **21**, 176 (1962).
[12] B. L. O'Dell, J. M. Vandenbelt, E. S. Bloom, and J. J. Pfiffner, *J. Am. Chem. Soc.* **69**, 250 (1947).

fractions 15–20. Dihydrohomofolate emerges in fractions 30–40. The method utilized is described below.

Oxidation of Tetrahydrohomofolate. Solutions of tetrahydrohomofolate in 0.001 M 2-mercaptoethanol, 0.001 M Tris, pH 7.2, oxidize to dihydro-homofolate on standing at room temperature for several hours.

The following evidence indicates that the material synthesized is di-hydrohomofolate. (1) The chemical procedures used for the reduction are the same as for the synthesis of dihydrofolate. (2) The UV-absorption spectrum is similar to that of dihydrofolate.[13] (3) It is reduced to tetra-hydrohomofolate in the presence of NADPH and dihydrofolate reductase.[2, 9]

In contrast to tetrahydrohomofolate, it supports the growth of *S. faecalis*, half-maximal growth occurring at 4×10^{-8} M.

Dihydrohomopteroate has been prepared by dithionite reduction and by oxidation of tetrahydrohomopteroate. It supports half-maximal growth of *S. faecalis* at 2×10^{-8} M. Its microbiological activity differs from tetrahydrohomopteroate, which is an inhibitor, and homopteroate, which is neither a growth factor nor an inhibitor.

Column Chromatography

Tetrahydrohomofolate can be separated from UV-absorbing impurities by gradient elution from diethylaminoethyl cellulose columns.[14] A 2 × 30 cm water-jacketed column is packed evenly with light tamping to a height of 10 cm, washed with 100 ml of 0.1 N KOH and 300 ml of H_2O. Ten milligrams of sodium tetrahydrohomofolate is dissolved in 5 ml of 0.005 M Tris HCl, pH 7.0, containing 0.2 M 2-mercaptoethanol and applied to the column. A gradient produced by having 200 ml of 0.005 M Tris HCl, pH 7.0, in the mixing chamber and 400 ml of 1.0 M Tris HCl, pH 7.0, in the reservoir is employed. Both solutions contain 0.2 M 2-mer-captoethanol. The flow rate is 2 ml/min. Five-milliliter fractions are collected. Ice-water is circulated through the jacket. Tetrahydrohomofolate is eluted in fractions 30–40; tetrahydrohomopteroate, in fractions 25–35. The fractions are identified by their UV-absorption spectra and their ability to inhibit the growth of *S. faecalis*.

Sodium tetrahydrohomofolate prepared from homofolic acid synthesized by the method of DeGraw *et al.*[6] shows negligible UV-absorbing impurities. Homofolic acid synthesized by the method of Kim *et al.*[8] yields tetra-hydrohomofolate showing two additional peaks, one at fractions 5–10, the other at fractions 25–30. Each of these peaks amounts to 5–10% of the total UV-absorbing material.

The method described does not separate dihydrohomofolate from tetra-

[13] A. J. Wahba and M. Friedkin, *J. Biol. Chem.* **236**, PCII (1961).
[14] C. K. Mathews and F. M. Huennekens, *J. Biol. Chem.* **235**, 3304 (1960).

hydrohomofolate, although the dihydro compound tends to be more concentrated in the later fractions. A procedure for separating these compounds has been described.[14a]

Microbiological Assay

Tetrahydrohomofolic acid and tetrahydrohomopteroic acid can be detected by their ability to inhibit the growth of *Streptococcus faecalis* (ATCC 8043).[2–4] Samples for assay are diluted in 0.6% potassium ascorbate pH 6.[15] The growth medium is that of Flynn *et al.*[16] supplemented with $2.2 \times 10^{-9} M$ folic acid; 2.5 ml of double strength medium plus 1.5 ml of H_2O are autoclaved for 5 minutes. The medium is cooled, then samples are added in 0.6% ascorbate and the volume is made to 5 ml with ascorbate. The inoculum is prepared by centrifuging 5 ml of an 18-hour nutrient broth culture (Difco B_{12} Inoculum Broth, 32 g/liter; yeast extract, 5 g/liter), suspending the pellet in 5 ml of 0.9% NaCl, and diluting it 1:100 in 0.9% NaCl. After the addition of one drop of this inoculum per tube, the assay medium is mixed, incubated for 15 hours at 37°, and the turbidity is measured. The concentration range between just detectable and complete inhibition for tetrahydrohomofolate is $0.1 \times 10^{-9} M$ to $0.5 \times 10^{-9} M$. The corresponding values for tetrahydrohomopteroate are $0.2 \times 10^{-9} M$ to $1.4 \times 10^{-9} M$.

The inhibitory potency of each of the two diastereoisomers produced by chemical reduction of homofolate to tetrahydrohomofolate has been determined.[16a] The diastereoisomer having a configuration at carbon 6 opposite to that found in naturally occurring tetrahydrofolate fully accounts for the inhibition obtained with the mixture of diastereoisomers.

Blood levels may be determined[17] after diluting 0.2 ml of freshly drawn blood in 9.8 ml of ascorbate. Blood does not interfere with the assay and actually stabilizes tetrahydrohomofolate. This solution can be kept at room temperature for at least 5 hours without losing activity. Activity is also retained for at least 1 week if the samples are stored frozen. In the absence of blood, tetrahydrohomofolate in 0.6% ascorbate loses 75% of its inhibitory ability after 15 hours at room temperature.

Urine levels are determined after the urine has been collected (24 hours or less) in a vessel (immersed in an ice bath) containing 15 ml of 10%

[14a] A. Nahas and M. Friedkin, *Cancer Research* **29**, 1937 (1969).

[15] H. A. Bakerman, *Anal. Biochem.* **2**, 558 (1961).

[16] L. M. Flynn, V. B. Williams, B. L. O'Dell, and A. G. Hogan, *Anal. Chem.* **23**, 180 (1951).

[16a] R. L. Kisliuk and Y. Gaumont, *Federation Proc.* **29**, 807 (1970).

[17] R. L. Kisliuk, M. Friedkin, V. Reid, E. J. Crawford, L. H. Schmidt, R. Rossan, D. Lewis, J. Harrison, and R. Sullivan, *J. Pharmacol. Exptl. Therap.* **159**, 416 (1968).

potassium ascorbate, pH 6.0, and enough mineral oil to cover the surface. Prior to assay the urine is sterilized by filtration through a Millipore filter.

The shape of the inhibition curve with material obtained from blood or urine (mouse[3] and monkey[17] have been tested) is the same as for the standard compound.[17a]

The technique of Bratton and Marshall has been applied to the assay of tetrahydrohomofolate in tissues.[18]

Molar Extinction Coefficients of Homofolic Acid Derivatives

Homofolic acid[6,7]
$\lambda_{max}^{pH\ 13}$ 255 nm (26,400), 281(19,450), 365(7,880)
Dihydrohomofolic acid
$\lambda_{max}^{pH\ 7.2}$ 282 nm (19,700)
$\lambda_{shoulder}^{pH\ 7.2}$ 300 nm (15,000)
$\lambda^{pH\ 7.2}$ 340 nm (4,700)
Tetrahydrohomofolic acid
$\lambda_{max}^{pH\ 7.2}$ 295 nm (20,500)
$\lambda^{pH\ 7.2}$ 340 nm (410)

The spectra of dihydrohomofolic acid and tetrahydrohomofolic acid were taken in 0.001 M Tris HCl, pH 7.2, containing 0.2 M 2-mercaptoethanol. The difference in the molar extinction value at 340 nm between dihydrohomofolic acid and tetrahydrohomofolic acid is 4290 as compared with 6400 for the corresponding folic acid derivatives. This difference must be taken into account when assaying thymidylate synthetase spectrophotometrically using tetrahydrohomofolate as a cofactor.[19]

Molar Extinction Coefficients of Homopteroic Acid Derivatives

Homopteroic acid[6]
$\lambda_{max}^{pH\ 13}$ 256 nm (26,900), 277(21,900), 365
                                                 (7,630)
Dihydrohomopteroic acid
$\lambda_{max}^{pH\ 7.2}$ 280 (21,500)
$\lambda_{shoulder}^{pH\ 7.2}$ 330 (5,500)

[17a] R. L. Kisliuk, M. Friedkin, E. J. Crawford, V. Reid, L. H. Schmidt, R. N. Rossan, D. Lewis, and R. Sullivan, in "Symposium on the Use of Subhuman Primates in Drug Evaluation" (H. Vagtborg, ed.), p. 23. University of Texas Press, Austin, Texas, 1968.

[18] L. C. Mishra and J. A. R. Mead, Federation Proc. 27, 659 (1968).

[19] R. L. Kisliuk, G. Strait, and E. J. Crawford, Abstr. Am. Chem. Soc. Div. Biol. Chem. 156, Abstr. 140 (1968).

Tetrahydrohomopteroic acid

$\lambda_{max}^{pH\ 7.2}$ 285 nm (20,000)

$\lambda^{pH\ 7.2}$ 340 nm (600)

The spectra of dihydrohomopteroic acid and tetrahydrohomopteroic acid were taken in 0.001 M Tris HCl, pH 7.2, containing 0.2 M 2-mercapto-ethanol.

[181] Synthesis of Biopterin, Neopterin, and Analogs

By H. REMBOLD and J. EDER

Principle

Polyhydroxyalkylpterins are prepared by condensation of 2,4,5-tri-amino-6-hydroxypyrimidine (I) with aldoses in the presence of hydrazine. The use of 5-deoxy-L-arabinose (II) gives rise to the formation of biopterin [2-amino-4-hydroxy-6-(L-*erythro*-1,2-dihydroxypropyl)pteridine]. It was found that biopterin in its reduced form is functioning as a cofactor in various mixed-function oxidase reactions.[1] As a byproduct of this condensa-tion, the isomeric 7-biopterin[2-amino-4-hydroxy-7-(L-*erythro*-1,2-dihydro-xypropyl)pteridine] is formed in an appreciable amount.

[1] S. Kaufman, *Ann. Rev. Biochem.* **36**, 171 (1967).

The separation and purification of the isomers (III) and (IV), which cannot be distinguished by paper chromatography, have been tried by various authors.[2] Their methods result only in isolating the mixture of the isomers from the crude condensation product or in an enrichment of either isomer. Separation and purification of biopterin and 7-biopterin is achieved by cation-exchange chromatography on phosphorylated cellulose (P-cellulose).[3]

Likewise, neopterin,[4] aminobiopterin,[5] and the optical isomers of biopterin[6] and neopterin[4] have been prepared in pure form by using the respective pentoses. Unlike the other purification procedures, this method has several advantages: The capacity of P-cellulose is high and therefore permits the use of small-volume columns; the retention volumes of the isomers exhibit a satisfactory difference (the specific retention volume = retention volume of the substance: volume of the column. This is about 13 for biopterin and 15 for 7-biopterin).

Procedure

Synthesis of Crude Biopterin

Hydrazine hydrate (119.6 mg of 96% purity; 2.3 millimoles) is weighed into a 50-ml round-bottom flask and heated with 110 mg of 5-deoxy-L-arabinose[6] (0.82 millimole) in 0.5 ml of water on a boiling water bath. After 15 minutes, 3.5 ml of water, 1.0 ml of glacial acetic acid, and 91.5 mg of 2,4,5-triamino-6-hydroxypyrimidine sulfate[7] (0.38 millimole) are added, and the solution is heated for a further 25 minutes on the water bath. The reaction flask must be shaken occasionally to achieve a rapid dissolution of the pyrimidine.

The crude product that precipitates after storage in the refrigerator overnight corresponds to a yield of 50–60% of the theoretical amount. It yields approximately 6 mg of biopterin (7% of theory) and 23 mg of 7-biopterin (26% of theory). Recoveries are strongly dependent upon the concentration and volume of the reaction mixture. Multiplication of the

---

[2] E. L. Patterson, R. Milstrey, and E. L. R. Stokstad, *J. Am. Chem. Soc.* **78**, 5868 (1956); M. Viscontini and H. Raschig, *Helv. Chim. Acta* **41**, 108 (1958); A. Butenandt and H. Rembold, *Z. Physiol. Chem.* **311**, 79 (1958); M. Viscontini, *Ind. Chim. Belge*, p. 1181 (1960); R. Tschesche, B. Hess, I. Ziegler, and H. Machleidt, *Ann.* **658**, 193 (1962).

[3] H. Rembold and H. Metzger, *Z. Physiol. Chem.* **329**, 291 (1962); *Ber.* **96**, 1395 (1963); H. Rembold, *in* "Pteridine Chemistry," p. 465. Pergamon Press, Oxford, 1964.

[4] H. Rembold and L. Buschmann, *Ber.* **96**, 1406 (1963).

[5] H. Rembold and J. Eder, *Tetrahedron* **23**, 1387 (1967).

[6] B. Green and H. Rembold, *Ber.* **99**, 2162 (1966).

[7] C. K. Cain, M. F. Mallette, and E. C. Taylor, *J. Am. Chem. Soc.* **68**, 1996 (1946).

above standard condensation results in very poor yields. Large amounts of biopterin therefore ought to be prepared by a suitable number of standard condensations.

Chromatography of Crude Biopterin. One hundred milligrams of crude biopterin is dissolved in 150 ml of water by heating; after cooling to room temperature, the solution is filtered. By means of a pipette the clear solution of biopterin is carefully applied to a P-cellulose column[8] of 24-mm diameter and 30-cm height. Subsequently the column is washed with several portions of warm water, and connected to a pump and a reservoir containing distilled water. Water desalted by ion-exchange resins always contains appreciable amounts of resin material; it is therefore advisable to use distilled water for elution. Under a UV lamp, the zone of pterins is visible as a blue fluorescent band on the top of the column. The effluent is regulated to a flow-rate of 8–10 drops per minute and collected in fractions of 10 or 25 ml each. The

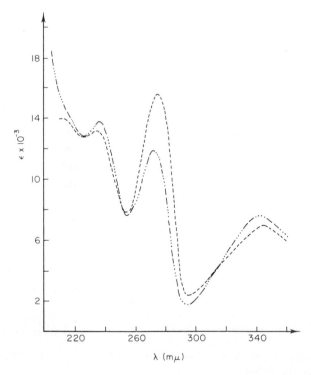

Fig. 1. Comparison of the molar extinctions in water: (———) biopterin; (— — —) 7-biopterin. Reproduced from H. Rembold and H. Metzger, *Ber.* **96,** 1395 (1963).

[8] See chapter by Rembold, this volume [178].

chromatography ought to be carried out in dim light to avoid decomposition by light. The elution pattern is recorded by measuring the UV absorption of the fractions at 272 nm; fractions with an optical density above 2.0 are diluted.

When the UV spectra of biopterin and 7-biopterin are compared in water (Fig. 1), there are striking differences between the molar extinction coefficients in the region of 270 nm. They permit the identification of the pure isomers already in the chromatographic fractions by comparing their UV extinctions and their retention volumes. The curve of the quotient that results by dividing the UV absorption at a wavelength with large differences by one with small differences is followed along the retention volume of biopterin. The quotient used is $E_{272}:E_{252}$; its value is 1.9–2.1 for biopterin and 1.4–1.5 for 7-biopterin. Hence, as soon as the first strongly blue fluorescent zone appears in the effluent, the quotient is determined.

Figure 2 exemplifies the chromatography of a 15-fold standard condensation. In the front peak, byproducts and decomposition products are

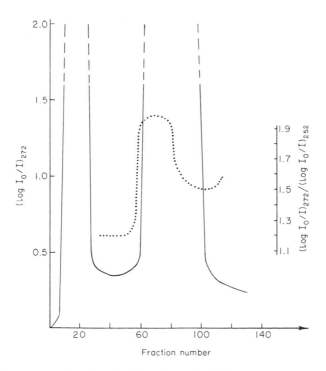

Fig. 2. Chromatography of crude biopterin on P-cellulose. (———) Optical density at 272 nm; (— — —) quotient $E_{272}:E_{252}$. Fifteenfold standard condensation; 5.6 × 27 cm column; 23-ml fractions. Reproduced from H. Rembold and H. Metzger, *Ber.* **96,** 1395 (1963).

eluted. Approximately between fraction 60 and 100 the mixture of the isomers appears in the effluent. The curve of the quotient shows that even the first chromatography of the crude product results in a good enrichment of the isomers.

The zone of biopterin with a quotient above 1.70 and the zone of 7-biopterin with a quotient between 1.70 and 1.50 are pooled and concentrated by the use of a rotary evaporator. The fractions are rechromatographed until a sufficiently stable quotient is reached. Fractions with constant quotients are pooled; zones with rising or falling quotients are collected for rechromatography. The solutions of the pure isomers are brought to dryness in a rotary evaporator. The residue is dissolved again in a minimum of boiling water, filtered, and stored at 0° for 24 hours. The colorless crystals are centrifuged and may be recrystallized from water, resuspended in water, and lyophilized to dryness.

Synthesis of the Optical Isomers of Biopterin. The three optical isomers of biopterin,[6] 6-D-*erythro*-, 6-L-*threo*-, and 6-D-*threo*-(1,2-dihydroxypropyl)-pterin, and their 7-isomers are prepared by condensation of 2,4,5-triamino-6-hydroxy-pyrimidine sulfate (I) with 5-deoxy-D-ribose, 5-deoxy-L-xylose, and 5-deoxy-D-xylose, respectively, according to the procedure described above. The synthesis of the deoxypentoses has been described.[6]

Synthesis of Neopterin

Neopterin has the configuration of a 2-amino-4-hydroxy-6-(D-*erythro*-1,2,3-trihydroxypropyl)pteridine.[4] It is obtained from compound (I) and D-ribose in analogy to the method described for biopterin.

The optical isomers of neopterin are prepared likewise by condensation of compound (I) with L-arabinose, L-xylose, and D-xylose, respectively.

Synthesis of Aminobiopterin and Aminoneopterin

The 4-amino analogs[5] of biopterin and neopterin are more water soluble and more basic than the 4-hydroxy compounds. Their preparation and purification requires a modification of the method described above, which makes the procedure more time-consuming.

Condensation. The crude condensation product is prepared according to the standard synthesis described for biopterin from 2,4,5,6-tetramino-pyrimidine sulfate[9] and 5-deoxy-L-arabinose. The pronounced tendency of 2,4,5,6-tetraminopyrimidine sulfate toward self-condensation can be met by using an excess of sugar and by dissolving the pyrimidine as quickly as possible.

Chromatography. The separation of the isomers from byproducts is

---

[9] J. Eder and H. Rembold, *Z. Anal. Chem.* **237**, 50 (1968).

achieved by chromatography on P-cellulose, as already described. The separation effect is not so good as in the case of biopterin. Hence, the packing of the columns has to be done very carefully, because inclined zones interfere with the separation effect.

The crude precipitate from 5 standard condensations is dissolved in water and applied to a column of 56 mm diameter and 270 mm height. The chromatography ought to be carried out in dim light. The column is washed with 500 ml of distilled water to elute a part of the byproducts. The 6- and 7-Aminobiopterin exhibit a yellow fluorescence and are eluted by using a gradient of formic acid. The column is connected to a closed mixing flask provided with a magnetic stirrer and containing 500 ml of water. The reservoir is successively supplied with batches of 500–1000 ml of dilute formic acid of the following N concentrations: 0.02, 0.03, 0.04, 0.05, 0.06.

The elution pattern is recorded by measuring the absorbance at 245 nm and 335 nm. The pure 6-isomer has a quotient $E_{245}:E_{335}$ of 1.60; the value for 7-aminobiopterin is 1.01. Each chromatography is divided into three zones: an enriched 6-isomer with a quotient above 1.50; a mixed fraction 1.50–1.30; and an enriched 7-isomer with a quotient below 1.30. The preparation of sufficient material with a constant quotient curve requires more rechromatographies than the preparation of biopterin. The fractions are pooled and concentrated in a rotary evaporator. The formic acid is removed by continuous extraction with ether. Sufficiently pure fractions are further concentrated, filtered, lyophilized to dryness, and finally recrystallized from n-propanol. The yellow crystals are recovered by centrifugation. This method of synthesis and purification is likewise applicable to the preparation of aminoneopterin.

Synthesis of the ^{14}C-Labeled Compounds

Syntheses have been described for biopterin-8a-^{14}C,[3] biopterin-2-^{14}C,[10] and neopterin-6,7,1′,2′,3′-^{14}C.[11] The synthesis is carried out by condensation of 2,4,5-triamino-6-hydroxypyrimidine sulfate-4-^{14}C or 2,4,5-triamino-6-hydroxypyrimidine sulfate-2-^{14}C[12] with 5-deoxy-L-arabinose or of (I) with uniformly labeled D-ribose according to the standard procedure. The separation of the isomers is achieved on P-cellulose, as described above.

Synthesis of 2,4,5-Triamino-6-hydroxypyrimidine Sulfate-4-^{14}C. Chloroacetic acid (203.5 mg; 2.15 millimoles) is weighed into a 100-ml round-bottom flask, together with 300 mg of water-free K_2CO_3; 1 ml of water is added slowly. When the development of CO_2 is finished, 130.2 mg of K^{14}CN (2

[10] H. Rembold and G. Hanser, *Z. Physiol. Chem.* **319**, 213 (1960).

[11] H. Rembold, H. Metzger, P. Sudershan, and W. Gutensohn, *Biochim. Biophys. Acta* **184**, 386 (1969).

[12] F. Weygand, H. J. Mann, and H. Simon, *Ber.* **85**, 463 (1952).

TABLE I

UV ABSORPTION OF THE PURE COMPOUNDS

	HCl, 0.1 N		NaOH, 0.1 N	
Pteridine[a,b]	λ_{max} (nm)	log ϵ_{max}	λ_{max} (nm)	log ϵ_{max}
Biopterin[a,b]	247	4.07	254	4.39
	320	3.93	363	3.90
7-Biopterin	245	3.96	251	4.35
	318	4.02	359	3.94
Aminobiopterin	243.5	4.18	224	4.05
	285.5	3.68	257.5	4.32
	336.5	3.98	(280)	(3.77)
	347	3.92	368	3.82
Amino-7-biopterin	241	3.40	224	3.93
	285	3.59	255	4.19
	333.5	3.96	(280)	3.46
	(345)	(3.91)	(365)	(3.80)

[a] The UV data for the respective optical isomers are identical.

[b] The isomeric trihydroxypropylpteridines have the same UV characteristics as the corresponding 1,2-dihydroxypropyl compounds.

millimoles; specific activity, 1 mCi/millimole) is added, and the mixture is heated on the water bath for 1 hour. The reaction mixture is brought to dryness on a rotary evaporator; subsequently the K salt of the cyanoacetic acid is dried over $CaCl_2$ for 2 days. The acid is esterified by refluxing for 1 hour with 0.8 ml of dimethyl sulfate (8.5 millimoles) in 4 ml of absolute methanol. After the addition of 292.8 mg of guanidine nitrate (2.4 millimoles), 4 ml of absolute methanol, and a solution of 0.3 g of Na in 3 ml

TABLE II

NUCLEAR MAGNETIC RESONANCE SPECTRA OF AMINOBIOPTERIN
AND AMINONEOPTERIN IN DMSO-D_6[a]

Compound	2-NH$_2$, 4-NH$_2$, resp.		1'-H	7-H	6-H for amino-7-biopterin only	3'-CH$_3$
Amino-biopterin	7.67 6.55	Broad, exchanging with deut.	4.50 d (1), J = 6 cps	8.71 s (1)	8.47 s (1)	1.12 d (3), J = 6 cps
Amino-neopterin	7.67 6.58	Broad, exchanging with deut.	4.71 d (1), J = 6 cps	8.75 s (1)	—	—

[a] Internal standard, TMS.

TABLE III

R_f Values, Specific Rotation Values, and Growth Activity in the *Crithidia* Test of Biopterin, Neopterin, and Analog

	R_f values[a] in							$[\alpha]_D^{25}$ in 0.1 N HCl	Half-optimum growth conc. (μg × 10^{-3}/ml of assay medium)
	a	b	c	d	e	f	g		
—(1,2-Dihydroxypropyl)pterin								(c = 0.2)	
6-L-*erythro* (biopterin)	0.34	0.28	0.30	0.28	0.32	0.62	0.59	−62°	0.006–0.015
6-D-*erythro*	0.34	0.28	0.30	0.28	0.32	0.62	0.59	+61°	0.7
6-L-*threo*	0.16	0.24	0.26	0.28	0.32	0.62	0.59	+95°	0.2
6-D-*threo*	0.16	0.24	0.26	0.28	0.32	0.62	0.59	−94°	0.5
7-L-*erythro*	0.39	0.28	0.30	0.28	0.32	0.62	0.59	−18°	
7-D-*erythro*	0.39	0.28	0.30	0.28	0.32	0.62	0.59	+7°	50
7-L-*threo*	0.24	0.24	0.26	0.28	0.32	0.62	0.59	+79°	
7-D-*threo*	0.24	0.24	0.26	0.28	0.32	0.62	0.59	−74°	
—(1,2,3-Trihydroxypropyl)pterin								(c = 0.3)	
6-D-*erythro* (neopterin)	—	—	0.21	0.13	0.21	0.60	0.55	+45°	10
6-L-*erythro*	—	—	0.21	0.13	0.21	0.60	0.55	−44°	0.03
6-D-*threo*	—	—	0.12	0.13	0.21	0.60	0.55	−92°	10
6-L-*threo*	—	—	0.12	0.13	0.21	0.60	0.55	+97°	2
7-L-*erythro*	—	—						−13°	
Aminobiopterin	—	—		0.28	0.33	0.63	0.36	(c = 0.16) −33°	0.01
Aminoneopterin	—	—		0.16	0.23	0.56	0.33	+23°	

[a] System a = thin-layer plates with silica gel G, applied by means of 0.1 M H$_3$BO$_3$; solvent isopropanol–5% aqueous H$_3$BO$_3$ (4:1). b = Paper, Schleicher and Schüll 2043 b; solvent, isopropanol–5% H$_3$BO$_3$ solution (6:1), descending. c = Isopropanol–10% H$_3$BO$_3$ solution (4:1), descending. d = n-Butanol–glacial acetic acid–water (20:3:7), ascending. e = n-Propanol–1% aqueous ammonia (2:1), ascending. f = 3% NH$_4$Cl solution, ascending. g = 4% Na citrate solution, ascending.

of absolute methanol (dropwise), the mixture is heated on the water bath for 2 hours. Five milliliters of water is added, the pH is adjusted to 4 by addition of conc. HCl, and the methanol is completely distilled off under vacuum. The residue is dissolved in 1 ml of water and warmed to 90°. An excess of conc. NaNO₂ solution is added, and after 5 minutes at 90° the pink precipitate of 2,4-diamino-6-hydroxy-5-nitrosopyrimidine-4-^{14}C is transferred by means of a pipette into a centrifuge tube. The sediment is washed with three 3-ml portions of water and finally dissolved in 3 ml of 10% NaOH. The solution is brought to 95°, and solid hydrosulfite (Na₂S₂O₄·2 H₂O) is added until the dark red solution becomes yellow. The temperature is kept at 95° for 5 minutes, then conc. HCl is added dropwise until pH 1 is reached. To remove sulfur, some charcoal is added. The hot solution is filtered by suction; 12 drops of conc. H₂SO₄ and 2 ml of methanol are added. After 12 hours in the refrigerator, 50.1 mg (11% of theory relative to KCN) of the radioactive pyrimidine is precipitated. The colorless needles are recovered by filtration and washed with methanol.

Properties and Criteria of Purity

The described compounds are characterized by their UV spectra (Table I); nuclear magnetic resonance data are given for the amino analogs (Table II). An extremely sensitive criterion of purity is the mentioned quotient of the UV extinctions at two suitable wavelengths.

Another useful criterion of purity is provided by paper chromatography. The behavior of the substances in several solvent systems is summarized in Table III. In addition, the specific rotation values and the growth activities found with the biopterin-requiring microorganism *Crithidia fasciculata*[13] are given. The degradation of the side chain to the corresponding carboxylic acids[8] by KMnO₄ and their separation by paper chromatography is another valuable and sensitive method, if it is necessary to prove that a pteridine is free of its isomer.

[13] H. N. Guttman, *in* "Pteridine Chemistry" (W. Pfleiderer and E. C. Taylor, eds.), p. 255. Pergamon Press, Oxford, 1964.

[182] Pterins and Folate Analogs

By M. VISCONTINI

The properties of the pterins vary considerably with the position and the type of substituent in the molecule, and it is very difficult to give a general method for the preparation of these products and of folate analogs. We believe that it is better to describe (a) the synthesis of some pterins, chosen for their particular character, (b) the hydrogenation of these pterins to 5,6,7,8-tetrahydro derivatives, and (c) the formation of the 7,8-dihydropterins by oxidation of the corresponding 5,6,7,8-tetrahydropterins, to make clear with these examples the difficult procedures that can be used to prepare these products, which are important in biochemistry.

Preparation of Pterins

The following three syntheses are described: (1) pterin, (2) monapterin and neopterin; (3) N^5-methyl-6,7-diphenyl-5,6-dihydropterin.

Synthesis of Pterin

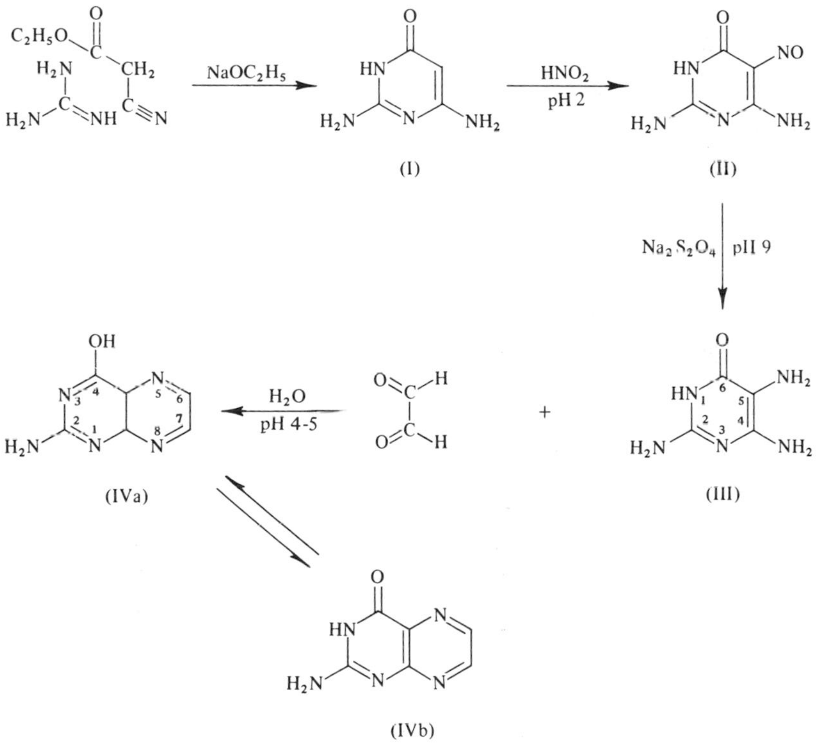

(I)   (II)

$Na_2S_2O_4$ | pH 9

(IVa)   (III)

(IVb)

Operation

Pterin is the trivial name for 2-amino-4-hydroxypteridine (IVa) or its tautomeric structure 2-amino-4-oxo-3,4-dihydropteridine (IVb). The best method for the preparation of this substance involves the condensation of the 2,4,5-triamino-6-oxo-1,6-dihydropyrimidine (III) with glyoxal.[1,2]

[1] C. K. Cain, M. F. Mallette, and E. C. Taylor, *J. Am. Chem. Soc.* **68,** 1998 (1946).

For the preparation of the pyrimidine (III), we use the old method of Traube,[1,3] which is modified to obtain directly the stable sulfate salt, as the basic pyrimidine (III) itself is very sensitive to air oxidation.

Procedure

2,4-Diamino-6-oxo-1,6-dihydropyrimidine (I). Thirty-five grams (1.52 moles) of sodium (note 1, p. 682) is added in portions to 1500 ml of absolute ethanol (note 2) contained in a 5-liter, round-bottomed, three-necked flask fitted with a reflux condenser, an efficient stirrer, and a dropping funnel. When the sodium has completely dissolved (ca. 2 hours), 110 g (1.2 moles) of finely powdered guanidine hydrochloride together with 135 g (1.2 moles) of ethyl cyanoacetate are poured into the flask (note 3). The flask is then electrically heated and vigorously stirred. A white precipitate of NaCl and of some sodium-enolate of the pyrimidine (I) is formed during the course of the reaction. The condensation is complete after heating under reflux for 3 hours.

2,4-Diamino-5-nitroso-6-oxo-1,6-dihydropyrimidine (II). The pyrimidine (I) is not directly isolated, but, after being cooled to room temperature, is brought into solution by adding 1000 ml of water to the reaction mixture (note 4). The solution is acidified with 100 ml of 10 N hydrochloric acid (note 5), and a solution of 100 g (1.2 moles) of sodium nitrite in 300 ml of water is slowly dropped into the acidified solution with vigorous stirring at room temperature. The solution becomes red, and the red-violet nitroso-pyrimidine (II) separates as a fine crystalline precipitate, insoluble in the reaction mixture (note 6). After 2 hours' stirring, the pyrimidine (II) is collected by suction on a Büchner funnel and is washed free from the faintly colored mother liquor with water.

2,4,5-Triamino-6-oxo-1,6-dihydropyrimidine (III) Sulfate. The wet nitrosopyrimidine (II) is placed in a clean 5-liter, round-bottomed, two-necked flask containing 3,000 ml of water. The suspension is heated at 70° with stirring. Then 5 N NaOH is slowly added until pH 9 is reached, and 500 g (2.37 mole) of sodium hydrosulfite·2 H_2O is cautiously added in small portions over 1.5–2 hours (note 7). The color of the mixture changes from red-violet to yellow (note 8). The solution is decolorized at 70° by stirring for a quarter of an hour with 10 g of charcoal; it is then filtered through a fluted paper in a heated filter. Before cooling, the filtrate is cautiously neutralized with 7 N H_2SO_4 (note 9), and the pyrimidine (III) sulfate soon separates as yellow or light brown needles. Sulfuric acid is added until the pH falls to 1–2; the suspension is cooled and left to stand

[2] J. H. Mowat, J. H. Boothe, B. L. Hutchings, E. L. R. Stokstad, C. W. Waller, R. B. Angier, J. Semb, D. B. Cosulich, and Y. SubbaRow, *J. Am. Chem. Soc.* **70**, 14 (1948).
[3] W. Traube, *Ber.* **33**, 1371 (1900).

in an ice bath for 5 hours. The sulfate is collected on a Büchner funnel and is washed with a solution of sodium hydrosulfite (1 g in 100 ml of water) (note 10). For purification, the wet sulfate is transferred to a 2-liter beaker containing 1500 ml of water and 2 g of sodium hydrosulfite. The suspension is heated to 70°, and a 10% NaOH solution is cautiously added to pH 8–9. The solution of the pyrimidine (III) is treated with 5 g of decolorizing carbon and is then filtered. The filtrate is cautiously acidified by stirring with a solution of 100 g of H_2SO_4 in 200 ml of water (pH 1–2). Light yellow needles are formed, and the suspension is cooled at 0° for 15 hours. The precipitate containing some Na_2SO_4 is washed by decantation with 500 ml of 0.1 N H_2SO_4. The supernatant solution is discarded, and the precipitate is washed by stirring with a further 500 ml of 0.1 N H_2SO_4; after the suspension has been allowed to stand for a few hours, the supernatant is discarded. The pure pyrimidine (III) sulfate is collected on a Büchner funnel, washed with water, then three times with 100 ml of ethanol and is dried in a vacuum desiccator over $CaCl_2$ for 2 days. The yield is 220 g of pyrimidine (III) sulfate monohydrate (78%).

2-Amino-4-oxo-3,4-dihydropteridine (IV). A 200-ml, two-necked flask placed on a heating magnetic stirrer is fitted with a reflux condenser and a bent glass tube arranged to introduce nitrogen gas not too near the bottom of the flask. One hundred ml of water, 2.57 g (10 millimoles) of pyrimidine (III) sulfate monohydrate, 4.08 g (30 millimoles) of $CH_3COOH \cdot 3H_2O$, and 3 ml of a glyoxal solution (30%) are introduced into the flask. The mixture is heated to the boiling point, and a slow stream of nitrogen is bubbled into the liquid with stirring. The sulfate goes into solution, and a yellow precipitate is slowly formed. After 1 hour of refluxing, the flask is cooled at 0°. After 10–15 hours, the precipitate of crude pterin (IV) is collected by suction filtration and is washed on the filter with 50 ml of water. The crude brown pterin (IV) is systematically purified as follows: The wet precipitate is dissolved in a minimal quantity of 0.5 N NaOH, the solution is filtered, if necessary, on a sintered-glass filter, and 20 ml of 10 N NaOH is added to the clear filtrate. The sodium enolate of the pterin (IV) precipitates and is collected by suction filtration on a sintered-glass filter and washed well with absolute ethanol (note 11).

The wet enolate is dissolved in 150 ml of water, and the solution is neutralized with 2 N HCl to pH 5. The pterin (IV) precipitates; it is collected by suction filtration and washed with water. The wet pterin is dissolved in a minimal quantity of 0.5 N HCl by stirring and gently heating (not over 80°). The acid solution is carefully neutralized with 2 N NH$_4$OH until a white turbidity appears (note 12). The solution is now decolorized at the boiling point by stirring for 10 minutes with a small quantity of active carbon (note 13) and is filtered through a fluted paper in a heated

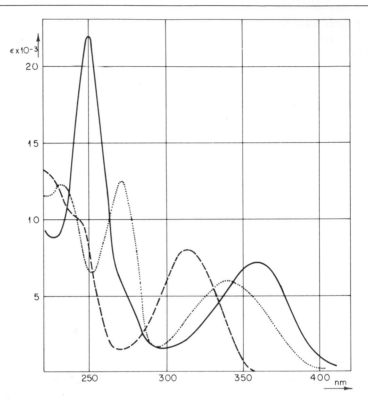

Fig. 1. UV spectra of pterin (IV). (. . . . .) 0.1 *N* HCl; (– – – –) pH 7; (————)
0.1 *N* NaOH.

filter. The warm clear solution is neutralized to pH 4 with 2 *N* NH$_4$OH
and is allowed to cool slowly. After 24 hours in the refrigerator, the microcrystalline pterin (II) is collected by suction filtration and washed on a
Büchner funnel with water, ethanol, and ether. The pterin is dried in a
vacuum desiccator over CaCl$_2$ for 2 days. The yield is 1.1 g of pterin (IV)
monohydrate (60%). The purity of the pterin must be verified by elemental
analysis (note 14), by paper chromatography with an authentic specimen,
and UV spectra (Fig. 1).

Notes

1. Sodium must be added rapidly (*caution!*) in order to make easier
its dissolution. The hydrogen evolution is very vigorous, and a stirrer
cannot be used because of the danger of explosion caused by an electric
spark.

2. A good grade of absolute ethanol should be used.

3. The commercial ethyl cyanoacetate should be purified by vacuum distillation (90°–95°, 10 mm).

4. The solution obtained should be clear and light yellow. A brown color indicates a bad condensation.

5. The yellow solution changes into a colorless one as soon as the sodium enolate is neutralized. The free pyrimidine (I) has, when pure, no absorption in visible light.

6. The nitrosation occurs best at pH 1–2, and the pH of the reaction should be continually followed. If the pH rises above 3, 10 N HCl should be added until the pH falls again to 1–2.

7. The reduction of the nitroso group to the amino group occurs best at pH 9, and the pH of the reaction should be continually followed. If the pH falls below 9, 5 N NaOH should be added until the pH rises again to 9.

8. If the yellow color is not obtained after the addition of 500 g of sodium hydrosulfite, some more reducing salt should be added until the red color of the suspension disappears.

9. This neutralization must be done under a hood because the decomposition of hydrosulfite and sulfite salts with H_2SO_4 evolves a great quantity of SO_2.

10. The pyrimidine (III) is very sensitive to air oxidation, giving red and brown condensation products. Sodium hydrosulfite is added to the washing water in order to prevent this oxidation.

11. The sodium enolate of the pterin (IV), soluble in water, is insoluble in concentrated solutions of NaOH (solubility product). If the precipitation is not complete after the addition of 10 N NaOH, it is necessary to add to the filtrate some pellets of NaOH with stirring and to collect the newly formed enolate by filtration.

12. The turbidity is formed by the precipitation of some impurities and some amorphous pterin (IV), which is insoluble in a neutral solution. The turbidity should not be too great, otherwise the loss of the product would be excessive during the purification. In this case, it is recommended that a few drops of 0.5 N HCl be added to the warm suspension in order to bring the bulk of the precipitate back into solution.

13. A large quantity of decolorizing carbon should be avoided, because pterin (IV) is easily adsorbed on carbon at a pH below 7. Generally, with a few milligrams of carbon, the brown solution turns to faint yellow.

14. It is very difficult to carry out the elemental analysis of pterin derivatives. Nitrogen values, especially, are seldom good; therefore it is recommended that the pterins be given to experienced microanalytical laboratories for elemental analysis.

Synthesis of L-Monapterin and D-Neopterin

Operation

L-Monapterin is the trivial name for the $(+)$-6-(L-*threo*-1′,2′,3′-trihydroxypropyl)pterin (V); and D-neopterin, the trivial name for $(+)$-6-(D-*erythro*-1′,2′,3′-trihydroxypropyl)pterin (VI).

(V)   (VI)

The method of preparation of these substances is based on the condensation of the 2,4,5-triamino-6-oxo-dihydropyrimidine (III) with a pentopyranose: L-xylose for monapterin and D-arabinose for neopterin. In order to avoid the formation of a mixture of 6- and 7-substituted derivatives, which always occurs with the free aldopentopyranoses, the condensation is carried out with a corresponding 1-deoxy-1-amino-ketopentose, for example, with the L-1-deoxy-1-benzylaminoxylulose (VII) for the synthesis of the L-monapterin, according to the scheme at top of the following page, and with D-1-deoxy-1-benzylamino-ribulose for the synthesis of the D-neopterin. This method of synthesis already has been published for the preparation of D-monapterin and L-neopterin.[4] Recently, two syntheses of biopterin have been published,[4a] as well as a new method to obtain monapterin and neopterin with yields better than 60%.[4b]

Procedure

N-Benzylamino-L-xyloside. One gram (6.7 millimoles) of L-xylose (note 1, p. 689) and 1 g (9.3 millimoles) of benzylamine (note 2) are introduced into a 100-ml beaker. The mixture is heated to 80° with stirring until the xylose has been dissolved. The faint yellow, viscous solution is cooled to 0°, and 10 ml of ethyl acetate and isopropanol (1:1) are added quickly to the cool solution. The mixture is stirred until crystallization of the N-benzylamino-L-xyloside occurs. The crystals are collected by suction filtration, washed with ethyl acetate and ether, and dried in a desiccator

---

[4] M. Viscontini and R. Provenzale, *Helv. Chim. Acta* **51**, 1495 (1968).

[4a] K. J. M. Andrews, W. E. Barber, and B. P. Tong, *J. Chem. Soc.* (C), p. 928 (1969); M. Viscontini and R. Provenzale, *Helv. Chim. Acta* **52**, 1225 (1969).

[4b] M. Viscontini, R. Provenzale, S. Ohlgart, and J. Mallevialle, *Helv. Chim. Acta* **53**, 1202 (1970).

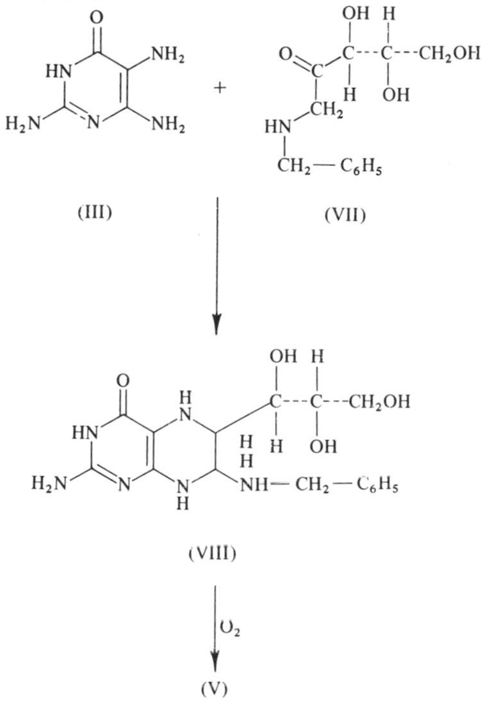

(III)   +   (VII)

(VIII)

$\downarrow$ O$_2$

(V)

over CaCl$_2$. The yield of the product, m.p. 90°, is 1.2 g (75%); $[\alpha]_D^{20} =$ + 50° ($c = 1$; CH$_3$OH).

1-Deoxy-1-benzylamino-L-xyloketose (VII) Oxalate. To the solution of 1.2 g (5 millimoles) of *N*-benzylamino-L-xyloside in a minimal quantity of absolute isopropanol at 20° is added 0.6 g (6.7 millimoles) of anhydrous oxalic acid. The mixture is gently heated with stirring to 80°, and crystals of the oxalate salt begin slowly to precipitate. After 5 minutes, the suspension is cooled to 0° and left at this temperature for 12 hours. The crystals are collected by suction filtration, washed with absolute ethanol and ether, and dried. The yield of the product, m.p. 130°–132°, is 1.5 g (90%); $[\alpha]_D^{20} =$ +0.5° ($c = 1$; CH$_3$OH) (note 3).

2,4,5-Triamino-6-oxo-1,6-dihydropyrimidine (III) Dihydrochloride.[5] Thirty-two grams (138 millimoles) of pyrimidine (III) sulfate (page 680) is poured into a 250-ml beaker containing a solution of 33 g (135 millimoles) of BaCl$_2$·2 H$_2$O, 0.5 g of sodium hydrosulfite, and 10 ml of 1 *N* HCl in 70 ml of water. The mixture is heated to 100° for 3 minutes and is filtered

[5] A. Albert, *J. Appl. Chem.* **3**, 521 (1953).

hot. The insoluble $BaSO_4$ is washed with 30 ml of hot water acidified with some drops of 1 N HCl, and the filtrate is decolorized, if necessary, by stirring at the boiling point for 5 minutes with a small quantity of charcoal. After filtration through a fluted paper in a heated filter, 50 ml of 10 N HCl is added to the clear solution, which is allowed to cool slowly and is then placed in a refrigerator for 15 hours. The faint yellow crystals of pyrimidine (III) dihydrochloride are collected by suction filtration on a Büchner filter, washed on the filter with ethanol and ether, and then dried in a desiccator over $CaCl_2$. The yield is 23.6 g (90%) of the dihydrochloride salt.

L-*Monapterin*. One and seven-hundredth grams (5 millimoles) of 2,4,5-triamino-6-oxo-1,6-dihydropyrimidine (III) dihydrochloride, 0.820 g (10 millimoles) of anhydrous sodium acetate, 200 ml of absolute methanol, and 5 drops of thioethanol are introduced in a 500-ml, three-necked, round-bottomed flask equipped with a reflux condenser, a gas inlet tube, and a magnetic stirrer (note 4). The mixture is boiled with stirring, while a slow stream of nitrogen is passed into the solution. After 15 minutes, NaCl precipitates. To the solution, 1.645 g (5 millimoles) of the xylulose (VII) oxalate is then added, and the mixture is boiled for 3 hours. After cooling, the insoluble inorganic salts are removed by filtration, and the solution of the tetrahydropterin VIII is evaporated at 30° under reduced pressure (12 mm) by rotatory evaporation. Twenty milliliters of a 0.03 N ammonium formate solution, adjusted to pH 7 with ammonia, is added to the solid, brown residue (note 5). The suspension is centrifuged, and the clear solution containing the pterins is chromatographied on a 4 × 30 cm column of Dowex 1-X4 (note 6). The solvent and substances should always be kept under a nitrogen atmosphere. The elution occurs with the same solution (0.03 N ammonium formate, pH 7) at the rate of 20 ml in 10 minutes. Fractions of 20 ml are collected. After oxidation with O_2, the fractions are monitored by paper chromatography with butanol–acetic acid–water (10:3:7) as solvent, and the fluorescent spots of pterins are located under UV light (Fig. 2). The first 15 fractions are brown and contain impurities; they are discarded. The monapterin appears in fractions 16–30, but fractions 27–30 also contain some 1'-deoxymonapterin and are, therefore, discarded. Fractions 16–26 are combined, brought to pH 5.6–6.5 with formic acid, and oxidized by bubbling O_2 through the solution for 12 hours. A portion of the pterins precipitates during the oxidation, and the solvent is removed under reduced pressure (12 mm) at 30° by rotatory evaporation. The dry, brown residue is washed with ethanol and ether, and is dissolved in 2–3 ml of an NH_4OH solution (1%). The suspension is centrifuged, and the clear solution is brought to pH 9–10 with formic acid; it is chromatographed on a 4 × 20 cm column of Dowex 1-X4 (note 6). The elution is accomplished with a buffer of 0.03 N ammonium formate,

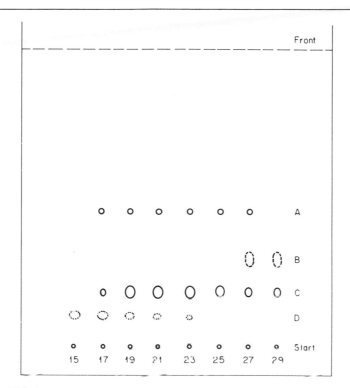

Fig. 2. Thin-layer chromatography of fractions 15–29 after oxidation with O_2. Solvent butanol-acetic acid-water (10:3:7). Spot A: pterin (IV); spot B: 1'-desoxymonapterin; spot C: L-monapterin (V); spot D: unknown impurity.

pH 8.5 (note 7). Initially the pterins remain at the top of the column, but a gradual separation into two fluorescent zones occurs. When this separation is reached, the pH of the buffer eluent is brought to 7–7.3 with a little formic acid to increase the velocity of the elution. The first fluorescent zone is discarded, and the second is continually monitored for monapterin by paper chromatography with butanol–acetic acid–water (10:3:7) as solvent. The fractions containing the pure monapterin are combined and concentrated to a small volume (2 ml) under reduced pressure (12 mm) at 30° by rotatory evaporation. To the reduced solution is added 15–20 ml of absolute ethanol. The monapterin precipitates, and the suspension is left in a refrigerator for 12 hours. After centrifugation, the crude monapterin is recrystallized from hot water, recovered by centrifugation, and finally washed with ethanol and ether, and dried. The yield is 130 mg (10%) of pure monapterin; $[\alpha]_D^{20} = +111° \pm 5°$ ($c = 0.2$; 0.1 N HCl) (note 8).

The purity of the monapterin should be verified by elemental analysis,

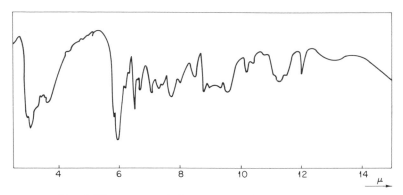

FIG. 3. IR spectrum of the L-monapterin measured in KBr.

paper chromatography with an authentic specimen, measurement of the specific optical rotation, and the infrared spectrum (Fig. 3).

N-Benzylamino-D-arabinoside.[6] Five grams (33.4 millimoles) of D-arabinose (note 9) and 5 g (46.5 millimoles) of benzylamine (note 2) are introduced into a 100-ml beaker. The mixture is heated to 80° with stirring until the arabinose has been dissolved (30 minutes). The faint yellow, viscous solution is then cooled to 0°, and ethyl acetate is added with stirring until crystallization commences. The crystals are collected by suction filtration, washed with ethyl acetate and ether, and dried in a desiccator over $CaCl_2$. The yield of the product, m.p. 115°–116°, is 7 g (90%); $[\alpha]_D^{25} = -4°$ ($c = 1$; CH_3OH).

1-Benzylamino-1-deoxy-d-ribulose Oxalate.[6] To the solution of 6 g (25 millimoles) of *N*-benzylamino-D-arabinoside in 200 ml of absolute dioxane at 20° is added a solution of 2.5 g (28 millimoles) of anhydrous oxalic acid in 50 ml of dioxane. A turbidity appears slowly in the mixture, and suddenly the contents of the flask solidifies in a crystalline mass. After 12 hours in the refrigerator the crystals are collected by suction filtration, washed on the Büchner funnel with isopropanol, and recrystallized from absolute ethanol. The yield of oxalate, m.p. 144°–145°, is 6 g (75%) $[\alpha]_D^{25} = +5°$ (note 3).

D-Neopterin. The synthesis is carried out identically to that of L-monapterin, taking *N*-benzylamino-D-ribulose oxalate as the starting material. The yield is 75 mg (6%) of pure neopterin, $[\alpha]_D^{20} = +51° \pm 3°$ ($c = 0.17$; 0.1 N HCl) (note 10), which is less soluble in water (50 mg/100 ml) than monapterin. The purity of the neopterin should be verified by elemental analysis, paper chromatography with an authentic specimen, and measurement of the specific optical rotation and the IR spectrum (Fig. 4).

[6] T. Neilson and H. C. S. Wood, *J. Chem. Soc.*, p. 44 (1962).

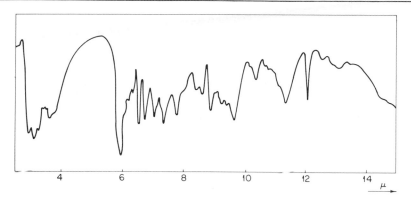

FIG. 4. IR spectrum of the D-neopterin measured in KBr.

Notes

1. L-Xylose should be pure ($[\alpha]_D^{20} = -19°$), and the purity should be checked by paper chromatography.

2. Commercial benzylamine should be purified by vacuum distillation (70°–71°, 10 mm).

3. As is the case for all linear sugar derivatives, the specific optical rotation of the aminoketoses is very low and is difficult to measure. The melting point may vary over some degrees and cannot be used as a criterion of purity. To check the purity of the substance, it is recommended that elemental analysis be carried out.

4. The free triamino-oxo-dihydropyrimidine (III), the tetrahydropterin (VIII), the monapterin, and the neopterin are all sensitive to oxidation and to UV light; therefore, it is recommended that some antioxidant (e.g., thioethanol, sulfite, hydrosulfite, isobutyl phenol) and a nitrogen gas stream be used during the chemical reactions. It is also necessary to work in a laboratory with red or yellow-brown light.

5. For the preparation of the buffer solution, 4.875 g (90 millimoles) of formic acid (85%) is dissolved in 3 liters of water. The pH of the solution is then brought to 7 by adding NH_4OH cautiously and checking it continually with a glass electrode.

6. A column of Dowex 1-X8 can be employed instead. The Dowex must be in the formate form, which is prepared as follows: 100 ml of formic acid (85%) in 200 ml of water is mixed with 20 g of NaOH in 200 ml of water. The column is washed with this solution and then with water until the pH of the effluent is 5. After this has been carried out, the column is washed with 100 ml of the buffer solution in which the substances are dissolved.

7. For the preparation of this buffer solution, 4.875 g (90 millimoles)

of formic acid (85%) is dissolved in 3 liters of water. The pH of the solution is brought to 8.5 by adding 3.8 ml (90 millimoles) of NH_4OH (40%), and is then adjusted with a little ammonia (glass electrode).

8. The (+)-L-monapterin was isolated for the first time from a culture of *Pseudomonas roseus fluorescens*.[7] The product is not very soluble in water (100 mg/100 ml). Specific optical rotation: $[\alpha]_D^{25} = +106° \pm 4°$ ($c = 94; 0.1\ N$ HCl). Synthetic and natural products are identical in every respect.

9. D-Arabinose should be pure ($[\alpha]_D^{20} = -104°$), and the purity should be checked by paper chromatography.

10. The (+)-neopterin was isolated for the first time from the pupae of bees,[8] but the specific optical rotation of the natural product has never been measured.

Synthesis of N^5-Methyl-6,7-diphenyl-5,6-dihydropterin

(IX)   (X)   (XI)

Operation

The interest of biochemists in N^5-methylpterins is steadily growing, since it has been shown that N^5-methyltetrahydrofolic acid (XIII) is one of the coenzymes of biological transmethylation.

One cannot make an N^5-methylpterin as a tertiary amine other than as a dihydro- or a tetrahydroderivative. N^5-Methyl-5,6-dihydrofolic acid has not yet been found in nature, but it has been described in the literature.[9] In contrast, N^5-methyl-5,6,7,8-tetrahydrofolic acid (XIII) was isolated in 1961 as a natural product[10,11] and has been synthesized by reduction of the methylenetetrahydrofolic acid (XII) with $NaBH_4$[11,12] and also with KBH_4.[9]

Some years ago we succeeded in obtaining in our laboratory an N^5-meth-

[7] M. Viscontini, M. Pouteau-Thouvenot, R. Bühler-Moor, and M. Schroeder, *Helv. Chim. Acta.* **47**, 1948 (1964); M. Viscontini and R. Bühler-Moor, *ibid.* **51**, 1548 (1968).

[8] H. Rembold and L. Buschmann, *Ann.* **662**, 72 (1963).

[9] V. S. Gupta and F. M. Huennekens, *Arch. Biochem. Biophys.* **120**, 712 (1967).

[10] W. Wilmanns, B. Rucker, and L. Jaenicke, *Z. Physiol. Chem.* **322**, 283 (1960); A. Larrabee and J. M. Buchanan, *Federation Proc.* **20**, 9 (1961).

[11] W. Sakami and I. Ukstins, *J. Biol. Chem.* **236**, PC50 (1961).

[12] L. Jaenicke, *Z. Physiol. Chem.* **326**, 168 (1961).

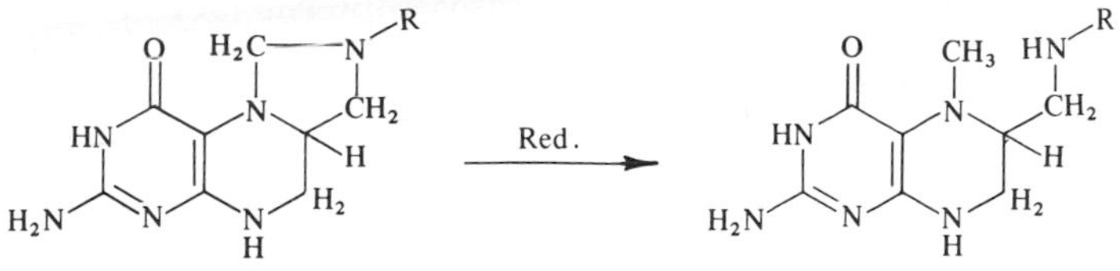

(XII)   (XIII)

yldihydropterin, namely N^5-methyl-6,7-diphenyl-5,6-dihydropterin (XI), by reaction of the desyl chloride (X) with 2,4-diamino-5-methylamino-6-oxo-1,6-dihydropyrimidine (IX) in an aqueous solution of ethanol.[13] We describe this synthesis in detail, since (a) this dihydropterin possesses special and very interesting properties, (b) it can be considered as a folate analog, and (c) the preparation is simple.

Procedure

2,4-Diamino-5-methylamino-6-oxo-1,6-dihydropyrimidine (IX) dihydrochloride,[14] (0.089 g, 0.384 millimole) 0.1 g (0.435 millimoles) of desyl chloride (X), 0.160 g (1.18 millimoles) of CH_3—$COONa·3$ H_2O and 10 ml of aqueous ethanol (25%) are introduced into a 25-ml, three-necked, round-bottomed flask equipped with a reflux condenser, a gas inlet tube, and a stirrer. The mixture is boiled with stirring while a slow stream of nitrogen is passed into the solution. The starting material dissolves slowly, and after 10 minutes the brown-red N^5-methyldihydropterin (XI) begins to precipitate. After 6 hours of refluxing, the red substance is collected by suction on a small Büchner funnel and is washed free from mineral salts with water.

The crude dihydropterin (XI) is placed in a 25-ml flask fitted with a reflux condenser, and is boiled twice with 10 ml of acetone (note 1, p. 691). The dihydropterin is collected each time by suction filtration, washed on a Büchner funnel with a little acetone, and finally dried over $CaCl_2$. The yield is 36 mg (28%) of the red N^5-methyldiphenyldihydropterin (XI) (note 2).

Notes

1. During the reaction some desyl chloride is hydrolyzed to benzoin which, on cooling, precipitates with the methyldihydropterin (XI). The benzoin is very soluble in hot acetone and can be extracted by boiling with this solvent.

[13] M. Viscontini and S. Huwyler, *Helv. Chim. Acta* **48**, 764 (1965).
[14] J. A. Haines, C. B. Reese, and A. R. Todd, *J. Chem. Soc.*, p. 5281 (1962); W. Pfleiderer and F. Sagi, *Ann.* **673**, 78 (1964).

2. The N^5-methyl-6,7-diphenyl-5,6-dihydropterin (XI) is insoluble in water, dilute acidic solutions, and nonpolar solvents. It is slightly soluble in methanol, and is very soluble in basic solutions (NaOH, KOH). When dry, the substance is very stable. However, in solution the dihydropterin (XI) tends to decompose like the other hydrogenated pterins, and therefore a recrystallization is not possible. To check the purity of the product, it is necessary to have its elemental analysis, and to measure its UV spectrum [in methanol: λ_{max} 269 nm ($\epsilon \times 10^{-3} = 25$), 423 nm (4.1), λ_{min} 229 nm (10.1), 362 nm (2.1)] and its nuclear magnetic resonance spectrum [in CF_3COOH with tetramethylsilane as internal standard; N^5-(C)H_3, singlet, 4.23 ppm, 3 protons; C^6H, singlet, 6.3 ppm, 1 proton; aromatic hydrogen atoms, multiplet, 7.2–7.8 ppm, 10 protons].

Preparation of 5,6,7,8-Tetrahydropterins

The following syntheses are described: (1) 5,6,7,8-tetrahydropterin; (2) tetrahydrofolic acid; (3) N^5-methyl-6,7-diphenyl-5,6,7,8-tetrahydropterin.

Synthesis of the 5,6,7,8-Tetrahydropterin

(IV)   (XIV)

Operation

It is well known since the discovery of the *Citrovorum* factor that hydrogenated pterins play an important role in the animal metabolism of carbon derivatives, and numerous syntheses have been attempted to obtain such products synthetically[15] by catalytic reduction of the corresponding pterins. It was soon known that pterins are able to add two molecules of hydrogen to the pyrazine ring at positions 5, 6, 7, and 8, and in 1958 we succeeded in obtaining for the first time a crystalline salt of the 5,6,7,8-tetrahydropterin.[16] At this time we carried out the reduction in

[15] B. L. Odell, J. M. Vandenbelt, E. S. Bloom, and J. Pfiffner, *J. Am. Chem. Soc.* **69**, 250 (1940); M. May, T. J. Bardos, F. L. Barger, M. Lansford, J. M. Ravel, G. L. Sutherland, and W. Shive, *ibid.* **73**, 3067 (1951); B. Roth, M. E. Hultquist, M. J. Fahrenbach, D. B. Cosulich, H. P. Broquist, J. A. Brockman, J. M. Smith, R. P. Parker, E. L. R. Stokstad, and T. H. Jukes, *ibid.* **74**, 3247 (1952).

[16] M. Viscontini and H. R. Weilenmann, *Helv. Chim. Acta* **41**, 2170 (1958).

basic solution with Pt as catalyst, but now we use CF_3COOH as solvent and Pt or Rd as catalyst.[17]

Procedure

5,6,7,8-Tetrahydropterin (XIV) Sulfate. In a semimicro apparatus for catalytic reduction (note 1, p. 695), 0.150 mg of platinum oxide is reduced to platinum black in 50 ml of trifluoroacetic acid by shaking at atmospheric pressure and room temperature. This reduction is complete within a few minutes. Then 0.3 g (1.85 millimoles) of pterin (IV) is dissolved in the trifluoroacetic solvent, and the shaking is continued until the theoretical amount of hydrogen (3.70 millimoles) has been absorbed. This reduction is generally finished within half an hour. The apparatus is quickly opened, and 2 ml of H_2SO_4 dissolved in 2 ml of absolute ethanol is immediately added to the solution (note 2). The tetrahydropterin (XIV) sulfate precipitates after a short time. The suspension is evaporated at 30° by rotatory evaporation. As soon as the sulfuric acid reaches a specific concentration in the solution, the sulfates dissolve and only platinum remains in suspension. To this suspension, 30–50 ml of a SO_2 solution (7%) is added, and the flask is gently warmed in order to maintain the tetrahydropterin sulfate in solution. The catalyst is filtered off, and the clear warm solution is decolorized with some active carbon; 5–10 ml of absolute ethanol and one drop of sulfuric acid are then added to the colorless solution. The solution is cooled, and the tetrahydropterin sulfate precipitates as thin white needles. After standing in the refrigerator for several hours, the needles are collected by centrifuging or by suction filtration. They are washed with ethanol and ether and dried in a vacuum desiccator over $CaCl_2$. The yield is 0.35 g (70%) of a product that is pure enough for use in biological experiments.

The recrystallization, being rather difficult, should be carried out as follows: One-tenth gram of the crude tetrahydropterin sulfate, 5 ml of water, one drop of sulfuric acid, and a few milligrams of sodium hydrosulfite are introduced in a 25-ml, three-necked, round-bottomed flask equipped with a reflux condenser, a gas inlet tube, and a magnetic stirrer. The mixture is heated with stirring while a slow stream of very pure nitrogen is passed into the solution. The tetrahydropterin sulfate dissolves slowly. More water is to be added if complete solution is not reached near the boiling point. The solution should be filtered while hot by suction after addition, if necessary, of some decolorizing charcoal. Absolute ethanol is then added to the colorless warm solution until the first turbidity appears. After some hours at room temperature and then in the refrigerator,

[17] A. Bobst and M. Viscontini, *Helv. Chim. Acta* **49**, 875 (1966).

the well formed crystals of the tetrahydropterin (XIV) sulfate are collected by suction filtration, washed with ethanol and ether, and dried in a vacuum desiccator over $CaCl_2$ (note 3).

The pure tetrahydropterin (XIV) sulfate has a typical UV spectrum (Fig. 5) resembling that of pure 2,4,5-triamino-6-oxo-1,6-dihydropyrimi-

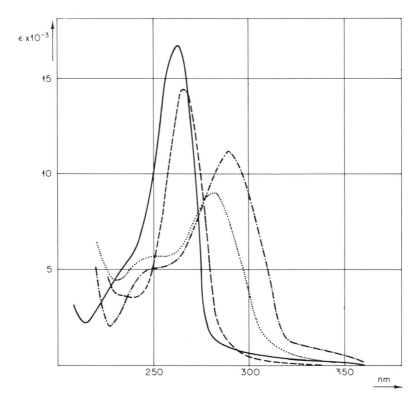

FIG. 5. UV spectra of tetrahydropterin (XIV). (————) 2.1 N HCl, (– – – – –) pH 3.5, (– · – · – · – ·) pH 8, (· · · · · · · ·) 0.1 N KOH.

dine (III) sulfate.[17] Its nuclear magnetic resonance (NMR) spectrum is quite different from that of the pterin (IV), as can be seen by the following data. NMR spectrum of the pterin (IV) (measured in CF_3COOH with tetramethylsilane as internal standard): N(1)-H and N(2′)-H_2, broad peak, 8.80 ppm, 3 protons; C(6)-H and C(7)-H, singlet, 9.12 ppm, 2 protons.[18] NMR spectrum of the tetrahydropterin (XIV) measured under the same conditions (directly after the reduction): C(6)-H_2 and C(7)-H_2,

[18] M. Viscontini, L. Merlini, and W. von Philipsborn, *Helv. Chim. Acta* **46,** 1181 (1963).

singlet, 3.93 ppm[17] (an estimate of the integration is not possible here). The signal of the vinyl protons has completely disappeared. One can also note the difference in the pK values of the two products (IV and XIV). pK values of the pterin (IV): 2.31 and 7.92[19]; 2.51 and 8.02[16]; pK values of the tetrahydropterin (XIV): 1.3, 5.6, and 10.6.[17] The new pK value of 5.6 is to be attributed to the N(5) atom, which becomes basic after the reduction (note 4).

Notes

1. The author uses an all-glass semimicro apparatus, but any other apparatus may be used without difficulty. The only condition is to have a recipient of glass where the reduction itself is carried out in order to be able to see the color of the solution and its fluorescence.

2. The 5,6,7,8-tetrahydropterin (XIV), like the other tetrahydropterins, is very sensitive to oxidation and to light; it should not stay too long in direct contact with them. The first step of the oxidation involves the removal of one of the two free electrons of the N-5 atom and the formation of a radical cation.[20,21] In order to avoid the loss of this electron, protons should be brought into the solution. They are immediately fixed to the lone electron pair of the N-5 atom, and the molecule is thus stabilized. A solution of stabilized tetrahydropterin is colorless and possesses no fluorescence. As soon as the oxidation of the tetrahydropterin begins, the solution becomes a faint red, and a blue fluorescence immediately appears.

3. The pure tetrahydropterin (XIV) sulfate in the crystalline state is very stable, even to the air, and may be kept in the dark without decomposition for many months. A tetrahydropterin hydrochloride can also be obtained and has been described.[17] It possesses the advantage of being much more soluble in water than the sulfate, but the crystals are not easily formed as they are somewhat hygroscopic, and they slowly decompose, even under vacuum.

4. The molecule of the triamino-oxopyrimidine (III) and that of the tetrahydropterin (XIV) are very similar. We have noted the similarity of both UV spectra. The similarity of the pK values is also striking: 1.3, 5.6, and 10.6 for the tetrahydropterin (XIV); 2.0, 5.1, and 10.1[22] for the triamino-oxopyrimidine (III). Both of them can also be considered as vinyl amides with the following limit structures:

[19] A. Albert, D. J. Brown, and G. Cheesemann, *J. Chem. Soc.*, p. 4219 (1952).
[20] M. Viscontini and T. Okada, *Helv. Chim. Acta* **50,** 1845 (1967).
[21] M. Viscontini, *Fortschr. Chem. Forsch.* **9,** 605 (1968).
[22] A. Pohland, E. H. Flynn, R. G. Jones, and W. Shive, *J. Am. Chem. Soc.* **73,** 3247 (1951).

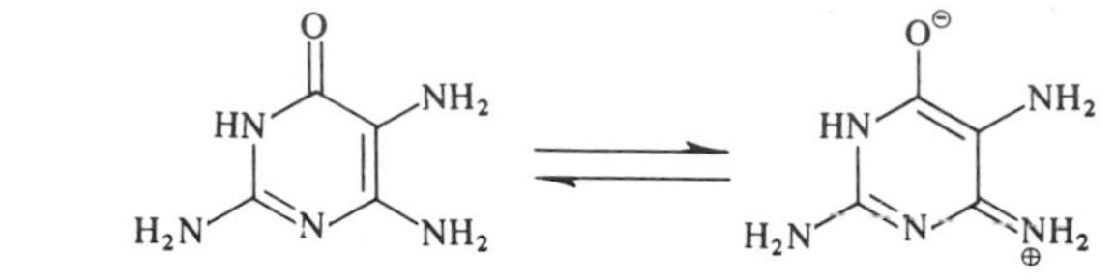

(III)   (IIIa)

(XIV)   (XIVa)

Thus the basic nitrogen atom should be N'(5) in the pyrimidine (III) and N(5) in the tetrahydropterin (XIV). The chemical properties of the products confirm entirely these assumptions.

Synthesis of 5,6,7,8-Tetrahydrofolic Acid

Operation

Synthetic 5,6,7,8-tetrahydrofolic acid was described very early in the history of pteridine chemistry,[15,22] but the product was not stabilized with a strong acid, and the UV spectra published at that time show that the different tetrahydrofolic acids obtained were not entirely pure. We describe below the preparation of the tetrahydrofolic acid as the sulfate, using the method given in the preceding section.

Procedure

Twenty-five milligrams of platinum oxide are reduced by shaking in 8.5 ml of trifluoroacetic acid at atmospheric pressure and room temperature. When this reduction is complete, 0.176 g (0.37 millimole) of folic acid is added to the solvent, and the shaking is continued until the theoretical amount of hydrogen (0.74 millimole) has been absorbed (0.5 hour).

Sulfate Salt. The platinum is filtered from the solution under an atmosphere of nitrogen, and then 0.050 g (0.51 millimole) of H_2SO_4 dissolved in 1 ml of absolute ethanol is added to the clear filtrate. After the solution has been allowed to stand half an hour at 0°, during which time a white crystalline precipitate is formed, the tetrahydrofolic acid sulfate is collected by centrifugation or by suction filtration on a sintered-glass filter. The salt is washed with absolute ethanol and ether, and is dried first in a vacuum desiccator over $CaCl_2$ and then for 4 hours at 110° under high vacuum (0.05 torr). The yield is 0.115 g (57%). The tetrahydrofolic acid sulfate

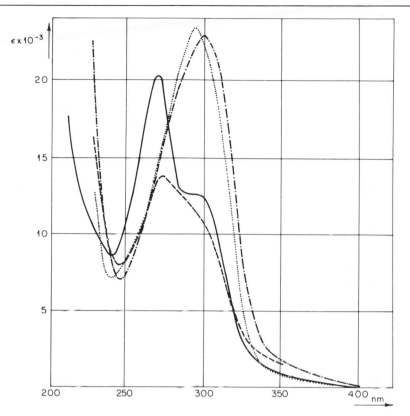

FIG. 6. UV spectra of tetrahydrofolic acid. (——— —) 2.1 N HCl, (– – – – – –) pH 3.5, (– · – · – · – · – · –) pH 8, (· · · · · · · · ·) 0.1 N KOH.

is not as stable as the sulfate of tetrahydropterin (XIV) and cannot be recrystallized. Its purity must be checked by elemental analysis and by its UV spectrum (Fig. 6).

Synthesis of the N^5-Methyl-6,7-diphenyl-5,6,7,8-tetrahydropterin

Operation

The preparation of N^5-methyltetrahydropterins is not easy, and the methods described in the literature[9,11,12] for the preparation of the N^5-methyltetrahydrofolic acid (XIII) give poor yields of the product. However, these substances are of great importance in biological methylation, and therefore we developed in our laboratory a method for preparing a pure N^5-methyltetrahydropterin (XV) by reduction of the N^5-methyl-6,7-diphenyl-5,6-dihydropterin (XI) with NaBH$_4$.[23]

[23] M. Viscontini and T. Okada, *Helv. Chim. Acta* **50**, 1492 (1967).

(XI)   (XV)

Procedure

Thirty-five milliliters of methanol, 3 ml of N NaOH, and 0.2 g (0.6 millimole) of methyldihydropterin (XI) are introduced in a 100-ml, round-bottomed, three-necked flask fitted with a dropping funnel, a magnetic stirrer, and a gas inlet tube. The nitrogen is slowly passed through the suspension, and 3 g (80 millimoles) of $NaBH_4$ dissolved in 10 ml of water is added dropwise with stirring at room temperature. Stirring is continued for 24 hours, during which time the methyldihydropterin goes entirely into solution, and the color of the solution changes from red to yellow.

After removal of the methanol on a rotatory evaporator, $1\ N$ H_2SO_4 is added slowly to the yellow solution in order to destroy the excess of $NaBH_4$ and to neutralize the reaction mixture. A white precipitate is formed which is collected by suction filtration and washed with water, ethanol, and ether. The yield is 0.11 g (50%). The product can be purified under a nitrogen atmosphere by dissolving in methanol, filtration, and finally precipitation by the addition of water. It is quite stable when dried in a vacuum desiccator over $CaCl_2$. Its UV spectrum is typical of a tetrahydropterin. The properties of this substance have been described in the literature.[20,23]

Preparation of 7,8-Dihydropterins

The following syntheses are described: (1) 6-methyl-7,8-dihydropterin; (2) 6,7-dimethyl-7,8-dihydropterin.

Synthesis of the 6-Methyl-7,8-dihydropterin

Operation

The first 7,8-dihydropterin described in the literature was the 6-methyl-7,8-dihydropterin,[24,25] but its preparation with the methods given is rather difficult. In our laboratory we prefer starting from the 6-methylpterin

---

[24] J. H. Boothe, C. W. Waller, E. L. R. Stokstad, B. L. Hutchings, J. H. Mowat, R. B. Angier, J. Semb, Y. SubbaRow, D. B. Cosulich, M. J. Fahrenbach, M. E. Hultquist, E. Kuh, E. H. Northey, D. R. Seeger, J. P. Sickels, and J. M. Smith, *J. Am. Chem. Soc.* **70**, 27 (1948).

[25] W. R. Boon and T. Leigh, *J. Chem. Soc.* 1951 p. 1497 (1951).

(XVII) and reducing it with zinc and sodium hydroxide,[24,26] according to the following scheme:

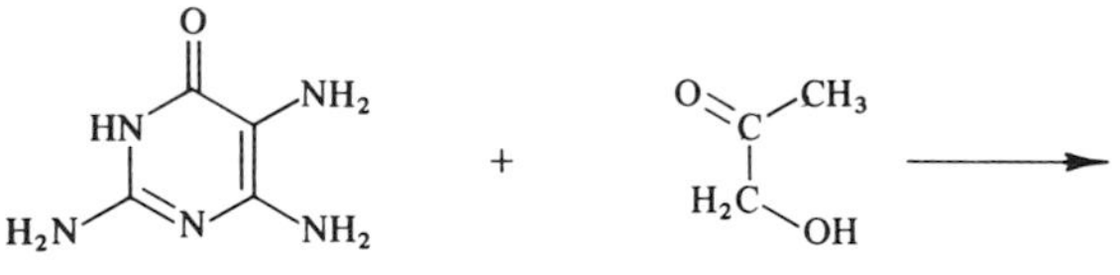

(III)   (XVI)

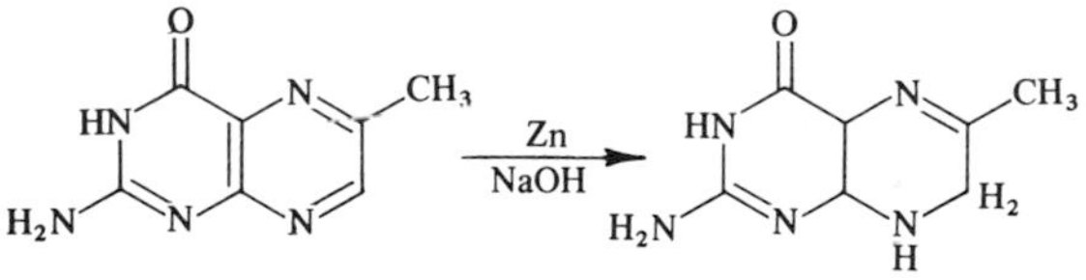

(XVII)   (XVIII)

Procedure

6-Methylpterin (XVII). A 2-liter, three-necked, round-bottomed flask placed over an efficient magnetic stirrer is fitted with a dropping funnel and a gas inlet tube. Four hundred milliliters of absolute methanol and 0.46 g (20 meq) of sodium are introduced into the flask with stirring. After the sodium has reacted, the solution is cooled to 0°, and 11.6 g (100 millimoles) of acetol acetate (note 1, p. 701) is introduced into the flask with stirring while nitrogen is slowly bubbled through the solution. After 15 minutes, the trans-acetylation is complete and the sodium methylate is neutralized by slowly adding 1.2 g (20 millimoles) of acetic acid to the reaction mixture.

An absolute methanolic solution (1100 ml) containing 10 drops of thioethanol (ethanol-1-thiol-2), 6.56 g (80 millimoles) of anhydrous sodium acetate and 10.7 g (50 millimoles) of 2,4,5-triamino-6-oxo-1,6-dihydro-pyrimidine (III) dihydrochloride are now added to the clear solution (note 2).

The flask is then securely stoppered and is allowed to stand at room temperature, and the faint yellow product (XVII) begins to precipitate. After 24 hours, oxygen is passed through the inlet tube until the precipitation of the 6-methylpterin is complete (note 3). The crude 6-methylpterin (XVII) is collected by suction filtration and is washed on a Büchner funnel with methanol, ethanol, and ether.

[26] W. Pfleiderer and H. Zondler, *Chem. Ber.* **99**, 3008 (1966).

After drying in a vacuum dessicator over $CaCl_2$, the yield is 7 g (70%) of pterin (XVII) monohydrate, which can be used without further purification for the preparation of the dihydropterin (XVIII) (note 4).

6-Methyl-7,8-dihydropterin (XVIII) Hydrochloride. Seven grams (35 millimoles) of the crude 6-methylpterin (XVII) monohydrate and 36 g of zinc powder are introduced into a 1-liter beaker containing 400 ml of 0.5 N NaOH. The mixture is stirred for 3 hours at room temperature. After removal of the remaining zinc by suction filtration, 60 ml of concentrated HCl is added to the clear filtrate, and the solution is stored under a nitrogen atmosphere in a refrigerator for 24 hours; during this time a yellow precipitate of crude dihydropterin (XVIII) hydrochloride is formed. The yellow crystals are collected by suction filtration and are washed on a Büchner funnel with ethanol and ether. The crystals are dried in a vacuum desiccator over $CaCl_2$. The yield is 4.2 g (63%) of a deep yellow product which is not sufficiently pure for biological experiments.

Purification is achieved by dissolving the product in a mixture of 850 ml of water and 20 ml of concentrated HCl, which is heated to the boiling point under a nitrogen atmosphere. The solution is then decolorized with

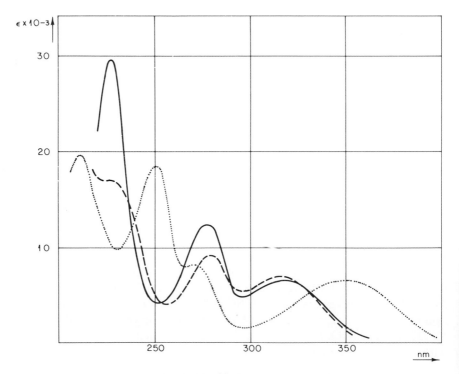

FIG. 7. UV spectra of 5,6-dimethyl-7,8-dihydropterin (XXV). ($\cdots\cdots$) 0.1 N HCl, ($-----$) pH 7, ($\underline{\qquad}$) 0.1 N NaOH.

the smallest quantity of active carbon and is filtered through paper in a heated filter. By cooling and storing for 24 hours in a refrigerator, 3.4 g (50%) of faintly yellow crystals is obtained.

As a hydrochloride salt, the dihydropterin (XVIII) is moderately stable, but it tends to oxidize to pterin (XVII) in the presence of oxygen. Its purity must be verified by elemental analysis, and by its very characteristic UV spectrum, which is identical with the UV spectrum of the 6,7-dimethyl-7,8-dihydropterin (XXV) (Fig. 7).

Notes

1. The commercial acetol acetate (acetoxyacetone) should be purified by vacuum distillation (70°–75°, 10 mm).

2. The solution contains free hydroxyacetone (XVI), which is rather unstable and undergoes polymerization easily. In order to avoid this problem, it is recommended that one proceed immediately with the next operation and work with a nitrogen atmosphere in the flask.

3. We believe that the condensation of the pterin occurs according to the following scheme[4,27]:

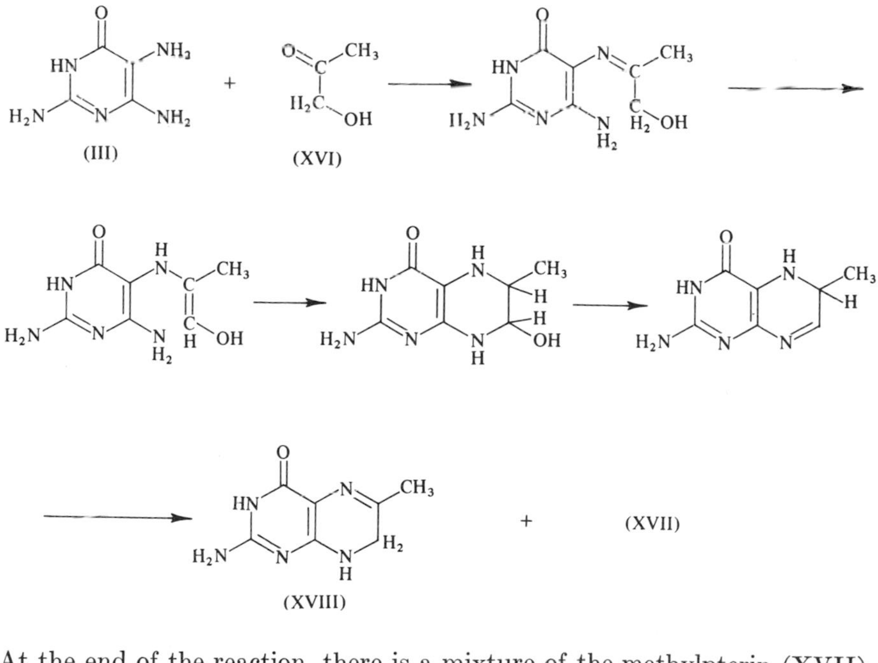

(III)   (XVI)

(XVII)

(XVIII)

At the end of the reaction, there is a mixture of the methylpterin (XVII) and the methyldihydropterin (XVIII) in the precipitate and in the solu-

[27] M. Viscontini and H. Leidner, *Helv. Chim. Acta* **51**, 1029 (1968).

tion, and it is necessary to oxidize the latter pterin to the former with oxygen.

4. The crude 6-methylpterin (XVII) monohydrate contains some isomeric 7-methylpterin; this can be easily shown by $KMnO_4$ oxidation. This small quantity of 7-methylpterin does not affect the subsequent reduction, and it is removed during the purification of the dihydropterin (XVIII). The UV spectrum of the product is very similar to the spectrum of the pterin (IV) (Fig. 1).

Preparation of the 6,7-Dimethyl-7,8-dihydropterin (XXV)

Operation

We have seen that the tetrahydropterins are very unstable substances that are easily oxidized, especially at pH 7. Studies of this oxidation have shown that intermediates are formed on contact with oxygen and that the reaction takes place according to the following scheme[20,21,28]:

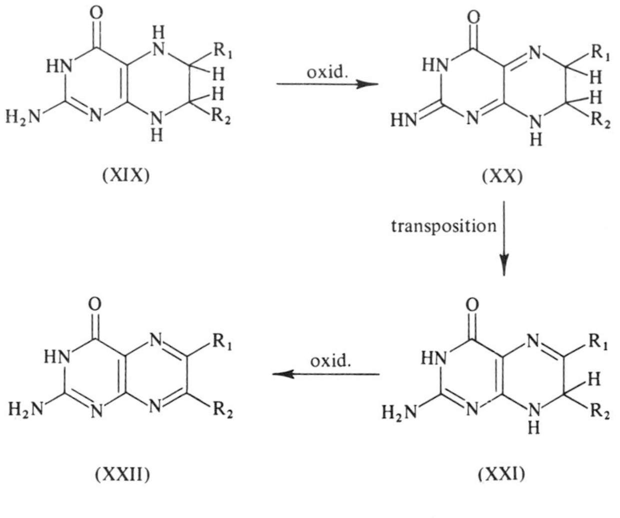

These products (XX) and (XXI) react too quickly to be isolated, but when R_1 and R_2 are alkyl groups, the 7,8-dihydropterin (XXI) is more stable and may be obtained in pure state. The first example of the preparation of a 7,8-dihydropterin by this method has been recently described[29] in the case of 6,7-dimethyl-7,8-dihydropterin.

[28] M. Viscontini and A. Bobst, Helv. Chim. Acta 48, 816 (1965).
[29] H. I. X. Mager, R. Addink, and W. Berends, Rec. Trav. Chim. 86, 833 (1967).

The author currently uses this method in his laboratory as well as the reduction of pterins with Zn and NaOH; it is described here as used to obtain the dihydropterin (XXV).

The dimethyltetrahydropterin (XXIV) is oxidized with O_2 to the dimethyldihydropterin (XXV), which, in the presence of diethylamine, gives an insoluble product. H. I. X. Mager et al. believe that this product is the diethylammonium salt (XXVI) of the dihydropterin (XXV).[29] The present author, however, considers that the insoluble product is the product (XXVII) formed by the addition of the diethylamine to the 5,6-double bond of the dihydropterin (XXV). Such additions are well known in the chemistry of the dihydropterins and have been studied in detail.[21] Whatever the structure of this insoluble product may be, it is easy to isolate and to decompose it with concentrated HCl, giving the desired 6,7-dimethyl-7,8-dihydropterin (XXV).

Procedure

6,7-Dimethylpterin (XXIII). This pterin is synthesized according to the method of Mager *et al.*[29]

6,7-Dimethyl-5,6,7,8-tetrahydropterin (XXIV) Sulfate. The reduction of the dimethylpterin (XXII) is conducted as described for the preparation of the tetrahydropterin sulfate (page 693).

Addition Product (XXVI) or (XXVII). Three hundred milliliters of diethylamine (note 1, p. 704), 13 ml of distilled water, and 1 g (3.4 millimoles) of 6,7-dimethyl-5,6,7,8-tetrahydropterin (XXIV) sulfate are introduced into a 750-ml two-necked, round-bottomed flask fitted with a gas inlet tube and an efficient magnetic stirrer. On stirring, the sulfate goes into solution. A stream of oxygen is now passed through the solution at room temperature (note 2). The faint yellow product (XXVI–XXVII) precipitates slowly. After 1 hour (note 3), the oxygen stream is replaced by nitrogen. The flask, full of nitrogen, is securely stoppered and is stored in a refrigerator for 24 hours. The free precipitate is collected by suction filtration, washed on a Büchner funnel with diethylamine and ether, and dried in a vacuum desiccator over $CaCl_2$ at room temperature.

6,7-Dimethyl-7,8-dihydropterin (XXV) Hydrochloride. Without further purification, the crude addition product (XXVI–XXVII) is dissolved in 5 ml of distilled water in the previously used flask (note 4). Fifty milliliters of ethanol and 3 ml of concentrated HCl are immediately added to this solution. The hydrochloride of the dihydropterin (XXV) precipitates as a fine microcrystalline substance. The mixture is stored in a refrigerator for 24 hours. The hydrochloride is then collected by suction filtration and is washed on a Büchner funnel with ethanol and ether, and dried in a vacuum desiccator over $CaCl_2$ at room temperature. The yield is 0.62 g (80%) of faint yellow crystals sufficiently pure for biological experiments (note 5). Like the 6-methyl-7,8-dihydropterin (XVIII), the dihydropterin (XXV) possesses a very characteristic UV spectrum (Fig. 7), which must be used for its identification.

Notes

1. The author uses diethylamine puriss. from Fluka AG, CH-9470 Buchs.

2. If the reaction is being attempted for the first time, it is recommended that the oxidation of the tetrahydropterin be monitored by paper chromatography using a 3% solution of sodium citrate as eluent. R_f of the dimethyltetrahydropterin (XXIV) is 0.6; of the dimethyl-7,8-dihydropterin (XXV), 0.5; of the dimethylpterin (XXIII), 0.33. Under ultraviolet

light (260 nm), the tetrahydropterin exhibits a dark spot, the dihydropterin a dark blue fluorescence, and the pterin a light blue fluorescence.

3. The oxygen stream (3–4 bubbles per second) is stopped as soon as the tetrahydropterin has completely disappeared. This generally occurs after 1 hour of oxidation. During this time the addition product (XXVI–XXVII) begins to precipitate on the inside of the flask, from which it is removed only with great difficulty. Only the crystals that precipitate later remain free in suspension.

4. It is recommended that one employ the flask used for the preparation of the addition product (XXVI–XXVII), because a great part of the product remains attached on its inner surface. In water, all the material immediately dissolves.

5. When dry, the dihydropterin hydrochloride is very stable and may be stored for weeks without decomposition. In contrast, its oxidation in aqueous solution is very rapid, especially at pH 7.

[183] Tetrahydrofolic Acid and Formaldehyde

By ROLAND G. KALLEN[1]

Tetrahydrofolic acid (THF) is the biologically active form of the vitamin, folic acid, and functions as a coenzyme in enzyme-catalyzed reactions involving one-carbon units at three oxidation levels. This communication is concerned with reactions at the intermediate (aldehyde) level of oxidation. Formaldehyde and glyoxalate react rapidly with THF in aqueous solution to produce imidazolidines: N^5,N^{10}-methylene-THF (MTHF) and the carboxylate derivative, respectively.[1a] Several review articles[2a–2d] and detailed reports on the preparation and characteristics of THF[3] and MTHF[4] have appeared previously.

[1] Supported by grants from the National Science Foundation and the National Institutes of Health (HD-01247, CA-16,912, and GM 13,777).

[1a] R. G. Kallen and W. P. Jencks, *J. Biol. Chem.* **241**, 5851 (1966).

[2a] M. Friedkin, *Ann. Rev. Biochem.* **32**, 185 (1963) and earlier reviews cited therein.

[2b] J. C. Rabinowitz *in* "The Enzymes" (P. D. Boyer, H. Lardy, and K. Myrbäck, eds.), 2nd ed., Vol. 2, p. 185. Academic Press, New York, 1960.

[2c] E. L. R. Stokstad and I. Koch, *Physiol. Rev.* **47**, 83 (1967).

[2d] R. L. Blakley, "The Biochemistry of Folic Acid and Related Pteridines," Wiley, New York, 1969.

[3] F. M. Huennekens, C. K. Matthews, and K. G. Scrimgeour, Vol. VI [113].

[4] F. M. Huennekens, P. P. K. Ho, and K. G. Scrimgeour, Vol. VI [114].

Properties of Tetrahydrofolic Acid and N^5,N^{10}-Methylenetetrahydrofolic Acid

Tetrahydrofolic Acid (THF)

THF is available commercially in solution containing 1.0 M 2-mercaptoethanol (Nutritional Biochemical Corp.) or as the solid containing two moles of acetic acid per mole of THF (mol. wt. 565.4)[5] (General Biochemicals, Sigma Chemical Corp.). Since 2-aminotetrahydropteridine-4-ones and to a significantly lesser extent N^5-substituted derivatives such as MTHF are exceeding oxygen labile,[6] the color of THF in all commercial preparations is tan. Recrystallization or an ether wash and precipitation from methanol-ether under anaerobic conditions[2b,7] can be utilized to obtain more highly purified preparations.

THF in solution decomposes to dihydrofolate (absorption maximum of 282 nm), xanthopterin and other compounds in the presence of air (oxygen) in reactions that appear to be catalyzed by light, acid, base, and heavy metal ions. Copper and iron appear to be especially effective metal ion catalysts.[6] The additional fact that THF is approximately isoionic and minimally soluble in the pH region from 2 to 5 has led to the selection of pH 5 to 6 as optimal for maintenance of stock THF solutions. Protection of solutions from light, the use of deionized water in conjunction with ethylenediaminetetraacetic acid (EDTA), 10^{-3} to 10^{-4} M, and maintenance of anaerobic conditions, for example, by bubbling argon continuously, are strongly recommended.

Although utilization of 2-mercaptoethanol (2-ME) to stabilize THF solutions[4] is widespread, whenever this technique is adopted in kinetic and equilibrium studies, the final 2-ME concentration must be on the order of 10^{-4} M or less, which is lower than is commonly recommended,[3] for the following reasons: (a) Hemithioacetals are formed from thiols and aldehydes in base-catalyzed reactions which are generally sufficiently rapid

[5] R. H. Himes and J. C. Rabinowitz, *J. Biol. Chem.* **237**, 2903 (1962).
[6] R. L. Blakley, *Biochem. J.* **65**, 331 (1957).
[7] R. G. Kallen and W. P. Jencks, *J. Biol. Chem.* **241**, 5845 (1966).

to compete successfully with imidazolidine formation for the available formaldehyde. For example, the equilibrium constant for hemithioacetal formation from 2-ME and formaldehyde is 620 M^{-1} at 25°, ionic strength 1.0 M, and accounts quantitatively for the inhibition of the rate of the reaction of THF with formaldehyde at pH values greater than 4.[1a] Significant inhibition is observed even at 2-ME concentrations as low as 2.7 × 10^{-3} M. (b) THF has been reported to complex directly with thiols[8] to form an N^5-substituted adduct, $HOCH_2CH_2S$-N^5-THF. Since ultraviolet spectral alterations are observed in other N^5-substituted THF derivatives when compared with THF, the failure to detect spectrophotometric evidence for the proposed complex in the course of experiments by previous workers leaves this question unsettled.

Stock solutions of THF have been obtained and maintained by the following procedure:

Solid THF is placed in serum bottles and sealed with tight-fitting

TABLE I

SUMMARY OF pK'_a VALUES AND SPECTROSCOPIC DATA ON TETRAHYDROFOLIC ACID[a]

		Tetrahydrofolate		
Dissociable group		pK'	λ_{max}[d] (mµ)	$\epsilon \times 10^{-3}$ [d] (M^{-1} cm^{-1})
	Amide	10.5	290	21.6[b]
	N^5	4.82	297	29.1
			220	31.4
			290	22.8
			270	25.4
			215	40.8
	Carboxyl (γ)	4.8[c]		
	Carboxyl (α)	3.5[c]		
	N^1	1.24		
			292	20.6
			267	15.4
	N^{10}	-1.25		
			265	20.6

decreasing pH (indicated along left margin with downward arrow)

[a] See Kallen and Jencks.[7] Table reprinted by permission of the *Journal of Biological Chemistry.*

[b] Solutions of pH 12 are unstable.

[c] Spectrophotometrically inert.

[d] ϵ = Molar absorptivity; λ = wavelength.

[8] S. F. Zakrzewski, *J. Biol. Chem.* **241**, 2957 (1966).

silicone-greased stoppers. The gaseous phase is exchanged 5–10 times, and the appropriate amounts of deaerated EDTA solution (final conc. 10^{-3} M) and alkali are introduced with a syringe to achieve a pH of 5–6; the preparation is shaken to dissolve the THF, the gaseous phase is exchanged another 5–10 times, and the solutions are stored frozen under slight positive pressure. Stock solutions of THF (0.03 M) are briefly thawed and refrozen for aliquot removal for the dilute solutions, which are made fresh daily. The latter solutions are maintained at about pH 5 in the presence of 10^{-4} M ethylenediaminetetraacetic acid with argon bubbling continuously. Final

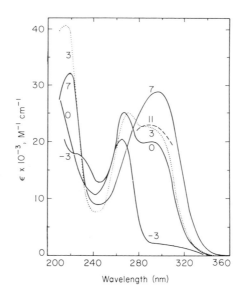

FIG. 1. Spectra of tetrahydrofolic acid between H_0 − 3.0 and pH 14. The lines represent regions in which spectra are independent of pH, that is, regions in which no spectrophotometrically detectable ionization is occurring. The pH was maintained with hydrochloric acid, sodium hydroxide, sodium formate, sodium acetate, and potassium phosphate buffers. For H_0 values see M. A. Paul and F. A. Long, *Chem. Rev.* **57**, 1 (1957). Reprinted by permission of the *Journal of Biological Chemistry*.

reaction mixtures are derived from solutions through which argon also is bubbled slowly and continuously. When further precautions prove necessary, particularly in the alkaline region, modified Thunberg cells of quartz (Pyrocell) are used for the study of THF-containing reaction mixtures.

Table I contains a summary of the pK'_a values, the spectral data by which they were obtained, and the groups to which the pK'_a values have been assigned.[7]

The pH dependence of the spectra of the THF is depicted in Fig. 1.

N^5,N^{10}-Methylenetetrahydrofolic Acid (MTHF)

Equilibria. The equilibrium constants for the formation of the various formaldehyde adducts of THF at 25°, ionic strength 1.0 M, water activity 1.0 are given in Table II.[1a,6,9]

As the pH decreases into the region of the protonation of the N^5 site of THF (pK'_a = 4.82), the formation of MTHF becomes less favorable; at pH 4.3, 22°, $K_{overall}$ (apparent) is reported to be $1.3 \times 10^4\ M^{-1}$,[10] and at pH 2.3, 25°, $K'_{overall}$ is estimated to be 785 M^{-1} from the pK'_a (3.21) of the N^5-group in MTHF (Scheme I) and $K'_{overall} = K_{overall}\ K'_{a_1}/K'_{a_2}$.[11]

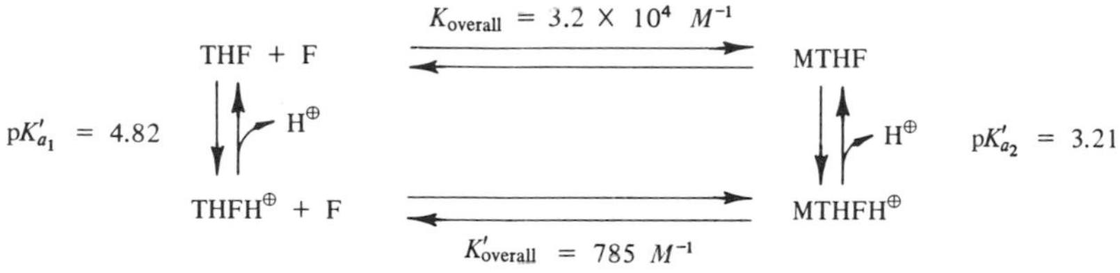

$$K_{overall} = 3.2 \times 10^4\ M^{-1}$$

THF + F $\longrightarrow$ MTHF

pK'_{a_1} = 4.82   H$^{\oplus}$

THFH$^{\oplus}$ + F $\longrightarrow$ MTHFH$^{\oplus}$

$$K'_{overall} = 785\ M^{-1}$$

H$^{\oplus}$ pK'_{a_2} = 3.21

SCHEME I

Kinetics. By making use of the spectral differences between THF and MTHF (Figs. 1 and 2), the kinetics of MTHF formation have been studied over the pH range 0–12.[1a]

The pH dependence of the rate of MTHF formation over much of this range is depicted in Fig. 3, which shows a bell-shaped profile with the maximum rate observed at about pH 6.

Scheme II (see footnote a, Table III) has been proposed to account for the available kinetic data on this reaction and postulates a change in rate-determining step at about pH 7 from general acid-catalyzed attack (step 1, Scheme II) in the acid region to general acid-catalyzed dehydration (step 2, Scheme II) in the alkaline region.[1a] The Brønsted alpha

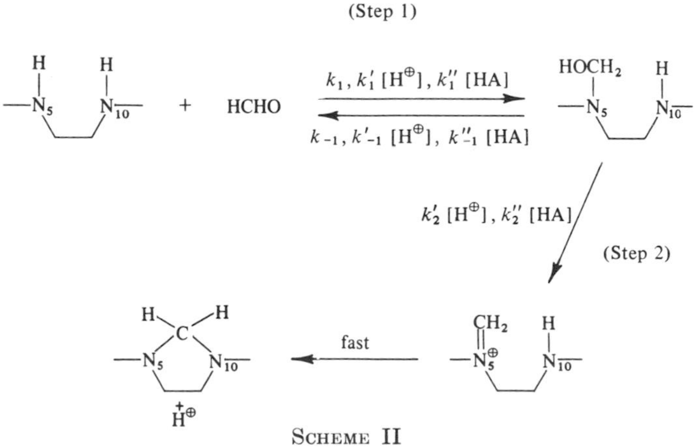

(Step 1)

SCHEME II

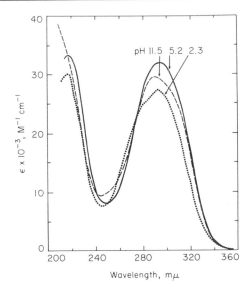

FIG. 2. Spectra of N^5,N^{10}-methylene-THF between pH 2 and 12. The lines represent regions in which spectra are independent of pH, that is, regions in which no spectrophotometrically detectable ionization is occurring except in the case of the pH 2.3 line, which is approximately equidistant from the pK'_a of 3.2 and the next lower pK'_a. The pH was maintained with hydrochloric acid, potassium hydroxide, and acetate buffers.

values for the attack step and dehydration steps are 0.20 and 0.75, respectively. The apparent rate constants are summarized in Table III.[1a] Nucleophilic catalysis of the rate of formation of MTHF by secondary amines has also been observed.[1a]

N^5,N^{10}-Methylene-THF in Enzyme Systems

Three enzyme-catalyzed hydroxymethylations and one enzyme-catalyzed methylation have been reported to involve THF and formaldehyde.[12,13a-13c] The reactions are: serine formation (serine hydroxymethylase[14]), deoxyhydroxymethylcytidylate formation (deoxycytidylate

[9] M. J. Osborn and F. M. Huennekens, *J. Biol. Chem.* **233**, 969 (1958).
[10] M. J. Osborn, P. T. Talbert, and F. M. Huennekens, *J. Am. Chem. Soc.* **82**, 4921 (1960).
[11] R. G. Kallen, unpublished data.
[12] L. Schirch and M. Mason, *J. Biol. Chem.* **237**, 2578 (1962); E. M. Wilson and E. E. Snell, *ibid.*, **237**, 3171 (1962); L. Schirch and W. T. Jenkins, *ibid.*, **239**, 3797, 3801 (1964).
[13a] Y. C. Yeh and G. R. Greenberg, *J. Biol. Chem.* **242**, 1307 (1967).
[13b] A. H. Alegria, F. M. Kahan, and J. Marmur, *Biochemistry* **7**, 3179 (1968).
[13c] M. I. S. Lomax and G. R. Greenberg, *J. Biol. Chem.* **242**, 109, 1302 (1967).
[14] L-Serine:tetrahydrofolate 5,10-hydroxymethyltransferase, EC 2.1.2.1.

TABLE II

APPARENT EQUILIBRIUM CONSTANTS FOR REACTION OF TETRAHYDROFOLIC ACID
AND FORMALDEHYDE[a] AT 25°, IONIC STRENGTH 1.0 M, WATER ACTIVITY 1.0

$$K_{\text{overall}} = \frac{\left[\text{structure: } \underset{N_5 \quad\quad N_{10}}{H_2C \text{ bridge}} \right]}{[\text{THF}][\text{F}]} \qquad 3.2 \times 10^4 \; M^{-1\,b}$$

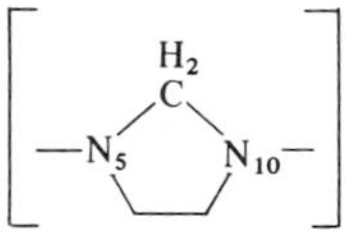

$$K'_{\text{overall}} = \frac{\left[\text{structure: } H\overset{\oplus}{N_5} \cdots N_{10}, \, H_2C \right]}{[\text{THFH}^{\oplus}][\text{F}]} \qquad 785 \; M^{-1}$$

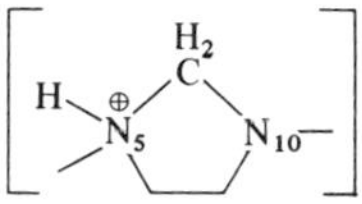

$$K_1 = \frac{[N_5\!-\!CH_2OH]}{[\text{THF}][\text{F}]} \qquad 32 \; M^{-1\,c}$$

$$K_2 = \frac{\left[\begin{array}{cc} OH & OH \\ | & | \\ CH_2 & CH_2 \\ | & | \\ N_5 & N_{10} \end{array} \right]}{\left[\begin{array}{c} H \\ | \\ O \\ | \\ CH_2 \quad H \\ | \quad\;\; | \\ N_5 \quad N_{10} \end{array} \right][\text{F}]} \qquad 2.0 \; M^{-1\,h}$$

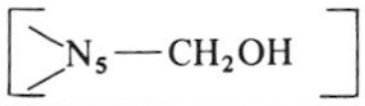

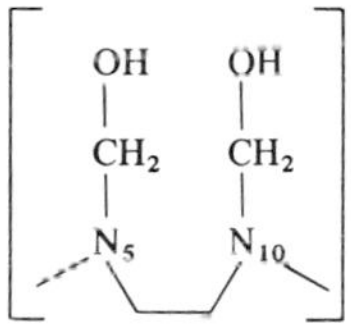

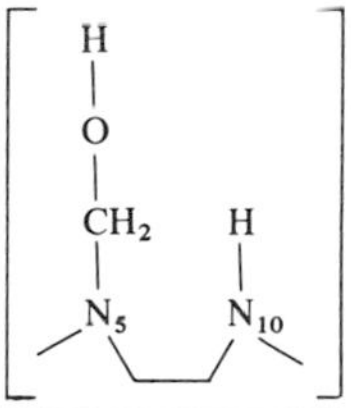

$$K_0/K_1 = \frac{\left[\underset{N_5 \quad\quad N_{10}}{H_2C \text{ bridge}} \right]}{[N_5\!-\!CH_2OH]} \qquad 10^3$$

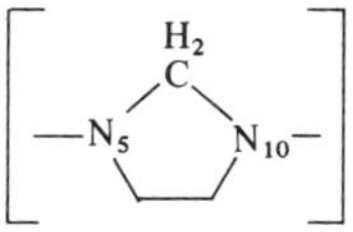

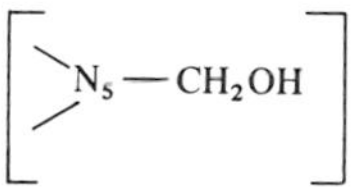

---

[a] F = formaldehyde hydrate. [c] THFH $^{\oplus}$ refers to the N[5] protonated species.

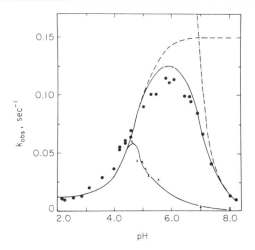

Fig. 3. Dependence on pH of the pseudo-first-order rate constants for the reaction of THF with 0.00167 M formaldehyde, ionic strength 1.0 M, and 25° in the presence ($\times$) and absence ($\bullet$) of 0.00267 M 2-mercaptoethanol. The rate constants are extrapolated to zero buffer concentration. The pH was maintained with hydrochloric acid, formate, acetate, phosphate, N-methylmorpholine, and triethylenediamine buffers. The lines at low (- - - - -) and high (— · —) pH are the calculated rates of the attack and dehydration steps, respectively, and the upper solid line is the calculated rate from the steady state rate equation (see Kallen and Jencks[1a]). Reprinted by permission of the *Journal of Biological Chemistry.*

hydroxymethylase[15]), deoxyhydroxymethyluridylate formation (deoxyuridylate hydroxymethylase[16]), and thymidylate formation (thymidylate synthetase).

There have been basically two mechanisms proposed for the formation of Mannich bases, which have been postulated to be intermediates in these reactions; the first involves S_N2 displacements on either N^5-hydroxymethyl-THF or MTHF[16a] (Paths 1a and 1b, Scheme III) and the second involves

[15] Deoxycytidylate:tetrahydrofolate 5,10-hydroxymethyltransferase.

[16] Deoxyuridylate:tetrahydrofolate 5,10-hydroxymethyltransferase.

[16a] L. Jaenicke, *in* "The Mechanism of Action of Water Soluble Vitamins," (A. V. S. Reuck and M. O'Connor, eds.), Ciba Symp. No. 11, p. 38. Little, Brown, Boston, Massachusetts, 1961; F. M. Huennekens, H. R. Whiteley, and M. J. Osborn, *J. Cell. Comp. Physiol.* **54,** Suppl. 1, p. 109 (1959).

[b] R. L. Blakley, *Biochem. J.* **65,** 331 (1957); M. J. Osborn and F. M. Huennekens, *J. Biol. Chem.* **233,** 969 (1958).

[c] Formaldehyde is hydrated to the extent of 99.9% or greater in aqueous solution [J. F. Walker, "Formaldehyde," 3rd ed., ACS Monograph Ser., Rheinhold, New York, 1964; R. P. Bell and P. G. Evans, *Proc. Roy. Soc.* **A291,** 297 (1966)]. Reprinted by permission of the *Journal of Biological Chemistry.*

TABLE III
APPARENT RATE CONSTANTS FOR REACTION OF TETRAHYDROFOLIC ACID AND FORMALDEHYDE[a] AT 25°, IONIC STRENGTH 1.0 M, WATER ACTIVITY 1.0[b]

k_1	$v = k_1$	$[\text{THF}][\text{F}]$	$90 \ M^{-1} \sec^{-1}$
k'_1	$v = k'_1$	$[\text{THF}][\text{F}][\text{H}^+]$	$4.8 \times 10^5 \ M^{-2} \sec^{-1}$
k''_1	$v = k''_1$	$[\text{THF}][\text{F}][\text{H}_2\text{PO}_4{}^{\ominus}]$	$2400 \ M^{-2} \sec^{-1}$
k_{-1}	$v = k_{-1}$	$[\text{N}_5\text{CH}_2\text{OH}]$	$2.8 \ \sec^{-1}$
k'_{-1}	$v = k'_{-1}$	$[\text{N}_5\text{CH}_2\text{OH}][\text{H}^+]$	$1.5 \times 10^4 \ M^{-1} \sec^{-1}$
k''_{-1}	$v = k''_{-1}$	$[\text{N}_5\text{CH}_2\text{OH}][\text{H}_2\text{PO}_4{}^{\ominus}]$	$75 \ M^{-1} \sec^{-1}$
k'_2	$v = k'_2$	$[\text{N}_5\text{CH}_2\text{OH}][\text{H}^+]$	$2.7 \times 10^7 \ M^{-1} \sec^{-1}$
k''_2	$v = k''_2$	$[\text{N}_5\text{CH}_2\text{OH}][\text{H}_2\text{PO}_4{}^{\ominus}]$	$42 \ M^{-1} \sec^{-1}$
k'_{2a}	$v = k'_{2a}$	$[\text{A}][\text{H}^+]$	$1.3 \times 10^9 \ M^{-1} \sec^{-1}$

$A = $ (THF anion)

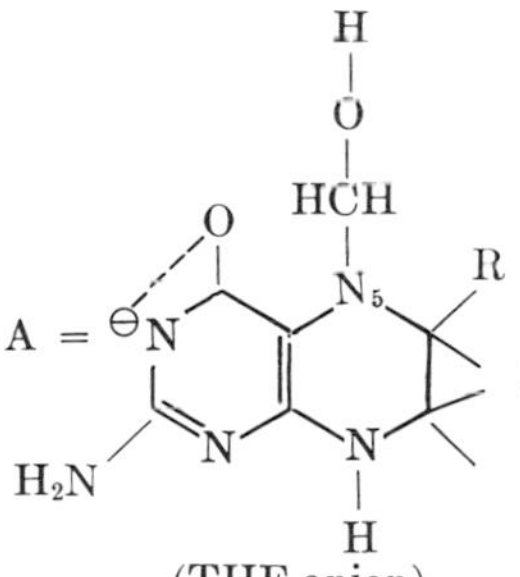

$F = $ formaldehyde hydrate

[a] Formaldehyde is hydrated to the extent of somewhat greater than 99.9% in aqueous solution (see references in footnote c, Table II). The apparent rate constants in this table have not been expressed in terms of unhydrated formaldehyde, which is in all probability the reactive species[1a]: This can be accomplished by multiplying the apparent rate constants by $(1 + K_H)$, where K_H is the hydration constant.

$$K_H = 2275 = \frac{[\text{F}]}{[\text{C}=\text{O}]} \quad \text{(water activity 1.0)}$$

[b] Reprinted by permission of the *Journal of Biological Chemistry*.

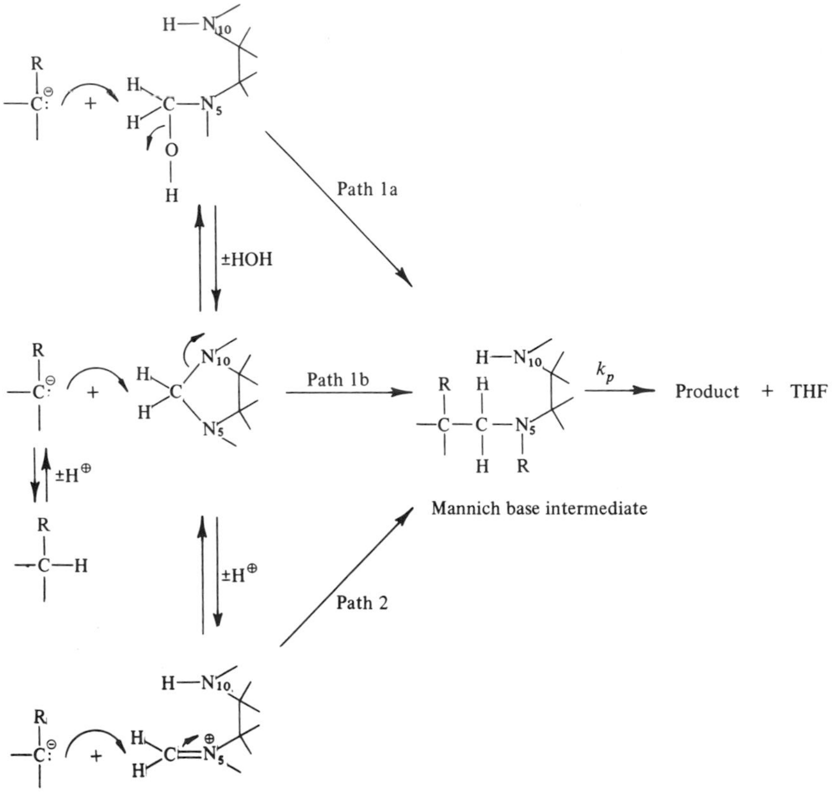

Scheme III

an elimination-addition sequence (Path 2, Scheme III).[16a,17] We favor an elimination-addition mechanism for formation of the Mannich base intermediate in the enzymatic reaction, rather than S_N2 displacement.[1a]

Proton exchange indicative of carbanion formation has been shown to occur with deoxycytidylate catalyzed by deoxycytidylate hydroxymethylase[13a] in the presence of all components except formaldehyde at a rate equal to that of the overall reaction. This has been interpreted to indicate that the formation of the deoxycytidylate carbanion is the rate-determining step. Alanine, a glycine analog, has been shown to undergo proton exchange catalyzed by serine hydroxymethylase in the presence of THF, and the concomitant spectrophotometric changes have been attributed to the formation of the carbanion.[12]

Thymidylate synthetase has been shown to catalyze the loss of tritium from the 5 position of tritium-labeled deoxyuridylate at a rate equal to

[17] L. Jaenicke and E. Brode, *Ann.* **624,** 120 (1959); E. Brode and L. Jaenicke, *Biochem. Z.* **332,** 259 (1960).

the overall rate of thymidylate synthesis. Since this tritium loss requires the presence of all the reaction components including formaldehyde,[13c] kinetic evidence for a carbanion intermediate has not yet been obtained. Among the possible interpretations of the data indicating tritium loss, the one proposed is that the rate-determining step in thymidylate synthesis is the further reaction of the Mannich base intermediate (step k_p in Scheme III).

For the hydroxymethylation reactions, the step following the formation of the Mannich base involves the replacement of THF by hydroxide in the Mannich base, while in the reaction catalyzed by thymidylate synthetase, a hydride or its equivalent from THF replaces THF from the Mannich base with the concomitant formation of dihydrofolate.[18] The nonenzymatic formation of thymine from a 5-thyminyl Mannich base in alkali has been reported.[19] As noted above, we favor elimination-addition mechanisms for such substitution reactions regenerating free coenzyme, but the activity of some serine hydroxymethylases[12] with substrate analogs that are α-substituted does not permit such a pathway in the case of those enzymes and requires another mechanism, for example, S_N2 displacement of THF from the Mannich base or perhaps a direct attack of the carbanion on formaldehyde itself without the formation of Mannich base intermediates.[19a] The stereospecific addition of formaldehyde to glycine to form serine provides evidence against the possibility that formaldehyde itself is the reactive species.[19b]

The suitability of THF as a cofactor in these one-carbon transfer reactions appears to reside in the following considerations:

1. The high association constant for MTHF formation lowers the concentration of free formaldehyde, a compound whose toxic properties have led to its use as an antimicrobial agent.

2. The large markedly asymmetric molecule with many functional groups provides the possibility of satisfying rather stringent stereochemical binding requirements for the proper positioning of the formaldehyde moiety at the enzyme active site.[20]

3. As THF is a secondary amine, a Schiff base (imine) at the N-5 or N-10 sites of THF would be cationic and a more reactive electrophile at all pH values.

4. The relatively acidic N-5 may make the release of THF more facile than the release of a more basic aliphatic secondary amine from Mannich

[18] E. J. Pastore and M. Friedkin, *J. Biol. Chem.* **237**, 3802 (1962).
[19] V. S. Gupta and F. M. Huennekens, *Federation Proc.* **24**, 541 (1965).
[19a] P. M. Jordan and M. Akhtar, *Biochem. J.* **116**, 277 (1967).
[19b] J.-F. Biellman and F. Schuber, *Biochem. Biophys. Res. Commun.* **27**, 517 (1967).
[20] It has been shown that l,L-THF is the active diastereoisomer in enzyme systems.[2d,3]

base intermediates, while the marked difference in basicity between N-10 and N-5 sites may make the formation of the N^5 imine from MTHF much more favorable on kinetic grounds.[21]

5. The redox properties of THF are such as to make the hydrogen replacement of the coenzyme in thymidylate synthesis favorable.

The report of an enzyme that catalyzes MTHF formation from THF and formaldehyde[23] has been regarded with some skepticism in the absence of additional information regarding the enzyme-catalyzed reaction.[2a,2d]

MTHF dehydrogenase[24] which catalyzes the NADP-linked interconversion of MTHF and N^5,N^{10}-methenyl-THF, is discussed elsewhere in this volume (see [172]).

Acknowledgment

We acknowledge permission received from the *Journal of Biological Chemistry* for the reproduction of several figures and tables.

[21] Suggested by the results of studies of N,N'-diphenylethylenediamines and related THF model compounds.[17,22]

[22] S. J. Benkovic, personal communication (1969).

[23] M. J. Osborn, E. N. Vercamer, P. T. Talbert, and F. M. Huennekens, *J. Am. Chem. Soc.* **79**, 6565 (1957).

[24] 5,10-Methylenetetrahydrofolate:NADP oxidoreductase (EC 1.5.1.5).

[184] The Nonenzymatic Hydrolysis of N^5,N^{10}-Methenyltetrahydrofolic Acid and Related Reactions[1]

By DWIGHT R. ROBINSON

The purpose of this article is to review some reactions of formyl derivatives of tetrahydrofolic acid (THF) as an aid to the preparation and use of these compounds in enzymatic reactions. The synthesis of these compounds and their enzyme-catalyzed reactions are reviewed elsewhere in this and previous volumes of this series and are not considered here. Attention is focused on the catalysis of the hydrolysis of mTHF to form N^{10}-fTHF, and some aspects of other reactions which lead to interconversion of formyl derivatives of THF are discussed.

[1] Abbreviations are N^5-fTHF, mTHF, N^{10}-fTHF, and N^5-formiminoTHF, which refer to the N^5-formyl, N^5,N^{10}-methenyl, N^{10}-formyl, and N^5-formimino derivatives of tetrahydrofolic acid. Publication No. 513 of the Lovett Memorial Unit for the Study of Diseases Causing Deformities.

Experimental Procedures

Synthetic N^5-fTHF is available commercially,[2] and the mTHF and N^{10}-fTHF derivatives are readily obtained from this material as described below. It should be noted that the synthetic N^5-fTHF is a racemic mixture of dl-stereoisomers because of the asymmetric 6-carbon atom. Enzymatic methods are used to synthesize the stereospecific l-isomers. Reactions involving formyl derivatives of THF are conveniently followed spectrophotometrically based on the absorbance of mTHF around 350 nm and that of N^5-fTHF and N^5-formiminoTHF at 285 nm (Table I).

TABLE I

ULTRAVIOLET ABSORBANCE OF FORMYL DERIVATIVES OF TETRAHYDROFOLIC ACID

Folate derivatives	λ_{max} (nm)	$\epsilon \times 10^{-4}$ (M^{-1} cm^{-1})	Acidity
mTHF[a]	345	2.60	1 N HCl
mTHF[b]	348	2.65	1 N HCl
mTHF[a]	352	2.50	0.01 N HCl
mTHF[a]	360	2.51	0.001 N HCl
N^{10}-fTHF[a,c]	258	~2.2[c]	pH 7–9
N^5-fTHF[d]	282	3.26	pH 13
N^5-fTHF[e]	286	—[e]	pH 8.4
N^5-FormiminoTHF[f]	285	3.54	pH 7.0

[a] J. C. Rabinowitz, Vol. VI [116].

[b] F. M. Huennekens, P. P. K. Ho, and K. C. Scrimgeour, Vol. VI [114].

[c] D. Robinson, unpublished data. Estimated assuming complete conversion of mTHF to N^{10}-fTHF at pH 9.0.

[d] F. M. Huennekens and M. J. Osborn, Advan. Enzymol. 21, (1959).

[e] D. Robinson, unpublished data. The absorbance at 286 nm at pH 8.4 is approximately 5% less than the absorbance at 282 nm of an equimolar sample in 0.1 N KOH.

[f] J. C. Rabinowitz, Vol. VI [115].

In common with other THF derivatives substituted on the N^5-nitrogen, N^5-fTHF, mTHF, and N^5-formiminoTHF are relatively stable to oxidation by air. The compound N^{10}-fTHF is readily oxidized by air to N^{10}-formyldihydrofolate and N^{10}-formylfolate, and may be protected from oxidation by the addition of thiols or ascorbate. Alternatively, reactions may be run under anaerobic conditions. Anaerobic Thunberg spectrophotometer cells, which fit into standard-sized spectrophotometer cell compartments,

[2] The synthetic compound, N^5-fTHF, may be obtained as the calcium salt from Lederle Laboratories, Pearl River, New York 10965. The material has been generously supplied to investigators by Lederle without charge.

may be obtained commercially.[3] In the author's laboratory larger Thunberg cells are used, and we have constructed simple light-tight aluminum boxes to fit spectrophotometer cell compartments, providing the necessary additional space. Solutions are gassed with argon through a gas-dispensing tube for several minutes before being transferred to the Thunberg cell. After assembling the cell, it is alternately evacuated and flushed with argon several times. Under these conditions dilute solutions of N^{10}-fTHF are stable to oxidation for periods of 24 hours or more.

Hydrolysis of mTHF

The hydrolysis of mTHF to form N^{10}-fTHF is a reasonably rapid reaction in neutral or slightly alkaline buffers. The equilibrium is pH dependent as indicated in Eq. (1), and the formation of N^{10}-fTHF proceeds essentially to completion at pH values above 7.0. The reaction can be shown to be reversible at lower pH values, if the N^{10}-fTHF is protected from oxidation.

$$
\underset{m\text{THF}}{\longrightarrow\!\overset{+}{N}\underset{\vphantom{/}}{\underset{}{}}\;N\!\longleftarrow} \;+\; H_2O \;\rightleftharpoons\; \underset{N^{10}\text{-}f\,\text{THF}}{\longrightarrow\!\underset{H}{N}\quad\underset{H\overset{}{C}=O}{N}\!\longleftarrow} \;+\; H^+ \tag{1}
$$

The hydrolysis of mTHF presents obvious difficulty in the study of enzymatic reactions of this compound. We will consider here the mechanism and catalysis of the nonenzymatic hydrolysis of mTHF, which may aid in selecting appropriate conditions for studies involving this compound. In addition, the mechanism of this reaction provides a model for the mechanisms of some enzymatic reactions of formyl derivatives of THF.

It has been known for several years that the rate of hydrolysis of mTHF at a given pH varies with different buffers. In 0.1 M buffers at pH 7.0, the reaction was shown to be more rapid in the presence of phosphate and certain amines than in maleate.[4] For this reason, several investigations of enzymatic reactions of mTHF have been carried out in maleate buffers. Recently the hydrolysis of mTHF has been investigated in more detail, and some of the conclusions of this work will be reviewed here.[5]

The hydrolysis of mTHF was followed in aqueous solutions at 25.0°,

[3] The anaerobic cells may be obtained from Pyrocell Manufacturing Co., 91 Carver Ave., Westwood, New Jersey.

[4] J. C. Rabinowitz, in "The Enzymes" (P. D. Boyer, H. Lardy, and K. Myrbäck, eds.), 2nd ed., Vol. 2, p. 185. Academic Press, New York, 1960.

[5] D. R. Robinson and W. P. Jencks, J. Am. Chem. Soc. **89**, 7098 (1967).

ionic strength 1.0 M, over the pH range 7–10. The reaction was followed spectrophotometrically by the disappearance of absorbance of mTHF. Good pseudo-first-order kinetics are observed if the reaction is followed anaerobically at 350–360 nm, near the absorption maximum of mTHF. In the presence of oxygen, the oxidation of N^{10}-fTHF leads to deviations from first-order kinetics at these wavelengths. First-order kinetics may be obtained aerobically at 390–400 nm, where none of the products of the reaction absorb appreciably.

It should be noted that the amide group in the pteridine ring of mTHF ionizes with a pK of 8.95, as determined spectrophotometrically. The kinetics show that the reactivity of mTHF can be accounted for by the protonated species over the pH range 7–10. No detectable reactivity of the conjugate base (I) is found under these conditions.

(I)

1. Buffer Catalysis of mTHF Hydrolysis. The rates of hydrolysis of mTHF were determined as a function of buffer concentration in 15 different buffers. Observed pseudo-first-order rate constants, k_{obs}, increase linearly with increasing buffer concentration at constant pH in dilute buffers. The slopes of plots of k_{obs} against total buffer concentration become steeper with increasing pH, indicating that the basic forms of the buffers are the catalytically active species. However, the catalytic effectiveness of the buffers increases with increasing pH to a greater extent than can be accounted for by the concentration of the basic forms of the buffers alone. This requires an hydroxide term in the rate law, which cannot be accounted for by an ionization of the basic forms of buffers, since some of these bases do not possess an ionizable proton. The buffer catalysis is described by the rate law of Eq. (2), where S^+ is the protonated form of the mTHF with respect to the pK of 8.95, and B is the basic form of the buffer.

$$v = k_1(S^+)(B) + k_2(S^+)(B)(OH^-) \qquad (2)$$

The solvent-catalyzed reaction also follows the rate law of Eq. (2), where B is hydroxide ion. Presumably a water term also exists, but it has not been demonstrated. It can be estimated that the value for the water term is <0.005 min^{-1}.

Some typical rate constants in dilute buffer solutions are given in

TABLE II

BUFFER CATALYSIS OF THE HYDROLYSIS OF N^5,N^{10}-METHENYLTETRAHYDROFOLATE
AT 25.0°, IONIC STRENGTH 1.0 M

Buffer	pK	Total buffer concentration $(M)^c$	pH	k_t(min^{-1})^a	k_0(min^{-1})^b
Maleate	5.6	0.10	7.23	0.012	<0.01
Phosphate	6.5	0.10	7.21	0.105	<0.01
Imidazole	7.2	0.05	7.38	0.50	<0.01
N-Methylmorpholine	7.8	0.05	7.85	0.22	0.010
Tris^d	8.3	0.05	8.27	0.43	0.03
Borate	9.0	0.05	9.02	0.37	0.14
Ethanolamine	9.8	0.05	9.20	2.13	0.18
Carbonate	9.8	0.05	9.55	4.0	0.36

^{a} Observed first-order rate constants. The half-times of the reactions may be obtained from the relation, $t_{\frac{1}{2}} = 0.693/k$.

^{b} First-order rate constants for the solvent-catalyzed reactions.

^{c} Rate constants increase linearly with increasing buffer concentrations at least up to the concentrations listed. Negative deviations are seen at higher buffer concentrations.

^{d} Tris(hydroxymethyl)aminomethane.

Table II. In almost every case, most of the observed rate constants, even in these dilute buffers, are accounted for by buffer catalysis; only a small fraction is accounted for by the solvent term, k_0. The reactivity of different buffers is generally related to their basicity. Brønsted plots of log k against pK have slopes (Brønsted β values) of approximately 0.5 and 0.3 for the k_1 and k_2 terms, respectively. Buffers differing widely in chemical structure and charge deviate from the Brønsted relationship by factors of less than 20 in every case, and usually much less. Even severely hindered buffers such as *sym*-collidine, N-methylmorpholine, and tris(hydroxymethyl)-aminomethane have normal reactivity. This behavior indicates that the bases are acting as classical general acid–base catalysts, not as covalent or nucleophilic catalysts. In contrast, nucleophilic reactions of carbonyl derivatives show a large sensitivity to the structure of the nucleophile, and nucleophiles of similar basicity may differ in reactivity by several orders of magnitude.[6] It should be noted also that there is no evidence for the formation of any product other than N^{10}-fTHF, even in the presence of compounds that are highly reactive nucleophiles, such as imidazole and hydrazine, toward carbonyl derivatives.

A consequence of these facts is that buffers of a given basicity differ in their ability to catalyze the hydrolysis of mTHF by relatively small factors. Therefore the rate of hydrolysis of mTHF in a given pH range cannot be dramatically reduced by the use of unreactive buffers. However,

[6] W. P. Jencks and J. Carriuolo, *J. Am. Chem. Soc.* **82**, 1778 (1960).

in attempting to decrease the rate of hydrolysis of mTHF, some advantage can be taken of the small differences in reactivity of different buffers. Imidazole and the k_1 term for carbonate are somewhat more reactive than expected for their basicity. Maleate has not been examined in detail, but the rate constant listed in Table II, for 0.05 M maleate buffer, suggests that it is relatively unreactive. Its poor buffering power at pH 7.0 and above may compensate somewhat for this advantage by requiring high maleate concentrations. Among more alkaline buffers, borate is relatively unreactive.

In summary, the following statements may serve as a guide for selecting conditions that will minimize the rate of nonenzymatic hydrolysis of mTHF. (1) The total buffer concentration should be as low as possible, preferably in the range of 0.01 M. It should be borne in mind that the hydrolysis of mTHF generates one equivalent of acid. (2) At a given pH and buffer concentration, different buffers have similar effects on mTHF hydrolysis, but borate and probably maleate are less reactive for their basicity than other buffers. At a given pH, the less basic of two buffers will generally be less reactive, but this advantage tends to be offset by the necessity of having a larger fraction of the buffer in the reactive free-base form. (3) The rate of hydrolysis of mTHF in buffer solutions increases markedly with increasing pH, which can be seen from the rate law [Eq. (2)]. The greater reactivity of buffers with increasing pH is in large part due to the k_3 terms in the rate law, and these terms are generally negligible at pH values near 7.0. The hydroxide ion-catalyzed reaction follows the same rate law, but the k_2 term is relatively unimportant below pH 9.0. At higher pH values (>9.0), the reactivity of mTHF is decreased by the ionization of mTHF to a less reactive form ($pK = 8.95$).

2. *Mechanisms of mTHF Hydrolysis.* The mechanism which has been proposed for the reaction is shown in Eq. (3). In this mechanism a tetrahedral intermediate (TH) is formed by the attack of solvent on mTHF, followed by the decomposition of the intermediate to form the product by general acid–base catalyzed pathways. It can be stated with some confidence that there is a tetrahedral intermediate (TH) on the reaction pathway, on the basis of the following evidence. First, there is a nonlinear dependence of the first-order rate constants on buffer concentration. It was pointed out above that plots of rate constants against buffer concentration were usually linear in dilute buffers (<0.1 M). At higher concen-

$$\begin{array}{ccc} m\text{THF} & \text{TH} & \text{N}^{10}\text{-}f\,\text{THF} \end{array}$$

trations, however, the plots deviate markedly *below* the values expected by extrapolating from data at low buffer concentrations. This behavior demands a change in rate law since it cannot be accounted for by the addition of terms to the rate law at low buffer concentrations. This change in rate law with changing buffer concentration requires a change in rate-determining step, and thus there must be an intermediate on the reaction pathway. This type of kinetic behavior has been considered previously by others as evidence for the presence of reactive intermediates in other reactions.[7] It should be pointed out that this type of kinetic behavior can also be accounted for by the formation of a complex of the buffer with the substrate, which is less reactive than the substrate alone. In the case of the hydrolysis of mTHF, evidence was presented that substantially rules out complexation as an explanation for these kinetics.

The second piece of evidence for the presence of the intermediate (TH) is the k_2 term in the rate law for the hydroxide-catalyzed reaction (B = OH⁻), in which the reaction is proportional to the second power of the hydroxide ion concentration. This is most reasonably accounted for by the participation of hydroxide ion, or its kinetic equivalent, in each of two steps on the reaction pathway. Finally, the hydrolysis of N,N'-diphenylimidazolinium chloride, which can be considered a simple model compound for mTHF, follows the same rate law as does mTHF [Eq. (2)].[8] More recently a tetrahedral intermediate has been directly observed spectrophotometrically during the alkaline hydrolysis of N,N'-diphenylimidazolinium chloride.[9]

The mechanism proposed for the hydrolysis of mTHF is summarized in Eq. (3). At low buffer concentrations, the neutral tetrahedral intermediate (TH) is formed from the attack of solvent on mTHF in a reversible preequilibrium step. The intermediate then breaks down to form the product through two general acid–base-catalyzed pathways. At high buffer concentrations, the first step becomes rate limiting. This step proceeds through direct attack of hydroxide ion on mTHF, or by the general base-catalyzed attack of water according to mechanism (II).

(II)

[7] See Robinson and Jencks,[5] footnote 8, for a list of pertinent references.
[8] D. R. Robinson and W. P. Jencks, *J. Am. Chem. Soc.* **89,** 7088 (1967).
[9] D. R. Robinson, *Tetrahedron Letters* **48,** 5007 (1968); *J. Am. Chem. Soc.* **92,** 3138 (1970).

Several possible mechanisms can be suggested for the second step of the reaction. Arguments have been given elsewhere for preferring the mechanisms shown below.[8] The k'_3 step corresponds to the k_1 term in the rate law and is interpreted as the general base-catalyzed breakdown of the protonated intermediate (III).

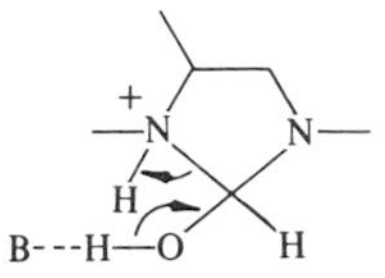

(III) 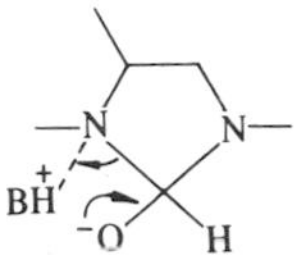   (IV)

The k'_4 step corresponds to the k_2 term in the rate law and proceeds according to mechanism (IV), the general acid-catalyzed breakdown of the anionic conjugate base of the intermediate. In both mechanisms, protonation of the leaving nitrogen atom either precedes or accompanies bond breaking in the transition state. This avoids the formation of the highly unstable nitrogen anion as an intermediate on the reaction pathway. The driving force for expulsion of the leaving group is the oxygen anion in mechanism (IV) which is the predominant mechanism at high pH. At lower pH values, mechanism (III) becomes important, and the driving force for expulsion of the leaving group is provided by removal of the hydroxyl proton by the general base.

Formation of mTHF from N^5-fTHF

The N^5-formyl derivative is stable at neutral and alkaline pH, and is converted to mTHF in acid solutions. The rate of the reaction increases with increasing acidity (Table III), but the kinetics of this reaction have not been studied in detail. Although both these compounds are relatively

TABLE III

RATES OF FORMATION of N^5,N^{10}-METHENYLTETRAHYDROFOLIC ACID
FROM N^5-FORMYLTETRAHYDROFOLIC ACID AT 25.0°[a]

(HCl) N	k_{obs} (min^{-1})
2.0	0.27
1.0	0.21
1.0[b]	0.25[b]
0.50	0.17
0.10	0.083

[a] Ionic strength maintained at 2.0 M with added KCl. The reactions were followed spectrophotometrically using the absorbance of mTHF at 400 nm.
[b] Ionic strength 1.0 M.

stable to oxidation, the solutions are usually gassed with nitrogen or argon. It has been reported that the formation of mTHF from N^5-fTHF does not proceed to completion at pH values above 2.0. The reverse reaction has not been demonstrated in acid solutions, and at neutral and alkaline pH values N^{10}-fTHF is the only hydrolysis product. Some of these observations are considered further below.

Reactions of N^{10}-fTHF

Under anaerobic conditions, N^{10}-fTHF is converted to N^5-fTHF at neutral pH after heating or prolonged standing. The conversion to N^5-fTHF is accelerated by alkaline pH, but in strong alkali, (e.g., $[OH^-] = 0.1\ N$), N^{10}-fTHF is readily hydrolyzed to give THF as the principal product.[10] As previously noted, the equilibrium between mTHF and N^{10}-fTHF favors the mTHF at pH values below 6.0.

N^5-Formiminotetrahydrofolic Acid

This compound has been prepared enzymatically by Rabinowitz, and some of its properties are described in a previous volume of this series.[11] It is relatively stable to oxidation by air and its ultraviolet spectrum is similar to that of N^5-fTHF (Table I). The formimino derivative is converted to mTHF in acid solutions but, unlike N^5-fTHF, it is unstable in neutral and alkaline solutions, decomposing with a half-life of about 1 hour between pH 5 and 9 at 37°.[11] The products of the latter reaction have not been characterized, but the compound might be expected to hydrolyze to give either N^5-fTHF, free THF, or both.

Relative Stability of Formyl Derivatives of THF

At neutral pH the negative free energy of hydrolysis of the formyl derivatives of THF decreases in the order, mTHF $> N^{10}$-fTHF $> N^5$-fTHF, based on considerations discussed below. The equilibrium constants were determined at room temperature or 25°. They include all ionic species of reactants and products and use the convention that the activity of water is equal to one.

The equilibrium constant for reaction (4), K_4, is $0.9 \times 10^{-6}\ M^{-1}$.[12]

$$m\text{THF}^+ + \text{H}_2\text{O} \rightleftharpoons N^{10}\text{-}f\text{THF} + \text{H}^+$$
$$K_4 = (N^{10}\text{-}f\text{THF})(\text{H}^+)/(m\text{THF}^+) \tag{4}$$

[10] M. May, T. J. Bardos, F. L. Barger, M. Lansford, J. M. Ravel, G. L. Sutherland, and W. Shive, *J. Am. Chem. Soc.* **73**, 3067 (1951).

[11] J. C. Rabinowitz, Vol. VI [115].

[12] L. D. Kay, M. J. Obsorn, Y. Hatefi, and F. M. Huennekens, *J. Biol. Chem.* **235**, 195 (1960).

This gives a value for the apparent equilibrium constant at pH 7.0, $K'_{4, \text{pH } 7.0}$, of 9.0, corresponding to an apparent standard free energy change, $\Delta F'_{4, \text{pH } 7.0}$, of -1.3 kcal/mole. The reaction therefore proceeds essentially to completion at pH values above 7.0.

The free energy of hydrolysis of N^{10}-fTHF has been calculated from the equilibrium constant of the formyltetrahydrofolate synthetase reaction, Eq. (5).[13]

$$\text{THF} + \text{formate} + \text{ATP} \rightleftharpoons N^{10}\text{-}f\text{THF} + \text{ADP} + \text{P}_i$$
$$K_5 = (N^{10}\text{-}f\text{THF})(\text{ADP})(\text{P}_i)/(\text{THF})(\text{formate})(\text{ATP}) \tag{5}$$

The value of $\Delta F'_{4, \text{pH } 8.0} = -2.1$ kcal/mole was determined for this reaction (5). From a value of -8 kcal/mole for the apparent standard free energy of hydrolysis of ATP at pH 8.0, the free energy of hydrolysis of N^{10}-fTHF, $\Delta F'_{5, \text{pH } 8.0}$, is calculated to be -6 kcal/mole, which would be approximately the same at pH 7.0.

The relative stability of N^5-fTHF remains uncertain. It is known that N^{10}-fTHF, if protected from oxidation, is converted to N^5-fTHF spontaneously at neutral or alkaline pH.[9] This indicates that the free energy of hydrolysis of N^5-fTHF is less negative than that of N^{10}-fTHF, but the equilibrium constant for this reaction has not been directly measured. A value of the equilibrium constant for the conversion of N^5-fTHF to mTHF has been reported, but the basis for this value may be questioned. The reaction was followed only in the direction of formation of mTHF in the pH range 2.1–3.5 with approximately equal concentrations of N^5-fTHF and mTHF reported at equilibrium at pH 2.7.[12] It was apparently not possible to follow the reaction in the opposite direction under these conditions, and the products present at equilibrium were not characterized. Therefore it may be questioned whether equilibrium was reached in these experiments.

The enzyme-catalyzed transformylation reaction (6):

$$N^5\text{-}f\text{THF} + \text{glutamate} \rightleftharpoons f\text{-glutamate} + \text{THF} \tag{6}$$

has an apparent equilibrium constant at pH 6.6 of approximately 0.1.[14] Huennekens has estimated a free energy of hydrolysis of ca. -2 kcal/mole for N^5-fTHF based on this equilibrium constant, assuming that the free energy of hydrolysis of N-formylglutamate was -3 to -4 kcal/mole.[15] In conclusion, it seems clear that the free energy of hydrolysis of N^5-fTHF is less negative than N^{10}-fTHF, but the exact value for N^5-fTHF is uncertain.

[13] R. H. Himes and J. C. Rabinowitz, *J. Biol. Chem.* **237**, 2903 (1962).

[14] M. Silverman, J. C. Keresztesy, G. T. Koval, and R. C. Gardiner, *J. Biol. Chem.* **226**, 83 (1957); M. Silverman, Vol. V [106].

[15] F. M. Huennekens, H. R. Whiteley, and M. J. Osborn, *J. Cellular Comp. Physiol.* **54**, Suppl. 1, 109 (1959).

[185] A New Preparation of Dihydrofolic Acid

By SIGMUND F. ZAKRZEWSKI and ANNETTE M. SANSONE

Preparation

Principle. Reduction of folic acid to dihydrofolic acid by sodium dithionite has been described.[1] The alternate method involving reduction of folic acid with zinc in alkaline solution[2] is described here. The product of this reaction is identical with that described by Blakley.[3] Use of the zinc reduction method is specially advantageous for preparation of tritium-labeled dihydrofolate. In this case the reaction is carried out in the presence of tritiated water. Of the total tritium incorporated, about 30% is at C-9 and the rest at C-7.[2] Starting with tritiated water of specific activity 90 mCi/ml, the yield of tritium incorporated is about 10%. Under comparable conditions the yield of tritium incorporation during the dithionite reduction was about 2.5%. In addition zinc reduction is applicable to derivatives of folate, e.g., 7-methylfolic acid, which cannot be reduced by dithionite.[2]

Reagents

Folic acid
NaOH, 1.0 N
Zinc dust
Sodium ascorbate solution, 50 mg/ml
HCl, 2 N
2-Mercaptoethanol
Acetone
Anhydrous ether

Procedure. Fifty milligrams of folic acid is dissolved in 1 ml of 1.0 N NaOH. One hundred milligrams of zinc dust is added, and the mixture is stirred mechanically for 30 minutes. Zinc is separated by filtration so as to allow the filtrate to flow directly into 9 ml of a solution of sodium ascorbate. The solution is then chilled in ice, and dihydrofolic acid is precipitated by the dropwise addition of 2 N HCl until the pH is about 2.8. The precipitate is separated by centrifugation, washed twice with 0.01 N HCl, twice with acetone, and finally with anhydrous ether. After drying *in vacuo*, 40–45 mg of dihydrofolic acid of about 85% purity is obtained. The purity was estimated from the ultraviolet spectrum (ϵ at 283 nm = 28.5 ×

[1] S. Futterman, Vol. VI [112].
[2] S. F. Zakrzewski and A. Sansone, *J. Biol. Chem.* **242**, 5661 (1967).
[3] R. L. Blakley, *Nature* **188**, 231 (1960).

$10^3 \ M^{-1} \ cm^{-1}$).[3] To obtain pure compound the crude material may be recrystallized as described by Blakley[3] or purified by ion-exchange chromatography on DEAE-cellulose as described below.

Purification Procedure

Reagents

DEAE-cellulose
Ammonium hydroxide, 1.5 N
2-Mercaptoethanol
HCl, 6 N
HCl, 1 N
Acetone
Anhydrous ether

Preparation of the Column. Eight grams of DEAE-cellulose (OH form) is slurried in 0.1 N NaOH. This slurry is poured into a chromatographic column of about 22 mm diameter. The column bed is allowed to settle by gravity, and 0.1 N NaOH is passed through it until the effluent is colorless. A paper disk is placed on top, and the cellulose is compressed with a plunger to the length of about 14–15 cm. The column is washed with H_2O until the pH of the effluent is that of distilled water, and then with a 1% solution of mercaptoethanol, just enough to saturate the column.

Chromatography. The crude dihydrofolic acid is suspended in 10 ml of 1% mercaptoethanol and dissolved by dropwise addition of 1.0 N NaOH. This solution is applied to the column and allowed to percolate into the column bed. The elution is carried out with a linear gradient of 500 ml of 1% mercaptoethanol in the mixing flask and 500 ml of 1.5 N NH_4OH containing 1% mercaptoethanol in the reservoir. Fractions of about 8 ml are collected. An aliquot of each fraction is diluted 10-fold with H_2O, and absorption at 280 nm is recorded. The main peak (dihydrofolic acid) appears after about 600 ml of the effluent is collected. The fractions containing the desired compound are pooled (about 150 ml) and chilled in ice. The solution is acidified first by dropwise addition of 6 N HCl to pH 7 and then using 1 N HCl to a final pH of 2.8. The precipitate is collected by centrifugation, washed twice with 0.01 N HCl, twice with acetone, and finally with anhydrous ether. After drying *in vacuo*, about 25 mg of pure dihydrofolic acid is obtained.

[186] A New Preparation of Tetrahydrofolic Acid

By Sigmund F. Zakrzewski and Annette M. Sansone

Principle. Preparation of tetrahydrofolic acid by catalytic hydrogenation of folic acid and by enzymatic reduction of dihydrofolic acid has been described.[1] Silverman and Noronka described reduction of folic acid to tetrahydrofolic acid with sodium dithionite.[2] Here a new method of isolation and purification of tetrahydrofolic acid from the crude reaction mixture is described.[3] This method is based on the observation that tetrahydrofolic acid, but not dihydrofolic and folic acid, can be eluted from a DEAE-cellulose acetate column with diluted acetic acid in the presence of mercaptoethanol. After evaporation of the effluent, a product is obtained that has been characterized as a complex between tetrahydrofolic acid and mercaptoethanol of approximate composition: $C_{19}H_{23}N_7O_6 \cdot HS—CH_2—CH_2—OH \cdot H_2O$ with an estimated molecular weight of 541.[3] This compound is slightly more stable than free tetrahydrofolic acid.[3] This method of chromatography described here is applicable to the synthetic as well as to the enzymatic preparation of tetrahydrofolic acid. In the case of the enzymatic preparation (especially when a crude dihydrofolic acid reductase is used as a source of the enzyme) the product of DEAE-cellulose chromatography might be contaminated with protein. Therefore it is recommended that the crude incubation mixture be passed through a Sephadex G-25 column prior to the purification on DEAE-cellulose.

Chemical Synthesis of Tetrahydrofolic Acid

Reagents

Folic acid
Ascorbate buffer, 0.5 M, pH 6.0
2-Mercaptoethanol
Sodium dithionite
NaOH, 1 N and 0.1 N
DEAE-cellulose
Acetate buffer, 1.0 M, pH 6
Acetic acid, 0.75 N
Anhydrous ether

[1] F. M. Huennekens, C. K. Mathews, and K. G. Scrimgeour, Vol. VI [113].
[2] M. Silverman and J. M. Noronka, *Biochem. Biophys. Res. Commun.* **4**, 180 (1961).
[3] S. F. Zakrzewski, *J. Biol. Chem.* **241**, 2957 (1966).

Preparation of the Column for Chromatography. Eight grams of DEAE-cellulose (OH-form) is slurried in 0.1 N NaOH. This slurry is poured into a chromatographic column of about 22 mm diameter. The column bed is allowed to settle by gravity, and 0.1 N NaOH solution is passed until the yellow color disappears from the effluent. A paper disk is placed on top, and the cellulose is compressed with a plunger to the length of about 14–15 cm. The column is then washed with H_2O until the pH of the effluent is that of distilled water. Subsequently, 1.0 M acetate buffer, pH 6, is passed through the column. There is initially an exchange of OH^- against acetate taking place, and the effluent becomes strongly alkaline. When the pH of the effluent comes to 6, the exchange is finished. To remove the excess of acetate the column is washed with 500 ml of H_2O and then with 1% solution of mercaptoethanol, just enough to saturate the column.

Procedure. Seventy milligrams of folic acid are suspended in 3 ml of water; 1.0 N NaOH is added dropwise until all folate has dissolved. This solution is then mixed with 6 ml of 0.5 M ascorbate buffer, pH 6.0. Subsequently 400 mg of solid $Na_2S_2O_4$ is added, and the mixture is incubated in a closed container at 75° for 90 minutes. This solution is applied to the column and allowed to percolate into the column bed.

Elution is carried out with a linear gradient of 500 ml of 1% mercaptoethanol in the mixing flask and 500 ml of 0.75 N acetic acid containing 1% mercaptoethanol in the reservoir. Fractions of about 8 ml are collected. Aliquots of each fraction are diluted 1:10, and absorption at 280 nm is recorded. There are three main components. The third one, which appears after about 250–275 ml of effluent is collected, represents tetrahydrofolic acid. The first two fractions of the tetrahydrofolate effluent are discarded, and the remaining are pooled (about 60 ml) and lyophilized. The residue is taken up in anhydrous ether, centrifuged, and washed once again with anhydrous ether. The yield is about 40–50 mg, purity about 85%.

By repetition of the chromatography described above, higher purity samples (about 95%) may be obtained. In this case tetrahydrofolic acid is suspended in 5 ml of 1% mercaptoethanol, and 1.0 N NaOH is added dropwise to dissolve the solid. This solution is applied to the column. The material appears after 350–400 ml of the effluent is collected.

The ultraviolet spectrum of tetrahydrofolic acid (at pH 7 in the presence of 0.1% mercaptoethanol) has a single broad peak with a maximum at 297 nm and extinction coefficient of 27×10^3 M^{-1} cm^{-1}. A maximum at 298 nm and extinction coefficient of 28×10^3 M^{-1} cm^{-1} were reported for tetrahydrofolic acid prepared by catalytic hydrogenation of folic acid.[1]

Estimation of Purity. The procedure for estimation of the purity of tetrahydrofolate preparations is based on the quantitative conversion of tetrahydrofolate to N^5,N^{10}-methenyltetrahydrofolate. A sample of about

0.5 mg is weighed accurately into a screw-cap tube; 0.5 ml of 97% formic acid is added, the cap is tightened, and the tube is immersed in a boiling water bath for 30 minutes. After cooling, the solution is diluted to 50 ml with 0.1 M HCl, and absorbance at 350 nm of this solution is determined. The purity of the preparation is calculated by Eq. (1).

$$\% \text{ purity} = \frac{0.102 \times \text{absorbance}}{\text{sample weight (in grams)}} \tag{1}$$

This equation is based on the molecular weight of tetrahydrofolic acid of 541 and ϵ_{350} for N^5,N^{10}-methenyltetrahydrofolic acid of $26.5 \times 10^3\ M^{-1}$ cm^{-1}.[4]

Enzymatic Synthesis of Tetrahydrofolic Acid
(5,6,7,8-Tetrahydrofolate:NADP Oxidoreductase, EC 1.5.1.3)

Reagents

Dihydrofolic acid
NADPH
Phosphate buffer, 0.01 M, pH 7
2-Mercaptoethanol
Sephadex G-25
Dialyzed extract of *Streptococcus faecalis*[5]

Procedure. Thirty milligrams of dihydrofolic acid and 43 mg of NADPH are dissolved in 150 ml of 0.01 M phosphate buffer, pH 7, containing 0.1% mercaptoethanol. The reaction is started by the addition of 10 ml of *S. faecalis* extract. Aliquots of 0.1 ml are withdrawn every 5 minutes and diluted with 1 ml of phosphate buffer; the absorbance of these solutions at 340 nm is determined. When there is no further change in absorbance (about 60 minutes), 1.5 ml of mercaptoethanol is added, and the solution is lyophilized. The residue is dissolved in 10 ml of 1% mercaptoethanol. The insoluble material is removed by centrifugation, and the clear supernatant is passed through a column (6 × 11 cm) of Sephadex G-25.[6] The column is washed with 1% mercaptoethanol until the yellow band is eluted.

---

[4] F. M. Huennekens, P. P. K. Ho, and K. G. Scrimgeour, Vol. VI [114].

[5] Packed cells of *Streptococcus faecalis* (8043) (7.5 g, wet weight) are disrupted in a Hughes press. The broken cells are suspended in 10 ml of 0.1 M phosphate buffer, pH 7, stirred with 1 mg of DNase at room temperature until the viscous material is broken down (about 15 minutes), and centrifuged. The supernatant is dialyzed overnight against 5 liters of 0.05 M phosphate buffer, pH 7.

[6] Sephadex G-25 is allowed to swell in water, and the suspension is poured into the column. After being allowed to settle by gravity, the Sephadex is washed with a 1% aqueous solution of 2-mercaptoethanol.

The bulk of protein is eluted first, and the yellow effluent (about 100 ml) represents low molecular weight material, only slightly contaminated by the protein. This fraction is lyophilized. The residue is redissolved in 10–15 ml of 1% mercaptoethanol and purified on a DEAE-cellulose column as described for the synthetic material. The yield is 20–25 mg.

[187] Preparation of Folinic Acid
(N^5-Formyltetrahydro Folic Acid)

By SIGMUND F. ZAKRZEWSKI and ANNETTE M. SANSONE

Preparation

Principle. Folinic acid is formed upon treatment of N^5,N^{10}-methenyltetrahydrofolic acid with alkali at elevated temperature.[1] The synthesis of N^5,N^{10}-methenyltetrahydrofolic acid by catalytic hydrogenation of folic acid in formic acid was described.[2] The synthesis of N^5,N^{10}-methenyltetrahydrofolic acid from tetrahydrofolic acid, its conversion to folinic acid and purification on DEAE-cellulose are described here. *dl*-L-Folinic acid is commercially available. The method reported here is therefore useful only when either *l*-L-folinic acid is desired or when preparation of radioactive material is being considered.

Reagents

    Tetrahydrofolic acid (prepared by any of the synthetic or enzymatic methods)[3,4]
    Formic acid, 97%
    Na_2CO_3 solution in 1% mercaptoethanol, 0.1 M
    NaOH, 5 N
    2-Mercaptoethanol
    NH_4OH, 2 N
    DEAE-cellulose
    Acetone
    Anhydrous ether

[1] B. Roth, M. E. Hultquist, M. J. Fahrenbach, D. B. Cosulick, H. P. Broquist, J. A. Brockman, Jr., J. M. Smith, Jr., R. B. Parker, E. L. R. Stokstad, and T. H. Jukes, *J. Am. Chem. Soc.* **74**, 3247 (1952).
[2] F. M. Huennekens, P. P. K. Ho, and K. G. Scrimgeour, Vol. VI [114].
[3] S. F. Zakrzewski and A. Sansone, this volume [186].
[4] F. M. Huennekens, C. K. Mathews, and K. G. Scrimgeour, Vol. VI [113].

Procedure. One hundred twenty milligrams of tetrahydrofolic acid are heated at 100° in 87 ml of 97% formic acid for 40 minutes. The solution is then evaporated to a syrup *in vacuo* (45°). Upon addition of acetone, a yellow precipitate forms. This is centrifuged, washed once with acetone and once with ether. After drying *in vacuo* 103 mg of crude N^5,N^{10}-methenyl-tetrahydrofolic acid is obtained.

Fifty milliliters of 0.1 M solution of Na_2CO_3 in 1% mercaptoethanol are adjusted to pH 12 with 5 N NaOH. This solution is placed in a screw-cap tube and preheated at 100° for 5 minutes. One hundred milligrams of N^5,N^{10}-methenyltetrahydrofolic acid are added, the cap is tightened, and heating at 100° is continued for 60 minutes. After cooling to room temperature, the pH is adjusted to 8, and the solution is applied to a DEAE-cellulose column in the OH⁻ form.[5] The column is eluted with a linear gradient of 500 ml of 1% mercaptoethanol in the mixing flask and 500 ml of 2 N NH_4OH containing 1% mercaptoethanol in the reservoir. Fractions of about 8 ml are collected. An aliquot of each fraction is diluted 10-fold with water, and absorption at 280 nm is recorded. There are several components, but the main one appears after about 200–250 ml of the effluent has passed through the column. Fractions containing folinic acid are pooled and lyophilized. The material left in the flask appears as a dried-out gel on the walls and cannot be removed from the flask in that form. About 10 ml of acetone is then added, and it is stirred vigorously with a magnetic stirrer while the position of the flask over the stirring motor is being changed constantly. A fine white precipitate of folinic acid forms. This suspension is transferred into a centrifuge tube with a pipette, and another portion of acetone is added. This is repeated until all material is removed from the flask. After centrifugation the precipitate is washed twice with anhydrous ether and dried *in vacuo*. The yield is 40–45 mg. The crude folinic acid is

TABLE I
ULTRAVIOLET SPECTRA IN 0.1 N NaOH

	Max		Min		
	λ	ε	λ	ε	Ratio, max/
Sample[a]	(nm)	$(M^{-1} cm^{-1})$	(nm)	$(M^{-1} cm^{-1})$	min
Leucovorin (Lederle)	282	37.9×10^3	248	18.6×10^3	2.03
Folinic acid (prepared above)	282	34.4×10^3	248	14.1×10^3	2.44

[a] All calculations are based on the following formulas: (1) calcium salt of "leucovorin" $C_{20}H_{21}O_7N_7 \cdot 4 H_2O$ Ca, molecular weight 583; (2) folinic acid (free acid prepared above) $C_{20}H_{23}O_7N_7 \cdot 3 H_2O$, molecular weight 527.

[5] S. F. Zakrzewski and A. Sansone, this volume [185].

TABLE II

ULTRAVIOLET SPECTRA AFTER STANDING FOR 90 MINUTES IN 0.1 N HCl
(N^5,N^{10}-METHENYLTETRAHYDROFOLATE)

Sample[a]	Max		Min			Minimum purity estimated from ϵ_{252}[b] (%)
	λ (nm)	ϵ (M^{-1} cm^{-1})	λ (nm)	ϵ (M^{-1} cm^{-1})	Ratio, max/min	
Leucovorin (Lederle)	352	23.9×10^3	302	10×10^3	2.39	90
Folinic acid (prepared above)	352	24.1×10^3	302	10×10^3	2.41	91

[a] See Table I.

[b] F. M. Huennekens, P. P. K. Ho, and K. G. Scrimgeour, Vol. VI [113].

then suspended in 5 ml of 1% mercaptoethanol solution; 1.0 N NaOH is added dropwise, just enough to dissolve the solid, and this solution is rechromatographed as described above. After lyophilization, treatment with acetone, washing with ether, and drying *in vacuo*, 35 mg of white powder are obtained.

Properties

To ascertain the purity of the above preparation some of its properties were compared with those of the calcium salt of "leucovorin" available from the Lederle division of American Cyanamid Company (Tables I and II).

On paper chromatography in 0.1 M phosphate buffer, pH 7, each compound appears as a single light-absorbing spot with an R_f: 0.73–0.74. In both preparations only traces of fluorescing materials are found.

[188] The Synthesis of N^5,N^{10}-Methenyltetrahydrofolic Acid

By PETER B. ROWE

N^5,N^{10}-Methenyltetrahydrofolic acid (anhydroleucovorin) is the specific formyl donor for the third step in purine biosynthesis, the introduction of the 8-carbon atom into the purine ring by formylation of glycinamide ribonucleotide to formylglycinamide ribonucleotide.[1,2]

[1] J. M. Buchanan and S. Hartman, *Advan. Enzymol.* **21**, 199 (1959).

[2] D. A. Goldthwait, R. A. Peabody, and G. R. Greenberg, *J. Am. Chem. Soc.* **76**, 5258 (1954).

This folic acid derivative is formed in acid solution (pH 1–2) from either N^5-formyl- or N^{10}-formyltetrahydrofolic acid by elimination of water[3]; neither of these is readily available.

An alternative procedure[4] consists in the platinum oxide-catalyzed reduction at pH 1–2 of N^{10}-formylfolic acid, produced by formylation of folic acid. A method has been developed for the direct synthesis, in high yield, of the pure N^5,N^{10}-methenyl derivative from commercial tetrahydrofolic acid.

Tetrahydrofolic acid (Sigma type III), 500 mg, is dissolved in 125 ml of 98% formic acid (Eastman Organic Chemicals) containing 2.0% (v/v) β-mercaptoethanol (Sigma) in a 500-ml round-bottom flask fitted with a reflux condenser. The apparatus is sealed from light with aluminum foil to minimize photodecomposition, and the flask is maintained at 60° for 3 hours in a heating mantle. Addition of thiol is required to prevent the oxidation of both the tetrahydrofolic acid and its N^{10}-formyl derivative, which is formed under these conditions. In the process, the N^{10}-formyl compound is dehydrated by the acid to form the N^5,N^{10}-methenyl derivative by a ring closure.

The cloudy yellow solution is filtered rapidly and lyophilized to dryness. The yield at this stage as calculated from an extinction coefficient of 2.5×10^4 (mole/liter)$^{-1}$ at 355 nm in 0.01 M HCl[5] is of the order of 80%.

Purification thereafter essentially follows that described by Huennekens.[6] A column 4×30 cm is packed under pressure with Whatman cellulose CF11 suspended in water. The column is washed with 1 liter of 0.1 M formic acid containing 0.01 M β-mercaptoethanol and is drained to near dryness. The lyophilized preparation, dissolved in 50 ml of the same acid–thiol solution, is adsorbed to the cellulose column. Elution is carried out with this acid–thiol solution. All effluent fractions exhibiting a value greater than 1.6 for the ratio of absorption at 355 nm to that at 280 nm ($E_{355}:E_{280}$) are pooled and lyophilized to dryness.

The N^5,N^{10}-methenyl derivative is crystallized from 0.1 M HCl–0.1 M β-mercaptoethanol according to Huennekens,[6] washed with absolute alcohol and ether, and stored in a vacuum desiccator at $-20°$.

This preparation, when examined by descending chromatography on

[3] M. T. May, J. Bardos, F. L. Barger, M. Lansford, J. M. Ravel, G. L. Sutherland, and W. Shive, *J. Am. Chem. Soc.* **73**, 3067 (1951).

[4] L. D. Kay, M. J. Osborn, Y. Hatefi, and F. M. Huennekens, *J. Biol. Chem.* **235**, 195 (1960).

[5] J. C. Rabinowitz, *in* "The Enzymes" (P. D. Boyer, H. Lardy, and K. Myrbäck, eds.), 2nd ed., Vol. 2, p. 185. Academic Press, New York, 1960.

[6] F. M. Huennekens, P. P. K. Ho, and K. G. Scrimgeour, Vol. VI [114].

Whatman No. 1 paper with a 1.0 M formic acid–0.01 M β-mercaptoethanol solvent system, demonstrates a single white fluorescent spot ($R_f = 0.50$) under ultraviolet light. The classical absorption spectrum is also seen with the absorption maximum further into the visible region than that of any of the other formyl derivatives.[7] In 1.0 M HCl, the absorption ratio at 348:305 nm ($E_{348}:E_{305}$) is 2.40, approaching the value of 2.46 quoted for the pure material.[6] The overall recovery, based on an extinction coefficient of 2.65×10^4 (mole/liter)$^{-1}$ at 348 nm in 1.0 M HCl, is of the order of 50%. The compound stoichiometrically formylates glycinamide ribonucleotide in the assay system[8] for formyl glycinamide ribonucleotide synthetase.

N^{10}-Formyltetrahydrofolic acid can be formed from the N^5,N^{10}-methenyl compound by allowing the latter to stand for 3 hours at 25° in the presence of 0.1 M Tris-Cl buffer, pH 8.0, containing 0.01 M β-mercaptoethanol. The N^{10}-formyl derivative is the formyl donor for the formylation of 5-amino-4-imidazole carboxamide ribotide.[9] Both the N^5,N^{10}-methenyl and N^{10}-formyl derivatives of tetrahydrofolic acid have been shown to be absolutely required for the biosynthesis of purines de novo in a partially purified enzyme system from pigeon liver.[10]

[7] E. L. R. Stokstad and J. Koch, *Physiol. Rev.* **47**, 88 (1967).
[8] L. Warren and J. M. Buchanan, *J. Biol. Chem.* **229**, 613 (1957).
[9] J. G. Flaks, M. J. Erwin, and J. M. Buchanan, *J. Biol. Chem.* **229**, 603 (1957).
[10] P. B. Rowe, unpublished data (1968).

[189] Preparation and Properties of Antigenic Vitamin and Coenzyme Derivatives

By JEAN-CLAUDE JATON and HANNA UNGAR-WARON

The primary biochemical role of folic acid (pteroylmonoglutamic acid, vitamin B_c) appears to be involved in the synthesis of nucleoproteins.[1] Although folic acid is not active as such in the mammalian organism, it is the precursor of the various coenzyme forms of the vitamin which participate in single-carbon transfer reactions. Its tetrahydro derivative, for example, serves as an acceptor of hydroxymethyl and formyl groups involved in the synthesis of compounds such as purines, pyrimidines, and certain amino acids like serine.[1] Because of the importance of folate co-

[1] E. L. R. Stokstad, *in* "The Vitamins" (W. H. Sebrell, Jr. and R. S. Harris, eds.), Vol. III, p. 89. Academic Press, New York, 1954.

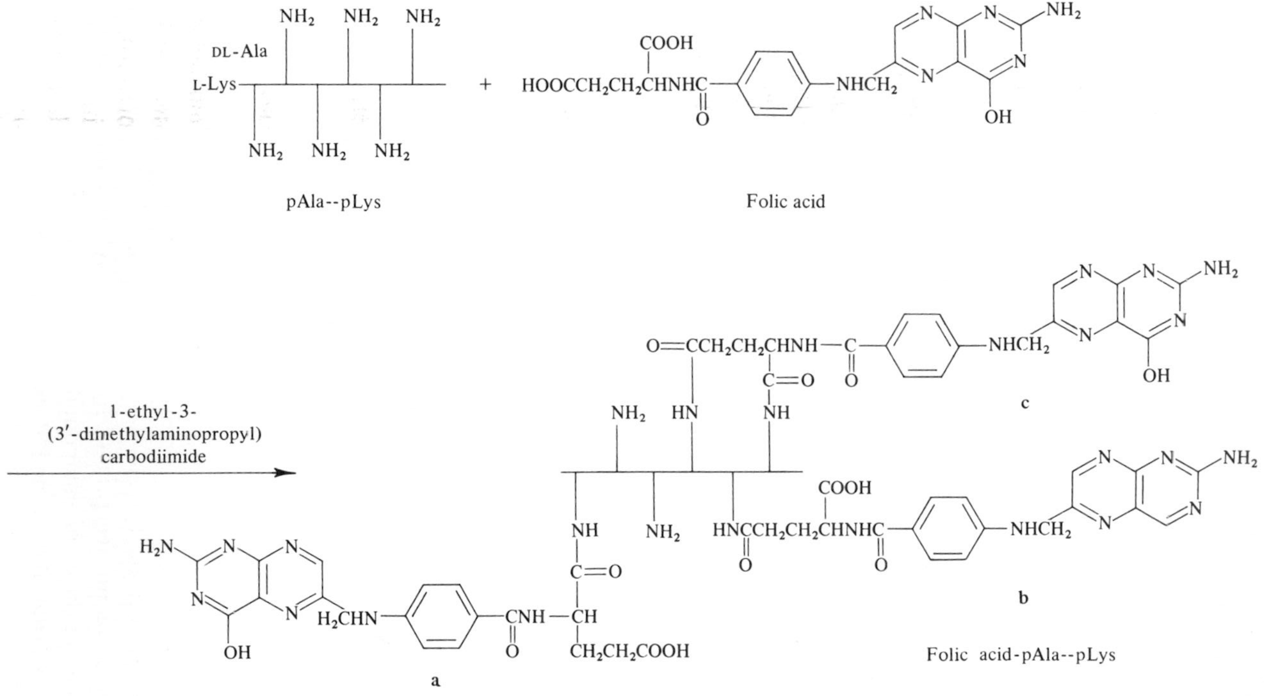

FIG. 1. Schematic representation of the synthesis of a folic acid multichain polypeptide conjugate, folic acid–polyDL-Ala–polyLLys. Possible coupling reactions of folic acid to the polymer through (a), the α-carboxyl; (b), the γ-carboxyl; or (c), both carboxylic groups. From J. C. Jaton and H. Ungar-Waron, *Arch. Biochem. Biophys.* **122**, 157 (1967).

enzymes in nearly all mammalian metabolic systems implying transfers of a one-carbon unit, it seems of interest to elucidate the antigenic properties of folic acid.

The latter is a small molecule which is not immunogenic per se; however, its chemical binding to high molecular weight "carriers" yields conjugates capable of eliciting antibodies specific to the folic acid moiety.[2,3]

Folic acid can be coupled by means of carbodiimide reagents to the terminal amino groups of synthetic multichain polypeptides.[3] The covalent attachment of folic acid to the essentially nonimmunogenic multichain poly-DL-alanyl–poly-L-lysine yields a water-soluble entirely synthetic antigen (Fig. 1), which elicits upon injection into rabbits, antibodies. The specificity of the immune response is mostly directed to the folic acid moiety, as apparent both from the weak cross-reaction of the antibodies with the carrier, multichain poly-DL-alanine, and from inhibition studies.[3]

Preparation of Folic Acid Conjugate of Multichain Poly-DL-alanine

The synthesis of folic acid-polyDLAla–polyLLys can be performed either in aqueous (Variant I) or nonaqueous media (Variant II). Multichain poly-DL-alanine is prepared by the method described in detail elsewhere.[4]

Variant I

1-Ethyl-3-(3'-dimethylaminopropyl)carbodiimide hydrochloride (Ott Chemical Co., Muskegon, Michigan), 0.9 millimole, is added to a mixture of polyDLAla–polyLLys (2 g) and folic acid (0.9 millimole) in 100 ml of 0.05 M sodium hydrogen carbonate adjusted to a final pH value of 8.3. The reaction mixture is stirred for 1 hour and then left overnight at room temperature. After 18 hours, the reaction mixture is dialyzed against several changes of 6 liters of 0.05 M sodium hydrogen carbonate, pH 8.3, and then gradually against a solution of lower salt content, and finally against distilled water. The conjugate is freeze dried and stored at −20°.

Variant II

Multichain poly-DL-alanine (2 g) is dissolved in 10 ml of distilled water and mixed with a solution of folic acid (0.9 millimole dissolved in 170 ml of dimethyl formamide upon heating). N,N'-Dicyclohexylcarbodiimide (DCC, 0.9 millimole) in 20 ml of dimethylformamide is then added, and the reaction mixture is left overnight at room temperature. The final

[2] R. Ricker and B. D. Stollar, *Biochemistry* **6**, 2001 (1967).
[3] J. C. Jaton and H. Ungar-Waron, *Arch. Biochem. Biophys.* **122**, 157 (1967).
[4] M. Sela and S. Fuchs, *in* "Methods in Immunology and Immunochemistry" (M. W. Chase and C. A. Williams, eds.), Vol. 1, p. 167. Academic Press, New York, 1967.

reaction mixture (200 ml) should not contain more than 5% water. After 18 hours, it is dialyzed against several changes of 6 liters of 0.05 M sodium hydrogen carbonate in order to remove excess folic acid and the dimethyl formamide. It is then dialyzed against distilled water. The precipitated material in the dialysis bags, because of the water-insoluble DCC and the urea derivatives produced, is filtered off and discarded; the filtrate is dialyzed again against distilled water, freeze dried, and stored at $-20°$.

The folic acid content of the polymers is determined from the ultraviolet spectrum of folic acid-polyDLAla–polyLLys in 0.1 N NaOH, as compared with that of folic acid in the range of 280–320 nm, and taking into consideration that multichain poly-DL-alanine itself does not contribute significantly to the absorbancy of the conjugates in the absorption spectrum range being analyzed.[3] The amount of folic acid attached, according to both methods of synthesis, is in the range of 9–11% (weight ratio) of the polymeric conjugate obtained. Molecular weights, derived from sedimentation and diffusion data, depend upon the molecular weight of the multichain poly-DL-alanine used as a carrier for the synthesis of the conjugates, and are usually in the range of 90,000–150,000.

Methods of Immunization

Groups of 5–8 rabbits are immunized with emulsions containing equal volumes of 2% antigen in 0.15 M NaCl solution and complete Freund's adjuvant. Upon each injection, the animals receive 10 mg of antigen administered intramuscularly into the thighs of hind legs. Three or four injections are given at 10-day intervals.

Alternatively, 2 injections of 10 mg of antigen each into multiple intradermal sites (about 15, distributed from the neck through the flanks) are given at a 15-day interval. For this purpose, an emulsified mixture composed of 1 ml of 1% antigen solution in 0.15 M NaCl and 1.5 ml of complete Freund's adjuvant is used.

After immune response is obtained, the rabbits are bled twice a week, and the antisera are pooled and kept frozen.

Isolation of Specific Antibodies to Folic Acid

The isolation of specific antibodies to folic acid can be achieved by using a specific insoluble immunoadsorbent prepared from bromoacetyl cellulose and folic acid-polyDLAla–polyLLys, in analogy to the immunoadsorbents described in detail by Robbins et al.[5] Antifolic acid sera (250 ml) are mixed with a conjugate of folic acid-polyDLAla–polyLLys to bromo-

[5] J. B. Robbins, J. Haimovich, and M. Sela, *Immunochemistry* **4**, 11 (1967).

acetyl cellulose (3 g)[6] and stirred overnight at 4°. The immunoadsorbent is spun down and washed several times with 0.15 M NaCl, until the absorbancy of the washing fluid at 280 nm is less than 0.1. Elution of specific antibodies from the immunoadsorbent can be carried out by resuspending the complex either in 0.1 N acetic acid (50 ml) or in a 0.02 M folic acid solution (50 ml) in 0.15 M NaCl adjusted at pH 7.3 with 0.05 N NaOH, and stirring at 37° for 1 hour. After centrifugation at 17,000 g for 15 minutes, the antibody solutions are exhaustively dialyzed against 0.15 M NaCl–0.02 M Tris-HCl buffer, pH 7.4, and a small amount of insoluble material formed during the dialysis is removed by centrifugation. Antibody solutions are then concentrated by vacuum dialysis to 10–20 mg/ml and stored at −20° prior to use.

[6] The optimal pH value for the physical binding of folic acid–polyDLAla–polyLLys to bromoacetyl cellulose is around 5.0 (0.1 M sodium phosphate–0.05 M citrate buffer). The procedure for the preparation of this particular conjugate is analogous to that described by J. C. Jaton, H. Ungar-Waron, and M. Sela, *European J. Biochem.* **2**, 106 (1967).

[190] Pteridine Synthesis in *Tetrahymena*[1]

By G. W. KIDDER and VIRGINIA C. DEWEY

Unlike the kinetoplastid flagellate *Crithidia fasciculata*,[1a] the ciliated protozoan *Tetrahymena pyriformis* does not synthesize biopterin, or any unconjugated pteridine from folic acid. It has been shown to synthesize its major pteridine, ciliapterin, entirely from the carbons and nitrogens of guanosine.[2] The procedures for isolating and characterizing this pteridine and demonstrating its precursors are given below.

Tetrahymena pyriformis W is grown in the defined medium given in Table I. Low-form culture flasks (2.5-liter) are best as the growth container, as they ensure adequate aeration, provided no more than 400 ml of medium per flask is employed. Incubation is carried out in total darkness, and all subsequent manipulations are performed in either red or very subdued light, as pteridines are subject to photolysis. After 5 days of growth, the total culture is made alkaline (pH 9–10) with ammonium hydroxide (which causes cell rupture), and the cell debris is removed by filtration on a Büchner funnel with a pad of filter aid (Celite). The filtrate (approximately

[1] Supported by research grants AM 01005 and CA 02924 from the National Institutes of Health, U.S. Public Health Service.
[1a] G. W. Kidder, V. C. Dewey, and H. Rembold, *Arch. Mikrobiol.* **59**, 180 (1967).
[2] G. W. Kidder and V. C. Dewey, *J. Biol. Chem.* **243**, 826 (1968).

TABLE I
COMPOSITION OF THE GROWTH MEDIUM

Component	Amount (μg/ml)	Component	Amount (μg/ml)
DL-Alanine	110	Glucose[a]	2500
L-Arginine	206	Na acetate	1000
L-Aspartic acid	122	$MgSO_4 \cdot 7H_2O$	100
Glycine	10	$Fe(NH_4)(SO_4)_2 \cdot 6H_2O$	25
L-Glutamic acid	233	$FeCl_3 \cdot 6H_2O$	1.25
L-Histidine	87	$MnCl_2 \cdot 4H_2O$	0.5
DL-Isoleucine	276	$ZnCl_2$	0.05
L-Leucine	344	$CaCl_2 \cdot 2H_2O$	50
L-Lysine	272	$CuCl_2 \cdot 2H_2O$	5
DL-Methionine	248	K_2HPO_4	1000
L-Phenylalanine	160	KH_2PO_4	1000
L-Proline	250	Purines and pyrimidines[b] as follows:	
DL-Serine	394	Guanine-2-^{14}C (50 μCi/400 ml) diluted with cold	
DL-Threonine	326	guanosine for a total of 30 μg/ml guanine	
L-Tryptophan	72	Adenylic acid	20 μg/ml
DL-Valine	162	Cytidylic acid	25 μg/ml
Ca pantothenate	0.10	Uracil	10 μg/ml
Nicotinamide	0.10		
Pyridoxal HCl	0.10		
Pyridoxamine	0.10		
Riboflavin	0.10		
Folic acid	0.01		
Thiamine HCl	1.00		
Biotin (free acid)	0.0005		
Choline Cl	1.0		
6-Thioctic acid	0.004		

[a] Autoclaved separately and added aseptically.

[b] When side chain precursors are being investigated, the adenylic acid is replaced with adenine (25 μg/ml); the guanine-2-^{14}C is replaced with guanosine-U-^{14}C (50 μCi/400 ml) plus unlabeled guanosine to give a concentration of 30 μg/ml. The cytidylic acid is omitted, and the concentration of the uracil is doubled.

400 ml) is applied to a column of Sephadex G-25 (fine),[3] 7 × 28 cm, and elution is carried out with 13 mM mercaptoethanol (20-ml fractions). The fractions are monitored spectrophotometrically at 275 nm; fluorometrically with a fluorometer having 365-nm primary and 450 nm secondary filters; and for radioactivity by streaking 50-μl aliquots from each fraction onto filter paper strips (1.5 in. wide) and running the dried strip through a 4π scanner. Fractions may also be monitored for biological (unconjugated pteridine) activity by adding 100 μl from each fraction to growth tubes containing pteridine-free medium. After sterilization and inoculation with

[3] V. C. Dewey and G. W. Kidder, *J. Chromatog.* **31**, 326 (1967).

pteridine-depleted *Crithidia* and incubation for 96 hours in the dark at 25°, growth is determined turbidimetrically.[4]

Fluorescent fractions containing biologically active pteridine are combined and placed on a longer (3.5 × 60 cm) Sephadex G-25 column and eluted as before. These gel filtrations are very advantageous because the pteridines are concentrated about 4-fold in a single passage[3] and are separated from the bulk of the ultraviolet absorbing materials in the original extract.

The fluorescent fractions from the second Sephadex column are combined, reduced to dryness by lyophilization, taken up in 8–10 ml of water, and chromatographed on a phosphocellulose (H[+] form) column (0.8 × 27 cm) with 13 mM mercaptoethanol as the eluent. Three-milliliter fractions are usually taken. The fluorescent fractions containing biologically active material are combined, dried by lyophilization, and again taken up in 8–10 ml of water. This material, nearly pure ciliapterin, is chromatographed on an anion-exchange column (either DEAE-cellulose or Dowex 1-X8 formate). The combined fluorescent fractions are again reduced to dryness by lyophilization, taken up in 1–2 ml of water, and applied as a streak to large filter paper (Whatman No. 1). Either 1-butanol–acetic acid–water (20:3:7) or 1-propanol–1% ammonia (2:1) is used as the solvent, and the chromatograms are developed in the descending manner. After drying in the dark, the fluorescent band is outlined under UV light, cut out, and eluted with water. Further purification (as judged by the ratio of absorbance at 272 nm–252 nm, by constant specific radioactivity and by constant biological activity in the *Crithidia* test[4]) may be carried out as above, but employing different solvent systems.

Degradations. All but one carbon of the side chain of the pteridines may be removed by alkaline permanganate oxidation, leaving the pterin carboxylate. The pteridine solution is made 0.1 N to NaOH, and small crystals of KMnO$_4$ are dissolved in the mixture in a hot (90°) water bath until the purple color of the permanganate persists. The excess permanganate is then destroyed with a few drops of ethanol, and the MnO$_2$ is removed by filtration. Pterin 6-carboxylate results when ciliapterin is oxidized in this way, and may be compared chromatographically, spectrophotometrically, and fluorometrically to authentic pterin 6-carboxylate, and its specific radioactivity can be determined.

Easily hydrolyzable bonds (such as phosphoester or glycosidic) are shown to be absent from ciliapterin by treating a solution with 2 N HCl for 1 hour at 100° and then cochromatographing the treated pteridine with the starting material.

In determining whether the side-chain carbons have been derived from

[4] V. C. Dewey and G. W. Kidder, this volume [174].

TABLE II
ORIGIN OF SIDE CHAIN OF CILIAPTERIN

Compound	Specific activity, average of two experiments (cpm/micromole)
Ciliapterin	30×10^5
Pterin-6-carboxylate-^{14}C	
Predicted if side chain not derived from GTP	30×10^5
Predicted if side chain derived from GTP (7/9)	23.4×10^5
Found	23.3×10^5

the ribose carbons of guanosine, as is the case in *Colias*,[5] or from some other precursor material in the medium, which has been claimed for *Anacystis*[6] and *Drosophila*,[7] it is necessary to grow *Tetrahymena* in a medium containing guanosine-U-^{14}C or guanosine monophosphate-U-^{14}C. No other ribosides

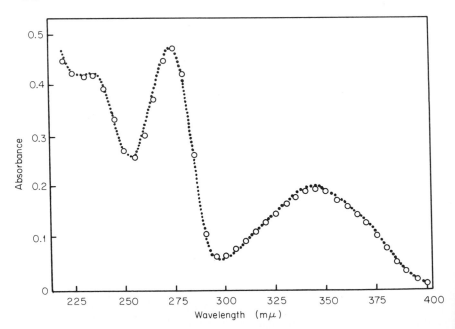

FIG. 1. Ultraviolet absorption spectrum of 6-biopterin (dotted line) and ciliapterin (open circles) at pH 7.

[5] W. B. Watt, *J. Biol. Chem.* **242**, 565 (1967).

[6] F. I. Maclean, H. S. Forrest, and J. Meyer, *Biochem. Biophys. Res. Commun.* **18**, 623 (1965).

[7] K. Sugiura and M. Goto, *Biochem. Biophys. Res. Commun.* **28**, 687 (1967).

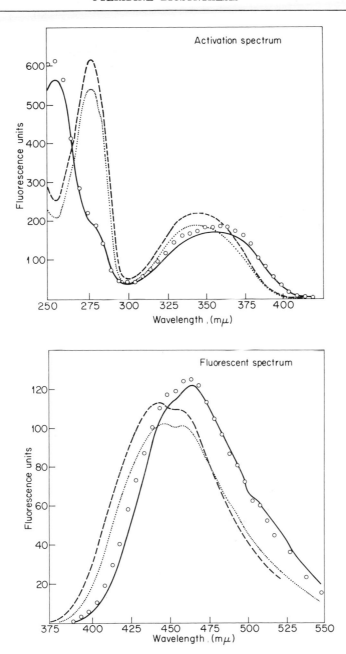

FIG. 2. Fluorescence and excitation spectra (corrected) of ciliapterin: neutral, dotted line; acid, dashed line; alkali, solid line. The open circles represent the spectra of 6-biopterin in alkali, for comparison.

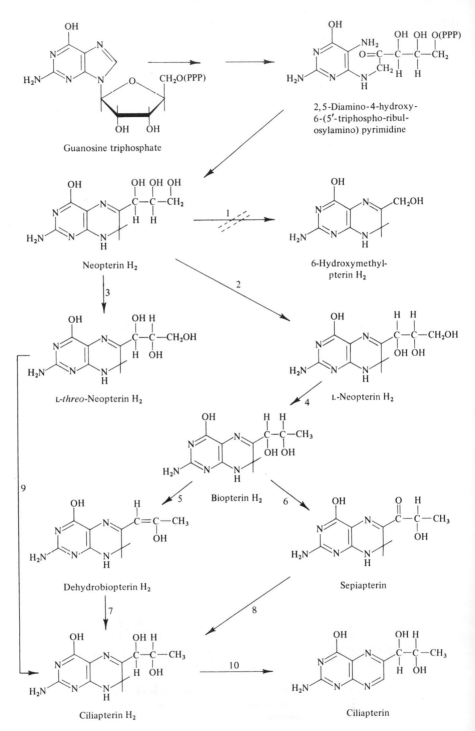

Guanosine triphosphate

2,5-Diamino-4-hydroxy-
6-(5'-triphospho-ribul-
osylamino) pyrimidine

Neopterin H$_2$

6-Hydroxymethyl-
pterin H$_2$

L-*threo*-Neopterin H$_2$

L-Neopterin H$_2$

Biopterin H$_2$

Dehydrobiopterin H$_2$

Sepiapterin

Ciliapterin H$_2$

Ciliapterin

or ribotides should be present, in order to preclude transribosylation. Accordingly, the pyrimidine requirement of the organism[8] is met by uracil. Adenine replaces adenylic acid, this to spare the guanine, which is an absolute dietary requirement for the organism.[8] Ciliapterin is isolated, purified, and its specific radioactivity determined by the methods described above. A sample of this ciliapterin is then subjected to alkaline permanganate oxidation, and the specific radioactivity of the resulting carboxylate determined. If the side chain carbons had been derived from precursors other than the guanosine ribose carbons then they would carry no label, and the specific radioactivities of the ciliapterin and the carboxylate derived from it would be the same. If, on the other hand, the side chain carbons are the carbons of the guanosine ribose, then the specific radioactivity of the carboxylate would be less than the original ciliapterin by 2/9. Table II shows that the latter conditions are met, and it can only be concluded that *Tetrahymena* forms its completed ciliapterin molecule, as *Colias*[5] does its sepiapterin molecule, by rearrangement of neopterin.

In order to determine whether the terminal carbon of the side chain of ciliapterin is reduced (—CH_3) or alcoholic (—CH_2OH), periodate oxidation is employed. The pteridinaldehyde is nonvolatile and remains in the reaction mixture, but the aldehyde(s) formed from the side chain would be formaldehyde if the side chain were the trihydroxypropyl, or acetaldehyde if the side chain were dihydroxypropyl with a terminal methyl carbon. Both formaldehyde and acetaldehyde are volatile and must be trapped, most readily as their hydrazones. A three-vessel (with ground-glass joints) train is used for this operation. The first and third vessels contain 2,4-dinitrophenylhydrazine (0.1 g/100 ml of 2 N HCl), the first to ensure against the entrance of air-borne aldehydes, and the third as a trap for the volatile aldehydes produced by the periodate oxidation of the ciliapterin. The third vessel is kept at 80° by being immersed in a water bath. The middle (second) vessel in the train contains acidified (0.1 N HCl) ciliapterin in 1.5 mM $NaIO_4$. Air is drawn through the vessels, and the oxidation is

---

[8] G. W. Kidder and V. C. Dewey, *Arch. Biochem.* **8**, 293 (1945).

Fɪɢ. 3. Scheme of reactions leading to the synthesis of ciliapterin in *Tetrahymena*. Reaction (1) is blocked in this organism, precluding the synthesis of folic acid. There is no evidence indicating which of the alternate pathways is operating: reaction (2), producing the ʟ-isomer of neopterin, which would then undergo reduction of the terminal carbon [reaction (4)] to form biopterin, or reaction (3) which would produce the *threo* compound from the *erythro*. Subsequently the terminal carbon would then be reduced [reaction (9)] to form dihydrociliapterin. If biopterin is indeed formed, dihydrociliapterin could be formed by dehydration [reaction (5)] and rehydration [reaction (7)]; or oxidation to sepiapterin [reaction (6)] and then reduction to dihydrociliapterin [reaction (8)]. Reaction (10) may not be biological.

allowed to continue for 2 hours. The hydrazone formed in the third vessel is extracted into ethyl acetate, the ethyl acetate is evaporated, and the hydrazone is taken up into carbon tetrachloride, again evaporated to dryness and finally taken up in ethyl acetate for chromatography and determination of absorption spectra. The spectrum obtained is compared with the spectra of authentic formaldehyde and acetaldehyde hydrazones under identical conditions. It is found that the correspondence with the acetaldehyde hydrazone is exact. Chromatography on paper together with formaldehyde and acetaldehyde hydrazones (all dissolved in ethyl acetate[9]) with methanol-saturated heptane as the solvent system and after tank equilibration according to Meigh,[10] demonstrated exact correspondence with the acetaldehyde hydrazone (R_f 0.25) and quite distinct from the formaldehyde hydrazone (R_f 0.14). The conclusion is that the terminal carbon on the side chain of ciliapterin is methyl rather than hydroxymethyl.

The structure of ciliapterin is, therefore, 2-amino-4-hydroxy-6-(*threo*-dihydroxypropyl)pteridine, or threobiopterin. The biological activity would indicate that it is probably of the L-configuration.

The ultraviolet spectrum of ciliapterin does not serve to differentiate it from biopterin[2] (Fig. 1). Fluorescence and excitation spectra (corrected) are presented in Fig. 2. A scheme of the reactions leading to the synthesis of ciliapterin by *Tetrahymena* has been suggested[2] and is presented in Fig. 3.

[9] G. W. Kidder and V. C. Dewey, *J. Chromatog.* **31,** 234 (1967).
[10] D. F. Meigh, *Nature* **170,** 579 (1952).

[191] Biosynthesis of Pteridines in *Drosophila melanogaster* and *Rana catesbeiana*

By MIKI GOTO and KATSURA SUGIURA

Using ^{14}C-labeled compounds, the biosynthesis of biopterin can be investigated in *Drosophila melanogaster*[1] and *Rana catesbeiana*.[2] It has been demonstrated that a guanine nucleotide, neopterin, 2-amino-4-hydroxy-6-hydroxymethylpteridine and reduced 2-amino-4-hydroxypteridine serve as precursors of biopterin and that 7,8-dihydroneopterin 3'-phosphate or its pyrophosphate is a direct precursor of biopterin.

[1] K. Sugiura and M. Goto, *Biochem. Biophys. Res. Commun.* **28,** 687 (1967).
[2] K. Sugiura and M. Goto, *J. Biochem.* **64,** 657 (1968).

Procedure

Chemical Syntheses of ^{14}C-Labeled Compounds

2,4,5-Triamino-6-hydroxypyrimidine-4-[14]C Sulfate Monohydrate[3,4]

A solution of 1.67 g of chloroacetic acid in 3.0 ml of water is neutralized with 1.30 g of anhydrous potassium carbonate, and the solution is made up to 5.0 ml with water. A 0.88-ml portion of the solution is added to 162 mg of potassium cyanide-[14]C (1.5 mCi/millimole) and heated 40 minutes at 95°; then water (0.2 ml) and absolute methanol (4 ml) are added. The solution is evaporated to dryness under nitrogen at 70°. Absolute methanol (4 ml) is again added and evaporated to dryness. The residue is dried under reduced pressure and methylated with dimethyl sulfate (0.57 ml) and dry dioxane (1.60 ml) by heating 1 hour at 95°–100°.

Guanidine nitrate (383 mg), absolute methanol (7.6 ml), and sodium methoxide solution (17.0 ml), prepared by dissolving 2.75 g of sodium in 100 ml of absolute methanol, are added to the reaction mixture. The mixture is then heated 1.5 hours at 75°–80°. Methanol is distilled off, and 12 ml of 2 N hydrochloric acid is added to the residue. After cooling to 30°, the resulting 2,4-diamino-6-hydroxypyrimidine-4-^{14}C in solution is nitrosated by slow addition of 400 mg of sodium nitrite dissolved in 10 ml of water. After heating to 80°, the mixture is allowed to cool. The bright rose 5-nitroso derivative is collected by centrifugation and washed twice with water (each, 10 ml). The wet nitrosopyrimidine is suspended in 6.0 ml of 2 N hydrochloric acid, 210 mg of iron powder are added, and it is then left 10 minutes at room temperature. The solution is heated 10 minutes at 50° and centrifuged. The precipitate is washed twice with a little 2 N hydrochloric acid (each, 3.0 ml, 50°). Concentrated sulfuric acid (0.40 ml) is added to the combined solution to precipitate the pyrimidine, and 21 ml of methanol to reduce the solubility. The needles of 2,4,5-triamino-6-hydroxypyrimidine-4-^{14}C sulfate monohydrate which form upon cooling are collected by centrifugation and washed with methanol and ether. The yield is 248 mg (39%, based on the radioactive cyanide).

2-Amino-4-hydroxypteridine-8a-[14]C[2]

2,4,5-Triamino-6-hydroxypyrimidine-4-^{14}C sulfate monohydrate (25.0 mg, 1.5 mC/millimole) is dissolved in 1.0 ml of 1 N hydrochloric acid, and the solution is heated to 70°–90°. Aqueous glyoxal solution (40%, 0.1 ml) is added, and the mixture is heated 1 hour at 70°–90°. The material is

[3] F. Korte and H. Barkemeyer, *Ber.* **90**, 392 (1957).
[4] F. Korte and H. Barkemeyer, *Ber.* **89**, 2400 (1956).

purified by chromatography on a Dowex 1-X8 column (3×14 cm, HCO_2^-). The column is eluted with formic acid (0.03 M) and ammonia, the pH of the developer being changed continuously from 9.2 to 7.4. The yield is 12 mg (75%).

7,8-Dihydro-2-amino-4-hydroxypteridine-8a-[14]C

2-Amino-4-hydroxypteridine-8a-[14]C (10 mg, 1.5 mCi/millimole) is dissolved in 1.0 ml of 0.1 N sodium hydroxide and hydrogenated over Adams' catalyst (15 mg). Hydrogenation ceases after 2 hours; the solution is no longer fluorescent. The solution is neutralized with phosphoric acid. The tetrahydro compound thus obtained is converted into the 7,8-dihydro form

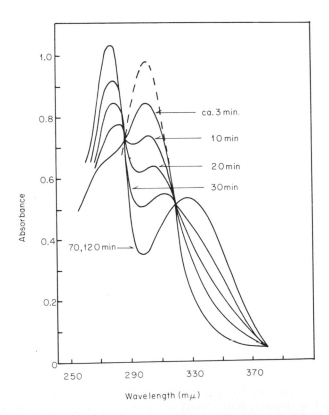

Fig. 1. UV spectral change of tetrahydro-2-amino-4-hydroxypteridine on air oxidation. 2-Amino-4-hydroxypteridine is hydrogenated to the tetrahydro compound, and the UV-absorption spectra are determined in 0.1 M phosphate buffer (pH 7.5) after 3, 10, 20, 30, 70, and 120 minutes. The broken line shows the supposed curve of the tetrahydro compound (0 minute). The spectrum at 70 and 120 minutes is that of the relatively stable 7,8-dihydro compound.

by autoxidation with air; a slight amount of 7,8-dihydroxanthopterin, which can be neglected, is produced. The UV spectral change of the tetrahydro compound under air oxidation in a phosphate buffer (pH 7.5) is shown in Fig. 1. The UV-absorption spectra of the dihydro compound obtained after 70 minutes of autoxidation are identical with those of the sodium salt of 7,8-dihydro-2-amino-4-hydroxypteridine, which is prepared from the bisulfite salt[5]; the 7,8-dihydro compound is comparatively stable and is used for the biosynthetic studies.

2-Amino-4-hydroxy-6-hydroxymethylpteridine-8a-$^{14}C^2$

The condensation is carried out by the procedure described by Forrest and Walker.[6] A mixture of dihydroxyacetone (0.35 g), water (2.0 ml), and hydrazine hydrate (0.23 ml, 80%) is stirred 30 minutes at room temperature. A 0.2-ml portion of the mixture is added to a mixture of 2,4,5-triamino-6-hydroxypyrimidine-4-^{14}C sulfate monohydrate (13.0 mg; 1.5 mCi/millimole), water (2.0 ml), boric acid (6.0 mg), and sodium acetate (9 mg, anhydrous). The resulting mixture is heated 2 hours under nitrogen at 90°. After cooling, the product is collected by centrifugation, washed with a little water and dried. The material is further purified by chromatography on a Dowex 1-X8 column (3 × 20 cm, HCO_2^-). The column is eluted with formic acid (0.03 M) and ammonia, the pH of the developer being changed continuously from 8.5 to 7.0. Ammonium formate is removed by sublimation under reduced pressure. This yield is 3.7 mg (38%) ($E_{253}:E_{362} = 3.2$).

2-Amino-4-hydroxy-6-(D-erythro-1',2',3'-trihydroxypropyl)pteridine-8a-^{14}C (Neopterin-8a-^{14}C)

The condensation is carried out by the procedure described by Weygand et al.[7] A mixture of D-ribose (750 mg) and p-toluidine (485 mg) in 0.03 N hydrochloric acid (0.80 ml) is heated under nitrogen 20 minutes at 95°–100°. Hydrazine hydrate (3.34 ml, 80%), acetic acid (1.68 ml), and water (10.9 ml) are added; the mixture is heated 1 hour at 95°–100°. The solution is extracted twice with 25 ml of ether to remove excess p-toluidine, and then acetic acid (3.34 ml) and 2,4,5-triamino-6-hydroxypyrimidine-4-^{14}C sulfate monohydrate (140 mg, 1.5 mCi/millimole) are added. The mixture is heated 1 hour at 90° under nitrogen. Paper chromatography of the reaction mixture shows three main radioactive spots: R_f 0.12 (76%), 0.21 (21%), 0.31 (3%); developer, 1-butanol–methanol–water (2:1:1); relative amounts, assayed by radioactivity, are given in parentheses. Neopterin

[5] A. Stuart, H. C. S. Wood, and D. Duncan, J. Chem. Soc. (C) p. 285 (1966).

[6] H. S. Forrest and J. Walker, J. Chem. Soc. p. 2077 (1949).

[7] F. Weygand, H. Simon, K. D. Keil, and H. Millauer, Ber. 97, 1002 (1964).

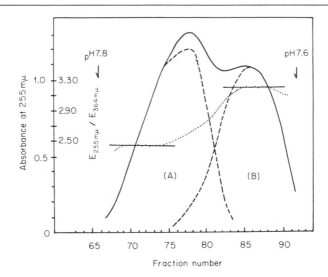

Fig. 2. Separation of neopterin and its 7-isomer with a Dowex 1-X8 column. Separation of neopterin from its 7-isomer is tried by chromatography using a Dowex 1-X8 column (25.0 × 4.0 cm); no perfect separation is, however, achieved. The column is eluted with 0.03 M formic acid and ammonia, the pH of the developer being changed continuously from 9.0 to 7.6. The flow rate is 400 ml per hour. Fractions of 100 ml are collected, and the desired fractions are selected on the basis of their typical ultraviolet absorption and the production of 2-amino-4-hydroxypteridine-6-carboxylic acid on oxidation ($KMnO_4$ or KIO_4) of a small sample. The absorptions given here are determined at pH 9.0 (a small amount of ammonia is added to the eluted solution). Fractions 85–92 contained pure neopterin.

is isolated from the reaction mixture by a chromatographic method using a cellulose column [25.0 × 8.0 cm; developer: 1-butanol–methanol–water (2:1:1)]. The blue fluorescent bands are eluted, and the eluate is evaporated to dryness under reduced pressure. The neopterin fraction (R_f 0.21) contained 55% of its 7-isomer; further purification is achieved by using a Dowex 1-X8 column (25 × 4 cm, HCO_2^-; developer: 0.03 M formic acid–ammonia, the pH value being changed continuously from 9.0 to 7.6). The elution curve is shown in Fig. 2. The fractions containing pure neopterin (Nos. 85–92) are combined; the solution is evaporated to dryness under reduced pressure and ammonium formate is removed by sublimation. The yield is 5.9 mg (4.3%) ($E_{255}:E_{364} = 3.2$). The improved methods for the synthesis of neopterin is described by Viscontini et al.[8]

7,8-Dihydro-2-amino-4-hydroxy-6-(D-erythro-1′,2′,3′-trihydroxypropyl)pteridine-8a-^{14}C (7,8-Dihydroneopterin-8a-^{14}C)

The reduction is carried out according to the procedure applied to folic acid.[9] Neopterin-8a-^{14}C (2.5 mg, 1.5 mCi/millimole) is dissolved in 3.0 ml

[8] M. Viscontini, R. Provenzale, S. Ohlgart, and J. Mallevialle, *Helv. Chim. Acta* **53**, 1202 (1970).

of hot water (60°), and the solution is cooled to room temperature. Sodium hydrosulfite (30 mg) is added, and the pH value of the solution is adjusted to 7–8 with 1 N sodium hydroxide. The solution is kept at room temperature for 45 minutes, then the product is purified on a Sephadex G-25 (fine) column (3 × 25 cm; developer: water).[10,11] The elution curve is given in Fig. 3. The fractions containing dihydroneopterin (Nos. 25–38) are combined and concentrated to dryness under reduced pressure below 40°. The UV spectra of the products are shown in Fig. 4.

2-Amino-4-hydroxy-6-(D-erythro-1′,2′,3′-trihydroxypropyl)pteridine 3′-phosphate-8a-¹⁴C (Neopterin 3′-phosphate-8a-¹⁴C)[11]

Pure neopterin-8a-¹⁴C (10 mg), obtained above and also from the reaction mixture (fractions Nos. 65–84, Fig. 2) by repetition of chromatography, is used for phosphorylation. A suspension of neopterin-8a-¹⁴C (10 mg, 1.5 mCi/millimole) in a pyridine solution of 2-cyanoethyl phosphate

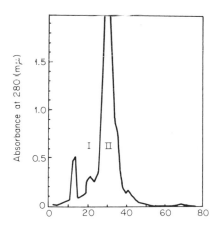

Fraction number

Fig. 3. Purification of 7,8-dihydroneopterin with a Sephadex G-25 (fine) column. Neopterin-8a-¹⁴C is reduced with aqueous sodium hydrosulfite at room temperature. The product is purified with a Sephadex G-25 (fine) column (3 × 25 cm); the column is eluted with water. The flow rate is 40 ml per hour, and fractions of 5.0 ml are collected. The desired fractions are selected on the basis of their typical ultraviolet absorption. Bands I and II are those of neopterin and 7,8-dihydroneopterin, respectively. 7,8-Dihydroneopterin on standing in acidic media spontaneously yields a green fluorescent compound (7,8-dihydro-2-amino-4-hydroxy-6-α,β-dihydroxypropionylpteridine), and a careful treatment is necessary.

[9] M. Friedkin, E. J. Crawford, and D. Misra, *Federation Proc.* **21**, 176 (1962).
[10] T. Fukushima and M. Akino, *Arch. Biochem. Biophys.* **128**, 1 (1968).
[11] V. C. Dewey and G. W. Kidder, *J. Chromatog.* **31**, 326 (1967).

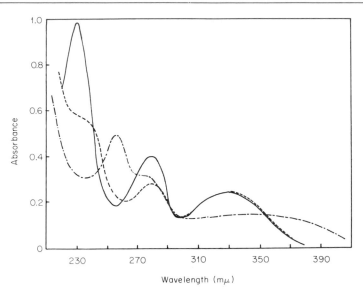

FIG. 4. UV-absorption spectra of 7,8-dihydroneopterin. (— — — —) In 0.1 N hydrochloric acid; (———) at pH 7.5 (in phosphate buffer); (- - - - -) in 0.1 N sodium hydroxide.

(1.5 ml, 0.1 M)[12] is concentrated to dryness under reduced pressure below 30°. The residue is taken up in 10 ml of anhydrous pyridine (dried over CaH_2), and the solution is concentrated to dryness under reduced pressure. This process is repeated. Then 20 ml of anhydrous pyridine and 60 mg of dicyclohexylcarbodiimide are added, and the mixture is refluxed for 1.5 hours. Water (2.5 ml) is added, and the solution is left 30 minutes at 25°. Water (5.0 ml) is again added, and pyridine is removed by concentrating the solution to dryness under reduced pressure. To the residue is added 10 ml of 10% acetic acid, and the solution is heated 1 hour under reflux. The reaction mixture is concentrated to dryness under reduced pressure, 10 ml of concentrated ammonia is added, and the solution is heated 60 minutes at 60°. The solution is concentrated under reduced pressure and filtered. The precipitate is washed with a little water. The combined filtrate and washings are placed on a 3.5 × 20.0 cm column of Dowex 1-X8 (Cl⁻, 200–400 mesh); a fluorescent band is adsorbed at the top of the column. The column is eluted with 0.02 N hydrochloric acid. The product is purified further by successive chromatography on cellulose columns (3.0 × 16.0 cm) using 2-propanol–2% ammonium acetate (1:1) and 2-propanol–1% ammonia (2:1); yield, 0.6 mg (4.5%). Neopterin is recovered in a 46%

[12] G. M. Tener, *J. Am. Chem. Soc.* **83**, 159 (1961).

yield. On treatment with alkaline phosphatase, the product gives neopterin quantitatively, and on oxidation with sodium metaperiodate for 20 hours (in 0.1 M aqueous sodium carbonate), 2-amino-4-hydroxypteridine-6-carboxylic acid is obtained in almost quantitative yield.

Radiochemical Purity of Synthetic Materials

The purity of the synthetic materials is confirmed by paper chromatography using five solvent systems: 1-butanol–acetic acid–water (4:1:1); 1-butanol–methanol–water (2:1:1); 2-propanol–2% ammonium acetate (1:1); 2-propanol–1% ammonia (2:1); and 3% ammonium chloride. The ascending method is employed, and no rigid control is kept of temperature or time of running, authentic materials always being run at the same time as controls. Radioactivity is detected with a paper chromatogram scanner; the position of compounds is also confirmed by observing the fluorescence using a Blak·Ray lamp (output maximum, 366 nm).

Feeding and Isolation of the Radioactive Compounds from D. melanogaster

The radioactive compound is suspended in a little water, and 4 ml of the medium (yeast 15%, agar 1%, water 76%, sugar 8%, and a little propionic acid, the whole heated 20 minutes at 120°) is added carefully. *D. melanogaster* larvae are allowed to develop on this food; the pteridines are extracted from the adult flies as follows: The flies are powdered with a mixture of 50% ethyl alcohol and 1% ammonia (1:1) in a mortar. The suspension is centrifuged, and the precipitate is extracted 3 more times with the same solvent. The combined extracts are evaporated to a small bulk under reduced pressure. Compounds are purified by paper chromatography in four solvent systems: 2-propanol–1% ammonia (2:1); 2-propanol–2% ammonium acetate (1:1); 1-butanol–acetic acid–water (4:1:1); and 3% ammonium chloride.

Incubation and Isolation of the Radioactive Compounds from the Skin of R. catesbeiana

The experimental procedures described here are the same as those used by Levy.[13] Skin, 5–7 cm in length, is removed from a number of tadpoles. The main pteridines in the skin are as follows (amounts present in 1 g of wet skins are in parentheses): biopterin (ca. 0.7 micromole/g); isoxanthopterin (ca. 0.8 micromole/g); 2-amino-4-hydroxypteridine-6-carboxylic acid (ca. 0.1 micromole/g); sepiapterin (trace); and 2-amino-4-hydroxypteridine (trace). These pteridines are not removed in advance in

[13] C. C. Levy, *J. Biol. Chem.* **239**, 560 (1964).

this experiment. The tissue (2–3 g, 30 heads) is washed with 20 ml of 0.1 M phosphate buffer (pH 7.5) and transferred into a flask containing the solution of radioactive compound in 10 ml of 0.1 M phosphate buffer (pH 7.5); the mixture is incubated anaerobically (under nitrogen) for 20 hours at 25°.

To isolate radioactive compounds from the skin, an equal amount of methanol is added to the incubation mixture, the skins are removed by centrifugation, and the tissue is resuspended in 50% methanol (20 ml), refluxed for 30 minutes, and centrifuged as before. The extraction procedure is repeated three times. The combined extracts are evaporated to a small bulk under reduced pressure. Methanol (10 ml, 50%) is added and the precipitate is removed by vacuum filtration (Celite). The precipitate is washed with 50% methanol (10 ml). The filtrate and washings are combined and concentrated under reduced pressure. Compounds are purified by paper chromatography in four solvent systems, in sequence, using preparative chromatographic papers (thickness: 4 mm) and then Whatman No. 1 papers; developer: 1-butanol–acetic acid–water (4:1:1); 2-propanol–1% ammonia (2:1); 2-propanol–2% ammonium acetate (1:1); and 3% ammonium chloride.

Determination of Radioactivity

When specific radioactivity is relatively high, the radioactive compounds are located with a paper chromatogram scanner; the position of compounds is also confirmed by observing the fluorescence using a Blak·Ray lamp (output maximum, 366 nm). When radioactivity is too low to be detected with a paper chromatogram scanner, the compounds are further purified by paper chromatography using Whatman No. 1 paper (40 × 30 cm) and the four solvent systems described above. Final chromatograms are cut into sections of 1 cm along vertical lines, and the material from each fraction is eluted with water. Each eluate is evaporated to dryness under reduced pressure below 30°, and the residue is dissolved in a little ammonia (1%). After measurement of the UV spectra, the solution is concentrated under reduced pressure, put on a stainless steel plate 15 mm in diameter, and evaporated to dryness under reduced pressure; the residue is counted. The position of the radioactive compounds is determined graphically both from radioactivity and UV absorptions. The purification procedures described above are repeated until a constant specific radioactivity is obtained.

Neopterin, biopterin, and 2-amino-4-hydroxy-6-hydroxymethylpteridine isolated from the skin of *R. catesbeiana* after incubation are oxidized to 2-amino-4-hydroxypteridine-6-carboxylic acid by heating the compounds with saturated potassium permanganate in 0.1 N sodium hydroxide

TABLE I

R_f Values and UV Absorption of Synthetic Pteridines

Compounds[a]	R_f values				UV absorption[c] (nm; $\epsilon \times 10^{-3}$)	Solvent
	A	B	C	D[b]		
Neopterin	0.14	0.40	0.55	0.66	255 (24.7); 364 (7.48)	1% ammonia
Biopterin	0.32	0.52	0.58	0.70	256 (22.4); 364 (7.00)	1% ammonia
P-6-CH₂OH	0.26	0.41	0.48	0.52	253 (22.6); 362 (7.00)	1% ammonia
AHP	0.36	0.43	0.48	0.52	251 (21.4); 360 (7.10)	1% ammonia
P-6-CO₂H	0.18	0.20	0.33	0.52	262 (20.6); 364 (9.13)	1% ammonia
Lumazine	0.38	0.51	0 52	0.63	252 (17.0); 365 (6.03)	0.1 N NaOH
Isoxanthopterin	0.20	0.27	0.33	0.32	252 (10.3); 339 (13.8)	1% ammonia
Neopterin phosphate	0.01	0.10	0.34	0.77	255 (24.7); 364 (7.48)	1% ammonia
Guanosine	0.22	0.45	0.56	0.61	—	—
Guanine	0.30	0.43	0.46	0.61	—	—

[a] P-6-CH₂OH = 2-amino-4-hydroxy-6-hydroxymethylpteridine; AHP = 2-amino-4-hydroxypteridine; P-6-CO₂H = 2-amino-4-hydroxypteridine-6-carboxylic acid.

[b] Solvents: (A) 1-butanol–acetic acid–water (4:1:1); (B) 2-propanol–1% ammonia (2:1); (C) 2-propanol–2% ammonium acetate (1:1); (D) 3% ammonium chloride.

[c] For UV absorption, see R. L. Blakley, "The Biochemistry of Folic Acid and Related Pteridines," p. 66. North-Holland, Amsterdam, 1969.

(95°–100°, 30 minutes). The specific radioactivity is again determined after chromatography in the four solvent systems listed above.

In the case of 2-amino-4-hydroxypteridine, this compound (1 micromole) after measurement of the specific radioactivity, is incubated in 1.0 ml of phosphate buffer (0.03 M, pH 8.0) with a milk xanthine oxidase preparation (0.2 unit) for 1 hour at 27°. Isoxanthopterin produced is purified by chromatographic methods, and the identity of the specific radioactivities is confirmed.

The R_f values and UV absorptions of the pteridines are listed in Table I.

Results and Metabolic Pathway

Drosophila melanogaster

The experimental results are summarized in Table II.

Guanosine 5'-monophosphate (GMP) is an effective precursor of biopterin and 2-amino-4-hydroxypteridine (AHP); the specific radioactivities of the products are approximately the same. Neopterin is also converted into biopterin and 2-amino-4-hydroxypteridine, and again the specific radioactivities of the products are essentially identical. 2-Amino-4-hydroxy-6-hydroxymethylpteridine (P-6-CH₂OH) is effectively converted into 2-amino-4-hydroxypteridine and, to a lesser extent, into bio-

TABLE II

BIOSYNTHESIS OF PTERIDINES IN *Drosophila melanogaster*[a]

Radioisotopes fed to flies				Pteridines isolated from flies		
	Specific radioactivity and amount		Species and number of larvae	Radiochemical data		
Compounds	mCi/mmole	mg		Compounds	Counts per minute	Amount ($\times10^{-8}$ mole)	Specific radioactivity ($\times10^6$ cpm/mmole)
GMP	293	0.068	(*ry²*) 1016	Biopterin	495.6 ± 5.2[b]	12.9	3.84
				AHP	437.4 ± 3.3	9.0	4.86
Neopterin	1.5	1.98	(+) 400	Neopterin	2006.3 ± 15.8	4.9	40.8
				Biopterin	5.21 ± 0.14	2.7	0.19
				AHP	3.85 ± 0.21	2.1	0.18
				Ixp	5.37 ± 0.19	3.7	0.15
Neopterin	1.5	1.85	(*ry²*) 604	Neopterin	4008 ± 21.8	3.5	114
				Biopterin	10.38 ± 0.32	6.3	0.16
				AHP	25.14 ± 0.39	14.8	0.17
P-6-CH₂OH	1.5	3.7	(*ry²*) 383	P-6-CH₂OH	7587 ± 43	23.6	32.1
				Biopterin	1.49 ± 0.19	2.5	0.058
				AHP	310.4 ± 2.64	11.3	2.75
				Lumazine	9.24 ± 0.12	—	—
AHP (reduced)	1.5	6.0	(*ry²*) 1062	Biopterin	10.86 ± 0.23	9.5	0.11
				AHP	13424 ± 57	16.9	79.4
				Lumazine	603.5 ± 6.1	4.9	12.3
AHP (reduced)	1.5	2.4	(+) 643	Biopterin	0.02 ± 0.10	2.5	(0.0008)
				AHP	3.35 ± 0.15	3.0	0.11
				Ixp	983.1 ± 7.5	19.8	4.97

[a] Abbreviations: GMP = guanosine 5'-monophosphate; P-6-CH₂OH = 2-amino-4-hydroxy-6-hydroxymethylpteridine; AHP = 2-amino-4-hydroxypteridine; Ixp = isoxanthopterin.

pterin (the specific radioactivity of biopterin is 2.1% of that of 2-amino-4-hydroxypteridine). Reduced 2-amino-4-hydroxypteridine is effectively converted into isoxanthopterin (Ixp) (wild-type flies) and lumazine (ry^2 flies), but to a lesser extent into biopterin. The results show that 2-amino-4-hydroxy-6-hydroxymethylpteridine is first converted into 2-amino-4-hydroxypteridine, which is then incorporated into biopterin.

The evidence presented here suggests that there exists a main catabolic pathway for neopterin, i.e., neopterin $\rightarrow$ P-6-CH$_2$OH $\rightarrow$ AHP $\rightarrow$ isoxanthopterin ($+$) or lumazine (ry^2).

A possible biosynthetic mechanism for biopterin, which is consistent with these results, is as follows: The glycerol side chain of neopterin is entirely replaced with an unknown C_3 fragment to give sepiapterin or biopterin; the interconversion of sepiapterin and biopterin is confirmed by different authors.[14,15] Reduced 2-amino-4-hydroxypteridine itself, however, probably is not a free intermediate in this enzymatic process, but it can act as a substrate in the biopterin-synthesizing enzyme system.

Rana catesbeiana

The experimental results are summarized in Table III.

Guanosine, 7,8-dihydroneopterin, and 7,8-dihydro-2-amino-4-hydroxypteridine (dihydro-AHP) are effective precursors of biopterin. Using nonradioactive pteridines as carrier compounds, it is demonstrated that guanosine is converted into neopterin and AHP; 7,8-dihydroneopterin into 2-amino-4-hydroxy-6-hydroxymethylpteridine (P-6-CH$_2$OH) and AHP; and 6,7-dihydro-AHP into lumazine and isoxanthopterin.

The evidence presented here suggests that there exists a main catabolic pathway for 7,8-dihydroneopterin: neopterin·2 H $\rightarrow$ P-6-CH$_2$OH·2 H $\rightarrow$ AHP·2 H $\rightarrow$ lumazine and isoxanthopterin. This pathway is observed also in *D. melanogaster*. There exists here the possibility for the direct conversion of neopterin·2H into AHP·2H with cleavage of a C_3-fragment, as addition of nonradioactive P-6-CH$_2$OH·2H (*inhibitor*) does not decrease the incorporation of dihydroneopterin-8a-^{14}C into AHP appreciably.

Both dihydro-AHP and dihydroneopterin function as precursors of biopterin. If the pathway involves the reaction sequence: neopterin·2H $\rightarrow$ P-6-CH$_2$OH·2H $\rightarrow$ AHP·2H $\rightarrow$ sepiapterin and biopterin, the incorporation of dihydroneopterin-8a-^{14}C into biopterin must be decreased by addition of the nonradioactive intermediates, i.e., AHP·2H and P-6-CH$_2$OH·2 H. However, the addition of such nonradioactive intermediates does not decrease the incorporation of dihydroneopterin-8a-^{14}C

[14] M. Matsubara, S. Katoh, M. Akino, and S. Kaufman, *Biochim. Biophys. Acta* **122**, 202 (1966).
[15] S. Kaufman, *J. Biol. Chem.* **242**, 3938 (1967).

TABLE III

BIOSYNTHESIS OF PTERIDINES IN *Rana catesbeiana*

Radioisotopes incubated with the skins	Specific radioactivity and amount	Added compounds (amount)	Pteridines isolated from the skins			
				Radiochemical data		
			Compounds	Counts per minute	Amount (micromoles)	Specific radioactivity (cpm/μmole)
Guanosine-^{14}C (U)	518 mCi/mmole 0.02 μmole	Dihydroneopterin (6.4 μmoles) and dihydro-AHP^d (6.1 μmoles)	Neopterin	11.1 ± 0.4^a	0.58	19
			Biopterin	8.5 ± 0.3	0.91	9
			AHP	17.3 ± 0.4	0.33	52
			Isoxanthopterin	—	—	—
			Lumazine	—	—	—
Neopterin 3′-phosphate-8a-^{14}C	1.5 mCi/mmole 2.1 μmole	ATP (5.1 μM) Mg^{2+} (20 μM)	Biopterin	108.7 ± 1.1	1.22	89^e
Dihydroneopterin-8a-^{14}C	1.5 mCi/mmole 3.7 μmole	—	Biopterin	57.6 ± 0.9	0.91	63

Substrate				Product			
Neopterin-8a-¹⁴C	1.5 mCi/mmole	4.5 μmole	—	Biopterin	23.6 ± 0.7	1.22	19
Dihydroneopterin-8a-¹⁴C	1.5 mCi/mmole	3.5 μmole	Dihydro-AHP (1.5 μM)	Biopterin	26.8 ± 0.5	0.44	61
				AHP	3847 ± 18	0.66	5800
Dihydroneopterin-8a-¹⁴Cᵇ	1.5 mCi/mmole	2.6 μmole	Dihydro-P-6-CH₂OHᵉ (2.9 μM) and dihydroAHP (2.3 μM)	P-6-CH₂OH	3608 ± 16	0.63	5700
				AHP	4588 ± 17	0.98	4700
				Biopterin	115.5 ± 1.5	1.04	110
				Isoxanthopterin	—	—	—
				Lumazine	—	—	—
Dihydro-AHP-8a-¹⁴Cᵇ	1.5 mCi/mmole	7.7 μmole	—	Isoxanthopterin	255.5 ± 4.7	0.02	13,000
				Lumazine	18,540 ± 79	0.16	115,000
				Biopterin	167.0 ± 1.6	0.63	260

ᵃ 1.5 mCi/μmole of the pteridines corresponds to 2.3×10^5 cpm/μmole by this gas flow counter. ± designates standard deviation due to a gas flow counter.

ᵇ Contrary to guanosine-¹⁴C (U), the pteridines-8a-¹⁴C are converted into only a limited number of the radioactive compounds; these are the pteridines described here.

ᶜ After incubation, it is found that about 85% of neopterin 3'-phosphate is hydrolyzed to neopterin; 15% remains as the phosphate.

ᵈ AHP = 2-amino-4-hydroxypteridine.

ᵉ P-6-CH₂OH = 2-amino-4-hydroxy-6-hydroxymethylpteridine.

into biopterin. This evidence suggests that both P-6-CH$_2$OH·2H and AHP·2H are not free intermediates in the biopterin and sepiapterin biosynthesis, addition of nonradioactive P-6-CH$_2$OH·2H rather accelerates the incorporation of dihydroneopterin-8a-^{14}C into biopterin.

Thus, a more possible direct precursor of biopterin and sepiapterin might be a 7,8-dihydroneopterin compound. The incorporation of neopterin, 7,8-dihydroneopterin, and neopterin 3'-phosphate into biopterin are compared. Neopterin 3'-phosphate is a more effective precursor than neopterin; the specific radioactivity of biopterin isolated from the skin incubated with neopterin 3'-phosphate is 4.5 times that of biopterin isolated from the skin incubated with neopterin. After incubation of neopterin 3'-phosphate with the tadpole skins, it is found that about 85% of the phosphate is hydrolyzed to neopterin; thus, actual incorporation of the phosphate into biopterin must be greater. 7,8-Dihydroneopterin is a more effective

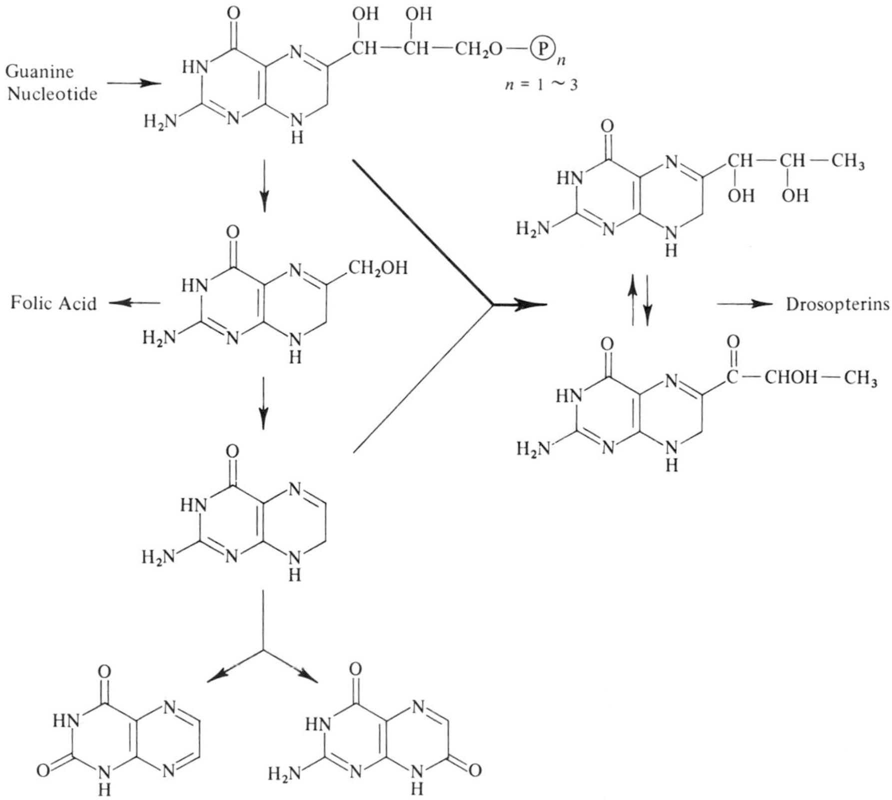

FIG. 5. Metabolic relationships among pteridines.

precursor of biopterin than neopterin itself, and this evidence suggests that the actual precursor of sepiapterin and biopterin may be 7,8-dihydro-neopterin 3'-phosphate or its pyrophosphate. Neopterin 3'-phosphate is actually isolated from *Escherichia coli*.[16]

Jones and Brown have reported the incorporation of 7,8-dihydro-neopterin into dihydro-P-6-CH₂OH and dihydropteroic acid in *E. coli*.[17] This work supports the postulation that dihydroneopterin 3'-phosphate or its pyrophosphate may be a key link between the biopterin group and the folic acid group.

The biosynthetic scheme shown in Fig. 5 may be postulated both in *D. melanogaster* and *R. catesbeiana*.

[16] M. Goto and H. S. Forrest, *Biochem. Biophys. Res. Commun.* **6**, 180 (1961).
[17] T. H. D. Jones and G. M. Brown, *J. Biol. Chem.* **242**, 3989 (1967).

[192] GTP Cyclohydrolase from *Escherichia coli*

By GENE M. BROWN

$$GTP \rightarrow formate + dihydroneopterin\ triphosphate$$

The first step in the biosynthetic pathway for the conversion of GTP to the pteridine portion of folic acid is the hydrolytic reaction that results in the elimination of carbon 8 of GTP as formate.[1-4] The enzyme that catalyzes this reaction has been termed "GTP cyclohydrolase."[2] Theoretical considerations suggest that the other product of this reaction should be a ribosylated derivative of 2,4,5-triamino-6-hydroxypyrimidine. However, Burg and Brown[2] have shown that, surprisingly, purified GTP cyclohydrolase catalyzes the conversion of GTP to dihydroneopterin[5] triphosphate. Thus, the enzyme is responsible for the formation of a product by a set of reactions that theoretically should include the following four steps: two hydrolytic reactions (for the removal from GTP of carbon 8 as formate); an Amadori rearrangement of the ribose moiety to a 1-deoxy-2-ketopentose triphosphate unit; and finally ring closure to yield dihydroneopterin triphosphate.[2] The observation that this is all accomplished

[1] A. W. Burg and G. M. Brown, *Biochim. Biophys. Acta* **117**, 275 (1966).
[2] A. W. Burg and G. M. Brown, *J. Biol. Chem.* **243**, 2349 (1968).
[3] T. Shiota and M. P. Palumbo, *J. Biol. Chem.* **240**, 4449 (1965).
[4] R. Dalal and J. S. Gots, *Biochem. Biophys. Res. Commun.* **20**, 509 (1965).
[5] Dihydroneopterin is the trivial name for 2-amino-4-hydroxy-6-(D-*erythro*-1',2',3'-trihydroxypropyl)-7,8-dihydropteridine.

by what analyzes as a single protein poses some interesting questions about the nature of such a protein.

Assay Method

Principle. The activity of GTP cyclohydrolase can be assessed most easily by measurement of the amount of formate released from GTP. Since crude extracts of *E. coli* contain formic dehydrogenase, the direct determination of formate when crude extracts are used as the source of enzyme is not possible. A procedure has therefore been devised[1] that results in the oxidation, either enzymatically or chemically, of formate to carbon dioxide. The latter compound can then be trapped as barium carbonate. A second more convenient assay has been developed for use with enzyme preparations that have been purified free from formic dehydrogenase.[2] The details of these methods are described below.

Reagents

    Tris buffer, 1 M, pH 8.5
    GTP-8-[14]C (300,000 cpm), 2 mM
    Sodium formate, 2 mM
    Trichloroacetic acid, 50%
    Formic acid, 1 N
    Acetic acid, 1 N
    Mercuric acetate, 2 mM
    Saturated solution of barium hydroxide
    Darco G-60 charcoal (Atlas Powder Co.)

Procedure for Method A.[1] Reaction mixtures are prepared to contain 0.1 ml of Tris buffer, 0.1 ml of GTP-8-^{14}C, 0.1 ml of sodium formate, enzyme preparation, and enough water to make a final volume of 1.0 ml. Incubation is for 30 minutes at 42°. To the incubated mixture are added 0.1 ml of 50% trichloroacetic acid, 0.1 ml of 1 N acetic acid, and 0.1 ml of mercuric acetate solution. The reaction mixture is then heated, and the evolved carbon dioxide is trapped by allowing the air to bubble through 5 ml of a saturated solution of barium hydroxide. The resulting barium carbonate is recovered by centrifugation and suspended in a small amount of water. An aliquot of this suspension is plated on a planchet, and the radioactivity of this material is measured with a Geiger counter.

Procedure for Method B.[2] A reaction mixture is prepared to contain the following components: 0.1 ml of Tris buffer; 0.1 ml of GTP-8-[14]C (30,000 cpm); enzyme preparation; and enough water to make a total volume of 0.5 ml. Incubation is for 30 minutes at 42°, after which 0.1 ml of 1 N formic acid is added to stop the reaction and to provide carrier nonradioactive formate. Darco G-60 charcoal (20 mg) is then added, and the suspension is mixed on a Vortex shaker in order to adsorb on the charcoal any unreacted GTP-8-[14]C. The mixture is centrifuged to remove the charcoal, and an aliquot (0.3 ml) of the supernatant fluid is removed and analyzed for radioactivity in a liquid scintillation counter. In this determination, radioactivity remaining in the charcoal-treated reaction mixture is directly proportional to the amount of formate produced. Control experiments show that under the conditions described no formate is adsorbed to the charcoal and that GTP is completely adsorbed. Burg and Brown[1] have presented proof that the 1-carbon compound produced in this enzymatic reaction is formate.

Purification Procedure[2]

Preparation of Extracts of E. coli. E. coli harvested in the late log phase is obtained frozen in 1% NaCl solution from General Biochemicals. The frozen cells are subjected to treatment with a Hughes press (Shandon Scientific Co.) to rupture the cells. The broken cells are suspended in 2–3 volumes of 0.05 M Tris buffer (pH 8.0) and treated with a small amount of DNase to reduce the viscosity of the suspension. The mixture is centrifuged at 105,000 g for 1 hour. The clear supernatant solution is called the "crude extract."

Treatment with RNase. To 435 ml of crude extract is added RNase (40 mg) and the solution is incubated for 30 minutes at 42° to degrade RNA which might otherwise interfere with subsequent fractionation steps. A portion of the material in the extract is rendered insoluble by this treatment and is removed by centrifugation and discarded.

Fractionation with Ammonium Sulfate. Enough of a saturated solution of ammonium sulfate is added to the RNase-treated extract to give a solution 31% saturated with the salt. The resulting precipitated protein is removed by centrifugation and discarded. To the supernatant solution is added additional ammonium sulfate to give a 41% saturated solution. The precipitate is recovered by centrifugation and is suspended in 30 ml of a buffer solution at pH 7.0 containing 5 mM phosphate and 5 mM EDTA. The material is subjected to dialysis overnight against 14 liters of the same buffer. A small amount of insoluble material is removed by centrifugation at 320,000 g for 30 minutes. The resulting supernatant fluid is called the "dialyzed extract." It is important to keep EDTA in the buffer to prevent a slow decay of enzyme activity which results in the absence of EDTA.

Fractionation on Sephadex. The dialyzed extract is reduced in volume to 15 ml by lyophilization. This solution is divided into 5 parts of 3 ml each for convenience of fractionation on Sephadex. Each 3-ml portion (124 mg of protein) is applied to a column (1.5 × 119 cm) of Sephadex G-200. A solution containing 5 mM phosphate and 5 mM EDTA (pH

7.0) is allowed to flow through the column, and fractions of 6 ml each are collected at a rate of 0.1 ml per minute at 4°. Each fraction is analyzed for GTP cyclohydrolase activity and for protein (protein measured either by the method of Lowry et al.[6] or by the method of Warburg and Christian[7]). The major portion of the protein is eluted from the column with the passage of 160–400 ml of eluting solution through the column, whereas the enzymatic activity is eluted faster, with the passage of 90–150 ml through the column. The fractions (15 through 25) that contain most of the enzymatic activity are combined and are called the "Sephadex eluate."

Fractionation on DEAE-Cellulose. The Sephadex eluate is applied to a column (2 × 31 cm) of DEAE-cellulose. The column is developed with a convex logarithmic concentration gradient of phosphate buffer, formed by placing 100 ml of 0.005 M phosphate buffer (pH 7.0) in the mixing chamber and 250 ml of 0.1 M phosphate buffer (pH 7.0) in the reservoir. Fractions of 10 ml each are collected at a rate of 0.5 ml per minute at 4°. Each fraction is analyzed for enzymatic activity. The protein concentration in each fraction is too low to be measured accurately. The fractions containing enzymatic activity are obtained when 100–175 ml of eluting solution has passed through the column. These fractions are combined, and the material is called the "DEAE-cellulose eluate." An aliquot of this material is reduced to a small volume by lyophilization, and protein is determined on this aliquot.

A summary of the purification achieved by the steps described above is given in the table. A unit of enzyme is defined as that amount of enzyme which catalyzes the production from GTP in 1 minute of 1 micromole of formate. It should be noted that although an apparent 2798-fold purification is achieved, this value is probably too high since the procedure from "crude extract" to "dialyzed extract" results in an overall yield of 400%, probably a result of the elimination from the extract of inhibitors. Thus, a more reasonable estimate of purification is approximately 700-fold.

Properties

Purified GTP cyclohydrolase (the DEAE-cellulose eluate) contains one major and four minor protein components, as judged by electrophoresis on polyacrylamide gel. The enzymatic activity for the release of formate and the formation of dihydroneopterin triphosphate from GTP are found only in the major component.

Neither GDP, GMP, guanosine, guanine, nor any other purine nucleotide can be used as substrate in place of GTP. The K_m for GTP is $2.2 \times 10^{-5}\ M$. No metal activator nor coenzyme stimulates the activity

[6] O. H. Lowry, N. J. Rosebrough, A. L. Farr, and R. J. Randall, *J. Biol. Chem.* **193,** 265 (1951).
[7] O. Warburg and W. Christian, *Biochem. Z.* **310,** 384 (1941).

PURIFICATION OF GTP CYCLOHYDROLASE FROM *Escherichia coli*

Enzyme preparation	Specific activity (units/mg, $\times 10^4$)	Overall yield (units)	Relative specific activity
Crude extract	0.08	0.30	1.0
RNase-treated extract	0.13	0.47	1.6
Dialyzed extract	2.14	1.20	25.8
Sephadex eluate	9.27	0.40	112
DEAE-cellulose eluate	232	0.37	2798

of the enzyme. The reaction proceeds optimally at pH 8.0 and at 42°. The activity of the enzyme is not affected by such sulfhydryl-binding agents as iodoacetamide and ethyleneimine. Hg^{2+}, Cu^{2+}, Mn^{2+}, and Fe^{2+}, all tested at 20 mM, inhibit the activity of the enzyme. Mg^{2+} neither stimulates nor inhibits. The activity of the purified enzyme can be maintained indefinitely by storage in the frozen state in the presence of 5 mM EDTA.

[193] Hydroxymethyldihydropteridine[1] Pyrophosphokinase and Dihydropteroate Synthetase from *Escherichia coli*

By DAVID P. RICHEY and GENE M. BROWN

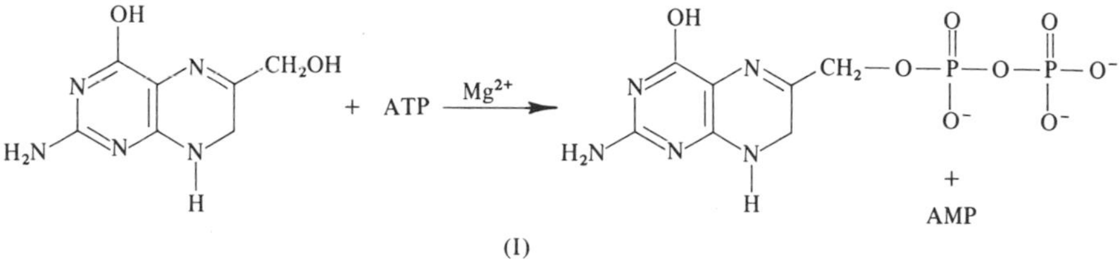

(I)

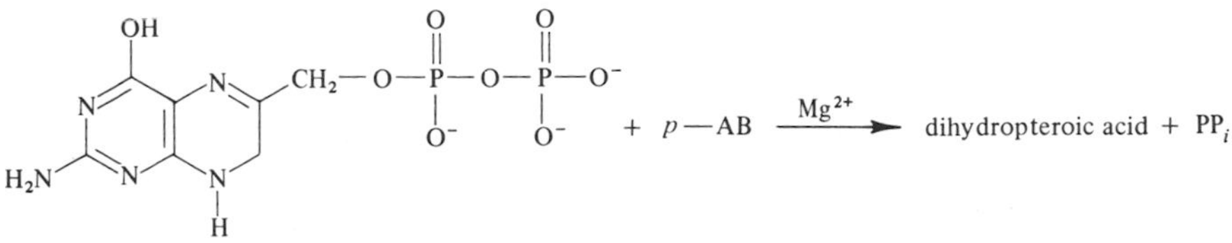

(II)

The synthesis of dihydropteroic acid from hydroxymethyldihydropteridine,[1] ATP, and p-AB[1] is catalyzed by two enzymes present in *Escherichia coli* and proceeds by the two steps shown above.[2,3] The participation of the pyrophosphorylated pteridine intermediate was suggested by the observation that cell-free systems from several different species of microorganisms use the chemically synthesized pyrophosphate compound for the synthesis of dihydropteroic acid in the absence of ATP.[2,4-6] Richey and Brown[3] have isolated an intermediate compound formed in the presence of a purified enzyme from *E. coli* and have shown that the intermediate is identical with the chemically synthesized pyrophosphate ester [shown as the product in (1) above] and that AMP is the other product of the reaction. They also have shown that during the second reaction [(2) above] the pyrophosphate group is released by the substitution of *p*-AB. The first enzyme has been named hydroxymethyldihydropteridine pyrophosphokinase and the second enzyme dihydropteroate synthetase.

Assays

Principle. Two kinds of assays can be used. Method A is used to measure only the activity of hydroxymethyldihydropteridine pyrophosphokinase, and Method B can be used to measure the activity of either the pyrophosphokinase or dihydropteroate synthetase.[3] If synthetase activity is to be measured, an excess of the pyrophosphokinase is added to the reaction mixture, and if pyrophosphokinase activity is to be determined, an excess of synthetase is supplied. In either case, the product that is determined is dihydropteroate (or pteroate, since no attempt is made to prevent oxidation during the procedure). This is accomplished by including *p*-AB-^{14}C in the reaction mixture and by separating the resulting ^{14}C-pteroate from unreacted *p*-AB-^{14}C by paper chromatography.

The alternative method (Method A) for measurement of the activity of hydroxymethyldihydropteridine pyrophosphokinase involves the addition of ^{14}C-hydroxymethyldihydropteridine to the reaction mixture and the determination of the amount of radioactive pyrophosphate ester formed. The radioactive product is separated from the residual unreacted substrate by paper chromatography. The ^{14}C-hydroxymethylpteridine is synthesized and reduced to the dihydro compound as described by Richey and Brown.[3]

---

[1] Abbreviations used are as follows: hydroxymethyldihydropteridine for 2-amino-4-hydroxy-6-hydroxymethyl-7,8-dihydropteridine; and *p*-AB for *p*-aminobenzoic acid.

[2] R. A. Weisman and G. M. Brown, *J. Biol. Chem.* **239**, 326 (1964).

[3] D. P. Richey and G. M. Brown, *J. Biol. Chem.* **244**, 1582 (1969).

[4] T. Shiota, M. N. Disraely, and M. P. McCann, *J. Biol. Chem.* **239**, 2259 (1964).

[5] T. Shiota, M. N. Disraely, and M. P. McCann, *Biochem. Biophys. Res. Commun.* **7**, 194 (1962).

[6] P. J. Ortiz and R. D. Hotchkiss, *Biochemistry* **5**, 67 (1966).

Reagents for Method A

Hydroxymethyldihydropteridine-^{14}C, 0.3 mM (4 × 10^5 cpm per micromole)

2-Mercaptoethanol, 1.0 M

Tris buffer, 1.0 M, pH 8.6

MgCl$_2$, 0.1 M

ATP, 0.05 M

EDTA, 0.25 M, pH 8.3

Procedure for Method A.[3] Reaction mixtures are prepared to contain the following components: 0.06 ml of ^{14}C-hydroxymethyldihydropteridine solution; 0.01 ml of mercaptoethanol; 0.02 ml of Tris buffer; 0.02 ml of MgCl$_2$ solution; 0.01 ml of ATP solution; the enzyme solution; and water to a final volume of 0.2 ml. Incubation is at 37° for 1 hour. The reactions are stopped by the addition of 0.02 ml of the EDTA solution. Each reaction mixture is evaporated to dryness over silica gel at 37°. Each residue is redissolved in 50 μl of water and applied quantitatively (in a 1.5 × 3.0 cm area) to Whatman 3 MM chromatography paper. Chromatograms are developed by the descending technique for 20 hours at 23° with 2-propanol–H$_2$O (4:1). The pteridine pyrophosphate product remains at the origin, whereas unreacted pteridine migrates with an R_f of approximately 0.3. Areas corresponding to the origins are cut as strips from the dried papers. Each strip is moistened with 0.5 ml of scintillator fluid, and the radioactivities of the strips are measured in a liquid scintillation counter.

Reagents for Method B. Reagents are the same as for Method A, except that hydroxymethyldihydropteridine (nonradioactive) is supplied at 0.33 mM and p-AB-^{14}C, 0.5 mM (2.8 × 10^6 cpm per micromole) is used. Also, a source of the pyrophosphokinase, uncontaminated with the synthetase, and a source of the synthetase, uncontaminated with the pyrophosphokinase, must be supplied. When this method was devised originally, the sources used for these enzymes were the preparations of Weisman and Brown.[2] These authors had separated the two enzymes by fractionation on DEAE-cellulose. Later, the method of separating the enzymes on Sephadex, devised by Richey and Brown[3] (this method is described in a later section of this article), was used to prepare the enzymes used in the assay procedure.

Procedure for Method B. Reaction mixtures are prepared to contain the following components: 0.03 ml hydroxymethyldihydropteridine solution; 0.01 ml of mercaptoethanol; 0.02 ml of Tris buffer; 0.02 ml of MgCl$_2$ solution; 0.01 ml of ATP solution; 0.01 ml of p-AB-^{14}C solution; a preparation of the complementary enzyme (provided in excess); the solution to be analyzed for either pyrophosphokinase or synthetase activity; and enough water to give a final volume of 0.2 ml. Reaction mixtures are incubated at 37° for 1 hour, after which the reactions are stopped by the addition of 0.02 ml of the EDTA solution. The reaction mixtures are then evaporated to dryness *in vacuo* over silica gel at 37°. Each reaction residue is dissolved in 50 μl of water and the material is applied quantitatively (each in a 1.5 × 3.0-cm area) to Whatman 3 MM chromatography paper. The chromatograms are developed by the descending technique with 0.1 M potassium phosphate buffer, pH 7.0, for 40 minutes at 23°. Neither dihydropteroate nor pteroate moves from the origin under these conditions, whereas unreacted p-AB migrates with an R_f value of 0.78. The material is present as pteroate on the chromatogram since the enzymatic product, dihydropteroate, becomes oxidized during the assay procedure. Areas corresponding to the origins of the developed chromatograms are cut as strips from the dried papers; each strip is moistened with 0.5 ml of scintillator fluid, and the radioactivities of the strips are measured in a liquid scintillation counter.

Purification of Hydroxymethyldihydropteridine Pyrophosphokinase[3]

Crude cell-free extracts of *E. coli* B are prepared as described for GTP cyclohydrolase[7] in this volume.

Treatment with RNase. Crude extract, 153 ml (87.5 mg of protein per milliliter), is incubated with 10 mg of RNase at 37° for 60 minutes and then dialyzed overnight against 20 liters of 0.01 M potassium phosphate buffer, pH 7.0. Insoluble material is removed by centrifugation, and the supernatant solution, 188 ml, is made 0.1 M with Tris buffer, pH 8.0.

Fractionation with Ammonium Sulfate. Enough of a saturated solution of ammonium sulfate is added to the RNase-treated extract to bring the solution to 20% saturation with the salt. Precipitated protein is removed by centrifugation at 40,000 g for 30 minutes and discarded. Additional ammonium sulfate solution is added to the supernatant fluid to give a solution 70% saturated with the salt. The precipitated protein is recovered by centrifugation and dissolved in 122 ml of 0.1 M Tris buffer, pH 8.0. This solution is dialyzed overnight against 14 liters of the same buffer and then subjected to centrifugation to remove a small amount of protein that precipitates during dialysis. The supernatant solution (182 ml, 51 mg of protein per milliliter) is then refractionated with ammonium sulfate and dialyzed in the manner described above to yield a 20–60% ammonium sulfate fraction containing 80 mg of protein per milliliter. The refractionation procedure removes a large quantity of undesirable inorganic pyrophosphatase activity, but retains all the recoverable dihydropteroate

[7] G. M. Brown, this volume [192].

synthesizing activity. The 20–60% fraction is used as the starting material for the purification of both the pyrophosphokinase and the synthetase.

Fractionation on Sephadex. Dialyzed 20–60% ammonium sulfate fraction containing 170 mg of protein in a volume of 2.0 ml is applied to a Sephadex G-100 column (2 × 73 cm) that has been equilibrated previously with 0.01 M potassium phosphate buffer, pH 7.0. The column is developed at 4° with the same buffer, and fractions of 3.5 ml are collected at a rate of 3.5 ml per hour. The fractions are analyzed for pyrophosphokinase and synthetase activities and for protein (protein is measured either by the method of Lowry *et al.*[8] or by the method of Warburg and Christian[9]). Although the major portion of the protein (more than 95%) is eluted with the passage of 60–120 ml of buffer through the column, the pyrophosphokinase is eluted much later, between 135 and 152 ml. Dihydropterate synthetase is completely separated from the pyrophosphokinase by this step. The protein eluting between 100 and 115 ml of buffer contains all the synthetase activity. The fractions containing the pyrophosphokinase activity are combined and referred to as the "Sephadex eluate."

Fractionation on DEAE-Cellulose. The Sephadex eluate (17 ml and 3.7 mg of protein) is applied to a column (1 × 17 cm) of DEAE-cellulose previously equilibrated with 0.01 M potassium phosphate buffer, pII 7.0, and the column is washed with 100 ml of the same buffer. Protein is eluted from the column with a linear gradient (0.01 M to 0.2 M; total volume, 200 ml) of potassium phosphate buffer, pH 7.0. Fractions of 3.6 ml each are collected at a flow rate of 27 ml per hour. The fractions that contain

TABLE I

PURIFICATION OF HYDROXYMETHYLDIHYDROPTERIDINE
PYROPHOSPHOKINASE FROM *Escherichia coli*

Enzyme preparation	Total activity (units, ×10³)	Specific activity (units/mg, ×10³)	Relative specific activity
Crude extract	981	0.0745	1.0
RNase-treated extract	1716	0.126	1.7
20–70% Ammonium sulfate fraction	1495	0.161	2.2
20–60% Ammonium sulfate fraction	966	0.161	2.2
Sephadex eluate	1053[a]	7.79	106
DEAE eluate	505[a]	29.95	413

[a] Corrected values to take into account that only a portion of the 20–60% ammonium sulfate fraction was applied to the Sephadex column.

[8] O. H. Lowry, N. J. Rosebrough, A. L. Farr, and R. J. Randall, *J. Biol. Chem.* **193**, 265 (1951).
[9] O. Warburg and W. Christian, *Biochem. Z.* **310**, 384 (1941).

pyrophosphokinase activity (eluted between 40 and 58 ml of the gradient) are combined and called the "DEAE eluate."

A summary of the purification achieved by the steps described above is given in Table I. A unit of enzyme activity is that amount of enzyme which catalyzes the synthesis of 1 micromole of product (the pyrophosphate ester) per minute.

Properties of the Pyrophosphokinase

Polyacrylamide gel electrophoresis of the DEAE eluate reveals the presence of only two significant protein bands, only one of which can be shown to possess hydroxymethyldihydropteridine pyrophosphokinase activity. The K_m values for both hydroxymethyldihydropteridine and for ATP are $1.5 \times 10^{-5}\ M$. Deoxy-ATP is used somewhat less effectively than ATP, but GTP, CTP, and UTP are not used at all. Magnesium ion is required with an optimum of 4 mM. Mn^{2+} can replace Mg^{2+} with a 10 mM optimum; Co^{2+} is somewhat less effective, but Ca^{2+}, Cu^{2+}, Zn^{2+}, and monovalent ions are totally ineffective. The enzyme exhibits activity between pH 7.5 and pH 10.8 with a maximum at pH 8.5. The molecular weight of the enzyme is estimated to be 15,000 from its behavior on Sephadex.[3] The enzyme is relatively heat stable; it retains 25% of its initial activity after incubation at 100° for 60 minutes. The activity of the enzyme is stable indefinitely upon storage at −20° as the Sephadex eluate, but the DEAE eluate loses its activity over a period of months at −20°.

Purification of Dihydropteroate Synthetase[3]

Those fractions from the Sephadex G-100 column (described above in the purification of the pyrophosphokinase) which contain dihydropteroate synthetase activity are combined (16.4 ml and 24 mg of protein) and applied to a DEAE-cellulose column (2 × 12 cm) previously equilibrated with 0.01 M potassium phosphate buffer, pH 7.0. The column is washed with 250 ml of the same buffer and protein is eluted at 4° with a linear gradient of 0.01 to 0.25 M potassium phosphate buffer (total gradient volume of 300 ml). Fractions of 3.9 ml are collected at a flow rate of 60 ml per hour. The synthetase is eluted between 195 and 250 ml of the gradient. The fractions containing enzyme activity are combined (54 ml) and reduced in volume to 3 ml by dialysis of the material against solid sucrose packed around the dialysis tubing.

This purification scheme is summarized in Table II. A unit of synthetase activity is that amount of enzyme which catalyzes the synthesis of 1 micromole of dihydropteroate per minute.

TABLE II

PURIFICATION OF DIHYDROPTEROATE SYNTHETASE FROM *Escherichia coli*

Enzyme preparation	Total activity (units, $\times 10^3$)	Specific activity (units/mg, $\times 10^3$)
Crude extract	3420	0.25
RNase-treated extract	6335	0.46
20–70% Ammonium sulfate fraction	4650	0.50
20–60% Ammonium sulfate fraction	4340	0.72
Sephadex eluate	4280	4.8
DEAE eluate	1653	6.5
Dialyzed DEAE eluate	1768	13.1

Properties of Dihydropteroate Synthetase

The K_m value for p-AB is $2.5 \times 10^{-6}\,M$. p-Aminobenzoylglutamate is an extremely poor inhibitor of p-AB utilization, having a K_i of about $10^{-3}\,M$. p-Aminohippuric acid and o- and m-aminobenzoic acid are totally ineffective as inhibitors of p-AB utilization. Dihydropteroate exhibits (competitive) product inhibition, but dihydrofolate and tetrahydropteroyl-triglutamate are not inhibitors. The other reaction product, inorganic pyrophosphate, does not inhibit the enzyme activity. Phosphate, arsenate, and fluoride also are without effect. Sulfathiazole and other sulfonamides exhibit typical inhibition of dihydropteroate synthesis as described previously by Brown.[10]

No reverse or exchange reactions are noted for this enzyme. The enzyme requires Mg^{2+} for activity. The pH optimum is 8.5 with 70% of maximal activity at pH 7.0 and at pH 10.5. The molecular weight of the enzyme is approximately 52,000. The enzyme is stable indefinitely when stored at $-20°$ in the presence of sucrose.

[10] G. M. Brown, *J. Biol. Chem.* **237**, 536 (1962).

[194] Dihydrofolate Synthetase

$$\text{Dihydropteroate} + \text{glutamate} \xrightarrow{\text{ATP}} \text{dihydrofolate}$$

By GENE M. BROWN

Assay Method

Principle. The method involves measurement of the formation of dihydrofolate. This can be accomplished by microbiological assay with

Lactobacillus casei (ATCC No. 7496). This microorganism needs folic acid as a growth factor, but folate can be replaced with various degrees of effectiveness by dihydrofolic acid, tetrahydrofolic acid, and pteroyldi- and pteroyltriglutamate (as well as the dihydro and tetrahydro forms of these compounds). Neither pteroic acid nor the reduced forms (dihydro and tetrahydro) of this compound can be used as a growth factor for this species of bacteria. It has been carefully determined that dihydrofolate is utilized in this assay 20% as effectively as folate.[1] Since, for the sake of convenience, folate is used as a standard in the assays, data have to be multiplied by a correction factor to determine the amount of dihydrofolate produced. Microbiological assays are performed as described by Herbert.[2] Aliquots of incubated reaction mixtures are prepared for assay by dilution in sterile sodium ascorbate, pH 7.0, and dispensed aseptically to assay tubes containing sterile medium (see Brown, Weisman, and Molnar[3] for details). This procedure is necessary to prevent the product, dihydrofolate, from being oxidized and destroyed. The dihydropteroic acid used as substrate was prepared from pteroic acid by reduction with sodium dithionite ($Na_2S_2O_4$) as described by Futterman[4] for the preparation of dihydrofolate from folate.

Reagents

Dihydropteroic acid, 0.25 mM
L-Glutamic acid, 20 mM
ATP, 50 mM
$MgCl_2$, 50 mM
Phosphate buffer (K salts), 2 M, pH 8.0
2-Mercaptoethanol, 0.67 M

Procedure. Reaction mixtures are prepared by mixing together the following components: 0.08 ml of phosphate buffer; 0.08 ml of ATP solution; 0.08 ml of mercaptoethanol; 0.08 ml of $MgCl_2$ solution; 0.08 ml of glutamic acid solution; 0.08 ml of dihydropteroic acid solution; and 0.1–0.2 ml of enzyme solution. Enough water is added to make a final volume of 0.8 ml. Reaction mixtures are incubated under nitrogen (to prevent oxidation of the substrate and the product) for 3 hours at 37°. Aliquots of 0.1 ml are then diluted in the appropriate amount of sterile sodium ascorbate, and the diluted samples are dispensed aseptically to individual micro-

[1] M. J. Griffin and G. M. Brown, *J. Biol. Chem.* **239**, 310 (1964).
[2] V. Herbert, *J. Clin. Invest.* **40**, 81 (1961).
[3] G. M. Brown, R. A. Weisman, and D. A. Molnar, *J. Biol. Chem.* **236**, 2534 (1961).
[4] S. Futterman, *J. Biol. Chem.* **228**, 1031 (1957).

biological assay tubes that have previously been prepared (see Herbert[2] for the details of the microbiological assay).

Definition of a Unit. A unit of enzyme is that amount that is required for the formation of 1 mμg of dihydrofolate under the conditions described above. Protein is determined by the method of Lowry *et al.*[5] with bovine serum albumin as the standard.

Purification Procedure[1]

Crude Extract. Crude extract of *Escherichia coli* is prepared by the same procedure described in a previous article of this volume.[6]

Treatment with RNase and Charcoal. Crude extracts of *E. coli* contain folic acid compounds (active in supporting the growth of *L. casei*) in large enough quantities so that synthesis of dihydrofolate can not be determined accurately. Therefore, a procedure is used for the removal of these materials from the extract. For this purpose the crude extract (125 ml, approximately 5 g of protein) is incubated with 3.75 mg of ribonuclease for 30 minutes at 37°. The suspension is then centrifuged to remove insoluble material. The supernatant solution is treated with Darco G-60 (Atlas Powder Co.) as follows. An amount of charcoal, equal in weight to the amount of protein present, is added to the solution, and the suspension is stirred at 2° for 15 minutes. The suspension is then centrifuged at 18,000 g to remove the charcoal, and the resulting supernatant solution is saved and will be referred to as the "crude extract."

Fractionation with Ammonium Sulfate. To the charcoal-treated extract is added, with stirring at 2°, enough of a saturated solution of ammonium sulfate (adjusted to pH 7.0 with NH_4OH) to make the resulting solution 63% saturated with the salt. The precipitated protein is recovered by centrifugation and dissolved in 110 ml of 0.01 M phosphate (potassium salts) buffer, pH 7.6. This solution is dialyzed for a total of 18 hours against three changes (6 hours each) of 14 liters of the same buffer. This dialyzed solution contains enzyme purified approximately 3-fold over the crude extract.

Fractionation with Calcium Phosphate Gel. Since dihydrofolate synthetase does not adsorb readily to calcium phosphate gel, a "negative gel step" is used. For this purpose, enough calcium phosphate gel (in 0.01 M potassium phosphate buffer, pH 7.6; this gel had been stored at 4° as a suspension, 44 mg/ml, in 0.01 M phosphate buffer, pH 6.8, for 6 months before use) is added with stirring (at 2°) to a 40-ml (800 mg of protein) portion of the 0–63% ammonium sulfate fraction to give a final concentra-

[5] O. H. Lowry, N. J. Rosebrough, A. L. Farr, and R. J. Randall, *J. Biol. Chem.* **193**, 265 (1951).

[6] G. M. Brown, this volume [192].

tion of 6 mg of gel per milligram of protein. The suspension is stirred for a total of 20 minutes and then centrifuged at 15,000 g for 10 minutes. The supernatant fluid contains most of the enzyme activity.

Fractionation on DEAE-Cellulose. The supernatant solution from the treatment with calcium phosphate gel is applied to a DEAE-cellulose column (6.3 × 26 cm) that is prepared by packing the DEAE-cellulose as a slurry in cold (4°) 0.05 M potassium phosphate buffer, pH 6.8. The ratio of weights of dry DEAE-cellulose to protein applied to the column is 75:1. After the protein solution is added, the column is washed with 250 ml (approximately 4 column volumes) of 0.05 M potassium phosphate buffer, pH 6.8. The column is then developed (at 4°) by a linear gradient elution procedure in which the concentration of potassium phosphate buffer (pH 6.8) is increased from 0.05 M initially to 0.4 M after a total of 500 ml of buffer has passed through the column. Fractions of 10 ml each are collected at a flow rate of 2.5 ml per minute. Each fraction is analyzed for protein and for enzyme activity. Two protein fractions are eluted: one with 0.15–0.18 M phosphate; and the other with 0.18–0.22 M phosphate. The first of these two fractions (eluted with 0.15 to 0.18 M phosphate) contains most of the enzyme activity but only a small part of the total protein eluted from the column.

A summary of the purification steps is given in the table. The overall purification achieved by the steps described is 30- to 35-fold. Material purified to this extent is referred to below as the "purified enzyme."

PURIFICATION OF DIHYDROFOLATE SYNTHETASE FROM EXTRACTS OF *Escherichia coli*[a]

Enzyme preparation	Enzyme activity (units/ml)	Protein (mg/ml)	Specific activity (units/mg protein)	Overall yield (%)	Overall purification (-fold)
Crude extract[b]	2500	45	56	—	—
0–63% Ammonium sulfate fraction	3200	20	160	128	2.9
Supernatant solution from calcium phosphate gel	500	1.4	360	56	6.5
Fraction from DEAE-cellulose	1000	0.6	1670	36	30

[a] Data from Griffin and Brown.[1]

[b] Crude extract that has been treated with RNase and charcoal.

Substrates

Neither pteroic acid nor tetrahydropteroic acid can replace dihydropteroic acid as substrate. The product formed from dihydropteroate can be shown to be dihydrofolic acid by bioautographic techniques.[1] Neither γ-glutamylglutamic acid nor γ-glutamyl-γ-glutamylglutamic acid can replace L-glutamic acid as substrate. Also, D-glutamate and pyrrolidone carboxylic acid cannot be used in place of L-glutamate. Other requirements for the reaction to take place are ATP, Mg^{2+}, and a monovalent cation that can be supplied either as K^+ or NH_4^+. ITP and GTP can be utilized only 60% and 35%, respectively, as effectively at ATP. CTP and UTP are totally ineffective. Rb^+ is used somewhat less effectively than K^+ or NH_4^+ for the monovalent cation requirement; Li^+, Na^+, and Cs^+ are without effect.

Under optimal conditions, 20% of the added dihydropteroate can be converted to dihydrofolate. The optimal concentration of glutamic acid needed is 2.0 mM, whereas maximal synthesis of product is achieved with 0.025 mM of the other substrate, dihydropteroate. The amount of ATP needed for maximal synthesis of product is reduced from 2.0 mM when crude extract is used as a source of enzyme to 0.5 mM when the purified enzyme is used. This probably reflects the removal during purification of adenosine triphosphatases.

Miscellaneous Properties

Dihydrofolate synthetase operates optimally at pH 9.0 with significant activity in the range from pH 8.0 to 10. The activity of the enzyme is not inhibited by 2 mM fluoride or 2 mM iodoacetate. p-Hydroxymercuribenzoate at 2 mM inhibits the formation of product by 70%.

Distribution of the Enzyme

Dihydrofolate synthetase has been found in a variety of microorganisms including the following diverse species[1]: E. coli, Corynebacterium sp., Saccharomyces cerevisiae, Bacillus megatherium, Neurospora crassa, Streptococcus faecalis (ATCC No. 8043), Mycobacterium phlei, and Mycobacterium avium.

[195] Dihydrofolic Reductase (5,6,7,8-Tetrahydrofolate: NADP Oxidoreductase, EC 1.5.1.3)

By B. G. STANLEY, G. E. NEAL, and D. C. WILLIAMS

$$\text{Dihydrofolate} + \text{TPNH} + \text{H}^+ \rightleftharpoons \text{tetrahydrofolate} + \text{TPN}^+ \tag{1}$$

Assay Method

Principle. Dihydrofolic reductase catalyzes the reversible reduction of dihydrofolate by TPNH, as shown in Eq. (1). At neutral pH the reaction normally proceeds in the forward direction, and the enzymatic activity can be assayed by measuring the decrease in absorbancy at 340 nm, which results from both the reduction of dihydrofolate and the oxidation of TPNH.[1] Considerable difficulties are encountered in attempts to assay the reverse reaction colorimetrically, using those techniques normally employed in assaying TPN+- or DPN+-dependent dehydrogenase enzymes.[1a] These are due mainly to the equilibrium lying relatively far to the right, and to the instability of solutions of tetrahydrofolate. Stabilization of the tetrahydrofolate solutions with antioxidants often results in nonenzymatic reactions with the chromogenic substrates.

The forward reaction, however, can be successfully assayed colorimetrically in the presence of the tetrazolium salt 3-(4,5-dimethylthiazolyl-2)-2,5-diphenyl tetrazolium bromide (MTT). The tetrahydrofolate produced by the enzymatic activity reduces the MTT to a formazan having a λ_{max} at 560 nm. The tetrahydrofolate is oxidized to reform dihydrofolate.

Assays Using Isolated Enzyme Preparations

Reagents

Tris-HCl buffer, 0.05 M, pH 7.5

Dihydrofolic acid, 20 μM. This is prepared fresh each day by the reduction of a solution of potassium folate by dithionite, according to the method of Futterman.[1b] The dihydrofolic acid is repeatedly washed in 5 mM HCl and then stored at 1°, suspended in 5 mM HCl under argon. It is dissolved at the required concentration in the Tris buffer immediately before use.

TPNH, 20 μM

MTT, 0.1 mg

Procedure. The above reagents are added to a glass cuvette, 1-cm optical path, in a total volume of 2 ml. Blank reactions, in which TPNH or dihydrofolate are omitted, are run at the same time. The reaction is started by addition of 0.01–0.1 ml of enzyme. The reaction may be followed for at least 5 minutes by direct measurement of the absorbancy at 560 nm. Subsequently, using highly active enzyme preparations, some divergence from a linear reaction rate is observed, which may be due to the insolubility of the formazan. Longer reaction periods can be accommodated by dissolving the formazan at the end of the experimental period by the addition of methanol to the assay tubes.

Using rat liver dihydrofolic reductase preparations, extensively purified by the method of Mathews *et al.*,[1] <5% of the activity of the complete

[1] C. K. Mathews, K. G. Scrimgeour, and F. M. Huennekens, Vol. VI [364].

[1a] A. G. E. Pearse, "Histochemistry," p. 908. Churchill, London, 1960.

[1b] S. Futterman, Vol. VI [112].

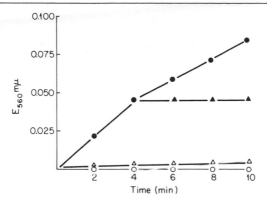

Fig. 1. Assay of dihydrofolic reductase using MTT. (●—●) complete system: 20 μM dihydrofolate, 20 μM TPNH, 0.05 M Tris (pH 7.5), 100 μg of MTT, 0.01 ml of highly purified dihydrofolic reductase, total volume 2 ml; (△—△) minus TPNH; (○—○) minus dihydrofolate; (▲—▲) complete system, 10^{-5} M methotrexate added after 4 minutes.

system is observed in either of the blanks in which dihydrofolate or TPNH are omitted. Sensitivity to addition of antifolate can further be used to confirm the extent to which the Δ absorbancy at 560 nm is due to dihydrofolic reductase activity. Care should be taken, when using methotrexate as the antifolate, to ensure the absence of an impurity capable of catalyzing the nonenzymatic reduction of MTT by TPNH.[2]

When less highly purified enzyme preparations are assayed, considerable color formation may be observed even in the absence of dihydrofolate, due to the presence of TPNH diaphorase activity. These blanks can be substantially reduced by addition of p-chloromercuribenzoate to the assay medium at a final concentration of 1.5 mM. Dihydrofolate reductase is either insensitive to, or actually stimulated by, this concentration of p-chloromercuribenzoate. Using the above conditions, the 560 nm assay method normally gives 50–70% of the Δ absorbancy observed using the same enzyme preparations and the 340 nm assay.

Assays Using Tissue Sections or Cell Suspensions

Reagents

Tris-HCl buffer, 0.05 M, pH 7.5
Dihydrofolic acid, 50 μM
TPNH, 80 μM
MTT, 0.2 mg
p-Chloromercuribenzoate, 2 mM

[2] B. G. Stanley, G. E. Neal, and D. C. Williams, *Biochem. Pharmacol.* **17**, 2228 (1968).

TABLE I

ASSAY OF DIHYDROFOLIC REDUCTASE IN RAT LIVER SECTIONS USING MTT

Sample	$E_{560m\mu}$/mg protein
Complete system[a]	0.131
Minus TPNH	0.030
Minus dihydrofolate	0.025
Complete system + 10^{-5} M methotrexate	0.064

[a] Complete system contained: 50 μM dihydrofolate, 80 μM TPNH, 0.05 M Tris (pH 7.4), 200 μg of MTT, 2 mM p-chloromercuribenzoate, and 20 sections (15 μ in thickness) containing 2.84 mg of protein, cut from fresh frozen liver. Total volume was 2 ml. Other systems varied as indicated with reference to the complete system.

Procedure. Twenty sections (15 μ in thickness), cut from freshly frozen rat liver using a cryostat, are used per assay tube. The sections are preincubated in 1 ml of 4 mM p-chloromercuribenzoate (dissolved at pH 8.5 and subsequently buffered with 0.05 M Tris at pH 7.5) for 30 minutes at room temperature. Methotrexate (10^{-5} M) is added to some of the tubes. Sections, after preincubation, are placed in the assay medium in a total volume of 2 ml. Methotrexate (at a final concentration of 10^{-5} M) is added to those assay tubes in which the sections were preincubated with methotrexate. Blanks from which TPNH or dihydrofolate are omitted are also used. Assay tubes are incubated at room temperature for 3 hours. The tissue is then spun down, the formazan is extracted into 3 ml of absolute methanol, and the color is determined at 560 nm (Table I).

The technique can also be successfully applied to intact ascites cells

TABLE II

ASSAY OF DIHYDROFOLIC REDUCTASE IN INTACT ASCITES CELLS USING MTT

	$E_{560m\mu}$/mg protein	
Sample	Ascites cells[a]	Ascites cells harvested 30 minutes after *in vivo* methotrexate injection[b]
Complete system	0.165	0.094
Complete system + 10^{-5} M methotrexate	0.098	0.096

[a] Complete system and incubation times were as for liver sections, but using 0.1 ml of dilute ascites cell suspension (containing 0.75 mg of protein) in place of the liver sections. The cells were maintained in an isotonic medium.

[b] Complete system was as for liver, but using 0.1 ml of ascites cell suspension, the cells having been harvested 30 minutes after intraperitoneal injection of 1 mg of methotrexate dissolved in 0.5 ml of physiological saline.

and can be used to demonstrate the inhibition of dihydrofolic reductase in intact cells following *in vivo* antifolate injections (Table II).

The method has been used to locate dihydrofolate reductase activity following electrophoresis on polyacrylamide gels,[3] or cellulose acetate membranes.[4]

[3] P. F. Nixon and R. L. Blakley, *J. Biol. Chem.* **243**, 4722 (1968).
[4] G. P. Mell, M. Martelli, J. Kirchner, and F. M. Huennekens, *Biochem. Biophys. Res. Commun.* **33**, 74 (1968).

[196] Small Molecule Inhibitors of Dihydrofolate Reductase

By Barbara Roth and James J. Burchall

Assay Method

The assay for dihydrofolate reductase activity is based on the decrease in absorbancy at 340 nm observed in the presence of dihydrofolate, NADPH, and enzyme. Protocols for the assay, properties of the enzyme, and methods of purification have been fully discussed in a previous essay in this series.[1] The assay system used in these studies contained 0.1 M phosphate buffer (Na_2HPO_4 and KH_2PO_4 in a molar ratio of 6:4), pH 7.0; 0.16 micromole of dihydrofolic acid; 0.24 micromole of NADPH; and 33 micromoles of 2-mercaptoethanol, in a final volume of 3.0 ml. Control reactions consisted of either the complete reaction without dihydrofolate or without NADPH, or more frequently, the complete reaction system with 10^{-5} M amethopterin.[2] Our assay system differs from that described in Mathews *et al.*[1] chiefly in our lower pH value (7.0 instead of 7.5) and our larger volume (3.0 ml instead of 1.5 ml), which allows more conveniently for the addition of candidate inhibitors. The sensitivity of the dihydrofolate reductase assay is improved when the system is coupled to the formate-activating enzyme,[1] and this method frequently is useful in detecting low levels of activity in blood and tissue. The sensitivity gained by the use of that method is unfortunately linked with a considerable loss of accuracy, and data obtained with inhibitors in such a system should be interpreted with great caution.

The initial information usually required of a new potential inhibitor of dihydrofolate reductase is that quantity of inhibitor required to reduce the rate of decrease in absorbancy at 340 nm to 50% of its control system without inhibitor. The level of enzyme used in these studies should not

[1] C. K. Mathews, K. G. Scrimgeour, and F. M. Huennekens, Vol. VI [364].
[2] J. J. Burchall and G. H. Hitchings, *Mol. Pharmacol.* **1**, 126 (1965).

exceed the range where enzyme concentration is proportional to substrate disappearance. In practice a quantity of enzyme giving a $\Delta OD_{340} = 0.005$ to 0.010/minute is usually sufficient. The level of inhibitor is then varied to obtain at least two levels of activity below or close to 50% and two levels above or close to 50%. The values are plotted on semilogarithmic paper and fitted through the 50% inhibition point by eye.

Frequently much additional information may be obtained by slight variation of the assay protocol. The relation of inhibition to pH may prove revealing, particularly if the pK_a of the inhibitor is known.[3] All inhibitors should be tested both with and without a 10-minute preincubation in the presence of the reductase. In this manner inhibitors of the pseudoirreversible type, such as methotrexate,[4] or those forming covalent bonds with the enzyme, such as Baker has designed,[5] may be differentiated from purely competitive inhibitors. The exact nature of the enzyme inhibitor is best determined by the standard double-reciprocal plot of 1/velocity against 1/substrate at various levels of the inhibitors.[6] Computer programs for processing enzyme kinetic data have been made available by Cleland.[7] When differences are seen in the level of inhibition as a function of preincubation time, the Ackermann-Potter plot[8] is an extremely useful method in distinguishing between reversible and pseudoirreversible inhibition.[4]

It should be emphasized that the 50% inhibition values do not consider the Michaelis constant (K_m) of dihydrofolate for the particular enzyme under investigation. These values may range from 1×10^{-4} M for the calf thymus reductase[9] to 5×10^{-7} M for the chicken liver reductase.[10] The inhibition constant (K_i) implicitly includes both terms, but in practice it is too laborious to determine for a large series of compounds. For compounds known to be competitive inhibitors, the use of the formula $K_i = (K_m \cdot I_{50})/S$ suggested by Baker[5] may be helpful.

Synthesis of Inhibitors

Inhibitors of this type are best exemplified by the 2,4-diamino-5-substituted or -5,6-condensed pyrimidines, diaminodihydrotriazines, and closely related analogs. Their chief difference from the direct analogs of

[3] B. Roth and J. Strelitz, *J. Org. Chem.* **34**, 821 (1969).

[4] W. C. Werkheiser, *J. Biol. Chem.* **236**, 888 (1961).

[5] B. R. Baker, "Design of Active-Site-Directed Irreversible Enzyme Inhibitors. The Organic Chemistry of the Enzymic Active-Site." Wiley, New York, 1967.

[6] H. Lineweaver and D. Burk, *J. Am. Chem. Soc.* **56**, 658 (1934).

[7] W. W. Cleland, *Nature* **198**, 463 (1963).

[8] W. W. Ackermann and V. R. Potter, *Proc. Soc. Exptl. Biol. Med.* **72**, 1 (1949).

[9] R. Nath and D. M. Greenberg, *Biochemistry* **1**, 435 (1962).

[10] M. J. Osborn, M. Freeman, and F. M. Huennekens, *Proc. Soc. Exptl. Biol. Med.* **97**, 429 (1958).

folic acid is that they lack the *p*-aminobenzoylglutamic acid moiety but have a smaller hydrophobic substituent in its place. Examples follow.

2,4-Diamino-5-(3',4',5'-trimethoxybenzyl)pyrimidine (trimethoprim) (III).[11] This compound is a broad spectrum antibacterial agent, which is active against gram-negative organisms, such as *Proteus vulgaris*.[12] It strongly potentiates the sulfonamides.[13] The compound is bound much more strongly to dihydrofolate reductase enzymes from bacterial, than from mammalian, sources (conc. for 50% inhibition for *E. coli* reductase, $5 \times 10^{-9}\ M$; for rat liver reductase, $2.6 \times 10^{-4}\ M$).[2] It is best synthesized by one of the following schemes.[14,15]

The preparation *via* (I) → (II) → (III) follows.

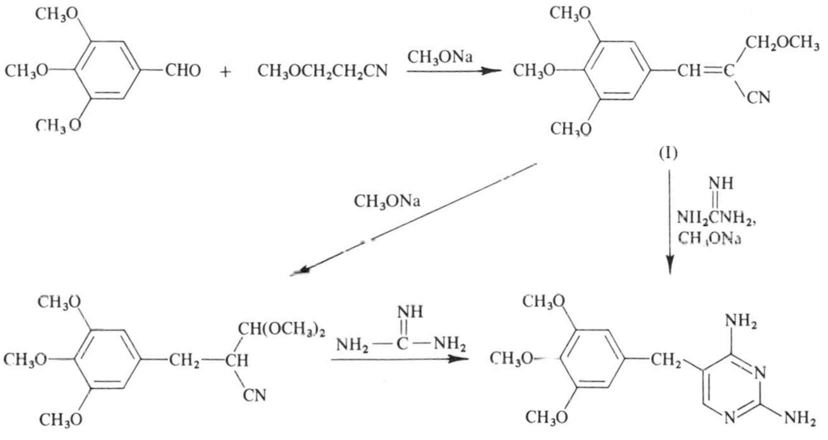

$$
\begin{array}{l}
\text{CH}_3\text{O—}\!\!\!\!\!\!\bigcirc\!\!\!\!\!\!\text{—CHO} \;+\; \text{CH}_3\text{OCH}_2\text{CH}_2\text{CN} \xrightarrow{\text{CH}_3\text{ONa}} \text{CH}_3\text{O—}\!\!\!\!\!\!\bigcirc\!\!\!\!\!\!\text{—CH=C}\!\!\begin{smallmatrix}\text{CH}_2\text{OCH}_3\\ \text{CN}\end{smallmatrix}
\end{array}
$$

(I)

$\xrightarrow[\text{CH}_3\text{ONa}]{}$

$$\xrightarrow[\substack{\text{NH}_2\text{CNH}_2,\\ \text{CH}_3\text{ONa}}]{\overset{\text{NH}}{\|}}$$

(II)   (III)

To a solution of 28 g (0.52 mole) of sodium methylate in 600 ml of anhydrous methanol is added 93.5 g (1.1 moles) of *β*-methoxypropionitrile and 196 g (1 mole) of 3,4,5-trimethoxybenzaldehyde. The mixture is refluxed for 4 hours, and is then chilled and diluted with 300 ml of cold water. The product separates and gradually crystallizes during 1 hour, still being chilled and stirred. The crude *3,4,5-trimethoxy-2'-methoxymethylcinnamonitrile* (I) is isolated by filtration, followed by washing well with very cold dilute methanol. The air-dried product weighs approximately

---

[11] U.S. Patent 2,909,522 (1959).
[12] B. Roth, E. A. Falco, G. H. Hitchings, and S. R. M. Bushby, *J. Med. Pharm. Chem.* **5**, 1103 (1962).
[13] S. R. M. Bushby and G. H. Hitchings, *Brit. J. Pharmacol. Chemotherap.* **33**, 72 (1968).
[14] P. Stenbuck, R. Baltzly, and H. M. Hood, *J. Org. Chem.* **28**, 1983 (1963); U.S. Patent 3,049,544 (1962).
[15] U.S. Patent 3,341,541 (1967).

180–185 g (ca. 70%); m.p. 78°–80°. This can be used in the next step without further purification. Recrystallization from methanol raises the melting point to 85°. UV: λ_{max}(EtOH), 232 nm (ϵ 15,140), 308 (15,060).

The crude product (184 g, 0.7 mole) is added to a solution of 78 g (1.44 mole) of sodium methylate in 525 ml of anhydrous methanol. The solution is heated under reflux for 24 hours, which causes it to turn brown. After cooling, the mixture is poured into 1750 ml of water. The precipitated oil is extracted repeatedly with benzene, after which the combined benzene extracts (900–1200 ml) are washed 3 times with water. The benzene is removed *in vacuo*, and the brown residual oil is distilled under vacuum; b.p., 215–225°/11 mm. The product, *2-(3',4',5'-trimethoxybenzyl)-3,3-dimethoxypropionitrile* (II), is obtained as a clear viscous oil which solidifies on standing; yield, 145 g (71%). After recrystallization from methanol, the substance melts at 69°–70°; n_D^{25}, 1.5190.

A methanolic guanidine solution is prepared as follows. Guanidine hydrochloride (110 g, 1.15 moles) is dissolved in 300 ml of warm methanol. To this is added a solution of 65 g (1.2 moles) of sodium methylate in 550 ml of methanol. The solution is rapidly filtered from salt with suction, and the precipitate is washed with an additional 100 ml of methanol, which is combined with the main filtrate. The 2-(3',4',5'-trimethoxybenzyl)-3,3-dimethoxypropionitrile (145 g, 0.49 mole) is added to this solution, and the mixture is refluxed for 2 hours. The reflux condenser is then removed, and the methanol is distilled from the solution at atmospheric pressure, using an oil bath heated to about 110°–120°. The residue crystallizes to a yellow mass. After cooling, it is slurried in water and the solid is isolated. The yield of crude *2,4-diamino-5-(3',4',5'-trimethoxybenzyl)pyrimidine* (III) is approximately 129 g (91%). The yellow discoloration can be removed by conversion of the product to the sulfate salt. A 100-g portion is added to 150 ml of 3 N sulfuric acid at 60° with stirring. The resultant solution is cooled to 5°–10° with stirring, and the precipitated sulfate is collected by vacuum filtration. The precipitate is washed twice with 50-ml portions of cold 3 N sulfuric acid, which removes colored material. It is then dissolved in 1 liter of hot water, treated with charcoal, and clarified with the aid of Hyflo-Supercel. The product is isolated from the clear colorless filtrate by the gradual addition of a solution of 100 g of sodium hydroxide in 200 ml of water, with chilling. The precipitate is isolated and washed thoroughly with water. The recovery is about 88% of material which melts at 200°–201°. If desired, the pyrimidine can be recrystallized from water or dilute ethanol or methanol. UV: λ_{max} (neutral species), sh 227.5 nm (ϵ 20,800), 287 (7250); λ_{max} (monocation), 269 (6060). The thermodynamic pK_a (20°) is 7.12 ± 0.03.[3]

Alternatively, the pyrimidine can be prepared in two steps, by treating the 3,4,5-trimethoxy-2'-methoxymethylcinnamonitrile directly with guanidine and sodium methylate.[14] However, the yields are somewhat inferior by this process.

2,4-Diamino-5-(p-chlorophenyl)-6-ethylpyrimidine (pyrimethamine). (VI).[16] Pyrimethamine is a highly potent antimalarial agent,[17,18] which owes its activity to the fact that it is a powerful inhibitor of dihydrofolate reductases from plasmodia. It exhibits a 50% inhibitory concentration of ca. 5×10^{-10} M against *Plasmodium berghei* reductase,[19] compared to 2.5 $\times 10^{-6}$ and 1.8×10^{-6} M for *E. coli* and human liver reductase enzymes, respectively. Preparative methods follow.[17,20,21]

A mixture of 151.5 g (1 mole) of *p*-chlorophenylacetonitrile and 306 g (3 moles) of ethyl propionate is added to a solution of 54 g (1 mole) of sodium methylate in 500 ml of absolute ethanol with stirring. The mixture is allowed to stand overnight and is then heated under reflux for 4 hours. The solvent is removed under vacuum, and the residue is poured into water and extracted with ether. The ether extract is discarded. The aqueous solution is neutralized with 1 N sulfuric acid, and the separated oil is extracted with ether. The ethereal extract is dried (Drierite or sodium sulfate), and the ether is removed. The residual oil crystallizes upon triturating with petroleum ether; weight, 170 g (82%). The product, *α-propionyl-α-(p-chlorophenyl)acetonitrile* (IV), may be purified by crystallization from ether–petroleum ether; m.p. 52°–54°.[17,20]

16 U.S. Patent 2,576,939 (1951).
17 P. B. Russell and G. H. Hitchings, *J. Am. Chem. Soc.* **73**, 3763 (1951).
18 G. H. Hitchings, *Clin. Pharmacol. Therap.* **1**, 570 (1960).
19 R. Ferone, J. J. Burchall, and G. H. Hitchings, *Federation Proc.* **27**, 390 (1968).
20 W. Logemann, L. Almirante, and L. Caprio, *Ber.* **87**, 435 (1954).
21 B. H. Chase and J. Walker, *J. Chem. Soc.* p. 3518 (1953).

The above nitrile (170 g, ca. 0.82 mole) is dissolved in 3 liters of toluene, and to this is added 67 g (0.9 mole) of i-butanol and 10 g of p-toluene-sulfonic acid. The mixture is heated vigorously under reflux in an apparatus containing a Dean-Stark trap for separation of water. Heating is continued until the theoretical amount of water separates (14.7 ml), which usually requires about 40–48 hours. The solution is cooled, washed with N sodium hydroxide and water, followed by removal of the toluene by distillation. The residual oil can be used directly in the next reaction, or alternatively, it can be distilled under high vacuum, to yield approximately 150 g (70%) of *3-isobutoxy-2-(p-chlorophenyl)pent-2-enonitrile* (V, R = i-butyl), boiling at 140°–141°/0.2 mm; n_D^{20}, 1.5587; infrared spectrum, 1600, 2210, 2850, 2935 cm^{-1}.[21]

An alternative procedure to the above is to convert the enol to its methyl ether, by the addition of a slight excess of ethereal diazomethane to an ethereal solution of (IV). The solution is allowed to stand overnight, followed by removal of the solvent. The crude product may be used directly in the next reaction.[17]

An ethanolic solution of guanidine is prepared from 56.5 g (0.6 mole) of guanidine hydrochloride plus 32.4 g (0.6 mole) of sodium methylate in 750 ml of absolute ethanol. To this is added 150 g (0.57 mole) of (V) (where R = i-butyl). The mixture is heated under reflux for 6 hours, and the solvent is removed *in vacuo*. The residue is extracted with dilute alkali, followed by water, and the solid is collected. The yield is approximately 127 g (90%) of *2,4-diamino-5-(p-chlorophenyl)-6-ethylpyrimidine* (VI). After recrystallization from dilute ethanol or i-butanol, the melting point is 235°–236°. UV: λ_{max} (neutral species), sh 220 mμ (ϵ 19,800), sh 248 (7450), 285° (8750); λ_{max} (cation), sh 222.5 (23,100), 272 (7320). The thermodynamic pK_a (20°) is 7.34 ± 0.04.[3]

1-(p-Butylphenyl)-2,2-dimethyl-4,6-diamino-1,2-dihydro-s-triazine Hydrochloride (VII). This compound is of interest as a dihydrofolate reductase inhibitor which is bound much more tightly to mammalian, than to bacterial, enzymes. Against *E. coli* reductase, the concentration required for 50% inhibition is 6.5×10^{-4} M, whereas against a guinea pig reductase the required concentration is 4×10^{-8} M.[2]

The preparation is as follows.

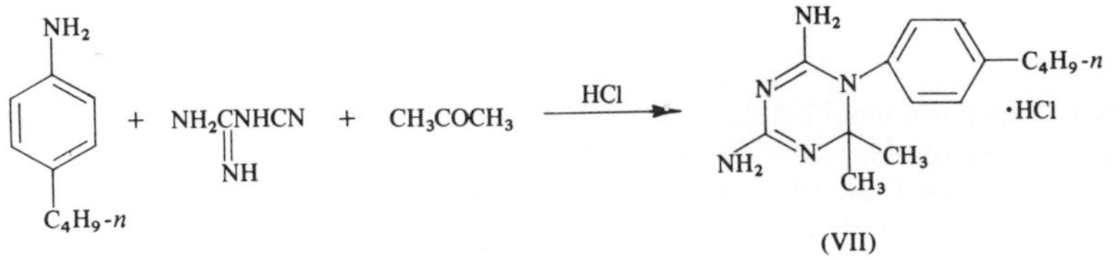

(VII)

A mixture of 149 g (1 mole) of p-(n-butyl)aniline, 90 g (1.07 mole) of dicyandiamide, 85 ml of concentrated HCl (1.02 mole), and 500 ml of acetone is heated under reflux for 18 hours. The mixture is chilled well, and the white crystalline product is isolated and washed well with cold acetone; weight, 135 g (44%). Additional material can be recovered from the filtrate upon evaporation, but it is of inferior quality. The product (VII) is purified by recrystallization from ethanol; m.p. 205°–207°.

2,4-Diamino-6-n-butylpyrido[2,3-d]pyrimidine (X).[22,23] The 2,4-diamino-pyrido[2,3-d]pyrimidines may be considered to have a closer analogy to the pteridine moiety of folic acid than do the 2,4-diamino-5-substituted pyrimidines, in that they contain a fused ring system which is similar to that of the pteridines. Such compounds are potent antibacterial agents. They do not exhibit the striking difference in binding between mammalian and bacterial enzymes which is seen with trimethoprim. However, some of them exhibit interesting differences in their binding to gram-positive *vs.* gram-negative organisms which are not seen with the latter compound. The title compound requires the following concentrations for 50% inhibition of enzymes from various sources: 50×10^{-8} for *E. coli* enzyme, 4×10^{-8} for *Staphylococcus aureus*, 95×10^{-8} for human liver enzyme.[4] These compounds strikingly potentiate the activity of the sulfonamides, and can consequently be used at low concentrations.[24] The preparation of the above derivative follows.

$(CH_3)_2NCH = \underset{\underset{C_4H_9\text{-}n}{|}}{C}CHO \;+\; COCl_2 \longrightarrow [(CH_3)_2N = CH - \underset{\underset{C_4H_9\text{-}n}{|}}{C} = CHCl]^+ \cdot Cl \;+\; CO_2$

(VIII)   (IX)

(IX) + [2,4-diamino-pyrimidine structure, H_2N and NH_2 groups] $\longrightarrow$ [n-C_4H_9-substituted pyrido[2,3-d]pyrimidine structure, NH_2 groups]

(X)

A solution of phosgene [CAUTION—*very toxic*] (13 g, 0.13 mole) in 100 ml of ethylene dichloride is added slowly to a cold solution of 2-butyl-3-dimethylaminoacrolein (VIII)[25] (20 g, 0.129 mole) in 50 ml of ethylene

[22] U.S. Patent 3,288,792 (1966).
[23] B. S. Hurlbert and B. F. Valenti, *J. Med. Chem.* **11**, 708 (1968).
[24] B. S. Hurlbert, R. Ferone, T. A. Herrmann, G. H. Hitchings, M. Barnett, and S. R. M. Bushby, *J. Med. Chem.* **11**, 711 (1968).
[25] Z. Arnold and F. Šorm, *Chem. Listy* **51**, 1082 (1957).

dichloride. When the evolution of carbon dioxide ceases, the solvent is removed *in vacuo*. A solution of 16.3 g (0.13 mole) of 2,4,6-triaminopyrimidine in 250 ml of absolute ethanol is then added to the residue (IX). This mixture is heated under reflux for 18 hours, made basic by adding 1 g of sodium methylate, and then heated 1 hour more. It is then cooled, and the precipitated pyridopyrimidine base is filtered, washed well with water, and recrystallized from 70% ethanol–water containing approximately an equimolar amount of hydrochloric acid. The crystalline product, *2,4-diamino-6-n-butylpyrido[2,3-d]pyrimidine hydrochloride* (X), weighs approximately 25.8 g (79%); m.p., 278°. The compound has a pK_a of 6.8. The base exhibits UV maxima at 248 nm (ϵ 19,900), 268 (9800) and 348 nm (6900). Maxima for the monocation (pH 2) are at 322 (8800) and 332 (sh) nm (7600).

[197] N^5-Formyltetrahydrofolic Acid Cyclodehydrase

By DAVID M. GREENBERG

$$N^5\text{-Formyl-FH}_4 + \text{ATP} \xrightarrow{\text{Mg}^{2+}} N^5,N^{10}\text{-methenyl-FH}_4 + \text{ADP} + \text{P}_i$$

This enzyme has been isolated from sheep liver acetone powders[1,2] and from extracts of *Micrococcus aerogenes*.[3]

Assay Method

Principle. The absorbance of the reaction product, 5,10-methenyl-FH$_4$, in the spectral region of 350–355 nm at pH 6.0 is used for the convenient assay of the enzyme.

Reagents

dl-N^5-formyl-FH$_4$, 2.5 mg/ml in 0.05 M sodium citrate buffer, pH 6.0
ATP, disodium salt, 20 micromoles/ml
MgSO$_4$, 20 micromoles/ml
Sodium citrate buffer, 0.05 M, pH 6.0[4]

The substrate dl-N^5-formyl-FH$_4$ obtained as the calcium salt of the commercial product (calcium leucovorin, American Cyanamid Corp.) is

[1] J. M. Peters and D. M. Greenberg, *J. Am. Chem. Soc.* **80**, 2719 (1958).
[2] L. K. Wynston and D. M. Greenberg, *Biochemistry* **4**, 1872 (1965).
[3] L. D. Kay, M. J. Osborn, G. Hatefi, and F. M. Huennekens, *J. Biol. Chem.* **235**, 195 (1960).
[4] G. Gomori, Vol. I [16].

purified by chromatography on a TEAE-cellulose column.[5] About 100 mg of the material dissolved in 10 ml of 5 mM ammonium acetate buffer, pH 5.3, is introduced onto the column (1.25 × 12 cm) and eluted stepwise with increasing concentrations of the buffer. The purified material emerges with 1.0 M ammonium acetate buffer in nine fractions of 8.0 ml each. Further purification is carried out by filtering the eluate through a bed of acid-washed charcoal (Darco G-60) formed on a bed of Celite in a glass cylinder (2.5 cm diameter) fitted with a sintered glass disk. The 5-formyl-FH_4 is eluted with 1500 ml of a mixture of butanol–ethanol–water (10:4:5 by volume) adjusted to pH 8.1 with ammonia. The solution is evaporated to dryness and sealed in ampoules for use.

Procedure. Incubation is performed by mixing 0.1 ml of 5-formyl-FH_4 solution (approx. 0.2 micromole), 2 micromoles of ATP (0.1 ml), 2 micromoles of $MgSO_4$ (0.1 ml), 100 micromoles of sodium citrate buffer, pH 6.0 (2.0 ml), and enzyme solution to make a total volume of 3.0 ml in a cuvette of 1-cm light path. The incubation is carried out at 30° in a Gilford recording spectrophotometer. (The Beckman instrument can be used and the recordings read visually if a recorder is not available.) Optical measurements are made at 343 nm. The 5,10-methenyl-FH_4 is comparatively stable in the citrate buffer, pH 6.0, with a half-life of 20 minutes at 30° and longer at lower temperatures.

Enzyme Unit. One unit of 5-formyl-FH_4 cyclodehydrase is defined as the amount of enzyme that will bring about a change of absorbancy of 0.001/unit/minute. The molar extinction coefficient has been estimated to be 23,600 based on the absorption ratio at 343 nm to 350 nm and the figure of 24,900 by Rabinowitz and Pricer at 350 nm in 0.17 M HCl.[6] A change in absorbancy of 0.001 corresponds to the production of 0.04 milli-micromoles of 5,10-methenyl-FH_4 per 3 ml. Specific activity of the enzyme is expressed in enzyme units per milligrams of protein. Protein concentrations were determined by the method of Lowry et al. as given by Layne[7] in crude solutions and by ultraviolet absorption in purified preparations.[7]

Purification Procedure

Fresh or freshly frozen livers are homogenized with acetone at −15° to prepare a highly stable acetone powder containing the enzyme. Approximately 130–150 g of liver are homogenized for 1 minute with 2 liters of cold acetone at medium speed in a Waring blender. The homogenate is filtered on a Büchner funnel and washed with 5 liters of cold acetone to remove most of the pigment. The powder is then dried on filter paper in a

[5] Procedure for purification of formyl-FH_4 was developed by Dr. Y.-C. Yeh.
[6] J. C. Rabinowitz and W. E. Pricer, Jr., *J. Biol. Chem.* **229**, 321 (1957).
[7] E. Layne, Vol. III [73].

vacuum desiccator. Enzyme activity is extracted from the acetone powder by stirring for 2 hours with cold 0.05 M sodium citrate buffer, pH 6.5, using 1500 ml of buffer per 100 g of powder. To the supernatant liquid obtained after centrifugation of the suspension at 10,000 g for 30 minutes, solid ammonium sulfate is added to reach 30% saturation (16.8 g/100 ml). The pH is maintained at 6.0–6.5 by the addition of dilute ammonium hydroxide. The precipitate that forms is removed by centrifugation, and more ammonium sulfate is added to reach 50% saturation (12.1 g/100 ml). After standing overnight at 4°, the suspension is centrifuged at 10,000 g for 30 minutes, and the supernatant liquid is discarded. The paste thus obtained retains enzyme activity indefinitely when frozen at −20°. Preliminary to further purification, salts are removed by dialysis against five changes of 4 liters each of distilled water. At this stage the specific activity of the enzyme preparation is 0.6–0.7.

Chromatography on CM-Cellulose. The CM-cellulose, purified by washing with sodium hydroxide and hydrochloric acid,[8] is packed into a 2.5 × 21 cm column and equilibrated to pH 5.4 with sodium citrate buffer and washed with 0.005 M buffer at the same pH.

The dialyzed protein is adjusted to pH 5.4 with citrate buffer, and the precipitate that forms is removed by centrifugation. About 200 ml of the protein solution containing 4–5 g of protein is introduced onto the CM-cellulose column. Chromatography is carried out in a cold room at 4°, and 15-ml fractions are collected at a flow rate of 3 ml/minute. The enzyme is eluted with 0.005 M sodium citrate buffer at pH 5.4. A peak zone remains at the top of the column, and the enzyme activity is eluted in a greenish zone which migrates with the solvent front.

Chromatography on DEAE-Cellulose. DEAE-cellulose purified as described for CM-cellulose is packed into a 2.0 × 30 cm column equilibrated to pH 7.5 and washed with 0.005 M potassium phosphate buffer. The active fraction from the CM-cellulose column is dialyzed against distilled water and 0.001 M potassium phosphate buffer, pH 7.5. The enzyme solution is concentrated on a rotary evaporator, and the concentrate is introduced onto the DEAE-cellulose column. Elution is begun with 0.005 M phosphate buffer and later changed to 0.035 M buffer. Ten-milliliter fractions are collected. The active enzyme fraction obtained assays about 11 units/mg. This preparation is relatively stable on storage at −20°. Most of the properties of the enzyme are determined on preparations of this activity.

Higher purification can be obtained by rechromatography on DEAE-cellulose. Samples of 70–155 mg of protein are introduced into a 2.0 × 23-cm column of DEAE-cellulose in 0.005 M phosphate buffer, pH 7.5.

---

[8] E. A. Peterson and H. A. Sober, *J. Am. Chem. Soc.* **78**, 751 (1956).

PURIFICATION OF SHEEP LIVER N^5-FORMYLTETRAHYDROFOLATE CYCLODEHYDRASE[a]

Step	Volume (ml)	Total protein (mg)	Total enzyme (units)	Specific activity	Yield (%)
Acetone powder extract	2640	55,440	4580	0.083	100
$(NH_4)_2SO_4$ treatment	210	5,580	3315	0.625	73
CM-cellulose	50	1,780	2840	1.60	62
First DEAE-cellulose	—	155	1558	10.7	34
Second DEAE-cellulose	—	18	596	33.1	13

[a] Sheep liver acetone powder (200 g) extracted with 3000 ml of 0.05 M sodium citrate buffer, pH 6.5, and centrifuged to remove insoluble material.

Elution is begun with 0.005 M buffer, then increased to 0.02 M and finally to 0.035 M, with the pH maintained at 7.5. The enzyme obtained on rechromatography, assaying 33–38 units/mg protein, is unstable and loses activity when concentrated by lyophilization. The table shows a typical preparation with the levels of enzyme activity and recovery at the several stages of purification.

Properties

pH Optimum. This was determined to be at pH 4.8.

Activators and Inhibitors. The cyclodehydrase is an SH enzyme. Its activity is inhibited by the usual SH reagents, and this inhibition can be reversed by glutathione. ATP and Mg^{2+} are required in the reaction.

Kinetic Properties. The Michaelis constant of N^5-formyl-FH_4 was estimated to be 1.4 $\times$ 10^{-4} at pH 6.0 and 30°. The Michaelis constant of ATP was similarly determined at an N^5-formyl-FH_4 concentration of 0.2 mM and found to be 4.5 $\times$ 10^{-4} M.

[198] N^5,N^{10}-Methenyltetrahydrofolate Cyclohydrolase

By DAVID M. GREENBERG

5,10-Methenyl-FH_4 + H_2O → 10-formyl-FH_4

This enzyme was observed in lyophilized cells of *Clostridium cylindrosporum* by Rabinowitz and Pricer.[1] It has been purified about 100-fold from beef liver by Lombrozo and Greenberg,[2] and some of its properties have been studied.

[1] J. C. Rabinowitz and W. E. Pricer, Jr., *J. Am. Chem. Soc.* **78**, 5702 (1956).
[2] L. Lombrozo and D. M. Greenberg, *Arch. Biochem. Biophys.* **118**, 297 (1967).

Assay Method

Reagents. dl-N^5-Formyl-FH$_4$, the calcium salt of the commercial product (calcium leucovorin), is purified by chromatography on a TEAE-cellulose column as described in the preceding article.[3] 5,10-Methenyl-FH$_4$ is formed daily from the purified 5-formyl-FH$_4$ by weighing 0.77 mg into a glass-stoppered test tube, displacing the air in the tube by N$_2$, and adding 1.0 ml of 0.1 N HCl. The solution is then allowed to stand at least 4 hours at room temperature to obtain complete conversion to 5,10-methenyl-FH$_4$. This preparation is comparatively stable for several days if kept frozen. 10-Formyl-FH$_4$ is prepared from 5,10-methenyl-FH$_4$ by adding 0.30 ml of 1.0 N KOH to a stock solution containing 0.90 ml of 5,10-methenyl-FH$_4$ and 0.15 ml of mercaptoethanol. The mixture is left standing 30 minutes at 25° to ensure complete conversion.

Enzyme Assay. The enzyme assays in the paper from which this is taken were performed with a Beckman D-K2 recording spectrophotometer. However, they can be carried out with any recording spectrophotometer, e.g., the Gilford instrument.

Enzyme measurements are made at 355 nm in a cuvette of 1-cm light path. To this is added 0.1 ml of 1 M potassium citrate buffer, pH 6.5, and 0.4 ml of the 5,10-methenyl-FH$_4$ solution. Water is added to bring the total volume to 1 ml. The rate of spontaneous decay of the substrate is recorded for 3–4 minutes, and the enzyme solution is then introduced (0.01–0.04 ml). Enzyme activity is estimated by subtracting the spontaneous decay rate from that obtained with added enzyme.

Enzyme Unit. This is defined as the amount of enzyme producing a change in absorbance of 1.0 unit per minute. Specific activity is expressed in enzyme units per milligram of protein. Protein concentrations are determined by the method of Lowry *et al.* as given in Layne[7] in crude solutions and by ultraviolet absorption in purified preparations.[4]

Purification Procedure

Purification of 5,10-methenyl-FH$_4$ cyclohydrolase is extremely difficult because of the marked instability of the enzyme. To preserve enzyme activity for a considerable period requires the presence of 25% glycerol and a reducing agent (mercaptoethanol).

Preparation of Homogenate. Two 100-g portions of freshly obtained sliced beef liver are minced and then homogenized in a Waring blendor for 1 minute each, in two 500-ml portions of a chilled buffer consisting of equal volumes of 0.25 M sucrose and 0.25 M Tris, pH 7.5. All operations

[3] D. M. Greenberg, this volume [197].
[4] E. Layne, Vol. III [73].

are carried out at 0°–4°. The homogenate is centrifuged for 30 minutes at 20,000 g, and the supernatant fluid is filtered through cheesecloth to remove fat and particulate material. The enzyme activity of this fraction is used as the basis of estimation of enzyme purification.

Ammonium Sulfate Fractionation. The bulk of the enzyme is precipitated at 50–60% $(NH_4)_2SO_4$ saturation in the first treatment and at 45–55% saturation in the second treatment. The first fraction is isolated by stirring in solid salt to 50% saturation, removing precipitated protein by centrifugation, and then adding more salt with stirring to 60% saturation. Stirring is continued for 30 minutes at each step, and the protein precipitate is isolated by centrifugation at 20,000 g for 30 minutes. The active fraction is dissolved in about 40 ml of 25% glycerol (v/v)–0.025 M Tris buffer, pH 7.5, and dialyzed overnight against three changes of buffer.

The dialyzate is centrifuged, and the $(NH_4)_2SO_4$ precipitation is repeated on the clear supernatant solution to obtain the 45–55% saturation fraction. The precipitate from this fraction is suspended in about 20 ml of the glycerol-Tris buffer and dialyzed as described above. Centrifugation is repeated to obtain a clear supernatant solution.

The second $(NH_4)_2SO_4$ fractionation does not appreciably increase the specific activity of the enzyme, but it is valuable in removing a great deal of contaminating pigment.

DEAE-Cellulose Column Chromatography. The chromatographic procedure has to be performed as rapidly as possible because of the increased instability of the enzyme on purification. A high flow rate for the column is achieved by removing the resin "fines" and by subjecting the column to an atmosphere of nitrogen at 3–4 psi.

The fractionated resin is placed in a jacketed column (5 × 55 cm) washed with 1 liter of 0.05 N NaCl and 2 liters of distilled water to clean the DEAE-cellulose and also to determine the bed volume. The column is then equilibrated with 2 liters of the 25% glycerol–0.025 M Tris buffer, pH 7.5. A coolant below 0° (usually at −5°) is pumped through the cooling jacket.

The enzyme solution from the second $(NH_4)_2SO_4$ fractionation is introduced into the column and eluted with the 25% glycerol–Tris buffer in which the concentration of Tris is increased stepwise to 0.025, 0.050, 0.075, and 0.1 M. Fifty-milliliter fractions are collected using a flow rate of 10 ml/minute after a quantity of buffer equivalent to the bed volume of the column has passed through. To each eluent fraction, 0.005–0.050 M mercaptoethanol is added as an additional protective agent. Up to 3.5 g of enzyme protein has been fractionated on the column with little decrease in resolution. A typical purification procedure is summarized in the table. The enrichment in activity was 100-fold with a recovery of about 11%.

PURIFICATION OF N^5,N^{10}-METHENYLTETRAHYDROFOLATE CYCLOHYDROLASE

Fraction	Volume (ml)	Activity (total units)	Total protein (mg)	Specific activity
Homogenate	1000	4410	37,800	0.12
Dialyzed 50–60% $(NH_4)_2SO_4$ (I)	67.5	4060	2,977	1.37
Dialyzed 45–55% $(NH_4)_2SO_4$ (II)	28	1195	899	1.33
DEAE-column eluate	50	543	45.2	12.00

Enzyme Properties

pH Optimum. This was determined by subtracting the spontaneous rate of hydrolysis of the 5,10-methenyl-FH$_4$ with respect to pH. Enzyme activity increased with increasing alkalinity from pH 6.0 on and reached a maximum value at pH 8.0. Beyond this value there is an abrupt loss of enzyme activity, probably due to denaturation of the enzyme.

Evidence for Sulfhydryl Nature of Enzyme. The 5,10-methenyl-FH$_4$ is inhibited by the SH reagents, PCMB, *o*-iodosobenzoate, and iodoacetate. It is protected to a considerable degree by mercaptoethanol and is partially reactivated (30–50%) by such reducing agents as sodium hydrosulfate, sodium bisulfate, and mercaptoethanol.

Inhibition of Metal Ions. Heavy metal ions are in general inhibitory to the enzyme. The greatest inhibitory effects were produced by Cu^{2+}, Cd^{2+}, Mg^{2+}, Fe^{2+}, and Fe^{3+}. The observed absorbance changes indicated that Fe^{2+} and Fe^{3+} interacted with the substrate as well as the enzyme protein.

Michaelis Equilibrium Constants. Values of K_m obtained were 2.5×10^{-4} M for 5,10-methenyl-FH$_4$ in the forward reaction and 1.0×10^{-4} M for 10-formyl-FH$_4$ in the reverse reaction. From these values and the maximum velocities in the forward and reverse reactions, an equilibrium constant of 1.05×10^{-8} was estimated by the Haldane equation[5] at pH 6.1 and 25°. This value compares favorably with the value of 2.4×10^{-8} estimated by Kay *et al.* for the equilibrium constant of the nonenzymatic reaction in the pH range 5.7–7.1.[6]

The 5,10-methenyl-FH$_4$ cyclohydrolase reaction is strongly inhibited by both of the substrates at higher concentrations.

[5] J. B. S. Haldane, "Enzymes," p. 80. Longmans, London, 1930.
[6] L. D. Kay, M. J. Osborn, Y. Hatefi, and F. M. Huennekens, *J. Biol. Chem.* **235**, 195 (1960).

[199] 10-Formyl Tetrahydrofolate:NADP Oxidoreductase

By CARL KUTZBACH and E. L. R. STOKSTAD

$$10\text{-CHO-}H_4\text{PteGlu} + NADP^+ + H_2O \rightarrow H_4\text{PteGlu} + CO_2 + NADPH + H^+ \quad (1)$$

Assay Methods

Principle. Assays based on measuring the formation of tetrahydrofolic acid at its absorption maximum at 300 nm or the consumption of 10-formyl tetrahydrofolate at 355 nm (after acidification) have been described[1-3] for the deacylase reaction[4] [Eq. (2)].

$$10\text{-CHO-}H_4\text{PteGlu} + H_2O \rightarrow H_4\text{PteGlu} + HCOOH \quad (2)$$

They can also be used for this enzyme if a substrate amount of NADP (0.2 mM) is added to the reaction mixture. The difference in molar extinction coefficient between the substrate and the product at 300 nm has been determined as 21×10^3 cm^2 mole^{-1}.[5] The small contribution of NADPH absorption at this wavelength can be corrected for by using a $\Delta\epsilon$ of 22.6, or, alternatively, the accumulation of NADPH can be prevented by the addition of 5 mM ketoglutarate, 5 mM NH$_4^+$ and an excess of glutamic dehydrogenase (EC 1.4.1.3). Although convenient to detect the presence of this enzyme or roughly follow a purification, these assays are not suited for more quantitative work, because a strong product inhibition by tetrahydrofolate makes reliable initial rate determinations impossible. This complication is overcome by coupling the assay with the formate:tetrahydrofolate ligase (ADP) reaction, Eq. (3), (EC 6.3.4.3). Thereby the product is continuously converted back to the substrate.

$$H_4\text{PteGlu} + HCOOH + ATP \rightarrow 10\text{-CHO-}H_4\text{PteGlu} + ADP + P_i \quad (3)$$
$$10\text{-CHO-}H_4\text{PteGlu} + NADP^+ + H_2O \rightarrow H_4\text{PteGlu} + CO_2 + NADPH + H^+ \quad (1)$$

$$\overline{HCOOH + ATP + NADP^+ + H_2O \rightarrow CO_2 + ADP + P_i + NADPH + H^+ \quad (4)}$$

The overall reaction, Eq. (4), has been measured either spectrophotometrically at the absorption band of NADPH (a) or by the formation of $^{14}CO_2$ from 10-^{14}CHO-H$_4$PteGlu and H^{14}COOH (b). Since only a catalytic amount

[1] F. M. Huennekens and K. G. Scrimgeour, Vol. VI [50].

[2] K. G. Scrimgeour and F. M. Huennekens, *in* "Hoppe-Seyler, Thierfelder: Handbuch der physiologisch- und pathologisch- chemischen Analyse," 10th ed., Vol. VI/B, p. 194. Springer, Berlin, 1966.

[3] H. R. Whiteley, *Comp. Biochem. Physiol.* **1**, 227 (1960).

[4] M. J. Osborn, Y. Hatefi, L. D. Kay, and F. M. Huennekens, *Biochim. Biophys. Acta* **26**, 208 (1957).

[5] C. Kutzbach and E. L. R. Stokstad, *Biochem. Biophys. Res. Commun.* **30**, 111 (1968).

TABLE I
COMPOSITION OF SPECTROPHOTOMETRIC ASSAY

Tris-HCl, pH 7.7	50 mM
2-Mercaptoethanol	100 mM
MgSO₄	30 mM
(±)-10-Formyl-H₄PteGlu[a]	0.07 mM
ATP	2.5 mM
NH₄-Formate	20 mM
NADP	0.2 mM
Formate: H₄PteGlu ligase[b]	5 IU

[a] Prepared fresh by alkalization from a stock solution of commercial 5-CHO-H₂Pte-Glu in 0.1 M HCl. See also D. Robinson, this volume [184].

[b] Prepared from *Clostridium cylindrosporum* according to R. H. Himes and J. C. Rabinowitz, Vol. VI [51] 5 IU ≅ 417 Units as defined by these authors.

of folate substrate is necessary for the overall reaction, it should be possible to do the assay with H₄PteGlu instead of 10-CHO-H₄PteGlu. If, however, chemically prepared (±)H₄PteGlu is used, only about 30% of the full activity is measured because of inhibition by the enzymatically inactive (+) isomer (see below).

Procedure a: Spectrophotometric. The components of the assay, except for NADP, are mixed to give the concentrations specified in Table I in a final volume of 1.0 ml. The enzyme sample is added, and the mixture is preincubated for 5 minutes at 30° in a cuvette with 1-cm lightpath. The reaction is then started by the addition of the NADP. Formation of NADPH is followed at 340 nm. The amount of enzyme is selected to give a change in extinction of about 0.1/minute.

Procedure b: Radioactive. The incubation is carried out at 30° with shaking under an atmosphere of nitrogen in Warburg vessels equipped with two sidearms and a center well. Contents of the vessel are shown in Table II. The reaction is started after temperature equilibration by addition of the enzyme sample from sidearm 1. The amount of enzyme is chosen to produce between 50 and 500 millimicromoles of $^{14}CO_2$. After 20 minutes, the reaction is terminated by the addition of the acid from sidearm 2, and the vessels are shaken for another 10 minutes to allow absorption of the CO_2 in the center well. The vessels are then opened, and the contents of the center well including the paper strip are transferred, using a Pasteur-type pipette, to counting vials with several washings of Bray's solution.[6] The vials are then filled to 10 ml with Bray's solution and counted. A blank value is obtained from an incubation without enzyme or without NADP.

[6] G. A. Bray, *Anal. Biochem.* **1,** 279 (1960).

TABLE II
COMPOSITION OF RADIOACTIVE ASSAY

Main Compartment	Total volume 2.0 ml with concentrations as in Table I, except:	
	(−)-10-^{14}CHO-H$_4$PteGlu[a]	0.1 mM
	NH$_4$formate-^{14}C[b]	10 mM
	NADP	1 mM
Side arm 1	Enzyme sample	
Side arm 2	0.2 ml of 2.5 N HClO$_4$	
Center well	A piece of filter paper soaked in 0.2 ml of phenylethylamine	

[a] Prepared by enzymatic synthesis essentially according to P. K. Ho and L. Jones, *Biochim. Biophys. Acta* **148,** 622 (1967).

[b] Of the same specific activity as the labeled folate; we used about 100,000 dpm/micromole.

Unit. Activities are calculated as international units (IU) (= micromoles · min^{-1} at 30°). For the radioactive assay the calculations are:

$$IU = \frac{(cpm_{sample} - cpm_{blank}) \times 100}{Counting\ Eff.\ (\%) \times 20\ (min) \times spec.\ act.\ of\ substrates\ (dpm/micromole)}$$

Comparison. Results obtained with procedures (a) and (b) are in good agreement and proportional to both the amount of enzyme and time; for assay (b), linearity has been checked up to 30 minutes. Assay (b) is preferable for measuring crude tissue extracts because of (1) its greater sensitivity and (2) its independence from NADPH oxidizing activities.

Purification Procedure

All operations were carried out at about 4° unless otherwise indicated. Buffer, if not otherwise given, is 0.05 M potassium phosphate (pH 7.2), 0.01 M 2-mercaptoethanol.

Step 1. Extraction. Frozen pig liver (500 g) is cut in small pieces and homogenized with 1.2 liters of buffer in a Waring blendor. The homogenate is adjusted to pH 6 with 4 M acetic acid with stirring and then centrifuged for 1 hour at 20,000 g; the sediment is discarded.

Step 2. Ammonium Sulfate Precipitation. Solid ammonium sulfate is added to 25% saturation with stirring. The precipitate is collected by centrifugation for 45 minutes at 20,000 g and discarded. More ammonium sulfate is added to the supernatant up to 42% saturation, and the mixture is centrifuged as before. The precipitate is dissolved in 300 ml of buffer and dialyzed overnight against 5 liters of buffer.

Step 3. Acetone Fractionation. The dialyzed solution is cooled to 0° and acetone cooled to −10° is added to 30% (v/v) with rapid stirring while the temperature is lowered to −5°. The precipitate is removed by centrifugation at $-5°$ for 10 minutes at 6000 g. More acetone is added to a concentration of 50% as before, while the temperature is lowered to $-10°$. This precipitate is collected by centrifugation as before and dissolved in 120 ml of buffer. Residual acetone should be removed by dialysis against 5 liters of buffer without delay to prevent losses of activity.

Step 4. *DEAE-Sephadex Chromatography.* A column is prepared from DEAE-Sephadex A-50 and equilibrated with a solution of 0.12 M KCl in buffer. KCl was added to the dialyzed enzyme solution to a concentration of 0.05 M and the solution is applied to the column with a flow rate of 100 ml/hour. The sample is followed by 0.12 M KCl in buffer. Most of the inactive protein is eluted under these conditions. The protein concentration of the effluent is recorded at 280 nm. When the absorption drops below half its maximal value, elution is continued with a linear gradient formed from 500 ml each of 0.12 M KCl and 0.5 M KCl in buffer; the effluent is collected in fractions. The enzyme activity is eluted approximately between 500 and 700 ml of effluent.

Step 5. *Ammonium Sulfate Precipitation.* Ammonium sulfate is added to the pooled active fractions up to 80% saturation. The precipitate is collected by centrifugation for 30 minutes at 20,000 g and suspended in a minimal volume of buffer. Residual ammonium sulfate is estimated by nesslerization and the concentration is adjusted to 45% by the addition of solid ammonium sulfate, if necessary. The precipitate is again collected by centrifugation for 30 minutes at 40,000 g and dissolved in a few milliliters of buffer.

Step 6. *Hydroxylapatite Chromatography.* A small column of hydroxylapatite,[7] approximately 1×3 cm, is equilibrated with 0.02 M potassium phosphate buffer, pH 7.2, 0.01 M 2-mercaptoethanol. The sample is dialyzed against 10 volumes of the same buffer and applied to the column. Stepwise elution is carried out with 0.05 M, 0.08 M, and 0.1 M buffers of the same pH with 2-mercaptoethanol added. Protein elution is recorded at 280 nm, and the buffer is changed when the elution of one peak is completed. Most of the activity is eluted with 0.08 M buffer; some additional material is eluted in the 0.1 M fraction.

Good reproducibility of this procedure has been obtained except for the last step, where results varied with different preparations of the adsorbent. A summary of a purification is given in Table III. The final product is not homogeneous during gel electrophoresis and gel chromatography; further purification, however, has not yet been attempted. A quantitative estimate of purity is not possible. The preparation is stable for several weeks when frozen at $-15°$ in the presence of 0.01 M 2-mercaptoethanol.

[7] O. Levin, Vol. V [2].

TABLE III
PURIFICATION OF 10-FORMYL-H$_2$PteGlu:NADP OXIDOREDUCTASE

Step	Volume (ml)	Units^b (IU)	Protein^a (mg)	Specific activity (IU/mg, ×10^{-3})	Yield (%)
1. Extract pH 6	1600	293	29,440	9.9	100
2. (NH$_4$)$_2$SO$_4$ fraction	330	231	8,250	28.0	79
3. Acetone fraction	150	176	3,470	50.7	60
4. DEAE-Sephadex	335	90	415	216.8	31
5. (NH$_4$)$_2$SO$_4$	6.5	63	134	469.0	22
6. Hydroxylapatite	5	32	33	980.0	11

^{a} Determined with the biuret method: G. Beisenherz, H. J. Boltze, Th. Bücher, R. Czok, K. H. Garbade, E. Meyer-Arendt, and G. Pfleiderer, Z. Naturforsch. 8b, 555 (1953).
^{b} Spectrophotometric assay (a).

Properties

A molecular weight of about 320,000 has been obtained by gel chromatography. The broad pH optimum centers at about pH 7.8, with a range of more than 60% activity extending from pH 6.5 to pH 8.5. Tris, imidazole, and phosphate buffers gave identical results; 40% higher activity was found in glycine buffer.

Specificity. The enzyme is specific for NADP and (−)-10-CHO-H$_4$-PteGlu. NAD (<5%), (+)-10-CHO-H$_4$PteGlu, and 5-CHO-H$_4$PteGlu do not react nor inhibit.

Michaelis Constants. K_m values have been obtained from Lineweaver-Burk plots of data obtained with assay (a): $K_{m(NADP)} = 3.5$ μM; $K_{m[(-)-10-CHO-H_4PteGlu]} = 8.2$ μM (the same value was obtained from experiments with enzymatically synthesized (−)-10-CHO-H$_4$PteGlu and the chemically prepared diastereoisomeric mixture, taking the active form as 50%).

Equilibrium. The reaction could not be reversed. From thermodynamic data an equilibrium constant of 1.6×10^8 can be calculated.

Inhibitors. The reaction is strongly inhibited by both isomers of (±)-H$_4$PteGlu. The presence of the inhibitor affects both slope (K_m) and intercept (V_{max}) in Lineweaver-Burk plots. K_i has roughly been estimated to be in the range of 1 μM for the (−) form and 10 μM for the (+) form of H$_4$PteGlu.[8] This high affinity for the inhibitor has the effect that a much larger K_m (50–70 μM) for 10-CHO-H$_4$PteGlu is obtained from initial rate measurement in an assay that is not coupled to formate:H$_4$PteGlu ligase, since 1 μM product accumulates before the progress of the reaction can be

[8] C. Kutzbach, unpublished data (1968).

observed. Intermediate K_m values are obtained if the amount of ligase added is insufficient.

5-CH$_3$-H$_4$PteGlu, 5-CHO-H$_4$PteGlu, folate, aminopterin, and tetra-hydroaminopterin did not inhibit up to $5 \times 10^{-4}\ M$. Inhibition by p-chloro-mercuribenzoate and iodoacetamide is reversed by the 2-mercaptoethanol included in the assay mixture.

Other Properties. In the absence of NADP, the enzyme catalyzes a hydrolytic cleavage of 10-CHO-H$_4$PteGlu to formate and H$_4$PteGlu at 15–30% of the rate of the oxidative reaction. This reaction is unaffected by the presence of $0.1\ M$ hydroxylamine, which strongly inhibits the NADP dependent reactions.

Distribution

The enzyme was first demonstrated in the livers of rats and pigs, and the stoichiometry of the reaction was established.[5] Smaller activities were assayed in rat kidney, heart, spleen, and chicken liver. No activity has so far been found in microorganisms. The enzyme is most probably identical to 10-CHO-H$_4$-PteGlu deacylase reported to catalyze the reaction shown in Eq. (2) with a catalytic requirement for NADP or NADPH.[4] No formation of formate from 10-CHO-H$_4$PteGlu could, however, be found in liver extracts when substrate amounts of NADP were present. Stoichiometric amounts of NADP are required for the formation of H$_4$PteGlu from 10-CHO-H$_4$PteGlu at maximal rate if the enzyme was purified to be free of NADPH-oxidizing activities.

Acknowledgment

This work was supported in part by a grant from the National Institutes of Health, No. AM 08171.

[200] Catabolism of Unconjugated Pteridines

By H. REMBOLD and W. GUTENSOHN

After the structure of biopterin had been elucidated by Patterson *et al.*[1] this compound was isolated from many natural sources. Tetrahydro-biopterin and some other unconjugated tetrahydropteridines[2-4] have

[1] E. L. Patterson, M. H. von Saltza, and E. L. R. Stokstad, *J. Am. Chem. Soc.* **78**, 5871 (1956).

[2] S. Kaufman, *Trans. N.Y. Acad. Sci.* **26**, 977 (1964).

[3] A. R. Brenneman and S. Kaufman, *Biochem. Biophys. Res. Commun.* **17**, 177 (1964).

[4] A. Tietz, M. Lindberg, and E. P. Kennedy, *J. Biol. Chem.* **239**, 4081 (1964).

cofactor activity in hydroxylating enzyme systems. Their chemical properties as well as their behavior in experiments *in vivo* and *in vitro* with the rat are described.

Analysis

All pteridines occur in reactions only in small quantities. No microbiological assays are available except the biopterin-specific *Crithidia* test. Therefore labeled compounds have to be used in general. Their synthesis[5] and their analysis[6] by ion-exchange chromatography are described in this volume.

Resorption, Distribution, and Excretion of Biopterin *in Vivo*

On an average the rat excretes 40–60 μg of biopterin per day in the urine. This quantity is not altered by feeding a biopterin-free diet.[7] After intraperitoneal injection of 14 μg of biopterin-8a-^{14}C (specific activity 1 mCi/millimole) dissolved in 1 ml of 0.9% NaCl, 84% of the radioactivity is excreted during the first 24 hours, a further 6% during the next day. In this case the total renal excretion is at least 90%, although the injected quantity corresponds to less than 50% of the normal daily biopterin excretion. The resorption is changed completely when the same quantity is fed by stomach tube (Table I).[8]

Biopterin reappears only in small quantities (10%) in the urine. Remarkably, the greater part of the isomeric 7-biopterin is not resorbed. The animals do not exhale measurable amounts of radioactive CO_2.

TABLE I

EXCRETION AS PERCENTAGE OF ADMINISTERED RADIOACTIVITY AFTER FEEDING
BIOPTERIN-8a-^{14}C AND 7-BIOPTERIN-8a-^{14}C TO RATS[8]

Day	Biopterin		7-Biopterin	
	Feces	Urine	Feces	Urine
1	3.7	3.7	43	4.7
2	1.4	0.3	4	0
	5.1	4.0	47	4.7
	Sum: 9.1		51.7	

[5] See this volume [181].
[6] See this volume [178].
[7] H. Kraut, W. Pabst, H. Rembold, and L. Wildemann, *Z. Physiol. Chem.* **332**, 101 (1963).
[8] H. Rembold and H. Metzger, *Z. Naturforsch.* **22b**, 827 (1967).

TABLE II

Crithidia ACTIVITY IN THE ORGANS OF THE RAT EQUIVALENT TO MICROGRAMS
OF BIOPTERIN PER GRAM FRESH WEIGHT[9]

Liver	6.0	Lung	0.9	Heart	0.2
Adrenals	4.8	Bone marrow	0.6	Serum	0.2
Spleen	3.0	Brain	0.2	Eyes	0.02
Kidney	1.2				

Microbiological analyses demonstrate an especially high *Crithidia* activity in liver, adrenals, and spleen (Table II).[9] Upon feeding 14 μg (7×10^4 cpm) of biopterin by stomach tube, low radioactivities are observed in liver, serum, and adrenals after 3 days. A trichloroacetic acid extract from a labeled liver was analyzed by chromatography. Ten percent of the isolated radioactivity is associated with biopterin, and small amounts with pterin-6-carboxylic acid.

On *incubation with rat liver homogenate* biopterin is stable. Small amounts are oxidized unspecifically by air to pterin-6-carboxylic acid. The same applies to neopterin and 6-hydroxymethylpterin.

Behavior of Tetrahydropteridines *in Vitro*

Hydrogenation. The device[10] shown in Fig. 1 is made up by the upper hydrogenation vessel, which can be stirred magnetically, and by the lower reservoir, into which the solution is filtered through a sintered-glass disk at the end of the reaction. The three gas inputs are connected with a mercury valve which is joined to a gas bomb. Before the reduction the device is flushed with hydrogen by opening the three input stopcocks and removing the output stoppers. The solution of the substance is brought into the hydrogenation vessel without interrupting the hydrogen stream. The stoppers are closed, and a slight pressure of hydrogen is maintained during the reduction. After the reaction the solution is brought to the vertical tube by inclining the vessel. By removing the stopper of the reservoir, it is pressed through the sintered glass. From the reservoir the solution can be pipetted out under streaming hydrogen nearly quantitatively.

For catalytic reduction 2 mg of solid PdO are first saturated with hydrogen in 1 ml of 0.2 N HCl. The catalyst is introduced through a glass tube 5 cm long with a little bulb on its end. After 20 minutes, 1 ml of a solution of the pteridine (0.5 mg/ml of water) is added under streaming hydrogen; it is reduced for 1 hour with stirring. The sample is neutralized by adding 36 mg of solid Tris and is filtered.

[9] H. Rembold, *in* "Pteridine Chemistry" (W. Pfleiderer and J. Taylor, eds.), p. 465. Pergamon Press, Oxford, 1964.
[10] H. Rembold and H. Metzger, *Z. Physiol. Chem.* **348,** 194 (1967).

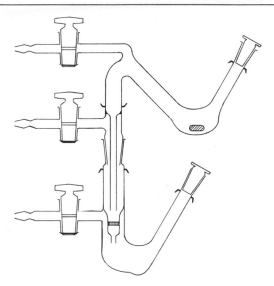

FIG. 1. Hydrogenation vessel, reduced. The original size of the vessel is 20 cm high. Reproduced from Reference 10.

To obtain the tetrahydropteridines in physiological saline for *in vivo* injections, the catalyst is saturated with hydrogen in 2 ml of 0.3 N HCl. The pteridine is added in 1 ml of 0.3 N NaOH, and after the reduction the sample is neutralized by another milliliter of 0.3 N NaOH. This method is generally recommended for pteridines which are only slightly soluble in water (e.g., pterin and lumazine) and can also be applied to larger amounts of substance (up to 10 micromoles). The tetrahydropteridines are prepared fresh for each experiment.

Incubation. Liver homogenates are prepared in a Potter-Elvehjem homogenizer (Teflon pestle) in 0.03 M Veronal-acetate buffer, pH 7.5, or in 0.1 M Tris-HCl, pH 7.5. The tetrahydropteridines (0.4–1.5 micromoles) are incubated in 2 ml of the homogenate at 37° under aerobic conditions and with shaking. The reaction is stopped by pouring the solution into 3 ml of boiling water. After short boiling, the denatured protein is centrifuged and the supernatant is analyzed by chromatography. This type of chromatography does not allow the reaction to be stopped by TCA.

Anaerobic incubations are done under a nitrogen atmosphere and are stopped by adding 1 ml of 0.2 N HCl. For the subsequent oxidation of hydrogenated products, 1 ml of an iodine solution is added, and the mixture is stirred under nitrogen for 2 hours at room temperature. To obtain the iodine solution, 200 mg of iodine are suspended in 4 ml of water, and potassium iodide is added until solution of the iodine is completed. The denatured protein is separated from the incubation mixture by centrifugation, and the solution is neutralized. Oxidation of tetrahydropteridines by air yields numerous compounds. Iodine on the contrary oxidizes hydrogenated pteridines to their nonhydrogenated derivatives under mild conditions.

Chromatography. Chromatographic analysis of the incubation mixtures is performed on Dowex 1 X8 (formate), Ecteola-cellulose (EC), and phosphocellulose (PC). The preparation of the columns has been described.[6] The following column dimensions are used: for Dowex 1-X8, 5 × 20 mm; for EC, 10 × 65 mm; and for PC, 17 × 170 mm. Immediately before analysis the Dowex and EC columns are washed once more with 600 ml and 1000 ml of water, respectively. This is necessary to obtain clear-cut separations. All compounds are applied to the columns in neutral aqueous solution. The analysis of a typical incubation reaction (tetrahydrobiopterin in 0.5% liver homogenate) is shown in Fig. 2.

Xanthopterin and 7,8-dihydroxanthopterin are not separated by chromatography on columns or on paper, but they can be distinguished by their UV spectra. Neopterin appears on PC in nearly the same fractions as biopterin. Pterin-6-carboxylic acid and 6,7-dihydroxylumazine are more acidic compounds. They are eluted from Dowex in zone II and can be purified further by rechromatography on EC. Isoxanthopterin as a less

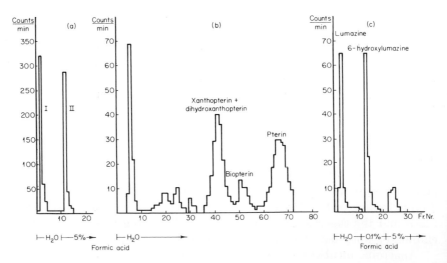

FIG. 2. Chromatographic analysis of an incubation reaction: 0.3 micromoles of tetrahydrobiopterin-2-^{14}C in 2 ml of 0.5% liver homogenate, 0.03 M Veronal-acetate, pH 7.5; temperature, 37°; incubation time, 30 minutes. (a) Separation on Dowex 1 X8 (formate). (b) Rechromatography of zone I from Dowex on phosphocellulose. (c) Rechromatography of zone II from Dowex on Ecteola cellulose. Reproduced from Reference 12a.

acidic substance appears in zone II from Dowex, but is eluted from EC with water. Identity and purity of the compounds ascribed to the radioactive peaks is checked by comparative paper chromatography in 4 systems.[6]

Incubations *in vitro* indicate 6-hydroxylumazine to be the main degradation product of tetrahydrobiopterin (or tetrahydroneopterin) in liver homogenates. The essential steps of this degradation are (1) cleavage of the side chain in position 6; (2) deamination of the amino group in position 2; (3) introduction of an oxygen function into position 6. Details and reaction mechanisms of this degradation scheme are discussed elsewhere.[11,12,12a]

Cleavage of the Side Chain. The irreversible cleavage of the side chain is a nonenzymatic reaction, which is preceded by an oxidation.

In the case of tetrahydroneopterin, the side chain can be trapped as glyceraldehyde. Ninety micrograms (corresponding to 5×10^5 cpm) of tetrahydroneopterin-6,7,1′,2′,3′-[14]C[5] and 3.3 mg of DL-glyceraldehyde are incubated for 2 hours (aerobically at 37°) in 2.5 ml of 0.1 M Tris-HCl, pH 7.5. According to a method of Neuberg,[13] a quantity of 2,4-dinitrophenylhydrazine equimolar to the glyceraldehyde dissolved in 4.3 ml of 2 N HCl is added. After standing for 30 minutes at 0°, the yellow derivative crystallizes. After filtration the precipitate is washed with diluted HCl and water and is recrystallized from 50% aqueous methanol to a constant radioactivity. Concentrations are determined spectrophotometrically in methanolic solution ($\epsilon_{359 \, m\mu} = 22,170$). Known amounts of methanolic solutions are evaporated to dryness. The residue is redissolved in a dioxane-scintillation mixture and its radioactivity is measured in a scintillation counter. Because of the yellow color of the solutions, quenching is high and must be determined by internal standards. A similar cleavage of the side chain of sepiapterin has been described by Nawa.[14]

Deamination. A protein fraction with deaminating enzyme activity can be obtained from rat liver.[11] A homogenate in 0.1 M citrate–phosphate buffer, pH 6.5, is centrifuged at 23,000 g. Upon an ammonium sulfate fractionation of the supernatant solution, the enzyme is precipitated between 50 and 60% saturation. By an intensive dialysis against 0.001 M citrate–phosphate buffer, pH 6.5, followed by a heat step (5 minutes at 60°) it can be further purified. A protein fraction obtained in this way is free from most of the liver xanthine oxidase. Since pterin is also a substrate

[11] H. Rembold and F. Simmersbach, *Biochim. Biophys. Acta* **184**, 589 (1969).
[12] H. Rembold and W. Gutensohn, *Biochem. Biophys. Res. Commun.* **31**, 837 (1968).
[12a] H. Rembold, H. Metzger, P. Sudershan, and W. Gutensohn, *Biochim. Biophys. Acta* **184**, 386 (1969).
[13] C. Neuberg and H. Collatz, *Biochem. Z.* **223**, 494 (1930).
[14] S. Nawa, *Bull. Chem. Soc. (Japan)* **33**, 1555 (1960).

TABLE III

PRODUCTS FROM *in Vitro* INCUBATIONS OF UNCONJUGATED PTERIDINES AND THEIR TETRAHYDRO DERIVATIVES

Incubation medium	Incubation conditions	Substrate	Biopterin	Neopterin	Pterin	7,8-Dihydroxanthopterin	Xanthopterin	Lumazine	6-Hydroxylumazine	Other products
Homogenate	Aerobic	Biopterin	+[a]	–	–	–	–	–	–	Pterin-6-carboxylic acid
Homogenate	Aerobic	Neopterin	–	+	–	–	–	–	–	Pterin-6-carboxylic acid
Buffer[b]	Aerobic	Tetrahydrobiopterin	4%[c]	–	36%	36%	–	–	–	–
Buffer[b]	Anaerobic + I₂-oxid.	Tetrahydroneopterin	–	46%	35%	–	–	–	–	–
0.5%[d] Homogenate	Aerobic	Tetrahydrobiopterin	5%	–	18%	13%	–	17%	22%	7-Hydroxylumazine, 4%
15% Homogenate	Aerobic	Tetrahydrobiopterin	42%	–	19%	9%	–	3%	5%	–
3% Homogenate	Anaerobic I₂-oxid.	Tetrahydrobiopterin	68%	–	5%	1%	–	–	–	–
0.5% Homogenate	Aerobic	Tetrahydroneopterin	–	+	+	+	–	+	+	–
Buffer	Aerobic	Tetrahydroneopterin	–	–	32%	\{ 52%	\}	–	–	–
Pure xanthine oxidase in buffer	Aerobic	Tetrahydroneopterin	–	–	–	\{ 81%	\}	–	–	Isoxanthopterin, 19%
0.5% Homogenate	Aerobic	Tetrahydropterin	–	–	+	+	–	–	+	Isoxanthopterin

Preparation	Conditions	Substrate							Products
0.5% Homogenate	Aerobic	Tetrahydro-6-hydroxy-methylpterin	+	—	—	—	—	—	—
0.5% Homogenate	Aerobic	Tetrahydro-pterin-6-carboxylic acid	+	—	—	—	—	—	—
Homogenate or pure xanthine oxidase	Aerobic	Pterin	—	—	—	—	+	—	Isoxantho-pterin
Homogenate or pure xanthine oxidase	Aerobic	Lumazine	—	+	—	—	—	—	7-Hydroxy-lumazine
Homogenate or pure xanthine oxidase	Aerobic	6-Hydroxy-lumazine	+	—	—	—	—	—	6,7-Dihydroxy-lumazine
Homogenate or pure xanthine oxidase	Aerobic	Tetrahydro-lumazine	+	+	—	—	—	—	—
Buffer	Aerobic	Tetrahydro-lumazine	+	+	—	—	—	—	—
Deaminase fraction	Aerobic	Pterin	—	+	—	—	+	—	—
Deaminase fraction	Aerobic	Isoxanthopterin	—	—	—	—	—	—	7-Hydroxy-lumazine
Deaminase fraction	Aerobic	Tetrahydro-pterin	—	+	—	—	+	—	—
Deaminase fraction + xanthine oxidase	Aerobic	Tetrahydro-pterin	—	—	—	—	—	—	Isoxantho-pterin + 6,7-dihydroxy lumazine[e]

(Continued)

TABLE III (*Continued*)

Incubation medium	Incubation conditions	Substrate	Products								
			Biopterin	Neopterin	Pterin	7,8-Dihydro-xanthopterin	Xanthopterin	Lumazine	6-Hydroxy-lumazine	Other products
Deaminase fraction + xanthine oxidase	Aerobic	Tetrahydro-neopterin	—	—	—	—	—	—	+	—

[a] Only those products are designated as + which have been isolated and unequivocally identified in respective experiments; they must not necessarily be the only products.

[b] 0.03 M Veronal-acetate, pH 7.5, or 0.1 M Tris-HCl, pH 7.5.

[c] Percent of the total regained radioactivity.

[d] Percent (w/v) corresponding to the liver fresh weight.

[e] From 6-hydroxylumazine, which is already formed during the incubation.

for xanthine oxidase and is turned over by this enzyme considerably faster than by the deaminase, the test described below can be reasonably applied only after separation of the xanthine oxidase.

The deaminase activity is tested as follows: 230 μg of pterin-2-^{14}C (corresponding to 10^5 cpm) are incubated in 4 ml of 0.03 M Veronal-acetate buffer, pH 6.5, at 37° for 7 minutes together with the protein fraction. The reaction is stopped by boiling the mixture. After cooling 0.05 mg of pure xanthine oxidase from milk (Boehringer) is added, and a further incubation follows for 30 minutes under the above conditions. After boiling the mixture once more, the denatured protein is centrifuged down and the supernatant solution is chromatographed on 5-ml EC columns. Isoxanthopterin is eluted with water and 7-hydroxylumazine with 1% formic acid. The 7-hydroxylumazine formed (determined by its radioactivity) is a measure for the activity of the deaminase. Controls are run with heat-denatured protein.

Xanthine Oxidase Reaction. The enzyme, which introduces the oxygen function into position 6, has been identified as rat liver xanthine oxidase.[12] Milk xanthine oxidase also catalyzes the reaction.

Tetrahydrolumazine (250 μg, 1.5 micromoles) is incubated together with the protein in 2 ml of 0.1 M Tris-HCl, pH 7.5, for 90 minutes at 37° under aerobic conditions. The reaction is stopped by adding 1 ml of 30% (w/v) TCA. The mixture is boiled for a short time and allowed to stand in the air for 24 hours. After separation of the precipitated protein the solution is brought to a volume of 4 ml. The yield of 6-hydroxylumazine is determined spectrophotometrically at 365 mμ ($\epsilon = 5500$).[15] The extinction of lumazine (from reoxidized substrate) at 365 mμ is negligible at low pH values. Controls are incubated in heat-denatured homogenates of the same protein concentration.

On changing from crude homogenates to enzyme fractions of high specific activity, the time of incubation has to be shortened considerably (from formerly 90 to 20 minutes). In that case, the test can be simplified by measuring the immediate enzyme product,[12] 7,8-dihydro-6-hydroxylumazine.

Tetrahydrolumazine (100 μg, 0.61 micromoles) is incubated in 2 ml of buffer as indicated above. The reaction is stopped by adding of 1 ml of 1 N HCl. The yield of enzyme product is calculated from the absorption of 300 nm ($\epsilon = 12,300$).[16] Zero time samples are used for correction.

Summary

A survey of all pteridine metabolites, which have been observed using various substrates and incubation conditions is given in Table III. The

[15] W. Pfleiderer, *Ber.* **90**, 2604 (1957).
[16] A. Albert and S. Matsuura, *J. Chem. Soc.* p. 2162 (1962).

table shows that all the tetrahydroderivatives are very unstable under aerobic conditions. They are degraded to a series of compounds for the most part unspecifically and by nonenzymatic reactions. The deamination, i.e., the transition from the pterin to the lumazine series, unequivocally is an enzymatic step. The participation of xanthine oxidase in the introduction of the oxygen function into position 6 of the pteridine ring (dihydroxanthopterin, xanthopterin, and 6-hydroxylumazine) has been clearly established.[11,12] But as is shown by the table, the same products are obtained by nonenzymatic reactions, although with lower yields.

[201] Prosthetic Group of an Alcohol Dehydrogenase of a Pseudomonad[1]

By C. ANTHONY

The prosthetic group of this dehydrogenase is included here as present evidence suggests that it may be a pteridine derivative. This evidence consists primarily of fluorescence spectra[2] and of data on cochromatography of the purified prosthetic group with known pteridines.[3] A definite identification has not been possible. Methods are given below for preparation and purification of the enzyme, together with procedures for preparation of the prosthetic group from the purified enzyme.

The Alcohol Dehydrogenase

The enzyme, which has been found only in bacteria capable of growth on methanol as sole source of carbon and energy, oxidizes methanol to the oxidation level of formaldehyde. When isolated, ammonia is required as activator and phenazine methosulfate as primary hydrogen acceptor; the natural hydrogen acceptor is unknown.[4,5]

Assay Method

The more rapid spectrophotometric assay using 2,6-dichlorophenol-indophenol as final hydrogen acceptor is given below. An alternative, more accurate, manometric assay,[5] using oxygen as final acceptor, may be used,

---

[1] The bulk of the work on this enzyme and its prosthetic group has been published elsewhere in collaboration with Dr. L. J. Zatman.

[2] C. Anthony and L. J. Zatman, *Biochem. J.* **104**, 960 (1967).

[3] C. Anthony, unpublished results.

[4] C. Anthony and L. J. Zatman, *Biochem. J.* **92**, 609 (1964).

[5] C. Anthony and L. J. Zatman, *Biochem. J.* **92**, 614 (1964).

but the spectrophotometric assay is sufficiently accurate for use during purification of the enzyme.

Reagents

Tris-HCl buffer, 0.6 M, pH 9.0
Methanol, 0.15 M
Ammonium chloride, 0.45 M
N-Methylphenazonium methosulfate (phenazine methosulfate), 0.033 M
2,6-Dichlorophenolindophenol, 0.0013 M

Procedure. The assay system in a 1-cm light path cuvette contains 0.5 ml of Tris buffer, 0.1 ml of methanol, 0.1 ml of ammonium chloride, 0.1 ml of phenazine methosulfate, and 0.1 ml of 2,6-dichlorophenolindophenol. The final volume in the cuvette is made to 3 ml with water, and the reference cuvette contains deionized water. Enzyme solution is blown in with a pipette, which may be used then for rapid mixing. The initial rate of dye reduction is taken as twice the change in E_{600} occurring between 15 and 45 seconds after addition of enzyme. The amount of enzyme is adjusted to give a rate of change of E_{600} of less than 0.6/minute.

Definition of Unit and Specific Activity. One unit of enzyme activity is defined as the amount of enzyme that produces a change in E_{600} of 0.01/minute between 15 and 45 seconds after addition of the enzyme; 570 of these units are equivalent to 1 standard unit as defined in Enzyme Nomenclature (1965).[6] A high rate of dye reduction is sometimes observed in the absence of substrate when using the spectrophotometric assay.[5] This is ignored and is not subtracted from the reduction measured in the presence of substrate. The specific activity is defined as the number of units per milligram of protein as determined by the method of Lowry *et al.*[7]

Purification Procedure

Growth of Bacteria[8] *and Preparation of Crude Extract. Pseudomonas* sp. M27[4] or *Pseudomonas* AMI[9,10] may be used, but *Pseudomonas* sp. M27 is

---

[6] Enzyme Nomenclature. "Recommendations (1964) of the International Union of Biochemistry on the Nomenclature and Classification of Enzymes." Elsevier, Amsterdam, 1965.

[7] O. H. Lowry, N. J. Rosebrough, A. L. Farr, and R. J. Randall, *J. Biol. Chem.* **193**, 265 (1951).

[8] The bacterial strains may be obtained from the National Collection of Industrial Bacteria, Torry Research Station, Aberdeen, Scotland, U.K. *Pseudomonas,* sp. M27 is catalog No. NC1B 9686. *Pseudomonas* AM1 is catalog No. NC1B 9133; it is also available as ATCC 14718.

[9] D. Peel and J. R. Quayle, *Biochem. J.* **81**, 465 (1961).

[10] P. A. Johnson and J. R. Quayle, *Biochem. J.* **93**, 281 (1964).

preferable for large-scale preparations as the enzyme is constitutive in this organism and gives more consistently a high specific activity. Stock cultures are maintained on methylamine-agar slopes.[9] The growth medium is a defined medium containing methanol + sodium lactate as carbon source. The bacteria are grown at 30° in well-aerated cultures and are harvested at the end of the exponential growth phase, which usually occurs 36–48 hours after inoculation. The inoculum is 5% of a late exponential phase culture. The yield is usually 2–3 g wet weight per liter. Harvested organisms may be stored at −20° for any length of time, and extracts may be prepared by means of Braun homogenizer, Hughes press, French pressure cell, or ultrasonic disintegrator. After cell breakage, crude extracts are prepared by centrifugation at 40,000 g for 1 hour at 2°. There are a number of procedures for purification of the enzyme,[5,11] but the best method for dealing with large quantities of material and for obtaining the purest product is given below.

Preparation of Pure Enzyme. The pH of crude extract is lowered very carefully with 1 N HCl to pH 4.0 at room temperature, and the heavy precipitate is removed by centrifugation at 40,000 g for 20 minutes. The pH of the supernatant liquid is raised to 6.0 with 1 N NaOH, and solid ammonium sulfate is added to give 65% saturation; after removal of the precipitate by centrifugation, ammonium sulfate is added to give 85% saturation. The precipitated enzyme is dissolved in 20 mM Tris-HCl buffer, pH 8.0, and passed through a column of DEAE-cellulose equilibrated with the same buffer. The enzyme is not adsorbed, but is eluted in the solvent front with 20 mM-Tris HCl buffer, pH 8.0. The pH of the pooled active fractions is lowered to 6.0, and solid ammonium sulfate is slowly added. Any precipitate that forms at less than 65% ammonium sulfate saturation is removed, and the 65–80% saturated fraction is dissolved in the minimum volume of 0.1 M Tris-HCl buffer, pH 8.0. At this stage the enzyme is dark brown. The enzyme solution is passed through a large column of Sephadex G-150 equilibrated with the same buffer; the total volume of the column should be about twenty times that of the sample. The main protein peak (as indicated by E_{280}) is collected and pooled. It should be golden yellow and completely separated from the red cytochrome c. If the absorption spectrum of the pooled fractions shows any absorption at 550 nm due to cytochrome c, they are concentrated with ammonium sulfate and passed through a similar column of Sephadex G-150. The pooled active fractions are dialyzed against 1000 volumes of deionized water at 2° before freeze-drying.

A summary of the purification procedure is given in the table.

[11] C. Anthony and L. J. Zatman, *Biochem. J.* **104,** 953 (1967).

PURIFICATION OF ALCOHOL DEHYDROGENASE FROM *Pseudomonas* SP. M27

Stage of purification	Volume (ml)	Total activity (units)	Specific activity (units/ mg)	Yield (%)	Purification (-fold)
Crude extract	415	1,050,000	43	100	—
pH 4.0 Supernatant	365	893,000	220	85	4.9
Ammonium sulfate fraction	34	650,000	326	62	7.6
Pooled fractions from DEAE-cellulose	34	495,000	384	47	8.9
Pooled fractions from Sephadex G-150	60	347,000	420	33	9.8

Purity and Stability of the Enzyme

Enzyme prepared as described above is at least 95% pure and is stable for at least 4 months in the freeze-dried state. Solutions of the enzyme in 20 mM Tris buffer, pH 8.0, lose about 15% activity after storage for 2 months at $-22°$ while the same solutions are stable at $2°$ for about a week.

Properties of the Enzyme

A full description of the enzyme including molecular weight, amino acid analysis, absorption spectrum, substrate specificity, and sensitivity to inhibitors has been published elsewhere.[2,5,11,12]

The Prosthetic Group

The prosthetic group is defined as the low molecular weight, green-fluorescent material produced from the nonfluorescent enzyme by boiling, or by lowering the pH to about 3, or by raising the pH to about 12. The prosthetic group has not been reversibly resolved from the enzyme; all attempts to do so either have failed to release the prosthetic group or have resulted in concomitant irreversible denaturation of the enzyme. A description of the methods used to prepare, purify, and assay this prosthetic group is given below.

Assay[2,3]

Because no assay that depends on reactivation of a resolved enzyme is possible, a method that involves measurement of fluorescence is used. The best ultraviolet source for detection of fluorescence is a lamp giving radiation at predominantly 366 nm; this is suitable for detection of fluorescent solutions or of fluorescent spots on chromatography or electrophoresis paper. Measurement of the prosthetic group is by means of a spectrophoto-

---

[12] C. Anthony and L. J. Zatman, *Biochem. J.* **96**, 808 (1965).

fluorimeter; the maximum excitation wavelength is at about 365 nm, and maximum fluorescence is at about 470 nm. These values are uncorrected for the instrument and vary with the pH value of the solution. The wavelengths for maximum excitation and fluorescence should be determined on the particular instrument to be used, and all solutions should be adjusted to the same pH value before measurement. There is relatively little fluorescence in alkaline solution.

Preparation[2,3]

Although the prosthetic group may be released from the enzyme by a variety of methods, the most straightforward is to boil the enzyme preparation (about 10 mg/ml) for 2 minutes. Details of the kinetics of liberation of prosthetic group from the enzyme have been published.[2] The supernatant liquid, after a brief centrifugation to remove denatured protein, contains all the fluorescent material. Ten milligrams of pure enzyme dissolved in 1 ml of 0.05 M phosphate buffer, pH 6.0, and treated in this way, yields a fluorescent supernatant solution which gives a reading of about 60% full-scale deflection using an Aminco-Bowman spectrophotofluorimeter with a xenon lamp and a type IP21 photomultiplier (sensitivity at 25; slit arrangement No. 4; meter multiplier at 0.3).

Purification[2,3]

The procedure described below removes from the supernatant liquid about half of the material which absorbs in the ultraviolet range of the spectrum (240–320 nm).

After boiling the enzyme, the supernatant liquid is passed through a small column of DEAE- or TEAE-cellulose equilibrated with 50 mM ammonium acetate buffer, pH 6.0. A column of about 1-cm diameter is suitable, and about 1 cm in height should be allowed for every 20 mg of pure enzyme originally used. The column is washed with the same buffer until no more UV-absorbing material is eluted and then with 0.5 M ammonium acetate buffer, pH 6.0, to remove any trace of blue-fluorescent material that may be present. The adsorbed green-fluorescent material, visible as a red-brown band, is eluted with a linear gradient of 0.7–1.2 M ammonium acetate buffer, pH 6.0. All the green-fluorescent fractions, which should have the same excitation/fluorescence characteristics, are pooled, diluted five times with water, and adsorbed on a small pad of DEAE- or TEAE-cellulose. This pad should be as small as possible, but large enough to adsorb all the fluorescent material, which is then eluted with the minimum volume (usually less than 1 ml) of 10 M ammonium acetate. This concentrated material is then freeze-dried to remove the ammonium acetate. The yield, as indicated by total (arbitrary) fluorescence units, is about 90%.

The procedure described here may also be used for preparation of prosthetic group from enzyme protein that has not been completely purified. The last and most difficult stage of the enzyme purification procedure is the removal of cytochrome c by gel filtration. When this last stage has been omitted, the preparation obtained after purification of the prosthetic group has not been seen to be different from that obtained when starting with pure enzyme. It should be noted, however, that this technique has not been used extensively, and although it saves considerable time and labor, it should be used with caution.

Properties[2,3]

At room temperature, the purified prosthetic group is freely soluble in water but not in methanol, ethanol, acetone, diethyl ether, light petroleum (b.p. 40°–60°), hexane, cyclohexane, or chloroform. The prosthetic group, prepared as described here, often appears to have two or more fluorescent species; the relative proportions of these depend on time of storage. Minor fluorescent species may be removed from the major green-fluorescent species by paper chromatography in 0.1 M phosphate buffer, pH 7.0, or by high-voltage electrophoresis in the same buffer; at pH 7.0 the prosthetic group is negatively charged. Excitation and fluorescence spectra are given elsewhere; the major component has a fluorescence maximum at about 470 nm with excitation maximum at about 365 nm at pH 7.0 (uncorrected values).

Addendum to Section VII

A Bioautographic Procedure for Detecting TPN, DPN, NMN, and NR[1]

By BERNARD WITHOLT

Bioassays are usually considerably more sensitive than chemical assays. Accordingly, a bioautographic technique based on the V factor requirement of *Hemophilus parainfluenza*, has[1a] been developed for several nicotinamide derivatives.

[1] This work was supported by grants from P-L Biochemicals Inc. and the American Cancer Society (P-77M) to Dr. N. O. Kaplan.
[1a] A. Lwoff and S. Lwoff, *Proc. Roy. Soc.* (London), *B*, **122**, 352 (1937).

Principle of the Method

A mixture containing NR, NMN, DPN, or TPN[2] is separated by chromatography or electrophoresis. The dried sheet or strip is applied to an agar layer seeded with *H. parainfluenza* and containing all the nutrients required by *H. parainfluenza* except the V factor. After removing the sheet or strip from the agar, the plate is incubated overnight at 30°–37°, and growth will occur wherever there is sufficient V factor. Since each of the above nicotinamide derivatives fulfills the V factor requirement,[3] growth zones develop corresponding to the separated spots on the sheet. The sensitivity of this method is approximately two orders of magnitude greater than that obtained by viewing the same sheet under UV light.

Materials

Chromatographic Solvent (CS). 2 M NH$_4$Cl in 0.12 M Na citrate pH 5.3/95% ethanol (1/3).

A = 32.1 gm NH$_4$Cl, 2.08 gm citric acid, 7.68 gm sodium citrate. Bring to 300 ml with water.

CS = prior to chromatography, mix one volume of A with three volumes of 95% ethanol.

Medium for H. parainfluenzae growth (B1). B = Add 3.7 gm Brain Heart Infusion (Difco) to 100 ml water in a 250 ml Erlenmeyer, with a metal cap (Bellco) or cotton plug. Autoclave at 121° for 40 minutes. D = 1 mg/ml DPN in water, filter sterilized. B1 = to 100 ml sterile B, add 0.1 ml D, using sterile technique.

Storage of H. parainfluenzae.[4] Cells are grown in B1 at 37°, to late log phase. Sterile glycerol is added to bring the culture to 15% with respect to glycerol. The culture is then distributed to small sterile capped vials or tubes in 2 ml aliquots, and stored at −70°. Such frozen cultures are viable for at least 6 months.

Preparation of Bioautographic Plates (BP). Frozen *H. parainfluenzae* stock (2 ml) is inoculated into 100 ml B1 in a 250 ml Erlenmeyer. The flask is shaken at 37°. Under these conditions the culture will be ready for use in 4–6 hours. A Pyrex baking tray (233, 13½″ × 8¾″ × 1¾″, DD–11) is covered with aluminum foil and autoclaved for 15 minutes at 121°. BP medium is prepared for each tray to be used: 3.75 gm agar (Difco), 5 gm Brain Heart Infusion (Difco), 250 ml water, and a magnetic stirring bar, in a 500 ml Erlenmeyer. This medium is autoclaved at 121° for 40 minutes.

---

[2] Abbreviations used: NR = nicotinamide ribose; NMN = nicotinamide mononucleotide; DPN = diphosphopyridine nucleotide; TPN = triphosphopyridine nucleotide.

[3] I. G. Leder and P. Handler, *J. Biol. Chem.* **189,** 889 (1951).

[4] J. W. Bendler and S. H. Goodgal, *Science* **162,** 464 (1968).

After autoclaving, the medium is allowed to equilibrate in a 43° incubator or water bath. This insures minimum killing of *H. parainfluenzae* while the medium is still liquid. However, BP medium cannot be left at 43° for more than 6–10 hours without some solidifying.

When the BP medium is at 43°, and *H. parainfluenzae* has reached the late log phase in B1 (the culture should be heavily turbid after 4–6 hours at 37°), 10 ml of the B1 culture is added to the BP medium over a magnetic stirrer. The stirrer is run so that there is adequate mixing without bubbling of the agar. After 1–2 minutes over the stirrer, the now seeded BP medium is poured into a sterile Pyrex tray. Before the agar hardens, the tray is slowly inclined in every direction, so that agar contacts the sides of the tray before setting. This insures adequate coverage of the entire tray and better adhesion of the agar. Bubbles on the surface can be removed by rapidly passing a flame over the surface before the agar hardens. Finally, the Pyrex tray is covered again, and stored at room temperature overnight. These plates should be used within 24 hours, because *H. parainfluenzae* dies in the absence of V factor. Storing at 0° does not improve the situation, and at 37° the death rate is considerably accelerated.[5]

Development of the Bioautographic Plate (BP)

On day one, samples are spotted onto Whatman #3 paper and eluted with CS by descending chromatography. This solvent separates nicotinamide, nicotinic acid, NR, NMN, DPN, and TPN with R_f values of 0.78, 0.73, 0.64, 0.46, 0.27, and 0.16, respectively. On day two, after overnight elution, the chromatogram is dried in an open hood. Simultaneously, *H. parainfluenzae* are grown in B1, BP medium is prepared, and plates are poured. On day three, the chromatogram is cut down to a maximum of $12'' \times 7\frac{1}{2}''$, encompassing all the spots of interest. It is heated to 70° for 30 minutes in an oven to sterilize the paper. The sheet is then carefully layered over the agar, starting at one end and slowly working towards the opposite end. A bent glass rod is used to eliminate air spaces between agar and paper. The plate is covered and turned over in a rapid, smooth movement to avoid collapsing or sliding of the agar. Reference points on the chromatogram (the four corners, or sample origins, for instance) are marked on the bottom of the Pyrex tray. The plate is turned over again, and after 10 minutes (from the initial application of sheet to agar) the chromatogram is removed from the agar and dried in the hood.

The plate is covered again and incubated at 30°–37°. Growth zones develop in 6–10 hours; usually the plates are left in the incubator overnight. Growth zones are recorded on day four. This can be done by dark field

---

[5] B. Witholt. (To be published).

photography or by delineating the zones on the back of the tray. Delineated zones are then traced on the original dried chromatogram.

Discussion

It is possible to detect as little as 100, 20, and 4 picomoles, respectively, of TPN, DPN, and NMN. The sensitivity to NR has not been measured. Whatman #3 paper or cellulose acetate (electrophoresis) is recommended because they are easily applied and removed from the agar. Thin layer chromatograms have been analyzed in this fashion but some of the thin layer inevitably adheres to the agar and sometimes alters *H. parainfluenzae* development.

Contamination of the plates should be avoided, because contaminants growing on the agar surface often excrete enough V factor to support the *H. parainfluenzae* below the surface. Thus, spurious growth zones can occur. The potential occurrence of various *H. parainfluenzae* inhibitors on the chromatogram should be borne in mind.

Author Index

Numbers in parentheses are reference numbers and indicate that an author's work is referred to although his name is not cited in the text.

Subject Index

A

nine) phosphoribosyltransferase, 197, 198, 203

6,7-Benzo-10-methylisoalloxazine, half-wave reduction potential, 486

N-Benzylamino-D-arabinoside, 688
conversion to 1-benzylamino-1-deoxy-D-ribulose oxalate, 688
synthesis, 688

1-Benzylamino-1-deoxy-D-ribulose oxalate
conversion to D-neopterin, 688
synthesis, 688

N-Benzylamino-L-xyloside
conversion to 1-deoxy-1-benzylamino-L-xyloketose oxalate, 685
synthesis, 684, 685

2-Benzylazinoriboflavin
substrate for liver flavokinase, 547
synthesis, 455, 456

Biopterin, see also Dihydrobiopterin, Tetrahydrobiopterin
biosynthesis, 746, 755–761
chromatography
column, 656, 658, 672, 674
paper, 655, 677
thin-layer, 677
isolation from bees, 660
resorption, distribution, and excretion in vivo, 799, 800
synthesis, 670–672, 675, 676

7-Biopterin
column chromatography, 672–674
synthesis, 670–672
ultraviolet absorption spectrum, 672, 676

Borohydride reduction of flavin and flavoproteins, 474–479
procedures, 474–477
properties of reduction products, 477–479

8-Bromoadenosine 5'-phosphoromorpholidate, use in synthesis of F8-bromoAD, 462, 463

1-(p-Butylphenyl)-2,2-dimethyl-4,6-diamino-1,2-dihydro-s-triazine hydrochloride, inhibition of dihydrofolate reductase and synthesis, 784, 785

Butyryl-CoA dehydrogenase, lack of sulfite complexing with pig liver enzyme, 473

C

8-Celluloseacetamido-7,10-dimethylisoalloxazine, see Flavin-cellulose compounds

7-Celluloseacetamido-10-(1'-D-ribityl)isoalloxazine, see Flavin-cellulose compounds

7-Chloro-10-diethylaminopropylisoalloxazine, half-wave reduction potential, 486

8-Chloro-10-diethylaminopropylisoalloxazine, half-wave reduction potential, 486

4-Chloro-2,6-dihydroxypyrimidine
conversion to 4-ribitylamino-5-nitroso-2,6-dihydropyrimidine, 518
formation from 4-chloro-2,6-dimethoxypyrimidine, 517

7-Chloro-8-methoxy-10-(ethyl-2'-N-pyrrolidino)isoalloxazine, half-wave reduction potential, 486

3-Chloro-4-methylacetanilide
nitration, 444
preparation, 444

3-Chloro-4-methyl-2-nitroacetanilide
deacetylation, 444
preparation, 444

3-Chloro-4-methyl-2-nitroaniline
conversion to 2,4-dichloro-5-nitrotoluene, 445
preparation, 444

7-Chloro-10-(1'-D-ribityl)isoalloxazine, see 7-Chlororiboflavin

7-Chlororiboflavin, substrate for liver flavokinase, 547

Ciliapterin, 739–746
fluorescence spectrum, 743
isolation and identification from Tetrahymena pyriformis, 739–746
ultraviolet spectrum, 742

Cinchomeronic acid, 112

Citrovorum factor, see N^5-Formyltetrahydrofolic acid

Clostridium acetobutyricum, source of riboflavin glycosidase and transglycosidase, 412

Clostridium cylindrosporum
source of methenyltetrahydrofolate cyclohydrolase, 789

N^5-Formyltetrahydrofolic acid cyclodehydrase, 786–789
assay, 786, 787
properties, 789
purification from liver, 787–789
Fusaric acid, 112

G

D-Galactoflavin, substrate for riboflavin
hydrolase, 573
D-Glucose, use of 1-^{3}H-compound for
preparation of NAD-nicotinamide-
^{3}H, 66, 67
Glucose dehydrogenase, use of beef liver
enzyme in assay for phenylalanine
hydroxylase cofactor, 601–603, 615–
617
Glucose oxidase
anaerobic photoreduction, 510, 511
borohydride reduction, 475
sulfite complex, 473
D-Glucose 6-phosphate, use in enzymatic
cycling of NADP and NADPH, 11–
16, 26, 27, 29, 30
L-Glutamate, 3–16
formation in enzymatic cycling of
NADP and NADPH, 11–16
use in catalytic assay for NAD, 3–11
L-Glutamic dehydrogenase
use in catalytic assay for NAD, 3–11
in enzymatic cycling of NADP and
NADPH, 11–16
in fluorometric recycling assay for
NAD, 28, 29
Glutaryl CoA, formation from 3-hydroxy-
kynurenine, 167
Glutathione reductase
borohydride reduction, 475
lack of sulfite complexing with yeast
enzyme, 473
D-Glyceraldehyde 3-phosphate, use in
catalytic assay for NAD, 3–11
Glyceraldehyde 3-phosphate dehydrogenase, use in catalytic assay for NAD,
3–11
DL-Glyceroflavin, substrate for riboflavin
hydrolase, 573
Glycolate oxidase
borohydride reduction, 475

purification over flavin-cellulose
column, 551, 552
sulfite complex, 473
Glyoxylate carboligase, lack of sulfite
complexing with E. coli enzyme, 473
GTP cyclohydrolase, 761–765
assay, 762, 763
properties, 764, 765
purification from E. coli, 763, 764
Guanosine 5'-phosphate, precursor of
pteridines, 755–758
in Drosophila melanogaster, 755, 756
in Rana catesbeiana, 757, 758
Guanosine triphosphate-dependent flavin
phosphate synthetase, 553–555
assay, 553
occurrence, 553
preparation from Rhizopus javanicus,
554, 555

H

Hemophilus parainfluenzae, microbiological assays for NADP, NAD,
NMN, and nicotinamide ribose, 813–
816
Homofolic acid
molar extinction coefficients, 669
paper chromatography, 664, 665
Homopteroic acid
molar extinction coefficients, 669
paper chromatography, 664, 665
Hydrazine, use in catalytic assay for
NAD, 3–11
3-Hydroxyanthranilate oxygenase, 165,
175
3-Hydroxyanthranilic acid, conversion to
α-amino-β-carboxymuconic ε-semialdehyde, 162–166
7-Hydroxybiopterin chromatography
column, 656, 658
paper, 655
4'-Hydroxybutylflavin, substrate for riboflavin hydrolase, 573
Hydroxyethylflavin, see 7,8-Dimethyl-
10-(2-hydroxyethyl)isoalloxazine
2-β-Hydroxyethyliminoriboflavin 5'-
phosphate, synthesis, 459–462
6'-Hydroxyhexylflavin, substrate for riboflavin hydrolase, 573

δ-Methylaminovaleric acid
  derivation from *N*-methyl-2-pyridone,
    103
  methylation to 5-trimethylaminovaleric
    acid, 104
7-Methyl-8-chlororiboflavin, synthesis,
    444–447
*N*³-Methyl-3,4-dihydrolumiflavin,
    formation from *N*³-methyllumiflavin,
    474–476
6-Methyl-7,8-dihydropterin, synthesis,
    698–702
7-Methyl-10-dimethylaminoethyliso-
    alloxazine, half-wave reduction
    potential, 485
*N*⁵-Methyl-6,7-diphenyl-5,6-dihydro-
    pterin, synthesis, 690–692
*N*⁵-Methyl-6,7-diphenyl-5,6,7,8-tetra-
    hydropterin, synthesis, 697, 698
2(or *α*-)-Methyleneglutaric acid
  derivation from 5-amino-4-carboxy-
    valeric acid, 82
  enzymatic conversion to dimethyl-
    maleic acid, 243, 244
  enzymatic formation from *α*-formyl-
    glutaric acid, 243
  reduction to 2-methylglutaric acid, 82
Methylenetetrahydrofolate dehydro-
    genase, 605–608
  purification from bakers' yeast, 606
    from *Clostridium cylindrosporum*,
    606, 607
  use in assays for *N*⁵,*N*¹⁰-methylene-
    tetrahydrofolate and tetrahydro-
    folate, 605–608
*N*⁵,*N*¹⁰-Methylenetetrahydrofolic acid
  determination with dehydrogenase,
    605–608
  enzymatic dehydrogenation to *N*⁵,*N*¹⁰-
    methenyltetrahydrofolate, 605, 606
  formation from tetrahydrofolate and
    formaldehyde, 709–716
  separation on DEAE-Sephadex, 663
  ultraviolet spectra, 710
7-Methyl-8-fluoro-10-(1'-ᴅ-ribityl) iso-
    alloxazine, *see* 7-Methyl-8-fluororibo-
    flavin
7-Methyl-8-fluororiboflavin, substrate for
    liver flavokinase, 547

2(or *α*-)-Methylglutaric acid
  derivation from 2-methyleneglutaric
    acid, 82
  oxidation to acetic acid, 83
6-Methyl-7-hydroxy-8-(1'-ᴅ-ribityl)
    lumazine
  bioproduct from 6,7-dimethyl-8-(1'-ᴅ-
    ribityl)lumazine, 515, 543
  chromatography
    column, 289, 524, 525
    paper, 525, 526
    thin-layer, 526, 527
  complex with riboflavin synthetase, 537
  fluorescence maximum, 289
  light absorption maximum, 289
  occurrence, 515, 516
10-Methylisoalloxazine, half-wave re-
    duction potential, 485
6-Methylisoxanthopterin, chro-
    matography
    column, 656, 658
    paper, 655
6-Methylisoxanthopterin-*O*-glucoside,
    column chromatography, 656
2-Methylmercaptoriboflavin, substrate
    for liver flavokinase, 547
*N*¹-Methylnicotinamide oxidase, 216–222
  assay, 216–219
  properties, 221, 222
  purification from rabbit liver, 219–221
7-Methyl-10-phenylisoalloxazine, half-
    wave reduction potential, 485
6-Methylpterin, 699–701
  reduction to 7,8-dihydro compound,
    700, 701
  synthesis, 699, 700
N-Methyl-2-pyridone
  conversion to δ-methylaminovaleric
    acid, 103, 104
  derivation from *N*-methyl-2-pyridone-
    5-carboxylic acid, 103
*N*¹-Methyl-2-pyridone-5-carboxamide,
    derivation from oxidation of *N*¹-
    methylnicotinamide, 216–219
*N*¹-Methyl-4-pyridone-3-carboxamide,
    derivation from oxidation of *N*¹-
    methylnicotinamide, 216–219
N-Methyl-2-pyridone-5-carboxylic acid
    decarboxylation, 103

properties, 134, 135
spectral identification, 22, 24
support of microbial growth, 134, 135
Nicotinic acid hydrolase, 235–239
    assay, 235, 236
    properties, 237–239
    purification from *Clostridium* sp., 236,
        237
Nicotinic acid mononucleotide
    enzymatic preparation from nicotinic
        acid adenine dinucleotide, 135, 136
    formation from 3-hydroxykynurenine,
        163–166
        from nicotinamide mononucleotide,
            192–197
        from quinolinic acid, 166
    preparation of ^{14}C-nicotinate-labeled
        compound, 56, 57
    spectral identification, 22, 136
Nicotinic acid ribonucleoside, enzymatic
    preparation from nicotinic acid ribo-
    nucleotide, 137
Nipecotic acid
    benzoylation, 81
    from hydrogenation of nicotinic acid,
        81
    oxidation, 84, 85
        of benzoyl derivative, 82
2-Nitro-4-methyl-5-chloro-N-D-ribityl-
    aniline
    conversion to 7-methyl-8-chlororibo-
        flavin, 446, 447
    preparation, 446
Nucleoside phosphorylase, 204–210
    assays during phosphorolysis of pyri-
        dine nucleosides
        radioactive nicotinate or nicotin-
            amide formed, 205, 206
        radioactive phosphate converted to
            ribose 1-phosphate, 206, 207
    properties, 209, 210
    purification from rat liver, 207–209
5'-Nucleotidase, snake venom enzyme
    used in hydrolysis of ribonucleotides
    of nicotinate and nicotinamide, 60
Nucleotide pyrophosphatase, potato
    enzyme used in hydrolysis of NAD,
        53

O

"Old yellow enzyme"
    borohydride reduction, 475
    lack of sulfite complexing, 473
Orotate-^{14}C, use in assay of 5-phospho-
    ribosyl 1-pyrophosphate, 31
Orotidine 5'-phosphate, intermediate in
    assay of phosphoribosyl pyrophos-
    phate, 31
Orotidine 5-phosphate decarboxylase, use
    in assay of phosphoribosyl pyrophos-
    phate, 31
Orotidine 5'-phosphate pyrophospho-
    rylase, use in assay of phosphoribo-
    syl pyrophosphate, 31
Oxynitrilase
    borohydride reduction, 475
    sulfite complex, 473

P

Pancreatic lipase, hydrolysis of ribo-
    flavin fatty acid esters, 451, 452
1,3,7,8,10-Pentamethylisoalloxazine, half-
    wave reduction potential, 486
Peracetylriboflavin, *see* 2',3',4',5'-Tetra-
    acetylriboflavin
Phenazine methosulfate, use in spectro-
    photometric recycling assay for NAD
    and NADP, 26, 27, 29, 30
L-Phenylalanine
    conversion of p-tritio to m-tritio com-
        pound catalyzed by phenylalanine
        hydroxylase, 601–603
    substrate for phenylalanine hydroxyl-
        ase, 600–605, 614–617
Phenylalanine hydroxylase
    use of enzyme from *Pseudomonas* sp.
        11299a to assay reduced L-threo-
        biopterin, 600–605
        from rat liver to assay reduced L-
            erythrobiopterin, 614–617
10-Phenylisoalloxazine, half-wave re-
    duction potential, 485
Phosphodiesterase from snake venom
    hydrolysis of deamido NAD, 135
    of NAD, 53, 60
6-Phospho-D-gluconate, formation in
    enzymatic cycling of NADP and
    NADPH, 12, 14, 15